Geometry and Computing

Volume 13

Geometric shapes belong to our every-day life, and modeling and optimization of such forms determine biological and industrial success. Similar to the digital revolution in image processing, which turned digital cameras and online video downloads into consumer products, nowadays we encounter a strong industrial need and scientific research on geometry processing technologies for 3D shapes.

Several disciplines are involved, many with their origins in mathematics, revived with computational emphasis within computer science, and motivated by applications in the sciences and engineering. Just to mention one example, the renewed interest in discrete differential geometry is motivated by the need for a theoretical foundation for geometry processing algorithms, which cannot be found in classical differential geometry.

Scope: This book series is devoted to new developments in geometry and computation and its applications. It provides a scientific resource library for education, research, and industry. The series constitutes a platform for publication of the latest research in mathematics and computer science on topics in this field.

- Discrete geometry
- Computational geometry
- Differential geometry
- Discrete differential geometry
- Computer graphics
- Geometry processing
- CAD/CAM
- Computer-aided geometric design
- Geometric topology
- Computational topology
- Statistical shape analysis
- Structural molecular biology
- Shape optimization
- Geometric data structures
- Geometric probability
- Geometric constraint solving
- Algebraic geometry
- Graph theory
- Physics-based modeling
- Kinematics
- Symbolic computation
- Approximation theory
- Scientific computing
- Computer vision

More information about this series at http://www.springer.com/series/7580

Jean Gallier • Jocelyn Quaintance

Differential Geometry and Lie Groups

A Second Course

 Springer

Jean Gallier
Department of Computer and Information
Science
University of Pennsylvania
Philadelphia, PA, USA

Jocelyn Quaintance
Department of Computer and Information
Science
University of Pennsylvania
Philadelphia, PA, USA

ISSN 1866-6795   ISSN 1866-6809   (electronic)
Geometry and Computing
ISBN 978-3-030-46049-5   ISBN 978-3-030-46047-1   (eBook)
https://doi.org/10.1007/978-3-030-46047-1

Mathematics Subject Classification: 58A10, 58C35, 33C55, 55R10, 55R25, 15A66

This Springer imprint is published by the registered company Springer Nature Switzerland AG
The registered company address is: Gewerbestrasse 11, 6330 Cham, Switzerland

To my daughter Mia, my wife Anne,

my son Philippe, and my daughter Sylvie.

To my parents Howard and Jane.

Preface

This book is written for a wide audience ranging from upper-undergraduate to advanced graduate students in mathematics, physics, and more broadly engineering students, especially in computer science. Basically, it covers topics which belong to a second course in differential geometry. The reader is expected to be familiar with the theory of manifolds and with some elements of Riemannian geometry, including connections, geodesics, and curvature. Some familiarity with the material presented in the following books is more than sufficient: Tu [105] (the first three chapters), Warner [109] (the first chapter and parts of Chapters 2–4), Do Carmo [37] (the first four chapters), Gallot, Hulin, Lafontaine [48] (the first two chapters and parts of Chapter 3), O'Neill [84] (Chapters 1 and 3), and Gallier and Quaintance [47], which contains all the preliminaries needed to read this book.

The goal of differential geometry is to study the geometry and the topology of manifolds using techniques involving differentiation in one way or another. The pillars of differential geometry are:

(1) Riemannian metrics,
(2) Connections,
(3) Geodesics, and
(4) Curvature.

There are many good books covering the above topics, and we also provided our own account (Gallier and Quaintance [47]). One of the goals of differential geometry is also to be able to generalize "calculus on $\mathbb{R}^n$" to spaces more general than $\mathbb{R}^n$, namely manifolds. We would like to differentiate functions $f : M \to \mathbb{R}$ defined on a manifold, optimize functions (find their minima or maxima), and also to integrate such functions, as well as compute areas and volumes of subspaces of our manifold.

The generalization of the notion of derivative of a function defined on a manifold is the notion of tangent map, and the notions of gradient and Hessian are easily generalized to manifolds equipped with a connection (or a Riemannian metric, which yields the Levi-Civita connection). However, the problem of defining the integral of a function whose domain is a manifold remains.

One of the main discoveries made at the beginning of the twentieth century by Poincaré and Élie Cartan is that the "right" approach to integration is to integrate *differential forms*, and not functions. To integrate a function f, we integrate the form $f\omega$, where ω is a *volume form* on the manifold M. The formalism of differential forms takes care of the process of the change of variables quite automatically and allows for a very clean statement of *Stokes' theorem*.

The theory of differential forms is one of the main tools in geometry and topology. This theory has a surprisingly large range of applications, and it also provides a relatively easy access to more advanced theories such as cohomology. For all these reasons, it is really an indispensable theory, and anyone with more than a passable interest in geometry should be familiar with it.

In this book, we discuss the following topics:

(1) Differential forms, including vector-valued differential forms and differential forms on Lie groups.
(2) An introduction to de Rham cohomology.
(3) Distributions and the Frobenius theorem.
(4) Integration on manifolds, starting with orientability, volume forms, and ending with Stokes' theorem on regular domains.
(5) Integration on Lie groups.
(6) Spherical harmonics and an introduction to the representations of compact Lie groups.
(7) Operators on Riemannian manifolds: Hodge Laplacian, Laplace–Beltrami Laplacian, and Bochner Laplacian.
(8) Fiber bundles, vector bundles, principal bundles, and metrics on bundles.
(9) Connections and curvature in vector bundles, culminating with an introduction to Pontrjagin classes, Chern classes, and the Euler class.
(10) Clifford algebras, Clifford groups, and the groups **Pin**(n), **Spin**(n), **Pin**(p, q), and **Spin**(p, q).

Topics (3)–(7) have more of an analytic than a geometric flavor. Topics (8) and (9) belong to the core of a second course on differential geometry. Clifford algebras and Clifford groups constitute a more algebraic topic. These can be viewed as a generalization of the quaternions. The groups **Spin**(n) are important because they are the universal covers of the groups **SO**(n).

Since this book is already quite long, we resolved ourselves, not without regrets, to omit many proofs. We feel that it is more important to motivate, demystify, and explain the reasons for introducing various concepts and to clarify the relationship between these notions rather than spelling out every proof in full detail. Whenever we omit a proof, we provide precise pointers to the literature.

We must acknowledge our debt to our main sources of inspiration: Bott and Tu [12], Bröcker and tom Dieck [18], Cartan [21], Chern [22], Chevalley [24], Dieudonné [31–33], do Carmo [37], Gallot, Hulin, Lafontaine [48], Hirzebruch [57], Knapp [66], Madsen and Tornehave [75], Milnor and Stasheff [78], Morimoto [81], Morita [82], Petersen [86], and Warner [109].

The chapters or sections marked with the symbol ⊛ contain material that is typically more specialized or more advanced, and they can be omitted upon first (or second) reading.

Acknowledgments We would like to thank Eugenio Calabi, Ching-Li Chai, Ted Chinburg, Chris Croke, Ron Donagi, Harry Gingold, H.W. Gould, Herman Gluck, David Harbater, Julia Hartmann, Jerry Kazdan, Alexander Kirillov, Florian Pop, Steve Shatz, Jim Stasheff, George Sparling, Doran Zeilberger, and Wolfgand Ziller for their encouragement, advice, inspiration, and for what they taught us. We also thank Christine Allen-Blanchette, Arthur Azevedo de Amorim, Kostas Daniilidis, Carlos Esteves, Spyridon Leonardos, Stephen Phillips, João Sedoc, Marcelo Siqueira, and Roberto Tron for reporting typos and for helpful comments.

Philadelphia, PA, USA   Jean Gallier
Philadelphia, PA, USA   Jocelyn Quaintance

Contents

Chapter 1
Introduction

This book covers topics which belong to a second course in differential geometry. Differential forms constitute the main tool needed to understand and prove many of the results presented in this book. Thus one need to have a solid understanding of differential forms, which turn out to be certain kinds of skew-symmetric (also called alternating) tensors. Differential forms have two main roles:

(1) To describe various systems of partial differential equations on manifolds.
(2) To define various geometric invariants reflecting the global structure of manifolds or bundles. Such invariants are obtained by integrating certain differential forms.

Differential forms can be combined using a notion of product called the wedge product, but what really gives power to the formalism of differential forms is the magical operation d of *exterior differentiation*. Given a form ω, we obtain another form $d\omega$, and remarkably, the following equation holds

$$dd\omega = 0.$$

As silly as it looks, the above equation lies at the core of the notion of cohomology, a powerful algebraic tool to understanding the topology of manifolds, and more generally of topological spaces.

Élie Cartan had many of the intuitions that led to the cohomology of differential forms, but it was Georges de Rham who defined it rigorously and proved some important theorems about it. It turns out that the notion of Laplacian can also be defined on differential forms using a device due to Hodge, and some important theorems can be obtained: the Hodge decomposition theorem, and Hodge's theorem about the isomorphism between the de Rham cohomology groups and the spaces of harmonic forms. Differential forms can also be used to define the notion of curvature of a connection on a certain type of manifold called a *vector bundle*.

© Springer Nature Switzerland AG 2020
J. Gallier, J. Quaintance, *Differential Geometry and Lie Groups*, Geometry and Computing 13, https://doi.org/10.1007/978-3-030-46047-1_1

Because differential forms are such a fundamental tool we made the (perhaps painful) decision to provide a fairly detailed exposition of tensors, starting with arbitrary tensors, and then specializing to symmetric and alternating tensors. In particular, we explain rather carefully the process of taking the dual of a tensor (of all three flavors). Tensors, symmetric tensors, tensor algebras, and symmetric algebras are discussed in Chapter 2. Alternating tensors and exterior algebras are discussed in Chapter 3. The Hodge $*$ operator is introduced, we discuss criteria for the decomposability of an alternating tensor in terms of hook operators, and we present the Grassmann-Plücker's equations.

We now give a preview of the topics discussed in this book.

Chapter 4 is devoted to a thorough presentation of differential forms, including vector-valued differential forms, differential forms on Lie Groups, and Maurer-Cartan forms. We also introduce de Rham cohomology.

Chapter 5 is a short chapter devoted to distributions and the Frobenius theorem. Distributions are a generalization of vector fields, and the issue is to understand when a distribution is integrable. The Frobenius theorem gives a necessary and sufficient condition for a distribution to have an integral manifold at every point. One version of the Frobenius theorem is stated in terms of vector fields, the second version in terms of differential forms.

The theory of integration on manifolds and Lie groups is presented in Chapter 6. We introduce the notion of orientation of a smooth manifold (of dimension n), volume forms, and then explain how to integrate a smooth n-form with compact support. We define densities which allow integrating n-forms even if the manifold is not orientable, but we do not go into the details of this theory. We define manifolds with boundary, and explain how to integrate forms on certain kinds of manifolds with boundaries called regular domains. We state and prove a version of the famous result known as *Stokes' theorem*. In the last section we discuss integrating functions on Riemannian manifolds or Lie groups.

The main theme of Chapter 7 is to generalize Fourier analysis on the circle to higher dimensional spheres. One of our goals is to understand the structure of the space $L^2(S^n)$ of real-valued square integrable functions on the sphere S^n, and its complex analog $L^2_{\mathbb{C}}(S^n)$. Both are Hilbert spaces if we equip them with suitable inner products. It turns out that each of $L^2(S^n)$ and $L^2_{\mathbb{C}}(S^n)$ contains a countable family of very nice finite-dimensional subspaces $\mathcal{H}_k(S^n)$ (and $\mathcal{H}^{\mathbb{C}}_k(S^n)$), where $\mathcal{H}_k(S^n)$ is the space of (real) *spherical harmonics* on S^n, that is, the restrictions of the harmonic homogeneous polynomials of degree k (in $n + 1$ real variables) to S^n (and similarly for $\mathcal{H}^{\mathbb{C}}_k(S^n)$); these polynomials satisfy the Laplace equation

$$\Delta P = 0,$$

where the operator Δ is the (Euclidean) *Laplacian*,

$$\Delta = \frac{\partial^2}{\partial x_1^2} + \cdots + \frac{\partial^2}{\partial x_{n+1}^2}.$$

Remarkably, each space $\mathcal{H}_k(S^n)$ (resp. $\mathcal{H}_k^{\mathbb{C}}(S^n)$) is the eigenspace of the Laplace-Beltrami operator Δ_{S^n} on S^n, a generalization to Riemannian manifolds of the standard Laplacian (in fact, $\mathcal{H}_k(S^n)$ is the eigenspace for the eigenvalue $-k(n+k-1)$). As a consequence, the spaces $\mathcal{H}_k(S^n)$ (resp. $\mathcal{H}_k^{\mathbb{C}}(S^n)$) are pairwise orthogonal. Furthermore (and this is where analysis comes in), the set of all finite linear combinations of elements in $\bigcup_{k=0}^{\infty} \mathcal{H}_k(S^n)$ (resp. $\bigcup_{k=0}^{\infty} \mathcal{H}_k^{\mathbb{C}}(S^n)$) is dense in $L^2(S^n)$ (resp. dense in $L_{\mathbb{C}}^2(S^n)$). These two facts imply the following fundamental result about the structure of the spaces $L^2(S^n)$ and $L_{\mathbb{C}}^2(S^n)$.

The family of spaces $\mathcal{H}_k(S^n)$ (resp. $\mathcal{H}_k^{\mathbb{C}}(S^n)$) yields a Hilbert space direct sum decomposition

$$L^2(S^n) = \bigoplus_{k=0}^{\infty} \mathcal{H}_k(S^n) \qquad (\text{resp.} \quad L_{\mathbb{C}}^2(S^n) = \bigoplus_{k=0}^{\infty} \mathcal{H}_k^{\mathbb{C}}(S^n)),$$

which means that the summands are closed, pairwise orthogonal, and that every $f \in L^2(S^n)$ (resp. $f \in L_{\mathbb{C}}^2(S^n)$) is the sum of a converging series

$$f = \sum_{k=0}^{\infty} f_k$$

in the L^2-norm, where the $f_k \in \mathcal{H}_k(S^n)$ (resp. $f_k \in \mathcal{H}_k^{\mathbb{C}}(S^n)$) are uniquely determined functions. Furthermore, given any orthonormal basis $(Y_k^1, \ldots, Y_k^{a_{k,n+1}})$ of $\mathcal{H}_k(S^n)$, we have

$$f_k = \sum_{m_k=1}^{a_{k,n+1}} c_{k,m_k} Y_k^{m_k}, \qquad \text{with} \quad c_{k,m_k} = \langle f, Y_k^{m_k} \rangle_{S^n}.$$

The coefficients c_{k,m_k} are "generalized" *Fourier coefficients* with respect to the Hilbert basis $\{Y_k^{m_k} \mid 1 \leq m_k \leq a_{k,n+1}, \ k \geq 0\}$; see Theorems 7.18 and 7.19.

When $n = 2$, the functions $Y_k^{m_k}$ correspond to the *spherical harmonics*, which are defined in terms of the Legendre functions. Along the way, we prove the famous Funk–Hecke formula.

The purpose of Section 7.9 is to generalize the results about the structure of the space of functions $L_{\mathbb{C}}^2(S^n)$ defined on the sphere S^n, especially the results of Sections 7.5 and 7.6 (such as Theorem 7.19, except part (3)), to homogeneous spaces G/K where G is a compact Lie group and K is a closed subgroup of G.

The first step is to consider the Hilbert space $L_{\mathbb{C}}^2(G)$ where G is a compact Lie group and to find a Hilbert sum decomposition of this space. The key to this generalization is the notion of (unitary) linear representation of the group G.

The result that we are alluding to is a famous theorem known as the *Peter–Weyl theorem* about unitary representations of compact Lie groups.

The Peter–Weyl theorem can be generalized to any representation $V \colon G \to$ $\mathrm{Aut}(E)$ of G into a separable Hilbert space E, and we obtain a Hilbert sum decomposition of E in terms of subspaces E_ρ of E.

The next step is to consider the subspace $L^2_{\mathbb{C}}(G/K)$ of $L^2_{\mathbb{C}}(G)$ consisting of the functions that are right-invariant under the action of K. These can be viewed as functions on the homogeneous space G/K. Again we obtain a Hilbert sum decomposition. It is also interesting to consider the subspace $L^2_{\mathbb{C}}(K\backslash G/K)$ of functions in $L^2_{\mathbb{C}}(G)$ consisting of the functions that are both left- and right-invariant under the action of K. The functions in $L^2_{\mathbb{C}}(K\backslash G/K)$ can be viewed as functions on the homogeneous space G/K that are invariant under the left action of K.

Convolution makes the space $L^2_{\mathbb{C}}(G)$ into a noncommutative algebra. Remarkably, it is possible to characterize when $L^2_{\mathbb{C}}(K\backslash G/K)$ is commutative (under convolution) in terms of a simple criterion about the irreducible representations of G. In this situation, (G, K) is a called a *Gelfand pair*.

When (G, K) is a Gelfand pair, it is possible to define a well-behaved notion of *Fourier transform* on $L^2_{\mathbb{C}}(K\backslash G/K)$. Gelfand pairs and the Fourier transform are briefly considered in Section 7.11.

Chapter 8 deals with various generalizations of the Laplacian to manifolds.

The Laplacian is a very important operator because it shows up in many of the equations used in physics to describe natural phenomena such as heat diffusion or wave propagation. Therefore, it is highly desirable to generalize the Laplacian to functions defined on a manifold. Furthermore, in the late 1930s, Georges de Rham (inspired by Élie Cartan) realized that it was fruitful to define a version of the Laplacian operating on differential forms, because of a fundamental and almost miraculous relationship between harmonics forms (those in the kernel of the Laplacian) and the de Rham cohomology groups on a (compact, orientable) smooth manifold. Indeed, as we will see in Section 8.6, for every cohomology group $H^k_{\mathrm{DR}}(M)$, every cohomology class $[\omega] \in H^k_{\mathrm{DR}}(M)$ is represented by a *unique harmonic k-form* ω; this is the Hodge theorem. The connection between analysis and topology lies deep and has many important consequences. For example, *Poincaré duality* follows as an "easy" consequence of the Hodge theorem.

Technically, the Hodge Laplacian can be defined on differential forms using the Hodge $*$ operator (Section 3.5). On functions, there is an alternate and equivalent definition of the Laplacian using only the covariant derivative and obtained by generalizing the notions of gradient and divergence to functions on manifolds.

Another version of the Laplacian on k-forms can be defined in terms of a generalization of the Levi-Civita connection $\nabla \colon \mathfrak{X}(M) \times \mathfrak{X}(M) \to \mathfrak{X}(M)$ to k-forms viewed as a linear map

$$\nabla \colon \mathcal{A}^k(M) \to \mathrm{Hom}_{C^\infty(M)}(\mathfrak{X}(M), \mathcal{A}^k(M)),$$

and in terms of a certain adjoint ∇^* of ∇, a linear map

$$\nabla^* \colon \mathrm{Hom}_{C^\infty(M)}(\mathfrak{X}(M), \mathcal{A}^k(M)) \to \mathcal{A}^k(M).$$

We obtain the *Bochner Laplacian* (or *connection Laplacian*) $\nabla^*\nabla$. Then it is natural to wonder how the Hodge Laplacian Δ differs from the connection Laplacian $\nabla^*\nabla$?

Remarkably, there is a formula known as *Weitzenböck's formula* (or *Bochner's formula*) of the form

$$\Delta = \nabla^*\nabla + C(R_\nabla),$$

where $C(R_\nabla)$ is a contraction of a version of the curvature tensor on differential forms (a fairly complicated term). In the case of one-forms,

$$\Delta = \nabla^*\nabla + \mathrm{Ric},$$

where Ric is a suitable version of the Ricci curvature operating on one-forms.

Weitzenböck-type formulae are at the root of the so-called "Bochner technique," which consists in exploiting curvature information to deduce topological information.

Chapter 9 is an introduction to bundle theory; we discuss fiber bundles, vector bundles, and principal bundles.

Intuitively, a *fiber bundle* over B is a family $E = (E_b)_{b \in B}$ of spaces E_b (fibers) indexed by B and varying smoothly as b moves in B, such that every E_b is diffeomorphic to some prespecified space F. The space E is called the total space, B the base space, and F the fiber. A way to define such a family is to specify a surjective map $\pi : E \to B$. We will assume that E, B, and F are smooth manifolds and that π is a smooth map. The type of bundles that we just described is too general and to develop a useful theory it is necessary to assume that locally, a bundle looks like a product. Technically, this is achieved by assuming that there is some open cover $\mathcal{U} = (U_\alpha)_{\alpha \in I}$ of B and that there is a family $(\varphi_\alpha)_{\alpha \in I}$ of diffeomorphisms

$$\varphi_\alpha : \pi^{-1}(U_\alpha) \to U_\alpha \times F.$$

Intuitively, above U_α, the open subset $\pi^{-1}(U_\alpha)$ looks like a product. The maps φ_α are called *local trivializations*.

The last important ingredient in the notion of a fiber bundle is the specification of the "twisting" of the bundle; that is, how the fiber $E_b = \pi^{-1}(b)$ gets twisted as b moves in the base space B. Technically, such twisting manifests itself on overlaps $U_\alpha \cap U_\beta \neq \emptyset$. It turns out that we can write

$$\varphi_\alpha \circ \varphi_\beta^{-1}(b, x) = (b, g_{\alpha\beta}(b)(x))$$

for all $b \in U_\alpha \cap U_\beta$ and all $x \in F$. The term $g_{\alpha\beta}(b)$ is a diffeomorphism of F. Then we require that the family of diffeomorphisms $g_{\alpha\beta}(b)$ belongs to a Lie group G, which is expressed by specifying that the maps $g_{\alpha\beta}$, called transitions maps, are maps

$$g_{\alpha\beta}: U_\alpha \cap U_\beta \to G.$$

The purpose of the group G, called the structure group, is to specify the "twisting" of the bundle.

Fiber bundles are defined in Section 9.1. The family of transition maps $g_{\alpha\beta}$ satisfies an important condition on nonempty overlaps $U_\alpha \cap U_\beta \cap U_\gamma$ called the *cocycle condition*:

$$g_{\alpha\beta}(b)g_{\beta\gamma}(b) = g_{\alpha\gamma}(b)$$

(where $g_{\alpha\beta}(b)$, $g_{\beta\gamma}(b)$, $g_{\alpha\gamma}(b) \in G$), for all α, β, γ such that $U_\alpha \cap U_\beta \cap U_\gamma \neq \emptyset$ and all $b \in U_\alpha \cap U_\beta \cap U_\gamma$.

In Section 9.2, following Hirzebruch [57] and Chern [22], we define bundle morphisms and the notion of equivalence of bundles over the same base. We show that two bundles (over the same base) are equivalent if and only if they are isomorphic.

In Section 9.3 we describe the construction of a fiber bundle with prescribed fiber F and structure group G from a base manifold, B, an open cover $\mathcal{U} = (U_\alpha)_{\alpha \in I}$ of B, and a family of maps $g_{\alpha\beta}: U_\alpha \cap U_\beta \to G$ satisfying the cocycle condition, called a *cocycle*. This construction is the basic tool for constructing new bundles from old ones.

Section 9.4 is devoted to a special kind of fiber bundle called *vector bundles*. A vector bundle is a fiber bundle for which the fiber is a finite-dimensional vector space V, and the structure group is a subgroup of the group of linear isomorphisms ($\mathbf{GL}(n, \mathbb{R})$ or $\mathbf{GL}(n, \mathbb{C})$, where $n = \dim V$). Typical examples of vector bundles are the tangent bundle TM and the cotangent bundle T^*M of a manifold M. We define maps of vector bundles and equivalence of vector bundles.

In Section 9.5 we describe various operations on vector bundles: Whitney sums, tensor products, tensor powers, exterior powers, symmetric powers, dual bundles, and $\mathcal{H}om$ bundles. We also define the complexification of a real vector bundle.

In Section 9.6 we discuss properties of the sections of a vector bundle ξ. We prove that the space of sections $\Gamma(\xi)$ is finitely generated projective $C^\infty(B)$-module.

Section 9.7 is devoted to the the covariant derivative of tensor fields and to the duality between vector fields and differential forms.

In Section 9.8 we explain how to give a vector bundle a Riemannian metric. This is achieved by supplying a smooth family $(\langle -, - \rangle_b)_{b \in B}$ of inner products on each fiber $\pi^{-1}(b)$ above $b \in B$. We describe the notion of reduction of the structure group and define orientable vector bundles.

In Section 9.9 we consider the special case of fiber bundles for which the fiber coincides with the structure group G, which acts on itself by left translations. Such fiber bundles are called *principal bundles*. It turns out that a principal bundle can be defined in terms of a free right action of Lie group on a smooth manifold. When principal bundles are defined in terms of free right actions, the notion of bundle morphism is also defined in terms of equivariant maps.

There are two constructions that allow us to reduce the study of fiber bundles to the study of principal bundles. Given a fiber bundle ξ with fiber F, we can construct a principal bundle $P(\xi)$ obtained by replacing the fiber F by the group G. Conversely, given a principal bundle ξ and an effective action of G on a manifold F, we can construct the fiber bundle $\xi[F]$ obtained by replacing G by F. The maps

$$\xi \mapsto \xi[F] \quad \text{and} \quad \xi \mapsto P(\xi)$$

induce a bijection between equivalence classes of principal G-bundles and fiber bundles (with structure group G). Furthermore, ξ is a trivial bundle iff $P(\xi)$ is a trivial bundle.

Section 9.10 is devoted to principal bundles that arise from proper and free actions of a Lie group. When the base space is a homogenous space, which means that it arises from a transitive action of a Lie group, then the total space is a principal bundle. There are many illustrations of this situation involving $\mathbf{SO}(n + 1)$ and $\mathbf{SU}(n + 1)$.

In Chapter 10 we discuss connections and curvature in vector bundles. In Section 10.2 we define connections on a vector bundle. This can be done in two equivalent ways. One of the two definitions is more abstract than the other because it involves a tensor product, but it is technically more convenient. This definition states that a connection on a vector bundle ξ, as an $\mathbb{R}$-linear map

$$\nabla : \Gamma(\xi) \to \mathcal{A}^1(B) \otimes_{C^\infty(B)} \Gamma(\xi) \tag{$*$}$$

that satisfies the "Leibniz rule"

$$\nabla(fs) = df \otimes s + f \nabla s,$$

with $s \in \Gamma(\xi)$ and $f \in C^\infty(B)$, where $\Gamma(\xi)$ and $\mathcal{A}^1(B)$ are treated as $C^\infty(B)$-modules. Here, $\mathcal{A}^1(B) = \Gamma(T^*B)$ is the space of 1-forms on B. Since there is an isomorphism

$$\mathcal{A}^1(B) \otimes_{C^\infty(B)} \Gamma(\xi) \cong \Gamma(T^*B \otimes \xi),$$

a connection can be defined equivalently as an $\mathbb{R}$-linear map

$$\nabla : \Gamma(\xi) \to \Gamma(T^*B \otimes \xi)$$

satisfying the Leibniz rule.

In Section 10.3 we show how a connection can be represented in a chart in terms of a certain matrix called a *connection matrix*. We prove that every vector bundle possesses a connection, and we give a formula describing how a connection matrix changes if we switch from one chart to another.

In Section 10.4 we define the notion of covariant derivative along a curve and parallel transport.

Section 10.5 is devoted to the very important concept of *curvature form* R^∇ of a connection ∇ on a vector bundle ξ. We show that the curvature form is a vector-valued two-form with values in $\Gamma(\mathcal{H}om(\xi, \xi))$. We also establish the relationship between R^∇ and the more familiar definition of the Riemannian curvature in terms of vector fields.

In Section 10.6 we show how the curvature form can be expressed in a chart in terms of a matrix of two-forms called a *curvature matrix*. The connection matrix and the curvature matrix are related by the *structure equation*. We also give a formula describing how a curvature matrix changes if we switch from one chart to another. Bianchi's identity gives an expression for the exterior derivative of the curvature matrix in terms of the curvature matrix itself and the connection matrix.

Section 10.8 deals with connections compatible with a metric and the Levi-Civita connection, which arise in the Riemannian geometry of manifolds. One way of characterizing the Levi-Civita connection involves defining the notion of connection on the dual bundle. This is achieved in Section 10.9.

Levi-Civita connections on the tangent bundle of a manifold are investigated in Section 10.10.

The purpose of Section 10.11 is to introduce the reader to *Pontrjagin Classes and Chern Classes*, which are fundamental invariants of real (resp. complex) vector bundles. Here we are dealing with one of the most sophisticated and beautiful parts of differential geometry.

A masterly exposition of the theory of characteristic classes is given in the classic book by Milnor and Stasheff [78]. Amazingly, the method of Chern and Weil using differential forms is quite accessible for someone who has reasonably good knowledge of differential forms and de Rham cohomology, as long as one is willing to gloss over various technical details. We give an introduction to characteristic classes using the method of Chern and Weil.

If ξ is a real orientable vector bundle of rank $2m$, and if ∇ is a metric connection on ξ, then it is possible to define a closed global form $\mathrm{eu}(R^\nabla)$, and its cohomology class $e(\xi)$ is called the *Euler class* of ξ. This is shown in Section 10.13. The Euler class $e(\xi)$ turns out to be a square root of the top Pontrjagin class $p_m(\xi)$ of ξ. A complex rank m vector bundle can be viewed as a real vector bundle of rank $2m$, which is always orientable. The Euler class $e(\xi)$ of this real vector bundle is equal to the top Chern class $c_m(\xi)$ of the complex vector bundle ξ.

The global form $\mathrm{eu}(R^\nabla)$ is defined in terms of a certain polynomial $\mathrm{Pf}(A)$ associated with a real skew-symmetric matrix A, which is a kind of square root of the determinant $\det(A)$. The polynomial $\mathrm{Pf}(A)$, called the *Pfaffian*, is defined in Section 10.12.

The culmination of this chapter is a statement of the generalization due to Chern of a classical theorem of Gauss and Bonnet. This theorem known as the *generalized Gauss–Bonnet formula* expresses the Euler characteristic $\chi(M)$ of an orientable, compact smooth manifold M of dimension $2m$ as

$$\chi(M) = \int_M \mathrm{eu}(R^\nabla),$$

where $\mathrm{eu}(R^\nabla)$ is the Euler form associated with the curvature form R^∇ of a metric connection ∇ on M.

The goal of Chapter 11 is to explain how rotations in $\mathbb{R}^n$ are induced by the action of a certain group $\mathbf{Spin}(n)$ on $\mathbb{R}^n$, in a way that generalizes the action of the unit complex numbers $\mathbf{U}(1)$ on $\mathbb{R}^2$, and the action of the unit quaternions $\mathbf{SU}(2)$ on $\mathbb{R}^3$ (*i.e.*, the action is defined in terms of multiplication in a larger algebra containing both the group $\mathbf{Spin}(n)$ and $\mathbb{R}^n$). The group $\mathbf{Spin}(n)$, called a *spinor group*, is defined as a certain subgroup of units of an algebra Cl_n, the *Clifford algebra* associated with $\mathbb{R}^n$.

Since the spinor groups are certain well-chosen subgroups of units of Clifford algebras, it is necessary to investigate Clifford algebras to get a firm understanding of spinor groups. This chapter provides a tutorial on *Clifford algebra* and the groups $\mathbf{Spin}$ and $\mathbf{Pin}$, including a study of the structure of the Clifford algebra $\mathrm{Cl}_{p,q}$ associated with a nondegenerate symmetric bilinear form of signature (p, q) and culminating in the beautiful 8-*periodicity theorem* of Élie Cartan and Raoul Bott (with proofs). We also explain when $\mathbf{Spin}(p, q)$ is a double-cover of $\mathbf{SO}(p, q)$.

Some preliminaries on algebras and tensor algebras are reviewed in Section 11.2.

In Section 11.3 we define Clifford algebras over the field $K = \mathbb{R}$. The Clifford groups (over $K = \mathbb{R}$) are defined in Section 11.4. In the second half of this section we restrict our attention to the real quadratic form $\Phi(x_1, \ldots, x_n) = -(x_1^2 + \cdots + x_n^2)$. The corresponding Clifford algebras are denoted Cl_n and the corresponding Clifford groups as Γ_n.

In Section 11.5 we define the groups $\mathbf{Pin}(n)$ and $\mathbf{Spin}(n)$ associated with the real quadratic form $\Phi(x_1, \ldots, x_n) = -(x_1^2 + \cdots + x_n^2)$. We prove that the maps $\rho \colon \mathbf{Pin}(n) \to \mathbf{O}(n)$ and $\rho \colon \mathbf{Spin}(n) \to \mathbf{SO}(n)$ are surjective with kernel $\{-1, 1\}$. We determine the groups $\mathbf{Spin}(n)$ for $n = 2, 3, 4$.

Section 11.6 is devoted to the Spin and Pin groups associated with the real nondegenerate quadratic form

$$\Phi(x_1, \ldots, x_{p+q}) = x_1^2 + \cdots + x_p^2 - (x_{p+1}^2 + \cdots + x_{p+q}^2).$$

We obtain Clifford algebras $\mathrm{Cl}_{p,q}$, Clifford groups $\Gamma_{p,q}$, and groups $\mathbf{Pin}(p, q)$ and $\mathbf{Spin}(p, q)$. We show that the maps $\rho \colon \mathbf{Pin}(p, q) \to \mathbf{O}(p, q)$ and $\rho \colon \mathbf{Spin}(p, q) \to \mathbf{SO}(p, q)$ are surjective with kernel $\{-1, 1\}$.

In Section 11.7 we show that the Lie groups $\mathbf{Pin}(p, q)$ and $\mathbf{Spin}(p, q)$ are double covers of $\mathbf{O}(p, q)$ and $\mathbf{SO}(p, q)$.

In Section 11.8 we prove an amazing result due to Élie Cartan and Raoul Bott, namely the 8-periodicity of the Clifford algebras $\mathrm{Cl}_{p,q}$. This result says that: for all $n \geq 0$, we have the following isomorphisms:

$$\mathrm{Cl}_{0,n+8} \cong \mathrm{Cl}_{0,n} \otimes \mathrm{Cl}_{0,8}$$

$$Cl_{n+8,0} \cong Cl_{n,0} \otimes Cl_{8,0}.$$

Furthermore,

$$Cl_{0,8} = Cl_{8,0} = \mathbb{R}(16),$$

the real algebra of 16×16 matrices.

Section 11.9 is devoted to the complex Clifford algebras $Cl(n, \mathbb{C})$. In this case, we have a 2-periodicity,

$$Cl(n + 2, \mathbb{C}) \cong Cl(n, \mathbb{C}) \otimes_{\mathbb{C}} Cl(2, \mathbb{C}),$$

with $Cl(2, \mathbb{C}) = \mathbb{C}(2)$, the complex algebra of 2×2 matrices.

Finally, in the last section, Section 11.10, we outline the theory of Clifford groups and of the Pin and Spin groups over any field K of characteristic $\neq 2$.

Chapter 2
Tensor Algebras and Symmetric Algebras

Tensors are creatures that we would prefer did not exist but keep showing up whenever multilinearity manifests itself.

One of the goals of differential geometry is to be able to generalize "calculus on $\mathbb{R}^n$" to spaces more general than $\mathbb{R}^n$, namely manifolds. We would like to differentiate functions $f : M \to \mathbb{R}$ defined on a manifold, optimize functions (find their minima or maxima), but also to integrate such functions, as well as compute areas and volumes of subspaces of our manifold.

The suitable notion of differentiation is the notion of tangent map, a linear notion. One of the main discoveries made at the beginning of the twentieth century by Poincaré and Élie Cartan is that the "right" approach to integration is to integrate *differential forms*, and not functions. To integrate a function f, we integrate the form $f\omega$, where ω is a *volume form* on the manifold M. The formalism of differential forms takes care of the process of the change of variables quite automatically, and allows for a very clean statement of *Stokes' formula*.

Differential forms can be combined using a notion of product called the wedge product, but what really gives power to the formalism of differential forms is the magical operation d of *exterior differentiation*. Given a form ω, we obtain another form $d\omega$, and remarkably, the following equation holds:

$$dd\omega = 0.$$

As silly as it looks, the above equation lies at the core of the notion of cohomology, a powerful algebraic tool to understanding the topology of manifolds, and more generally of topological spaces.

Élie Cartan had many of the intuitions that led to the cohomology of differential forms, but it was Georges de Rham who defined it rigorously and proved some important theorems about it. It turns out that the notion of Laplacian can also be

---

The original version of this chapter was revised. The correction to this chapter is available at https://doi.org/10.1007/978-3-030-46047-1_12

© Springer Nature Switzerland AG 2020, corrected publication 2020
J. Gallier, J. Quaintance, *Differential Geometry and Lie Groups*, Geometry and Computing 13, https://doi.org/10.1007/978-3-030-46047-1_2

defined on differential forms using a device due to Hodge, and some important theorems can be obtained: the Hodge decomposition theorem and Hodge's theorem about the isomorphism between the de Rham cohomology groups and the spaces of harmonic forms.

To understand all this, one needs to learn about differential forms, which turn out to be certain kinds of skew-symmetric (also called alternating) tensors.

If one's only goal is to define differential forms, then it is possible to take some short cuts and to avoid introducing the general notion of a tensor. However, tensors that are not necessarily skew-symmetric arise naturally, such as the curvature tensor, and in the theory of vector bundles, general tensor products are needed.

Consequently, we made the (perhaps painful) decision to provide a fairly detailed exposition of tensors, starting with arbitrary tensors, and then specializing to symmetric and alternating tensors. In particular, we explain rather carefully the process of taking the dual of a tensor (of all three flavors).

We refrained from following the approach in which a tensor is defined as a multilinear map defined on a product of dual spaces, because it seems very artificial and confusing (certainly to us). This approach relies on duality results that only hold in finite dimension, and consequently unnecessarily restricts the theory of tensors to finite-dimensional spaces. We also feel that it is important to begin with a coordinate-free approach. Bases can be chosen for computations, but tensor algebra should not be reduced to raising or lowering indices.

Readers who feel that they are familiar with tensors could skip this chapter and the next. They can come back to them "by need."

We begin by defining tensor products of vector spaces over a field and then we investigate some basic properties of these tensors, in particular the existence of bases and duality. After this we investigate special kinds of tensors, namely symmetric tensors and skew-symmetric tensors. Tensor products of modules over a commutative ring with identity will be discussed very briefly. They show up naturally when we consider the space of sections of a tensor product of vector bundles.

Given a linear map $f : E \to F$ (where E and F are two vector spaces over a field K), we know that if we have a basis $(u_i)_{i \in I}$ for E, then f is completely determined by its values $f(u_i)$ on the basis vectors. For a multilinear map $f : E^n \to F$, we don't know if there is such a nice property but it would certainly be very useful.

In many respects tensor products allow us to define multilinear maps in terms of their action on a suitable basis. The crucial idea is to *linearize*, that is, to create a new vector space $E^{\otimes n}$ such that the multilinear map $f : E^n \to F$ is turned into a *linear map* $f_\otimes : E^{\otimes n} \to F$ which is equivalent to f in a strong sense. If in addition, f is symmetric, then we can define a symmetric tensor power $\mathrm{Sym}^n(E)$, and every symmetric multilinear map $f : E^n \to F$ is turned into a *linear map* $f_\odot : \mathrm{Sym}^n(E) \to F$ which is equivalent to f in a strong sense. Similarly, if f is alternating, then we can define a skew-symmetric tensor power $\bigwedge^n(E)$, and every alternating multilinear map is turned into a *linear map* $f_\wedge : \bigwedge^n(E) \to F$ which is equivalent to f in a strong sense.

Tensor products can be defined in various ways, some more abstract than others. We try to stay down to earth, without excess.

In Section 2.1, we review some facts about dual spaces and pairings. In particular, we show that an inner product on a finite-dimensional vector space E induces a canonical isomorphism between E and its dual space E^*. Pairings will be used to deal with dual spaces of tensors. We also show that there is a canonical isomorphism between the vector space of bilinear forms on E and the vector space of linear maps from E to itself.

Tensor products are defined in Section 2.2. Given two vector spaces E_1 and E_2 over a field K, the *tensor product* $E_1 \otimes E_2$ is defined by a universal mapping property: it is a vector space with an injection $i_\otimes : E_1 \times E_2 \to E_1 \otimes E_2$, such that for every vector space F and every bilinear map $f : E_1 \times E_2 \to F$, there is a *unique linear map* $f_\otimes : E_1 \otimes E_2 \to F$ such that

$$f = f_\otimes \circ i_\otimes,$$

as illustrated in the following diagram:

$$E_1 \times E_2 \xrightarrow{\ i_\otimes\ } E_1 \otimes E_2$$

$$f \searrow \qquad \downarrow f_\otimes$$

$$F$$

We prove that the above universal mapping property defines $E_1 \otimes E_2$ up to isomorphism, and then we prove its existence by constructing a suitable quotient of the free vector space $K^{(E_1 \otimes E_2)}$ generated by $E_1 \otimes E_2$. The generalization to any finite number of vector spaces $E_1, \ldots, E_n$ is immediate.

The universal mapping property of the tensor product yields an isomorphism between the vector space of linear maps $\mathrm{Hom}(E_1 \otimes \cdots \otimes E_n, F)$ and the vector space of multilinear maps $\mathrm{Hom}(E_1, \ldots, E_n; F)$. We show that tensor product is functorial, which means that given two linear maps $f : E \to F$ and $g : E' \to F'$, there is a linear map $f \otimes g : E \otimes F \to E' \otimes F'$.

In Section 2.3, we show how to construct a basis for the tensor product $E_1 \otimes \cdots \otimes E_n$ from bases for the spaces $E_1, \ldots, E_n$.

In Section 2.4, we prove some basic isomorphisms involving tensor products. One of these isomorphisms states that $\mathrm{Hom}(E \otimes F, G)$ is isomorphic to $\mathrm{Hom}(E, \mathrm{Hom}(F, G))$.

Section 2.5 deals with duality for tensor products. It is a very important section which needs to be thoroughly understood in order to study vector bundles. The main isomorphisms state that if $E_1, \ldots, E_n$ are finite dimensional, then

$$(E_1 \otimes \cdots \otimes E_n)^* \cong E_1^* \otimes \cdots \otimes E_n^* \cong \mathrm{Hom}(E_1, \ldots, E_n; K).$$

The second isomorphism arises from the pairing μ defined on generators by

$$\mu(v_1^* \otimes \cdots \otimes v_n^*)(u_1, \ldots, u_n) = v_1^*(u_1) \cdots v_n^*(u_n).$$

We also prove that if either E or F is finite dimensional, then there is a canonical isomorphism between $E^* \otimes F$ and $\text{Hom}(E, F)$.

In Section 2.6 we define the *tensor algebra* $T(V)$. This is the direct sum of the tensor powers $V^{\otimes m} = \underbrace{V \otimes \cdots \otimes V}_{m}$,

$$T(V) = \bigoplus_{m \geq 0} V^{\otimes m}.$$

In addition to being a vector space, $T(V)$ is equipped with an associative multiplication, $\otimes$. The tensor algebra $T(V)$ satisfies a universal mapping property with respect to (associative) algebras.

We also define the tensor algebras $T^{r,s}(V)$ and the tensor algebra $T^{\bullet,\bullet}(V)$. We define contraction operations. If E and F are algebras, we show how to make $E \otimes F$ into an algebra.

In Section 2.7 to turn to the special case of symmetric tensor powers, which correspond to *symmetric multilinear maps* $\varphi \colon E^n \to F$. There are multilinear maps that are invariant under permutation of its arguments.

Given a vector space E over a field K, for any $n \geq 1$, the *symmetric tensor power* $S^n(E)$ is defined by a universal mapping property: it is a vector space with an injection $i_\odot \colon E^n \to S^n(E)$, such that for every vector space F and every symmetric multilinear map $f \colon E^n \to F$, there is a *unique linear map* $f_\odot \colon S^n(E) \to F$ such that

$$f = f_\odot \circ i_\odot,$$

as illustrated in the following diagram:

$$
\begin{array}{ccc}
E^n & \xrightarrow{\ i_\odot\ } & S^n(E) \\
 & {\scriptstyle f} \searrow & \downarrow {\scriptstyle f_\odot} \\
 & & F
\end{array}
$$

We prove that the above universal mapping property defines $S^n(E)$ up to isomorphism, and then we prove its existence by constructing the quotient of the tensor power $E^{\otimes n}$ by the subspace C of $E^{\otimes n}$ generated by the vectors of the form

$$u_1 \otimes \cdots \otimes u_n - u_{\sigma(1)} \otimes \cdots \otimes u_{\sigma(n)},$$

for all $u_i \in E$, and all permutations $\sigma \colon \{1, \ldots, n\} \to \{1, \ldots, n\}$. As a corollary, there is an isomorphism between the vector space of linear maps $\text{Hom}(S^n(E), F)$ and the vector space of symmetric multilinear maps $\text{Sym}^n(E; F)$. We also show that given two linear maps $f, g \colon E \to E'$, there is a linear map $f \odot g \colon S^2(E) \to S^2(E')$.

A basic isomorphism involving the symmetric power of a direct sum is shown at the end of this section.

In Section 2.8, we show how to construct a basis of the tensor power $S^n(E)$ from a basis of E. This involves multisets.

Section 2.9 is devoted to duality in symmetric powers. There is a nondegenerate pairing $S^n(E^*) \times S^n(E) \longrightarrow K$ defined on generators as follows:

$$(v_1^* \odot \cdots \odot v_n^*, u_1 \odot \cdots \odot u_n) \mapsto \sum_{\sigma \in \mathfrak{S}_n} v_{\sigma(1)}^*(u_1) \cdots v_{\sigma(n)}^*(u_n).$$

As a consequence, if E is finite dimensional and if K is a field of characteristic 0, we have canonical isomorphisms

$$(S^n(E))^* \cong S^n(E^*) \cong \mathrm{Sym}^n(E; K).$$

The symmetric tensor power $S^n(E)$ is also naturally embedded in $E^{\otimes n}$.

In Section 2.10 we define symmetric tensor algebras. As in the case of tensors, we can pack together all the symmetric powers $S^n(V)$ into an algebra. Given a vector space V, the space

$$S(V) = \bigoplus_{m \geq 0} S^m(V),$$

is called the *symmetric tensor algebra of* V. The symmetric tensor algebra $S(V)$ satisfies a universal mapping property with respect to commutative algebras.

We conclude with Section 2.11 which gives a quick introduction to tensor products of modules over a commutative ring. Such tensor products arise because vector fields and differential forms on a smooth manifold are modules over the ring of smooth functions $C^\infty(M)$. Except for the results about bases and duality, most other results still hold for these more general tensors. We introduce *projective modules*, which behave better under duality. Projective modules will show up when dealing with vector bundles.

2.1   Linear Algebra Preliminaries: Dual Spaces and Pairings

We assume that we are dealing with vector spaces over a field K. As usual the *dual space E^** of a vector space E is defined by $E^* = \mathrm{Hom}(E, K)$. The dual space E^* is the vector space consisting of all linear maps $\omega \colon E \to K$ with values in the field K.

A problem that comes up often is to decide when a space E is isomorphic to the dual F^* of some other space F (possibly equal to E). The notion of pairing due to Pontrjagin provides a very clean criterion.

Definition 2.1. Given two vector spaces E and F over a field K, a map $\langle -, - \rangle : E \times F \to K$ is a *nondegenerate pairing* iff it is bilinear and iff $\langle u, v \rangle = 0$ for all $v \in F$ implies $u = 0$, and $\langle u, v \rangle = 0$ for all $u \in E$ implies $v = 0$. A nondegenerate pairing induces two linear maps $\varphi : E \to F^*$ and $\psi : F \to E^*$ defined such that for all $u \in E$ and all $v \in F$, $\varphi(u)$ is the linear form in F^* and $\psi(v)$ is the linear form in E^* given by

$$\varphi(u)(y) = \langle u, y \rangle \quad \text{for all } y \in F$$

$$\psi(v)(x) = \langle x, v \rangle \quad \text{for all } x \in E.$$

Schematically, $\varphi(u) = \langle u, - \rangle$ and $\psi(v) = \langle -, v \rangle$.

Proposition 2.1. *For every nondegenerate pairing $\langle -, - \rangle : E \times F \to K$, the induced maps $\varphi : E \to F^*$ and $\psi : F \to E^*$ are linear and injective. Furthermore, if E and F are finite dimensional, then $\varphi : E \to F^*$ and $\psi : F \to E^*$ are bijective.*

Proof. The maps $\varphi : E \to F^*$ and $\psi : F \to E^*$ are linear because $u, v \mapsto \langle u, v \rangle$ is bilinear. Assume that $\varphi(u) = 0$. This means that $\varphi(u)(y) = \langle u, y \rangle = 0$ for all $y \in F$, and as our pairing is nondegenerate, we must have $u = 0$. Similarly, ψ is injective. If E and F are finite dimensional, then $\dim(E) = \dim(E^*)$ and $\dim(F) = \dim(F^*)$. However, the injectivity of φ and ψ implies that that $\dim(E) \le \dim(F^*)$ and $\dim(F) \le \dim(E^*)$. Consequently $\dim(E) \le \dim(F)$ and $\dim(F) \le \dim(E)$, so $\dim(E) = \dim(F)$. Therefore, $\dim(E) = \dim(F^*)$ and φ is bijective (and similarly $\dim(F) = \dim(E^*)$ and ψ is bijective). $\qquad\square$

Proposition 2.1 shows that when E and F are finite dimensional, a nondegenerate pairing induces *canonical isomorphisms* $\varphi : E \to F^*$ and $\psi : F \to E^*$; that is, isomorphisms that do not depend on the choice of bases. An important special case is the case where $E = F$ and we have an inner product (a symmetric, positive definite bilinear form) on E.

Remark. When we use the term "canonical isomorphism," we mean that such an isomorphism is defined independently of any choice of bases. For example, if E is a finite-dimensional vector space and $(e_1, \ldots, e_n)$ is any basis of E, we have the dual basis $(e_1^*, \ldots, e_n^*)$ of E^* (where, $e_i^*(e_j) = \delta_{i\,j}$), and thus the map $e_i \mapsto e_i^*$ is an isomorphism between E and E^*. This isomorphism is *not* canonical.

On the other hand, if $\langle -, - \rangle$ is an inner product on E, then Proposition 2.1 shows that the nondegenerate pairing $\langle -, - \rangle$ on $E \times E$ induces a canonical isomorphism between E and E^*. This isomorphism is often denoted $\flat : E \to E^*$, and we usually write $u^\flat$ for $\flat(u)$, with $u \in E$. Schematically, $u^\flat = \langle u, - \rangle$. The inverse of $\flat$ is denoted $\sharp : E^* \to E$, and given any linear form $\omega \in E^*$, we usually write $\omega^\sharp$ for $\sharp(\omega)$. Schematically, $\omega = \langle \omega^\sharp, - \rangle$.

Given any basis, $(e_1, \ldots, e_n)$ of E (not necessarily orthonormal), let (g_{ij}) be the $n \times n$-matrix given by $g_{ij} = \langle e_i, e_j \rangle$ (the *Gram* matrix of the inner product). Recall that the *dual basis* $(e_1^*, \ldots, e_n^*)$ of E^* consists of the coordinate forms $e_i^* \in E^*$, which are characterized by the following properties:

$$e_i^*(e_j) = \delta_{ij}, \quad 1 \leq i, j \leq n.$$

The inverse of the Gram matrix (g_{ij}) is often denoted by (g^{ij}) (by raising the indices).

The tradition of raising and lowering indices is pervasive in the literature on tensors. It is indeed useful to have some notational convention to distinguish between vectors and linear forms (also called *one-forms* or *covectors*). The usual convention is that coordinates of vectors are written using superscripts, as in $u = \sum_{i=1}^{n} u^i e_i$, and coordinates of one-forms are written using subscripts, as in $\omega = \sum_{i=1}^{n} \omega_i e_i^*$. Actually, since vectors are indexed with subscripts, one-forms are indexed with superscripts, so e_i^* should be written as e^i.

The motivation is that summation signs can then be omitted, according to the *Einstein summation convention*. According to this convention, whenever a summation variable (such as i) appears as both a subscript and a superscript in an expression, it is assumed that it is involved in a summation. For example the sum $\sum_{i=1}^{n} u^i e_i$ is abbreviated as

$$u^i e_i,$$

and the sum $\sum_{i=1}^{n} \omega_i e^i$ is abbreviated as

$$\omega_i e^i.$$

In this text we will not use the Einstein summation convention, which we find somewhat confusing, and we will also write e_i^* instead of e^i.

The maps $\flat$ and $\sharp$ can be described explicitly in terms of the Gram matrix of the inner product and its inverse.

Proposition 2.2. *For any vector space E, given a basis $(e_1, \ldots, e_n)$ for E and its dual basis $(e_1^*, \ldots, e_n^*)$ for E^*, for any inner product $\langle -, - \rangle$ on E, if (g_{ij}) is its Gram matrix, with $g_{ij} = \langle e_i, e_j \rangle$, and (g^{ij}) is its inverse, then for every vector $u = \sum_{j=1}^{n} u^j e_j \in E$ and every one-form $\omega = \sum_{i=1}^{n} \omega_i e_i^* \in E^*$, we have*

$$u^\flat = \sum_{i=1}^{n} \omega_i e_i^*, \quad \text{with} \quad \omega_i = \sum_{j=1}^{n} g_{ij} u^j,$$

and

$$\omega^\sharp = \sum_{j=1}^{n} (\omega^\sharp)^j e_j, \quad \text{with} \quad (\omega^\sharp)^i = \sum_{j=1}^{n} g^{ij} \omega_j.$$

Proof. For every $u = \sum_{j=1}^{n} u^j e_j$, since $u^\flat(v) = \langle u, v \rangle$ for all $v \in E$, we have

$$u^\flat(e_i) = \langle u, e_i \rangle = \left\langle \sum_{j=1}^{n} u^j e_j, e_i \right\rangle = \sum_{j=1}^{n} u^j \langle e_j, e_i \rangle = \sum_{j=1}^{n} g_{ij} u^j,$$

so we get

$$u^\flat = \sum_{i=1}^{n} \omega_i e_i^*, \quad \text{with} \quad \omega_i = \sum_{j=1}^{n} g_{ij} u^j.$$

If we write $\omega \in E^*$ as $\omega = \sum_{i=1}^{n} \omega_i e_i^*$ and $\omega^\sharp \in E$ as $\omega^\sharp = \sum_{j=1}^{n} (\omega^\sharp)^j e_j$, since

$$\omega_i = \omega(e_i) = \langle \omega^\sharp, e_i \rangle = \sum_{j=1}^{n} (\omega^\sharp)^j g_{ij}, \quad 1 \leq i \leq n,$$

we get

$$(\omega^\sharp)^i = \sum_{j=1}^{n} g^{ij} \omega_j,$$

where (g^{ij}) is the inverse of the matrix (g_{ij}). $\qquad\square$

The map $\flat$ has the effect of lowering (flattening!) indices, and the map $\sharp$ has the effect of raising (sharpening!) indices.

Here is an explicit example of Proposition 2.2. Let (e_1, e_2) be a basis of E such that

$$\langle e_1, e_1 \rangle = 1, \qquad \langle e_1, e_2 \rangle = 2, \qquad \langle e_2, e_2 \rangle = 5.$$

Then

$$g = \begin{pmatrix} 1 & 2 \\ 2 & 5 \end{pmatrix}, \qquad g^{-1} = \begin{pmatrix} 5 & -2 \\ -2 & 1 \end{pmatrix}.$$

Set $u = u^1 e_1 + u^2 e_2$ and observe that

$$u^\flat(e_1) = \langle u^1 e_1 + u^2 e_2, e_1 \rangle = \langle e_1, e_1 \rangle u^1 + \langle e_2, e_1 \rangle u^2 = g_{11} u^1 + g_{12} u^2 = u^1 + 2u^2$$

$$u^\flat(e_2) = \langle u^1 e_1 + u^2 e_2, e_2 \rangle = \langle e_1, e_2 \rangle u^1 + \langle e_2, e_2 \rangle u^2 = g_{21} u^1 + g_{22} u^2 = 2u^1 + 5u^2,$$

which in turn implies that

$$u^\flat = \omega_1 e_1^* + \omega_2 e_2^* = u^\flat(e_1)e_1^* + u^\flat(e_2)e_2^* = (u^1 + 2u^2)e_1^* + (2u^1 + 5u^2)e_2^*.$$

Given $\omega = \omega_1 e_1^* + \omega_2 e_2^*$, we calculate $\omega^\sharp = (\omega^\sharp)^1 e_1 + (\omega^\sharp)^2 e_2$ from the following two linear equalities:

$$\omega_1 = \omega(e_1) = \langle \omega^\sharp, e_1 \rangle = \langle (\omega^\sharp)^1 e_1 + (\omega^\sharp)^2 e_2, e_1 \rangle$$
$$= \langle e_1, e_1 \rangle (\omega^\sharp)^1 + \langle e_2, e_1 \rangle (\omega^\sharp)^2 = (\omega^\sharp)^1 + 2(\omega^\sharp)^2 = g_{11}(\omega^\sharp)^1 + g_{12}(\omega^\sharp)^2$$
$$\omega_2 = \omega(e_2) = \langle \omega^\sharp, e_2 \rangle = \langle (\omega^\sharp)^1 e_1 + (\omega^\sharp)^2 e_2, e_2 \rangle$$
$$= \langle e_1, e_2 \rangle (\omega^\sharp)^1 + \langle e_2, e_2 \rangle (\omega^\sharp)^2 = 2(\omega^\sharp)^1 + 5(\omega^\sharp)^2 = g_{21}(\omega^\sharp)^1 + g_{22}(\omega^\sharp)^2.$$

These equalities are concisely written as

$$\begin{pmatrix} \omega_1 \\ \omega_2 \end{pmatrix} = \begin{pmatrix} 1 & 2 \\ 2 & 5 \end{pmatrix} \begin{pmatrix} (\omega^\sharp)^1 \\ (\omega^\sharp)^2 \end{pmatrix} = g \begin{pmatrix} (\omega^\sharp)^1 \\ (\omega^\sharp)^2 \end{pmatrix}.$$

Then

$$\begin{pmatrix} (\omega^\sharp)^1 \\ (\omega^\sharp)^2 \end{pmatrix} = g^{-1} \begin{pmatrix} \omega_1 \\ \omega_2 \end{pmatrix} = \begin{pmatrix} 5 & -2 \\ -2 & 1 \end{pmatrix} \begin{pmatrix} \omega_1 \\ \omega_2 \end{pmatrix},$$

which in turn implies

$$(\omega^\sharp)^1 = 5\omega_1 - 2\omega_2, \qquad (\omega^\sharp)^2 = -2\omega_1 + \omega_2,$$

i.e.

$$\omega^\sharp = (5\omega_1 - 2\omega_2)e_1 + (-2\omega_1 + \omega_2)e_2.$$

The inner product $\langle -, - \rangle$ on E induces an inner product on E^* denoted $\langle -, - \rangle_{E^*}$, and given by

$$\langle \omega_1, \omega_2 \rangle_{E^*} = \langle \omega_1^\sharp, \omega_2^\sharp \rangle, \qquad \text{for all } \omega_1, \omega_2 \in E^*.$$

Then we have

$$\langle u^\flat, v^\flat \rangle_{E^*} = \langle (u^\flat)^\sharp, (v^\flat)^\sharp \rangle = \langle u, v \rangle \quad \text{for all} \quad u, v \in E.$$

If $(e_1, \ldots, e_n)$ is a basis of E and $g_{ij} = \langle e_i, e_j \rangle$, as

$$(e_i^*)^\sharp = \sum_{k=1}^n g^{ik} e_k,$$

an easy computation shows that

$$\langle e_i^*, e_j^* \rangle_{E^*} = \langle (e_i^*)^\sharp, (e_j^*)^\sharp \rangle = g^{ij};$$

that is, in the basis $(e_1^*, \ldots, e_n^*)$, the inner product on E^* is represented by the matrix (g^{ij}), the inverse of the matrix (g_{ij}).

The inner product on a finite vector space also yields a canonical isomorphism between the space $\mathrm{Hom}(E, E; K)$ of bilinear forms on E, and the space $\mathrm{Hom}(E, E)$ of linear maps from E to itself. Using this isomorphism, we can define the trace of a bilinear form in an intrinsic manner. This technique is used in differential geometry, for example, to define the divergence of a differential one-form.

Proposition 2.3. *If $\langle -, - \rangle$ is an inner product on a finite vector space E (over a field, K), then for every bilinear form $f : E \times E \to K$, there is a unique linear map $f^\sharp : E \to E$ such that*

$$f(u, v) = \langle f^\sharp(u), v \rangle, \quad \text{for all } u, v \in E.$$

The map $f \mapsto f^\sharp$ is a linear isomorphism between $\mathrm{Hom}(E, E; K)$ and $\mathrm{Hom}(E, E)$.

Proof. For every $g \in \mathrm{Hom}(E, E)$, the map given by

$$f(u, v) = \langle g(u), v \rangle, \quad u, v \in E$$

is clearly bilinear. It is also clear that the above defines a linear map from $\mathrm{Hom}(E, E)$ to $\mathrm{Hom}(E, E; K)$. This map is injective, because if $f(u, v) = 0$ for all $u, v \in E$, as $\langle -, - \rangle$ is an inner product, we get $g(u) = 0$ for all $u \in E$. Furthermore, both spaces $\mathrm{Hom}(E, E)$ and $\mathrm{Hom}(E, E; K)$ have the same dimension, so our linear map is an isomorphism. $\square$

If $(e_1, \ldots, e_n)$ is an orthonormal basis of E, then we check immediately that the trace of a linear map g (which is independent of the choice of a basis) is given by

$$\mathrm{tr}(g) = \sum_{i=1}^{n} \langle g(e_i), e_i \rangle,$$

where $n = \dim(E)$.

Definition 2.2. We define the *trace of the bilinear form* f by

$$\mathrm{tr}(f) = \mathrm{tr}(f^\sharp).$$

From Proposition 2.3, $\mathrm{tr}(f)$ is given by

$$\mathrm{tr}(f) = \sum_{i=1}^{n} f(e_i, e_i),$$

for any orthonormal basis $(e_1, \ldots, e_n)$ of E. We can also check directly that the above expression is independent of the choice of an orthonormal basis.

We demonstrate how to calculate $\mathrm{tr}(f)$ where $f : \mathbb{R}^2 \times \mathbb{R}^2 \to \mathbb{R}$ with $f((x_1, y_1), (x_2, y_2)) = x_1 x_2 + 2x_2 y_1 + 3x_1 y_2 - y_1 y_2$. Under the standard basis for $\mathbb{R}^2$, the bilinear form f is represented as

$$\begin{pmatrix} x_1 & y_1 \end{pmatrix} \begin{pmatrix} 1 & 3 \\ 2 & -1 \end{pmatrix} \begin{pmatrix} x_2 \\ y_2 \end{pmatrix}.$$

This matrix representation shows that

$$f^{\natural} = \begin{pmatrix} 1 & 3 \\ 2 & -1 \end{pmatrix}^{\top} = \begin{pmatrix} 1 & 2 \\ 3 & -1 \end{pmatrix},$$

and hence

$$\mathrm{tr}(f) = \mathrm{tr}(f^{\natural}) = \mathrm{tr} \begin{pmatrix} 1 & 2 \\ 3 & -1 \end{pmatrix} = 0.$$

We will also need the following proposition to show that various families are linearly independent.

Proposition 2.4. *Let E and F be two nontrivial vector spaces and let $(u_i)_{i \in I}$ be any family of vectors $u_i \in E$. The family $(u_i)_{i \in I}$ is linearly independent iff for every family $(v_i)_{i \in I}$ of vectors $v_i \in F$, there is some linear map $f : E \to F$ so that $f(u_i) = v_i$ for all $i \in I$.*

Proof. Left as an exercise. □

2.2 Tensor Products

First we define tensor products, and then we prove their existence and uniqueness up to isomorphism.

Definition 2.3. Let K be a given field, and let $E_1, \ldots, E_n$ be $n \geq 2$ given vector spaces. For any vector space F, a map $f : E_1 \times \cdots \times E_n \to F$ is *multilinear* iff it is linear in each of its argument; that is,

$$f(u_1, \ldots u_{i_1}, v + w, u_{i+1}, \ldots, u_n) = f(u_1, \ldots u_{i_1}, v, u_{i+1}, \ldots, u_n)$$
$$+ f(u_1, \ldots u_{i_1}, w, u_{i+1}, \ldots, u_n)$$
$$f(u_1, \ldots u_{i_1}, \lambda v, u_{i+1}, \ldots, u_n) = \lambda f(u_1, \ldots u_{i_1}, v, u_{i+1}, \ldots, u_n),$$

for all $u_j \in E_j$ ($j \neq i$), all $v, w \in E_i$, and all $\lambda \in K$, for $i = 1 \ldots, n$.

The set of multilinear maps as above forms a vector space denoted $L(E_1, \ldots, E_n; F)$ or $\mathrm{Hom}(E_1, \ldots, E_n; F)$. When $n = 1$, we have the vector space of linear maps $L(E, F)$ (also denoted $\mathrm{Hom}(E, F)$). (To be very precise, we write $\mathrm{Hom}_K(E_1, \ldots, E_n; F)$ and $\mathrm{Hom}_K(E, F)$.)

Definition 2.4. A *tensor product* of $n \geq 2$ vector spaces $E_1, \ldots, E_n$ is a vector space T together with a multilinear map $\varphi \colon E_1 \times \cdots \times E_n \to T$, such that for every vector space F and for every multilinear map $f \colon E_1 \times \cdots \times E_n \to F$, there is a unique linear map $f_\otimes \colon T \to F$ with

$$f(u_1, \ldots, u_n) = f_\otimes(\varphi(u_1, \ldots, u_n)),$$

for all $u_1 \in E_1, \ldots, u_n \in E_n$, or for short

$$f = f_\otimes \circ \varphi.$$

Equivalently, there is a unique linear map $f_\otimes$ such that the following diagram commutes.

$$
\begin{array}{ccc}
E_1 \times \cdots \times E_n & \xrightarrow{\ \varphi\ } & T \\
 & {}_{f}\searrow & \big\downarrow{}^{f_\otimes} \\
 & & F
\end{array}
$$

The above property is called the *universal mapping property* of the tensor product (T, φ).

We show that any two tensor products (T_1, φ_1) and (T_2, φ_2) for $E_1, \ldots, E_n$, are isomorphic.

Proposition 2.5. *Given any two tensor products (T_1, φ_1) and (T_2, φ_2) for $E_1, \ldots, E_n$, there is an isomorphism $h \colon T_1 \to T_2$ such that*

$$\varphi_2 = h \circ \varphi_1.$$

Proof. Focusing on (T_1, φ_1), we have a multilinear map $\varphi_2 \colon E_1 \times \cdots \times E_n \to T_2$, and thus there is a unique linear map $(\varphi_2)_\otimes \colon T_1 \to T_2$ with

$$\varphi_2 = (\varphi_2)_\otimes \circ \varphi_1$$

as illustrated by the following commutative diagram.

$$
\begin{array}{ccc}
E_1 \times \cdots \times E_n & \xrightarrow{\ \varphi_1\ } & T_1 \\
 & {}_{\varphi_2}\searrow & \big\downarrow{}^{(\varphi_2)_\otimes} \\
 & & T_2
\end{array}
$$

Similarly, focusing now on (T_2, φ_2), we have a multilinear map $\varphi_1 \colon E_1 \times \cdots \times E_n \to T_1$, and thus there is a unique linear map $(\varphi_1)_\otimes \colon T_2 \to T_1$ with

$$\varphi_1 = (\varphi_1)_\otimes \circ \varphi_2$$

as illustrated by the following commutative diagram.

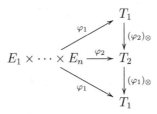

Putting these diagrams together, we obtain the commutative diagrams

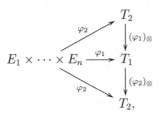

and

$$
\begin{array}{c}
\end{array}
$$

which means that

$$\varphi_1 = (\varphi_1)_\otimes \circ (\varphi_2)_\otimes \circ \varphi_1 \quad \text{and} \quad \varphi_2 = (\varphi_2)_\otimes \circ (\varphi_1)_\otimes \circ \varphi_2.$$

On the other hand, focusing on (T_1, φ_1), we have a multilinear map $\varphi_1 \colon E_1 \times \cdots \times E_n \to T_1$, but the unique linear map $h \colon T_1 \to T_1$ with

$$\varphi_1 = h \circ \varphi_1$$

is $h = \mathrm{id}$, as illustrated by the following commutative diagram

$$E_1 \times \cdots \times E_n \xrightarrow{\varphi_1} T_1$$
$$\searrow_{\varphi_1} \quad \downarrow \text{id}$$
$$T_1,$$

and since $(\varphi_1)_\otimes \circ (\varphi_2)_\otimes$ is linear as a composition of linear maps, we must have

$$(\varphi_1)_\otimes \circ (\varphi_2)_\otimes = \text{id}.$$

Similarly, we have the commutative diagram

$$E_1 \times \cdots \times E_n \xrightarrow{\varphi_2} T_2$$
$$\searrow_{\varphi_2} \quad \downarrow \text{id}$$
$$T_2,$$

and we must have

$$(\varphi_2)_\otimes \circ (\varphi_1)_\otimes = \text{id}.$$

This shows that $(\varphi_1)_\otimes$ and $(\varphi_2)_\otimes$ are inverse linear maps, and thus, $(\varphi_2)_\otimes \colon T_1 \to T_2$ is an isomorphism between T_1 and T_2.   □

Now that we have shown that tensor products are unique up to isomorphism, we give a construction that produces them. Tensor products are obtained from free vector spaces by a quotient process, so let us begin by describing the construction of the free vector space generated by a set.

For simplicity assume that our set I is finite, say

$$I = \{\heartsuit, \diamondsuit, \spadesuit, \clubsuit\}.$$

The construction works for any field K (and in fact for any commutative ring A, in which case we obtain the free A-module generated by I). Assume that $K = \mathbb{R}$. The *free vector space generated by* I is the set of all formal linear combinations of the form

$$a\heartsuit + b\diamondsuit + c\spadesuit + d\clubsuit,$$

with $a, b, c, d \in \mathbb{R}$. It is assumed that the order of the terms does not matter. For example,

$$2\heartsuit - 5\diamondsuit + 3\spadesuit = -5\diamondsuit + 2\heartsuit + 3\spadesuit.$$

Addition and multiplication by a scalar are defined as follows:

$$(a_1\heartsuit + b_1\diamondsuit + c_1\spadesuit + d_1\clubsuit) + (a_2\heartsuit + b_2\diamondsuit + c_2\spadesuit + d_2\clubsuit)$$
$$= (a_1 + a_2)\heartsuit + (b_1 + b_2)\diamondsuit + (c_1 + c_2)\spadesuit + (d_1 + d_2)\clubsuit,$$

and

$$\alpha \cdot (a\heartsuit + b\diamondsuit + c\spadesuit + d\clubsuit) = \alpha a\heartsuit + \alpha b\diamondsuit + \alpha c\spadesuit + \alpha d\clubsuit,$$

for all $a, b, c, d, \alpha \in \mathbb{R}$. With these operations, it is immediately verified that we obtain a vector space denoted $\mathbb{R}^{(I)}$. The set I can be viewed as embedded in $\mathbb{R}^{(I)}$ by the injection ι given by

$$\iota(\heartsuit) = 1\heartsuit, \quad \iota(\diamondsuit) = 1\diamondsuit, \quad \iota(\spadesuit) = 1\spadesuit, \quad \iota(\clubsuit) = 1\clubsuit.$$

Thus, $\mathbb{R}^{(I)}$ can be viewed as the vector space with the special basis $I = \{\heartsuit, \diamondsuit, \spadesuit, \clubsuit\}$. In our case, $\mathbb{R}^{(I)}$ is isomorphic to $\mathbb{R}^4$.

The exact same construction works for any field K, and we obtain a vector space denoted by $K^{(I)}$ and an injection $\iota \colon I \to K^{(I)}$.

The main reason why the free vector space $K^{(I)}$ over a set I is interesting is that it satisfies a *universal mapping property*. This means that for every vector space F (over the field K), any function $h \colon I \to F$, where F is *considered just a set*, has a unique linear extension $\overline{h} \colon K^{(I)} \to F$. By extension, we mean that $\overline{h}(i) = h(i)$ for all $i \in I$, or more rigorously that $h = \overline{h} \circ \iota$.

For example, if $I = \{\heartsuit, \diamondsuit, \spadesuit, \clubsuit\}$, $K = \mathbb{R}$, and $F = \mathbb{R}^3$, the function h given by

$$h(\heartsuit) = (1, 1, 1), \quad h(\diamondsuit) = (1, 1, 0), \quad h(\spadesuit) = (1, 0, 0), \quad h(\clubsuit) = (0, 0 - 1)$$

has a unique linear extension $\overline{h} \colon \mathbb{R}^{(I)} \to \mathbb{R}^3$ to the free vector space $\mathbb{R}^{(I)}$, given by

$$\overline{h}(a\heartsuit + b\diamondsuit + c\spadesuit + d\clubsuit) = a\overline{h}(\heartsuit) + b\overline{h}(\diamondsuit) + c\overline{h}(\spadesuit) + d\overline{h}(\clubsuit)$$
$$= ah(\heartsuit) + bh(\diamondsuit) + ch(\spadesuit) + dh(\clubsuit)$$
$$= a(1, 1, 1) + b(1, 1, 0) + c(1, 0, 0) + d(0, 0, -1)$$
$$= (a + b + c, a + b, a - d).$$

To generalize the construction of a free vector space to infinite sets I, we observe that the formal linear combination $a\heartsuit + b\diamondsuit + c\spadesuit + d\clubsuit$ can be viewed as the function $f \colon I \to \mathbb{R}$ given by

$$f(\heartsuit) = a, \quad f(\diamondsuit) = b, \quad f(\spadesuit) = c, \quad f(\clubsuit) = d,$$

where $a, b, c, d \in \mathbb{R}$. More generally, we can replace $\mathbb{R}$ by any field K. If I is finite, then the set of all such functions is a vector space under pointwise addition and pointwise scalar multiplication. If I is infinite, since addition and scalar multiplication only make sense for finite vectors, we require that our functions $f : I \rightarrow K$ take the value 0 except for possibly finitely many arguments. We can think of such functions as an infinite sequences $(f_i)_{i \in I}$ of elements f_i of K indexed by I, with only finitely many nonzero f_i. The formalization of this construction goes as follows.

Given any set I viewed as an index set, let $K^{(I)}$ be the set of all functions $f : I \rightarrow K$ such that $f(i) \neq 0$ only for finitely many $i \in I$. As usual, denote such a function by $(f_i)_{i \in I}$; it is a family of finite support. We make $K^{(I)}$ into a vector space by defining addition and scalar multiplication by

$$(f_i) + (g_i) = (f_i + g_i)$$
$$\lambda(f_i) = (\lambda f_i).$$

The family $(e_i)_{i \in I}$ is defined such that $(e_i)_j = 0$ if $j \neq i$ and $(e_i)_i = 1$. It is a basis of the vector space $K^{(I)}$, so that every $w \in K^{(I)}$ can be uniquely written as a finite linear combination of the e_i. There is also an injection $\iota : I \rightarrow K^{(I)}$ such that $\iota(i) = e_i$ for every $i \in I$. Furthermore, it is easy to show that for any vector space F, and for any function $h : I \rightarrow F$, there is a unique linear map $\overline{h} : K^{(I)} \rightarrow F$ such that $h = \overline{h} \circ \iota$, as in the following diagram.

Definition 2.5. The vector space $(K^{(I)}, \iota)$ constructed as above from a set I is called the *free vector space generated by I* (or over I). The commutativity of the above diagram is called the *universal mapping property* of the free vector space $(K^{(I)}, \iota)$ over I.

Using the proof technique of Proposition 2.5, it is not hard to prove that any two vector spaces satisfying the above universal mapping property are isomorphic.

We can now return to the construction of tensor products. For simplicity consider two vector spaces E_1 and E_2. Whatever $E_1 \otimes E_2$ and $\varphi : E_1 \times E_2 \rightarrow E_1 \otimes E_2$ are, since φ is supposed to be bilinear, we must have

$$\varphi(u_1 + u_2, v_1) = \varphi(u_1, v_1) + \varphi(u_2, v_1)$$
$$\varphi(u_1, v_1 + v_2) = \varphi(u_1, v_1) + \varphi(u_1, v_2)$$
$$\varphi(\lambda u_1, v_1) = \lambda \varphi(u_1, v_1)$$
$$\varphi(u_1, \mu v_1) = \mu \varphi(u_1, v_1)$$

for all $u_1, u_2 \in E_1$, all $v_1, v_2 \in E_2$, and all $\lambda, \mu \in K$. Since $E_1 \otimes E_2$ must satisfy the universal mapping property of Definition 2.4, we may want to define $E_1 \otimes E_2$ as the free vector space $K^{(E_1 \times E_2)}$ generated by $I = E_1 \times E_2$ and let φ be the injection of $E_1 \times E_2$ into $K^{(E_1 \times E_2)}$. The problem is that in $K^{(E_1 \times E_2)}$, vectors such that

$$(u_1 + u_2, v_1) \quad and \quad (u_1, v_1) + (u_2, v_2)$$

are different, when they should really be the same, since φ is bilinear. Since $K^{(E_1 \times E_2)}$ is free, there are no relations among the generators and this vector space is too big for our purpose.

The remedy is simple: take the quotient of the free vector space $K^{(E_1 \times E_2)}$ by the subspace N generated by the vectors of the form

$$(u_1 + u_2, v_1) - (u_1, v_1) - (u_2, v_1)$$

$$(u_1, v_1 + v_2) - (u_1, v_1) - (u_1, v_2)$$

$$(\lambda u_1, v_1) - \lambda(u_1, v_1)$$

$$(u_1, \mu v_1) - \mu(u_1, v_1).$$

Then, if we let $E_1 \otimes E_2$ be the quotient space $K^{(E_1 \times E_2)}/N$ and let φ be the quotient map, this forces φ to be bilinear. Checking that $(K^{(E_1 \times E_2)}/N, \varphi)$ satisfies the universal mapping property is straightforward. Here is the detailed construction.

Theorem 2.6. *Given $n \geq 2$ vector spaces $E_1, \ldots, E_n$, a tensor product $(E_1 \otimes \cdots \otimes E_n, \varphi)$ for $E_1, \ldots, E_n$ can be constructed. Furthermore, denoting $\varphi(u_1, \ldots, u_n)$ as $u_1 \otimes \cdots \otimes u_n$, the tensor product $E_1 \otimes \cdots \otimes E_n$ is generated by the vectors $u_1 \otimes \cdots \otimes u_n$, where $u_1 \in E_1, \ldots, u_n \in E_n$, and for every multilinear map $f : E_1 \times \cdots \times E_n \to F$, the unique linear map $f_\otimes : E_1 \otimes \cdots \otimes E_n \to F$ such that $f = f_\otimes \circ \varphi$ is defined by*

$$f_\otimes(u_1 \otimes \cdots \otimes u_n) = f(u_1, \ldots, u_n)$$

on the generators $u_1 \otimes \cdots \otimes u_n$ of $E_1 \otimes \cdots \otimes E_n$.

Proof. First we apply the construction of a free vector space to the Cartesian product $I = E_1 \times \cdots \times E_n$, obtaining the free vector space $M = K^{(I)}$ on $I = E_1 \times \cdots \times E_n$. Since every basis generator $e_i \in M$ is uniquely associated with some n-tuple $i = (u_1, \ldots, u_n) \in E_1 \times \cdots \times E_n$, we denote e_i by $(u_1, \ldots, u_n)$.

Next let N be the subspace of M generated by the vectors of the following type:

$$(u_1, \ldots, u_i + v_i, \ldots, u_n) - (u_1, \ldots, u_i, \ldots, u_n) - (u_1, \ldots, v_i, \ldots, u_n),$$

$$(u_1, \ldots, \lambda u_i, \ldots, u_n) - \lambda(u_1, \ldots, u_i, \ldots, u_n).$$

We let $E_1 \otimes \cdots \otimes E_n$ be the quotient M/N of the free vector space M by N, $\pi : M \to M/N$ be the quotient map, and set

$$\varphi = \pi \circ \iota.$$

By construction, φ is multilinear, and since π is surjective and the $\iota(i) = e_i$ generate M, the fact that each i is of the form $i = (u_1, \ldots, u_n) \in E_1 \times \cdots \times E_n$ implies that $\varphi(u_1, \ldots, u_n)$ generate M/N. Thus, if we denote $\varphi(u_1, \ldots, u_n)$ as $u_1 \otimes \cdots \otimes u_n$, the space $E_1 \otimes \cdots \otimes E_n$ is generated by the vectors $u_1 \otimes \cdots \otimes u_n$, with $u_i \in E_i$.

It remains to show that $(E_1 \otimes \cdots \otimes E_n, \varphi)$ satisfies the universal mapping property. To this end, we begin by proving there is a map h such that $f = h \circ \varphi$. Since $M = K^{(E_1 \times \cdots \times E_n)}$ is free on $I = E_1 \times \cdots \times E_n$, there is a unique linear map $\overline{f} : K^{(E_1 \times \cdots \times E_n)} \to F$, such that

$$f = \overline{f} \circ \iota,$$

as in the diagram below.

$$E_1 \times \cdots \times E_n \xrightarrow{\ \iota\ } K^{(E_1 \times \cdots \times E_n)} = M$$

$$\xrightarrow[\ f\ ]{} \quad \downarrow{\overline{f}}$$

$$F$$

Because f is multilinear, note that we must have $\overline{f}(w) = 0$ for every $w \in N$; for example, on the generator

$$(u_1, \ldots, u_i + v_i, \ldots, u_n) - (u_1, \ldots, u_i, \ldots, u_n) - (u_1, \ldots, v_i, \ldots, u_n)$$

we have

$$\overline{f}((u_1, \ldots, u_i + v_i, \ldots, u_n) - (u_1, \ldots, u_i, \ldots, u_n) - (u_1, \ldots, v_i, \ldots, u_n))$$

$$= f(u_1, \ldots, u_i + v_i, \ldots, u_n) - f(u_1, \ldots, u_i, \ldots, u_n) - f(u_1, \ldots, v_i, \ldots, u_n)$$

$$= f(u_1, \ldots, u_i, \ldots, u_n) + f(u_1, \ldots, v_i, \ldots, u_n) - f(u_1, \ldots, u_i, \ldots, u_n)$$

$$- f(u_1, \ldots, v_i, \ldots, u_n)$$

$$= 0.$$

But then, $\overline{f} : M \to F$ factors through M/N, which means that there is a unique linear map $h : M/N \to F$ such that $\overline{f} = h \circ \pi$ making the following diagram commute:

by defining $h([z]) = \overline{f}(z)$ for every $z \in M$, where $[z]$ denotes the equivalence class in M/N of $z \in M$. Indeed, the fact that $\overline{f}$ vanishes on N ensures that h is well defined on M/N, and it is clearly linear by definition. Since $f = \overline{f} \circ \iota$, from the equation $\overline{f} = h \circ \pi$, by composing on the right with ι, we obtain

$$f = \overline{f} \circ \iota = h \circ \pi \circ \iota = h \circ \varphi,$$

as in the following commutative diagram.

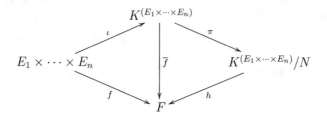

We now prove the uniqueness of h. For any linear map $f_{\otimes} \colon E_1 \otimes \cdots \otimes E_n \to F$ such that $f = f_{\otimes} \circ \varphi$, since the vectors $u_1 \otimes \cdots \otimes u_n$ generate $E_1 \otimes \cdots \otimes E_n$ and since $\varphi(u_1, \ldots, u_n) = u_1 \otimes \cdots \otimes u_n$, the map $f_{\otimes}$ is uniquely defined by

$$f_{\otimes}(u_1 \otimes \cdots \otimes u_n) = f(u_1, \ldots, u_n).$$

Since $f = h \circ \varphi$, the map h is unique, and we let $f_{\otimes} = h$.   □

The map φ from $E_1 \times \cdots \times E_n$ to $E_1 \otimes \cdots \otimes E_n$ is often denoted by $\iota_{\otimes}$, so that

$$\iota_{\otimes}(u_1, \ldots, u_n) = u_1 \otimes \cdots \otimes u_n.$$

What is important about Theorem 2.6 is not so much the construction itself but the fact that it produces a tensor product with the universal mapping property with respect to multilinear maps. Indeed, Theorem 2.6 yields a canonical isomorphism

$$\mathrm{L}(E_1 \otimes \cdots \otimes E_n, F) \cong \mathrm{L}(E_1, \ldots, E_n; F)$$

between the vector space of linear maps $\mathrm{L}(E_1 \otimes \cdots \otimes E_n, F)$, and the vector space of multilinear maps $\mathrm{L}(E_1, \ldots, E_n; F)$, *via* the linear map $- \circ \varphi$ defined by

$$h \mapsto h \circ \varphi,$$

where $h \in L(E_1 \otimes \cdots \otimes E_n, F)$. Indeed, $h \circ \varphi$ is clearly multilinear, and since by Theorem 2.6, for every multilinear map $f \in L(E_1, \ldots, E_n; F)$, there is a unique linear map $f_\otimes \in L(E_1 \otimes \cdots \otimes E_n, F)$ such that $f = f_\otimes \circ \varphi$, the map $- \circ \varphi$ is bijective. As a matter of fact, its inverse is the map

$$f \mapsto f_\otimes.$$

We record this fact as the following proposition.

Proposition 2.7. *Given a tensor product* $(E_1 \otimes \cdots \otimes E_n, \varphi)$, *the linear map* $h \mapsto h \circ \varphi$ *is a canonical isomorphism*

$$L(E_1 \otimes \cdots \otimes E_n, F) \cong L(E_1, \ldots, E_n; F)$$

between the vector space of linear maps $L(E_1 \otimes \cdots \otimes E_n, F)$, *and the vector space of multilinear maps* $L(E_1, \ldots, E_n; F)$.

Using the "Hom" notation, the above canonical isomorphism is written as

$$\text{Hom}(E_1 \otimes \cdots \otimes E_n, F) \cong \text{Hom}(E_1, \ldots, E_n; F).$$

Remarks.

(1) To be very precise, since the tensor product depends on the field K, we should subscript the symbol $\otimes$ with K and write

$$E_1 \otimes_K \cdots \otimes_K E_n.$$

However, we often omit the subscript K unless confusion may arise.

(2) For $F = K$, the base field, Proposition 2.7 yields a canonical isomorphism between the vector space $L(E_1 \otimes \cdots \otimes E_n, K)$, and the vector space of multilinear forms $L(E_1, \ldots, E_n; K)$. However, $L(E_1 \otimes \cdots \otimes E_n, K)$ is the dual space $(E_1 \otimes \cdots \otimes E_n)^*$, and thus the vector space of multilinear forms $L(E_1, \ldots, E_n; K)$ is canonically isomorphic to $(E_1 \otimes \cdots \otimes E_n)^*$.

Since this isomorphism is used often, we record it as the following proposition.

Proposition 2.8. *Given a tensor product* $E_1 \otimes \cdots \otimes E_n$, *there is a canonical isomorphism*

$$L(E_1, \ldots, E_n; K) \cong (E_1 \otimes \cdots \otimes E_n)^*$$

between the vector space of multilinear maps $L(E_1, \ldots, E_n; K)$ *and the dual* $(E_1 \otimes \cdots \otimes E_n)^*$ *of the tensor product* $E_1 \otimes \cdots \otimes E_n$.

The fact that the map $\varphi \colon E_1 \times \cdots \times E_n \to E_1 \otimes \cdots \otimes E_n$ is multilinear, can also be expressed as follows:

$$u_1 \otimes \cdots \otimes (v_i + w_i) \otimes \cdots \otimes u_n = (u_1 \otimes \cdots \otimes v_i \otimes \cdots \otimes u_n)$$
$$+ (u_1 \otimes \cdots \otimes w_i \otimes \cdots \otimes u_n),$$
$$u_1 \otimes \cdots \otimes (\lambda u_i) \otimes \cdots \otimes u_n = \lambda(u_1 \otimes \cdots \otimes u_i \otimes \cdots \otimes u_n).$$

Of course, this is just what we wanted!

Definition 2.6. Tensors in $E_1 \otimes \cdots \otimes E_n$ are called *n-tensors*, and tensors of the form $u_1 \otimes \cdots \otimes u_n$, where $u_i \in E_i$ are called *simple (or decomposable) n-tensors*. Those *n*-tensors that are not simple are often called *compound n-tensors*.

Not only do tensor products act on spaces, but they also act on linear maps (they are functors).

Proposition 2.9. *Given two linear maps* $f\colon E \to E'$ *and* $g\colon F \to F'$, *there is a unique linear map*

$$f \otimes g\colon E \otimes F \to E' \otimes F'$$

such that

$$(f \otimes g)(u \otimes v) = f(u) \otimes g(v),$$

for all $u \in E$ *and all* $v \in F$.

Proof. We can define $h\colon E \times F \to E' \otimes F'$ by

$$h(u, v) = f(u) \otimes g(v).$$

It is immediately verified that h is bilinear, and thus it induces a unique linear map

$$f \otimes g\colon E \otimes F \to E' \otimes F'$$

making the following diagram commutes:

$$E \times F \xrightarrow{\iota_\otimes} E \otimes F$$
$$h \searrow \qquad \downarrow f \otimes g$$
$$E' \otimes F',$$

such that $(f \otimes g)(u \otimes v) = f(u) \otimes g(v)$, for all $u \in E$ and all $v \in F$. □

Definition 2.7. The linear map $f \otimes g\colon E \otimes F \to E' \otimes F'$ given by Proposition 2.9 is called the *tensor product* of $f\colon E \to E'$ and $g\colon F \to F'$.

Another way to define $f \otimes g$ proceeds as follows. Given two linear maps $f: E \to E'$ and $g: F \to F'$, the map $f \times g$ is the linear map from $E \times F$ to $E' \times F'$ given by

$$(f \times g)(u, v) = (f(u), g(v)), \quad \text{for all } u \in E \text{ and all } v \in F.$$

Then the map h in the proof of Proposition 2.9 is given by $h = \iota'_{\otimes} \circ (f \times g)$, and $f \otimes g$ is the unique linear map making the following diagram commute.

$$
\begin{array}{ccc}
E \times F & \xrightarrow{\iota_{\otimes}} & E \otimes F \\
{\scriptstyle f \times g}\downarrow & & \downarrow{\scriptstyle f \otimes g} \\
E' \times F' & \xrightarrow[\iota'_{\otimes}]{} & E' \otimes F'
\end{array}
$$

Remark. The notation $f \otimes g$ is potentially ambiguous, because $\mathrm{Hom}(E, F)$ and $\mathrm{Hom}(E', F')$ are vector spaces, so we can form the tensor product $\mathrm{Hom}(E, F) \otimes \mathrm{Hom}(E', F')$ which contains elements also denoted $f \otimes g$. To avoid confusion, the first kind of tensor product of linear maps defined in Proposition 2.9 (which yields a linear map in $\mathrm{Hom}(E \otimes F, E' \otimes F')$) can be denoted by $T(f, g)$. If we denote the tensor product $E \otimes F$ by $T(E, F)$, this notation makes it clearer that T is a bifunctor. If E, E' and F, F' are finite dimensional, by picking bases it is not hard to show that the map induced by $f \otimes g \mapsto T(f, g)$ is an isomorphism

$$\mathrm{Hom}(E, F) \otimes \mathrm{Hom}(E', F') \cong \mathrm{Hom}(E \otimes F, E' \otimes F').$$

Proposition 2.10. *Suppose we have linear maps* $f: E \to E'$, $g: F \to F'$, $f': E' \to E''$, *and* $g': F' \to F''$. *Then the following identity holds:*

$$(f' \circ f) \otimes (g' \circ g) = (f' \otimes g') \circ (f \otimes g). \tag{$*$}$$

Proof. We have the commutative diagram

$$
\begin{array}{ccc}
E \times F & \xrightarrow{\iota_{\otimes}} & E \otimes F \\
{\scriptstyle f \times g}\downarrow & & \downarrow{\scriptstyle f \otimes g} \\
E' \times F' & \xrightarrow{\iota'_{\otimes}} & E' \otimes F' \\
{\scriptstyle f' \times g'}\downarrow & & \downarrow{\scriptstyle f' \otimes g'} \\
E'' \times F'' & \xrightarrow[\iota''_{\otimes}]{} & E'' \otimes F'',
\end{array}
$$

and thus the commutative diagram.

$$\begin{array}{ccc} E \times F & \xrightarrow{\iota_\otimes} & E \otimes F \\ {\scriptstyle (f' \times g') \circ (f \times g)} \downarrow & & \downarrow {\scriptstyle (f' \otimes g') \circ (f \otimes g)} \\ E'' \times F'' & \xrightarrow[\iota''_\otimes]{} & E'' \otimes F'' \end{array}$$

We also have the commutative diagram.

$$\begin{array}{ccc} E \times F & \xrightarrow{\iota_\otimes} & E \otimes F \\ {\scriptstyle (f' \circ f) \times (g' \circ g)} \downarrow & & \downarrow {\scriptstyle (f' \circ f) \otimes (g' \circ g)} \\ E'' \times F'' & \xrightarrow[\iota''_\otimes]{} & E'' \otimes F''. \end{array}$$

Since we immediately verify that

$$(f' \circ f) \times (g' \circ g) = (f' \times g') \circ (f \times g),$$

by uniqueness of the map between $E \otimes F$ and $E'' \otimes F''$ in the above diagram, we conclude that

$$(f' \circ f) \otimes (g' \circ g) = (f' \otimes g') \circ (f \otimes g),$$

as claimed.   □

The above formula (∗) yields the following useful fact.

Proposition 2.11. *If $f \colon E \to E'$ and $g \colon F \to F'$ are isomorphism, then $f \otimes g \colon E \otimes F \to E' \otimes F'$ is also an isomorphism.*

Proof. If $f^{-1} \colon E' \to E$ is the inverse of $f \colon E \to E'$ and $g^{-1} \colon F' \to F$ is the inverse of $g \colon F \to F'$, then $f^{-1} \otimes g^{-1} \colon E' \otimes F' \to E \otimes F$ is the inverse of $f \otimes g \colon E \otimes F \to E' \otimes F'$, which is shown as follows:

$$\begin{aligned} (f \otimes g) \circ (f^{-1} \otimes g^{-1}) &= (f \circ f^{-1}) \otimes (g \circ g^{-1}) \\ &= \mathrm{id}_{E'} \otimes \mathrm{id}_{F'} \\ &= \mathrm{id}_{E' \otimes F'}, \end{aligned}$$

and

$$\begin{aligned} (f^{-1} \otimes g^{-1}) \circ (f \otimes g) &= (f^{-1} \circ f) \otimes (g^{-1} \circ g) \\ &= \mathrm{id}_E \otimes \mathrm{id}_F \\ &= \mathrm{id}_{E \otimes F}. \end{aligned}$$

Therefore, $f \otimes g \colon E \otimes F \to E' \otimes F'$ is an isomorphism. $\qquad\qquad\square$

The generalization to the tensor product $f_1 \otimes \cdots \otimes f_n$ of $n \geq 3$ linear maps $f_i \colon E_i \to F_i$ is immediate, and left to the reader.

2.3 Bases of Tensor Products

We showed that $E_1 \otimes \cdots \otimes E_n$ is generated by the vectors of the form $u_1 \otimes \cdots \otimes u_n$. However, these vectors are not linearly independent. This situation can be fixed when considering bases.

To explain the idea of the proof, consider the case when we have two spaces E and F both of dimension 3. Given a basis (e_1, e_2, e_3) of E and a basis (f_1, f_2, f_3) of F, we would like to prove that

$$e_1 \otimes f_1, \ e_1 \otimes f_2, \ e_1 \otimes f_3, \ e_2 \otimes f_1, \ e_2 \otimes f_2, \ e_2 \otimes f_3, \ e_3 \otimes f_1, \ e_3 \otimes f_2, \ e_3 \otimes f_3$$

are linearly independent. To prove this, it suffices to show that for any vector space G, if $w_{11}, w_{12}, w_{13}, w_{21}, w_{22}, w_{23}, w_{31}, w_{32}, w_{33}$ are any vectors in G, then there is a bilinear map $h \colon E \times F \to G$ such that

$$h(e_i, e_j) = w_{ij}, \quad 1 \leq i, j \leq 3.$$

Because h yields a unique linear map $h_\otimes \colon E \otimes F \to G$ such that

$$h_\otimes(e_i \otimes e_j) = w_{ij}, \quad 1 \leq i, j \leq 3,$$

and by Proposition 2.4, the vectors

$$e_1 \otimes f_1, \ e_1 \otimes f_2, \ e_1 \otimes f_3, \ e_2 \otimes f_1, \ e_2 \otimes f_2, \ e_2 \otimes f_3, \ e_3 \otimes f_1, \ e_3 \otimes f_2, \ e_3 \otimes f_3$$

are linearly independent. This suggests understanding how a bilinear function $f \colon E \times F \to G$ is expressed in terms of its values $f(e_i, f_j)$ on the basis vectors (e_1, e_2, e_3) and (f_1, f_2, f_3), and this can be done easily. Using bilinearity we obtain

$$f(u_1 e_1 + u_2 e_2 + u_3 e_3, v_1 f_1 + v_2 f_2 + v_3 f_3)$$
$$= u_1 v_1 f(e_1, f_1) + u_1 v_2 f(e_1, f_2) + u_1 v_3 f(e_1, f_3)$$
$$+ u_2 v_1 f(e_2, f_1) + u_2 v_2 f(e_2, f_2) + u_2 v_3 f(e_2, f_3)$$
$$+ u_3 v_1 f(e_3, f_1) + u_3 v_2 f(e_3, f_2) + u_3 v_3 f(e_3, f_3).$$

Therefore, given $w_{11}, w_{12}, w_{13}, w_{21}, w_{22}, w_{23}, w_{31}, w_{32}, w_{33} \in G$, the function h given by

$$h(u_1e_1 + u_2e_2 + u_3e_3, v_1f_1 + v_2f_2 + v_3f_3) = u_1v_1w_{11} + u_1v_2w_{12} + u_1v_3w_{13}$$
$$+ u_2v_1w_{21} + u_2v_2w_{22} + u_2v_3w_{23}$$
$$+ u_3v_1w_{31} + u_3v_2w_{33} + u_3v_3w_{33}$$

is clearly bilinear, and by construction $h(e_i, f_j) = w_{ij}$, so it does the job.

The generalization of this argument to any number of vector spaces of any dimension (even infinite) is straightforward.

Proposition 2.12. *Given $n \geq 2$ vector spaces $E_1, \ldots, E_n$, if $(u_i^k)_{i \in I_k}$ is a basis for E_k, $1 \leq k \leq n$, then the family of vectors*

$$(u_{i_1}^1 \otimes \cdots \otimes u_{i_n}^n)_{(i_1,\ldots,i_n) \in I_1 \times \ldots \times I_n}$$

is a basis of the tensor product $E_1 \otimes \cdots \otimes E_n$.

Proof. For each k, $1 \leq k \leq n$, every $v^k \in E_k$ can be written uniquely as

$$v^k = \sum_{j \in I_k} v_j^k u_j^k,$$

for some family of scalars $(v_j^k)_{j \in I_k}$. Let F be any nontrivial vector space. We show that for every family

$$(w_{i_1,\ldots,i_n})_{(i_1,\ldots,i_n) \in I_1 \times \ldots \times I_n},$$

of vectors in F, there is some linear map $h \colon E_1 \otimes \cdots \otimes E_n \to F$ such that

$$h(u_{i_1}^1 \otimes \cdots \otimes u_{i_n}^n) = w_{i_1,\ldots,i_n}.$$

Then by Proposition 2.4, it follows that

$$(u_{i_1}^1 \otimes \cdots \otimes u_{i_n}^n)_{(i_1,\ldots,i_n) \in I_1 \times \ldots \times I_n}$$

is linearly independent. However, since $(u_i^k)_{i \in I_k}$ is a basis for E_k, the $u_{i_1}^1 \otimes \cdots \otimes u_{i_n}^n$ also generate $E_1 \otimes \cdots \otimes E_n$, and thus, they form a basis of $E_1 \otimes \cdots \otimes E_n$.

We define the function $f \colon E_1 \times \cdots \times E_n \to F$ as follows: For any n nonempty finite subsets $J_1, \ldots, J_n$ such that $J_k \subseteq I_k$ for $k = 1, \ldots, n$,

$$f\left(\sum_{j_1 \in J_1} v_{j_1}^1 u_{j_1}^1, \ldots, \sum_{j_n \in J_n} v_{j_n}^n u_{j_n}^n\right) = \sum_{j_1 \in J_1, \ldots, j_n \in J_n} v_{j_1}^1 \cdots v_{j_n}^n w_{j_1,\ldots,j_n}.$$

It is immediately verified that f is multilinear. By the universal mapping property of the tensor product, the linear map $f_\otimes \colon E_1 \otimes \cdots \otimes E_n \to F$ such that $f = f_\otimes \circ \varphi$ is the desired map h. $\square$

In particular, when each I_k is finite and of size $m_k = \dim(E_k)$, we see that the dimension of the tensor product $E_1 \otimes \cdots \otimes E_n$ is $m_1 \cdots m_n$. As a corollary of Proposition 2.12, if $(u_i^k)_{i \in I_k}$ is a basis for E_k, $1 \leq k \leq n$, then every tensor $z \in E_1 \otimes \cdots \otimes E_n$ can be written in a unique way as

$$z = \sum_{(i_1, \ldots, i_n) \, \in \, I_1 \times \ldots \times I_n} \lambda_{i_1, \ldots, i_n} \, u_{i_1}^1 \otimes \cdots \otimes u_{i_n}^n,$$

for some unique family of scalars $\lambda_{i_1, \ldots, i_n} \in K$, all zero except for a finite number.

2.4   Some Useful Isomorphisms for Tensor Products

Proposition 2.13. *Given three vector spaces E, F, G, there exist unique canonical isomorphisms*

(1) $E \otimes F \cong F \otimes E$
(2) $(E \otimes F) \otimes G \cong E \otimes (F \otimes G) \cong E \otimes F \otimes G$
(3) $(E \oplus F) \otimes G \cong (E \otimes G) \oplus (F \otimes G)$
(4) $K \otimes E \cong E$
      such that respectively

> *(a)* $u \otimes v \mapsto v \otimes u$
> *(b)* $(u \otimes v) \otimes w \mapsto u \otimes (v \otimes w) \mapsto u \otimes v \otimes w$
> *(c)* $(u, v) \otimes w \mapsto (u \otimes w, v \otimes w)$
> *(d)* $\lambda \otimes u \mapsto \lambda u.$

Proof. Except for (3), these isomorphisms are proved using the universal mapping property of tensor products.

(1) The map from $E \times F$ to $F \otimes E$ given by $(u, v) \mapsto v \otimes u$ is clearly bilinear, thus it induces a unique linear $\alpha : E \otimes F \to F \otimes E$ making the following diagram commute

$$E \times F \xrightarrow{\iota_\otimes} E \otimes F$$
$$\searrow \qquad \downarrow{\alpha}$$
$$F \otimes E,$$

such that

$$\alpha(u \otimes v) = v \otimes u, \quad \text{for all } u \in E \text{ and all } v \in F.$$

Similarly, the map from $F \times E$ to $E \otimes F$ given by $(v, u) \mapsto u \otimes v$ is clearly bilinear, thus it induces a unique linear $\beta \colon F \otimes E \;\to\; E \otimes F$ making the following diagram commute

$$
\begin{array}{ccc}
F \times E & \xrightarrow{\;\iota_\otimes\;} & F \otimes E \\
 & \searrow & \downarrow{\scriptstyle \beta} \\
 & & E \otimes F,
\end{array}
$$

such that

$$\beta(v \otimes u) = u \otimes v, \quad \text{for all } u \in E \text{ and all } v \in F.$$

It is immediately verified that

$$(\beta \circ \alpha)(u \otimes v) = u \otimes v \quad \text{and} \quad (\alpha \circ \beta)(v \otimes u) = v \otimes u$$

for all $u \in E$ and all $v \in F$. Since the tensors of the form $u \otimes v$ span $E \otimes F$ and similarly the tensors of the form $v \otimes u$ span $F \otimes E$, the map $\beta \circ \alpha$ is actually the identity on $E \otimes F$, and similarly $\alpha \circ \beta$ is the identity on $F \otimes E$, so α and β are isomorphisms.

(2) Fix some $w \in G$. The map

$$(u, v) \mapsto u \otimes v \otimes w$$

from $E \times F$ to $E \otimes F \otimes G$ is bilinear, and thus there is a linear map $f_w \colon E \otimes F \to E \otimes F \otimes G$ making the following diagram commute

$$
\begin{array}{ccc}
E \times F & \xrightarrow{\;\iota_\otimes\;} & E \otimes F \\
 & \searrow & \downarrow{\scriptstyle f_w} \\
 & & E \otimes F \otimes G,
\end{array}
$$

with $f_w(u \otimes v) = u \otimes v \otimes w$.

Next consider the map

$$(z, w) \mapsto f_w(z),$$

from $(E \otimes F) \times G$ into $E \otimes F \otimes G$. It is easily seen to be bilinear, and thus it induces a linear map $f \colon (E \otimes F) \otimes G \to E \otimes F \otimes G$ making the following diagram commute

$$(E \otimes F) \times G \xrightarrow{\iota_\otimes} (E \otimes F) \otimes G$$
$$\downarrow f$$
$$E \otimes F \otimes G,$$

with $f((u \otimes v) \otimes w) = u \otimes v \otimes w$.

Also consider the map

$$(u, v, w) \mapsto (u \otimes v) \otimes w$$

from $E \times F \times G$ to $(E \otimes F) \otimes G$. It is trilinear, and thus there is a linear map $g \colon E \otimes F \otimes G \to (E \otimes F) \otimes G$ making the following diagram commute

$$E \times F \times G \xrightarrow{\iota_\otimes} E \otimes F \otimes G$$
$$\downarrow g$$
$$(E \otimes F) \otimes G,$$

with $g(u \otimes v \otimes w) = (u \otimes v) \otimes w$. Clearly, $f \circ g$ and $g \circ f$ are identity maps, and thus f and g are isomorphisms. The other case is similar.

(3) Given a fixed vector space G, for any two vector spaces M and N and every linear map $f \colon M \to N$, let $\tau_G(f) = f \otimes \mathrm{id}_G$ be the unique linear map making the following diagram commute.

$$
\begin{array}{ccc}
M \times G & \xrightarrow{\iota_{M\otimes}} & M \otimes G \\
{\scriptstyle f \times \mathrm{id}_G} \downarrow & & \downarrow {\scriptstyle f \otimes \mathrm{id}_G} \\
N \times G & \xrightarrow{\iota_{N\otimes}} & N \otimes G
\end{array}
$$

The identity $(*)$ proved in Proposition 2.10 shows that if $g \colon N \to P$ is another linear map, then

$$\tau_G(g) \circ \tau_G(f) = (g \otimes \mathrm{id}_G) \circ (f \otimes \mathrm{id}_G)$$
$$= (g \circ f) \otimes (\mathrm{id}_G \circ \mathrm{id}_G)$$
$$= (g \circ f) \otimes \mathrm{id}_G = \tau_G(g \circ f).$$

Clearly, $\tau_G(0) = 0$, and a direct computation on generators also shows that

$$\tau_G(\mathrm{id}_M) = (\mathrm{id}_M \otimes \mathrm{id}_G) = \mathrm{id}_{M\otimes G},$$

and that if $f' \colon M \to N$ is another linear map, then

$$\tau_G(f + f') = \tau_G(f) + \tau_G(f').$$

In fancy terms, τ_G is a functor. Now, if $E \oplus F$ is a direct sum, it is a standard fact of linear algebra that if $\pi_E : E \oplus F \to E$ and $\pi_F : E \oplus F \to F$ are the projection maps, then

$$\pi_E \circ \pi_E = \pi_E \qquad\qquad \pi_F \circ \pi_F = \pi_F \qquad\qquad \pi_E \circ \pi_F = 0$$

$$\pi_F \circ \pi_E = 0 \qquad\qquad \pi_E + \pi_F = \mathrm{id}_{E \oplus F}.$$

If we apply τ_G to these identities, we get

$$\tau_G(\pi_E) \circ \tau_G(\pi_E) = \tau_G(\pi_E) \qquad\qquad \tau_G(\pi_F) \circ \tau_G(\pi_F) = \tau_G(\pi_F)$$

$$\tau_G(\pi_E) \circ \tau_G(\pi_F) = 0 \qquad\qquad \tau_G(\pi_F) \circ \tau_G(\pi_E) = 0$$

$$\tau_G(\pi_E) + \tau_G(\pi_F) = \mathrm{id}_{(E \oplus F) \otimes G}.$$

Observe that $\tau_G(\pi_E) = \pi_E \otimes \mathrm{id}_G$ is a map from $(E \oplus F) \otimes G$ onto $E \otimes G$ and that $\tau_G(\pi_F) = \pi_F \otimes \mathrm{id}_G$ is a map from $(E \oplus F) \otimes G$ onto $F \otimes G$, and by linear algebra, the above equations mean that we have a direct sum

$$(E \otimes G) \oplus (F \otimes G) \cong (E \oplus F) \otimes G.$$

(4) We have the linear map $\epsilon : E \to K \otimes E$ given by

$$\epsilon(u) = 1 \otimes u, \quad \text{for all } u \in E.$$

The map $(\lambda, u) \mapsto \lambda u$ from $K \times E$ to E is bilinear, so it induces a unique linear map $\eta : K \otimes E \to E$ making the following diagram commute

such that $\eta(\lambda \otimes u) = \lambda u$, for all $\lambda \in K$ and all $u \in E$. We have

$$(\eta \circ \epsilon)(u) = \eta(1 \otimes u) = 1u = u,$$

and

$$(\epsilon \circ \eta)(\lambda \otimes u) = \epsilon(\lambda u) = 1 \otimes (\lambda u) = \lambda(1 \otimes u) = \lambda \otimes u,$$

which shows that both $\epsilon \circ \eta$ and $\eta \circ \epsilon$ are the identity, so ϵ and η are isomorphisms. $\square$

Remark. The isomorphism (3) can be generalized to finite and even arbitrary direct sums $\bigoplus_{i \in I} E_i$ of vector spaces (where I is an arbitrary nonempty index set). We have an isomorphism

$$\left(\bigoplus_{i \in I} E_i \right) \otimes G \cong \bigoplus_{i \in I} (E_i \otimes G).$$

This isomorphism (with isomorphism (1)) can be used to give another proof of Proposition 2.12 (see Bertin [11], Chapter 4, Section 1) or Lang [67], Chapter XVI, Section 2).

Proposition 2.14. *Given any three vector spaces E, F, G, we have the canonical isomorphism*

$$\mathrm{Hom}(E, F; G) \cong \mathrm{Hom}(E, \mathrm{Hom}(F, G)).$$

Proof. Any bilinear map $f \colon E \times F \to G$ gives the linear map $\varphi(f) \in \mathrm{Hom}(E, \mathrm{Hom}(F, G))$, where $\varphi(f)(u)$ is the linear map in $\mathrm{Hom}(F, G)$ given by

$$\varphi(f)(u)(v) = f(u, v).$$

Conversely, given a linear map $g \in \mathrm{Hom}(E, \mathrm{Hom}(F, G))$, we get the bilinear map $\psi(g)$ given by

$$\psi(g)(u, v) = g(u)(v),$$

and it is clear that φ and ψ and mutual inverses.   □

Since by Proposition 2.7 there is a canonical isomorphism

$$\mathrm{Hom}(E \otimes F, G) \cong \mathrm{Hom}(E, F; G),$$

together with the isomorphism

$$\mathrm{Hom}(E, F; G) \cong \mathrm{Hom}(E, \mathrm{Hom}(F, G))$$

given by Proposition 2.14, we obtain the important corollary:

Proposition 2.15. *For any three vector spaces E, F, G, we have the canonical isomorphism*

$$\mathrm{Hom}(E \otimes F, G) \cong \mathrm{Hom}(E, \mathrm{Hom}(F, G)).$$

2.5   Duality for Tensor Products

In this section all vector spaces are assumed to have *finite dimension*, unless specified otherwise. Let us now see how tensor products behave under duality. For this, we define a pairing between $E_1^* \otimes \cdots \otimes E_n^*$ and $E_1 \otimes \cdots \otimes E_n$ as follows: For any fixed $(v_1^*, \ldots, v_n^*) \in E_1^* \times \cdots \times E_n^*$, we have the multilinear map

$$l_{v_1^*, \ldots, v_n^*} \colon (u_1, \ldots, u_n) \mapsto v_1^*(u_1) \cdots v_n^*(u_n)$$

from $E_1 \times \cdots \times E_n$ to K. The map $l_{v_1^*, \ldots, v_n^*}$ extends uniquely to a linear map $L_{v_1^*, \ldots, v_n^*} \colon E_1 \otimes \cdots \otimes E_n \longrightarrow K$ making the following diagram commute.

$$
\begin{array}{ccc}
E_1 \times \cdots \times E_n & \xrightarrow{\iota_\otimes} & E_1 \otimes \cdots \otimes E_n \\
& \searrow{\scriptstyle l_{v_1^*, \ldots, v_n^*}} & \downarrow{\scriptstyle L_{v_1^*, \ldots, v_n^*}} \\
& & K
\end{array}
$$

We also have the multilinear map

$$(v_1^*, \ldots, v_n^*) \mapsto L_{v_1^*, \ldots, v_n^*}$$

from $E_1^* \times \cdots \times E_n^*$ to $\mathrm{Hom}(E_1 \otimes \cdots \otimes E_n, K)$, which extends to a unique linear map L from $E_1^* \otimes \cdots \otimes E_n^*$ to $\mathrm{Hom}(E_1 \otimes \cdots \otimes E_n, K)$ making the following diagram commute.

$$
\begin{array}{ccc}
E_1^* \times \cdots \times E_n^* & \xrightarrow{\iota_\otimes} & E_1^* \otimes \cdots \otimes E_n^* \\
& \searrow{\scriptstyle L_{v_1^*, \ldots, v_n^*}} & \downarrow{\scriptstyle L} \\
& & \mathrm{Hom}(E_1 \otimes \cdots \otimes E_n; K)
\end{array}
$$

However, in view of the isomorphism

$$\mathrm{Hom}(U \otimes V, W) \cong \mathrm{Hom}(U, \mathrm{Hom}(V, W))$$

given by Proposition 2.15, with $U = E_1^* \otimes \cdots \otimes E_n^*$, $V = E_1 \otimes \cdots \otimes E_n$ and $W = K$, we can view L as a linear map

$$L \colon (E_1^* \otimes \cdots \otimes E_n^*) \otimes (E_1 \otimes \cdots \otimes E_n) \to K,$$

which corresponds to a bilinear map

$$\langle -, - \rangle \colon (E_1^* \otimes \cdots \otimes E_n^*) \times (E_1 \otimes \cdots \otimes E_n) \longrightarrow K, \qquad (\dagger\dagger)$$

via the isomorphism $(U \otimes V)^* \cong \mathrm{Hom}(U, V; K)$ given by Proposition 2.8. This pairing is given explicitly on generators by

$$\langle v_1^* \otimes \cdots \otimes v_n^*, u_1 \ldots, u_n \rangle = v_1^*(u_1) \cdots v_n^*(u_n).$$

This pairing is nondegenerate, as proved below.

Proof. If $(e_1^1, \ldots, e_{m_1}^1), \ldots, (e_1^n, \ldots, e_{m_n}^n)$ are bases for $E_1, \ldots, E_n$, then for every basis element $(e_{i_1}^1)^* \otimes \cdots \otimes (e_{i_n}^n)^*$ of $E_1^* \otimes \cdots \otimes E_n^*$, and any basis element $e_{j_1}^1 \otimes \cdots \otimes e_{j_n}^n$ of $E_1 \otimes \cdots \otimes E_n$, we have

$$\langle (e_{i_1}^1)^* \otimes \cdots \otimes (e_{i_n}^n)^*, e_{j_1}^1 \otimes \cdots \otimes e_{j_n}^n \rangle = \delta_{i_1\, j_1} \cdots \delta_{i_n\, j_n},$$

where $\delta_{i\, j}$ is *Kronecker delta*, defined such that $\delta_{i\, j} = 1$ if $i = j$, and 0 otherwise. Given any $\alpha \in E_1^* \otimes \cdots \otimes E_n^*$, assume that $\langle \alpha, \beta \rangle = 0$ for all $\beta \in E_1 \otimes \cdots \otimes E_n$. The vector α is a finite linear combination $\alpha = \sum \lambda_{i_1,\ldots,i_n} (e_{i_1}^1)^* \otimes \cdots \otimes (e_{i_n}^n)^*$, for some unique $\lambda_{i_1,\ldots,i_n} \in K$. If we choose $\beta = e_{i_1}^1 \otimes \cdots \otimes e_{i_n}^n$, then we get

$$0 = \langle \alpha, e_{i_1}^1 \otimes \cdots \otimes e_{i_n}^n \rangle = \left\langle \sum \lambda_{i_1,\ldots,i_n} (e_{i_1}^1)^* \otimes \cdots \otimes (e_{i_n}^n)^*, e_{i_1}^1 \otimes \cdots \otimes e_{i_n}^n \right\rangle$$

$$= \sum \lambda_{i_1,\ldots,i_n} \langle (e_{i_1}^1)^* \otimes \cdots \otimes (e_{i_n}^n)^*, e_{i_1}^1 \otimes \cdots \otimes e_{i_n}^n \rangle$$

$$= \lambda_{i_1,\ldots,i_n}.$$

Therefore, $\alpha = 0$,

Conversely, given any $\beta \in E_1 \otimes \cdots \otimes E_n$, assume that $\langle \alpha, \beta \rangle = 0$, for all $\alpha \in E_1^* \otimes \cdots \otimes E_n^*$. The vector β is a finite linear combination $\beta = \sum \lambda_{i_1,\ldots,i_n} e_{i_1}^1 \otimes \cdots \otimes e_{i_n}^n$, for some unique $\lambda_{i_1,\ldots,i_n} \in K$. If we choose $\alpha = (e_{i_1}^1)^* \otimes \cdots \otimes (e_{i_n}^n)^*$, then we get

$$0 = \langle (e_{i_1}^1)^* \otimes \cdots \otimes (e_{i_n}^n)^*, \beta \rangle = \left\langle (e_{i_1}^1)^* \otimes \cdots \otimes (e_{i_n}^n)^*, \sum \lambda_{i_1,\ldots,i_n} e_{i_1}^1 \otimes \cdots \otimes e_{i_n}^n \right\rangle$$

$$= \sum \lambda_{i_1,\ldots,i_n} \langle (e_{i_1}^1)^* \otimes \cdots \otimes (e_{i_n}^n)^*, e_{i_1}^1 \otimes \cdots \otimes e_{i_n}^n \rangle$$

$$= \lambda_{i_1,\ldots,i_n}.$$

Therefore, $\beta = 0$.   □

By Proposition 2.1,[1] we have a canonical isomorphism

$$(E_1 \otimes \cdots \otimes E_n)^* \cong E_1^* \otimes \cdots \otimes E_n^*.$$

Here is our main proposition about duality of tensor products.

Proposition 2.16. *We have canonical isomorphisms*

$$(E_1 \otimes \cdots \otimes E_n)^* \cong E_1^* \otimes \cdots \otimes E_n^*,$$

---

[1]This is where the assumption that our spaces are finite dimensional is used.

and

$$\mu \colon E_1^* \otimes \cdots \otimes E_n^* \cong \mathrm{Hom}(E_1, \ldots, E_n; K).$$

Proof. The second isomorphism follows from the isomorphism $(E_1 \otimes \cdots \otimes E_n)^* \cong E_1^* \otimes \cdots \otimes E_n^*$ together with the isomorphism $\mathrm{Hom}(E_1, \ldots, E_n; K) \cong (E_1 \otimes \cdots \otimes E_n)^*$ given by Proposition 2.8. $\square$

Remarks.

1. The isomorphism $\mu \colon E_1^* \otimes \cdots \otimes E_n^* \cong \mathrm{Hom}(E_1, \ldots, E_n; K)$ can be described explicitly as the linear extension to $E_1^* \otimes \cdots \otimes E_n^*$ of the map given by

$$\mu(v_1^* \otimes \cdots \otimes v_n^*)(u_1 \ldots, u_n) = v_1^*(u_1) \cdots v_n^*(u_n).$$

2. The canonical isomorphism of Proposition 2.16 holds under more general conditions. Namely, that K is a commutative ring with identity and that the E_i are finitely generated projective K-modules (see Definition 2.23). See Bourbaki [14] (Chapter III, §11, Section 5, Proposition 7).

We prove another useful canonical isomorphism that allows us to treat linear maps as tensors.

Let E and F be two vector spaces and let $\alpha \colon E^* \times F \to \mathrm{Hom}(E, F)$ be the map defined such that

$$\alpha(u^*, f)(x) = u^*(x) f,$$

for all $u^* \in E^*$, $f \in F$, and $x \in E$. This map is clearly bilinear, and thus it induces a linear map $\alpha_\otimes \colon E^* \otimes F \to \mathrm{Hom}(E, F)$ making the following diagram commute

$$
\begin{array}{ccc}
E^* \times F & \xrightarrow{\ \iota_\otimes\ } & E^* \otimes F \\
 & {\scriptstyle \alpha} \searrow & \downarrow {\scriptstyle \alpha_\otimes} \\
 & & \mathrm{Hom}(E, F),
\end{array}
$$

such that

$$\alpha_\otimes(u^* \otimes f)(x) = u^*(x) f.$$

Proposition 2.17. *If E and F are vector spaces (not necessarily finite dimensional), then the following properties hold:*

(1) The linear map $\alpha_\otimes \colon E^* \otimes F \to \mathrm{Hom}(E, F)$ is injective.

(2) *If E is finite dimensional, then $\alpha_\otimes \colon E^* \otimes F \to \operatorname{Hom}(E, F)$ is a canonical isomorphism.*

(3) *If F is finite dimensional, then $\alpha_\otimes \colon E^* \otimes F \to \operatorname{Hom}(E, F)$ is a canonical isomorphism.*

Proof.

(1) Let $(e_i^*)_{i \in I}$ be a basis of E^* and let $(f_j)_{j \in J}$ be a basis of F. Then we know that $(e_i^* \otimes f_j)_{i \in I, j \in J}$ is a basis of $E^* \otimes F$. To prove that $\alpha_\otimes$ is injective, let us show that its kernel is reduced to (0). For any vector

$$\omega = \sum_{i \in I', j \in J'} \lambda_{ij}\, e_i^* \otimes f_j$$

in $E^* \otimes F$, with I' and J' some finite sets, assume that $\alpha_\otimes(\omega) = 0$. This means that for every $x \in E$, we have $\alpha_\otimes(\omega)(x) = 0$; that is,

$$\sum_{i \in I', j \in J'} \alpha_\otimes(\lambda_{ij}\, e_i^* \otimes f_j)(x) = \sum_{j \in J'} \left(\sum_{i \in I'} \lambda_{ij} e_i^*(x) \right) f_j = 0.$$

Since $(f_j)_{j \in J}$ is a basis of F, for every $j \in J'$, we must have

$$\sum_{i \in I'} \lambda_{ij} e_i^*(x) = 0, \quad \text{for all } x \in E.$$

But then $(e_i^*)_{i \in I'}$ would be linearly dependent, contradicting the fact that $(e_i^*)_{i \in I}$ is a basis of E^*, so we must have

$$\lambda_{ij} = 0, \quad \text{for all } i \in I' \text{ and all } j \in J',$$

which shows that $\omega = 0$. Therefore, $\alpha_\otimes$ is injective.

(2) Let $(e_j)_{1 \le j \le n}$ be a finite basis of E, and as usual, let $e_j^* \in E^*$ be the linear form defined by

$$e_j^*(e_k) = \delta_{j,k},$$

where $\delta_{j,k} = 1$ iff $j = k$ and 0 otherwise. We know that $(e_j^*)_{1 \le j \le n}$ is a basis of E^* (this is where we use the finite dimension of E). For any linear map $f \in \operatorname{Hom}(E, F)$, for every $x = x_1 e_1 + \cdots + x_n e_n \in E$, we have

$$f(x) = f(x_1 e_1 + \cdots + x_n e_n) = x_1 f(e_1) + \cdots + x_n f(e_n)$$
$$= e_1^*(x) f(e_1) + \cdots + e_n^*(x) f(e_n).$$

Consequently, every linear map $f \in \operatorname{Hom}(E, F)$ can be expressed as

$$f(x) = e_1^*(x)f_1 + \cdots + e_n^*(x)f_n,$$

for some $f_i \in F$. Furthermore, if we apply f to e_i, we get $f(e_i) = f_i$, so the f_i are unique. Observe that

$$(\alpha_{\otimes}(e_1^* \otimes f_1 + \cdots + e_n^* \otimes f_n))(x) = \sum_{i=1}^{n}(\alpha_{\otimes}(e_i^* \otimes f_i))(x) = \sum_{i=1}^{n}e_i^*(x)f_i.$$

Thus, $\alpha_{\otimes}$ is surjective, so $\alpha_{\otimes}$ is a bijection.

(3) Let $(f_1, \ldots, f_m)$ be a finite basis of F, and let $(f_1^*, \ldots, f_m^*)$ be its dual basis. Given any linear map $h \colon E \to F$, for all $u \in E$, since $f_i^*(f_j) = \delta_{ij}$, we have

$$h(u) = \sum_{i=1}^{m} f_i^*(h(u))f_i.$$

If

$$h(u) = \sum_{j=1}^{m} v_j^*(u)f_j \quad \text{for all } u \in E \tag{$*$}$$

for some linear forms $(v_1^*, \ldots, v_m^*) \in (E^*)^m$, then

$$f_i^*(h(u)) = \sum_{j=1}^{m} v_j^*(u)f_i^*(f_j) = v_i^*(u) \quad \text{for all } u \in E,$$

which shows that $v_i^* = f_i^* \circ h$ for $i = 1, \ldots, m$. This means that h has a unique expression in terms of linear forms as in $(*)$. Define the map α from $(E^*)^m$ to $\mathrm{Hom}(E, F)$ by

$$\alpha(v_1^*, \ldots, v_m^*)(u) = \sum_{j=1}^{m} v_j^*(u)f_j \quad \text{for all } u \in E.$$

This map is linear. For any $h \in \mathrm{Hom}(E, F)$, we showed earlier that the expression of h in $(*)$ is unique, thus α is an isomorphism. Similarly, $E^* \otimes F$ is isomorphic to $(E^*)^m$. Any tensor $\omega \in E^* \otimes F$ can be written as a linear combination

$$\sum_{k=1}^{p} u_k^* \otimes y_k$$

for some $u_k^* \in E^*$ and some $y_k \in F$, and since $(f_1, \ldots, f_m)$ is a basis of F, each y_k can be written as a linear combination of $(f_1, \ldots, f_m)$, so ω can be expressed as

$$\omega = \sum_{i=1}^{m} v_i^* \otimes f_i, \tag{$\dagger$}$$

for some linear forms $v_i^* \in E^*$ which are linear combinations of the u_k^*. If we pick a basis $(w_i^*)_{i \in I}$ for E^*, then we know that the family $(w_i^* \otimes f_j)_{i \in I, 1 \leq j \leq m}$ is a basis of $E^* \otimes F$, and this implies that the v_i^* in ($\dagger$) are unique. Define the linear map β from $(E^*)^m$ to $E^* \otimes F$ by

$$\beta(v_1^*, \ldots, v_m^*) = \sum_{i=1}^{m} v_i^* \otimes f_i.$$

Since every tensor $\omega \in E^* \otimes F$ can be written in a unique way as in ($\dagger$), this map is an isomorphism. $\qquad \square$

Note that in Proposition 2.17, we have an isomorphism if either E or F has finite dimension. The following proposition allows us to view a multilinear as a tensor product.

Proposition 2.18. *If the $E_1, \ldots E_n$ are finite-dimensional vector spaces and F is any vector space, then we have the canonical isomorphism*

$$\mathrm{Hom}(E_1, \ldots, E_n; F) \cong E_1^* \otimes \cdots \otimes E_n^* \otimes F.$$

Proof. In view of the canonical isomorphism

$$\mathrm{Hom}(E_1, \ldots, E_n; F) \cong \mathrm{Hom}(E_1 \otimes \cdots \otimes E_n, F)$$

given by Proposition 2.7 and the canonical isomorphism $(E_1 \otimes \cdots \otimes E_n)^* \cong E_1^* \otimes \cdots \otimes E_n^*$ given by Proposition 2.16, if the E_i's are finite dimensional, then Proposition 2.17 yields the canonical isomorphism

$$\mathrm{Hom}(E_1, \ldots, E_n; F) \cong E_1^* \otimes \cdots \otimes E_n^* \otimes F,$$

as claimed. $\qquad \square$

2.6 Tensor Algebras

Our goal is to define a vector space $T(V)$ obtained by taking the direct sum of the tensor products

$$\underbrace{V \otimes \cdots \otimes V}_{m},$$

and to define a multiplication operation on $T(V)$ which makes $T(V)$ into an algebraic structure called an algebra. The algebra $T(V)$ satisfies a universal property stated in Proposition 2.19, which makes it the "free algebra" generated by the vector space V.

Definition 2.8. The tensor product

$$\underbrace{V \otimes \cdots \otimes V}_{m}$$

is also denoted as

$$\overset{m}{\bigotimes} V \quad \text{or} \quad V^{\otimes m}$$

and is called the *m-th tensor power of* V (with $V^{\otimes 1} = V$, and $V^{\otimes 0} = K$).

We can pack all the tensor powers of V into the "big" vector space

$$T(V) = \bigoplus_{m \geq 0} V^{\otimes m},$$

denoted $T^{\bullet}(V)$ or $\bigotimes V$ to avoid confusion with the tangent bundle.

This is an interesting object because we can define a multiplication operation on it which makes it into an *algebra*.

When V is of finite dimension n, we can pick some basis $(e_1 \ldots, e_n)$ of V, and then every tensor $\omega \in T(V)$ can be expressed as a linear combination of terms of the form $e_{i_1} \otimes \cdots \otimes e_{i_k}$, where $(i_1, \ldots, i_k)$ is any sequence of elements from the set $\{1, \ldots, n\}$. We can think of the tensors $e_{i_1} \otimes \cdots \otimes e_{i_k}$ as monomials in the noncommuting variables $e_1, \ldots, e_n$. Thus the space $T(V)$ corresponds to the algebra of polynomials with coefficients in K in n *noncommuting variables*.

Let us review the definition of an algebra over a field. Let K denote any (commutative) field, although for our purposes, we may assume that $K = \mathbb{R}$ (and occasionally, $K = \mathbb{C}$). Since we will only be dealing with associative algebras with a multiplicative unit, we only define algebras of this kind.

Definition 2.9. Given a field K, a *K-algebra* is a K-vector space A together with a bilinear operation $\cdot : A \times A \to A$, called *multiplication*, which makes A into a ring with unity 1 (or 1_A, when we want to be very precise). This means that $\cdot$ is associative and that there is a multiplicative identity element 1 so that $1 \cdot a = a \cdot 1 = a$, for all $a \in A$. Given two K-algebras A and B, a *K-algebra homomorphism* $h: A \to B$ is a linear map that is also a ring homomorphism, with $h(1_A) = 1_B$; that is,

$$h(a_1 \cdot a_2) = h(a_1) \cdot h(a_2) \quad \text{for all } a_1, a_2 \in A$$

$$h(1_A) = 1_B.$$

The set of K-algebra homomorphisms between A and B is denoted $\mathrm{Hom}_{\mathrm{alg}}(A, B)$.

For example, the ring $\mathrm{M}_n(K)$ of all $n \times n$ matrices over a field K is a K-algebra.

There is an obvious notion of ideal of a K-algebra.

Definition 2.10. Let A be a K-algebra. An *ideal* $\mathfrak{A} \subseteq A$ is a linear subspace of A that is also a two-sided ideal with respect to multiplication in A; this means that for all $a \in \mathfrak{A}$ and all $\alpha, \beta \in A$, we have $\alpha a \beta \in \mathfrak{A}$.

If the field K is understood, we usually simply say an algebra instead of a K-algebra.

We would like to define a multiplication operation on $T(V)$ which makes it into a K-algebra. As

$$T(V) = \bigoplus_{i \geq 0} V^{\otimes i},$$

for every $i \geq 0$, there is a natural injection $\iota_n \colon V^{\otimes n} \to T(V)$, and in particular, an injection $\iota_0 \colon K \to T(V)$. The multiplicative unit $\mathbf{1}$ of $T(V)$ is the image $\iota_0(1)$ in $T(V)$ of the unit 1 of the field K. Since every $v \in T(V)$ can be expressed as a finite sum

$$v = \iota_{n_1}(v_1) + \cdots + \iota_{n_k}(v_k),$$

where $v_i \in V^{\otimes n_i}$ and the n_i are natural numbers with $n_i \neq n_j$ if $i \neq j$, to define multiplication in $T(V)$, using bilinearity, it is enough to define multiplication operations $\cdot \colon V^{\otimes m} \times V^{\otimes n} \longrightarrow V^{\otimes(m+n)}$, which, using the isomorphisms $V^{\otimes n} \cong \iota_n(V^{\otimes n})$, yield multiplication operations $\cdot \colon \iota_m(V^{\otimes m}) \times \iota_n(V^{\otimes n}) \longrightarrow \iota_{m+n}(V^{\otimes(m+n)})$. First, for $\omega_1 \in V^{\otimes m}$ and $\omega_2 \in V^{\otimes n}$, we let

$$\omega_1 \cdot \omega_2 = \omega_1 \otimes \omega_2.$$

This defines a bilinear map so it defines a multiplication $V^{\otimes m} \times V^{\otimes n} \longrightarrow V^{\otimes m} \otimes V^{\otimes n}$. This is not quite what we want, but there is a canonical isomorphism

$$V^{\otimes m} \otimes V^{\otimes n} \cong V^{\otimes(m+n)}$$

which yields the desired multiplication $\cdot \colon V^{\otimes m} \times V^{\otimes n} \longrightarrow V^{\otimes(m+n)}$.

The isomorphism $V^{\otimes m} \otimes V^{\otimes n} \cong V^{\otimes(m+n)}$ can be established by induction using the isomorphism $(E \otimes F) \otimes G \cong E \otimes F \otimes G$. First we prove by induction on $m \geq 2$ that

$$V^{\otimes(m-1)} \otimes V \cong V^{\otimes m},$$

and then by induction on $n \geq 1$ than

$$V^{\otimes m} \otimes V^{\otimes n} \cong V^{\otimes(m+n)}.$$

In summary the multiplication $V^{\otimes m} \times V^{\otimes n} \longrightarrow V^{\otimes(m+n)}$ is defined so that

$$(v_1 \otimes \cdots \otimes v_m) \cdot (w_1 \otimes \cdots \otimes w_n) = v_1 \otimes \cdots \otimes v_m \otimes w_1 \otimes \cdots \otimes w_n.$$

(This has to be made rigorous by using isomorphisms involving the associativity of tensor products, for details, see Jacobson [61], Section 3.9, or Bertin [11], Chapter 4, Section 2.)

Definition 2.11.   Given a K-vector space V (not necessarily finite dimensional), the vector space

$$T(V) = \bigoplus_{m \geq 0} V^{\otimes m}$$

denoted $T^\bullet(V)$ or $\bigotimes V$ equipped with the multiplication operations $V^{\otimes m} \times V^{\otimes n} \longrightarrow V^{\otimes(m+n)}$ defined above is called the *tensor algebra of* V

Remark.   It is important to note that multiplication in $T(V)$ is **not** commutative. Also, in all rigor, the unit **1** of $T(V)$ is **not equal** to 1, the unit of the field K. However, in view of the injection $\iota_0 \colon K \to T(V)$, for the sake of notational simplicity, we will denote **1** by 1. More generally, in view of the injections $\iota_n \colon V^{\otimes n} \to T(V)$, we identify elements of $V^{\otimes n}$ with their images in $T(V)$.

The algebra $T(V)$ satisfies a universal mapping property which shows that it is unique up to isomorphism. For simplicity of notation, let $i \colon V \to T(V)$ be the natural injection of V into $T(V)$.

Proposition 2.19.   *Given any K-algebra A, for any linear map $f \colon V \to A$, there is a unique K-algebra homomorphism $\overline{f} \colon T(V) \to A$ so that*

$$f = \overline{f} \circ i,$$

as in the diagram below.

Proof. Left an exercise (use Theorem 2.6). A proof can be found in Knapp [66] (Appendix A, Proposition A.14) or Bertin [11] (Chapter 4, Theorem 2.4).   □

Proposition 2.19 implies that there is a natural isomorphism

$$\mathrm{Hom_{alg}}(T(V), A) \cong \mathrm{Hom}(V, A),$$

where the algebra A on the right-hand side is viewed as a vector space. Proposition 2.19 also has the following corollary.

Proposition 2.20. *Given a linear map $h: V_1 \rightarrow V_2$ between two vectors spaces V_1, V_2 over a field K, there is a unique K-algebra homomorphism $\otimes h: T(V_1) \rightarrow T(V_2)$ making the following diagram commute.*

$$
\begin{array}{ccc}
V_1 & \xrightarrow{\ i_1\ } & T(V_1) \\
{\scriptstyle h}\downarrow & & \downarrow{\scriptstyle \otimes h} \\
V_2 & \xrightarrow{\ i_2\ } & T(V_2).
\end{array}
$$

Most algebras of interest arise as well-chosen quotients of the tensor algebra $T(V)$. This is true for the *exterior algebra* $\bigwedge(V)$ (also called *Grassmann algebra*), where we take the quotient of $T(V)$ modulo the ideal generated by all elements of the form $v \otimes v$, where $v \in V$, and for the *symmetric algebra* Sym(V), where we take the quotient of $T(V)$ modulo the ideal generated by all elements of the form $v \otimes w - w \otimes v$, where $v, w \in V$.

Algebras such as $T(V)$ are graded in the sense that there is a sequence of subspaces $V^{\otimes n} \subseteq T(V)$ such that

$$T(V) = \bigoplus_{k \geq 0} V^{\otimes n},$$

and the multiplication $\otimes$ behaves well w.r.t. the grading, *i.e.*, $\otimes: V^{\otimes m} \times V^{\otimes n} \rightarrow V^{\otimes(m+n)}$.

Definition 2.12. A K-algebra E is said to be a *graded algebra* iff there is a sequence of subspaces $E^n \subseteq E$ such that

$$E = \bigoplus_{k \geq 0} E^n,$$

(with $E^0 = K$) and the multiplication $\cdot$ respects the grading; that is, $\cdot: E^m \times E^n \rightarrow E^{m+n}$. Elements in E^n are called *homogeneous elements of rank (or degree) n*.

In differential geometry and in physics it is necessary to consider slightly more general tensors.

Definition 2.13. Given a vector space V, for any pair of nonnegative integers (r, s), the *tensor space* $T^{r,s}(V)$ *of type* (r, s) is the tensor product

$$T^{r,s}(V) = V^{\otimes r} \otimes (V^*)^{\otimes s} = \underbrace{V \otimes \cdots \otimes V}_{r} \otimes \underbrace{V^* \otimes \cdots \otimes V^*}_{s},$$

with $T^{0,0}(V) = K$. We also define the *tensor algebra* $T^{\bullet,\bullet}(V)$ as the direct sum (coproduct)

$$T^{\bullet,\bullet}(V) = \bigoplus_{r,s \geq 0} T^{r,s}(V).$$

Tensors in $T^{r,s}(V)$ are called *homogeneous of degree* (r, s).

Note that tensors in $T^{r,0}(V)$ are just our "old tensors" in $V^{\otimes r}$. We make $T^{\bullet,\bullet}(V)$ into an algebra by defining multiplication operations

$$T^{r_1,s_1}(V) \times T^{r_2,s_2}(V) \longrightarrow T^{r_1+r_2,s_1+s_2}(V)$$

in the usual way, namely: For $u = u_1 \otimes \cdots \otimes u_{r_1} \otimes u_1^* \otimes \cdots \otimes u_{s_1}^*$ and $v = v_1 \otimes \cdots \otimes v_{r_2} \otimes v_1^* \otimes \cdots \otimes v_{s_2}^*$, let

$$u \otimes v = u_1 \otimes \cdots \otimes u_{r_1} \otimes v_1 \otimes \cdots \otimes v_{r_2} \otimes u_1^* \otimes \cdots \otimes u_{s_1}^* \otimes v_1^* \otimes \cdots \otimes v_{s_2}^*.$$

Denote by $\mathrm{Hom}(V^r, (V^*)^s; W)$ the vector space of all multilinear maps from $V^r \times (V^*)^s$ to W. Then we have the universal mapping property which asserts that there is a canonical isomorphism

$$\mathrm{Hom}(T^{r,s}(V), W) \cong \mathrm{Hom}(V^r, (V^*)^s; W).$$

In particular,

$$(T^{r,s}(V))^* \cong \mathrm{Hom}(V^r, (V^*)^s; K).$$

For finite-dimensional vector spaces, the duality of Section 2.5 is also easily extended to the tensor spaces $T^{r,s}(V)$. We define the pairing

$$T^{r,s}(V^*) \times T^{r,s}(V) \longrightarrow K$$

as follows: if

$$v^* = v_1^* \otimes \cdots \otimes v_r^* \otimes u_{r+1} \otimes \cdots \otimes u_{r+s} \in T^{r,s}(V^*)$$

and

$$u = u_1 \otimes \cdots \otimes u_r \otimes v_{r+1}^* \otimes \cdots \otimes v_{r+s}^* \in T^{r,s}(V),$$

then

$$(v^*, u) = v_1^*(u_1) \cdots v_{r+s}^*(u_{r+s}).$$

This is a nondegenerate pairing, and thus we get a canonical isomorphism

$$(T^{r,s}(V))^* \cong T^{r,s}(V^*).$$

Consequently, we get a canonical isomorphism

$$T^{r,s}(V^*) \cong \mathrm{Hom}(V^r, (V^*)^s; K).$$

We summarize these results in the following proposition.

Proposition 2.21. *Let V be a vector space and let*

$$T^{r,s}(V) = V^{\otimes r} \otimes (V^*)^{\otimes s} = \underbrace{V \otimes \cdots \otimes V}_{r} \otimes \underbrace{V^* \otimes \cdots \otimes V^*}_{s}.$$

We have the canonical isomorphisms

$$(T^{r,s}(V))^* \cong T^{r,s}(V^*),$$

and

$$T^{r,s}(V^*) \cong \mathrm{Hom}(V^r, (V^*)^s; K).$$

Remark. The tensor spaces $T^{r,s}(V)$ are also denoted $T_s^r(V)$. A tensor $\alpha \in T^{r,s}(V)$ is said to be *contravariant* in the first r arguments and *covariant* in the last s arguments. This terminology refers to the way tensors behave under coordinate changes. Given a basis $(e_1, \ldots, e_n)$ of V, if $(e_1^*, \ldots, e_n^*)$ denotes the dual basis, then every tensor $\alpha \in T^{r,s}(V)$ is given by an expression of the form

$$\alpha = \sum_{\substack{i_1, \ldots, i_r \\ j_1, \ldots, j_s}} a_{j_1, \ldots, j_s}^{i_1, \ldots, i_r} e_{i_1} \otimes \cdots \otimes e_{i_r} \otimes e_{j_1}^* \otimes \cdots \otimes e_{j_s}^*.$$

The tradition in classical tensor notation is to use lower indices on vectors and upper indices on linear forms and in accordance to *Einstein summation convention* (or *Einstein notation*) the position of the indices on the coefficients is reversed. *Einstein summation convention* (already encountered in Section 2.1) is to assume that a summation is performed for all values of every index that appears simultaneously once as an upper index and once as a lower index. According to this convention, the tensor α above is written as

$$\alpha = a^{i_1,\ldots,i_r}_{j_1,\ldots,j_s} e_{i_1} \otimes \cdots \otimes e_{i_r} \otimes e^{j_1} \otimes \cdots \otimes e^{j_s}.$$

An older view of tensors is that they are multidimensional arrays of coefficients,

$$\left(a^{i_1,\ldots,i_r}_{j_1,\ldots,j_s} \right),$$

subject to the rules for changes of bases.

Another operation on general tensors, contraction, is useful in differential geometry.

Definition 2.14. For all $r, s \geq 1$, the *contraction* $c_{i,j} \colon T^{r,s}(V) \to T^{r-1,s-1}(V)$, with $1 \leq i \leq r$ and $1 \leq j \leq s$, is the linear map defined on generators by

$$c_{i,j}(u_1 \otimes \cdots \otimes u_r \otimes v_1^* \otimes \cdots \otimes v_s^*)$$
$$= v_j^*(u_i)\, u_1 \otimes \cdots \otimes \widehat{u_i} \otimes \cdots \otimes u_r \otimes v_1^* \otimes \cdots \otimes \widehat{v_j^*} \otimes \cdots \otimes v_s^*,$$

where the hat over an argument means that it should be omitted.

Let us figure out what is $c_{1,1} \colon T^{1,1}(V) \to \mathbb{R}$, that is $c_{1,1} \colon V \otimes V^* \to \mathbb{R}$. If $(e_1, \ldots, e_n)$ is a basis of V and $(e_1^*, \ldots, e_n^*)$ is the dual basis, by Proposition 2.17 every $h \in V \otimes V^* \cong \mathrm{Hom}(V, V)$ can be expressed as

$$h = \sum_{i,j=1}^{n} a_{ij}\, e_i \otimes e_j^*.$$

As

$$c_{1,1}(e_i \otimes e_j^*) = \delta_{i,j},$$

we get

$$c_{1,1}(h) = \sum_{i=1}^{n} a_{ii} = \mathrm{tr}(h),$$

where $\mathrm{tr}(h)$ is the *trace* of h, where h is viewed as the linear map given by the matrix, (a_{ij}). Actually, since $c_{1,1}$ is defined independently of any basis, $c_{1,1}$ provides an intrinsic definition of the trace of a linear map $h \in \mathrm{Hom}(V, V)$.

Remark. Using the Einstein summation convention, if

$$\alpha = a^{i_1,\ldots,i_r}_{j_1,\ldots,j_s} e_{i_1} \otimes \cdots \otimes e_{i_r} \otimes e^{j_1} \otimes \cdots \otimes e^{j_s},$$

then

$$c_{k,l}(\alpha) = a^{i_1,\ldots,i_{k-1},i_{k+1}\ldots,i_r}_{j_1,\ldots,j_{l-1},j_{l+1},\ldots,j_s} e_{i_1} \otimes \cdots \otimes \widehat{e_{i_k}} \otimes \cdots \otimes e_{i_r} \otimes e^{j_1} \otimes \cdots \otimes \widehat{e^{j_l}} \otimes \cdots \otimes e^{j_s}.$$

If E and F are two K-algebras, we know that their tensor product $E \otimes F$ exists as a vector space. We can make $E \otimes F$ into an algebra as well. Indeed, we have the multilinear map

$$E \times F \times E \times F \longrightarrow E \otimes F$$

given by $(a, b, c, d) \mapsto (ac) \otimes (bd)$, where ac is the product of a and c in E and bd is the product of b and d in F. By the universal mapping property, we get a linear map,

$$E \otimes F \otimes E \otimes F \longrightarrow E \otimes F.$$

Using the isomorphism

$$E \otimes F \otimes E \otimes F \cong (E \otimes F) \otimes (E \otimes F),$$

we get a linear map

$$(E \otimes F) \otimes (E \otimes F) \longrightarrow E \otimes F,$$

and thus a bilinear map,

$$(E \otimes F) \times (E \otimes F) \longrightarrow E \otimes F$$

which is our multiplication operation in $E \otimes F$. This multiplication is determined by

$$(a \otimes b) \cdot (c \otimes d) = (ac) \otimes (bd).$$

In summary we have the following proposition.

Proposition 2.22. *Given two K-algebra E and F, the operation on $E \otimes F$ defined on generators by*

$$(a \otimes b) \cdot (c \otimes d) = (ac) \otimes (bd)$$

makes $E \otimes F$ into a K-algebra.

We now turn to symmetric tensors.

2.7 Symmetric Tensor Powers

Our goal is to come up with a notion of tensor product that will allow us to treat symmetric multilinear maps as linear maps. Note that we have to restrict ourselves to a *single* vector space E, rather than n vector spaces $E_1, \ldots, E_n$, so that symmetry makes sense.

Definition 2.15. A multilinear map $f \colon E^n \to F$ is *symmetric* iff

$$f(u_{\sigma(1)}, \ldots, u_{\sigma(n)}) = f(u_1, \ldots, u_n),$$

for all $u_i \in E$ and all permutations, $\sigma \colon \{1, \ldots, n\} \to \{1, \ldots, n\}$. The group of permutations on $\{1, \ldots, n\}$ (the *symmetric group*) is denoted $\mathfrak{S}_n$. The vector space of all symmetric multilinear maps $f \colon E^n \to F$ is denoted by $\mathrm{Sym}^n(E; F)$ or $\mathrm{Hom}_{\mathrm{symlin}}(E^n, F)$. Note that $\mathrm{Sym}^1(E; F) = \mathrm{Hom}(E, F)$.

We could proceed directly as in Theorem 2.6 and construct symmetric tensor products from scratch. However, since we already have the notion of a tensor product, there is a more economical method. First we define symmetric tensor powers.

Definition 2.16. An *n-th symmetric tensor power* of a vector space E, where $n \geq 1$, is a vector space S together with a symmetric multilinear map $\varphi \colon E^n \to S$ such that, for every vector space F and for every symmetric multilinear map $f \colon E^n \to F$, there is a unique linear map $f_\odot \colon S \to F$, with

$$f(u_1, \ldots, u_n) = f_\odot(\varphi(u_1, \ldots, u_n)),$$

for all $u_1, \ldots, u_n \in E$, or for short

$$f = f_\odot \circ \varphi.$$

Equivalently, there is a unique linear map $f_\odot$ such that the following diagram commutes.

The above property is called the *universal mapping property* of the symmetric tensor power (S, φ).

We next show that any two symmetric n-th tensor powers (S_1, φ_1) and (S_2, φ_2) for E are isomorphic.

Proposition 2.23. *Given any two symmetric n-th tensor powers (S_1, φ_1) and (S_2, φ_2) for E, there is an isomorphism $h \colon S_1 \to S_2$ such that*

$$\varphi_2 = h \circ \varphi_1.$$

Proof. Replace tensor product by n-th symmetric tensor power in the proof of Proposition 2.5.   □

We now give a construction that produces a symmetric n-th tensor power of a vector space E.

Theorem 2.24. *Given a vector space E, a symmetric n-th tensor power $(S^n(E), \varphi)$ for E can be constructed $(n \geq 1)$. Furthermore, denoting $\varphi(u_1, \ldots, u_n)$ as $u_1 \odot \cdots \odot u_n$, the symmetric tensor power $S^n(E)$ is generated by the vectors $u_1 \odot \cdots \odot u_n$, where $u_1, \ldots, u_n \in E$, and for every symmetric multilinear map $f \colon E^n \to F$, the unique linear map $f_\odot \colon S^n(E) \to F$ such that $f = f_\odot \circ \varphi$ is defined by*

$$f_\odot(u_1 \odot \cdots \odot u_n) = f(u_1, \ldots, u_n)$$

on the generators $u_1 \odot \cdots \odot u_n$ of $S^n(E)$.

Proof. The tensor power $E^{\otimes n}$ is too big, and thus we define an appropriate quotient. Let C be the subspace of $E^{\otimes n}$ generated by the vectors of the form

$$u_1 \otimes \cdots \otimes u_n - u_{\sigma(1)} \otimes \cdots \otimes u_{\sigma(n)},$$

for all $u_i \in E$, and all permutations $\sigma \colon \{1, \ldots, n\} \to \{1, \ldots, n\}$. We claim that the quotient space $(E^{\otimes n})/C$ does the job.

Let $p \colon E^{\otimes n} \to (E^{\otimes n})/C$ be the quotient map, and let $\varphi \colon E^n \to (E^{\otimes n})/C$ be the map given by

$$\varphi = p \circ \varphi_0,$$

where $\varphi_0 \colon E^n \to E^{\otimes n}$ is the injection given by $\varphi_0(u_1, \ldots, u_n) = u_1 \otimes \cdots \otimes u_n$.

Let us denote $\varphi(u_1, \ldots, u_n)$ as $u_1 \odot \cdots \odot u_n$. It is clear that φ is symmetric. Since the vectors $u_1 \otimes \cdots \otimes u_n$ generate $E^{\otimes n}$, and p is surjective, the vectors $u_1 \odot \cdots \odot u_n$ generate $(E^{\otimes n})/C$.

It remains to show that $((E^{\otimes n})/C, \varphi)$ satisfies the universal mapping property. To this end we begin by proving that there is a map h such that $f = h \circ \varphi$. Given any symmetric multilinear map $f \colon E^n \to F$, by Theorem 2.6 there is a linear map $f_\otimes \colon E^{\otimes n} \to F$ such that $f = f_\otimes \circ \varphi_0$, as in the diagram below.

However, since f is symmetric, we have $f_\otimes(z) = 0$ for every $z \in C$. Thus, we get an induced linear map $h\colon (E^{\otimes n})/C \to F$ making the following diagram commute.

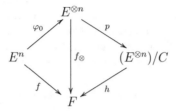

If we define $h([z]) = f_\otimes(z)$ for every $z \in E^{\otimes n}$, where $[z]$ is the equivalence class in $(E^{\otimes n})/C$ of $z \in E^{\otimes n}$, the above diagram shows that $f = h \circ p \circ \varphi_0 = h \circ \varphi$. We now prove the uniqueness of h. For any linear map $f_\odot\colon (E^{\otimes n})/C \to F$ such that $f = f_\odot \circ \varphi$, since $\varphi(u_1, \ldots, u_n) = u_1 \odot \cdots \odot u_n$ and the vectors $u_1 \odot \cdots \odot u_n$ generate $(E^{\otimes n})/C$, the map $f_\odot$ is uniquely defined by

$$f_\odot(u_1 \odot \cdots \odot u_n) = f(u_1, \ldots, u_n).$$

Since $f = h \circ \varphi$, the map h is unique, and we let $f_\odot = h$. Thus, $S^n(E) = (E^{\otimes n})/C$ and φ constitute a symmetric n-th tensor power of E.   □

The map φ from E^n to $S^n(E)$ is often denoted $\iota_\odot$, so that

$$\iota_\odot(u_1, \ldots, u_n) = u_1 \odot \cdots \odot u_n.$$

Again, the actual construction is not important. What is important is that the symmetric n-th power has the universal mapping property with respect to symmetric multilinear maps.

Remark. The notation $\odot$ for the commutative multiplication of symmetric tensor powers is not standard. Another notation commonly used is $\cdot$. We often abbreviate "symmetric tensor power" as "symmetric power." The symmetric power $S^n(E)$ is also denoted $\mathrm{Sym}^n E$ but we prefer to use the notation Sym to denote spaces of symmetric multilinear maps. To be consistent with the use of $\odot$, we could have used the notation $\odot^n E$. Clearly, $S^1(E) \cong E$ and it is convenient to set $S^0(E) = K$.

The fact that the map $\varphi\colon E^n \to S^n(E)$ is symmetric and multilinear can also be expressed as follows:

$$u_1 \odot \cdots \odot (v_i + w_i) \odot \cdots \odot u_n = (u_1 \odot \cdots \odot v_i \odot \cdots \odot u_n)$$
$$+ (u_1 \odot \cdots \odot w_i \odot \cdots \odot u_n),$$
$$u_1 \odot \cdots \odot (\lambda u_i) \odot \cdots \odot u_n = \lambda(u_1 \odot \cdots \odot u_i \odot \cdots \odot u_n),$$
$$u_{\sigma(1)} \odot \cdots \odot u_{\sigma(n)} = u_1 \odot \cdots \odot u_n,$$

for all permutations $\sigma \in \mathfrak{S}_n$.

The last identity shows that the "operation" $\odot$ is commutative. This allows us to view the symmetric tensor $u_1 \odot \cdots \odot u_n$ as an object called a multiset.

Given a set A, a multiset with elements from A is a generalization of the concept of a set that allows multiple instances of elements from A to occur. For example, if $A = \{a, b, c, d\}$, the following are multisets:

$$M_1 = \{a, a, b\}, \quad M_2 = \{a, a, b, b, c\}, \quad M_3 = \{a, a, b, b, c, d, d, d\}.$$

Here is another way to represent multisets as tables showing the multiplicities of the elements in the multiset:

$$M_1 = \begin{pmatrix} a\ b\ c\ d \\ 2\ 1\ 0\ 0 \end{pmatrix}, \quad M_2 = \begin{pmatrix} a\ b\ c\ d \\ 2\ 2\ 1\ 0 \end{pmatrix}, \quad M_3 = \begin{pmatrix} a\ b\ c\ d \\ 2\ 2\ 1\ 3 \end{pmatrix}.$$

The above are just graphs of functions from the set $A = \{a, b, c, d\}$ to $\mathbb{N}$. This suggests the following definition.

Definition 2.17. A finite *multiset* M over a set A is a function $M \colon A \to \mathbb{N}$ such that $M(a) \neq 0$ for finitely many $a \in A$. The *multiplicity* of an element $a \in A$ in M is $M(a)$. The set of all multisets over A is denoted by $\mathbb{N}^{(A)}$, and we let $\mathrm{dom}(M) = \{a \in A \mid M(a) \neq 0\}$, which is a finite set. The set $\mathrm{dom}(M)$ is the set of elements in A that actually occur in M. For any multiset $M \in \mathbb{N}^{(A)}$, note that $\sum_{a \in A} M(a)$ makes sense, since $\sum_{a \in A} M(a) = \sum_{a \in \mathrm{dom}(A)} M(a)$, and $\mathrm{dom}(M)$ is finite; this sum is the total number of elements in the multiset A and is called the *size* of M. Let $|M| = \sum_{a \in A} M(a)$.

Going back to our symmetric tensors, we can view the tensors of the form $u_1 \odot \cdots \odot u_n$ as multisets of size n over the set E.

Theorem 2.24 implies the following proposition.

Proposition 2.25. *There is a canonical isomorphism*

$$\mathrm{Hom}(S^n(E), F) \cong \mathrm{Sym}^n(E; F)$$

between the vector space of linear maps $\mathrm{Hom}(S^n(E), F)$ *and the vector space of symmetric multilinear maps* $\mathrm{Sym}^n(E; F)$ *given by the linear map* $- \circ \varphi$ *defined by* $h \mapsto h \circ \varphi$, *with* $h \in \mathrm{Hom}(S^n(E), F)$.

Proof. The map $h \circ \varphi$ is clearly symmetric multilinear. By Theorem 2.24, for every symmetric multilinear map $f \in \mathrm{Sym}^n(E; F)$ there is a unique linear map $f_\odot \in \mathrm{Hom}(\mathrm{S}^n(E), F)$ such that $f = f_\odot \circ \varphi$, so the map $- \circ \varphi$ is bijective. Its inverse is the map $f \mapsto f_\odot$. $\square$

In particular, when $F = K$, we get the following important fact.

Proposition 2.26. *There is a canonical isomorphism*

$$\left(\mathrm{S}^n(E)\right)^* \cong \mathrm{Sym}^n(E; K).$$

Definition 2.18. Symmetric tensors in $\mathrm{S}^n(E)$ are called *symmetric n-tensors*, and tensors of the form $u_1 \odot \cdots \odot u_n$, where $u_i \in E$, are called *simple (or decomposable) symmetric n-tensors*. Those symmetric n-tensors that are not simple are often called *compound symmetric n-tensors*.

Given a linear map $f \colon E \to E'$, since the map $\iota'_\odot \circ (f \times f)$ is bilinear and symmetric, there is a unique linear map $f \odot f \colon \mathrm{S}^2(E) \to \mathrm{S}^2(E')$ making the following diagram commute.

$$
\begin{array}{ccc}
E^2 & \xrightarrow{\ \iota_\odot\ } & \mathrm{S}^2(E) \\
{\scriptstyle f \times f}\Big\downarrow & & \Big\downarrow{\scriptstyle f \odot f} \\
(E')^2 & \xrightarrow[\ \iota'_\odot\ ]{} & \mathrm{S}^2(E').
\end{array}
$$

Observe that $f \odot f$ is determined by

$$(f \odot f)(u \odot v) = f(u) \odot f(v).$$

Proposition 2.27. *Given any two linear maps $f \colon E \to E'$ and $f' \colon E' \to E''$, we have*

$$(f' \circ f) \odot (f' \circ f) = (f' \odot f') \circ (f \odot f).$$

By using the proof techniques of Proposition 2.13 (3), we can show the following property of symmetric tensor products.

Proposition 2.28. *We have the following isomorphism:*

$$\mathrm{S}^n(E \oplus F) \cong \bigoplus_{k=0}^{n} \mathrm{S}^k(E) \otimes \mathrm{S}^{n-k}(F).$$

The generalization to the symmetric tensor product $f \odot \cdots \odot f$ of $n \geq 3$ copies of the linear map $f \colon E \to E'$ is immediate, and left to the reader.

2.8   Bases of Symmetric Powers

The vectors $u_1 \odot \cdots \odot u_m$ where $u_1, \ldots, u_m \in E$ generate $S^m(E)$, but they are not linearly independent. We will prove a version of Proposition 2.12 for symmetric tensor powers using multisets.

Recall that a (finite) multiset over a set I is a function $M \colon I \to \mathbb{N}$, such that $M(i) \neq 0$ for finitely many $i \in I$. The set of all multisets over I is denoted as $\mathbb{N}^{(I)}$ and we let $\mathrm{dom}(M) = \{i \in I \mid M(i) \neq 0\}$, the finite set of elements in I that actually occur in M. The size of the multiset M is $|M| = \sum_{a \in A} M(a)$.

To explain the idea of the proof, consider the case when $m = 2$ and E has dimension 3. Given a basis (e_1, e_2, e_3) of E, we would like to prove that

$$e_1 \odot e_1, \quad e_1 \odot e_2, \quad e_1 \odot e_3, \quad e_2 \odot e_2, \quad e_2 \odot e_3, \quad e_3 \odot e_3$$

are linearly independent. To prove this, it suffices to show that for any vector space F, if $w_{11}, w_{12}, w_{13}, w_{22}, w_{23}, w_{33}$ are any vectors in F, then there is a symmetric bilinear map $h \colon E^2 \to F$ such that

$$h(e_i, e_j) = w_{ij}, \quad 1 \leq i \leq j \leq 3.$$

Because h yields a unique linear map $h_\odot \colon S^2(E) \to F$ such that

$$h_\odot(e_i \odot e_j) = w_{ij}, \quad 1 \leq i \leq j \leq 3,$$

by Proposition 2.4, the vectors

$$e_1 \odot e_1, \quad e_1 \odot e_2, \quad e_1 \odot e_3, \quad e_2 \odot e_2, \quad e_2 \odot e_3, \quad e_3 \odot e_3$$

are linearly independent. This suggests understanding how a symmetric bilinear function $f \colon E^2 \to F$ is expressed in terms of its values $f(e_i, e_j)$ on the basis vectors (e_1, e_2, e_3), and this can be done easily. Using bilinearity and symmetry, we obtain

$$\begin{aligned}
f(u_1 e_1 &+ u_2 e_2 + u_3 e_3, v_1 e_1 + v_2 e_2 + v_3 e_3) \\
&= u_1 v_1 f(e_1, e_1) + (u_1 v_2 + u_2 v_1) f(e_1, e_2) \\
&\quad + (u_1 v_3 + u_3 v_1) f(e_1, e_3) + u_2 v_2 f(e_2, e_2) \\
&\quad + (u_2 v_3 + u_3 v_2) f(e_2, e_3) + u_3 v_3 f(e_3, e_3).
\end{aligned}$$

Therefore, given $w_{11}, w_{12}, w_{13}, w_{22}, w_{23}, w_{33} \in F$, the function h given by

$$h(u_1 e_1 + u_2 e_2 + u_3 e_3, v_1 e_1 + v_2 e_2 + v_3 e_3) = u_1 v_1 w_{11} + (u_1 v_2 + u_2 v_1) w_{12}$$
$$+ (u_1 v_3 + u_3 v_1) w_{13} + u_2 v_2 w_{22}$$
$$+ (u_2 v_3 + u_3 v_2) w_{23} + u_3 v_3 w_{33}$$

is clearly bilinear symmetric, and by construction $h(e_i, e_j) = w_{ij}$, so it does the job.

The generalization of this argument to any $m \geq 2$ and to a space E of any dimension (even infinite) is conceptually clear, but notationally messy. If $\dim(E) = n$ and if $(e_1, \ldots, e_n)$ is a basis of E, for any m vectors $v_j = \sum_{i=1}^{n} u_{i,j} e_i$ in E, for any symmetric multilinear map $f \colon E^m \to F$, we have

$$f(v_1, \ldots, v_m) = \sum_{\substack{k_1 + \cdots + k_n = m}} \sum_{\substack{I_1 \cup \cdots \cup I_n = \{1, \ldots, m\} \\ I_i \cap I_j = \emptyset,\, i \neq j,\, |I_j| = k_j}}$$

$$\left(\prod_{i_1 \in I_1} u_{1,i_1} \right) \cdots \left(\prod_{i_n \in I_n} u_{n,i_n} \right) f(\underbrace{e_1, \ldots, e_1}_{k_1}, \ldots, \underbrace{e_n, \ldots, e_n}_{k_n}).$$

Definition 2.19. Given any set J of $n \geq 1$ elements, say $J = \{j_1, \ldots, j_n\}$, and given any $m \geq 2$, for any sequence $(k_1 \ldots, k_n)$ of natural numbers $k_i \in \mathbb{N}$ such that $k_1 + \cdots + k_n = m$, the multiset M of size m

$$M = \{\underbrace{j_1, \ldots, j_1}_{k_1}, \underbrace{j_2, \ldots, j_2}_{k_2}, \ldots, \underbrace{j_n, \ldots, j_n}_{k_n}\}$$

is denoted by $M(m, J, k_1, \ldots, k_n)$. Note that $M(j_i) = k_i$, for $i = 1, \ldots, n$. Given any $k \geq 1$, and any $u \in E$, we denote $\underbrace{u \odot \cdots \odot u}_{k}$ as $u^{\odot k}$.

We can now prove the following proposition.

Proposition 2.29. *Given a vector space E, if $(e_i)_{i \in I}$ is a basis for E, then the family of vectors*

$$\left(e_{i_1}^{\odot M(i_1)} \odot \cdots \odot e_{i_k}^{\odot M(i_k)} \right)_{\substack{M \in \mathbb{N}^{(I)},\ |M| = m, \\ \{i_1, \ldots, i_k\} = dom(M)}}$$

is a basis of the symmetric m-th tensor power $S^m(E)$.

Proof. The proof is very similar to that of Proposition 2.12. First assume that E has finite dimension n. In this case $I = \{1, \ldots, n\}$, and any multiset $M \in \mathbb{N}^{(I)}$ of size $|M| = m$ is of the form $M(m, \{1, \ldots, n\}, k_1, \ldots, k_n)$, with $k_i = M(i)$ and $k_1 + \cdots + k_n = m$.

For any nontrivial vector space F, for any family of vectors

$$(w_M)_{M \in \mathbb{N}^{(I)}, \, |M| = m},$$

we show the existence of a symmetric multilinear map $h \colon S^m(E) \to F$, such that for every $M \in \mathbb{N}^{(I)}$ with $|M| = m$, we have

$$h(e_{i_1}^{\odot M(i_1)} \odot \cdots \odot e_{i_k}^{\odot M(i_k)}) = w_M,$$

where $\{i_1, \ldots, i_k\} = \mathrm{dom}(M)$. We define the map $f \colon E^m \to F$ as follows: for any m vectors $v_1, \ldots, v_m \in E$ we can write $v_k = \sum_{i=1}^n u_{i,k} e_i$ for $k = 1, \ldots, m$ and we set

$$f(v_1, \ldots, v_m) = \sum_{k_1 + \cdots + k_n = m} \sum_{\substack{I_1 \cup \cdots \cup I_n = \{1, \ldots, m\} \\ I_i \cap I_j = \emptyset, \, i \neq j, \, |I_j| = k_j}}$$

$$\left(\prod_{i_1 \in I_1} u_{1, i_1} \right) \cdots \left(\prod_{i_n \in I_n} u_{n, i_n} \right) w_{M(m, \{1, \ldots, n\}, k_1, \ldots, k_n)}.$$

It is not difficult to verify that f is symmetric and multilinear. By the universal mapping property of the symmetric tensor product, the linear map $f_\odot \colon S^m(E) \to F$ such that $f = f_\odot \circ \varphi$ is the desired map h. Then by Proposition 2.4, it follows that the family

$$\left(e_{i_1}^{\odot M(i_1)} \odot \cdots \odot e_{i_k}^{\odot M(i_k)} \right)_{\substack{M \in \mathbb{N}^{(I)}, \, |M| = m, \\ \{i_1, \ldots, i_k\} = \mathrm{dom}(M)}}$$

is linearly independent. Using the commutativity of $\odot$, we can also show that these vectors generate $S^m(E)$, and thus, they form a basis for $S^m(E)$.

If I is infinite dimensional, then for any m vectors $v_1, \ldots, v_m \in F$ there is a finite subset J of I such that $v_k = \sum_{j \in J} u_{j,k} e_j$ for $k = 1, \ldots, m$, and if we write $n = |J|$, then the formula for $f(v_1, \ldots, v_m)$ is obtained by replacing the set $\{1, \ldots, n\}$ by J. The details are left as an exercise. $\qquad\square$

As a consequence, when I is finite, say of size $p = \dim(E)$, the dimension of $S^m(E)$ is the number of finite multisets $(j_1, \ldots, j_p)$, such that $j_1 + \cdots + j_p = m$, $j_k \geq 0$. We leave as an exercise to show that this number is $\binom{p+m-1}{m}$. Thus, if $\dim(E) = p$, then the dimension of $S^m(E)$ is $\binom{p+m-1}{m}$. Compare with the dimension of $E^{\otimes m}$, which is p^m. In particular, when $p = 2$, the dimension of $S^m(E)$ is $m + 1$. This can also be seen directly.

Remark. The number $\binom{p+m-1}{m}$ is also the number of homogeneous monomials

$$X_1^{j_1} \cdots X_p^{j_p}$$

of total degree m in p variables (we have $j_1 + \cdots + j_p = m$). This is not a coincidence! Given a vector space E and a basis $(e_i)_{i \in I}$ for E, Proposition 2.29 shows that every symmetric tensor $z \in S^m(E)$ can be written in a unique way as

$$z = \sum_{\substack{M \in \mathbb{N}^{(I)} \\ \sum_{i \in I} M(i) = m \\ \{i_1, \dots, i_k\} = \mathrm{dom}(M)}} \lambda_M \, e_{i_1}^{\odot M(i_1)} \odot \cdots \odot e_{i_k}^{\odot M(i_k)},$$

for some unique family of scalars $\lambda_M \in K$, all zero except for a finite number.

This looks like a homogeneous polynomial of total degree m, where the monomials of total degree m are the symmetric tensors

$$e_{i_1}^{\odot M(i_1)} \odot \cdots \odot e_{i_k}^{\odot M(i_k)}$$

in the "indeterminates" e_i, where $i \in I$ (recall that $M(i_1) + \cdots + M(i_k) = m$) and implies that polynomials can be defined in terms of symmetric tensors.

2.9 Duality for Symmetric Powers

In this section all vector spaces are assumed to have *finite dimension over a field of characteristic zero*. We define a nondegenerate pairing $S^n(E^*) \times S^n(E) \longrightarrow K$ as follows: Consider the multilinear map

$$(E^*)^n \times E^n \longrightarrow K$$

given by

$$(v_1^*, \dots, v_n^*, u_1, \dots, u_n) \mapsto \sum_{\sigma \in \mathfrak{S}_n} v_{\sigma(1)}^*(u_1) \cdots v_{\sigma(n)}^*(u_n).$$

Note that the expression on the right-hand side is "almost" the determinant $\det(v_j^*(u_i))$, except that the sign $\mathrm{sgn}(\sigma)$ is missing (where $\mathrm{sgn}(\sigma)$ is the signature of the permutation σ; that is, the parity of the number of transpositions into which σ can be factored). Such an expression is called a *permanent*.

It can be verified that this expression is symmetric w.r.t. the u_i's and also w.r.t. the v_j^*. For any fixed $(v_1^*, \ldots, v_n^*) \in (E^*)^n$, we get a symmetric multilinear map

$$l_{v_1^*, \ldots, v_n^*} : (u_1, \ldots, u_n) \mapsto \sum_{\sigma \in \mathfrak{S}_n} v_{\sigma(1)}^*(u_1) \cdots v_{\sigma(n)}^*(u_n)$$

from E^n to K. The map $l_{v_1^*, \ldots, v_n^*}$ extends uniquely to a linear map $L_{v_1^*, \ldots, v_n^*} : S^n(E) \to K$ making the following diagram commute:

$$
\begin{array}{ccc}
E^n & \xrightarrow{\iota_\odot} & S^n(E) \\
 & \searrow{\scriptstyle l_{v_1^*, \ldots, v_n^*}} & \downarrow{\scriptstyle L_{v_1^*, \ldots, v_n^*}} \\
 & & K.
\end{array}
$$

We also have the symmetric multilinear map

$$(v_1^*, \ldots, v_n^*) \mapsto L_{v_1^*, \ldots, v_n^*}$$

from $(E^*)^n$ to $\mathrm{Hom}(S^n(E), K)$, which extends to a linear map L from $S^n(E^*)$ to $\mathrm{Hom}(S^n(E), K)$ making the following diagram commute:

$$
\begin{array}{ccc}
(E^*)^n & \xrightarrow{\iota_{\odot^*}} & S^n(E^*) \\
 & \searrow & \downarrow{\scriptstyle L} \\
 & & \mathrm{Hom}(S^n(E), K).
\end{array}
$$

However, in view of the isomorphism

$$\mathrm{Hom}(U \otimes V, W) \cong \mathrm{Hom}(U, \mathrm{Hom}(V, W)),$$

with $U = S^n(E^*)$, $V = S^n(E)$ and $W = K$, we can view L as a linear map

$$L : S^n(E^*) \otimes S^n(E) \longrightarrow K,$$

which by Proposition 2.8 corresponds to a bilinear map

$$\langle -, - \rangle : S^n(E^*) \times S^n(E) \longrightarrow K. \qquad (*)$$

This pairing is given explicitly on generators by

$$\langle v_1^* \odot \cdots \odot v_n^*, u_1 \odot \cdots \odot u_n \rangle = \sum_{\sigma \in \mathfrak{S}_n} v_{\sigma(1)}^*(u_1) \cdots v_{\sigma(n)}^*(u_n).$$

Now this pairing in nondegenerate. This can be shown using bases.[2] If $(e_1, \ldots, e_m)$ is a basis of E, then for every basis element $(e_{i_1}^*)^{\odot n_1} \odot \cdots \odot (e_{i_k}^*)^{\odot n_k}$ of $S^n(E^*)$, with $n_1 + \cdots + n_k = n$, we have

$$\langle (e_{i_1}^*)^{\odot n_1} \odot \cdots \odot (e_{i_k}^*)^{\odot n_k}, e_{i_1}^{\odot n_1} \odot \cdots \odot e_{i_k}^{\odot n_k} \rangle = n_1! \cdots n_k!,$$

and

$$\langle (e_{i_1}^*)^{\odot n_1} \odot \cdots \odot (e_{i_k}^*)^{\odot n_k}, e_{j_1} \odot \cdots \odot e_{j_n} \rangle = 0$$

if $(j_1 \ldots, j_n) \neq (\underbrace{i_1, \ldots, i_1}_{n_1}, \ldots, \underbrace{i_k, \ldots, i_k}_{n_k})$.

If the field K has characteristic zero, then $n_1! \cdots n_k! \neq 0$. We leave the details as an exercise to the reader. Therefore we get a canonical isomorphism

$$(S^n(E))^* \cong S^n(E^*).$$

The following proposition summarizes the duality properties of symmetric powers.

Proposition 2.30. *Assume the field K has characteristic zero. We have the canonical isomorphisms*

$$(S^n(E))^* \cong S^n(E^*)$$

and

$$S^n(E^*) \cong \mathrm{Sym}^n(E; K) = \mathrm{Hom}_{\mathrm{symlin}}(E^n, K),$$

which allows us to interpret symmetric tensors over E^* as symmetric multilinear maps.

Proof. The isomorphism

$$\mu \colon S^n(E^*) \cong \mathrm{Sym}^n(E; K)$$

---

[2]This is where the assumptions that we are in finite dimension and that the field has characteristic zero are used.

follows from the isomorphisms $(S^n(E))^* \cong S^n(E^*)$ and $(S^n(E))^* \cong \mathrm{Sym}^n(E; K)$ given by Proposition 2.26. $\square$

Remarks.

1. The isomorphism $\mu\colon S^n(E^*) \cong \mathrm{Sym}^n(E; K)$ discussed above can be described explicitly as the linear extension of the map given by

$$\mu(v_1^* \odot \cdots \odot v_n^*)(u_1 \odot \cdots \odot u_n) = \sum_{\sigma \in \mathfrak{S}_n} v_{\sigma(1)}^*(u_1) \cdots v_{\sigma(n)}^*(u_n).$$

   If $(e_1, \ldots, e_m)$ is a basis of E, then for every basis element $(e_{i_1}^*)^{\odot n_1} \odot \cdots \odot (e_{i_k}^*)^{\odot n_k}$ of $S^n(E^*)$, with $n_1 + \cdots + n_k = n$, we have

$$\mu((e_{i_1}^*)^{\odot n_1} \odot \cdots \odot (e_{i_k}^*)^{\odot n_k})(\underbrace{e_{i_1}, \ldots, e_{i_1}}_{n_1}, \ldots, \underbrace{e_{i_k}, \ldots, e_{i_k}}_{n_k}) = n_1! \cdots n_k!,$$

   If the field K has positive characteristic, then it is possible that $n_1! \cdots n_k! = 0$, and this is why we required K to be of characteristic 0 in order for Proposition 2.30 to hold.

2. The canonical isomorphism of Proposition 2.30 holds under more general conditions. Namely, that K is a commutative algebra with identity over $\mathbb{Q}$, and that the E is a finitely generated projective K-module (see Definition 2.23). See Bourbaki [14] (Chapter III, § 11, Section 5, Proposition 8).

   The map from E^n to $S^n(E)$ given by $(u_1, \ldots, u_n) \mapsto u_1 \odot \cdots \odot u_n$ yields a surjection $\pi\colon E^{\otimes n} \to S^n(E)$. Because we are dealing with vector spaces, this map has some section; that is, there is some injection $\beta\colon S^n(E) \to E^{\otimes n}$ with $\pi \circ \beta = \mathrm{id}$. Since our field K has characteristic 0, there is a special injection having a natural definition involving a symmetrization process defined as follows: For every permutation σ, we have the map $r_\sigma\colon E^n \to E^{\otimes n}$ given by

$$r_\sigma(u_1, \ldots, u_n) = u_{\sigma(1)} \otimes \cdots \otimes u_{\sigma(n)}.$$

As r_σ is clearly multilinear, r_σ extends to a linear map $(r_\sigma)_\otimes\colon E^{\otimes n} \to E^{\otimes n}$ making the following diagram commute

$$
\begin{array}{ccc}
E^n & \xrightarrow{\iota_\otimes} & E^{\otimes n} \\
 & \searrow{\scriptstyle r_\sigma} & \downarrow{\scriptstyle (r_\sigma)_\otimes} \\
 & & E^{\otimes n},
\end{array}
$$

and we get a map $\mathfrak{S}_n \times E^{\otimes n} \longrightarrow E^{\otimes n}$, namely

$$\sigma \cdot z = (r_\sigma)_\otimes(z).$$

It is immediately checked that this is a left action of the symmetric group $\mathfrak{S}_n$ on $E^{\otimes n}$, and the tensors $z \in E^{\otimes n}$ such that

$$\sigma \cdot z = z, \quad \text{for all} \quad \sigma \in \mathfrak{S}_n$$

are called *symmetrized* tensors.

We define the map $\eta \colon E^n \to E^{\otimes n}$ by

$$\eta(u_1, \ldots, u_n) = \frac{1}{n!} \sum_{\sigma \in \mathfrak{S}_n} \sigma \cdot (u_1 \otimes \cdots \otimes u_n) = \frac{1}{n!} \sum_{\sigma \in \mathfrak{S}_n} u_{\sigma(1)} \otimes \cdots \otimes u_{\sigma(n)}.$$

As the right-hand side is clearly symmetric, we get a linear map $\eta_\odot \colon S^n(E) \to E^{\otimes n}$ making the following diagram commute.

$$
\begin{array}{ccc}
E^n & \xrightarrow{\ \iota_\odot\ } & S^n(E) \\
& \eta \searrow & \downarrow \eta_\odot \\
& & E^{\otimes n}
\end{array}
$$

Clearly, $\eta_\odot(S^n(E))$ is the set of symmetrized tensors in $E^{\otimes n}$. If we consider the map $S = \eta_\odot \circ \pi \colon E^{\otimes n} \longrightarrow E^{\otimes n}$, where π is the surjection $\pi \colon E^{\otimes n} \to S^n(E)$, it is easy to check that $S \circ S = S$. Therefore, S is a projection, and by linear algebra, we know that

$$E^{\otimes n} = S(E^{\otimes n}) \oplus \operatorname{Ker} S = \eta_\odot(S^n(E)) \oplus \operatorname{Ker} S.$$

It turns out that $\operatorname{Ker} S = E^{\otimes n} \cap \mathfrak{I} = \operatorname{Ker} \pi$, where $\mathfrak{I}$ is the two-sided ideal of $T(E)$ generated by all tensors of the form $u \otimes v - v \otimes u \in E^{\otimes 2}$ (for example, see Knapp [66], Appendix A). Therefore, $\eta_\odot$ is injective,

$$E^{\otimes n} = \eta_\odot(S^n(E)) \oplus (E^{\otimes n} \cap \mathfrak{I}) = \eta_\odot(S^n(E)) \oplus \operatorname{Ker} \pi,$$

and the symmetric tensor power $S^n(E)$ is naturally embedded into $E^{\otimes n}$.

2.10   Symmetric Algebras

As in the case of tensors, we can pack together all the symmetric powers $S^n(V)$ into an algebra.

Definition 2.20. Given a vector space V, the space

$$S(V) = \bigoplus_{m \geq 0} S^m(V),$$

is called the *symmetric tensor algebra of* V.

We could adapt what we did in Section 2.6 for general tensor powers to symmetric tensors but since we already have the algebra $T(V)$, we can proceed faster. If $\mathfrak{I}$ is the two-sided ideal generated by all tensors of the form $u \otimes v - v \otimes u \in V^{\otimes 2}$, we set

$$\mathbf{S}^\bullet(V) = T(V)/\mathfrak{I}.$$

Observe that since the ideal $\mathfrak{I}$ is generated by elements in $V^{\otimes 2}$, every tensor in $\mathfrak{I}$ is a linear combination of tensors of the form $\omega_1 \otimes (u \otimes v - v \otimes u) \otimes \omega_2$, with $\omega_1 \in V^{\otimes n_1}$ and $\omega_2 \in V^{\otimes n_2}$ for some $n_1, n_2 \in \mathbb{N}$, which implies that

$$\mathfrak{I} = \bigoplus_{m \geq 0} (\mathfrak{I} \cap V^{\otimes m}).$$

Then, $\mathbf{S}^\bullet(V)$ automatically inherits a multiplication operation which is commutative, and since $T(V)$ is graded, that is

$$T(V) = \bigoplus_{m \geq 0} V^{\otimes m},$$

we have

$$\mathbf{S}^\bullet(V) = \bigoplus_{m \geq 0} V^{\otimes m}/(\mathfrak{I} \cap V^{\otimes m}).$$

However, it is easy to check that

$$\mathbf{S}^m(V) \cong V^{\otimes m}/(\mathfrak{I} \cap V^{\otimes m}),$$

so

$$\mathbf{S}^\bullet(V) \cong \mathbf{S}(V).$$

When V is of finite dimension n, $\mathbf{S}(V)$ corresponds to *the algebra of polynomials with coefficients in K in n variables* (this can be seen from Proposition 2.29). When V is of infinite dimension and $(u_i)_{i \in I}$ is a basis of V, the algebra $\mathbf{S}(V)$ corresponds to the algebra of polynomials in infinitely many variables in I. What's nice about the symmetric tensor algebra $\mathbf{S}(V)$ is that it provides an intrinsic definition of a polynomial algebra in any set of I variables.

It is also easy to see that $\mathbf{S}(V)$ satisfies the following universal mapping property.

Proposition 2.31. *Given any commutative K-algebra A, for any linear map $f: V \to A$, there is a unique K-algebra homomorphism $\overline{f}: \mathbf{S}(V) \to A$ so that*

$$f = \overline{f} \circ i,$$

as in the diagram below.

Remark. If E is finite dimensional, recall the isomorphism $\mu \colon S^n(E^*) \longrightarrow \mathrm{Sym}^n(E; K)$ defined as the linear extension of the map given by

$$\mu(v_1^* \odot \cdots \odot v_n^*)(u_1, \ldots, u_n) = \sum_{\sigma \in \mathfrak{S}_n} v_{\sigma(1)}^*(u_1) \cdots v_{\sigma(n)}^*(u_n).$$

Now we have also a multiplication operation $S^m(E^*) \times S^n(E^*) \longrightarrow S^{m+n}(E^*)$. The following question then arises:

Can we define a multiplication $\mathrm{Sym}^m(E; K) \times \mathrm{Sym}^n(E; K) \longrightarrow \mathrm{Sym}^{m+n}(E; K)$ directly on symmetric multilinear forms, so that the following diagram commutes?

$$
\begin{array}{ccc}
S^m(E^*) \times S^n(E^*) & \xrightarrow{\ \ \odot\ \ } & S^{m+n}(E^*) \\
\downarrow{\scriptstyle \mu_m \times \mu_n} & & \downarrow{\scriptstyle \mu_{m+n}} \\
\mathrm{Sym}^m(E; K) \times \mathrm{Sym}^n(E; K) & \longrightarrow & \mathrm{Sym}^{m+n}(E; K)
\end{array}
$$

The answer is *yes*! The solution is to define this multiplication such that for $f \in \mathrm{Sym}^m(E; K)$ and $g \in \mathrm{Sym}^n(E; K)$,

$$(f \cdot g)(u_1, \ldots, u_{m+n})$$

$$= \sum_{\sigma \in \mathrm{shuffle}(m,n)} f(u_{\sigma(1)}, \ldots, u_{\sigma(m)}) g(u_{\sigma(m+1)}, \ldots, u_{\sigma(m+n)}), \qquad (*)$$

where $\mathrm{shuffle}(m, n)$ consists of all (m, n)-"shuffles;" that is, permutations σ of $\{1, \ldots m + n\}$ such that $\sigma(1) < \cdots < \sigma(m)$ and $\sigma(m + 1) < \cdots < \sigma(m + n)$. Observe that a (m, n)-shuffle is completely determined by the sequence $\sigma(1) < \cdots < \sigma(m)$.

For example, suppose $m = 2$ and $n = 1$. Given $v_1^*, v_2^*, v_3^* \in E^*$, the multiplication structure on $S(E^*)$ implies that $(v_1^* \odot v_2^*) \cdot v_3^* = v_1^* \odot v_2^* \odot v_3^* \in S^3(E^*)$. Furthermore, for $u_1, u_2, u_3, \in E$,

$$\mu_3(v_1^* \odot v_2^* \odot v_3^*)(u_1, u_2, u_3) = \sum_{\sigma \in \mathfrak{S}_3} v_{\sigma(1)}^*(u_1) v_{\sigma(2)}^*(u_2) v_{\sigma(3)}^*(u_3)$$

$$= v_1^*(u_1) v_2^*(u_2) v_3^*(u_3) + v_1^*(u_1) v_3^*(u_2) v_2^*(u_3)$$

$$+ v_2^*(u_1) v_1^*(u_2) v_3^*(u_3) + v_2^*(u_1) v_3^*(u_2) v_1^*(u_3)$$

$$+ v_3^*(u_1) v_1^*(u_2) v_2^*(u_3) + v_3^*(u_1) v_2^*(u_2) v_1^*(u_3).$$

Now the $(2, 1)$- shuffles of $\{1, 2, 3\}$ are the following three permutations, namely

$$\begin{pmatrix} 1\ 2\ 3 \\ 1\ 2\ 3 \end{pmatrix}, \quad \begin{pmatrix} 1\ 2\ 3 \\ 1\ 3\ 2 \end{pmatrix}, \quad \begin{pmatrix} 1\ 2\ 3 \\ 2\ 3\ 1 \end{pmatrix}.$$

If $f \cong \mu_2(v_1^* \odot v_2^*)$ and $g \cong \mu_1(v_3^*)$, then $(*)$ implies that

$$(f \cdot g)(u_1, u_2, u_3) = \sum_{\sigma \in \text{shuffle}(2,1)} f(u_{\sigma(1)}, u_{\sigma(2)}) g(u_{\sigma(3)})$$

$$= f(u_1, u_2) g(u_3) + f(u_1, u_3) g(u_2) + f(u_2, u_3) g(u_1)$$

$$= \mu_2(v_1^* \odot v_2^*)(u_1, u_2) \mu_1(v_3^*)(u_3)$$

$$+ \mu_2(v_1^* \odot v_2^*)(u_1, u_3) \mu_1(v_3^*)(u_2)$$

$$+ \mu_2(v_1^* \odot v_2^*)(u_2, u_3) \mu_1(v_3^*)(u_1)$$

$$= (v_1^*(u_1) v_2^*(u_2) + v_2^*(u_1) v_1^*(u_2)) v_3^*(u_3)$$

$$+ (v_1^*(u_1) v_2^*(u_3) + v_2^*(u_1) v_1^*(u_3)) v_3^*(u_2)$$

$$+ (v_1^*(u_2) v_2^*(u_3) + v_2^*(u_2) v_1^*(u_3)) v_3^*(u_1)$$

$$= \mu_3(v_1^* \odot v_2^* \odot v_3^*)(u_1, u_2, u_3).$$

We leave it as an exercise for the reader to verify Equation $(*)$ for arbitrary nonnegative integers m and n.

Another useful canonical isomorphism (of K-algebras) is given below.

Proposition 2.32. *For any two vector spaces E and F, there is a canonical isomorphism (of K-algebras)*

$$S(E \oplus F) \cong S(E) \otimes S(F).$$

2.11   Tensor Products of Modules Over a Commutative Ring

This section provides some background on modules which is needed for Section 9.8 about metrics on vector bundles and for Chapter 10 on connections and curvature on vector bundles. What happens is that given a manifold M, the space $\mathfrak{X}(M)$ of vector fields on M and the space $\mathcal{A}^p(M)$ of differential p-forms on M are vector spaces, but vector fields and p-forms can also be multiplied by smooth functions in $C^\infty(M)$. This operation is a left action of $C^\infty(M)$ which satisfies all the axioms of the scalar multiplication in a vector space, but since $C^\infty(M)$ is not a field, the resulting structure is not a vector space. Instead it is a module, a more general notion.

Definition 2.21. If R is a commutative ring with identity (say 1), a *module over R* (or *R-module*) is an abelian group M with a scalar multiplication $\cdot : R \times M \to M$ such that all the axioms of a vector space are satisfied.

At first glance, a module does not seem any different from a vector space, but the lack of multiplicative inverses in R has drastic consequences, one being that unlike vector spaces, modules are generally not free; that is, have no bases. Furthermore, a module may have *torsion elements*, that is, elements $m \in M$ such that $\lambda \cdot m = 0$, even though $m \neq 0$ and $\lambda \neq 0$. For example, for any nonzero integer $n \in \mathbb{Z}$, the $\mathbb{Z}$-module $\mathbb{Z}/n\mathbb{Z}$ has no basis and $n \cdot \overline{m} = 0$ for all $\overline{m} \in \mathbb{Z}/n\mathbb{Z}$. Similarly, $\mathbb{Q}$ as a $\mathbb{Z}$-module has no basis. In fact, any two distinct nonzero elements p_1/q_1 and p_2/q_2 are linearly dependent, since

$$(p_2 q_1)\left(\frac{p_1}{q_1}\right) - (p_1 q_2)\left(\frac{p_2}{q_2}\right) = 0.$$

Nevertheless, it is possible to define tensor products of modules over a ring, just as in Section 2.2, and the results of that section continue to hold. The results of Section 2.4 also continue to hold since they are based on the universal mapping property. However, the results of Section 2.3 on bases generally fail, except for free modules. Similarly, the results of Section 2.5 on duality generally fail. Tensor algebras can be defined for modules, as in Section 2.6. Symmetric tensor and alternating tensors can be defined for modules, but again, results involving bases generally fail.

Tensor products of modules have some unexpected properties. For example, if p and q are relatively prime integers, then

$$\mathbb{Z}/p\mathbb{Z} \otimes_\mathbb{Z} \mathbb{Z}/q\mathbb{Z} = (0).$$

This is because, by Bezout's identity, there are $a, b \in \mathbb{Z}$ such that

$$ap + bq = 1,$$

so, for all $x \in \mathbb{Z}/p\mathbb{Z}$ and all $y \in \mathbb{Z}/q\mathbb{Z}$, we have

$$
\begin{aligned}
x \otimes y &= ap(x \otimes y) + bq(x \otimes y) \\
&= a(px \otimes y) + b(x \otimes qy) \\
&= a(0 \otimes y) + b(x \otimes 0) \\
&= 0.
\end{aligned}
$$

It is possible to salvage certain properties of tensor products holding for vector spaces by restricting the class of modules under consideration. For example, *projective modules* have a pretty good behavior w.r.t. tensor products.

Definition 2.22. A free R-module F is a module that has a basis (*i.e.*, there is a family $(e_i)_{i \in I}$ of linearly independent vectors in F that span F).

Projective modules generalize free modules. They have many equivalent characterizations. Here is one that is best suited for our needs.

Definition 2.23. An R-module P is *projective* if it is a summand of a free module; that is, if there is a free R-module F, and some R-module Q, so that

$$F = P \oplus Q.$$

For example, we show in Section 9.8 that the space $\Gamma(\xi)$ of global sections of a vector bundle ξ over a base manifold B is a finitely generated $C^\infty(B)$-projective module.

Given any R-module M, we let $M^* = \mathrm{Hom}_R(M, R)$ be its *dual*. We have the following proposition.

Proposition 2.33. *For any finitely generated projective R-module P and any R-module Q, we have the isomorphisms:*

$$P^{**} \cong P$$

$$\mathrm{Hom}_R(P, Q) \cong P^* \otimes_R Q.$$

Proof Sketch. We only consider the second isomorphism. Since P is projective, we have some R-modules P_1, F with

$$P \oplus P_1 = F,$$

where F is some free module. We know that for any R-modules U, V, W, we have

$$\mathrm{Hom}_R(U \oplus V, W) \cong \mathrm{Hom}_R(U, W) \prod \mathrm{Hom}_R(V, W)$$

$$\cong \mathrm{Hom}_R(U, W) \oplus \mathrm{Hom}_R(V, W),$$

so

$$P^* \oplus P_1^* \cong F^*, \qquad \mathrm{Hom}_R(P, Q) \oplus \mathrm{Hom}_R(P_1, Q) \cong \mathrm{Hom}_R(F, Q).$$

By tensoring with Q and using the fact that tensor distributes w.r.t. coproducts, we get

$$(P^* \otimes_R Q) \oplus (P_1^* \otimes Q) \cong (P^* \oplus P_1^*) \otimes_R Q \cong F^* \otimes_R Q.$$

Now, the proof of Proposition 2.17 goes through because F is free and finitely generated. This implies

$$F^* \otimes Q \cong \text{Hom}(F, Q),$$

so

$$\alpha_\otimes : (P^* \otimes_R Q) \oplus (P_1^* \otimes Q) \cong F^* \otimes_R Q \longrightarrow \text{Hom}_R(F, Q)$$
$$\cong \text{Hom}_R(P, Q) \oplus \text{Hom}_R(P_1, Q)$$

is an isomorphism, and as $\alpha_\otimes$ maps $P^* \otimes_R Q$ to $\text{Hom}_R(P, Q)$, it yields an isomorphism between these two spaces.   □

The isomorphism $\alpha_\otimes : P^* \otimes_R Q \cong \text{Hom}_R(P, Q)$ of Proposition 2.33 is still given by

$$\alpha_\otimes(u^* \otimes f)(x) = u^*(x)f, \qquad u^* \in P^*, \ f \in Q, \ x \in P.$$

It is convenient to introduce the *evaluation map* $\text{Ev}_x : P^* \otimes_R Q \to Q$ defined for every $x \in P$ by

$$\text{Ev}_x(u^* \otimes f) = u^*(x)f, \qquad u^* \in P^*, \ f \in Q.$$

In Section 10.5 we will need to consider a slightly weaker version of the universal mapping property of tensor products. The situation is this: We have a commutative R-algebra S, where R is a field (or even a commutative ring), we have two R-modules U and V, and moreover, U is a right S-module and V is a left S-module. In Section 10.5, this corresponds to $R = \mathbb{R}$, $S = C^\infty(B)$, $U = \mathcal{A}^i(B)$, and $V = \Gamma(\xi)$, where ξ is a vector bundle. Then we can form the tensor product $U \otimes_R V$, and we let $U \otimes_S V$ be the quotient module $(U \otimes_R V)/W$, where W is the submodule of $U \otimes_R V$ generated by the elements of the form

$$us \otimes_R v - u \otimes_R sv.$$

As S is commutative, we can make $U \otimes_S V$ into an S-module by defining the action of S *via*

$$s(u \otimes_S v) = us \otimes_S v.$$

It is verified that this S-module is isomorphic to the tensor product of U and V as S-modules, and the following universal mapping property holds:

Proposition 2.34. *For every R-bilinear map $f : U \times V \to Z$, if f satisfies the property*

$$f(us, v) = f(u, sv), \qquad \text{for all } u \in U, \ v \in V, \ s \in S,$$

then f induces a unique R-linear map $\widehat{f} \colon U \otimes_S V \to Z$ such that

$$f(u, v) = \widehat{f}(u \otimes_S v), \qquad \text{for all } u \in U, \ v \in V.$$

Note that the linear map $\widehat{f} \colon U \otimes_S V \to Z$ is *only R-linear*; it is *not S-linear* in general.

2.12 Problems

Problem 2.1. Prove Proposition 2.4.

Problem 2.2. Given two linear maps $f \colon E \to E'$ and $g \colon F \to F'$, we defined the unique linear map

$$f \otimes g \colon E \otimes F \to E' \otimes F'$$

by

$$(f \otimes g)(u \otimes v) = f(u) \otimes g(v),$$

for all $u \in E$ and all $v \in F$. See Proposition 2.9. Thus $f \otimes g \in \mathrm{Hom}(E \otimes F, E' \otimes F')$. If we denote the tensor product $E \otimes F$ by $T(E, F)$, and we assume that E, E' and F, F' are finite dimensional, pick bases and show that the map induced by $f \otimes g \mapsto T(f, g)$ is an isomorphism

$$\mathrm{Hom}(E, F) \otimes \mathrm{Hom}(E', F') \cong \mathrm{Hom}(E \otimes F, E' \otimes F').$$

Problem 2.3. Adjust the proof of Proposition 2.13 (2) to show that

$$E \otimes (F \otimes G) \cong E \otimes F \otimes G,$$

whenever E, F, and G are arbitrary vector spaces.

Problem 2.4. Given a fixed vector space G, for any two vector spaces M and N and every linear map $f \colon M \to N$, we defined $\tau_G(f) = f \otimes \mathrm{id}_G$ to be the unique linear map making the following diagram commute.

$$
\begin{array}{ccc}
M \times G & \xrightarrow{\ \iota_{M\otimes}\ } & M \otimes G \\
{\scriptstyle f \times \mathrm{id}_G}\downarrow & & \downarrow{\scriptstyle f \otimes \mathrm{id}_G} \\
N \times G & \xrightarrow[\ \iota_{N\otimes}\ ]{} & N \otimes G
\end{array}
$$

See the proof of Proposition 2.13 (3). Show that

(1) $\tau_G(0) = 0$,
(2) $\tau_G(\mathrm{id}_M) = (\mathrm{id}_M \otimes \mathrm{id}_G) = \mathrm{id}_{M \otimes G}$,
(3) If $f' \colon M \to N$ is another linear map, then $\tau_G(f + f') = \tau_G(f) + \tau_G(f')$.

Problem 2.5. Induct on $m \geq 2$ to prove the canonical isomorphism

$$V^{\otimes m} \otimes V^{\otimes n} \cong V^{\otimes (m+n)}.$$

Use this isomorphism to show that $\cdot \colon V^{\otimes m} \times V^{\otimes n} \longrightarrow V^{\otimes (m+n)}$ defined as

$$(v_1 \otimes \cdots \otimes v_m) \cdot (w_1 \otimes \cdots \otimes w_n) = v_1 \otimes \cdots \otimes v_m \otimes w_1 \otimes \cdots \otimes w_n.$$

induces a multiplication on $T(V)$.
Hint. See Jacobson [61], Section 3.9, or Bertin [11], Chapter 4, Section 2.

Problem 2.6. Prove Proposition 2.19.
Hint. See Knapp [66] (Appendix A, Proposition A.14) or Bertin [11] (Chapter 4, Theorem 2.4).

Problem 2.7. Given linear maps $f \colon E \to E'$ and $f' \colon E' \to E''$, show that

$$(f' \circ f) \odot (f' \circ f) = (f' \odot f') \circ (f \odot f).$$

Problem 2.8. Complete the proof of Proposition 2.29 for the case of an infinite-dimensional vector space E.

Problem 2.9. Let I be a finite index set of cardinality p. Let m be a nonnegative integer. Show that the number of multisets over I with cardinality m is $\binom{p+m-1}{m}$.

Problem 2.10. Prove Proposition 2.28.

Problem 2.11. Using bases, show that the bilinear map at $(*)$ in Section 2.9 produces a nondegenerate pairing.

Problem 2.12. Let $\mathfrak{I}$ be the two-sided ideal generated by all tensors of the form $u \otimes v - v \otimes u \in V^{\otimes 2}$. Prove that $\mathrm{S}^m(V) \cong V^{\otimes m}/(\mathfrak{I} \cap V^{\otimes m})$.

Problem 2.13. Verify Equation $(*)$ of Section 2.10 for arbitrary nonnegative integers m and n.

Problem 2.14. Let P be a finitely generated projective R-module. Recall that $P^* = \mathrm{Hom}_R(P, R)$. Show that $P^{**} \cong P$.

Problem 2.15. Let S be a commutative R-algebra, where R is a commutative ring. Suppose we have R-modules U and V, where U is a right S-module and V is a left S-module. We form the tensor product $U \otimes_R V$, and we let $U \otimes_S V$ be the quotient module $(U \otimes_R V)/W$, where W is the submodule of $U \otimes_R V$ generated by the elements of the form

$$us \otimes_R v - u \otimes_R sv.$$

As S is commutative, we can make $U \otimes_S V$ into an S-module by defining the action of S via

$$s(u \otimes_S v) = us \otimes_S v.$$

Verify this S-module is isomorphic to the tensor product of U and V as S-modules.

Problem 2.16. Prove Proposition 2.34.

Chapter 3
Exterior Tensor Powers and Exterior Algebras

In this chapter we consider *alternating* (also called *skew-symmetric*) multilinear maps and *exterior tensor powers* (also called *alternating tensor powers*), denoted $\bigwedge^n(E)$. In many respects alternating multilinear maps and exterior tensor powers can be treated much like symmetric tensor powers, except that $\text{sgn}(\sigma)$ needs to be inserted in front of the formulae valid for symmetric powers.

Roughly speaking, we are now in the world of determinants rather than in the world of permanents. However, there are also some fundamental differences, one of which being that the exterior tensor power $\bigwedge^n(E)$ is the trivial vector space (0) when E is finite dimensional and $n > \dim(E)$. This chapter provides the firm foundations for understanding differential forms.

In Section 3.1 we define the *exterior powers* of a vector space E. This time, instead of dealing with symmetric multilinear maps, we deal with *alternating multilinear maps*, which are multilinear maps $f \colon E^n \to F$ such that $f(u_1, \ldots, u_n) = 0$ whenever two adjacent arguments are identical. This implies that $f(u_1, \ldots, u_n) = 0$ whenever any two arguments are identical, and that $f(\ldots, u_i, u_{i+1}, \ldots) = -f(\ldots, u_{i+1}, u_i, \ldots)$.

Given a vector space E over a field K, for any $n \geq 1$, the *exterior tensor power* $\bigwedge^n(E)$ is defined by a universal mapping property: it is a vector space with an injection $i_\wedge \colon E^n \to \bigwedge^n(E)$, such that for every vector space F and every alternating multilinear map $f \colon E^n \to F$, there is a *unique linear map* $f_\wedge \colon \bigwedge^n(E) \to F$ such that

$$f = f_\wedge \circ i_\wedge,$$

as illustrated in the following diagram:

The original version of this chapter was revised. The correction to this chapter is available at https://doi.org/10.1007/978-3-030-46047-1_12

We prove that the above universal mapping property defines $\bigwedge^n(E)$ up to isomorphism, and then we prove its existence by constructing the quotient

$$\bigwedge^n(E) = E^{\otimes n}/(\mathfrak{I}_a \cap E^{\otimes n}),$$

where $\mathfrak{I}_a$ is the two-sided ideal of the tensor algebra $T(E)$ generated by all tensors of the form $u \otimes u \in E^{\otimes 2}$. As a corollary, there is an isomorphism

$$\mathrm{Hom}(\bigwedge^n(E), F) \cong \mathrm{Alt}^n(E; F)$$

between the vector space of linear maps $\mathrm{Hom}(\bigwedge^n(E), F)$ and the vector space of alternating multilinear maps $\mathrm{Alt}^n(E; F)$. A new phenomenon that arises with exterior tensor powers is that if E has dimension n, then $\bigwedge^k(E) = (0)$ for all $k > n$.

Given any two linear maps $f, g \colon E \to E'$, there is a linear map $f \wedge g \colon \bigwedge^2(E) \to \bigwedge^2(E')$.

A basic isomorphism involving the exterior power of a direct sum is shown at the end of this section.

In Section 3.2 we show how to construct a basis of $\bigwedge^k(E)$ from a basis of E ($1 \le k \le n$). If E has dimension n and if $(e_1, \dots, e_n)$ is a basis of E, for any finite sequence $I = (i_1, \dots, i_k)$ with $1 \le i_1 < i_2 < \cdots < i_k \le n$, if we write

$$e_I = e_{i_1} \wedge \cdots \wedge e_{i_k},$$

then the family of all the e_I is a basis of $\bigwedge^k(E)$. Thus $\bigwedge^k(E)$ has dimension $\binom{n}{k}$.

Section 3.3 is devoted to duality in exterior powers. There is a nondegenerate pairing

$$\langle -, - \rangle \colon \bigwedge^n(E^*) \times \bigwedge^n(E) \longrightarrow K$$

defined in terms of generators by

$$\langle v_1^* \wedge \cdots \wedge v_n^*, u_1 \wedge \cdots \wedge u_n \rangle = \det(v_j^*(u_i)).$$

As a consequence, if E is finite dimensional, we have canonical isomorphisms

$$(\bigwedge^n(E))^* \cong \bigwedge^n(E^*) \cong \text{Alt}^n(E; K).$$

The exterior tensor power $\bigwedge^k(E)$ is naturally embedded in $E^{\otimes n}$ (if K has characteristic 0).

In Section 3.4 we define exterior algebras (or Grassmann algebras). As in the case of symmetric tensors, we can pack together all the exterior powers $\bigwedge^n(V)$ into an algebra. Given any vector space V, the vector space

$$\bigwedge(V) = \bigoplus_{m \geq 0}^m \bigwedge^m(V)$$

is an algebra called the *exterior algebra (or Grassmann algebra) of V*. The exterior algebra satisfies a universal mapping condition.

If we define

$$\text{Alt}(E) = \bigoplus_{n \geq 0} \text{Alt}^n(E; K),$$

then this is an algebra under a combinatorial definition of the wedge operation, and this algebra is isomorphic to $\bigwedge(E^*)$.

In Section 3.5, we introduce the *Hodge ∗-operator*. Given a vector space V of dimension n with an inner product $\langle -, - \rangle$, for some chosen orientation of V, for each k such that $1 \leq k \leq n$, there is an isomorphism $*$ from $\bigwedge^k(V)$ to $\bigwedge^{n-k}(V)$. The Hodge $*$ operator can be extended to an isomorphism of $\bigwedge(V)$. It is the main tool used to define a generalization of the Laplacian (the *Hodge Laplacian*) to a smooth manifold.

The next three sections are somewhat more technical. They deal with some contraction operators called *left hooks* and *right hooks*. The motivation comes from the problem of understanding when a tensor $\alpha \in \bigwedge^k(E)$ is *decomposable*. An arbitrary tensor $\alpha \in \bigwedge^k(E)$ is a linear combination of tensors of the form $u_1 \wedge \cdots \wedge u_k$, called *decomposable*. The issue is to find criteria for decomposability. This is not as obvious as it looks. For example, we have

$$e_1 \wedge e_2 + e_1 \wedge e_3 + e_2 \wedge e_3 = (e_1 + e_2) \wedge (e_2 + e_3),$$

where the tensor on the right is clearly decomposable, but the tensor on the left does not look decomposable at first glance. Criteria for testing decomposability using left hooks are given in Section 3.7.

Say $\dim(E) = n$. Using our nonsingular pairing

$$\langle -, - \rangle \colon \bigwedge^p E^* \times \bigwedge^p E \longrightarrow K \qquad (1 \leq p \leq n)$$

defined on generators by

$$\langle u_1^* \wedge \cdots \wedge u_p^*, v_1 \wedge \cdots \wedge u_p \rangle = \det(u_i^*(v_j)),$$

in Section 3.6 we define various contraction operations (partial evaluation operators)

$$\lrcorner : \overset{p}{\bigwedge} E \times \overset{p+q}{\bigwedge} E^* \longrightarrow \overset{q}{\bigwedge} E^*, \quad \lrcorner : \overset{p}{\bigwedge} E^* \times \overset{p+q}{\bigwedge} E \longrightarrow \overset{q}{\bigwedge} E \quad \text{left hook}$$

and

$$\llcorner : \overset{p+q}{\bigwedge} E^* \times \overset{p}{\bigwedge} E \longrightarrow \overset{q}{\bigwedge} E^*, \quad \llcorner : \overset{p+q}{\bigwedge} E \times \overset{p}{\bigwedge} E^* \longrightarrow \overset{q}{\bigwedge} E \quad \text{right hook.}$$

These left and right hooks also have combinatorial definitions in terms of the basis vectors e_I and e_J^*. The right hooks can be expressed in terms of the left hooks. Left and right hooks induce isomorphisms $\gamma : \bigwedge^p E \to \bigwedge^{n-p} E^*$ and $\delta : \bigwedge^p E^* \to \bigwedge^{n-p} E$.

A criterion for testing decomposability in terms of left hooks is presented in Section 3.7.

In Section 3.8, based on the criterion established in Section 3.7, we derive a criterion for testing decomposability in terms of equations known as the *Grassmann-Plücker's equations*. We also show that the Grassmannian manifold $G(k, n)$ can be embedded as an algebraic variety into $\mathbb{RP}^{\binom{n}{k}-1}$ defined by equations of degree 2.

Section 3.9 discusses vector-valued alternating forms. The purpose of this section is to present the technical background needed for Sections 4.5 and 4.6 on vector-valued differential forms, in particular in the case of Lie groups where differential forms taking their values in a Lie algebra arise naturally.

Given a finite-dimensional vector space E and any vector space F, there is an isomorphism

$$\mu_F : \left(\overset{n}{\bigwedge} (E^*) \right) \otimes F \longrightarrow \text{Alt}^n(E; F)$$

defined on generators by

$$\mu_F((v_1^* \wedge \cdots \wedge v_n^*) \otimes f)(u_1, \ldots, u_n) = (\det(v_j^*(u_i))) f,$$

with $v_1^*, \ldots, v_n^* \in E^*, u_1, \ldots, u_n \in E$, and $f \in F$. We also discuss a generalization of the wedge product.

3.1 Exterior Tensor Powers

As in the case of symmetric tensor powers, since we already have the tensor algebra $T(V)$, we can proceed rather quickly. But first let us review some basic definitions and facts.

Definition 3.1. Let $f\colon E^n \to F$ be a multilinear map. We say that f *alternating* iff for all $u_i \in E$, $f(u_1, \ldots, u_n) = 0$ whenever $u_i = u_{i+1}$, for some i with $1 \le i \le n - 1$; that is, $f(u_1, \ldots, u_n) = 0$ whenever two adjacent arguments are identical. We say that f is *skew-symmetric* (or *anti-symmetric*) iff

$$f(u_{\sigma(1)}, \ldots, u_{\sigma(n)}) = \mathrm{sgn}(\sigma) f(u_1, \ldots, u_n),$$

for every permutation $\sigma \in \mathfrak{S}_n$, and all $u_i \in E$.

For $n = 1$, we agree that every linear map $f\colon E \to F$ is alternating. The vector space of all multilinear alternating maps $f\colon E^n \to F$ is denoted $\mathrm{Alt}^n(E; F)$. Note that $\mathrm{Alt}^1(E; F) = \mathrm{Hom}(E, F)$. The following basic proposition shows the relationship between alternation and skew-symmetry.

Proposition 3.1. *Let $f\colon E^n \to F$ be a multilinear map. If f is alternating, then the following properties hold:*

(1) For all i, with $1 \le i \le n - 1$,

$$f(\ldots, u_i, u_{i+1}, \ldots) = -f(\ldots, u_{i+1}, u_i, \ldots).$$

(2) For every permutation $\sigma \in \mathfrak{S}_n$,

$$f(u_{\sigma(1)}, \ldots, u_{\sigma(n)}) = \mathrm{sgn}(\sigma) f(u_1, \ldots, u_n).$$

(3) For all i, j, with $1 \le i < j \le n$,

$$f(\ldots, u_i, \ldots u_j, \ldots) = 0 \quad \text{whenever } u_i = u_j.$$

Moreover, if our field K has characteristic different from 2, then every skew-symmetric multilinear map is alternating.

Proof.

(1) By multilinearity applied twice, we have

$$f(\ldots, u_i + u_{i+1}, u_i + u_{i+1}, \ldots) = f(\ldots, u_i, u_i, \ldots) + f(\ldots, u_i, u_{i+1}, \ldots)$$
$$+ f(\ldots, u_{i+1}, u_i, \ldots) + f(\ldots, u_{i+1}, u_{i+1}, \ldots).$$

Since f is alternating, we get

$$0 = f(\ldots, u_i, u_{i+1}, \ldots) + f(\ldots, u_{i+1}, u_i, \ldots);$$

that is, $f(\ldots, u_i, u_{i+1}, \ldots) = -f(\ldots, u_{i+1}, u_i, \ldots)$.

(2) Clearly, the symmetric group, $\mathfrak{S}_n$, acts on $\mathrm{Alt}^n(E; F)$ on the left, *via*

$$\sigma \cdot f(u_1, \ldots, u_n) = f(u_{\sigma(1)}, \ldots, u_{\sigma(n)}).$$

Consequently, as $\mathfrak{S}_n$ is generated by the transpositions (permutations that swap exactly two elements), since for a transposition, (2) is simply (1), we deduce (2) by induction on the number of transpositions in σ.

(3) There is a permutation σ that sends u_i and u_j respectively to u_1 and u_2. By hypothesis $u_i = u_j$, so we have $u_{\sigma(1)} = u_{\sigma(2)}$, and as f is alternating we have

$$f(u_{\sigma(1)}, \ldots, u_{\sigma(n)}) = 0.$$

However, by (2),

$$f(u_1, \ldots, u_n) = \mathrm{sgn}(\sigma) f(u_{\sigma(1)}, \ldots, u_{\sigma(n)}) = 0.$$

Now when f is skew-symmetric, if σ is the transposition swapping u_i and $u_{i+1} = u_i$, as $\mathrm{sgn}(\sigma) = -1$, we get

$$f(\ldots, u_i, u_i, \ldots) = -f(\ldots, u_i, u_i, \ldots),$$

so that

$$2f(\ldots, u_i, u_i, \ldots) = 0,$$

and in every characteristic except 2, we conclude that $f(\ldots, u_i, u_i, \ldots) = 0$, namely f is alternating. $\square$

Proposition 3.1 shows that in every field of characteristic different from 2, alternating and skew-symmetric multilinear maps are identical. Using Proposition 3.1 we easily deduce the following crucial fact.

Proposition 3.2. *Let $f\colon E^n \to F$ be an alternating multilinear map. For any families of vectors, $(u_1, \ldots, u_n)$ and $(v_1, \ldots, v_n)$, with $u_i, v_i \in E$, if*

$$v_j = \sum_{i=1}^{n} a_{ij} u_i, \qquad 1 \le j \le n,$$

then

$$f(v_1, \ldots, v_n) = \left(\sum_{\sigma \in \mathfrak{S}_n} \text{sgn}(\sigma)\, a_{\sigma(1),1} \cdots a_{\sigma(n),n} \right) f(u_1, \ldots, u_n)$$

$$= \det(A) f(u_1, \ldots, u_n),$$

where A is the $n \times n$ matrix, $A = (a_{ij})$.

Proof. Use Property (ii) of Proposition 3.1.   □

We are now ready to define and construct exterior tensor powers.

Definition 3.2. An *n-th exterior tensor power* of a vector space E, where $n \geq 1$, is a vector space A together with an alternating multilinear map $\varphi \colon E^n \to A$, such that for every vector space F and for every alternating multilinear map $f \colon E^n \to F$, there is a unique linear map $f_\wedge \colon A \to F$ with

$$f(u_1, \ldots, u_n) = f_\wedge(\varphi(u_1, \ldots, u_n)),$$

for all $u_1, \ldots, u_n \in E$, or for short

$$f = f_\wedge \circ \varphi.$$

Equivalently, there is a unique linear map $f_\wedge$ such that the following diagram commutes:

The above property is called the *universal mapping property* of the exterior tensor power (A, φ).

We now show that any two *n*-th exterior tensor powers (A_1, φ_1) and (A_2, φ_2) for E are isomorphic.

Proposition 3.3. *Given any two n-th exterior tensor powers (A_1, φ_1) and (A_2, φ_2) for E, there is an isomorphism $h \colon A_1 \to A_2$ such that*

$$\varphi_2 = h \circ \varphi_1.$$

Proof. Replace tensor product by *n*-th exterior tensor power in the proof of Proposition 2.5.   □

We next give a construction that produces an *n*-th exterior tensor power of a vector space E.

Theorem 3.4. *Given a vector space E, an n-th exterior tensor power $(\bigwedge^n(E), \varphi)$ for E can be constructed ($n \geq 1$). Furthermore, denoting $\varphi(u_1, \ldots, u_n)$ as $u_1 \wedge \cdots \wedge u_n$, the exterior tensor power $\bigwedge^n(E)$ is generated by the vectors $u_1 \wedge \cdots \wedge u_n$, where $u_1, \ldots, u_n \in E$, and for every alternating multilinear map $f: E^n \to F$, the unique linear map $f_\wedge: \bigwedge^n(E) \to F$ such that $f = f_\wedge \circ \varphi$ is defined by*

$$f_\wedge(u_1 \wedge \cdots \wedge u_n) = f(u_1, \ldots, u_n)$$

on the generators $u_1 \wedge \cdots \wedge u_n$ of $\bigwedge^n(E)$.

Proof Sketch. We can give a quick proof using the tensor algebra $T(E)$. Let $\mathfrak{I}_a$ be the two-sided ideal of $T(E)$ generated by all tensors of the form $u \otimes u \in E^{\otimes 2}$. Then let

$$\bigwedge^n(E) = E^{\otimes n}/(\mathfrak{I}_a \cap E^{\otimes n})$$

and let π be the projection $\pi: E^{\otimes n} \to \bigwedge^n(E)$. If we let $u_1 \wedge \cdots \wedge u_n = \pi(u_1 \otimes \cdots \otimes u_n)$, it is easy to check that $(\bigwedge^n(E), \wedge)$ satisfies the conditions of Theorem 3.4. $\square$

Remark. We can also define

$$\bigwedge(E) = T(E)/\mathfrak{I}_a = \bigoplus_{n \geq 0} \bigwedge^n(E),$$

the *exterior algebra* of E. This is the skew-symmetric counterpart of $S(E)$, and we will study it a little later.

For simplicity of notation, we may write $\bigwedge^n E$ for $\bigwedge^n(E)$. We also abbreviate "exterior tensor power" as "exterior power." Clearly, $\bigwedge^1(E) \cong E$, and it is convenient to set $\bigwedge^0(E) = K$.

The fact that the map $\varphi: E^n \to \bigwedge^n(E)$ is alternating and multilinear can also be expressed as follows:

$$u_1 \wedge \cdots \wedge (u_i + v_i) \wedge \cdots \wedge u_n = (u_1 \wedge \cdots \wedge u_i \wedge \cdots \wedge u_n)$$
$$+ (u_1 \wedge \cdots \wedge v_i \wedge \cdots \wedge u_n),$$
$$u_1 \wedge \cdots \wedge (\lambda u_i) \wedge \cdots \wedge u_n = \lambda(u_1 \wedge \cdots \wedge u_i \wedge \cdots \wedge u_n),$$
$$u_{\sigma(1)} \wedge \cdots \wedge u_{\sigma(n)} = \text{sgn}(\sigma)\, u_1 \wedge \cdots \wedge u_n,$$

for all $\sigma \in \mathfrak{S}_n$.

The map φ from E^n to $\bigwedge^n(E)$ is often denoted $\iota_\wedge$, so that

$$\iota_\wedge(u_1, \ldots, u_n) = u_1 \wedge \cdots \wedge u_n.$$

Theorem 3.4 implies the following result.

Proposition 3.5. *There is a canonical isomorphism*

$$\mathrm{Hom}(\bigwedge^n(E), F) \cong \mathrm{Alt}^n(E; F)$$

between the vector space of linear maps $\mathrm{Hom}(\bigwedge^n(E), F)$ *and the vector space of alternating multilinear maps* $\mathrm{Alt}^n(E; F)$, *given by the linear map* $- \circ \varphi$ *defined by* $\mapsto h \circ \varphi$, *with* $h \in \mathrm{Hom}(\bigwedge^n(E), F)$. *In particular, when* $F = K$, *we get a canonical isomorphism*

$$\left(\bigwedge^n(E)\right)^* \cong \mathrm{Alt}^n(E; K).$$

Definition 3.3. Tensors $\alpha \in \bigwedge^n(E)$ are called *alternating n-tensors* or *alternating tensors of degree n* and we write $\deg(\alpha) = n$. Tensors of the form $u_1 \wedge \cdots \wedge u_n$, where $u_i \in E$, are called *simple (or decomposable) alternating n-tensors*. Those alternating n-tensors that are not simple are often called *compound alternating n-tensors*. Simple tensors $u_1 \wedge \cdots \wedge u_n \in \bigwedge^n(E)$ are also called *n-vectors* and tensors in $\bigwedge^n(E^*)$ are often called *(alternating) n-forms*.

Given a linear map $f \colon E \to E'$, since the map $\iota'_\wedge \circ (f \times f)$ is bilinear and alternating, there is a unique linear map $f \wedge f \colon \bigwedge^2(E) \to \bigwedge^2(E')$ making the following diagram commute:

$$
\begin{array}{ccc}
E^2 & \xrightarrow{\iota_\wedge} & \bigwedge^2(E) \\
{\scriptstyle f \times f}\downarrow & & \downarrow{\scriptstyle f \wedge f} \\
(E')^2 & \xrightarrow{\iota'_\wedge} & \bigwedge^2(E').
\end{array}
$$

The map $f \wedge f \colon \bigwedge^2(E) \to \bigwedge^2(E')$ is determined by

$$(f \wedge f)(u \wedge v) = f(u) \wedge f(v).$$

Proposition 3.6. *Given any two linear maps* $f \colon E \to E'$ *and* $f' \colon E' \to E''$, *we have*

$$(f' \circ f) \wedge (f' \circ f) = (f' \wedge f') \circ (f \wedge f).$$

The generalization to the alternating product $f \wedge \cdots \wedge f$ of $n \geq 3$ copies of the linear map $f \colon E \to E'$ is immediate, and left to the reader.

We can show the following property of the exterior tensor product, using the proof technique of Proposition 2.13.

Proposition 3.7. *We have the following isomorphism:*

$$\overset{n}{\bigwedge}(E \oplus F) \cong \bigoplus_{k=0}^{n} \overset{k}{\bigwedge}(E) \otimes \overset{n-k}{\bigwedge}(F).$$

3.2　Bases of Exterior Powers

Definition 3.4. Let E be any vector space. For any basis $(u_i)_{i \in \Sigma}$ for E, we assume that some total ordering $\leq$ on the index set Σ has been chosen. Call the pair $((u_i)_{i \in \Sigma}, \leq)$ an *ordered basis*. Then for any nonempty finite subset $I \subseteq \Sigma$, let

$$u_I = u_{i_1} \wedge \cdots \wedge u_{i_m},$$

where $I = \{i_1, \ldots, i_m\}$, with $i_1 < \cdots < i_m$.

Since $\bigwedge^n(E)$ is generated by the tensors of the form $v_1 \wedge \cdots \wedge v_n$, with $v_i \in E$, in view of skew-symmetry, it is clear that the tensors u_I with $|I| = n$ generate $\bigwedge^n(E)$ (where $((u_i)_{i \in \Sigma}, \leq)$ is an ordered basis). Actually they form a basis. To gain an intuitive understanding of this statement, let $m = 2$ and E be a 3-dimensional vector space lexicographically ordered basis $\{e_1, e_2, e_3\}$. We claim that

$$e_1 \wedge e_2, \qquad e_1 \wedge e_3, \qquad e_2 \wedge e_3$$

form a basis for $\bigwedge^2(E)$ since they not only generate $\bigwedge^2(E)$ but are linearly independent. The linear independence is argued as follows: given any vector space F, if w_{12}, w_{13}, w_{23} are any vectors in F, there is an alternating bilinear map $h \colon E^2 \to F$ such that

$$h(e_1, e_2) = w_{12}, \qquad h(e_1, e_3) = w_{13}, \qquad h(e_2, e_3) = w_{23}.$$

Because h yields a unique linear map $h_\wedge \colon \bigwedge^2 E \to F$ such that

$$h_\wedge(e_i \wedge e_j) = w_{ij}, \qquad 1 \leq i < j \leq 3,$$

by Proposition 2.4, the vectors

$$e_1 \wedge e_2, \qquad e_1 \wedge e_3, \qquad e_2 \wedge e_3$$

are linearly independent. This suggests understanding how an alternating bilinear function $f\colon E^2 \to F$ is expressed in terms of its values $f(e_i, e_j)$ on the basis vectors (e_1, e_2, e_3). Using bilinearity and alternation, we obtain

$$f(u_1e_1 + u_2e_2 + u_3e_3, v_1e_1 + v_2e_2 + v_3e_3) = (u_1v_2 - u_2v_1)f(e_1, e_2)$$
$$+ (u_1v_3 - u_3v_1)f(e_1, e_3)$$
$$+ (u_2v_3 - u_3v_2)f(e_2, e_3).$$

Therefore, given $w_{12}, w_{13}, w_{23} \in F$, the function h given by

$$h(u_1e_1+u_2e_2+u_3e_3, v_1e_1+v_2e_2+v_3e_3)=(u_1v_2-u_2v_1)w_{12}$$
$$+ (u_1v_3-u_3v_1)w_{13}$$
$$+ (u_2v_3 - u_3v_2)w_{23}$$

is clearly bilinear and alternating, and by construction $h(e_i, e_j) = w_{ij}$, with $1 \leq i < j \leq 3$ does the job.

We now prove the assertion that tensors u_I with $|I| = n$ generate $\bigwedge^n(E)$ for arbitrary n.

Proposition 3.8. *Given any vector space E, if E has finite dimension $d = \dim(E)$, then for all $n > d$, the exterior power $\bigwedge^n(E)$ is trivial; that is $\bigwedge^n(E) = (0)$. If $n \leq d$ or if E is infinite dimensional, then for every ordered basis $((u_i)_{i \in \Sigma}, \leq)$, the family (u_I) is basis of $\bigwedge^n(E)$, where I ranges over finite nonempty subsets of Σ of size $|I| = n$.*

Proof. First assume that E has finite dimension $d = \dim(E)$ and that $n > d$. We know that $\bigwedge^n(E)$ is generated by the tensors of the form $v_1 \wedge \cdots \wedge v_n$, with $v_i \in E$. If $u_1, \ldots, u_d$ is a basis of E, as every v_i is a linear combination of the u_j, when we expand $v_1 \wedge \cdots \wedge v_n$ using multilinearity, we get a linear combination of the form

$$v_1 \wedge \cdots \wedge v_n = \sum_{(j_1,\ldots,j_n)} \lambda_{(j_1,\ldots,j_n)} \, u_{j_1} \wedge \cdots \wedge u_{j_n},$$

where each $(j_1, \ldots, j_n)$ is some sequence of integers $j_k \in \{1, \ldots, d\}$. As $n > d$, each sequence $(j_1, \ldots, j_n)$ must contain two identical elements. By alternation, $u_{j_1} \wedge \cdots \wedge u_{j_n} = 0$, and so $v_1 \wedge \cdots \wedge v_n = 0$. It follows that $\bigwedge^n(E) = (0)$.

Now assume that either $\dim(E) = d$ and $n \leq d$, or that E is infinite dimensional. The argument below shows that the u_I are nonzero and linearly independent. As usual, let $u_i^* \in E^*$ be the linear form given by

$$u_i^*(u_j) = \delta_{ij}.$$

For any nonempty subset $I = \{i_1, \ldots, i_n\} \subseteq \Sigma$ with $i_1 < \cdots < i_n$, for any n vectors $v_1, \ldots, v_n \in E$, let

$$l_I(v_1, \ldots, v_n) = \det(u_{i_j}^*(v_k)) = \begin{vmatrix} u_{i_1}^*(v_1) & \cdots & u_{i_1}^*(v_n) \\ \vdots & \ddots & \vdots \\ u_{i_n}^*(v_1) & \cdots & u_{i_n}^*(v_n) \end{vmatrix}.$$

If we let the n-tuple $(v_1, \ldots, v_n)$ vary we obtain a map l_I from E^n to K, and it is easy to check that this map is alternating multilinear. Thus l_I induces a unique linear map $L_I \colon \bigwedge^n(E) \to K$ making the following diagram commute.

$$
\begin{array}{ccc}
E^n & \xrightarrow{\ l_\wedge\ } & \bigwedge^n(E) \\
 & \searrow{\scriptstyle l_I} & \downarrow{\scriptstyle L_I} \\
 & & K
\end{array}
$$

Observe that for any nonempty finite subset $J \subseteq \Sigma$ with $|J| = n$, we have

$$L_I(u_J) = \begin{cases} 1 & \text{if } I = J \\ 0 & \text{if } I \neq J. \end{cases}$$

Note that when $\dim(E) = d$ and $n \leq d$, or when E is infinite dimensional, the forms $u_{i_1}^*, \ldots, u_{i_n}^*$ are all distinct, so the above does hold. Since $L_I(u_I) = 1$, we conclude that $u_I \neq 0$. If we have a linear combination

$$\sum_I \lambda_I u_I = 0,$$

where the above sum is finite and involves nonempty finite subset $I \subseteq \Sigma$ with $|I| = n$, for every such I, when we apply L_I we get $\lambda_I = 0$, proving linear independence. $\qquad\square$

As a corollary, if E is finite dimensional, say $\dim(E) = d$, and if $1 \leq n \leq d$, then we have

$$\dim(\overset{n}{\bigwedge}(E)) = \binom{n}{d},$$

and if $n > d$, then $\dim(\bigwedge^n(E)) = 0$.

Remark. When $n = 0$, if we set $u_\emptyset = 1$, then $(u_\emptyset) = (1)$ is a basis of $\bigwedge^0(V) = K$.

It follows from Proposition 3.8 that the family $(u_I)_I$ where $I \subseteq \Sigma$ ranges over finite subsets of Σ is a basis of $\bigwedge(V) = \bigoplus_{n \geq 0} \bigwedge^n(V)$.

As a corollary of Proposition 3.8 we obtain the following useful criterion for linear independence.

Proposition 3.9. *For any vector space E, the vectors $u_1, \ldots, u_n \in E$ are linearly independent iff $u_1 \wedge \cdots \wedge u_n \neq 0$.*

Proof. If $u_1 \wedge \cdots \wedge u_n \neq 0$, then $u_1, \ldots, u_n$ must be linearly independent. Otherwise, some u_i would be a linear combination of the other u_j's (with $j \neq i$), and then, as in the proof of Proposition 3.8, $u_1 \wedge \cdots \wedge u_n$ would be a linear combination of wedges in which two vectors are identical, and thus zero.

Conversely, assume that $u_1, \ldots, u_n$ are linearly independent. Then we have the linear forms $u_i^* \in E^*$ such that

$$u_i^*(u_j) = \delta_{i,j} \qquad 1 \leq i, j \leq n.$$

As in the proof of Proposition 3.8, we have a linear map $L_{u_1,\ldots,u_n} \colon \bigwedge^n(E) \to K$ given by

$$L_{u_1,\ldots,u_n}(v_1 \wedge \cdots \wedge v_n) = \det(u_j^*(v_i)) = \begin{vmatrix} u_1^*(v_1) & \cdots & u_1^*(v_n) \\ \vdots & \ddots & \vdots \\ u_n^*(v_1) & \cdots & u_n^*(v_n) \end{vmatrix},$$

for all $v_1 \wedge \cdots \wedge v_n \in \bigwedge^n(E)$. As $L_{u_1,\ldots,u_n}(u_1 \wedge \cdots \wedge u_n) = 1$, we conclude that $u_1 \wedge \cdots \wedge u_n \neq 0$. $\qquad \square$

Proposition 3.9 shows that *geometrically every nonzero wedge $u_1 \wedge \cdots \wedge u_n$ corresponds to some oriented version of an n-dimensional subspace of E.*

3.3 Duality for Exterior Powers

In this section *all vector spaces are assumed to have finite dimension.* We define a nondegenerate pairing $\bigwedge^n(E^*) \times \bigwedge^n(E) \longrightarrow K$ as follows: Consider the multilinear map

$$(E^*)^n \times E^n \longrightarrow K$$

given by

$$(v_1^*, \ldots, v_n^*, u_1, \ldots, u_n) \mapsto \sum_{\sigma \in \mathfrak{S}_n} \mathrm{sgn}(\sigma) \, v_{\sigma(1)}^*(u_1) \cdots v_{\sigma(n)}^*(u_n) = \det(v_j^*(u_i))$$

$$= \begin{vmatrix} v_1^*(u_1) & \cdots & v_1^*(u_n) \\ \vdots & \ddots & \vdots \\ v_n^*(u_1) & \cdots & v_n^*(u_n) \end{vmatrix}.$$

It is easily checked that this expression is alternating w.r.t. the u_i's and also w.r.t. the v_j^*. For any fixed $(v_1^*, \ldots, v_n^*) \in (E^*)^n$, we get an alternating multilinear map

$$l_{v_1^*,\ldots,v_n^*} : (u_1, \ldots, u_n) \mapsto \det(v_j^*(u_i))$$

from E^n to K. The map $l_{v_1^*,\ldots,v_n^*}$ extends uniquely to a linear map $L_{v_1^*,\ldots,v_n^*} : \bigwedge^n(E) \to K$ making the following diagram commute:

$$
\begin{array}{ccc}
E^n & \xrightarrow{\;\iota_\wedge\;} & \bigwedge^n(E) \\
 & \searrow_{l_{v_1^*,\ldots,v_n^*}} & \downarrow{L_{v_1^*,\ldots,v_n^*}} \\
 & & K.
\end{array}
$$

We also have the alternating multilinear map

$$(v_1^*, \ldots, v_n^*) \mapsto L_{v_1^*,\ldots,v_n^*}$$

from $(E^*)^n$ to $\mathrm{Hom}(\bigwedge^n(E), K)$, which extends to a linear map L from $\bigwedge^n(E^*)$ to $\mathrm{Hom}(\bigwedge^n(E), K)$ making the following diagram commute:

$$
\begin{array}{ccc}
(E^*)^n & \xrightarrow{\;\iota_{\wedge^*}\;} & \bigwedge^n(E^*) \\
 & \searrow & \downarrow{L} \\
 & & \mathrm{Hom}(\bigwedge^n(E), K).
\end{array}
$$

However, in view of the isomorphism

$$\mathrm{Hom}(U \otimes V, W) \cong \mathrm{Hom}(U, \mathrm{Hom}(V, W)),$$

with $U = \bigwedge^n(E^*)$, $V = \bigwedge^n(E)$ and $W = K$, we can view L as a linear map

$$L : \bigwedge^n(E^*) \otimes \bigwedge^n(E) \longrightarrow K,$$

which by Proposition 2.8 corresponds to a bilinear map

$$\langle -, - \rangle : \bigwedge^n(E^*) \times \bigwedge^n(E) \longrightarrow K. \tag{$*$}$$

This pairing is given explicitly in terms of generators by

$$\langle v_1^* \wedge \cdots \wedge v_n^*, u_1 \wedge \cdots \wedge u_n \rangle = \det(v_j^*(u_i)).$$

Now this pairing in nondegenerate. This can be shown using bases. Given any basis $(e_1, \ldots, e_m)$ of E, for every basis element $e_{i_1}^* \wedge \cdots \wedge e_{i_n}^*$ of $\bigwedge^n(E^*)$ (with $1 \leq i_1 < \cdots < i_n \leq m$), we have

$$\langle e_{i_1}^* \wedge \cdots \wedge e_{i_n}^*, e_{j_1} \wedge \cdots \wedge e_{j_n} \rangle = \begin{cases} 1 & \text{if } (j_1, \ldots, j_n) = (i_1, \ldots, i_n) \\ 0 & \text{otherwise.} \end{cases}$$

We leave the details as an exercise to the reader. As a consequence we get the following canonical isomorphisms.

Proposition 3.10. *There is a canonical isomorphism*

$$(\bigwedge^n(E))^* \cong \bigwedge^n(E^*).$$

There is also a canonical isomorphism

$$\mu \colon \bigwedge^n(E^*) \cong \operatorname{Alt}^n(E; K)$$

which allows us to interpret alternating tensors over E^* as alternating multilinear maps.

Proof. The second isomorphism follows from the canonical isomorphism $(\bigwedge^n(E))^* \cong \bigwedge^n(E^*)$ and the canonical isomorphism $(\bigwedge^n(E))^* \cong \operatorname{Alt}^n(E; K)$ given by Proposition 3.5. $\qquad\square$

Remarks.

1. The isomorphism $\mu \colon \bigwedge^n(E^*) \cong \operatorname{Alt}^n(E; K)$ discussed above can be described explicitly as the linear extension of the map given by

$$\mu(v_1^* \wedge \cdots \wedge v_n^*)(u_1, \ldots, u_n) = \det(v_j^*(u_i)).$$

2. The canonical isomorphism of Proposition 3.10 holds under more general conditions. Namely, that K is a commutative ring with identity and that E is a finitely generated projective K-module (see Definition 2.22). See Bourbaki [14] (Chapter III, §11, Section 5, Proposition 7).
3. Variants of our isomorphism μ are found in the literature. For example, there is a version μ', where

$$\mu' = \frac{1}{n!}\mu,$$

with the factor $\frac{1}{n!}$ added in front of the determinant. Each version has its own merits and inconveniences. Morita [82] uses μ' because it is more convenient than μ when dealing with characteristic classes. On the other hand, μ' may not be defined for a field with positive characteristic, and when using μ', some extra factor is needed in defining the wedge operation of alternating multilinear forms (see Section 3.4) and for exterior differentiation. The version μ is the one adopted by Warner [109], Knapp [66], Fulton and Harris [45], and Cartan [20, 21].

If $f: E \to F$ is any linear map, by transposition we get a linear map $f^\top: F^* \to E^*$ given by

$$f^\top(v^*) = v^* \circ f, \qquad v^* \in F^*.$$

Consequently, we have

$$f^\top(v^*)(u) = v^*(f(u)), \qquad \text{for all } u \in E \text{ and all } v^* \in F^*.$$

For any $p \geq 1$, the map

$$(u_1, \ldots, u_p) \mapsto f(u_1) \wedge \cdots \wedge f(u_p)$$

from E^p to $\bigwedge^p F$ is multilinear alternating, so it induces a unique linear map $\bigwedge^p f: \bigwedge^p E \to \bigwedge^p F$ making the following diagram commute

and defined on generators by

$$\left(\overset{p}{\bigwedge} f \right)(u_1 \wedge \cdots \wedge u_p) = f(u_1) \wedge \cdots \wedge f(u_p).$$

Combining $\bigwedge^p$ and duality, we get a linear map $\bigwedge^p f^\top: \bigwedge^p F^* \to \bigwedge^p E^*$ defined on generators by

$$\left(\overset{p}{\bigwedge} f^\top \right)(v_1^* \wedge \cdots \wedge v_p^*) = f^\top(v_1^*) \wedge \cdots \wedge f^\top(v_p^*).$$

Proposition 3.11. *If $f: E \to F$ is any linear map between two finite-dimensional vector spaces E and F, then*

$$\mu\left(\left(\bigwedge^p f^\top\right)(\omega)\right)(u_1, \ldots, u_p) = \mu(\omega)(f(u_1), \ldots, f(u_p)),$$

$$\omega \in \bigwedge^p F^*, \quad u_1, \ldots, u_p \in E.$$

Proof. It is enough to prove the formula on generators. By definition of μ, we have

$$\mu\left(\left(\bigwedge^p f^\top\right)(v_1^* \wedge \cdots \wedge v_p^*)\right)(u_1, \ldots, u_p) = \mu(f^\top(v_1^*) \wedge \cdots \wedge f^\top(v_p^*))(u_1, \ldots, u_p)$$

$$= \det(f^\top(v_j^*)(u_i))$$

$$= \det(v_j^*(f(u_i)))$$

$$= \mu(v_1^* \wedge \cdots \wedge v_p^*)(f(u_1), \ldots, f(u_p)),$$

as claimed. $\qquad\square$

Remark. The map $\bigwedge^p f^\top$ is often denoted f^*, although this is an ambiguous notation since p is dropped. Proposition 3.11 gives us the behavior of $\bigwedge^p f^\top$ under the identification of $\bigwedge^p E^*$ and $\mathrm{Alt}^p(E; K)$ *via* the isomorphism μ.

As in the case of symmetric powers, the map from E^n to $\bigwedge^n(E)$ given by $(u_1, \ldots, u_n) \mapsto u_1 \wedge \cdots \wedge u_n$ yields a surjection $\pi: E^{\otimes n} \to \bigwedge^n(E)$. Now this map has some section, so there is some injection $\beta: \bigwedge^n(E) \to E^{\otimes n}$ with $\pi \circ \beta = \mathrm{id}$. As we saw in Proposition 3.10 there is a canonical isomorphism

$$\left(\bigwedge^n(E)\right)^* \cong \bigwedge^n(E^*)$$

for any field K, even of positive characteristic. However, if our field K has characteristic 0, then there is a special injection having a natural definition involving an antisymmetrization process.

Recall, from Section 2.9 that we have a left action of the symmetric group $\mathfrak{S}_n$ on $E^{\otimes n}$.

The tensors $z \in E^{\otimes n}$ such that

$$\sigma \cdot z = \mathrm{sgn}(\sigma)\, z, \quad \text{for all} \quad \sigma \in \mathfrak{S}_n$$

are called *antisymmetrized* tensors. We define the map $\eta: E^n \to E^{\otimes n}$ by

$$\eta(u_1, \ldots, u_n) = \frac{1}{n!} \sum_{\sigma \in \mathfrak{S}_n} \mathrm{sgn}(\sigma)\, u_{\sigma(1)} \otimes \cdots \otimes u_{\sigma(n)}.^{[1]}$$

As the right-hand side is an alternating map, we get a unique linear map $\bigwedge^n \eta \colon \bigwedge^n(E) \to E^{\otimes n}$ making the following diagram commute.

$$
\begin{array}{ccc}
E^n & \xrightarrow{\iota_\wedge} & \bigwedge^n(E) \\
 & \searrow{\scriptstyle \eta} & \downarrow{\scriptstyle \bigwedge^n \eta} \\
 & & E^{\otimes n}.
\end{array}
$$

Clearly, $\bigwedge^n \eta(\bigwedge^n(E))$ is the set of antisymmetrized tensors in $E^{\otimes n}$. If we consider the map $A = (\bigwedge^n \eta) \circ \pi \colon E^{\otimes n} \longrightarrow E^{\otimes n}$, it is easy to check that $A \circ A = A$. Therefore, A is a projection, and by linear algebra, we know that

$$E^{\otimes n} = A(E^{\otimes n}) \oplus \mathrm{Ker}\, A = \bigwedge^n \eta(\bigwedge^n(E)) \oplus \mathrm{Ker}\, A.$$

It turns out that $\mathrm{Ker}\, A = E^{\otimes n} \cap \mathfrak{I}_a = \mathrm{Ker}\, \pi$, where $\mathfrak{I}_a$ is the two-sided ideal of $T(E)$ generated by all tensors of the form $u \otimes u \in E^{\otimes 2}$ (for example, see Knapp [66], Appendix A). Therefore, $\bigwedge^n \eta$ is injective,

$$E^{\otimes n} = \bigwedge^n \eta(\bigwedge^n(E)) \oplus (E^{\otimes n} \cap \mathfrak{I}_a) = \bigwedge^n \eta(\bigwedge^n(E)) \oplus \mathrm{Ker}\, \pi,$$

and the exterior tensor power $\bigwedge^n(E)$ is naturally embedded into $E^{\otimes n}$.

3.4   Exterior Algebras

As in the case of symmetric tensors, we can pack together all the exterior powers $\bigwedge^n(V)$ into an algebra.

Definition 3.5. Given any vector space V, the vector space

$$\bigwedge(V) = \bigoplus_{m \geq 0} \bigwedge^m(V)$$

is called the *exterior algebra (or Grassmann algebra) of V.*

---

[1] It is the division by $n!$ that requires the field to have characteristic zero.

   To make $\bigwedge(V)$ into an algebra, we mimic the procedure used for symmetric powers. If $\mathfrak{I}_a$ is the two-sided ideal generated by all tensors of the form $u \otimes u \in V^{\otimes 2}$, we set

$$\overset{\bullet}{\bigwedge}(V) = T(V)/\mathfrak{I}_a.$$

Then $\bigwedge^\bullet(V)$ automatically inherits a multiplication operation, called *wedge product*, and since $T(V)$ is graded, that is

$$T(V) = \bigoplus_{m \geq 0} V^{\otimes m},$$

we have

$$\overset{\bullet}{\bigwedge}(V) = \bigoplus_{m \geq 0} V^{\otimes m}/(\mathfrak{I}_a \cap V^{\otimes m}).$$

However, it is easy to check that

$$\overset{m}{\bigwedge}(V) \cong V^{\otimes m}/(\mathfrak{I}_a \cap V^{\otimes m}),$$

so

$$\overset{\bullet}{\bigwedge}(V) \cong \bigwedge(V).$$

When V has finite dimension d, we actually have a finite direct sum (coproduct)

$$\bigwedge(V) = \bigoplus_{m=0}^{d} \overset{m}{\bigwedge}(V),$$

and since each $\bigwedge^m(V)$ has dimension $\binom{d}{m}$, we deduce that

$$\dim(\bigwedge(V)) = 2^d = 2^{\dim(V)}.$$

   The multiplication, $\wedge \colon \bigwedge^m(V) \times \bigwedge^n(V) \to \bigwedge^{m+n}(V)$, is skew-symmetric in the following precise sense:

Proposition 3.12. *For all $\alpha \in \bigwedge^m(V)$ and all $\beta \in \bigwedge^n(V)$, we have*

$$\beta \wedge \alpha = (-1)^{mn} \alpha \wedge \beta.$$

Proof. Since $v \wedge u = -u \wedge v$ for all $u, v \in V$, Proposition 3.12 follows by induction.
□

Since $\alpha \wedge \alpha = 0$ for every *simple* (also called *decomposable*) tensor $\alpha = u_1 \wedge \cdots \wedge u_n$, it seems natural to infer that $\alpha \wedge \alpha = 0$ for *every* tensor $\alpha \in \bigwedge(V)$. If we consider the case where $\dim(V) \leq 3$, we can indeed prove the above assertion. However, if $\dim(V) \geq 4$, the above fact is generally false! For example, when $\dim(V) = 4$, if (u_1, u_2, u_3, u_4) is a basis for V, for $\alpha = u_1 \wedge u_2 + u_3 \wedge u_4$, we check that

$$\alpha \wedge \alpha = 2 u_1 \wedge u_2 \wedge u_3 \wedge u_4,$$

which is nonzero. However, if $\alpha \in \bigwedge^m E$ with m odd, since m^2 is also odd, we have

$$\alpha \wedge \alpha = (-1)^{m^2} \alpha \wedge \alpha = -\alpha \wedge \alpha,$$

so indeed $\alpha \wedge \alpha = 0$ (if K is not a field of characteristic 2).

The above discussion suggests that it might be useful to know when an alternating tensor is simple (decomposable). We will show in Section 3.6 that for tensors $\alpha \in \bigwedge^2(V)$, $\alpha \wedge \alpha = 0$ iff α is simple.

A general criterion for decomposability can be given in terms of some operations known as *left hook* and *right hook* (also called *interior products*); see Section 3.6.

It is easy to see that $\bigwedge(V)$ satisfies the following universal mapping property.

Proposition 3.13. *Given any K-algebra A, for any linear map $f: V \to A$, if $(f(v))^2 = 0$ for all $v \in V$, then there is a unique K-algebra homomorphism $\overline{f}: \bigwedge(V) \to A$ so that*

$$f = \overline{f} \circ i,$$

as in the diagram below.

$$
\begin{array}{ccc}
V & \xrightarrow{\ i\ } & \bigwedge(V) \\
& {}_{f}\searrow & \downarrow{\overline{f}} \\
& & A
\end{array}
$$

When E is finite dimensional, recall the isomorphism $\mu: \bigwedge^n(E^*) \longrightarrow \mathrm{Alt}^n(E; K)$, defined as the linear extension of the map given by

$$\mu(v_1^* \wedge \cdots \wedge v_n^*)(u_1, \ldots, u_n) = \det(v_j^*(u_i)).$$

Now, we have also a multiplication operation $\bigwedge^m(E^*) \times \bigwedge^n(E^*) \longrightarrow \bigwedge^{m+n}(E^*)$. The following question then arises:

Can we define a multiplication $\text{Alt}^m(E; K) \times \text{Alt}^n(E; K) \longrightarrow \text{Alt}^{m+n}(E; K)$ directly on alternating multilinear forms, so that the following diagram commutes?

$$
\begin{array}{ccc}
\bigwedge^m(E^*) \times \bigwedge^n(E^*) & \xrightarrow{\ \wedge\ } & \bigwedge^{m+n}(E^*) \\
\big\downarrow{\scriptstyle \mu_m \times \mu_n} & & \big\downarrow{\scriptstyle \mu_{m+n}} \\
\text{Alt}^m(E; K) \times \text{Alt}^n(E; K) & \xrightarrow{\ \wedge\ } & \text{Alt}^{m+n}(E; K)
\end{array}
$$

As in the symmetric case, the answer is *yes*! The solution is to define this multiplication such that, for $f \in \text{Alt}^m(E; K)$ and $g \in \text{Alt}^n(E; K)$,

$$(f \wedge g)(u_1, \ldots, u_{m+n})$$

$$= \sum_{\sigma \in \text{shuffle}(m,n)} \text{sgn}(\sigma)\, f(u_{\sigma(1)}, \ldots, u_{\sigma(m)}) g(u_{\sigma(m+1)}, \ldots, u_{\sigma(m+n)}),$$

$$(**)$$

where $\text{shuffle}(m, n)$ consists of all (m, n)-"shuffles;" that is, permutations σ of $\{1, \ldots m + n\}$ such that $\sigma(1) < \cdots < \sigma(m)$ and $\sigma(m + 1) < \cdots < \sigma(m + n)$. For example, when $m = n = 1$, we have

$$(f \wedge g)(u, v) = f(u)g(v) - g(u)f(v).$$

When $m = 1$ and $n \geq 2$, check that

$$(f \wedge g)(u_1, \ldots, u_{m+1}) = \sum_{i=1}^{m+1} (-1)^{i-1} f(u_i) g(u_1, \ldots, \widehat{u_i}, \ldots, u_{m+1}),$$

where the hat over the argument u_i means that it should be omitted.

Here is another explicit example. Suppose $m = 2$ and $n = 1$. Given $v_1^*, v_2^*, v_3^* \in E^*$, the multiplication structure on $\bigwedge(E^*)$ implies that $(v_1^* \wedge v_2^*) \cdot v_3^* = v_1^* \wedge v_2^* \wedge v_3^* \in \bigwedge^3(E^*)$. Furthermore, for $u_1, u_2, u_3, \in E$,

$$\mu_3(v_1^* \wedge v_2^* \wedge v_3^*)(u_1, u_2, u_3) = \sum_{\sigma \in \mathfrak{S}_3} \text{sgn}(\sigma) v_{\sigma(1)}^*(u_1) v_{\sigma(2)}^*(u_2) v_{\sigma(3)}^*(u_3)$$

$$= v_1^*(u_1)v_2^*(u_2)v_3^*(u_3) - v_1^*(u_1)v_3^*(u_2)v_2^*(u_3)$$

$$- v_2^*(u_1)v_1^*(u_2)v_3^*(u_3) + v_2^*(u_1)v_3^*(u_2)v_1^*(u_3)$$

$$+ v_3^*(u_1)v_1^*(u_2)v_2^*(u_3) - v_3^*(u_1)v_2^*(u_2)v_1^*(u_3).$$

Now the $(2, 1)$- shuffles of $\{1, 2, 3\}$ are the following three permutations, namely

$$\begin{pmatrix} 1\ 2\ 3 \\ 1\ 2\ 3 \end{pmatrix}, \quad \begin{pmatrix} 1\ 2\ 3 \\ 1\ 3\ 2 \end{pmatrix}, \quad \begin{pmatrix} 1\ 2\ 3 \\ 2\ 3\ 1 \end{pmatrix}.$$

If $f \cong \mu_2(v_1^* \wedge v_2^*)$ and $g \cong \mu_1(v_3^*)$, then (∗∗) implies that

$$(f \cdot g)(u_1, u_2, u_3) = \sum_{\sigma \in \text{shuffle}(2,1)} \text{sgn}(\sigma) f(u_{\sigma(1)}, u_{\sigma(2)}) g(u_{\sigma(3)})$$

$$= f(u_1, u_2)g(u_3) - f(u_1, u_3)g(u_2) + f(u_2, u_3)g(u_1)$$

$$= \mu_2(v_1^* \wedge v_2^*)(u_1, u_2)\mu_1(v_3^*)(u_3)$$

$$- \mu_2(v_1^* \wedge v_2^*)(u_1, u_3)\mu_1(v_3^*)(u_2)$$

$$+ \mu_2(v_1^* \wedge v_2^*)(u_2, u_3)\mu_1(v_3^*)(u_1)$$

$$= (v_1^*(u_1)v_2^*(u_2) - v_2^*(u_1)v_1^*(u_2))v_3^*(u_3)$$

$$- (v_1^*(u_1)v_2^*(u_3) - v_2^*(u_1)v_1^*(u_3))v_3^*(u_2)$$

$$+ (v_1^*(u_2)v_2^*(u_3) - v_2^*(u_2)v_1^*(u_3))v_3^*(u_1)$$

$$= \mu_3(v_1^* \wedge v_2^* \wedge v_3^*)(u_1, u_2, u_3).$$

As a result of all this, the direct sum

$$\text{Alt}(E) = \bigoplus_{n \geq 0} \text{Alt}^n(E; K)$$

is an algebra under the above multiplication, and this algebra is isomorphic to $\bigwedge(E^*)$. For the record we state the following.

Proposition 3.14. *When E is finite dimensional, the maps $\mu \colon \bigwedge^n(E^*) \longrightarrow \text{Alt}^n(E; K)$ induced by the linear extensions of the maps given by*

$$\mu(v_1^* \wedge \cdots \wedge v_n^*)(u_1, \ldots, u_n) = \det(v_j^*(u_i))$$

yield a canonical isomorphism of algebras $\mu \colon \bigwedge(E^*) \longrightarrow \text{Alt}(E)$, where the multiplication in $\text{Alt}(E)$ is defined by the maps $\wedge \colon \text{Alt}^m(E; K) \times \text{Alt}^n(E; K) \longrightarrow \text{Alt}^{m+n}(E; K)$, with

$$(f \wedge g)(u_1, \ldots, u_{m+n})$$
$$= \sum_{\sigma \in \text{shuffle}(m,n)} \text{sgn}(\sigma) \, f(u_{\sigma(1)}, \ldots, u_{\sigma(m)})(u_{\sigma(m+1)}, \ldots, u_{\sigma(m+n)}),$$

where $\text{shuffle}(m, n)$ consists of all (m, n)-"shuffles," that is, permutations σ of $\{1, \ldots m + n\}$ such that $\sigma(1) < \cdots < \sigma(m)$ and $\sigma(m + 1) < \cdots < \sigma(m + n)$.

Remark. The algebra $\bigwedge(E)$ is a graded algebra. Given two graded algebras E and F, we can make a new tensor product $E \widehat{\otimes} F$, where $E \widehat{\otimes} F$ is equal to $E \otimes F$ as a vector space, but with a skew-commutative multiplication given by

$$(a \otimes b) \wedge (c \otimes d) = (-1)^{\deg(b)\deg(c)}(ac) \otimes (bd),$$

where $a \in E^m, b \in F^p, c \in E^n, d \in F^q$. Then, it can be shown that

$$\bigwedge(E \oplus F) \cong \bigwedge(E) \,\widehat{\otimes}\, \bigwedge(F).$$

3.5   The Hodge ∗-Operator

In order to define a generalization of the Laplacian that applies to differential forms on a Riemannian manifold, we need to define isomorphisms

$$\overset{k}{\bigwedge} V \longrightarrow \overset{n-k}{\bigwedge} V,$$

for any Euclidean vector space V of dimension n and any k, with $0 \leq k \leq n$. If $\langle -, - \rangle$ denotes the inner product on V, we define an inner product on $\bigwedge^k V$, denoted $\langle -, - \rangle_\wedge$, by setting

$$\langle u_1 \wedge \cdots \wedge u_k, v_1 \wedge \cdots \wedge v_k \rangle_\wedge = \det(\langle u_i, v_j \rangle),$$

for all $u_i, v_i \in V$, and extending $\langle -, - \rangle_\wedge$ by bilinearity.

It is easy to show that if $(e_1, \ldots, e_n)$ is an orthonormal basis of V, then the basis of $\bigwedge^k V$ consisting of the e_I (where $I = \{i_1, \ldots, i_k\}$, with $1 \leq i_1 < \cdots < i_k \leq n$) is an orthonormal basis of $\bigwedge^k V$. Since the inner product on V induces an inner product on V^* (recall that $\langle \omega_1, \omega_2 \rangle = \langle \omega_1^\sharp, \omega_2^\sharp \rangle$, for all $\omega_1, \omega_2 \in V^*$), we also get an inner product on $\bigwedge^k V^*$.

Definition 3.6. An *orientation* of a vector space V of dimension n is given by the choice of some basis $(e_1, \ldots, e_n)$. We say that a basis $(u_1, \ldots, u_n)$ of V is *positively oriented* iff $\det(u_1, \ldots, u_n) > 0$ (where $\det(u_1, \ldots, u_n)$ denotes the determinant of the matrix whose jth column consists of the coordinates of u_j over the basis $(e_1, \ldots, e_n)$), otherwise it is *negatively oriented*. An *oriented vector space* is a vector space V together with an orientation of V.

If V is oriented by the basis $(e_1, \ldots, e_n)$, then V^* is oriented by the dual basis $(e_1^*, \ldots, e_n^*)$. If σ is any permutation of $\{1, \ldots, n\}$, then the basis $(e_{\sigma(1)}, \ldots, e_{\sigma(n)})$ has positive orientation iff the signature $\mathrm{sgn}(\sigma)$ of the permutation σ is even.

If V is an oriented vector space of dimension n, then we can define a linear isomorphism

$$*: \overset{k}{\bigwedge} V \to \overset{n-k}{\bigwedge} V,$$

called the *Hodge ∗-operator*. The existence of this operator is guaranteed by the following proposition.

Proposition 3.15. *Let V be any oriented Euclidean vector space whose orientation is given by some chosen orthonormal basis $(e_1, \ldots, e_n)$. For any alternating tensor $\alpha \in \bigwedge^k V$, there is a unique alternating tensor $*\alpha \in \bigwedge^{n-k} V$ such that*

$$\alpha \wedge \beta = \langle *\alpha, \beta \rangle_\wedge \, e_1 \wedge \cdots \wedge e_n$$

for all $\beta \in \bigwedge^{n-k} V$. The alternating tensor $*\alpha$ is independent of the choice of the positive orthonormal basis $(e_1, \ldots, e_n)$.

Proof. Since $\bigwedge^n V$ has dimension 1, the alternating tensor $e_1 \wedge \cdots \wedge e_n$ is a basis of $\bigwedge^n V$. It follows that for any fixed $\alpha \in \bigwedge^k V$, the linear map λ_α from $\bigwedge^{n-k} V$ to $\bigwedge^n V$ given by

$$\lambda_\alpha(\beta) = \alpha \wedge \beta$$

is of the form

$$\lambda_\alpha(\beta) = f_\alpha(\beta) \, e_1 \wedge \cdots \wedge e_n$$

for some linear form $f_\alpha \in \left(\bigwedge^{n-k} V \right)^*$. But then, by the duality induced by the inner product $\langle -, - \rangle$ on $\bigwedge^{n-k} V$, there is a unique vector $*\alpha \in \bigwedge^{n-k} V$ such that

$$f_\lambda(\beta) = \langle *\alpha, \beta \rangle_\wedge \quad \text{for all } \beta \in \overset{n-k}{\bigwedge} V,$$

which implies that

$$\alpha \wedge \beta = \lambda_\alpha(\beta) = f_\alpha(\beta) \, e_1 \wedge \cdots \wedge e_n = \langle *\alpha, \beta \rangle_\wedge \, e_1 \wedge \cdots \wedge e_n,$$

as claimed. If $(e'_1, \ldots, e'_n)$ is any other positively oriented orthonormal basis, by Proposition 3.2, $e'_1 \wedge \cdots \wedge e'_n = \det(P) \, e_1 \wedge \cdots \wedge e_n = e_1 \wedge \cdots \wedge e_n$, since $\det(P) = 1$ where P is the change of basis from $(e_1, \ldots, e_n)$ to $(e'_1, \ldots, e'_n)$ and both bases are positively oriented. ☐

Definition 3.7. The operator $*$ from $\bigwedge^k V$ to $\bigwedge^{n-k} V$ defined by Proposition 3.15 is called the *Hodge ∗-operator*.

Observe that the Hodge ∗-operator is linear.

The Hodge ∗-operator is defined in terms of the orthonormal basis elements of $\bigwedge V$ as follows: For any increasing sequence $(i_1, \ldots, i_k)$ of elements $i_p \in \{1, \ldots, n\}$, if $(j_1, \ldots, j_{n-k})$ is the increasing sequence of elements $j_q \in \{1, \ldots, n\}$ such that

$$\{i_1, \ldots, i_k\} \cup \{j_1, \ldots, j_{n-k}\} = \{1, \ldots, n\},$$

then

$$*(e_{i_1} \wedge \cdots \wedge e_{i_k}) = \operatorname{sign}(i_1, \ldots i_k, j_1, \ldots, j_{n-k})\, e_{j_1} \wedge \cdots \wedge e_{j_{n-k}}.$$

In particular, for $k = 0$ and $k = n$, we have

$$*(1) = e_1 \wedge \cdots \wedge e_n$$

$$*(e_1 \wedge \cdots \wedge e_n) = 1.$$

For example, if $n = 3$, we have

$$*e_1 = e_2 \wedge e_3$$

$$*e_2 = -e_1 \wedge e_3$$

$$*e_3 = e_1 \wedge e_2$$

$$*(e_1 \wedge e_2) = e_3$$

$$*(e_1 \wedge e_3) = -e_2$$

$$*(e_2 \wedge e_3) = e_1.$$

The Hodge ∗-operators $*\colon \bigwedge^k V \to \bigwedge^{n-k} V$ induce a linear map $*\colon \bigwedge(V) \to \bigwedge(V)$. We also have Hodge ∗-operators $*\colon \bigwedge^k V^* \to \bigwedge^{n-k} V^*$.

The following proposition shows that the linear map $*\colon \bigwedge(V) \to \bigwedge(V)$ is an isomorphism.

Proposition 3.16. *If V is any oriented vector space of dimension n, for every k with $0 \le k \le n$, we have*

(i) $** = (-id)^{k(n-k)}$.
(ii) $\langle x, y \rangle_\wedge = *(x \wedge *y) = *(y \wedge *x)$, for all $x, y \in \bigwedge^k V$.

Proof.

(1) Let $(e_i)_{i=1}^n$ is an orthonormal basis of V. It is enough to check the identity on basis elements. We have

$$*(e_{i_1} \wedge \cdots \wedge e_{i_k}) = \operatorname{sign}(i_1, \ldots i_k, j_1, \ldots, j_{n-k})\, e_{j_1} \wedge \cdots \wedge e_{j_{n-k}}$$

and

$$**(e_{i_1} \wedge \cdots \wedge e_{i_k}) = \operatorname{sign}(i_1, \ldots i_k, j_1, \ldots, j_{n-k}) *(e_{j_1} \wedge \cdots \wedge e_{j_{n-k}})$$

$$= \operatorname{sign}(i_1, \ldots i_k, j_1, \ldots, j_{n-k}) \operatorname{sign}(j_1, \ldots, j_{n-k}, i_1, \ldots i_k)$$

$$e_{i_1} \wedge \cdots \wedge e_{i_k}.$$

It is easy to see that

$$\text{sign}(i_1, \ldots i_k, j_1, \ldots, j_{n-k}) \, \text{sign}(j_1, \ldots, j_{n-k}, i_1, \ldots i_k) = (-1)^{k(n-k)},$$

which yields

$$**(e_{i_1} \wedge \cdots \wedge e_{i_k}) = (-1)^{k(n-k)} \, e_{i_1} \wedge \cdots \wedge e_{i_k},$$

as claimed.

(ii) These identities are easily checked on basis elements; see Jost [62], Chapter 2, Lemma 2.1.1. In particular let

$$x = e_{i_1} \wedge \cdots \wedge e_{i_k}, \qquad y = e_{i_j} \wedge \cdots \wedge e_{i_j}, \qquad x, y \in \bigwedge^k V,$$

where $(e_i)_{i=1}^n$ is an orthonormal basis of V. If $x \neq y$, $\langle x, y \rangle_\wedge = 0$ since there is some e_{i_p} of x not equal to any e_{j_q} of y by the orthonormality of the basis, this means the pth row of $(\langle e_{i_l}, e_{j_s} \rangle)$ consists entirely of zeroes. Also $x \neq y$ implies that $y \wedge *x = 0$ since

$$*x = \text{sign}(i_1, \ldots i_k, l_1, \ldots, l_{n-k}) e_{l_1} \wedge \cdots \wedge e_{l_{n-k}},$$

where e_{l_s} is the same as some e_p in y. A similar argument shows that if $x \neq y$, $x \wedge *y = 0$. So now assume $x = y$. Then

$$*(e_{i_1} \wedge \cdots \wedge e_{i_k} \wedge *(e_{i_1} \wedge \cdots \wedge e_{i_k})) = *(e_1 \wedge e_2 \cdots \wedge e_n)$$
$$= 1 = \langle x, x \rangle_\wedge.$$

$\square$

In Section 8.2 we will need to express $*(1)$ in terms of any basis (not necessarily orthonormal) of V.

Proposition 3.17. *If V is any finite-dimensional oriented vector space, for any basis $(v_1, \ldots, v_n)$ of V, we have*

$$*(1) = \frac{1}{\sqrt{\det(\langle v_i, v_j \rangle)}} \, v_1 \wedge \cdots \wedge v_n.$$

Proof. If $(e_1, \ldots, e_n)$ is an orthonormal basis of V and $(v_1, \ldots, v_n)$ is any other basis of V, then

$$\langle v_1 \wedge \cdots \wedge v_n, v_1 \wedge \cdots \wedge v_n \rangle_\wedge = \det(\langle v_i, v_j \rangle),$$

and since

$$v_1 \wedge \cdots \wedge v_n = \det(A)\, e_1 \wedge \cdots \wedge e_n$$

where A is the matrix expressing the v_j in terms of the e_i, we have

$$\langle v_1 \wedge \cdots \wedge v_n, v_1 \wedge \cdots \wedge v_n \rangle_\wedge = \det(A)^2 \langle e_1 \wedge \cdots \wedge e_n, e_1 \wedge \cdots \wedge e_n \rangle = \det(A)^2.$$

As a consequence, $\det(A) = \sqrt{\det(\langle v_i, v_j \rangle)}$, and

$$v_1 \wedge \cdots \wedge v_n = \sqrt{\det(\langle v_i, v_j \rangle)}\, e_1 \wedge \cdots \wedge e_n,$$

from which it follows that

$$*(1) = \frac{1}{\sqrt{\det(\langle v_i, v_j \rangle)}}\, v_1 \wedge \cdots \wedge v_n$$

(see Jost [62], Chapter 2, Lemma 2.1.3). □

3.6 Left and Right Hooks ⊛

The motivation for defining left hooks and right hook comes from the problem of understanding when a tensor $\alpha \in \bigwedge^k(E)$ is *decomposable*. An arbitrary tensor $\alpha \in \bigwedge^k(E)$ is a linear combination of tensors of the form $u_1 \wedge \cdots \wedge u_k$, called *decomposable*. The issue is to find criteria for decomposability. Criteria for testing decomposability using left hooks are given in Section 3.7.

In this section *all vector spaces are assumed to have finite dimension*. Say $\dim(E) = n$. Using our nonsingular pairing

$$\langle -, - \rangle \colon \bigwedge^p E^* \times \bigwedge^p E \longrightarrow K \qquad (1 \leq p \leq n)$$

defined on generators by

$$\langle u_1^* \wedge \cdots \wedge u_p^*, v_1 \wedge \cdots \wedge u_p \rangle = \det(u_i^*(v_j)),$$

we define various contraction operations (partial evaluation operators)

$$\lrcorner \colon \bigwedge^p E \times \bigwedge^{p+q} E^* \longrightarrow \bigwedge^q E^* \qquad \text{(left hook)}$$

and

$$\lrcorner : \overset{p+q}{\bigwedge} E^* \times \overset{p}{\bigwedge} E \longrightarrow \overset{q}{\bigwedge} E^* \qquad \text{(right hook)},$$

as well as the versions obtained by replacing E by E^* and E^{**} by E. We begin with the *left interior product or left hook,* $\lrcorner$.

Let $u \in \bigwedge^p E$. For any q such that $p + q \leq n$, multiplication on the right by u is a linear map

$$\wedge_R(u) : \overset{q}{\bigwedge} E \longrightarrow \overset{p+q}{\bigwedge} E$$

given by

$$v \mapsto v \wedge u$$

where $v \in \bigwedge^q E$. The transpose of $\wedge_R(u)$ yields a linear map

$$(\wedge_R(u))^\top : \left(\overset{p+q}{\bigwedge} E \right)^* \longrightarrow \left(\overset{q}{\bigwedge} E \right)^*,$$

which, using the isomorphisms $\left(\bigwedge^{p+q} E \right)^* \cong \bigwedge^{p+q} E^*$ and $\left(\bigwedge^q E \right)^* \cong \bigwedge^q E^*$, can be viewed as a map

$$(\wedge_R(u))^\top : \overset{p+q}{\bigwedge} E^* \longrightarrow \overset{q}{\bigwedge} E^*$$

given by

$$z^* \mapsto z^* \circ \wedge_R(u),$$

where $z^* \in \bigwedge^{p+q} E^*$. We denote $z^* \circ \wedge_R(u)$ by $u \lrcorner z^*$. In terms of our pairing, the adjoint $u \lrcorner$ of $\wedge_R(u)$ defined by

$$\langle u \lrcorner z^*, v \rangle = \langle z^*, \wedge_R(u)(v) \rangle;$$

this in turn leads to the following definition.

Definition 3.8. Let $u \in \bigwedge^p E$ and $z^* \in \bigwedge^{p+q} E^*$. We define $u \lrcorner z^* \in \bigwedge^q E^*$ to be q-vector uniquely determined by

$$\langle u \lrcorner z^*, v \rangle = \langle z^*, v \wedge u \rangle, \quad \text{for all } v \in \overset{q}{\bigwedge} E.$$

Remark. Note that to be precise the operator

$$\lrcorner : \overset{p}{\bigwedge} E \times \overset{p+q}{\bigwedge} E^* \longrightarrow \overset{q}{\bigwedge} E^*$$

depends of p, q, so we really defined a family of operators $\lrcorner_{p,q}$. This family of operators $\lrcorner_{p,q}$ induces a map

$$\lrcorner : \bigwedge E \times \bigwedge E^* \longrightarrow \bigwedge E^*,$$

with

$$\lrcorner_{p,q} : \overset{p}{\bigwedge} E \times \overset{p+q}{\bigwedge} E^* \longrightarrow \overset{q}{\bigwedge} E^*$$

as defined before. The common practice is to omit the subscripts of $\lrcorner$.

It is immediately verified that

$$(u \wedge v) \lrcorner z^* = u \lrcorner (v \lrcorner z^*),$$

for all $u \in \bigwedge^k E, v \in \bigwedge^{p-k} E, z^* \in \bigwedge^{p+q} E^*$ since

$$\langle (u \wedge v) \lrcorner z^*, w \rangle = \langle z^*, w \wedge u \wedge v \rangle = \langle v \lrcorner z^*, w \wedge u \rangle = \langle u \lrcorner (v \lrcorner z^*), w \rangle,$$

whenever $w \in \bigwedge^q E$. This means that

$$\lrcorner : \bigwedge E \times \bigwedge E^* \longrightarrow \bigwedge E^*$$

is a left action of the (noncommutative) ring $\bigwedge E$ with multiplication $\wedge$ on $\bigwedge E^*$, which makes $\bigwedge E^*$ into a left $\bigwedge E$-module.

By interchanging E and E^* and using the isomorphism

$$\left(\overset{k}{\bigwedge} F \right)^* \cong \overset{k}{\bigwedge} F^*,$$

we can also define some maps

$$\lrcorner : \overset{p}{\bigwedge} E^* \times \overset{p+q}{\bigwedge} E \longrightarrow \overset{q}{\bigwedge} E,$$

and make the following definition.

Definition 3.9. Let $u^* \in \bigwedge^p E^*$, and $z \in \bigwedge^{p+q} E$. We define $u^* \lrcorner z \in \bigwedge^q$ as the q-vector uniquely defined by

$$\langle v^* \wedge u^*, z \rangle = \langle v^*, u^* \lrcorner z \rangle, \quad \text{for all } v^* \in \bigwedge^q E^*.$$

As for the previous version, we have a family of operators $\lrcorner_{p,q}$ which define an operator

$$\lrcorner : \bigwedge E^* \times \bigwedge E \longrightarrow \bigwedge E.$$

We easily verify that

$$(u^* \wedge v^*) \lrcorner z = u^* \lrcorner (v^* \lrcorner z),$$

whenever $u^* \in \bigwedge^k E^*$, $v^* \in \bigwedge^{p-k} E^*$, and $z \in \bigwedge^{p+q} E$; so this version of $\lrcorner$ is a left action of the ring $\bigwedge E^*$ on $\bigwedge E$ which makes $\bigwedge E$ into a left $\bigwedge E^*$-module.

In order to proceed any further we need some combinatorial properties of the basis of $\bigwedge^p E$ constructed from a basis $(e_1, \ldots, e_n)$ of E. Recall that for any (nonempty) subset $I \subseteq \{1, \ldots, n\}$, we let

$$e_I = e_{i_1} \wedge \cdots \wedge e_{i_p},$$

where $I = \{i_1, \ldots, i_p\}$ with $i_1 < \cdots < i_p$. We also let $e_\emptyset = 1$.

Given any two nonempty subsets $H, L \subseteq \{1, \ldots, n\}$ both listed in increasing order, say $H = \{h_1 < \ldots < h_p\}$ and $L = \{\ell_1 < \ldots < \ell_q\}$, if H and L are disjoint, let $H \cup L$ be union of H and L considered as the ordered sequence

$$(h_1, \ldots, h_p, \ell_1, \ldots, \ell_q).$$

Then let

$$\rho_{H,L} = \begin{cases} 0 & \text{if } H \cap L \neq \emptyset, \\ (-1)^\nu & \text{if } H \cap L = \emptyset, \end{cases}$$

where

$$\nu = |\{(h, l) \mid (h, l) \in H \times L, \; h > l\}|.$$

Observe that when $H \cap L = \emptyset$, $|H| = p$, and $|L| = q$, the number ν is the number of inversions of the sequence

$$(h_1, \cdots, h_p, \ell_1, \cdots, \ell_q),$$

where an inversion is a pair (h_i, ℓ_j) such that $h_i > \ell_j$.

Unless $p + q = n$, the function whose graph is given by

$$\begin{pmatrix} 1 & \cdots & p & p+1 & \cdots & p+q \\ h_1 & \cdots & h_p & \ell_1 & \cdots & \ell_q \end{pmatrix}$$

is **not** a permutation of $\{1, \ldots, n\}$. We can view v as a slight generalization of the notion of the number of inversions of a permutation.

Proposition 3.18. *For any basis* $(e_1, \ldots, e_n)$ *of* E *the following properties hold:*

(1) If $H \cap L = \emptyset$, $|H| = p$, *and* $|L| = q$, *then*

$$\rho_{H,L}\rho_{L,H} = (-1)^v(-1)^{pq-v} = (-1)^{pq}.$$

(2) For $H, L \subseteq \{1, \ldots, m\}$ *listed in increasing order, we have*

$$e_H \wedge e_L = \rho_{H,L}e_{H \cup L}.$$

Similarly,

$$e_H^* \wedge e_L^* = \rho_{H,L}e_{H \cup L}^*.$$

(3) For the left hook

$$\lrcorner : \overset{p}{\bigwedge} E \times \overset{p+q}{\bigwedge} E^* \longrightarrow \overset{q}{\bigwedge} E^*,$$

we have

$$e_H \lrcorner e_L^* = 0 \quad \text{if } H \nsubseteq L$$

$$e_H \lrcorner e_L^* = \rho_{L-H,H}e_{L-H}^* \quad \text{if } H \subseteq L.$$

(4) For the left hook

$$\lrcorner : \overset{p}{\bigwedge} E^* \times \overset{p+q}{\bigwedge} E \longrightarrow \overset{q}{\bigwedge} E,$$

we have

$$e_H^* \lrcorner e_L = 0 \quad \text{if } H \nsubseteq L$$

$$e_H^* \lrcorner e_L = \rho_{L-H,H}e_{L-H} \quad \text{if } H \subseteq L.$$

Proof. These are proved in Bourbaki [14] (Chapter III, §11, Section 11), but the proofs of (3) and (4) are very concise. We elaborate on the proofs of (2) and (4), the proof of (3) being similar.

In (2) if $H \cap L \neq \emptyset$, then $e_H \wedge e_L$ contains some vector twice and so $e_H \wedge e_L = 0$. Otherwise, $e_H \wedge e_L$ consists of

$$e_{h_1} \wedge \cdots \wedge e_{h_p} \wedge e_{\ell_1} \wedge \cdots \wedge e_{\ell_q},$$

and to order the sequence of indices in increasing order we need to transpose any two indices (h_i, ℓ_j) corresponding to an inversion, which yields $\rho_{H,L} e_{H \cup L}$.

Let us now consider (4). We have $|L| = p + q$ and $|H| = p$, and the q-vector $e_H^* \lrcorner e_L$ is characterized by

$$\langle v^*, e_H^* \lrcorner e_L \rangle = \langle v^* \wedge e_H^*, e_L \rangle$$

for all $v^* \in \bigwedge^q E^*$. There are two cases.

Case 1. $H \not\subseteq L$. If so, no matter what $v^* \in \bigwedge^q E^*$ is, since H contains some index h not in L, the hth row $(e_h^*(e_{\ell_1}), \ldots, e_h^*(e_{\ell_{p+q}}))$ of the determinant $\langle v^* \wedge e_H^*, e_L \rangle$ must be zero, so $\langle v^* \wedge e_H^*, e_L \rangle = 0$ for all $v^* \in \bigwedge^q E^*$, and since the pairing is nongenerate, we must have $e_H^* \lrcorner e_L = 0$.

Case 2. $H \subseteq L$. In this case, for $v^* = e_{L-H}^*$, by (2) we have

$$\langle e_{L-H}^*, e_H^* \lrcorner e_L \rangle = \langle e_{L-H}^* \wedge e_H^*, e_L \rangle = \langle \rho_{L-H,H} e_L^*, e_L \rangle = \rho_{L-H,H},$$

which yields

$$\langle e_{L-H}^*, e_H^* \lrcorner e_L \rangle = \rho_{L-H,H}.$$

The q-vector $e_H^* \lrcorner e_L$ can be written as a linear combination $e_H^* \lrcorner e_L = \sum_J \lambda_J e_J$ with $|J| = q$ so

$$\langle e_{L-H}^*, e_H^* \lrcorner e_L \rangle = \sum_J \lambda_J \langle e_{L-H}^*, e_J \rangle.$$

By definition of the pairing, $\langle e_{L-H}^*, e_J \rangle = 0$ unless $J = L - H$, which means that

$$\langle e_{L-H}^*, e_H^* \lrcorner e_L \rangle = \lambda_{L-H} \langle e_{L-H}^*, e_{L-H} \rangle = \lambda_{L-H},$$

so $\lambda_{L-H} = \rho_{L-H,H}$, as claimed.   □

Using Proposition 3.18, we have the following.

Proposition 3.19. *For the left hook*

$$\lrcorner : E \times \overset{q+1}{\bigwedge} E^* \longrightarrow \overset{q}{\bigwedge} E^*,$$

for every $u \in E$, $x^* \in \bigwedge^{q+1-s} E^*$, and $y^* \in \bigwedge^s E^*$, we have

$$u \lrcorner (x^* \wedge y^*) = (-1)^s (u \lrcorner x^*) \wedge y^* + x^* \wedge (u \lrcorner y^*).$$

Proof. We can prove the above identity assuming that x^* and y^* are of the form e_I^* and e_J^* using Proposition 3.18 and leave the details as an exercise for the reader.   □

Thus, $\lrcorner : E \times \bigwedge^{q+1} E^* \longrightarrow \bigwedge^q E^*$ is almost an anti-derivation, except that the sign $(-1)^s$ is applied to the wrong factor.

We have a similar identity for the other version of the left hook

$$\lrcorner : E^* \times \overset{q+1}{\bigwedge} E \longrightarrow \overset{q}{\bigwedge} E,$$

namely

$$u^* \lrcorner (x \wedge y) = (-1)^s (u^* \lrcorner x) \wedge y + x \wedge (u^* \lrcorner y)$$

for every $u^* \in E^*$, $x \in \bigwedge^{q+1-s} E$, and $y \in \bigwedge^s E$.

An application of this formula when $q = 3$ and $s = 2$ yields an interesting equation. In this case, $u^* \in E^*$ and $x, y \in \bigwedge^2 E$, so we get

$$u^* \lrcorner (x \wedge y) = (u^* \lrcorner x) \wedge y + x \wedge (u^* \lrcorner y).$$

In particular, for $x = y$, since $x \in \bigwedge^2 E$ and $u^* \lrcorner x \in E$, Proposition 3.12 implies that $(u^* \lrcorner x) \wedge x = x \wedge (u^* \lrcorner x)$, and we obtain

$$u^* \lrcorner (x \wedge x) = 2((u^* \lrcorner x) \wedge x). \tag{†}$$

As a consequence, $(u^* \lrcorner x) \wedge x = 0$ iff $u^* \lrcorner (x \wedge x) = 0$. We will use this identity together with Proposition 3.25 to prove that a 2-vector $x \in \bigwedge^2 E$ is decomposable iff $x \wedge x = 0$.

It is also possible to define a *right interior product or right hook* ∟, using multiplication on the left rather than multiplication on the right. Then we use the maps

$$\llcorner : \overset{p+q}{\bigwedge} E^* \times \overset{p}{\bigwedge} E \longrightarrow \overset{q}{\bigwedge} E^*$$

to make the following definition.

Definition 3.10. Let $u \in \bigwedge^p E$ and $z^* \in \bigwedge^{p+q} E^*$. We define $z^* \llcorner u \in \bigwedge^q E^*$ to be the q-vector uniquely defined as

$$\langle z^* \llcorner u, v \rangle = \langle z^*, u \wedge v \rangle, \qquad \text{for all } v \in \overset{q}{\bigwedge} E.$$

This time we can prove that

$$z^* \llcorner (u \wedge v) = (z^* \llcorner u) \llcorner v,$$

so the family of operators $\llcorner_{p,q}$ defines a right action

$$\llcorner : \bigwedge E^* \times \bigwedge E \longrightarrow \bigwedge E^*$$

of the ring $\bigwedge E$ on $\bigwedge E^*$ which makes $\bigwedge E^*$ into a right $\bigwedge E$-module.
Similarly, we have maps

$$\llcorner : \overset{p+q}{\bigwedge} E \times \overset{p}{\bigwedge} E^* \longrightarrow \overset{q}{\bigwedge} E,$$

which in turn leads to the following dual formation of the right hook.

Definition 3.11. Let $u^* \in \bigwedge^p E^*$ and $z \in \bigwedge^{p+q} E$. We define $z \llcorner u^* \in \bigwedge^q$ to be the q-vector uniquely defined by

$$\langle u^* \wedge v^*, z \rangle = \langle v^*, z \llcorner u^* \rangle, \qquad \text{for all } v^* \in \overset{q}{\bigwedge} E^*.$$

We can prove that

$$z \llcorner (u^* \wedge v^*) = (z \llcorner u^*) \llcorner v^*,$$

so the family of operators $\llcorner_{p,q}$ defines a right action

$$\llcorner : \bigwedge E \times \bigwedge E^* \longrightarrow \bigwedge E$$

of the ring $\bigwedge E^*$ on $\bigwedge E$ which makes $\bigwedge E$ into a right $\bigwedge E^*$-module.
Since the left hook $\lrcorner : \bigwedge^p E \times \bigwedge^{p+q} E^* \longrightarrow \bigwedge^q E^*$ is defined by

$$\langle u \lrcorner z^*, v \rangle = \langle z^*, v \wedge u \rangle, \quad \text{for all } u \in \overset{p}{\bigwedge} E, v \in \overset{q}{\bigwedge} E \text{ and } z^* \in \overset{p+q}{\bigwedge} E^*,$$

the right hook

$$\llcorner : \overset{p+q}{\bigwedge} E^* \times \overset{p}{\bigwedge} E \longrightarrow \overset{q}{\bigwedge} E^*$$

by

$$\langle z^* \llcorner u, v \rangle = \langle z^*, u \wedge v \rangle, \quad \text{for all } u \in \bigwedge^p E, v \in \bigwedge^q E, \text{ and } z^* \in \bigwedge^{p+q} E^*,$$

and $v \wedge u = (-1)^{pq} u \wedge v$, we conclude that

$$z^* \llcorner u = (-1)^{pq} u \lrcorner z^*.$$

Similarly, since

$$\langle v^* \wedge u^*, z \rangle = \langle v^*, u^* \lrcorner z \rangle, \quad \text{for all } u^* \in \bigwedge^p E^*, v^* \in \bigwedge^q E^* \text{ and } z \in \bigwedge^{p+q} E$$

$$\langle u^* \wedge v^*, z \rangle = \langle v^*, z \llcorner u^* \rangle, \quad \text{for all } u^* \in \bigwedge^p E^*, v^* \in \bigwedge^q E^*, \text{ and } z \in \bigwedge^{p+q} E,$$

and $v^* \wedge u^* = (-1)^{pq} u^* \wedge v^*$, we have

$$z \llcorner u^* = (-1)^{pq} u^* \lrcorner z.$$

We summarize the above facts in the following proposition.

Proposition 3.20. *The following identities hold:*

$$z^* \llcorner u = (-1)^{pq} u \lrcorner z^* \quad \text{for all } u \in \bigwedge^p E \text{ and all } z^* \in \bigwedge^{p+q} E^*$$

$$z \llcorner u^* = (-1)^{pq} u^* \lrcorner z \quad \text{for all } u^* \in \bigwedge^p E^* \text{ and all } z \in \bigwedge^{p+q} E.$$

Therefore the left and right hooks are not independent, and in fact each one determines the other. As a consequence, we can restrict our attention to only one of the hooks, for example the left hook, but there are a few situations where it is nice to use both, for example in Proposition 3.23.

A version of Proposition 3.18 holds for right hooks, but beware that the indices in $\rho_{L-H,H}$ are permuted. This permutation has to do with the fact that the left hook and the right hook are related *via* a sign factor.

Proposition 3.21. *For any basis $(e_1, \ldots, e_n)$ of E the following properties hold:*

(1) For the right hook

$$\llcorner : \bigwedge^{p+q} E \times \bigwedge^p E^* \longrightarrow \bigwedge^q E$$

we have

$$e_L \lrcorner e_H^* = 0 \quad \text{if } H \nsubseteq L$$

$$e_L \lrcorner e_H^* = \rho_{H,L-H} e_{L-H} \quad \text{if } H \subseteq L.$$

(2) For the right hook

$$\lrcorner : \overset{p+q}{\bigwedge} E^* \times \overset{p}{\bigwedge} E \longrightarrow \overset{q}{\bigwedge} E^*$$

we have

$$e_L^* \lrcorner e_H = 0 \quad \text{if } H \nsubseteq L$$

$$e_L^* \lrcorner e_H = \rho_{H,L-H} e_{L-H}^* \quad \text{if } H \subseteq L.$$

Remark. Our definition of left hooks as left actions $\lrcorner : \bigwedge^p E \times \bigwedge^{p+q} E^* \longrightarrow \bigwedge^q E^*$ and $\lrcorner : \bigwedge^p E^* \times \bigwedge^{p+q} E \longrightarrow \bigwedge^q E$ and right hooks as right actions $\lrcorner : \bigwedge^{p+q} E^* \times \bigwedge^p E \longrightarrow \bigwedge^q E^*$ and $\lrcorner : \bigwedge^{p+q} E \times \bigwedge^p E^* \longrightarrow \bigwedge^q E$ is identical to the definition found in Fulton and Harris [45] (Appendix B). However, the reader should be aware that this is not a universally accepted notation. In fact, the left hook $u^* \lrcorner z$ defined in Bourbaki [14] is our right hook $z \lrcorner u^*$, up to the sign $(-1)^{p(p-1)/2}$. This has to do with the fact that Bourbaki uses a different pairing which also involves an extra sign, namely

$$\langle v^*, u^* \lrcorner z \rangle = (-1)^{p(p-1)/2} \langle u^* \wedge v^*, z \rangle.$$

One of the side-effects of this choice is that Bourbaki's version of Formula (4) of Proposition 3.18 (Bourbaki [14], Chapter III, page 168) is

$$e_H^* \lrcorner e_L = 0 \quad \text{if } H \nsubseteq L$$

$$e_H^* \lrcorner e_L = (-1)^{p(p-1)/2} \rho_{H,L-H} e_{L-H} \quad \text{if } H \subseteq L,$$

where $|H| = p$ and $|L| = p+q$. This correspond to Formula (1) of Proposition 3.21 up to the sign factor $(-1)^{p(p-1)/2}$, which we find horribly confusing. Curiously, an older edition of Bourbaki (1958) uses the same pairing as Fulton and Harris [45]. The reason (and the advantage) for this change of sign convention is not clear to us.

We also have the following version of Proposition 3.19 for the right hook.

Proposition 3.22. *For the right hook*

$$\lrcorner : \overset{q+1}{\bigwedge} E^* \times E \longrightarrow \overset{q}{\bigwedge} E^*,$$

for every $u \in E$, $x^* \in \bigwedge^r E^*$, and $y^* \in \bigwedge^{q+1-r} E^*$, we have

$$(x^* \wedge y^*) \llcorner u = (x^* \llcorner u) \wedge y^* + (-1)^r x^* \wedge (y^* \llcorner u).$$

Proof. A proof involving determinants can be found in Warner [109], Chapter 2.

□

Thus, $\llcorner : \bigwedge^{q+1} E^* \times E \longrightarrow \bigwedge^q E^*$ is an anti-derivation. A similar formula holds for the the right hook $\llcorner : \bigwedge^{q+1} E \times E^* \longrightarrow \bigwedge^q E$, namely

$$(x \wedge y) \llcorner u^* = (x \llcorner u^*) \wedge y + (-1)^r x \wedge (y \llcorner u^*),$$

for every $u^* \in E, \in \bigwedge^r E$, and $y \in \bigwedge^{q+1-r} E$. This formula is used by Shafarevitch [99] to define a hook, but beware that Shafarevitch uses the left hook notation $u^* \lrcorner x$ rather than the right hook notation. Shafarevitch uses the terminology *convolution*, which seems very unfortunate.

For $u \in E$, the right hook $z^* \llcorner u$ is also denoted $i(u)z^*$, and called *insertion operator* or *interior product*. This operator plays an important role in differential geometry.

Definition 3.12. Let $u \in E$ and $z^* \in \bigwedge^{n+1}(E^*)$. If we view z^* as an alternating multilinear map in $\mathrm{Alt}^{n+1}(E; K)$, then we define $i(u)z^* \in \mathrm{Alt}^n(E; K)$ as given by

$$(i(u)z^*)(v_1, \ldots, v_n) = z^*(u, v_1, \ldots, v_n).$$

Using the left hook $\lrcorner$ and the right hook $\llcorner$ we can define two linear maps $\gamma : \bigwedge^p E \to \bigwedge^{n-p} E^*$ and $\delta : \bigwedge^p E^* \to \bigwedge^{n-p} E$ as follows:

Definition 3.13. For any basis $(e_1, \ldots, e_n)$ of E, if we let $M = \{1, \ldots, n\}$, $e = e_1 \wedge \cdots \wedge e_n$, and $e^* = e_1^* \wedge \cdots \wedge e_n^*$, define $\gamma : \bigwedge^p E \to \bigwedge^{n-p} E^*$ and $\delta : \bigwedge^p E^* \to \bigwedge^{n-p} E$ as

$$\gamma(u) = u \lrcorner e^* \quad \text{and} \quad \delta(v^*) = e \llcorner v^*,$$

for all $u \in \bigwedge^p E$ and all $v^* \in \bigwedge^p E^*$.

Proposition 3.23. *The linear maps $\gamma : \bigwedge^p E \to \bigwedge^{n-p} E^*$ and $\delta : \bigwedge^p E^* \to \bigwedge^{n-p} E$ are isomorphism, and $\gamma^{-1} = \delta$. The isomorphisms γ and δ map decomposable vectors to decomposable vectors. Furthermore, if $z \in \bigwedge^p E$ is decomposable, say $z = u_1 \wedge \cdots \wedge u_p$ for some $u_i \in E$, then $\gamma(z) = v_1^* \wedge \cdots \wedge v_{n-p}^*$ for some $v_j^* \in E^*$, and $v_j^*(u_i) = 0$ for all i, j. A similar property holds for $v^* \in \bigwedge^p E^*$ and $\delta(v^*)$. If $(e_1', \ldots, e_n')$ is any other basis of E and $\gamma' : \bigwedge^p E \to \bigwedge^{n-p} E^*$ and $\delta' : \bigwedge^p E^* \to \bigwedge^{n-p} E$ are the corresponding isomorphisms, then $\gamma' = \lambda \gamma$ and $\delta' = \lambda^{-1}\delta$ for some nonzero $\lambda \in K$.*

Proof. Using Propositions 3.18 and 3.21, for any subset $J \subseteq \{1, \ldots, n\} = M$ such that $|J| = p$, we have

$$\gamma(e_J) = e_J \lrcorner e^* = \rho_{M-J,J} e^*_{M-J} \quad \text{and} \quad \delta(e^*_{M-J}) = e \llcorner e^*_{M-J} = \rho_{M-J,J} e_J.$$

Thus,

$$\delta \circ \gamma(e_J) = \rho_{M-J,J} \rho_{M-J,J} e_J = e_J,$$

since $\rho_{M-J,J} = \pm 1$. A similar result holds for $\gamma \circ \delta$. This implies that

$$\delta \circ \gamma = \mathrm{id} \quad \text{and} \quad \gamma \circ \delta = \mathrm{id}.$$

Thus, γ and δ are inverse isomorphisms.

If $z \in \bigwedge^p E$ is decomposable, then $z = u_1 \wedge \cdots \wedge u_p$ where $u_1, \ldots, u_p$ are linearly independent since $z \neq 0$, and we can pick a basis of E of the form $(u_1, \ldots, u_n)$. Then the above formulae show that

$$\gamma(z) = \pm u^*_{p+1} \wedge \cdots \wedge u^*_n.$$

Since $(u^*_1, \ldots, u^*_n)$ is the dual basis of $(u_1, \ldots, u_n)$, we have $u^*_i(u_j) = \delta_{ij}$, If $(e'_1, \ldots, e'_n)$ is any other basis of E, because $\bigwedge^n E$ has dimension 1, we have

$$e'_1 \wedge \cdots \wedge e'_n = \lambda e_1 \wedge \cdots \wedge e_n$$

for some nonzero $\lambda \in K$, and the rest is trivial.   □

Applying Proposition 3.23 to the case where $p = n - 1$, the isomorphism $\gamma \colon \bigwedge^{n-1} E \to \bigwedge^1 E^*$ maps indecomposable vectors in $\bigwedge^{n-1} E$ to indecomposable vectors in $\bigwedge^1 E^* = E^*$. But every vector in E^* is decomposable, so every vector in $\bigwedge^{n-1} E$ is decomposable.

Corollary 3.24. *If E is a finite-dimensional vector space, then every vector in $\bigwedge^{n-1} E$ is decomposable.*

3.7 Testing Decomposability ⊛

We are now ready to tackle the problem of finding criteria for decomposability. Such criteria will use the left hook. Once again, in this section *all vector spaces are assumed to have finite dimension*. But before stating our criteria, we need a few preliminary results.

Proposition 3.25. *Given $z \in \bigwedge^p E$ with $z \neq 0$, the smallest vector space $W \subseteq E$ such that $z \in \bigwedge^p W$ is generated by the vectors of the form*

$$u^* \lrcorner z, \qquad with \ u^* \in \bigwedge^{p-1} E^*.$$

Proof. First let W be any subspace such that $z \in \bigwedge^p(W)$ and let $(e_1, \ldots, e_r, e_{r+1}, \ldots, e_n)$ be a basis of E such that $(e_1, \ldots, e_r)$ is a basis of W. Then, $u^* = \sum_I \lambda_I e_I^*$, where $I \subseteq \{1, \ldots, n\}$ and $|I| = p - 1$, and $z = \sum_J \mu_J e_J$, where $J \subseteq \{1, \ldots, r\}$ and $|J| = p \leq r$. It follows immediately from the formula of Proposition 3.18 (4), namely

$$e_I^* \lrcorner e_J = \rho_{J-I,J} e_{J-I},$$

that $u^* \lrcorner z \in W$, since $J - I \subseteq \{1, \ldots, r\}$.

Next we prove that if W is the smallest subspace of E such that $z \in \bigwedge^p(W)$, then W is generated by the vectors of the form $u^* \lrcorner z$, where $u^* \in \bigwedge^{p-1} E^*$. Suppose not. Then the vectors $u^* \lrcorner z$ with $u^* \in \bigwedge^{p-1} E^*$ span a proper subspace U of W. We prove that for every subspace W' of W with $\dim(W') = \dim(W) - 1 = r - 1$, it is not possible that $u^* \lrcorner z \in W'$ for all $u^* \in \bigwedge^{p-1} E^*$. But then, as U is a proper subspace of W, it is contained in some subspace W' with $\dim(W') = r - 1$, and we have a contradiction.

Let $w \in W - W'$ and pick a basis of W formed by a basis $(e_1, \ldots, e_{r-1})$ of W' and w. Any $z \in \bigwedge^p(W)$ can be written as $z = z' + w \wedge z''$, where $z' \in \bigwedge^p W'$ and $z'' \in \bigwedge^{p-1} W'$, and since W is the smallest subspace containing z, we have $z'' \neq 0$. Consequently, if we write $z'' = \sum_I \lambda_I e_I$ in terms of the basis $(e_1, \ldots, e_{r-1})$ of W', there is some e_I, with $I \subseteq \{1, \ldots, r - 1\}$ and $|I| = p - 1$, so that the coefficient λ_I is nonzero. Now, using any basis of E containing $(e_1, \ldots, e_{r-1}, w)$, by Proposition 3.18 (4), we see that

$$e_I^* \lrcorner (w \wedge e_I) = \lambda w, \qquad \lambda = \pm 1.$$

It follows that

$$e_I^* \lrcorner z = e_I^* \lrcorner (z' + w \wedge z'') = e_I^* \lrcorner z' + e_I^* \lrcorner (w \wedge z'') = e_I^* \lrcorner z' + \lambda \lambda_I w,$$

with $e_I^* \lrcorner z' \in W'$, which shows that $e_I^* \lrcorner z \notin W'$. Therefore, W is indeed generated by the vectors of the form $u^* \lrcorner z$, where $u^* \in \bigwedge^{p-1} E^*$. □

To help understand Proposition 3.25, let E be the vector space with basis $\{e_1, e_2, e_3, e_4\}$ and $z = e_1 \wedge e_2 + e_2 \wedge e_3$. Note that $z \in \bigwedge^2 E$. To find the smallest vector space $W \subseteq E$ such that $z \in \bigwedge^2 W$, we calculate $u^* \lrcorner z$, where $u^* \in \bigwedge^1 E^*$. The multilinearity of $\lrcorner$ implies it is enough to calculate $u^* \lrcorner z$ for $u^* \in \{e_1^*, e_2^*, e_3^*, e_4^*\}$. Proposition 3.18 (4) implies that

$$e_1^* \lrcorner z = e_1^* \lrcorner (e_1 \wedge e_2 + e_2 \wedge e_3) = e_1^* \lrcorner e_1 \wedge e_2 = -e_2$$

$$e_2^* \lrcorner z = e_2^* \lrcorner (e_1 \wedge e_2 + e_2 \wedge e_3) = e_1 - e_3$$

$$e_3^* \lrcorner z = e_3^* \lrcorner (e_1 \wedge e_2 + e_2 \wedge e_3) = e_3^* \lrcorner e_2 \wedge e_3 = e_2$$
$$e_4^* \lrcorner z = e_4^* \lrcorner (e_1 \wedge e_2 + e_2 \wedge e_3) = 0.$$

Thus W is the two-dimensional vector space generated by the basis $\{e_2, e_1 - e_3\}$. This is not surprising since $z = -e_2 \wedge (e_1 - e_3)$ and is in fact decomposable. As this example demonstrates, the action of the left hook provides a way of extracting a basis of W from z.

Proposition 3.25 implies the following corollary.

Corollary 3.26. *Any nonzero* $z \in \bigwedge^p E$ *is decomposable iff the smallest subspace W of E such that $z \in \bigwedge^p W$ has dimension p. Furthermore, if $z = u_1 \wedge \cdots \wedge u_p$ is decomposable, then $(u_1, \ldots, u_p)$ is a basis of the smallest subspace W of E such that $z \in \bigwedge^p W$*

Proof. If $\dim(W) = p$, then for any basis $(e_1, \ldots, e_p)$ of W we know that $\bigwedge^p W$ has $e_1 \wedge \cdots \wedge e_p$ has a basis, and thus has dimension 1. Since $z \in \bigwedge^p W$, we have $z = \lambda e_1 \wedge \cdots \wedge e_p$ for some nonzero λ, so z is decomposable.

Conversely assume that $z \in \bigwedge^p W$ is nonzero and decomposable. Then, $z = u_1 \wedge \cdots \wedge u_p$, and since $z \neq 0$, by Proposition 3.9 $(u_1, \ldots, u_p)$ are linearly independent. Then for any $v_i^* = u_1^* \wedge \cdots u_{i-1}^* \wedge u_{i+1}^* \wedge \cdots \wedge u_p^*$ (where u_i^* is omitted), we have

$$v_i^* \lrcorner z = (u_1^* \wedge \cdots u_{i-1}^* \wedge u_{i+1}^* \wedge \cdots \wedge u_p^*) \lrcorner (u_1 \wedge \cdots \wedge u_p) = \pm u_i,$$

so by Proposition 3.25 we have $u_i \in W$ for $i = 1, \ldots, p$. This shows that $\dim(W) \geq p$, but since $z = u_1 \wedge \cdots \wedge u_p$, we have $\dim(W) = p$, which means that $(u_1, \ldots, u_p)$ is a basis of W. $\square$

Finally we are ready to state and prove the criterion for decomposability with respect to left hooks.

Proposition 3.27. *Any nonzero* $z \in \bigwedge^p E$ *is decomposable iff*

$$(u^* \lrcorner z) \wedge z = 0, \qquad \text{for all } u^* \in \bigwedge^{p-1} E^*.$$

Proof. First assume that $z \in \bigwedge^p E$ is decomposable. If so, by Corollary 3.26, the smallest subspace W of E such that $z \in \bigwedge^p W$ has dimension p, so we have $z = e_1 \wedge \cdots \wedge e_p$ where $e_1, \ldots, e_p$ form a basis of W. By Proposition 3.25, for every $u^* \in \bigwedge^{p-1} E^*$, we have $u^* \lrcorner z \in W$, so each $u^* \lrcorner z$ is a linear combination of the e_i's, say

$$u^* \lrcorner z = \alpha_1 e_1 + \cdots + \alpha_p e_p,$$

and

$$(u^* \lrcorner z) \wedge z = \sum_{i=1}^{p} \alpha_i e_i \wedge e_1 \wedge \cdots \wedge e_i \wedge \cdots \wedge e_p = 0.$$

Now assume that $(u^* \lrcorner z) \wedge z = 0$ for all $u^* \in \bigwedge^{p-1} E^*$, and that $\dim(W) = m > p$, where W is the smallest subspace of E such that $z \in \bigwedge^p W$. If $e_1, \ldots, e_m$ is a basis of W, then we have $z = \sum_I \lambda_I e_I$, where $I \subseteq \{1, \ldots, m\}$ and $|I| = p$. Recall that $z \neq 0$, and so, some λ_I is nonzero. By Proposition 3.25, each e_i can be written as $u^* \lrcorner z$ for some $u^* \in \bigwedge^{p-1} E^*$, and since $(u^* \lrcorner z) \wedge z = 0$ for all $u^* \in \bigwedge^{p-1} E^*$, we get

$$e_j \wedge z = 0 \quad \text{for} \quad j = 1, \ldots, m.$$

By wedging $z = \sum_I \lambda_I e_I$ with each e_j, as $m > p$, we deduce $\lambda_I = 0$ for all I, so $z = 0$, a contradiction. Therefore, $m = p$ and Corollary 3.26 implies that z is decomposable. □

As a corollary of Proposition 3.27 we obtain the following fact that we stated earlier without proof.

Proposition 3.28. *Given any vector space E of dimension n, a vector $x \in \bigwedge^2 E$ is decomposable iff $x \wedge x = 0$.*

Proof. Recall that as an application of Proposition 3.19 we proved the formula (†), namely

$$u^* \lrcorner (x \wedge x) = 2((u^* \lrcorner x) \wedge x)$$

for all $x \in \bigwedge^2 E$ and all $u^* \in E^*$. As a consequence, $(u^* \lrcorner x) \wedge x = 0$ iff $u^* \lrcorner (x \wedge x) = 0$. By Proposition 3.27, the 2-vector x is decomposable iff $u^* \lrcorner (x \wedge x) = 0$ for all $u^* \in E^*$ iff $x \wedge x = 0$. Therefore, a 2-vector x is decomposable iff $x \wedge x = 0$. □

As an application of Proposition 3.28, assume that $\dim(E) = 3$ and that (e_1, e_2, e_3) is a basis of E. Then any 2-vector $x \in \bigwedge^2 E$ is of the form

$$x = \alpha e_1 \wedge e_2 + \beta e_1 \wedge e_3 + \gamma e_2 \wedge e_3.$$

We have

$$x \wedge x = (\alpha e_1 \wedge e_2 + \beta e_1 \wedge e_3 + \gamma e_2 \wedge e_3) \wedge (\alpha e_1 \wedge e_2 + \beta e_1 \wedge e_3 + \gamma e_2 \wedge e_3) = 0,$$

because all the terms involved are of the form $c\, e_{i_1} \wedge e_{i_2} \wedge e_{i_3} \wedge e_{i_4}$ with $i_1, i_2, i_3, i_4 \in \{1, 2, 3\}$, and so at least two of these indices are identical. Therefore, every 2-vector $x = \alpha e_1 \wedge e_2 + \beta e_1 \wedge e_3 + \gamma e_2 \wedge e_3$ is decomposable, although this is not obvious at first glance. For example,

$$e_1 \wedge e_2 + e_1 \wedge e_3 + e_2 \wedge e_3 = (e_1 + e_2) \wedge (e_2 + e_3).$$

We now show that Proposition 3.27 yields an equational criterion for the decomposability of an alternating tensor $z \in \bigwedge^p E$.

3.8   The Grassmann-Plücker's Equations and Grassmannian Manifolds ⊛

Let E be a vector space of dimensions n, let $(e_1, \ldots, e_n)$ be a basis of E, and let $(e_1^*, \ldots, e_n^*)$ be its dual basis. Our objective is to determine whether a nonzero vector $z \in \bigwedge^p E$ is decomposable, in terms of equations.

We follow an argument adapted from Bourbaki [14] (Chapter III, §11, Section 13). By Proposition 3.27, the vector z is decomposable iff $(u^* \lrcorner z) \wedge z = 0$ for all $u^* \in \bigwedge^{p-1} E^*$. We can let u^* range over a basis of $\bigwedge^{p-1} E^*$, and then the conditions are

$$(e_H^* \lrcorner z) \wedge z = 0$$

for all $H \subseteq \{1, \ldots, n\}$, with $|H| = p - 1$. Since $(e_H^* \lrcorner z) \wedge z \in \bigwedge^{p+1} E$, this is equivalent to

$$\langle e_J^*, (e_H^* \lrcorner z) \wedge z \rangle = 0$$

for all $H, J \subseteq \{1, \ldots, n\}$, with $|H| = p - 1$ and $|J| = p + 1$. Then, for all $I, I' \subseteq \{1, \ldots, n\}$ with $|I| = |I'| = p$, Formulae (2) and (4) of Proposition 3.18 show that

$$\langle e_J^*, (e_H^* \lrcorner e_I) \wedge e_{I'} \rangle = 0,$$

unless there is some $i \in \{1, \ldots, n\}$ such that

$$I - H = \{i\}, \quad J - I' = \{i\}.$$

In this case, $I = H \cup \{i\}$ and $I' = J - \{i\}$, and using Formulae (2) and (4) of Proposition 3.18, we have

$$\langle e_J^*, (e_H^* \lrcorner e_{H \cup \{i\}}) \wedge e_{J-\{i\}} \rangle = \langle e_J^*, \rho_{\{i\},H} e_i \wedge e_{J-\{i\}} \rangle = \langle e_J^*, \rho_{\{i\},H} \rho_{\{i\},J-\{i\}} e_J \rangle$$

$$= \rho_{\{i\},H} \rho_{\{i\},J-\{i\}}.$$

If we let

$$\epsilon_{i,J,H} = \rho_{\{i\},H} \rho_{\{i\},J-\{i\}},$$

we have $\epsilon_{i,J,H} = +1$ if the parity of the number of $j \in J$ such that $j < i$ is the same as the parity of the number of $h \in H$ such that $h < i$, and $\epsilon_{i,J,H} = -1$ otherwise.

Finally we obtain the following criterion in terms of quadratic equations (*Plücker's equations*) for the decomposability of an alternating tensor.

Proposition 3.29 (Grassmann-Plücker's Equations). *For* $z = \sum_I \lambda_I e_I \in \bigwedge^p E$, *the conditions for* $z \neq 0$ *to be decomposable are*

$$\sum_{i \in J - H} \epsilon_{i,J,H} \lambda_{H \cup \{i\}} \lambda_{J - \{i\}} = 0,$$

with $\epsilon_{i,J,H} = \rho_{\{i\},H} \rho_{\{i\},J-\{i\}}$, *for all* $H, J \subseteq \{1, \ldots, n\}$ *such that* $|H| = p - 1$, $|J| = p + 1$, *and all* $i \in J - H$.

Using the above criterion, it is a good exercise to reprove that if $\dim(E) = n$, then every tensor in $\bigwedge^{n-1}(E)$ is decomposable. We already proved this fact as a corollary of Proposition 3.23.

Given any $z = \sum_I \lambda_I e_I \in \bigwedge^p E$ where $\dim(E) = n$, the family of scalars (λ_I) (with $I = \{i_1 < \cdots < i_p\} \subseteq \{1, \ldots, n\}$ listed in increasing order) is called the *Plücker coordinates* of z. The Grassmann-Plücker's equations give necessary and sufficient conditions for any nonzero z to be decomposable.

For example, when $\dim(E) = n = 4$ and $p = 2$, these equations reduce to the single equation

$$\lambda_{12}\lambda_{34} - \lambda_{13}\lambda_{24} + \lambda_{14}\lambda_{23} = 0.$$

However, it should be noted that the equations given by Proposition 3.29 are not independent in general.

We are now in the position to prove that the Grassmannian $G(p, n)$ can be embedded in the projective space $\mathbb{RP}^{\binom{n}{p}-1}$.

For any $n \geq 1$ and any k with $1 \leq p \leq n$, recall that the Grassmannian $G(p, n)$ is the set of all linear p-dimensional subspaces of $\mathbb{R}^n$ (also called *p-planes*). Any p-dimensional subspace U of $\mathbb{R}^n$ is spanned by p linearly independent vectors $u_1, \ldots, u_p$ in $\mathbb{R}^n$; write $U = \operatorname{span}(u_1, \ldots, u_k)$. By Proposition 3.9, $(u_1, \ldots, u_p)$ are linearly independent iff $u_1 \wedge \cdots \wedge u_p \neq 0$. If $(v_1, \ldots, v_p)$ are any other linearly independent vectors spanning U, then we have

$$v_j = \sum_{i=1}^{p} a_{ij} u_i, \quad 1 \leq j \leq p,$$

for some $a_{ij} \in \mathbb{R}$, and by Proposition 3.2

$$v_1 \wedge \cdots \wedge v_p = \det(A)\, u_1 \wedge \cdots \wedge u_p,$$

where $A = (a_{ij})$. As a consequence, we can define a map $i_G : G(p, n) \to \mathbb{RP}^{\binom{n}{p}-1}$ such that for any k-plane U, for any basis $(u_1, \ldots, u_p)$ of U,

$$i_G(U) = [u_1 \wedge \cdots \wedge u_p],$$

the point of $\mathbb{RP}^{\binom{n}{p}-1}$ given by the one-dimensional subspace of $\mathbb{R}^{\binom{n}{p}}$ spanned by $u_1 \wedge \cdots \wedge u_p$.

Proposition 3.30. *The map* $i_G : G(p, n) \to \mathbb{RP}^{\binom{n}{p}-1}$ *is injective.*

Proof. Let U and V be any two p-planes and assume that $i_G(U) = i_G(V)$. This means that there is a basis $(u_1, \ldots, u_p)$ of U and a basis $(v_1, \ldots, v_p)$ of V such that

$$v_1 \wedge \cdots \wedge v_p = c\, u_1 \wedge \cdots \wedge u_p$$

for some nonzero $c \in \mathbb{R}$. The above implies that the smallest subspaces W and W' of $\mathbb{R}^n$ such that $u_1 \wedge \cdots \wedge u_p \in \bigwedge^p W$ and $v_1 \wedge \cdots \wedge v_p \in \bigwedge^p W'$ are identical, so $W = W'$. By Corollary 3.26, this smallest subspace W has both $(u_1, \ldots, u_p)$ and $(v_1, \ldots, v_p)$ as bases, so the v_j are linear combinations of the u_i (and vice versa), and $U = V$. $\qquad\square$

Since any nonzero $z \in \bigwedge^p \mathbb{R}^n$ can be uniquely written as

$$z = \sum_I \lambda_I e_I$$

in terms of its Plücker coordinates (λ_I), every point of $\mathbb{RP}^{\binom{n}{p}-1}$ is defined by the Plücker coordinates (λ_I) viewed as homogeneous coordinates. The points of $\mathbb{RP}^{\binom{n}{p}-1}$ corresponding to one-dimensional spaces associated with decomposable alternating p-tensors are the points whose coordinates satisfy the Grassmann-Plücker's equations of Proposition 3.29. Therefore, the map i_G embeds the Grassmannian $G(p, n)$ as an algebraic variety in $\mathbb{RP}^{\binom{n}{p}-1}$ defined by equations of degree 2.

We can replace the field $\mathbb{R}$ by $\mathbb{C}$ in the above reasoning and we obtain an embedding of the complex Grassmannian $G_{\mathbb{C}}(p, n)$ as an algebraic variety in $\mathbb{CP}^{\binom{n}{p}-1}$ defined by equations of degree 2.

In particular, if $n = 4$ and $p = 2$, the equation

$$\lambda_{12}\lambda_{34} - \lambda_{13}\lambda_{24} + \lambda_{14}\lambda_{23} = 0$$

is the homogeneous equation of a quadric in $\mathbb{CP}^5$ known as the *Klein quadric*. The points on this quadric are in one-to-one correspondence with the lines in $\mathbb{CP}^3$.

There is also a simple algebraic criterion to decide whether the smallest subspaces U and V associated with two nonzero decomposable vectors $u_1 \wedge \cdots \wedge u_p$ and $v_1 \wedge \cdots \wedge v_q$ have a nontrivial intersection.

Proposition 3.31. *Let E be any n-dimensional vector space over a field K, and let U and V be the smallest subspaces of E associated with two nonzero decomposable* *vectors* $u = u_1 \wedge \cdots \wedge u_p \in \bigwedge^p U$ *and* $v = v_1 \wedge \cdots \wedge v_q \in \bigwedge^q V$. *The following properties hold:*

(1) We have $U \cap V = (0)$ *iff* $u \wedge v \neq 0$.
(2) If $U \cap V = (0)$, *then* $U + V$ *is the least subspace associated with* $u \wedge v$.

Proof. Assume $U \cap V = (0)$. We know by Corollary 3.26 that $(u_1, \ldots, u_p)$ is a basis of U and $(v_1, \ldots, v_q)$ is a basis of V. Since $U \cap V = (0)$, $(u_1, \ldots, u_p, v_1, \ldots, v_q)$ is a basis of $U + V$, and by Proposition 3.9, we have

$$u \wedge v = u_1 \wedge \cdots \wedge u_p \wedge v_1 \wedge \cdots \wedge v_q \neq 0.$$

This also proves (2).

Conversely, assume that $\dim(U \cap V) \geq 1$. Pick a basis $(w_1, \ldots, w_r)$ of $W = U \cap V$, and extend this basis to a basis $(w_1, \ldots, w_r, w_{r+1}, \ldots, w_p)$ of U and to a basis $(w_1, \ldots, w_r, w_{p+1}, \ldots, w_{p+q-r})$ of V. By Corollary 3.26, $(u_1, \ldots, u_p)$ is also basis of U, so

$$u_1 \wedge \cdots \wedge u_p = a \, w_1 \wedge \cdots \wedge w_r \wedge w_{r+1} \wedge \cdots \wedge w_p$$

for some $a \in K$, and $(v_1, \ldots, v_q)$ is also basis of V, so

$$v_1 \wedge \cdots \wedge v_q = b \, w_1 \cdots \wedge w_r \wedge w_{p+1} \wedge \cdots \wedge w_{p+q-r}$$

for some $b \in K$, and thus

$$u \wedge v = u_1 \wedge \cdots \wedge u_p \wedge v_1 \wedge \cdots \wedge v_q = 0$$

since it contains some repeated w_i, with $1 \leq i \leq r$.   □

As an application of Proposition 3.31, consider two projective lines D_1 and D_2 in $\mathbb{RP}^3$, which means that D_1 and D_2 correspond to two 2-planes in $\mathbb{R}^4$, and thus by Proposition 3.30, to two points in $\mathbb{RP}^{\binom{4}{2}-1} = \mathbb{RP}^5$. These two points correspond to the 2-vectors

$$z = a_{1,2}e_1 \wedge e_2 + a_{1,3}e_1 \wedge e_3 + a_{1,4}e_1 \wedge e_4 + a_{2,3}e_2 \wedge e_3 + a_{2,4}e_2 \wedge e_4 + a_{3,4}e_3 \wedge e_4$$

and

$$z' = a'_{1,2}e_1 \wedge e_2 + a'_{1,3}e_1 \wedge e_3 + a'_{1,4}e_1 \wedge e_4 + a'_{2,3}e_2 \wedge e_3 + a'_{2,4}e_2 \wedge e_4 + a'_{3,4}e_3 \wedge e_4$$

whose Plücker coordinates (where $a_{i,j} = \lambda_{ij}$) satisfy the equation

$$\lambda_{12}\lambda_{34} - \lambda_{13}\lambda_{24} + \lambda_{14}\lambda_{23} = 0$$

of the Klein quadric, and D_1 and D_2 intersect iff $z \wedge z' = 0$ iff

$$a_{1,2}a'_{3,4} - a_{1,3}a'_{3,4} + a_{1,4}a'_{2,3} + a_{2,3}a'_{1,4} - a_{2,4}a'_{1,3} + a_{3,4}a'_{1,2} = 0.$$

Observe that for D_1 fixed, this is a linear condition. This fact is very helpful for solving problems involving intersections of lines. A famous problem is to find how many lines in $\mathbb{RP}^3$ meet four given lines in general position. The answer is at most 2.

3.9   Vector-Valued Alternating Forms

The purpose of this section is to present the technical background needed for Sections 4.5 and 4.6 on vector-valued differential forms, in particular in the case of Lie groups where differential forms taking their values in a Lie algebra arise naturally.

In this section the vector space E is assumed to have *finite dimension*. We know that there is a canonical isomorphism $\bigwedge^n(E^*) \cong \text{Alt}^n(E; K)$ between alternating n-forms and alternating multilinear maps. As in the case of general tensors, the isomorphisms provided by Propositions 3.5, 2.17, and 3.10, namely

$$\text{Alt}^n(E; F) \cong \text{Hom}\left(\bigwedge^n(E), F\right)$$

$$\text{Hom}\left(\bigwedge^n(E), F\right) \cong \left(\bigwedge^n(E)\right)^* \otimes F$$

$$\left(\bigwedge^n(E)\right)^* \cong \bigwedge^n(E^*)$$

yield a canonical isomorphism

$$\text{Alt}^n(E; F) \cong \left(\bigwedge^n(E^*)\right) \otimes F$$

which we record as a corollary.

Corollary 3.32. *For any finite-dimensional vector space E and any vector space F, we have a canonical isomorphism*

$$\text{Alt}^n(E; F) \cong \left(\bigwedge^n(E^*)\right) \otimes F.$$

Note that F may have infinite dimension. This isomorphism allows us to view the tensors in $\bigwedge^n(E^*) \otimes F$ as *vector-valued alternating forms*, a point of view that is useful in differential geometry. If $(f_1, \ldots, f_r)$ is a basis of F, every tensor $\omega \in \bigwedge^n(E^*) \otimes F$ can be written as some linear combination

$$\omega = \sum_{i=1}^{r} \alpha_i \otimes f_i,$$

with $\alpha_i \in \bigwedge^n (E^*)$. We also let

$$\bigwedge(E; F) = \bigoplus_{n=0}^{n} \left(\bigwedge^n (E^*) \right) \otimes F = \left(\bigwedge(E) \right) \otimes F.$$

Given three vector spaces, F, G, H, if we have some bilinear map $\Phi \colon F \times G \to H$, then we can define a multiplication operation

$$\wedge_\Phi \colon \bigwedge(E; F) \times \bigwedge(E; G) \to \bigwedge(E; H)$$

as follows: For every pair (m, n), we define the multiplication

$$\wedge_\Phi \colon \left(\left(\bigwedge^m (E^*) \right) \otimes F \right) \times \left(\left(\bigwedge^n (E^*) \right) \otimes G \right) \longrightarrow \left(\bigwedge^{m+n} (E^*) \right) \otimes H$$

by

$$\omega \wedge_\Phi \eta = (\alpha \otimes f) \wedge_\Phi (\beta \otimes g) = (\alpha \wedge \beta) \otimes \Phi(f, g).$$

As in Section 3.4 (following H. Cartan [21]), we can also define a multiplication

$$\wedge_\Phi \colon \mathrm{Alt}^m (E; F) \times \mathrm{Alt}^n (E; G) \longrightarrow \mathrm{Alt}^{m+n} (E; H)$$

directly on alternating multilinear maps as follows: For $f \in \mathrm{Alt}^m (E; F)$ and $g \in \mathrm{Alt}^n (E; G)$,

$$(f \wedge_\Phi g)(u_1, \ldots, u_{m+n})$$

$$= \sum_{\sigma \in \mathrm{shuffle}(m,n)} \mathrm{sgn}(\sigma) \, \Phi\Big(f(u_{\sigma(1)}, \ldots, u_{\sigma(m)}), g(u_{\sigma(m+1)}, \ldots, u_{\sigma(m+n)}) \Big),$$

where $\mathrm{shuffle}(m, n)$ consists of all (m, n)-"shuffles;" that is, permutations σ of $\{1, \ldots m + n\}$ such that $\sigma(1) < \cdots < \sigma(m)$ and $\sigma(m + 1) < \cdots < \sigma(m + n)$.

A special case of interest is the case where $F = G = H$ is a Lie algebra and $\Phi(a, b) = [a, b]$ is the Lie bracket of F. In this case, using a basis $(f_1, \ldots, f_r)$ of F, if we write $\omega = \sum_i \alpha_i \otimes f_i$ and $\eta = \sum_j \beta_j \otimes f_j$, we have

$$\omega \wedge_\Phi \eta = [\omega, \eta] = \sum_{i,j} \alpha_i \wedge \beta_j \otimes [f_i, f_j].$$

It is customary to denote $\omega \wedge_\Phi \eta$ by $[\omega, \eta]$ (unfortunately, the bracket notation is overloaded). Consequently,

$$[\eta, \omega] = (-1)^{mn+1}[\omega, \eta].$$

In general not much can be said about $\wedge_\Phi$, unless Φ has some additional properties. In particular, $\wedge_\Phi$ is generally not associative.

We now use vector-valued alternating forms to generalize both the μ map of Proposition 3.14 and generalize Proposition 2.17 by defining the map

$$\mu_F : \left(\bigwedge^n (E^*) \right) \otimes F \longrightarrow \mathrm{Alt}^n (E; F)$$

on generators by

$$\mu_F((v_1^* \wedge \cdots \wedge v_n^*) \otimes f)(u_1, \ldots, u_n) = (\det(v_j^*(u_i)))f,$$

with $v_1^*, \ldots, v_n^* \in E^*$, $u_1, \ldots, u_n \in E$, and $f \in F$.

Proposition 3.33. *The map*

$$\mu_F : \left(\bigwedge^n (E^*) \right) \otimes F \longrightarrow \mathrm{Alt}^n (E; F)$$

defined as above is a canonical isomorphism for every $n \geq 0$. Furthermore, given any three vector spaces, F, G, H, and any bilinear map $\Phi : F \times G \to H$, for all $\omega \in \left(\bigwedge^n (E^*) \right) \otimes F$ and all $\eta \in \left(\bigwedge^n (E^*) \right) \otimes G$,

$$\mu_H(\omega \wedge_\Phi \eta) = \mu_F(\omega) \wedge_\Phi \mu_G(\eta).$$

Proof. Since we already know that $\left(\bigwedge^n (E^*) \right) \otimes F$ and $\mathrm{Alt}^n (E; F)$ are isomorphic, it is enough to show that μ_F maps some basis of $\left(\bigwedge^n (E^*) \right) \otimes F$ to linearly independent elements. Pick some bases $(e_1, \ldots, e_p)$ in E and $(f_j)_{j \in J}$ in F. Then we know that the vectors $e_I^* \otimes f_j$, where $I \subseteq \{1, \ldots, p\}$ and $|I| = n$, form a basis of $\left(\bigwedge^n (E^*) \right) \otimes F$. If we have a linear dependence

$$\sum_{I,j} \lambda_{I,j} \mu_F(e_I^* \otimes f_j) = 0,$$

applying the above combination to each $(e_{i_1}, \ldots, e_{i_n})$ $(I = \{i_1, \ldots, i_n\}, i_1 < \cdots < i_n)$, we get the linear combination

$$\sum_j \lambda_{I,j} f_j = 0,$$

and by linear independence of the f_j's, we get $\lambda_{I,j} = 0$ for all I and all j. Therefore, the $\mu_F(e_I^* \otimes f_j)$ are linearly independent, and we are done. The second part of the proposition is checked using a simple computation. $\square$

The following proposition will be useful in dealing with vector-valued differential forms.

Proposition 3.34. *If $(e_1, \ldots, e_p)$ is any basis of E, then every element $\omega \in \left(\bigwedge^n (E^*) \right) \otimes F$ can be written in a unique way as*

$$\omega = \sum_I e_I^* \otimes f_I, \qquad f_I \in F,$$

where the e_I^* are defined as in Section 3.2.

Proof. Since, by Proposition 3.8, the e_I^* form a basis of $\bigwedge^n (E^*)$, elements of the form $e_I^* \otimes f$ span $\left(\bigwedge^n (E^*) \right) \otimes F$. Now if we apply $\mu_F(\omega)$ to $(e_{i_1}, \ldots, e_{i_n})$, where $I = \{i_1, \ldots, i_n\} \subseteq \{1, \ldots, p\}$, we get

$$\mu_F(\omega)(e_{i_1}, \ldots, e_{i_n}) = \mu_F(e_I^* \otimes f_I)(e_{i_1}, \ldots, e_{i_n}) = f_I.$$

Therefore, the f_I are uniquely determined by f. $\square$

Proposition 3.34 can also be formulated in terms of alternating multilinear maps, a fact that will be useful to deal with differential forms.

Corollary 3.35. *Define the product $\cdot \colon \mathrm{Alt}^n(E; \mathbb{R}) \times F \to \mathrm{Alt}^n(E; F)$ as follows: For all $\omega \in \mathrm{Alt}^n(E; \mathbb{R})$ and all $f \in F$,*

$$(\omega \cdot f)(u_1, \ldots, u_n) = \omega(u_1, \ldots, u_n) f,$$

for all $u_1, \ldots, u_n \in E$. Then for every $\omega \in \left(\bigwedge^n (E^*) \right) \otimes F$ of the form

$$\omega = u_1^* \wedge \cdots \wedge u_n^* \otimes f,$$

we have

$$\mu_F(u_1^* \wedge \cdots \wedge u_n^* \otimes f) = \mu_F(u_1^* \wedge \cdots \wedge u_n^*) \cdot f.$$

Then Proposition 3.34 yields the following result.

Proposition 3.36. *If $(e_1, \ldots, e_p)$ is any basis of E, then every element $\omega \in \mathrm{Alt}^n(E; F)$ can be written in a unique way as*

$$\omega = \sum_I e_I^* \cdot f_I, \qquad f_I \in F,$$

where the e_I^* are defined as in Section 3.2.

3.10   Problems

Problem 3.1. Complete the induction argument used in the proof of Proposition 3.1 (2).

Problem 3.2. Prove Proposition 3.2.

Problem 3.3. Prove Proposition 3.7.

Problem 3.4. Show that the pairing given by $(*)$ in Section 3.3 is nondegenerate.

Problem 3.5. Let $\mathfrak{I}_a$ be the two-sided ideal generated by all tensors of the form $u \otimes u \in V^{\otimes 2}$. Prove that

$$\bigwedge^m(V) \cong V^{\otimes m}/(\mathfrak{I}_a \cap V^{\otimes m}).$$

Problem 3.6. Complete the induction proof of Proposition 3.12.

Problem 3.7. Prove the following lemma: If V is a vector space with $\dim(V) \leq 3$, then $\alpha \wedge \alpha = 0$ whenever $\alpha \in \bigwedge(V)$.

Problem 3.8. Prove Proposition 3.13.

Problem 3.9. Given two graded algebras E and F, define $E \,\widehat{\otimes}\, F$ to be the vector space $E \otimes F$, but with a skew-commutative multiplication given by

$$(a \otimes b) \wedge (c \otimes d) = (-1)^{\deg(b)\deg(c)}(ac) \otimes (bd),$$

where $a \in E^m, b \in F^p, c \in E^n, d \in F^q$. Show that

$$\bigwedge(E \oplus F) \cong \bigwedge(E) \,\widehat{\otimes}\, \bigwedge(F).$$

Problem 3.10. If $\langle -, - \rangle$ denotes the inner product on V, recall that we defined an inner product on $\bigwedge^k V$, also denoted $\langle -, - \rangle$, by setting

$$\langle u_1 \wedge \cdots \wedge u_k, v_1 \wedge \cdots \wedge v_k \rangle = \det(\langle u_i, v_j \rangle),$$

for all $u_i, v_i \in V$, and extending $\langle -, - \rangle$ by bilinearity.

Show that if $(e_1, \ldots, e_n)$ is an orthonormal basis of V, then the basis of $\bigwedge^k V$ consisting of the e_I (where $I = \{i_1, \ldots, i_k\}$, with $1 \leq i_1 < \cdots < i_k \leq n$) is also an orthonormal basis of $\bigwedge^k V$.

Problem 3.11. Show that

$$(u^* \wedge v^*) \lrcorner z = u^* \lrcorner (v^* \lrcorner z),$$

whenever $u^* \in \bigwedge^k E^*$, $v^* \in \bigwedge^{p-k} E^*$, and $z \in \bigwedge^{p+q} E$.

Problem 3.12. Prove Statement (3) of Proposition 3.18.

Problem 3.13. Prove Proposition 3.19.
Also prove the identity

$$u^* \lrcorner (x \wedge y) = (-1)^s (u^* \lrcorner x) \wedge y + x \wedge (u^* \lrcorner y),$$

where $u^* \in E^*$, $x \in \bigwedge^{q+1-s} E$, and $y \in \bigwedge^s E$.

Problem 3.14. Using the Grassmann-Plücker's equations prove that if $\dim(E) = n$, then every tensor in $\bigwedge^{n-1}(E)$ is decomposable.

Problem 3.15. Recall that the map

$$\mu_F : \left(\bigwedge^n (E^*) \right) \otimes F \longrightarrow \mathrm{Alt}^n(E; F)$$

is defined on generators by

$$\mu_F((v_1^* \wedge \cdots \wedge v_n^*) \otimes f)(u_1, \ldots, u_n) = (\det(v_j^*(u_i)))f,$$

with $v_1^*, \ldots, v_n^* \in E^*$, $u_1, \ldots, u_n \in E$, and $f \in F$.
Given any three vector spaces, F, G, H, and any bilinear map $\Phi : F \times G \to H$, for all $\omega \in \left(\bigwedge^n(E^*) \right) \otimes F$ and all $\eta \in \left(\bigwedge^n(E^*) \right) \otimes G$ prove that

$$\mu_H(\omega \wedge_\Phi \eta) = \mu_F(\omega) \wedge_\Phi \mu_G(\eta).$$

Chapter 4
Differential Forms

The theory of differential forms is one of the main tools in geometry and topology. This theory has a surprisingly large range of applications, and it also provides a relatively easy access to more advanced theories such as cohomology. For all these reasons, it is really an indispensable theory, and anyone with more than a passible interest in geometry should be familiar with it.

The theory of differential forms was initiated by Poincaré and further elaborated by Élie Cartan at the end of the nineteenth century. Differential forms have two main roles:

(1) To describe various systems of partial differential equations on manifolds.
(2) To define various geometric invariants reflecting the global structure of manifolds or bundles. Such invariants are obtained by integrating certain differential forms.

As we will see shortly, as soon as one tries to define integration on higher-dimensional objects, such as manifolds, one realizes that it is not functions that are integrated, but instead differential forms. Furthermore, as by magic, the algebra of differential forms handles changes of variables automatically and yields a neat form of "Stokes formula."

We begin with differential forms defined on an open subset U of $\mathbb{R}^n$. A *p-form* is any smooth function $\omega \colon U \to \bigwedge^p (\mathbb{R}^n)^*$ taking as values alternating tensors in the exterior power $\bigwedge^p (\mathbb{R}^n)^*$. The set of all p-forms on U is a vector space denoted $\mathcal{A}^p(U)$. The vector space $\mathcal{A}^*(U) = \bigoplus_{p \geq 0} \mathcal{A}^p(U)$ is the set of *differential forms on U*.

Proposition 3.14 shows that for every finite-dimensional vector space E, there are isomorphisms

$$\mu \colon \bigwedge^n (E^*) \longrightarrow \mathrm{Alt}^n(E; \mathbb{R}),$$

The original version of this chapter was revised. The correction to this chapter is available at https://doi.org/10.1007/978-3-030-46047-1_12

J. Gallier, J. Quaintance, *Differential Geometry and Lie Groups*, Geometry and Computing 13, https://doi.org/10.1007/978-3-030-46047-1_4

and these yield a canonical isomorphism of algebras $\mu \colon \bigwedge(E^*) \longrightarrow \operatorname{Alt}(E)$, where

$$\operatorname{Alt}(E) = \bigoplus_{n \geq 0} \operatorname{Alt}^n(E; \mathbb{R}),$$

and where $\operatorname{Alt}^n(E; \mathbb{R})$ is the vector space of real valued alternating multilinear maps on E^n.

In view of these isomorphisms, *we will identify ω and $\mu(\omega)$ for any $\omega \in \bigwedge^n(E^*)$, and we will write $\omega(u_1, \ldots, u_n)$ as an abbreviation for $\mu(\omega)(u_1, \ldots, u_n)$.*

Because $\operatorname{Alt}(\mathbb{R}^n)$ is an algebra under the wedge product, differential forms also have a wedge product, and thus $\mathcal{A}^*(U)$ is an algebra with the wedge product $\wedge$ on forms.

However, the power of differential forms stems from the *exterior differential*

$$d \colon \mathcal{A}^p(U) \to \mathcal{A}^{p+1}(U),$$

which is a skew-symmetric version of the usual differentiation operator. In Section 4.1 we prove some basic properties of the wedge product and of the exterior differential d. One of the most crucial properties of d is that the composition $\mathcal{A}^p(U) \xrightarrow{d} \mathcal{A}^{p+1}(U) \xrightarrow{d} \mathcal{A}^{p+2}(U)$ is identically zero; that is

$$d \circ d = 0,$$

which is an abbreviation for $d^{p+1} \circ d^p = 0$.

We explain that in $\mathbb{R}^3$, the notions of gradient, curl, and divergence arise naturally from the exterior differential d.

When is there a smooth field (P, Q, R) (in $\mathbb{R}^3$) whose curl is given by a prescribed smooth field (A, B, C)? Equivalently, when is there a 1-form $\omega = P\,dx + Q\,dy + R\,dz$ such that

$$d\omega = \eta = A\,dy \wedge dz + B\,dz \wedge dx + C\,dx \wedge dy?$$

Because $d \circ d = 0$, it is necessary that $d\eta = 0$; that is, (A, B, C) must have zero divergence. However, this condition is not sufficient in general; it depends on the topology of U.

More generally, we say that a differential p-form ω is *closed* if $d\omega = 0$ and *exact* if $\omega = d\eta$ for some $(p-1)$-form η. Since $d \circ d = 0$, every exact form is closed, but the converse is false in general. The purpose of de Rham cohomology is to measure the failure of a differential forms to be exact in terms of certain abelian groups (in fact, algebras).

The diagram (a *cochain complex*)

$$\mathcal{A}^0(U) \xrightarrow{d} \mathcal{A}^1(U) \longrightarrow \cdots \longrightarrow \mathcal{A}^{p-1}(U) \xrightarrow{d} \mathcal{A}^p(U) \xrightarrow{d} \mathcal{A}^{p+1}(U) \longrightarrow \cdots$$

is called the *de Rham complex* of U.

For every $p \geq 0$, let

$$Z^p(U) = \{\omega \in \mathcal{A}^p(U) \mid d\omega = 0\} = \operatorname{Ker} d \colon \mathcal{A}^p(U) \longrightarrow \mathcal{A}^{p+1}(U)$$

be the vector space of closed p-forms, also called p-cocycles, and for every $p \geq 1$, let

$$B^p(U) = \{\omega \in \mathcal{A}^p(U) \mid \exists \eta \in \mathcal{A}^{p-1}(U), \omega = d\eta\} = \operatorname{Im} d \colon \mathcal{A}^{p-1}(U) \longrightarrow \mathcal{A}^p(U)$$

be the vector space of exact p-forms, also called p-coboundaries. Set $B^0(U) = (0)$. Forms in $\mathcal{A}^p(U)$ are also called p-cochains. As $B^p(U) \subseteq Z^p(U)$ for every $p \geq 0$, we define the pth de Rham cohomology group of U as the quotient space

$$H_{\mathrm{DR}}^p(U) = Z^p(U)/B^p(U);$$

The real vector space $H_{\mathrm{DR}}^\bullet(U) = \bigoplus_{p \geq 0} H_{\mathrm{DR}}^p(U)$ is called the de Rham cohomology algebra of U.

The de Rham cohomology groups will be generalized to smooth manifolds in Section 4.3. They are important invariants of a manifold (which means that diffeomorphic manifolds have isomorphic cohomology groups).

In Section 4.2 we consider the behavior of differential forms under smooth maps $\varphi \colon U \to V$. Any such map induces a map $\varphi^* \colon \mathcal{A}^p(V) \to \mathcal{A}^p(U)$ on differential p-forms called a *pullback* (notice the reversal of U and V). Note that φ need not be a diffeomorphism, which is one of the technical advantages of forms over vector fields. We state various properties of the behavior of wedge products and the exterior differential d under pullback. In particular,

$$d\varphi^*(\omega) = \varphi^*(d\omega).$$

This property shows that a map $\varphi \colon U \to V$ induces a map $H_{\mathrm{DR}}^\bullet(\varphi) \colon H_{\mathrm{DR}}^\bullet(V) \to H_{\mathrm{DR}}^\bullet(U)$ on cohomology.

We state a fundamental result known as the *Poincaré lemma*, which says that the de Rham cohomology of a star-shaped open subset of $\mathbb{R}^n$ vanishes for $p \geq 1$, and that $H^0(U) = \mathbb{R}$. Thus every closed p-form on such a domain is exact ($p \geq 1$).

In Section 4.3 we generalize differential forms to smooth manifolds. Having defined differential forms on open subsets of $\mathbb{R}^n$, this is not a difficult task.

Technically, the set $\mathcal{A}^k(M)$ of *smooth differential k-forms* on M is the set of smooth sections $\Gamma(M, \bigwedge^k T^*M)$ of the bundle $\bigwedge^k T^*M$, and the set $\mathcal{A}^*(M)$ of all *smooth differential forms* on M is the set of smooth sections $\Gamma(M, \bigwedge T^*M)$ of the bundle $\bigwedge T^*M$.

These definitions are quite abstract, so we explain how p-forms are defined locally in terms of charts. Wedge products, pullbacks, and the exterior differential

$$d \colon \mathcal{A}^k(M) \to \mathcal{A}^{k+1}(M).$$

are defined. As in the case of open subsets of $\mathbb{R}^n$, we have

$$d \circ d = 0,$$

and d commutes with pullbacks. As a consequence, we have the de Rham complex

$$\mathcal{A}^0(M) \xrightarrow{d} \mathcal{A}^1(M) \longrightarrow \cdots \longrightarrow \mathcal{A}^{k-1}(M) \xrightarrow{d} \mathcal{A}^k(M) \xrightarrow{d} \mathcal{A}^{k+1}(M) \longrightarrow \cdots,$$

and we can define the *cohomology groups* $H^k_{\mathrm{DR}}(M)$ and the graded *cohomology algebra* $H^\bullet_{\mathrm{DR}}(M)$.

Another important property of the exterior differential d is that it is a *local operator*, which means that the value of $d\omega$ at p only depends of the values of ω near p. As a consequence, we obtain a characterization of the operator d; see Theorem 4.14.

Smooth differential forms can also be defined in terms of alternating $C^\infty(M)$-multilinear maps on smooth vector fields. This approach also yields a global formula for the exterior derivative $d\omega(X_1, \ldots, X_{k+1})$ of a k-form ω applied to $k + 1$ vector fields $X_1, \ldots, X_{k+1}$. This formula is not very useful for computing $d\omega$ at a given point p since it requires vector fields as input, but it is quite useful in theoretical investigations.

Let $\omega \in \mathcal{A}^k(M)$ be any smooth k-form on M. Then ω induces an alternating multilinear map

$$\omega \colon \underbrace{\mathfrak{X}(M) \times \cdots \times \mathfrak{X}(M)}_{k} \longrightarrow C^\infty(M)$$

as follows: for any k smooth vector fields $X_1, \ldots, X_k \in \mathfrak{X}(M)$,

$$\omega(X_1, \ldots, X_k)(p) = \omega_p(X_1(p), \ldots, X_k(p)).$$

This map is obviously alternating and $\mathbb{R}$-linear, but it is also $C^\infty(M)$-linear.

Let M be a smooth manifold. It is shown in Proposition 4.15 that for every $k \geq 0$, there is an isomorphism between the space of k-forms $\mathcal{A}^k(M)$ and the space $\mathrm{Alt}^k_{C^\infty(M)}(\mathfrak{X}(M))$ of alternating $C^\infty(M)$-multilinear maps on smooth vector fields. That is,

$$\mathcal{A}^k(M) \cong \mathrm{Alt}^k_{C^\infty(M)}(\mathfrak{X}(M)),$$

viewed as $C^\infty(M)$-modules. Then Proposition 4.16 gives an expression for $d\omega(X_1, \ldots, X_{k+1})$ (where $X_1, \ldots, X_{k+1}$ are vector fields) in terms of the X_i and some of their Lie brackets.

Section 4.4 is a technical section devoted to *Lie derivatives* of differential forms. We prove various properties about the interaction of Lie derivatives with the wedge operator and the exterior differential d. In particular, we prove *Cartan's formula*, which expresses the Lie derivative of a differential form in terms of d and an operator $i(X): \mathcal{A}^k(M) \rightarrow \mathcal{A}^{k-1}(M)$ called an *insertion operator*, where X is a vector field. We also generalize Lie derivatives to tensors.

In Section 4.5 we show how differential forms can be generalized so that they take values in any vector space F, rather than just $\mathbb{R}$. Vector-valued differential forms are needed in the theory of Lie groups and to define connections and curvature on vector bundles; see Chapter 10.

For simplicity, assume that U is an open subset of $\mathbb{R}^n$. Then it is natural to define differential forms with values in F as smooth maps $\omega: U \rightarrow \text{Alt}^p(\mathbb{R}^n; F)$, where $\text{Alt}^p(\mathbb{R}^n; F)$ denotes the vector space of alternating multilinear linear maps with values in F. The vector space of all p-forms on U with values in F is denoted $\mathcal{A}^p(U; F)$, and the vector space $\mathcal{A}^*(U; F) = \bigoplus_{p \geq 0} \mathcal{A}^p(U; F)$ is the set of *differential forms on U with values in F*.

There is no difficulty in defining *exterior differential* $d: \mathcal{A}^p(U; F) \rightarrow \mathcal{A}^{p+1}(U; F)$, and it can be shown that $d \circ d = 0$. The pullback of a form in $\mathcal{A}^p(V; F)$ along a smooth map $\varphi: U \rightarrow V$ is defined as before. The major difference is that there is no longer an obvious notion of wedge product. To define such an operation we need a bilinear form $\Phi: F \times G \rightarrow H$, where F, G, H are some vector spaces. Then we can define a wedge product

$$\wedge_\Phi: \mathcal{A}^p(U; F) \times \mathcal{A}^q(U; G) \rightarrow \mathcal{A}^{p+q}(U; H).$$

Such a wedge product is not associative in general, and not much can be said about it unless Φ has some additional properties. In general, unlike the case where $F = \mathbb{R}$, there is no nice formula for $d(\omega \wedge_\Phi \eta)$, unless F, G, H are finite dimensional. The case where $F = H = G = \mathfrak{g}$ where $\mathfrak{g}$ is a Lie algebra and $\Phi(a, b) = [a, b]$ is of particular interest.

The generalization of vector-valued differential forms to manifolds is no problem, except that some results involving the wedge product fail for the same reason that they fail in the case of forms on open subsets of $\mathbb{R}^n$.

In Section 4.6 we discuss left-invariant one-forms on a Lie group G. They form a space isomorphic to the dual $\mathfrak{g}^*$ of the Lie algebra $\mathfrak{g}$ of G. We prove the Maurer–Cartan equations in two versions, the second one involving a $\mathfrak{g}$-valued one-form ω_{MC} called the *Maurer–Cartan form*.

Our main goal is to define differential forms on manifolds, but we begin with differential forms on open subsets of $\mathbb{R}^n$ in order to build up intuition.

4.1 Differential Forms on Subsets of $\mathbb{R}^n$ and de Rham Cohomology

Differential forms are smooth functions on open subsets U of $\mathbb{R}^n$, taking as values alternating tensors in some exterior power $\bigwedge^p(\mathbb{R}^n)^*$.

Definition 4.1. Given any open subset U of $\mathbb{R}^n$, a smooth *differential p-form on U*, for short a *p-form on U*, is any smooth function $\omega \colon U \to \bigwedge^p(\mathbb{R}^n)^*$. The vector space of all p-forms on U is denoted $\mathcal{A}^p(U)$. The vector space $\mathcal{A}^*(U) = \bigoplus_{p \geq 0} \mathcal{A}^p(U)$ is the set of *differential forms on U*.

Observe that $\mathcal{A}^0(U) = C^\infty(U, \mathbb{R})$, the vector space of smooth functions on U, and $\mathcal{A}^1(U) = C^\infty(U, (\mathbb{R}^n)^*)$, the set of smooth functions from U to the set of linear forms on $\mathbb{R}^n$. Also, $\mathcal{A}^p(U) = (0)$ for $p > n$.

Remark. The space $\mathcal{A}^*(U)$ is also denoted $\mathcal{A}^\bullet(U)$. Other authors use $\Omega^p(U)$ instead of $\mathcal{A}^p(U)$, but we prefer to reserve Ω^p for holomorphic forms.

Recall from Sections 3.3 and 3.4, in particular Proposition 3.14, that for every finite-dimensional vector space E, the isomorphisms $\mu \colon \bigwedge^n(E^*) \longrightarrow \mathrm{Alt}^n(E; \mathbb{R})$ induced by the linear extensions of the maps given by

$$\mu(v_1^* \wedge \cdots \wedge v_n^*)(u_1, \ldots, u_n) = \begin{vmatrix} v_1^*(u_1) & \cdots & v_1^*(u_n) \\ \vdots & \ddots & \vdots \\ v_n^*(u_1) & \cdots & v_n^*(u_n) \end{vmatrix} = \det(v_j^*(u_i))$$

yield a canonical isomorphism of algebras $\mu \colon \bigwedge(E^*) \longrightarrow \mathrm{Alt}(E)$, where

$$\mathrm{Alt}(E) = \bigoplus_{n \geq 0} \mathrm{Alt}^n(E; \mathbb{R}),$$

and where $\mathrm{Alt}^n(E; \mathbb{R})$ is the vector space of real valued alternating multilinear maps on E^n. Recall that multiplication on alternating multilinear forms is defined such that, for $f \in \mathrm{Alt}^m(E; K)$ and $g \in \mathrm{Alt}^n(E; K)$,

$$(f \wedge g)(u_1, \ldots, u_{m+n}) = \sum_{\sigma \in \mathrm{shuffle}(m,n)} \mathrm{sgn}(\sigma) \, f(u_{\sigma(1)}, \ldots, u_{\sigma(m)}) g(u_{\sigma(m+1)}, \ldots, u_{\sigma(m+n)}),$$

$$(**)$$

where $\mathrm{shuffle}(m, n)$ consists of all (m, n)-"shuffles;" that is, permutations σ of $\{1, \ldots m + n\}$ such that $\sigma(1) < \cdots < \sigma(m)$ and $\sigma(m + 1) < \cdots < \sigma(m + n)$. The isomorphism μ has the property that

$$\mu(\omega \wedge \eta) = \mu(\omega) \wedge \mu(\eta), \quad \omega, \eta \in \bigwedge(E^*),$$

where the wedge operation on the left is the wedge on the exterior algebra $\bigwedge(E^*)$, and the wedge on the right is the multiplication on $\mathrm{Alt}(E)$ defined in $(**)$.

In view of these isomorphisms, *we will identify ω and $\mu(\omega)$ for any $\omega \in \bigwedge^n(E^*)$, and we will write $\omega(u_1, \ldots, u_n)$ as an abbreviation for $\mu(\omega)(u_1, \ldots, u_n)$.*

Because $\mathrm{Alt}(\mathbb{R}^n)$ is an algebra under the wedge product, differential forms also have a wedge product. However, the power of differential forms stems from the *exterior differential d*, which is a skew-symmetric version of the usual differentiation operator.

Recall from Section 3.2 that if $(e_1, \ldots, e_n)$ is any basis of $\mathbb{R}^n$ and $(e_1^*, \ldots, e_n^*)$ is its dual basis, then the alternating tensors

$$e_I^* = e_{i_1}^* \wedge \cdots \wedge e_{i_p}^*$$

form basis of $\bigwedge^p(\mathbb{R}^n)^*$, where $I = \{i_1, \ldots, i_p\} \subseteq \{1, \ldots, n\}$, with $i_1 < \cdots < i_p$. Thus, with respect to the basis $(e_1, \ldots, e_n)$, every p-form ω can be uniquely written as

$$\omega(x) = \sum_I f_I(x)\, e_{i_1}^* \wedge \cdots \wedge e_{i_p}^* = \sum_I f_I(x)\, e_I^* \qquad x \in U,$$

where each f_I is a smooth function on U. For example, if $U = \mathbb{R}^2 - \{0\}$, then

$$\omega(x, y) = \frac{-y}{x^2 + y^2}\, e_1^* + \frac{x}{x^2 + y^2}\, e_2^*$$

is a 1-form on U (with $e_1 = (1, 0)$ and $e_2 = (0, 1)$).

We often write ω_x instead of $\omega(x)$. Now, not only is $\mathcal{A}^*(U)$ a vector space, it is also an algebra.

Definition 4.2. The *wedge product* on $\mathcal{A}^*(U)$ is defined as follows: For all $p, q \geq 0$, the wedge product $\wedge : \mathcal{A}^p(U) \times \mathcal{A}^q(U) \to \mathcal{A}^{p+q}(U)$ is given by

$$(\omega \wedge \eta)_x = \omega_x \wedge \eta_x, \qquad x \in U.$$

For example, if ω and η are one-forms, then

$$(\omega \wedge \eta)_x(u, v) = \omega_x(u)\eta_x(v) - \omega_x(v)\eta_x(u).$$

In particular, if $U \subseteq \mathbb{R}^3$ and $\omega_x = a_1 e_1^* + a_3 e_3^*$ and $\eta_x = b_1 e_1^* + b_2 e_2^*$, for $u = (u_1, u_2, u_3) \in \mathbb{R}^3$ and $v = (v_1, v_2, v_3) \in \mathbb{R}^3$, the preceding line implies

$$\omega_x(u)\eta_x(v) - \omega_x(v)\eta_x(u)$$

$$= \big(a_1 e_1^*(u) + a_3 e_3^*(u)\big)\big(b_1 e_1^*(v) + b_2 e_2^*(v)\big)$$

$$\quad - \big(a_1 e_1^*(v) + a_3 e_3^*(v)\big)\big(b_1 e_1^*(u) + b_2 e_2^*(u)\big)$$

$$= (a_1 u_1 + a_3 u_3)(b_1 v_1 + b_2 v_2) - (a_1 v_1 + a_3 v_3)(b_1 u_1 + b_2 u_2)$$

$$= a_1 b_2(u_1 v_2 - v_1 u_2) - a_3 b_1(u_1 v_3 - v_1 u_3) - a_3 b_2(u_2 v_3 - u_3 v_2)$$

$$= a_1 b_2 \begin{vmatrix} e_1^*(u) & e_1^*(v) \\ e_2^*(u) & e_2^*(v) \end{vmatrix} - a_3 b_1 \begin{vmatrix} e_1^*(u) & e_1^*(v) \\ e_3^*(u) & e_3^*(v) \end{vmatrix} - a_3 b_2 \begin{vmatrix} e_2^*(u) & e_2^*(v) \\ e_3^*(u) & e_3^*(v) \end{vmatrix}$$

$$= (a_1b_2e_1^* \wedge e_2^* - a_3b_1e_1^* \wedge e_3^* - a_3b_2e_2^* \wedge e_3^*)(u, v)$$
$$= (a_1b_1e_1^* \wedge e_1^* + a_1b_2e_1^* \wedge e_2^* + a_3b_1e_3^* \wedge e_1^* + a_3b_2e_3^* \wedge e_2^*)(u, v)$$
$$= ((a_1e_1^* + a_3e_3^*) \wedge (b_1e_1^* + b_2e_2^*))(u, v)$$
$$= (\omega \wedge \eta)_x(u, v),$$

since $e_i^* \wedge e_i^* = 0$ and $e_i^* \wedge e_j^* = -e_j^* \wedge e_i^*$ for all $1 \le i < j \le 3$.

For $f \in \mathcal{A}^0(U) = C^\infty(U, \mathbb{R})$ and $\omega \in \mathcal{A}^p(U)$, we have $f \wedge \omega = f\omega$. Thus, the algebra $\mathcal{A}^*(U)$ is also a $C^\infty(U, \mathbb{R})$-module,

Proposition 3.12 immediately yields the following.

Proposition 4.1. *For all forms* $\omega \in \mathcal{A}^p(U)$ *and* $\eta \in \mathcal{A}^q(U)$, *we have*

$$\eta \wedge \omega = (-1)^{pq} \omega \wedge \eta.$$

We now come to the crucial operation of exterior differentiation. First recall that if $f : U \to V$ is a smooth function from $U \subseteq \mathbb{R}^n$ to a (finite-dimensional) normed vector space V, the derivative $f' : U \to \mathrm{Hom}(\mathbb{R}^n, V)$ of f (also denoted Df) is a function with domain U, with $f'(x)$ a linear map in $\mathrm{Hom}(\mathbb{R}^n, V)$ for every $x \in U$, such that if $(e_1, \dots, e_n)$ is the canonical basis of $\mathbb{R}^n$, $(u_1, \dots, u_m)$ is a basis of V, and if $f(x) = f_1(x)u_1 + \dots + f_m(x)u_m$, then

$$f'(x)(y_1e_1 + \dots + y_ne_n) = \sum_{i=1}^m \left(\sum_{j=1}^n \frac{\partial f_i}{\partial x_j}(x) \, y_j \right) u_i.$$

The $m \times n$ matrix

$$\left(\frac{\partial f_i}{\partial x_j}(x) \right)$$

is the *Jacobian matrix* of f at x, and if we write

$$z_1u_1 + \dots + z_mu_m = f'(x)(y_1e_1 + \dots + y_ne_n),$$

then in matrix form, we have

$$\begin{pmatrix} z_1 \\ \vdots \\ z_m \end{pmatrix} = \begin{pmatrix} \frac{\partial f_1}{\partial x_1}(x) & \cdots & \frac{\partial f_1}{\partial x_n}(x) \\ \vdots & \ddots & \vdots \\ \frac{\partial f_m}{\partial x_1}(x) & \cdots & \frac{\partial f_m}{\partial x_n}(x) \end{pmatrix} \begin{pmatrix} y_1 \\ \vdots \\ y_n \end{pmatrix}.$$

We also write $f_x'(u)$ for $f'(x)(u)$. Observe that since a p-form is a smooth map $\omega : U \to \bigwedge^p (\mathbb{R}^n)^*$, its derivative is a map

$$\omega': U \to \operatorname{Hom}\left(\mathbb{R}^n, \overset{p}{\bigwedge}(\mathbb{R}^n)^*\right)$$

such that ω'_x is a linear map from $\mathbb{R}^n$ to $\bigwedge^p(\mathbb{R}^n)^*$ for every $x \in U$. By the isomorphism $\bigwedge^p(\mathbb{R}^n)^* \cong \operatorname{Alt}^p(\mathbb{R}^n; \mathbb{R})$, we can view ω'_x as a linear map $\omega_x: \mathbb{R}^n \to \operatorname{Alt}^p(\mathbb{R}^n; \mathbb{R})$, or equivalently as a multilinear form $\omega'_x: (\mathbb{R}^n)^{p+1} \to \mathbb{R}$ which is alternating in its last p arguments. The exterior derivative $(d\omega)_x$ is obtained by making ω'_x into an alternating map in all of its $p+1$ arguments.

To make things more concrete, let us pick a basis $(e_1, \ldots, e_n)$ of $\mathbb{R}^n$, so that the $\binom{n}{p}$ tensors e_I^* form a basis of $\bigwedge^p(\mathbb{R}^n)^*$, where I is any subset $I = \{i_1, \ldots, i_p\} \subseteq \{1, \ldots, n\}$ such that $i_1 < \cdots < i_p$. Then every p-form ω can be uniquely written as

$$\omega_x = \sum_I f_I(x)\, e_I^* \qquad x \in U,$$

where each f_I is a smooth function on U, and for any $v = (v_1, \ldots, v_n) \in \mathbb{R}^n$,

$$\omega'_x(v) = \sum_I f'_I(x)(v)\, e_I^* = \sum_I \sum_{j=1}^{n} \frac{\partial f_I}{\partial x_j}(x)\, v_j\, e_I^* = \sum_I (\operatorname{grad}(f_I)_x \cdot v) e_I^*,$$

where $\cdot$ is the standard Euclidean inner product.

Remark. Observe that ω'_x is given by the $\binom{n}{p} \times n$ Jacobian matrix

$$\left(\frac{\partial f_I}{\partial x_j}(x)\right)$$

and that the product of the Ith row of the above matrix by v

$$\left(\frac{\partial f_I}{\partial x_1}(x) \cdots \frac{\partial f_I}{\partial x_n}(x)\right) \begin{pmatrix} v_1 \\ \vdots \\ v_n \end{pmatrix}$$

gives the coefficient $\operatorname{grad}(f_I)_x \cdot v$ of e_I^*.

Definition 4.3. For every $p \geq 0$, the *exterior differential* $d: \mathcal{A}^p(U) \to \mathcal{A}^{p+1}(U)$ is given by

$$(d\omega)_x(u_1, \ldots, u_{p+1}) = \sum_{i=1}^{p+1} (-1)^{i-1} \omega'_x(u_i)(u_1, \ldots, \widehat{u_i}, \ldots, u_{p+1}),$$

for all $\omega \in \mathcal{A}^p(U)$, all $x \in U$, and all $u_1, \ldots, u_{p+1} \in \mathbb{R}^n$, where the hat over the argument u_i means that it should be omitted.

In terms of a basis $(e_1, \ldots, e_n)$ of $\mathbb{R}^n$, if $\omega_x = \sum_I f_I(x)\, e_I^*$, then

$$(d\omega)_x(u_1, \ldots, u_{p+1}) = \sum_{i=1}^{p+1} (-1)^{i-1} \sum_I f_I'(x)(u_i)\, e_I^*(u_1, \ldots, \widehat{u_i}, \ldots, u_{p+1})$$

$$= \sum_{i=1}^{p+1} (-1)^{i-1} \sum_I (\operatorname{grad}(f_I)_x \cdot u_i) e_I^*(u_1, \ldots, \widehat{u_i}, \ldots, u_{p+1}).$$

One should check that $(d\omega)_x$ is indeed alternating, but this is easy. If necessary to avoid confusion, we write $d^p \colon \mathcal{A}^p(U) \to \mathcal{A}^{p+1}(U)$ instead of $d \colon \mathcal{A}^p(U) \to \mathcal{A}^{p+1}(U)$.

Remark. Definition 4.3 is the definition adopted by Cartan [20, 21][1] and Madsen and Tornehave [75]. Some authors use a different approach often using Propositions 4.2 and 4.3 as a starting point, but we find the approach using Definition 4.3 more direct. Furthermore, this approach extends immediately to the case of vector-valued forms.

For any smooth function, $f \in \mathcal{A}^0(U) = C^\infty(U, \mathbb{R})$, we get

$$df_x(u) = f_x'(u).$$

Therefore, for smooth functions, *the exterior differential df coincides with the usual derivative f'* (we identify $\bigwedge^1(\mathbb{R}^n)^*$ and $(\mathbb{R}^n)^*$). For any 1-form $\omega \in \mathcal{A}^1(U)$, we have

$$d\omega_x(u, v) = \omega_x'(u)(v) - \omega_x'(v)(u).$$

It follows that the map

$$(u, v) \mapsto \omega_x'(u)(v)$$

is symmetric iff $d\omega = 0$.

For a concrete example of exterior differentiation, consider

$$\omega_{(x,y)} = \frac{-y}{x^2 + y^2}\, e_1^* + \frac{x}{x^2 + y^2}\, e_2^* = f_1(x, y)e_1^* + f_2(x, y)e_2^*.$$

Since

$$\operatorname{grad}(f_1)_{(x,y)}^\top = \left(\frac{2xy}{(x^2 + y^2)^2} \quad \frac{y^2 - x^2}{(x^2 + y^2)^2} \right)$$

---

[1] We warn the reader that a few typos have crept up in the English translation, Cartan [21], of the original version Cartan [20].

$$\operatorname{grad}(f_2)_{(x,y)}^{\top} = \left(\frac{y^2 - x^2}{(x^2 + y^2)^2} \quad \frac{-2xy}{(x^2 + y^2)^2} \right),$$

if we write $u_1 = \begin{pmatrix} u_{11} \\ u_{12} \end{pmatrix}$ and $u_2 = \begin{pmatrix} u_{21} \\ u_{22} \end{pmatrix}$, then we have

$$\omega'_{(x,y)}(u_1)(u_2) = (\operatorname{grad}(f_1)_{(x,y)} \cdot u_1)e_1^*(u_2) + (\operatorname{grad}(f_2)_{(x,y)} \cdot u_1)e_2^*(u_2)$$

$$= \left(\frac{2xy}{(x^2 + y^2)^2} \quad \frac{y^2 - x^2}{(x^2 + y^2)^2} \right) \begin{pmatrix} u_{11} \\ u_{12} \end{pmatrix} e_1^* \begin{pmatrix} u_{21} \\ u_{22} \end{pmatrix}$$

$$+ \left(\frac{y^2 - x^2}{(x^2 + y^2)^2} \quad \frac{-2xy}{(x^2 + y^2)^2} \right) \begin{pmatrix} u_{11} \\ u_{12} \end{pmatrix} e_2^* \begin{pmatrix} u_{21} \\ u_{22} \end{pmatrix}$$

$$= \frac{2xy(u_{11}u_{21} - u_{12}u_{22}) + (y^2 - x^2)(u_{12}u_{21} + u_{11}u_{22})}{(x^2 + y^2)^2}.$$

A similar computation shows that

$$\omega'_{(x,y)}(u_2)(u_1) = \frac{2xy(u_{11}u_{21} - u_{12}u_{22}) + (y^2 - x^2)(u_{12}u_{21} + u_{11}u_{22})}{(x^2 + y^2)^2}$$

$$= \omega'_{(x,y)}(u_1)(u_2),$$

and so

$$d\omega_{(x,y)}(u_1, u_2) = \omega'_{(x,y)}(u_1)(u_2) - \omega'_{(x,y)}(u_2)(u_1) = 0.$$

Therefore $d\omega_{(x,y)} = 0$ for all $(x, y) \in U$, that is, $d\omega = 0$.

The following observation is quite trivial but it will simplify notation: On $\mathbb{R}^n$, we have the projection function $pr_i : \mathbb{R}^n \to \mathbb{R}$ with $pr_i(u_1, \dots, u_n) = u_i$. Note that $pr_i = e_i^*$, where $(e_1, \dots, e_n)$ is the canonical basis of $\mathbb{R}^n$. Let $x_i : U \to \mathbb{R}$ be the restriction of pr_i to U. Then note that x_i' is the constant map given by

$$x_i'(x) = pr_i, \qquad x \in U.$$

It follows that $dx_i = x_i'$ is the constant function with value $pr_i = e_i^*$. Now, since every p-form ω can be uniquely expressed as

$$\omega_x = \sum_I f_I(x) \, e_{i_1}^* \wedge \cdots \wedge e_{i_p}^* = \sum_I f_I(x)e_I^*, \qquad x \in U,$$

using Definition 4.2, we see immediately that ω can be uniquely written in the form

$$\omega = \sum_I f_I(x)\, dx_{i_1} \wedge \cdots \wedge dx_{i_p}, \qquad\qquad (*1)$$

where the f_I are smooth functions on U.

Observe that for $f \in \mathcal{A}^0(U) = C^\infty(U, \mathbb{R})$, we have

$$df_x = \sum_{i=1}^n \frac{\partial f}{\partial x_i}(x)\, e_i^* \quad \text{and} \quad df = \sum_{i=1}^n \frac{\partial f}{\partial x_i}\, dx_i.$$

Proposition 4.2. *For every p form $\omega \in \mathcal{A}^p(U)$ with $\omega = f\, dx_{i_1} \wedge \cdots \wedge dx_{i_p}$, we have*

$$d\omega = df \wedge dx_{i_1} \wedge \cdots \wedge dx_{i_p}.$$

Proof. Recall that $\omega_x = f e_{i_1}^* \wedge \cdots \wedge e_{i_p}^* = f e_I^*$, so

$$\omega_x'(u) = f_x'(u) e_I^* = df_x(u) e_I^*,$$

and by Definition 4.3, we get

$$d\omega_x(u_1, \ldots, u_{p+1}) = \sum_{i=1}^{p+1} (-1)^{i-1} df_x(u_i) e_I^*(u_1, \ldots, \widehat{u_i}, \ldots, u_{p+1})$$

$$= (df_x \wedge e_I^*)(u_1, \ldots, u_{p+1}),$$

where the last equation is an instance of the equation stated just before Proposition 3.14. $\square$

In practice we use Proposition 4.2 to compute $d\omega$. For example, if we take the previous example of

$$\omega = \frac{-y}{x^2 + y^2}\, dx + \frac{x}{x^2 + y^2}\, dy,$$

Proposition 4.2 implies that

$$d\omega = d\left(\frac{-y}{x^2 + y^2}\right) \wedge dx + d\left(\frac{x}{x^2 + y^2}\right) \wedge dy$$

$$= \left(\frac{2xy}{(x^2+y^2)^2}\, dx + \frac{y^2 - x^2}{(x^2+y^2)^2}\, dy\right) \wedge dx + \left(\frac{y^2 - x^2}{(x^2+y^2)^2}\, dx - \frac{2xy}{(x^2+y^2)^2}\, dy\right) \wedge dy$$

$$= \frac{y^2 - x^2}{(x^2+y^2)^2}\, dy \wedge dx + \frac{y^2 - x^2}{(x^2+y^2)^2}\, dx \wedge dy = 0.$$

We can now prove the following.

Proposition 4.3. *For all $\omega \in \mathcal{A}^p(U)$ and all $\eta \in \mathcal{A}^q(U)$,*

$$d(\omega \wedge \eta) = d\omega \wedge \eta + (-1)^p \omega \wedge d\eta.$$

Proof. In view of the unique representation $(*)$, it is enough to prove the proposition when $\omega = f e_I^*$ and $\eta = g e_J^*$. In this case, as $\omega \wedge \eta = fg\, e_I^* \wedge e_J^*$, by Proposition 4.2 we have

$$
\begin{aligned}
d(\omega \wedge \eta) &= d(fg) \wedge e_I^* \wedge e_J^* \\
&= ((df)g + f(dg)) \wedge e_I^* \wedge e_J^* \\
&= (df)g \wedge e_I^* \wedge e_J^* + f(dg) \wedge e_I^* \wedge e_J^* \\
&= df \wedge e_I^* \wedge g e_J^* + (-1)^p f e_I^* \wedge dg \wedge e_J^* \\
&= d\omega \wedge \eta + (-1)^p \omega \wedge d\eta
\end{aligned}
$$

since by Proposition 4.2, $d\omega = df \wedge e_I^*$ and $d\eta = g_J \wedge e_J^*$.   □

We say that d is an *anti-derivation of degree* -1.

Finally, here is the crucial and almost magical property of d.

Proposition 4.4. *For every $p \geq 0$, the composition $\mathcal{A}^p(U) \xrightarrow{d} \mathcal{A}^{p+1}(U) \xrightarrow{d} \mathcal{A}^{p+2}(U)$ is identically zero; that is*

$$d \circ d = 0,$$

which is an abbreviation for $d^{p+1} \circ d^p = 0$.

Proof. It is enough to prove the proposition when $\omega = f e_I^*$. We have

$$d\omega_x = df_x \wedge e_I^* = \frac{\partial f}{\partial x_1}(x)\, e_1^* \wedge e_I^* + \cdots + \frac{\partial f}{\partial x_n}(x)\, e_n^* \wedge e_I^*.$$

As $e_i^* \wedge e_j^* = -e_j^* \wedge e_i^*$ and $e_i^* \wedge e_i^* = 0$, we get

$$
\begin{aligned}
(d \circ d)\omega &= \sum_{i,j=1}^{n} \frac{\partial^2 f}{\partial x_i \partial x_j}(x)\, e_i^* \wedge e_j^* \wedge e_I^* \\
&= \sum_{i<j} \left(\frac{\partial^2 f}{\partial x_i \partial x_j}(x) - \frac{\partial^2 f}{\partial x_j \partial x_i}(x) \right) e_i^* \wedge e_j^* \wedge e_I^* = 0,
\end{aligned}
$$

since partial derivatives commute (as f is smooth).   □

It turns out that Propositions 4.3 and 4.4 together with the fact that d coincides with the derivative on $\mathcal{A}^0(U)$ characterize the differential d.

Theorem 4.5. *There is a unique linear map* $d\colon \mathcal{A}^*(U) \to \mathcal{A}^*(U)$ *with* $d = (d^p)$ *and* $d^p\colon \mathcal{A}^p(U) \to \mathcal{A}^{p+1}(U)$ *for every* $p \geq 0$, *such that*

(1) $df = f'$, *for every* $f \in \mathcal{A}^0(U) = C^\infty(U, \mathbb{R})$.
(2) $d \circ d = 0$.
(3) *For every* $\omega \in \mathcal{A}^p(U)$ *and every* $\eta \in \mathcal{A}^q(U)$,

$$d(\omega \wedge \eta) = d\omega \wedge \eta + (-1)^p \omega \wedge d\eta.$$

Proof. Existence has already been shown, so we only have to prove uniqueness. Let δ be another linear map satisfying Conditions (1)–(3). By (1), $df = \delta f = f'$ if $f \in \mathcal{A}^0(U)$. In particular, this holds when $f = x_i$, with $x_i\colon U \to \mathbb{R}$ the restriction of pr_i to U. In this case, we know that $\delta x_i = e_i^*$, the constant function $e_i^* = pr_i$. By (2), $\delta e_i^* = 0$. Using (3), we get $\delta e_I^* = 0$ for every nonempty subset $I \subseteq \{1, \ldots, n\}$. If $\omega = f e_I^*$, by (3), we get

$$\delta\omega = \delta f \wedge e_I^* + f \wedge \delta e_I^* = \delta f \wedge e_I^* = df \wedge e_I^* = d\omega.$$

Finally, since every differential form is a linear combination of special forms $f_I e_I^*$, we conclude that $\delta = d$. $\square$

Propositions 4.2, 4.3, and 4.4 can be summarized by saying that $\mathcal{A}^*(U)$ together with the product $\wedge$ and the differential d is a *differential graded algebra*. As $\mathcal{A}^*(U) = \bigoplus_{p \geq 0} \mathcal{A}^p(U)$ and $d^p\colon \mathcal{A}^p(U) \to \mathcal{A}^{p+1}(U)$, we can view $d = (d^p)$ as a linear map $d\colon \mathcal{A}^*(U) \to \mathcal{A}^*(U)$ such that

$$d \circ d = 0.$$

Let us consider one more example. Assume $n = 3$ and consider any function $f \in \mathcal{A}^0(U)$. We have

$$df = \frac{\partial f}{\partial x}\, dx + \frac{\partial f}{\partial y}\, dy + \frac{\partial f}{\partial z}\, dz,$$

and the vector

$$\left(\frac{\partial f}{\partial x}, \ \frac{\partial f}{\partial y}, \ \frac{\partial f}{\partial z} \right)$$

is the *gradient* of f. Next let

$$\omega = P\, dx + Q\, dy + R\, dz$$

be a 1-form on some open $U \subseteq \mathbb{R}^3$. An easy calculation yields

$$d\omega = dP \wedge dx + dQ \wedge dy + dR \wedge dz$$

$$= \left(\frac{\partial P}{\partial x} dx + \frac{\partial P}{\partial y} dy + \frac{\partial P}{\partial z} dz \right) \wedge dx + \left(\frac{\partial Q}{\partial x} dx + \frac{\partial Q}{\partial y} dy + \frac{\partial Q}{\partial z} dz \right) \wedge dy$$

$$+ \left(\frac{\partial R}{\partial x} dx + \frac{\partial R}{\partial y} dy + \frac{\partial R}{\partial z} dz \right) \wedge dz$$

$$= \frac{\partial P}{\partial y} dy \wedge dx + \frac{\partial P}{\partial z} dz \wedge dx + \frac{\partial Q}{\partial x} dx \wedge dy + \frac{\partial Q}{\partial z} dz \wedge dy$$

$$+ \frac{\partial R}{\partial x} dx \wedge dz + \frac{\partial R}{\partial y} dy \wedge dz$$

$$= \left(\frac{\partial R}{\partial y} - \frac{\partial Q}{\partial z} \right) dy \wedge dz + \left(\frac{\partial P}{\partial z} - \frac{\partial R}{\partial x} \right) dz \wedge dx + \left(\frac{\partial Q}{\partial x} - \frac{\partial P}{\partial y} \right) dx \wedge dy.$$

The vector field given by

$$\left(\frac{\partial R}{\partial y} - \frac{\partial Q}{\partial z}, \quad \frac{\partial P}{\partial z} - \frac{\partial R}{\partial x}, \quad \frac{\partial Q}{\partial x} - \frac{\partial P}{\partial y} \right)$$

is the *curl* of the vector field given by (P, Q, R). Now if

$$\eta = A dy \wedge dz + B dz \wedge dx + C dx \wedge dy$$

is a 2-form on $\mathbb{R}^3$, we get

$$d\eta = dA \wedge dy \wedge dz + dB \wedge dz \wedge dx + dC \wedge dx \wedge dy$$

$$= \left(\frac{\partial A}{\partial x} dx + \frac{\partial A}{\partial y} dy + \frac{\partial A}{\partial z} dz \right) \wedge dy \wedge dz$$

$$+ \left(\frac{\partial B}{\partial x} dx + \frac{\partial B}{\partial y} dy + \frac{\partial B}{\partial z} dz \right) \wedge dz \wedge dx$$

$$+ \left(\frac{\partial C}{\partial x} dx + \frac{\partial C}{\partial y} dy + \frac{\partial C}{\partial z} dz \right) \wedge dx \wedge dy$$

$$= \frac{\partial A}{\partial x} dx \wedge dy \wedge dz + \frac{\partial B}{\partial y} dy \wedge dz \wedge dx + \frac{\partial C}{\partial z} dz \wedge dx \wedge dy$$

$$= \left(\frac{\partial A}{\partial x} + \frac{\partial B}{\partial y} + \frac{\partial C}{\partial z} \right) dx \wedge dy \wedge dz.$$

The real number

$$\frac{\partial A}{\partial x} + \frac{\partial B}{\partial y} + \frac{\partial C}{\partial z}$$

is called the *divergence* of the vector field (A, B, C).

When is there a smooth field (P, Q, R) whose curl is given by a prescribed smooth field (A, B, C)? Equivalently, when is there a 1-form $\omega = Pdx + Qdy + Rdz$ such that

$$d\omega = \eta = Ady \wedge dz + Bdz \wedge dx + Cdx \wedge dy?$$

By Proposition 4.4 it is necessary that $d\eta = 0$; that is, (A, B, C) has zero divergence. However, this condition is not sufficient in general; it depends on the topology of U. If U is *star-like*, Poincaré's Lemma (to be considered shortly) says that this condition is sufficient.

Definition 4.4. The diagram

$$\mathcal{A}^0(U) \xrightarrow{d} \mathcal{A}^1(U) \longrightarrow \cdots \longrightarrow \mathcal{A}^{p-1}(U) \xrightarrow{d} \mathcal{A}^p(U) \xrightarrow{d} \mathcal{A}^{p+1}(U) \longrightarrow \cdots$$

is called the *de Rham complex* of U. It is a *cochain complex*.

Definition 4.5. A differential form ω is *closed* iff $d\omega = 0$; *exact* iff $\omega = d\eta$ for some differential form η. For every $p \geq 0$, let

$$Z^p(U) = \{\omega \in \mathcal{A}^p(U) \mid d\omega = 0\} = \operatorname{Ker} d \colon \mathcal{A}^p(U) \longrightarrow \mathcal{A}^{p+1}(U)$$

be the vector space of closed p-forms, also called *p-cocycles*, and for every $p \geq 1$, let

$$B^p(U) = \{\omega \in \mathcal{A}^p(U) \mid \exists \eta \in \mathcal{A}^{p-1}(U), \omega = d\eta\} = \operatorname{Im} d \colon \mathcal{A}^{p-1}(U) \longrightarrow \mathcal{A}^p(U)$$

be the vector space of exact p-forms, also called *p-coboundaries*. Set $B^0(U) = (0)$. Forms in $\mathcal{A}^p(U)$ are also called *p-cochains*. As $B^p(U) \subseteq Z^p(U)$ (by Proposition 4.4), for every $p \geq 0$, we define the pth *de Rham cohomology group* of U as the quotient space

$$H^p_{\mathrm{DR}}(U) = Z^p(U)/B^p(U);$$

This is an abelian group under addition of cosets. An element of $H^p_{\mathrm{DR}}(U)$ is called a *cohomology class* and is denoted $[\omega]$, where $\omega \in Z^p(U)$ is a cocycle. The real vector space $H^\bullet_{\mathrm{DR}}(U) = \bigoplus_{p \geq 0} H^p_{\mathrm{DR}}(U)$ is called the *de Rham cohomology algebra* of U. We also define the vector spaces $Z^*(U)$ and $B^*(U)$ by

$$Z^*(U) = \bigoplus_{p \geq 0} Z^p(U) \quad \text{and} \quad B^*(U) = \bigoplus_{p \geq 0} B^p(U).$$

We often drop the subscript DR and write $H^p(U)$ for $H^p_{\mathrm{DR}}(U)$ (resp. $H^\bullet(U)$ for $H^\bullet_{\mathrm{DR}}(U)$), when no confusion arises. Proposition 4.4 shows that every exact form is closed, but the converse is false in general. Measuring the extent to which closed forms are not exact is the object of *de Rham cohomology*.

For example, if we consider the form

$$
\omega_{(x,y)} = \frac{-y}{x^2 + y^2}\, dx + \frac{x}{x^2 + y^2}\, dy,
$$

on $U = \mathbb{R}^2 - \{0\}$, we have $d\omega = 0$. Yet, it is not hard to show (using integration, see Madsen and Tornehave [75], Chapter 1) that there is no smooth function f on U such that $df = \omega$. Thus, ω is a closed form which is not exact. This is because U is punctured.

Observe that $H^0(U) = Z^0(U) = \{f \in C^\infty(U, \mathbb{R}) \mid df = 0\}$; that is, $H^0(U)$ is the space of locally constant functions on U, equivalently, the space of functions that are constant on the connected components of U. Thus, the cardinality of $H^0(U)$ gives the number of connected components of U. For a large class of open sets (for example, open sets that can be covered by finitely many convex sets), the cohomology groups $H^p(U)$ are finite dimensional.

Now, $\mathcal{A}^*(U)$ is a graded algebra with multiplication $\wedge$.

Proposition 4.6. *The vector space $Z^*(U)$ is a subalgebra of $\mathcal{A}^*(U)$, and $B^*(U)$ is an ideal in $Z^*(U)$.*

Proof. The vector space $Z^*(U)$ is a subalgebra of $\mathcal{A}^*(U)$, because

$$
d(\omega \wedge \eta) = d\omega \wedge \eta + (-1)^p \omega \wedge d\eta,
$$

so $d\omega = 0$ and $d\eta = 0$ imply $d(\omega \wedge \eta) = 0$. The vector space $B^*(U)$ is an ideal in $Z^*(U)$, because if $\omega = d\eta$ and $d\tau = 0$, then

$$
d(\eta \wedge \tau) = d\eta \wedge \tau + (-1)^{p-1}\eta \wedge d\tau = \omega \wedge \tau,
$$

with $\eta \in \mathcal{A}^{p-1}(U)$. $\square$

Therefore, $H^\bullet_{\mathrm{DR}} = Z^*(U)/B^*(U)$ inherits a graded algebra structure from $\mathcal{A}^*(U)$. Explicitly, the multiplication in $H^\bullet_{\mathrm{DR}}$ is given by

$$
[\omega][\eta] = [\omega \wedge \eta].
$$

We now consider the action of smooth maps $\varphi \colon U \to U'$ on differential forms in $\mathcal{A}^*(U')$. We will see that φ induces a map from $\mathcal{A}^*(U')$ to $\mathcal{A}^*(U)$ called a *pull-back map. This corresponds to a change of variables.*

4.2   Pull-Back of Differential Forms

Recall Proposition 3.11 which states that if $f \colon E \to F$ is any linear map between two finite-dimensional vector spaces E and F, then

$$\mu\Big(\Big(\overset{p}{\bigwedge} f^\top\Big)(\omega)\Big)(u_1, \ldots, u_p) = \mu(\omega)(f(u_1), \ldots, f(u_p)), \qquad \omega \in \overset{p}{\bigwedge} F^*, \ u_1, \ldots, u_p \in E.$$

We apply this proposition with $E = \mathbb{R}^n$, $F = \mathbb{R}^m$, and $f = \varphi'_x$ ($x \in U$), and get

$$\mu\Big(\Big(\overset{p}{\bigwedge}(\varphi'_x)^\top\Big)(\omega_{\varphi(x)})\Big)(u_1, \ldots, u_p) = \mu(\omega_{\varphi(x)})(\varphi'_x(u_1), \ldots, \varphi'_x(u_p)), \qquad \omega \in \mathcal{A}^p(V), \ u_i \in \mathbb{R}^n.$$

This gives us the behavior of $\bigwedge^p(\varphi'_x)^\top$ under the identification of $\bigwedge^p(\mathbb{R})^*$ and $\mathrm{Alt}^n(\mathbb{R}^n; \mathbb{R})$ *via* the isomorphism μ. Consequently, denoting $\bigwedge^p(\varphi'_x)^\top$ by φ^*, we make the following definition:

Definition 4.6. Let $U \subseteq \mathbb{R}^n$ and $V \subseteq \mathbb{R}^m$ be two open subsets. For every smooth map $\varphi \colon U \to V$, for every $p \geq 0$, we define the map $\varphi^* \colon \mathcal{A}^p(V) \to \mathcal{A}^p(U)$ by

$$\varphi^*(\omega)_x(u_1, \ldots, u_p) = \omega_{\varphi(x)}(\varphi'_x(u_1), \ldots, \varphi'_x(u_p)),$$

for all $\omega \in \mathcal{A}^p(V)$, all $x \in U$, and all $u_1, \ldots, u_p \in \mathbb{R}^n$. We say that $\varphi^*(\omega)$ (for short, $\varphi^*\omega$) is the *pull-back* of ω by lφ.

As φ is smooth, $\varphi^*\omega$ is a smooth p-form on U. The maps $\varphi^* \colon \mathcal{A}^p(V) \to \mathcal{A}^p(U)$ induce a map also denoted $\varphi^* \colon \mathcal{A}^*(V) \to \mathcal{A}^*(U)$. Using the chain rule we obtain the following result.

Proposition 4.7. *The following identities hold:*

$$\mathrm{id}^* = \mathrm{id},$$
$$(\psi \circ \varphi)^* = \varphi^* \circ \psi^*.$$

Let $U = [0, 1] \times [0, 1] \subset \mathbb{R}^2$ and let $V = \mathbb{R}^3$. Define $\varphi \colon Q \to \mathbb{R}^3$ as $\varphi(u, v) = (\varphi_1(u, v), \varphi_2(u, v), \varphi_3(u, v)) = (x, y, z)$ where

$$x = u + v, \qquad y = u - v, \qquad z = uv.$$

Let $w = x dy \wedge dz + y dx \wedge dz$ be a 2-form in V. Clearly

$$\varphi'_{(u,v)} = \begin{pmatrix} 1 & 1 \\ 1 & -1 \\ v & u \end{pmatrix}.$$

Set $u_1 = \begin{pmatrix} u_{11} \\ u_{12} \end{pmatrix}$ and $u_2 = \begin{pmatrix} u_{21} \\ u_{22} \end{pmatrix}$. Definition 4.6 implies that the pull back of ω into U is

$$\varphi^*(\omega)_{(u,v)}(u_1, u_2) = \omega_{\varphi(u,v)}(\varphi'_{(u,v)}(u_1), \varphi'_{(u,v)}(u_2))$$

$$= \omega_{\varphi(u,v)}\left(\left(\begin{pmatrix} 1 & 1 \\ 1 & -1 \\ v & u \end{pmatrix}\begin{pmatrix} u_{11} \\ u_{12} \end{pmatrix}\right), \left(\begin{pmatrix} 1 & 1 \\ 1 & -1 \\ v & u \end{pmatrix}\begin{pmatrix} u_{21} \\ u_{22} \end{pmatrix}\right)\right)$$

$$= \omega_{\varphi(u,v)}\left(\left(\begin{matrix} u_{11} + u_{12} \\ u_{11} - u_{12} \\ vu_{11} + uu_{12} \end{matrix}\right), \left(\begin{matrix} u_{21} + u_{22} \\ u_{21} - u_{22} \\ vu_{21} + uu_{22} \end{matrix}\right)\right)$$

$$= (u + v)dy \wedge dz \left(\left(\begin{matrix} u_{11} + u_{12} \\ u_{11} - u_{12} \\ vu_{11} + uu_{12} \end{matrix}\right), \left(\begin{matrix} u_{21} + u_{22} \\ u_{21} - u_{22} \\ vu_{21} + uu_{22} \end{matrix}\right)\right)$$

$$+ (u - v)dx \wedge dz \left(\left(\begin{matrix} u_{11} + u_{12} \\ u_{11} - u_{12} \\ vu_{11} + uu_{12} \end{matrix}\right), \left(\begin{matrix} u_{21} + u_{22} \\ u_{21} - u_{22} \\ vu_{21} + uu_{22} \end{matrix}\right)\right)$$

$$= (u + v)\begin{vmatrix} u_{11} - u_{12} & u_{21} - u_{22} \\ vu_{11} + uu_{12} & vu_{21} + uu_{22} \end{vmatrix} + (u - v)\begin{vmatrix} u_{11} + u_{12} & u_{21} + u_{22} \\ vu_{11} + uu_{12} & vu_{21} + uu_{22} \end{vmatrix}$$

$$= (u + v)(u + v)(u_{11}u_{22} - u_{21}u_{12}) + (u - v)(u - v)(u_{11}u_{22} - u_{21}u_{12})$$

$$= (u + v)(u + v)\begin{vmatrix} u_{11} & u_{21} \\ u_{12} & u_{22} \end{vmatrix} + (u - v)(u - v)\begin{vmatrix} u_{11} & u_{21} \\ u_{12} & u_{22} \end{vmatrix}$$

$$= (u + v)(u + v)du \wedge dv(u_1, u_2) + (u - v)(u - v)du \wedge dv(u_1, u_2)$$

$$= 2(u^2 + v^2)du \wedge dv(u_1, u_2).$$

As the preceding example demonstrates, Definition 4.6 is not convenient for computations, so it is desirable to derive rules that yield a recursive definition of the pull-back.

The first rule has to do with the constant form $\omega = e_i^*$.

Proposition 4.8. *We have*

$$\varphi^* e_i^* = d\varphi_i, \quad with \; \varphi_i = pr_i \circ \varphi.$$

Proof. We have $\varphi_x = (\varphi_1)_x e_1 + \cdots + (\varphi_m)_x e_m$ for all $x \in U$, $\varphi_x'(u) = (\varphi_1)_x'(u)e_1 + \cdots + (\varphi_m)_x'(u)e_m$, and

$$(\varphi_i)_x'(u) = \sum_{l=1}^{n} \frac{\partial \varphi_i}{\partial x_l}(x)\, u_l = \sum_{l=1}^{n} \frac{\partial \varphi_i}{\partial x_l}(x)\, e_l^*(u),$$

so

$$\varphi^*(e_i^*)_x(u) = e_i^*(\varphi_x'(u))$$

$$= e_i^*((\varphi_1)_x'(u)e_1 + \cdots + (\varphi_m)_x'(u)e_m)$$

$$= (\varphi_i)_x'(u)$$

$$= \sum_{l=1}^{n} \frac{\partial \varphi_i}{\partial x_l}(x)\, e_l^*(u) = d(\varphi_i)_x(u),$$

as claimed.   □

The next proposition shows that the pull-back behaves well with respect to the wedge and the exterior derivative and provides the rest of the computational rules necessary for efficiently computing a pull-back.

Proposition 4.9. *Let $U \subseteq \mathbb{R}^n$ and $V \subseteq \mathbb{R}^m$ be two open sets and let $\varphi \colon U \to V$ be a smooth map. Then*

(i) $\varphi^*(\omega \wedge \eta) = \varphi^*\omega \wedge \varphi^*\eta$, *for all $\omega \in \mathcal{A}^p(V)$ and all $\eta \in \mathcal{A}^q(V)$.*
(ii) $\varphi^*(f) = f \circ \varphi$, *for all $f \in \mathcal{A}^0(V)$.*
(iii) $d\varphi^*(\omega) = \varphi^*(d\omega)$, *for all $\omega \in \mathcal{A}^p(V)$; that is, the following diagram commutes for all $p \geq 0$:*

$$
\begin{array}{ccc}
\mathcal{A}^k(N) & \xrightarrow{\;\varphi^*\;} & \mathcal{A}^k(M) \\
\downarrow{\scriptstyle d} & & \downarrow{\scriptstyle d} \\
\mathcal{A}^{k+1}(N) & \xrightarrow{\;\varphi^*\;} & \mathcal{A}^{k+1}(M)
\end{array}
$$

Proof.

(i) (See Madsen and Tornehave [75], Chapter 3.) For any $x \in U$ and any vectors $u_1, \ldots, u_{p+q} \in \mathbb{R}^n$ (with $p, q \geq 1$), we have

$$\varphi^*(\omega \wedge \eta)_x(u_1, \ldots, u_{p+q})$$

$$= (\omega \wedge \eta)_{\varphi(x)}(\varphi_x'(u_1), \ldots, \varphi_x'(u_{p+q}))$$

$$= \sum_{\sigma \in \mathrm{shuffle}(p,q)} \mathrm{sgn}(\sigma)\, \omega_{\varphi(x)}(\varphi_x'(u_{\sigma(1)}), \ldots, \varphi_x'(u_{\sigma(p)}))$$

$$\eta_{\varphi(x)}(\varphi_x'(u_{\sigma(p+1)}), \ldots, \varphi_x'(u_{\sigma(p+q)}))$$

$$= \sum_{\sigma \in \mathrm{shuffle}(p,q)} \mathrm{sgn}(\sigma)\, \varphi^*(\omega)_x(u_{\sigma(1)}, \ldots, u_{\sigma(p)})$$

$$\varphi^*(\eta)_x(u_{\sigma(p+1)}, \ldots, u_{\sigma(p+q)})$$

$$= (\varphi^*(\omega)_x \wedge \varphi^*(\eta)_x)(u_1, \ldots, u_{p+q}).$$

If $p = 0$ or $q = 0$, the proof is similar but simpler. We leave it as an exercise to the reader.

(ii) If $f \in \mathcal{A}^0(V) = C^\infty(V)$, by definition $\varphi^*(f)_x = f(\varphi(x))$, which means that $\varphi^*(f) = f \circ \varphi$.

First we prove (iii) in the case $\omega \in \mathcal{A}^0(V)$. Using (i) and (ii) and the fact that $\varphi^* e_i^* = d\varphi_i$, since

$$df = \sum_{k=1}^m \frac{\partial f}{\partial x_k} e_k^*,$$

we have

$$\varphi^*(df) = \sum_{k=1}^m \varphi^* \left(\frac{\partial f}{\partial x_k} \right) \wedge \varphi^*(e_k^*)$$

$$= \sum_{k=1}^m \left(\frac{\partial f}{\partial x_k} \circ \varphi \right) \wedge \left(\sum_{l=1}^n \frac{\partial \varphi_k}{\partial x_l} e_l^* \right)$$

$$= \sum_{k=1}^m \sum_{l=1}^n \left(\frac{\partial f}{\partial x_k} \circ \varphi \right) \left(\frac{\partial \varphi_k}{\partial x_l} \right) e_l^*$$

$$= \sum_{l=1}^n \left(\sum_{k=1}^m \left(\frac{\partial f}{\partial x_k} \circ \varphi \right) \frac{\partial \varphi_k}{\partial x_l} \right) e_l^*$$

$$= \sum_{l=1}^n \frac{\partial (f \circ \varphi)}{\partial x_l} e_l^*$$

$$= d(f \circ \varphi) = d(\varphi^*(f)).$$

For the case where $\omega = f e_I^*$, we know by Proposition 4.2 that $d\omega = df \wedge e_I^*$. We claim that

$$d\varphi^*(e_I^*) = 0.$$

To prove this first we show by induction on p that

$$d\varphi^*(e_I^*) = d(\varphi^*(e_{i_1}^* \wedge \cdots \wedge e_{i_p}^*)) = d(\varphi^*(e_{i_1}^*) \wedge \cdots \wedge \varphi^*(e_{i_p}^*))$$

$$= \sum_{k=1}^p (-1)^{k-1} \varphi^*(e_{i_1}^*) \wedge \cdots \wedge d(\varphi^*(e_{i_k}^*)) \wedge \cdots \wedge \varphi^*(e_{i_p}^*).$$

The base case $p = 1$ is trivial. Assuming that the induction hypothesis holds for any $p \geq 1$, with $I = \{i_1 < i_2 < \cdots < i_{p+1}\}$, using Proposition 4.3, we have

$$d\varphi^*(e_I^*) = d(\varphi^*(e_{i_1}^*) \wedge \varphi^*(e_{i_2}^*) \wedge \cdots \wedge \varphi^*(e_{i_{p+1}}^*))$$

$$= d(\varphi^*(e_{i_1}^*)) \wedge \varphi^*(e_{i_2}^*) \wedge \cdots \wedge \varphi^*(e_{i_{p+1}}^*)$$
$$+ (-1)^1 \varphi^*(e_{i_1}^*) \wedge d(\varphi^*(e_{i_2}^*) \wedge \cdots \wedge \varphi^*(e_{i_{p+1}}^*))$$

$$= d(\varphi^*(e_{i_1}^*)) \wedge \varphi^*(e_{i_2}^*) \wedge \cdots \wedge \varphi^*(e_{i_{p+1}}^*)$$
$$- \varphi^*(e_{i_1}^*) \wedge \left(\sum_{k=2}^{p+1} (-1)^{k-2} \varphi^*(e_{i_2}^*) \wedge \cdots \wedge d(\varphi^*(e_{i_k}^*)) \wedge \cdots \wedge \varphi^*(e_{i_{p+1}}^*) \right)$$

$$= d(\varphi^*(e_{i_1}^*)) \wedge \varphi^*(e_{i_2}^*) \wedge \cdots \wedge \varphi^*(e_{i_{p+1}}^*)$$
$$+ \sum_{k=2}^{p+1} (-1)^{k-1} \varphi^*(e_{i_1}^*) \wedge \varphi^*(e_{i_2}^*) \wedge \cdots \wedge d(\varphi^*(e_{i_k}^*)) \wedge \cdots \wedge \varphi^*(e_{i_{p+1}}^*)$$

$$= \sum_{k=1}^{p+1} (-1)^{k-1} \varphi^*(e_{i_1}^*) \wedge \cdots \wedge d(\varphi^*(e_{i_k}^*)) \wedge \cdots \wedge \varphi^*(e_{i_{p+1}}^*),$$

establishing the induction hypothesis.

As a consequence of the above equation, we have

$$d\varphi^*(e_I^*) = d(\varphi^*(e_{i_1}^*) \wedge \cdots \wedge \varphi^*(e_{i_p}^*))$$

$$= \sum_{k=1}^{p} (-1)^{k-1} \varphi^*(e_{i_1}^*) \wedge \cdots \wedge d(\varphi^*(e_{i_k}^*)) \wedge \cdots \wedge \varphi^*(e_{i_p}^*) = 0,$$

since $\varphi^*(e_{i_k}^*) = d\varphi_{i_k}$ and $d \circ d = 0$. Consequently, Proposition 4.3 implies that

$$d(\varphi^*(f) \wedge \varphi^*(e_I^*)) = d(\varphi^* f) \wedge \varphi^*(e_I^*).$$

Then we have

$$\varphi^*(d\omega) = \varphi^*(df) \wedge \varphi^*(e_I^*) = d(\varphi^* f) \wedge \varphi^*(e_I^*)$$
$$= d(\varphi^*(f) \wedge \varphi^*(e_I^*)) = d(\varphi^*(f e_I^*)) = d(\varphi^* \omega).$$

Since every differential form is a linear combination of special forms $f e_I^*$, we are done.   □

We use Proposition 4.9 to recompute the pull-back of $w = x \, dy \wedge dz + y \, dx \wedge dz$. Recall $Q = [0, 1] \times [0, 1] \subset \mathbb{R}^2$ and $\varphi : U \to \mathbb{R}^3$ was defined via

$$x = u + v, \qquad y = u - v, \qquad z = uv.$$

Proposition 4.9 implies that

$$\varphi^*(\omega) = (u+v)\varphi^*(dy) \wedge \varphi^*(dz) + (u-v)\varphi^*(dx) \wedge \varphi^*(dz)$$

$$= (u+v)d(\varphi^*y) \wedge d(\varphi^*z) + (u-v)d(\varphi^*x) \wedge d(\varphi^*z)$$

$$= (u+v)d(u-v) \wedge d(uv) + (u-v)d(u+v) \wedge d(uv)$$

$$= (u+v)(du-dv) \wedge (vdu+udv) + (u-v)(du+dv) \wedge (vdu+udv)$$

$$= 2(u^2+v^2)du \wedge dv.$$

We may generalize the techniques of the preceding calculation by using Proposition 4.9 to compute $\varphi^*\omega$ where $\varphi \colon U \to V$ is a smooth map between two open subsets U and V of $\mathbb{R}^n$ and $\omega = f\,dy_1 \wedge \cdots \wedge dy_n$ is a p-form on V. We can write $\varphi = (\varphi_1, \ldots, \varphi_n)$ with $\varphi_i \colon U \to \mathbb{R}$. By Proposition 4.9, we have

$$\varphi^*\omega = \varphi^*(f)\varphi^*(dy_1) \wedge \cdots \wedge \varphi^*(dy_n)$$

$$= \varphi^*(f)d(\varphi^*y_1) \wedge \cdots \wedge d(\varphi^*y_n)$$

$$= (f \circ \varphi)d(\varphi^*y_1) \wedge \cdots \wedge d(\varphi^*y_n).$$

However, $\varphi^*y_i = \varphi_i$ so we have

$$\varphi^*\omega = (f \circ \varphi)d\varphi_1 \wedge \cdots \wedge d\varphi_n.$$

For any $x \in U$, since

$$d(\varphi_i)_x = \sum_{j=1}^{n} \frac{\partial \varphi_i}{\partial x_j}(x)\,dx_j$$

we get

$$d\varphi_1 \wedge \cdots \wedge d\varphi_n = \det\left(\frac{\partial \varphi_i}{\partial x_j}(x) \right) dx_1 \wedge \cdots \wedge dx_n = J(\varphi)_x\, dx_1 \wedge \cdots \wedge dx_n,$$

where

$$J(\varphi)_x = \det\left(\frac{\partial \varphi_i}{\partial x_j}(x) \right)$$

is the Jacobian of φ at $x \in U$. It follows that

$$(\varphi^*\omega)_x = \varphi^*(f\,dy_1 \wedge \cdots \wedge dy_n)_x = f(\varphi(x))J(\varphi)_x\, dx_1 \wedge \cdots \wedge dx_n.$$

The fact that d and pull-back commutes is an important fact. It allows us to show that a map $\varphi \colon U \to V$ induces a map $H^\bullet(\varphi) \colon H^\bullet(V) \to H^\bullet(U)$ on cohomology, and it is crucial in generalizing the exterior differential to manifolds.

To a smooth map $\varphi\colon U \to V$, we associate the map $H^p(\varphi)\colon H^p(V) \to H^p(U)$ given by

$$H^p(\varphi)([\omega]) = [\varphi^*(\omega)].$$

This map is well defined, because if we pick any representative $\omega + d\eta$ in the cohomology class $[\omega]$ specified by the closed form ω, then

$$d\varphi^*\omega = \varphi^*d\omega = 0,$$

so $\varphi^*\omega$ is closed, and

$$\varphi^*(\omega + d\eta) = \varphi^*\omega + \varphi^*(d\eta) = \varphi^*\omega + d\varphi^*\eta,$$

which shows that $H^p(\varphi)([\omega])$ is well defined. It is also clear that

$$H^{p+q}(\varphi)([\omega][\eta]) = H^p(\varphi)([\omega])H^q(\varphi)([\eta]),$$

which means that $H^\bullet(\varphi)$ is a homomorphism of graded algebras. We often denote $H^\bullet(\varphi)$ by φ^*.

We conclude this section by stating without proof an important result known as the *Poincaré Lemma*. Recall that a subset $S \subseteq \mathbb{R}^n$ is *star-shaped* iff there is some point $c \in S$ such that for every point $x \in S$, the closed line segment $[c, x]$ joining c and x is entirely contained in S.

Theorem 4.10 (Poincaré's Lemma). *If $U \subseteq \mathbb{R}^n$ is any star-shaped open set, then we have $H^p(U) = (0)$ for $p > 0$ and $H^0(U) = \mathbb{R}$. Thus, for every $p \geq 1$, every closed form $\omega \in \mathcal{A}^p(U)$ is exact.*

Sketch of Proof. Pick c so that U is star-shaped w.r.t. c and let $g\colon U \to U$ be the constant function with value c. Then we see that

$$g^*\omega = \begin{cases} 0 & \text{if } \omega \in \mathcal{A}^p(U), \text{ with } p \geq 1, \\ \omega(c) & \text{if } \omega \in \mathcal{A}^0(U), \end{cases}$$

where $\omega(c)$ denotes the constant function with value $\omega(c)$. The trick is to find a family of linear maps $h^p\colon \mathcal{A}^p(U) \to \mathcal{A}^{p-1}(U)$, for $p \geq 1$, with $h^0 = 0$, such that

$$d \circ h^p + h^{p+1} \circ d = \text{id} - g^*, \qquad p > 0,$$

called a *chain homotopy*. Indeed, if $\omega \in \mathcal{A}^p(U)$ is closed and $p \geq 1$, we get $dh^p\omega = \omega$, so ω is exact, and if $p = 0$ we get $h^1 d\omega = 0 = \omega - \omega(c)$, so ω is constant. It remains to find the h^p, which is not obvious. A construction of these maps can be found in Madsen and Tornehave [75] (Chapter 3), Warner [109] (Chapter 4), Cartan [21] (Section 2), and Morita [82] (Chapter 3). □

In Section 4.3, we promote differential forms to manifolds. As preparation, note that every open subset $U \subseteq \mathbb{R}^n$ is a manifold, and that for every $x \in U$, the tangent space $T_x U$ to U at x is canonically isomorphic to $\mathbb{R}^n$. It follows that the tangent bundle TU and the cotangent bundle T^*U are trivial, namely $TU \cong U \times \mathbb{R}^n$ and $T^*U \cong U \times (\mathbb{R}^n)^*$, so the bundle

$$\bigwedge^k T^*U \cong U \times \bigwedge^k (\mathbb{R}^n)^*$$

is also trivial. Consequently, we can view $\mathcal{A}^k(U)$ as the set of smooth sections of the vector bundle $\bigwedge^k T^*(U)$. The generalization to manifolds is then to define the space of differential p-forms on a manifold M as the space of smooth sections of the bundle $\bigwedge^k T^*M$.

4.3 Differential Forms on Manifolds

Let M be any smooth manifold of dimension n. We define the vector bundle $\bigwedge T^*M$ as the direct sum bundle

$$\bigwedge T^*M = \bigoplus_{k=0}^{n} \bigwedge^k T^*M;$$

see Section 9.5 for details.

Recall that a smooth section of the bundle $\bigwedge^k T^*M$ is a smooth function $\omega \colon M \to \bigwedge^k T^*M$ such that $\omega(p) \in \bigwedge^k T_p^*M$ for all $p \in M$.

Definition 4.7. Let M be any smooth manifold of dimension n. The set $\mathcal{A}^k(M)$ of *smooth differential k-forms* on M is the set of smooth sections $\Gamma(M, \bigwedge^k T^*M)$ of the bundle $\bigwedge^k T^*M$, and the set $\mathcal{A}^*(M)$ of all *smooth differential forms* on M is the set of smooth sections $\Gamma(M, \bigwedge T^*M)$ of the bundle $\bigwedge T^*M$.

Observe that $\mathcal{A}^0(M) \cong C^\infty(M, \mathbb{R})$, the set of smooth functions on M, since the bundle $\bigwedge^0 T^*M$ is isomorphic to $M \times \mathbb{R}$, and smooth sections of $M \times \mathbb{R}$ are just graphs of smooth functions on M. We also write $C^\infty(M)$ for $C^\infty(M, \mathbb{R})$. If $\omega \in \mathcal{A}^*(M)$, we often write ω_p for $\omega(p)$.

Definition 4.7 is quite abstract, and it is important to get a more down-to-earth feeling by taking a local view of differential forms, namely with respect to a chart. So let (U, φ) be a local chart on M, with $\varphi \colon U \to \mathbb{R}^n$, and let $x_i = pr_i \circ \varphi$, the ith local coordinate ($1 \leq i \leq n$); see Tu [105] (Chapter 3, §8) or Gallier and Quaintance [47]. Recall that for any $p \in U$, the vectors

$$\left(\frac{\partial}{\partial x_1} \right)_p, \ldots, \left(\frac{\partial}{\partial x_x} \right)_p$$

form a basis of the tangent space T_pM. Furthermore, the linear forms $(dx_1)_p, \ldots, (dx_n)_p$ form a basis of T_p^*M, (where $(dx_i)_p$, the differential of x_i at p, is identified with the linear form such that $df_p(v) = v(\mathbf{f})$, for every smooth function f on U and every $v \in T_pM$). Consequently, locally on U, every k-form $\omega \in \mathcal{A}^k(M)$ can be written uniquely as

$$\omega_p = \sum_I f_I(p) dx_{i_1} \wedge \cdots \wedge dx_{i_k} = \sum_I f_I(p) dx_I, \quad p \in U,$$

where $I = \{i_1, \ldots, i_k\} \subseteq \{1, \ldots, n\}$, with $i_1 < \ldots < i_k$ and $dx_I = dx_{i_1} \wedge \cdots \wedge dx_{i_k}$. Furthermore, each f_I is a smooth function on U.

Remark. We define the set of smooth (r, s)-*tensor fields* as the set $\Gamma(M, T^{r,s}(M))$ of smooth sections of the tensor bundle $T^{r,s}(M) = T^{\otimes r}M \otimes (T^*M)^{\otimes s}$. Then locally in a chart (U, φ), every tensor field $\omega \in \Gamma(M, T^{r,s}(M))$ can be written uniquely as

$$\omega = \sum f^{i_1, \ldots, i_r}_{j_1, \ldots, j_s} \left(\frac{\partial}{\partial x_{i_1}} \right) \otimes \cdots \otimes \left(\frac{\partial}{\partial x_{i_r}} \right) \otimes dx_{j_1} \otimes \cdots \otimes dx_{j_s}.$$

The operations on the algebra $\bigwedge T^*M$ yield operations on differential forms using pointwise definitions. If $\omega, \eta \in \mathcal{A}^*(M)$ and $\lambda \in \mathbb{R}$, then for every $x \in M$,

$$(\omega + \eta)_x = \omega_x + \eta_x$$

$$(\lambda \omega)_x = \lambda \omega_x$$

$$(\omega \wedge \eta)_x = \omega_x \wedge \eta_x.$$

Actually, it is necessary to check that the resulting forms are smooth, but this is easily done using charts. When $f \in \mathcal{A}^0(M)$, we write $f\omega$ instead of $f \wedge \omega$. It follows that $\mathcal{A}^*(M)$ is a graded real algebra and a $C^\infty(M)$-module.

Proposition 4.1 generalizes immediately to manifolds.

Proposition 4.11. *For all forms $\omega \in \mathcal{A}^r(M)$ and $\eta \in \mathcal{A}^s(M)$, we have*

$$\eta \wedge \omega = (-1)^{pq} \omega \wedge \eta.$$

For any smooth map $\varphi \colon M \to N$ between two manifolds M and N, we have the differential map $d\varphi \colon TM \to TN$, also a smooth map, and for every $p \in M$, the map $d\varphi_p \colon T_pM \to T_{\varphi(p)}N$ is linear. As in Section 4.1, Proposition 3.11 gives us the formula

$$\mu\left(\left(\bigwedge^k (d\varphi_p)^\top \right) (\omega_{\varphi(p)}) \right) (u_1, \ldots, u_k)$$

$$= \mu(\omega_{\varphi(p)})(d\varphi_p(u_1), \ldots, d\varphi_p(u_k)), \quad \omega \in \mathcal{A}^k(N),$$

for all $u_1, \ldots, u_k \in T_pM$. This gives us the behavior of $\bigwedge^k (d\varphi_p)^\top$ under the identification of $\bigwedge^k T_p^*M$ and $\text{Alt}^k(T_pM; \mathbb{R})$ *via* the isomorphism μ. Here is the extension of Definition 4.6 to differential forms on a manifold.

Definition 4.8. For any smooth map $\varphi \colon M \to N$ between two smooth manifolds M and N, for every $k \geq 0$, we define the map $\varphi^* \colon \mathcal{A}^k(N) \to \mathcal{A}^k(M)$ by

$$\varphi^*(\omega)_p(u_1, \ldots, u_k) = \omega_{\varphi(p)}(d\varphi_p(u_1), \ldots, d\varphi_p(u_k)),$$

for all $\omega \in \mathcal{A}^k(N)$, all $p \in M$, and all $u_1, \ldots, u_k \in T_pM$. We say that $\varphi^*(\omega)$ (for short, $\varphi^*\omega$) is the *pull-back of ω by φ.*

The maps $\varphi^* \colon \mathcal{A}^k(N) \to \mathcal{A}^k(M)$ induce a map also denoted $\varphi^* \colon \mathcal{A}^*(N) \to \mathcal{A}^*(M)$. Using the chain rule, we check immediately that

$$\text{id}^* = \text{id},$$

$$(\psi \circ \varphi)^* = \varphi^* \circ \psi^*.$$

We need to check that $\varphi^*\omega$ is smooth, and for this it is enough to check it locally on a chart (U, ψ). For any chart (V, θ) on N such that $\varphi(U) \subseteq V$, on V we know that $\omega \in \mathcal{A}^k(N)$ can be written uniquely as

$$\omega = \sum_I f_I dx_{i_1} \wedge \cdots \wedge dx_{i_k},$$

with f_I smooth on V, and it is easy to see (using the definition) that locally on U we have

$$\varphi^*\omega = \sum_I (f_I \circ \varphi) d(x_{i_1} \circ \varphi) \wedge \cdots \wedge d(x_{i_k} \circ \varphi), \tag{$\dagger$}$$

which is smooth.

In the special case of $M = \mathbb{R}^n$, $\varphi \colon M \to N$ is a parametrization of N, and ($\dagger$) is what we use to efficiently calculate the pull-back of ω on the embedded manifold N. For example, let $M = \{(\theta, \varphi) : 0 < \theta < \pi,\ 0 < \varphi < 2\pi\} \subset \mathbb{R}^2$, $N = S^2$ and $\psi \colon M \to N$ the parametrization of S^2 given by

$$x = \sin\theta\cos\varphi, \qquad y = \sin\theta\sin\varphi, \qquad z = \cos\theta.$$

See Figure 4.1. Let $w = x\,dy$ be a form on S^2. The pull-back of ω into M is calculated via ($\dagger$) as

$$\psi^*w = \sin\theta\cos\varphi\,d(\sin\theta\sin\varphi)$$

$$= \sin\theta\cos\varphi(\cos\theta\sin\varphi\,d\theta + \sin\theta\cos\varphi\,d\varphi),$$

Fig. 4.1 The spherical
coordinates of S^2.

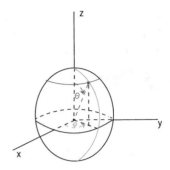

where we applied Proposition 4.2 since $M \subset \mathbb{R}^2$.

Remark. The fact that the pull-back of differential forms makes sense for arbitrary smooth maps $\varphi: M \to N$, and not just diffeomorphisms, is a major technical superiority of forms over vector fields.

The next step is to define d on $\mathcal{A}^*(M)$. There are several ways to proceed, but since we already considered the special case where M is an open subset of $\mathbb{R}^n$, we proceed using charts.

Given a smooth manifold M of dimension n, let (U, φ) be any chart on M. For any $\omega \in \mathcal{A}^k(M)$ and any $p \in U$, define $(d\omega)_p$ as follows: If $k = 0$, that is $\omega \in C^\infty(M)$, let

$$(d\omega)_p = d\omega_p, \qquad \text{the differential of } \omega \text{ at } p,$$

and if $k \geq 1$, let

$$(d\omega)_p = \varphi^*\left(d((\varphi^{-1})^*\omega)_{\varphi(p)}\right)_p,$$

where d is the exterior differential on $\mathcal{A}^k(\varphi(U))$. More explicitly, $(d\omega)_p$ is given by

$$(d\omega)_p(u_1, \ldots, u_{k+1}) = d((\varphi^{-1})^*\omega)_{\varphi(p)}(d\varphi_p(u_1), \ldots, d\varphi_p(u_{k+1})), \qquad (**)$$

for every $p \in U$ and all $u_1, \ldots, u_{k+1} \in T_pM$. Observe that the above formula is still valid when $k = 0$ if we interpret the symbols d in $d((\varphi^{-1})^*\omega)_{\varphi(p)} = d(\omega \circ \varphi^{-1})_{\varphi(p)}$ as the differential.

Since $\varphi^{-1}: \varphi(U) \to U$ is map whose domain is an open subset $W = \varphi(U)$ of $\mathbb{R}^n$, the form $(\varphi^{-1})^*\omega$ is a differential form in $\mathcal{A}^*(W)$, so $d((\varphi^{-1})^*\omega)$ is well defined.

The formula at Line (**) encapsulates the following "natural" three step procedure:

Step 1:   Take the form ω on the manifold M and precompose ω with the parameterization φ^{-1} so that $(\varphi^{-1})^*\omega$ is now a form in U, a subset of $\mathbb{R}^m$, where m is the dimension of M.

Step 2:   Differentiate $(\varphi^{-1})^*\omega$ via Proposition 4.2.

Step 3:   Compose the result of Step 2 with the chart map φ and pull the differential
form on U back into M.

We need to check that the definition at Line (∗∗) does not depend on the chart
(U, φ).

Proof. For any other chart (V, ψ), with $U \cap V \neq \emptyset$, the map $\theta = \psi \circ \varphi^{-1}$ is
a diffeomorphism between the two open subsets $\varphi(U \cap V)$ and $\psi(U \cap V)$, and
$\psi = \theta \circ \varphi$. Let $x = \varphi(p)$ and $y = \psi(p)$. We need to check that

$$d((\varphi^{-1})^*\omega)_x(d\varphi_p(u_1), \ldots, d\varphi_p(u_{k+1})) = d((\psi^{-1})^*\omega)_y(d\psi_p(u_1), \ldots, d\psi_p(u_{k+1})),$$

for every $p \in U \cap V$ and all $u_1, \ldots, u_{k+1} \in T_pM$. However, $y = \psi(p) = \theta(\varphi(p)) = \theta(x)$, so

$$d((\psi^{-1})^*\omega)_y(d\psi_p(u_1), \ldots, d\psi_p(u_{k+1}))$$
$$= d((\varphi^{-1} \circ \theta^{-1})^*\omega)_{\theta(x)}(d(\theta \circ \varphi)_p(u_1), \ldots, d(\theta \circ \varphi)_p(u_{k+1})).$$

Since

$$(\varphi^{-1} \circ \theta^{-1})^* = (\theta^{-1})^* \circ (\varphi^{-1})^*$$

and, by Proposition 4.9 (iii),

$$d(((\theta^{-1})^* \circ (\varphi^{-1})^*)\omega) = d((\theta^{-1})^*((\varphi^{-1})^*\omega)) = (\theta^{-1})^*(d((\varphi^{-1})^*\omega)),$$

we get

$$d((\varphi^{-1} \circ \theta^{-1})^*\omega)_{\theta(x)}(d(\theta \circ \varphi)_p(u_1), \ldots, d(\theta \circ \varphi)_p(u_{k+1}))$$
$$= (\theta^{-1})^*(d((\varphi^{-1})^*\omega))_{\theta(x)}(d(\theta \circ \varphi)_p(u_1), \ldots, d(\theta \circ \varphi)_p(u_{k+1})).$$

Then by Definition 4.8, we obtain

$$(\theta^{-1})^*(d((\varphi^{-1})^*\omega))_{\theta(x)}(d(\theta \circ \varphi)_p(u_1), \ldots, d(\theta \circ \varphi)_p(u_{k+1}))$$
$$= d((\varphi^{-1})^*\omega)_x((d\theta^{-1})_{\theta(x)}(d(\theta \circ \varphi)_p(u_1)), \ldots, (d\theta^{-1})_{\theta(x)}(d(\theta \circ \varphi)_p(u_{k+1}))).$$

As $(d\theta^{-1})_{\theta(x)}(d(\theta \circ \varphi)_p(u_i)) = d(\theta^{-1} \circ (\theta \circ \varphi))_p(u_i) = d\varphi_p(u_i)$, by the chain
rule, we obtain

$$d((\psi^{-1})^*\omega)_{\theta(x)}(d\psi_p(u_1), \ldots, d\psi_p(u_{k+1})) = d((\varphi^{-1})^*\omega)_x(d\varphi_p(u_1), \ldots, d\varphi_p(u_{k+1})),$$

as desired.   □

Observe that $(d\omega)_p$ is smooth on U, and as our definition of $(d\omega)_p$ does not depend on the choice of a chart, the forms $(d\omega) \upharpoonright U$ agree on overlaps and yield a differential form $d\omega$ defined on the whole of M. Thus we can make the following definition:

Definition 4.9. If M is any smooth manifold, there is a linear map $d: \mathcal{A}^k(M) \to \mathcal{A}^{k+1}(M)$ for every $k \geq 0$, such that for every $\omega \in \mathcal{A}^k(M)$, for every chart (U, φ), for every $p \in U$, if $k = 0$, that is $\omega \in C^\infty(M)$, then

$$(d\omega)_p = d\omega_p, \qquad \text{the differential of } \omega \text{ at } p,$$

else if $k \geq 1$, then

$$(d\omega)_p = \varphi^* \big(d((\varphi^{-1})^* \omega)_{\varphi(p)} \big)_p,$$

where d is the exterior differential on $\mathcal{A}^k(\varphi(U))$ from Definition 4.3. We obtain a linear map $d: \mathcal{A}^*(M) \to \mathcal{A}^*(M)$ called *exterior differentiation*.

To explicitly demonstrate Definition 4.9, we return to our previous example of $\psi : M \to S^2$ and $\omega = xdy$ considered as a one form on S^2. Note that

$$\psi(\theta, \varphi) = (\sin\theta \cos\varphi, \sin\theta \sin\varphi, \cos\theta)$$

is a parameterization of the S^2 and hence

$$\psi^{-1}(x, y, z) = (\cos^{-1}(z), \tan^{-1}(y/x))$$

provides the structure of a chart on S^2. We already found that the pull-back of ω into M is

$$\psi^*\omega = \sin\theta \cos\varphi \cos\theta \sin\varphi \, d\theta + \sin\theta \cos\varphi \sin\theta \cos\varphi \, d\varphi.$$

Proposition 4.2 is now applied $\psi^*\omega$ to give us

$$d\psi^*\omega = d(\sin\theta \cos\varphi \cos\theta \sin\varphi \, d\theta) + d(\sin\theta \cos\varphi \sin\theta \cos\varphi \, d\varphi)$$

$$= d(\sin\theta \cos\varphi \cos\theta \sin\varphi) \wedge d\theta + d(\sin\theta \cos\varphi \sin\theta \cos\varphi) \wedge d\varphi$$

$$= \frac{\partial}{\partial\varphi}(\sin\theta \cos\varphi \cos\theta \sin\varphi)d\varphi \wedge d\theta + \frac{\partial}{\partial\theta}(\sin\theta \cos\varphi \sin\theta \cos\varphi)d\theta \wedge d\varphi$$

$$= \sin\theta \cos\theta(-\sin^2\varphi + \cos^2\varphi)d\varphi \wedge d\theta + 2\sin\theta \cos\theta \cos^2\varphi \, d\theta \wedge d\varphi$$

$$= \sin\theta \cos\theta(\sin^2\varphi + \cos^2\varphi)d\theta \wedge d\varphi$$

$$= \sin\theta \cos\theta \, d\theta \wedge d\varphi.$$

It just remains to compose $d\psi^*\omega$ with ψ^{-1} to obtain

$$d\omega = (\psi^{-1})^*(d\psi^*\omega) = z\sqrt{1-z^2}d(\cos^{-1}z) \wedge d(\tan^{-1}y/x)$$

$$= z\sqrt{1-z^2}\left(-\frac{1}{\sqrt{1-z^2}}dz\right) \wedge \left(\frac{-\frac{y}{x^2}}{1+\frac{y^2}{x^2}}dx + \frac{\frac{1}{x}}{1+\frac{y^2}{x^2}}dy\right)$$

$$= -z\,dz \wedge \left(-\frac{y}{x^2+y^2}dx + \frac{x}{x^2+y^2}dy\right)$$

$$= \frac{zy}{x^2+y^2}dz \wedge dx - \frac{zx}{x^2+y^2}dz \wedge dy.$$

Since $x^2 + y^2 + z^2 = 1$, we obtain the constraint

$$x\,dx + y\,dy + z\,dz = 0,$$

which implies that $-z\,dz = x\,dx + y\,dy$. Then we find that $d\omega$ is equivalent to

$$d\omega = -z\,dz \wedge \left(-\frac{y}{x^2+y^2}dx + \frac{x}{x^2+y^2}dy\right)$$

$$= (x\,dx + y\,dy) \wedge \left(-\frac{y}{x^2+y^2}dx + \frac{x}{x^2+y^2}dy\right)$$

$$= \frac{x^2}{x^2+y^2}dx \wedge dy + \frac{y^2}{x^2+y^2}dx \wedge dy = dx \wedge dy,$$

where we interpret $dx \wedge dy$ as the restriction of 2-form in $\mathbb{R}^3$ to S^2, i.e., $dx \wedge dy|_{S^2}$ is defined as

$$(dx \wedge dy|_{S^2})_p(v) = (dx \wedge dy)_p(v), \qquad p \in S^2, \quad v \in T_pS^2.$$

Propositions 4.3, 4.4, and 4.9 generalize to manifolds.

Proposition 4.12. *Let M and N be smooth manifolds and let $\varphi \colon M \to N$ be a smooth map.*

(1) For all $\omega \in \mathcal{A}^r(M)$ and all $\eta \in \mathcal{A}^s(M)$,

$$d(\omega \wedge \eta) = d\omega \wedge \eta + (-1)^r \omega \wedge d\eta.$$

(2) For every $k \geq 0$, the composition $\mathcal{A}^k(M) \xrightarrow{d} \mathcal{A}^{k+1}(M) \xrightarrow{d} \mathcal{A}^{k+2}(M)$ is identically zero; that is,

$$d \circ d = 0.$$

(3) $\varphi^*(\omega \wedge \eta) = \varphi^*\omega \wedge \varphi^*\eta$, for all $\omega \in \mathcal{A}^r(N)$ and all $\eta \in \mathcal{A}^s(N)$.

(4) $\varphi^*(f) = f \circ \varphi$, for all $f \in \mathcal{A}^0(N)$.

(5) $d\varphi^*(\omega) = \varphi^*(d\omega)$, for all $\omega \in \mathcal{A}^k(N)$; that is, the following diagram commutes for all $k \geq 0$.

$$
\begin{array}{ccc}
\mathcal{A}^p(V; F) & \xrightarrow{\varphi^*} & \mathcal{A}^p(U; F) \\
\downarrow{\scriptstyle d} & & \downarrow{\scriptstyle d} \\
\mathcal{A}^{p+1}(V; F) & \xrightarrow{\varphi^*} & \mathcal{A}^{p+1}(U; F)
\end{array}
$$

Proof. It is enough to prove these properties in a chart (U, φ), which is easy. We only check (2). We have

$$
\begin{aligned}
(d(d\omega))_p &= d\big(\varphi^*\big(d((\varphi^{-1})^*\omega)_{\varphi(p)}\big)\big)_p \\
&= \varphi^*\Big[d\Big((\varphi^{-1})^*\big(\varphi^*\big(d((\varphi^{-1})^*\omega)_{\varphi(p)}\big)\big)\Big)_{\varphi(p)}\Big]_p \\
&= \varphi^*\Big[d\big(d((\varphi^{-1})^*\omega)_{\varphi(p)}\big)_{\varphi(p)}\Big]_p \\
&= 0,
\end{aligned}
$$

as $(\varphi^{-1})^* \circ \varphi^* = (\varphi \circ \varphi^{-1})^* = \mathrm{id}^* = \mathrm{id}$ and $d \circ d = 0$ on forms in $\mathcal{A}^k(\varphi(U))$, with $\varphi(U) \subseteq \mathbb{R}^n$. $\square$

As a consequence, Definition 4.5 of the de Rham cohomology generalizes to manifolds.

Definition 4.10. For every manifold M, we have the de Rham complex

$$
\mathcal{A}^0(M) \xrightarrow{d} \mathcal{A}^1(M) \longrightarrow \cdots \longrightarrow \mathcal{A}^{k-1}(M) \xrightarrow{d} \mathcal{A}^k(M) \xrightarrow{d} \mathcal{A}^{k+1}(M) \longrightarrow \cdots,
$$

and we can define the *cohomology groups* $H_{\mathrm{DR}}^k(M)$ and the graded *cohomology algebra* $H_{\mathrm{DR}}^\bullet(M)$. For every $k \geq 0$, let

$$
Z^k(M) = \{\omega \in \mathcal{A}^k(M) \mid d\omega = 0\} = \mathrm{Ker}\, d\colon \mathcal{A}^k(M) \longrightarrow \mathcal{A}^{k+1}(M)
$$

be the vector space of closed k-forms, and for every $k \geq 1$, let

$$
B^k(M) = \{\omega \in \mathcal{A}^k(M) \mid \exists \eta \in \mathcal{A}^{k-1}(M), \omega = d\eta\} = \mathrm{Im}\, d\colon \mathcal{A}^{k-1}(M) \longrightarrow \mathcal{A}^k(M)
$$

be the vector space of exact k-forms, with $B^0(M) = (0)$. Then, for every $k \geq 0$, we define the kth *de Rham cohomology group* of M as the quotient space

$$
H_{\mathrm{DR}}^k(M) = Z^k(M)/B^k(M).
$$

This is an abelian group under addition of cosets. The real vector space $H^\bullet_{\mathrm{DR}}(M) = \bigoplus_{k \geq 0} H^k_{\mathrm{DR}}(M)$ is called the *de Rham cohomology algebra* of M. We often drop the subscript DR when no confusion arises. Every smooth map $\varphi \colon M \to N$ between two manifolds induces an algebra map $\varphi^* \colon H^\bullet(N) \to H^\bullet(M)$.

Another important property of the exterior differential is that it is a *local operator*, which means that the value of $d\omega$ at p only depends of the values of ω near p. Not all operators are local. For example, the operator $I : C^\infty([a,b]) \to C^\infty([a,b])$ given by

$$I(f) = \int_a^b f(t)\,dt,$$

where $I(f)$ is the constant function on $[a,b]$, is not local since for any point $p \in [a,b]$, the calculation of $I(f)$ requires evaluating f over $[a,b]$.

More generally, we have the following definition.

Definition 4.11. A linear map $D \colon \mathcal{A}^*(M) \to \mathcal{A}^*(M)$ is a *local operator* if for all $k \geq 0$, for any nonempty open subset $U \subseteq M$ and for any two k-forms $\omega, \eta \in \mathcal{A}^k(M)$, if $\omega \restriction U = \eta \restriction U$, then $(D\omega) \restriction U = (D\eta) \restriction U$. Since D is linear, the above condition is equivalent to saying that for any k-form $\omega \in \mathcal{A}^k(M)$, if $\omega \restriction U = 0$, then $(D\omega) \restriction U = 0$.

Since Property (1) of Proposition 4.12 comes up a lot, we introduce the following definition.

Definition 4.12. Given any smooth manifold M, a linear map $D \colon \mathcal{A}^*(M) \to \mathcal{A}^*(M)$ is called an *antiderivation* if for all $r, s \geq 0$, for all $\omega \in \mathcal{A}^r(M)$ and all $\eta \in \mathcal{A}^s(M)$,

$$D(\omega \wedge \eta) = D\omega \wedge \eta + (-1)^r \omega \wedge D\eta.$$

The antiderivation is of *degree* $m \in \mathbb{Z}$ if $D \colon \mathcal{A}^p(M) \to \mathcal{A}^{p+m}(M)$ for all p such that $p + m \geq 0$.

By Proposition 4.12, exterior differentiation $d \colon \mathcal{A}^*(M) \to \mathcal{A}^*(M)$ is an antiderivation of degree 1.

Proposition 4.13. *Let M be a smooth manifold. Any linear antiderivation $D \colon \mathcal{A}^*(M) \to \mathcal{A}^*(M)$ is a local operator.*

Proof. By linearity, it is enough to show that if $\omega \restriction U = 0$, then $(D\omega) \restriction U = 0$. There is an apparent problem, which is that although ω is zero on U, it may not be zero outside U, so it is not obvious that we can conclude that $D\omega$ is zero on U. The crucial ingredient to circumvent this difficulty is the existence of "bump functions;" see Tu [105] (Chapter 3, §8) or Morita [82] (Chapter 1, Section 13(b)). By Lemma 1.28 of Morita applied to the constant function with value 1, for every $p \in U$, there some open subset $V \subseteq U$ containing p and a smooth function

$f \colon M \to \mathbb{R}$ such that supp $f \subseteq U$ and $f \equiv 1$ on V. Consequently, $f\omega$ is a smooth differential form which is identically zero, and since D is an antiderivation

$$D(f\omega) = Df \wedge \omega + f D\omega,$$

which, evaluated at p yields

$$0 = Df_p \wedge \omega_p + 1 D\omega_p = Df_p \wedge 0 + 1 D\omega_p = D\omega_p;$$

that is, $D\omega_p = 0$, as claimed. $\square$

Remark. If $D \colon \mathcal{A}^*(M) \to \mathcal{A}^*(M)$ is a linear map which is a *derivation*, which means that

$$D(\omega \wedge \eta) = D\omega \wedge \eta + \omega \wedge D\eta$$

for all $\omega \in \mathcal{A}^r(M)$ and all $\eta \in \mathcal{A}^s(M)$, then the proof of Proposition 4.13 still works and shows that D is also a local operator.

By Proposition 4.13, exterior differentiation $d \colon \mathcal{A}^*(M) \to \mathcal{A}^*(M)$ is a local operator. As in the case of differential forms on $\mathbb{R}^n$, the operator d is uniquely determined by the properties of Theorem 4.5.

Theorem 4.14. *Let M be a smooth manifold. There is a unique linear operator* $d \colon \mathcal{A}^*(M) \to \mathcal{A}^*(M)$, *with $d = (d^k)$ and $d^k \colon \mathcal{A}^k(M) \to \mathcal{A}^{k+1}(M)$ for every* $k \geq 0$, *such that*

(1) $(df)_p = df_p$, where df_p is the differential of f at $p \in M$ for every $f \in \mathcal{A}^0(M) = C^\infty(M)$.
(2) $d \circ d = 0$.
(3) For every $\omega \in \mathcal{A}^r(M)$ and every $\eta \in \mathcal{A}^s(M)$,

$$d(\omega \wedge \eta) = d\omega \wedge \eta + (-1)^r \omega \wedge d\eta.$$

Furthermore, any linear operator d satisfying (1)–(3) is a local operator.

Proof. Existence has already been established.

Let $D \colon \mathcal{A}^*(M) \to \mathcal{A}^*(M)$ be any linear operator satisfying (1)–(3). We need to prove that $D = d$ where d is defined in Definition 4.9. For any $k \geq 0$, pick any $\omega \in \mathcal{A}^k(M)$. For every $p \in M$, we need to prove that $(D\omega)_p = (d\omega)_p$. Let (U, φ) be any chart with $p \in U$, and let $x_i = pr_i \circ \varphi$ be the corresponding local coordinate maps. We know that $\omega \in \mathcal{A}^k(M)$ can be written uniquely as

$$\omega_q = \sum_I f_I(q) dx_{i_1} \wedge \cdots \wedge dx_{i_k} \qquad q \in U.$$

Using a bump function, there is some open subset V of U with $p \in V$ and some functions $\widetilde{f}_I$, and $\widetilde{x}_{i_1}, \ldots, \widetilde{x}_{i_k}$ defined on M and agreeing with $f_I, x_{i_1}, \ldots, x_{i_k}$ on

V. If we define

$$\widetilde{\omega} = \sum_I \widetilde{f_I} d\widetilde{x}_{i_1} \wedge \cdots \wedge d\widetilde{x}_{i_k}$$

then $\widetilde{\omega}$ is defined for all $p \in M$ and

$$\omega \restriction V = \widetilde{\omega} \restriction V.$$

By Proposition 4.13, since D is a linear map satisfying (3), it is a local operator so

$$D\omega \restriction V = D\widetilde{\omega} \restriction V.$$

Since D satisfies (1), we have $D\widetilde{x}_{i_j} = d\widetilde{x}_{i_j}$. Then at p, by linearity, we have

$$
\begin{aligned}
(D\omega)_p = (D\widetilde{\omega})_p &= \left(D\left(\sum_I \widetilde{f_I} d\widetilde{x}_{i_1} \wedge \cdots \wedge d\widetilde{x}_{i_k} \right) \right)_p \\
&= \left(D\left(\sum_I \widetilde{f_I} D\widetilde{x}_{i_1} \wedge \cdots \wedge D\widetilde{x}_{i_k} \right) \right)_p \\
&= \sum_I D\left(\widetilde{f_I} D\widetilde{x}_{i_1} \wedge \cdots \wedge D\widetilde{x}_{i_k} \right)_p.
\end{aligned}
$$

As in the proof of Proposition 4.9(iii), we can show by induction that

$$D(D\widetilde{x}_{i_1} \wedge \cdots \wedge D\widetilde{x}_{i_k}) = \sum_{j=1}^{k} (-1)^{j-1} D\widetilde{x}_{i_1} \wedge \cdots \wedge DD\widetilde{x}_{i_j} \wedge \cdots \wedge D\widetilde{x}_{i_k}$$

and since by (2) $DD\widetilde{x}_{i_j} = 0$, we have

$$D(D\widetilde{x}_{i_1} \wedge \cdots \wedge D\widetilde{x}_{i_k}) = 0.$$

Then, using the above, by (3) and (1), we get

$$
\begin{aligned}
\sum_I D\left(\widetilde{f_I} D\widetilde{x}_{i_1} \wedge \cdots \wedge D\widetilde{x}_{i_k} \right)_p &= \left(\sum_I D\widetilde{f_I} \wedge D\widetilde{x}_{i_1} \wedge \cdots \wedge D\widetilde{x}_{i_k} \right)_p \\
&\quad + \left(\sum_I \widetilde{f_I} \wedge D(D\widetilde{x}_{i_1} \wedge \cdots \wedge D\widetilde{x}_{i_k}) \right)_p \\
&= \left(\sum_I d\widetilde{f_I} \wedge d\widetilde{x}_{i_1} \wedge \cdots \wedge d\widetilde{x}_{i_k} \right)_p
\end{aligned}
$$

$$= \left(\sum_I df_I \wedge dx_{i_1} \wedge \cdots \wedge dx_{i_k} \right)_p$$

$$= (d\omega)_p.$$

Therefore $(D\omega)_p = (d\omega)_p$, which proves that $D = d$.   □

Remark. A closer look at the proof of Theorem 4.14 shows that it is enough to assume $DD\omega = 0$ on forms $\omega \in \mathcal{A}^0(M) = C^\infty(M)$.

Smooth differential forms can also be defined in terms of alternating $C^\infty(M)$-multilinear maps on smooth vector fields. This approach also yields a global formula for the exterior derivative $d\omega(X_1, \ldots, X_{k+1})$ of a k-form ω applied to $k + 1$ vector fields $X_1, \ldots, X_{k+1}$. This formula is not very useful for computing $d\omega$ at a given point p since it requires vector fields as input, but it is quite useful in theoretical investigations.

Let $\omega \in \mathcal{A}^k(M)$ be any smooth k-form on M. Then ω induces an alternating multilinear map

$$\omega \colon \underbrace{\mathfrak{X}(M) \times \cdots \times \mathfrak{X}(M)}_{k} \longrightarrow C^\infty(M)$$

as follows: For any k smooth vector fields $X_1, \ldots, X_k \in \mathfrak{X}(M)$,

$$\omega(X_1, \ldots, X_k)(p) = \omega_p(X_1(p), \ldots, X_k(p)).$$

This map is obviously alternating and $\mathbb{R}$-linear, but it is also $C^\infty(M)$-linear, since for every $f \in C^\infty(M)$,

$$\omega(X_1, \ldots, fX_i, \ldots X_k)(p) = \omega_p(X_1(p), \ldots, f(p)X_i(p), \ldots, X_k(p))$$
$$= f(p)\omega_p(X_1(p), \ldots, X_i(p), \ldots, X_k(p))$$
$$= (f\omega)_p(X_1(p), \ldots, X_i(p), \ldots, X_k(p)).$$

(Recall, that the set of smooth vector fields $\mathfrak{X}(M)$ is a real vector space and a $C^\infty(M)$-module.)

Interestingly, every alternating $C^\infty(M)$-multilinear map on smooth vector fields determines a differential form. This is because $\omega(X_1, \ldots, X_k)(p)$ only depends on the values of $X_1, \ldots, X_k$ at p.

Proposition 4.15. *Let M be a smooth manifold. For every $k \geq 0$, there is an isomorphism between the space of k-forms $\mathcal{A}^k(M)$ and the space $\mathrm{Alt}^k_{C^\infty(M)}(\mathfrak{X}(M))$ of alternating $C^\infty(M)$-multilinear maps on smooth vector fields. That is,*

$$\mathcal{A}^k(M) \cong \mathrm{Alt}^k_{C^\infty(M)}(\mathfrak{X}(M)),$$

viewed as $C^\infty(M)$-modules.

Proof. We follow the proof in O'Neill [84] (Chapter 2, Lemma 3 and Proposition 2). Let $\Phi \colon \underbrace{\mathfrak{X}(M) \times \cdots \times \mathfrak{X}(M)}_{k} \longrightarrow C^{\infty}(M)$ be an alternating $C^{\infty}(M)$-multilinear map. First we prove that for any vector fields $X_1, \ldots, X_k$ and $Y_1, \ldots, Y_k$, for every $p \in M$, if $X_i(p) = Y_i(p)$, then

$$\Phi(X_1, \ldots, X_k)(p) = \Phi(Y_1, \ldots, Y_k)(p).$$

Observe that

$$\begin{aligned}
\Phi(X_1, &\ldots, X_k) - \Phi(Y_1, \ldots, Y_k) \\
&= \Phi(X_1 - Y_1, X_2, \ldots, X_k) + \Phi(Y_1, X_2 - Y_2, X_3, \ldots, X_k) \\
&\quad + \Phi(Y_1, Y_2, X_3 - Y_3, \ldots, X_k) + \cdots \\
&\quad + \Phi(Y_1, \ldots, Y_{k-2}, X_{k-1} - Y_{k-1}, X_k) \\
&\quad + \Phi(Y_1, \ldots, Y_{k-1}, X_k - Y_k).
\end{aligned}$$

As a consequence, it is enough to prove that if $X_i(p) = 0$ for some i, then

$$\Phi(X_1, \ldots, X_k)(p) = 0.$$

Without loss of generality, assume $i = 1$. In any local chart (U, φ) near p, we can write

$$X_1 = \sum_{i=1}^{n} f_i \frac{\partial}{\partial x_i},$$

and as $X_i(p) = 0$, we have $f_i(p) = 0$ for $i = 1, \ldots, n$. Since the expression on the right-hand side is only defined on U, we extend it using a bump function once again. There is some open subset $V \subseteq U$ containing p and a smooth function $h \colon M \to \mathbb{R}$ such that $\operatorname{supp} h \subseteq U$ and $h \equiv 1$ on V. Then we let $h_i = h f_i$, a smooth function on M, $Y_i = h \frac{\partial}{\partial x_i}$, a smooth vector field on M, and we have $h_i \restriction V = f_i \restriction V$ and $Y_i \restriction V = \frac{\partial}{\partial x_i} \restriction V$. Now, since $h^2 = 1$ on V, it is obvious that

$$X_1 = \sum_{i=1}^{n} h_i Y_i + (1 - h^2) X_1$$

on V, so as Φ is $C^{\infty}(M)$-multilinear, $h_i(p) = 0$ and $h(p) = 1$, we get

$$\Phi(X_1, X_2, \ldots, X_k)(p) = \Phi\left(\sum_{i=1}^{n} h_i Y_i + (1 - h^2) X_1, X_2, \ldots, X_k \right)(p)$$

$$= \sum_{i=1}^{n} h_i(p)\Phi(Y_i, X_2, \ldots, X_k)(p) + (1 - h^2(p))\Phi(X_1, X_2, \ldots, X_k)(p) = 0,$$

as claimed.

Next we show that Φ induces a smooth differential form. For every $p \in M$, for any $u_1, \ldots, u_k \in T_pM$, we can pick smooth functions f_i equal to 1 near p and 0 outside some open near p, so that we get smooth vector fields $X_1, \ldots, X_k$ with $X_k(p) = u_k$. We set

$$\omega_p(u_1, \ldots, u_k) = \Phi(X_1, \ldots, X_k)(p).$$

As we proved that $\Phi(X_1, \ldots, X_k)(p)$ only depends on $X_1(p) = u_1, \ldots, X_k(p) = u_k$, the function ω_p is well defined, and it is easy to check that it is smooth. Therefore, the map $\Phi \mapsto \omega$ just defined is indeed an isomorphism. □

Remarks.

(1) The space $\mathrm{Hom}_{C^\infty(M)}(\mathfrak{X}(M), C^\infty(M))$ of all $C^\infty(M)$-linear maps $\mathfrak{X}(M) \longrightarrow C^\infty(M)$ is also a $C^\infty(M)$-module, called the *dual* of $\mathfrak{X}(M)$, and sometimes denoted $\mathfrak{X}^*(M)$. Proposition 4.15 shows that as $C^\infty(M)$-modules,

$$\mathcal{A}^1(M) \cong \mathrm{Hom}_{C^\infty(M)}(\mathfrak{X}(M), C^\infty(M)) = \mathfrak{X}^*(M).$$

(2) A result analogous to Proposition 4.15 holds for tensor fields. Indeed, there is an isomorphism between the set of tensor fields $\Gamma(M, T^{r,s}(M))$, and the set of $C^\infty(M)$-multilinear maps

$$\Phi\colon \underbrace{\mathcal{A}^1(M) \times \cdots \times \mathcal{A}^1(M)}_{r} \times \underbrace{\mathfrak{X}(M) \times \cdots \times \mathfrak{X}(M)}_{s} \longrightarrow C^\infty(M),$$

where $\mathcal{A}^1(M)$ and $\mathfrak{X}(M)$ are $C^\infty(M)$-modules.

Recall that for any function $f \in C^\infty(M)$ and every vector field $X \in \mathfrak{X}(M)$, the Lie derivative $X[f]$ (or $X(f)$) of f w.r.t. X is defined so that

$$X[f]_p = df_p(X(p));$$

Also recall the notion of the *Lie bracket* $[X, Y]$ of two vector fields; see Warner [109] (Chapter 1), Morita [82] (Chapter 1, Section 1.4), and Gallier and Quaintance [47]. The interpretation of differential forms as $C^\infty(M)$-multilinear maps given by Proposition 4.15 yields the following formula for $(d\omega)(X_1, \ldots, X_{k+1})$, where the X_i are vector fields:

Proposition 4.16. *Let M be a smooth manifold. For every k-form $\omega \in \mathcal{A}^k(M)$, we have*

$$(d\omega)(X_1, \ldots, X_{k+1}) = \sum_{i=1}^{k+1} (-1)^{i-1} X_i[\omega(X_1, \ldots, \widehat{X_i}, \ldots, X_{k+1})]$$

$$+ \sum_{i<j} (-1)^{i+j} \omega([X_i, X_j], X_1, \ldots, \widehat{X_i}, \ldots, \widehat{X_j}, \ldots, X_{k+1})],$$

for all vector fields, $X_1, \ldots, X_{k+1} \in \mathfrak{X}(M)$:

Proof Sketch. First one checks that the right-hand side of the formula in Proposition 4.16 is alternating and $C^\infty(M)$-multilinear. For this, use Proposition 5.3 from Chapter 0 of Do Carmo [37], or see Gallier and Quaintance [47]. Consequently, by Proposition 4.15, this expression defines a $(k + 1)$-form. Secondly, it is enough to check that both sides of the equation agree on charts (U, φ). We know in a chart that ω can be written uniquely as

$$\omega_p = \sum_I f_I(p) dx_{i_1} \wedge \cdots \wedge dx_{i_k} \qquad p \in U.$$

Also, as differential forms are $C^\infty(M)$-multilinear, it is enough to consider vector fields of the form $X_i = \frac{\partial}{\partial x_{j_i}}$. However, for such vector fields, $[X_i, X_j] = 0$, and then it is a simple matter to check that the equation holds. For more details, see Morita [82] (Chapter 2). $\qquad\qquad\qquad\qquad\qquad\qquad\qquad\qquad\qquad\qquad\qquad\qquad\qquad\square$

In particular, when $k = 1$, Proposition 4.16 yields the often used formula:

Corollary 4.17. *The following formula holds:*

$$d\omega(X, Y) = X[\omega(Y)] - Y[\omega(X)] - \omega([X, Y]).$$

There are other ways of proving the formula of Proposition 4.16, for instance, using Lie derivatives.

Before considering the Lie derivative $L_X\omega$ of differential forms, we define interior multiplication by a vector field, $i(X)(\omega)$. We will see shortly that there is a relationship between L_X, $i(X)$, and d, known as *Cartan's Formula.*

Definition 4.13. Let M be a smooth manifold. For every vector field $X \in \mathfrak{X}(M)$, for all $k \geq 1$, there is a linear map $i(X) : \mathcal{A}^k(M) \to \mathcal{A}^{k-1}(M)$ defined so that, for all $\omega \in \mathcal{A}^k(M)$, for all $p \in M$, for all $u_1, \ldots, u_{k-1} \in T_pM$,

$$(i(X)\omega)_p(u_1, \ldots, u_{k-1}) = \omega_p(X_p, u_1, \ldots, u_{k-1}).$$

Obviously, $i(X)$ is $C^\infty(M)$-linear in X, namely

$$i(fX)\omega = fi(X)\omega, \qquad i(X)(f\omega) = fi(X)\omega, \qquad f \in C^\infty(M), \ \omega \in \mathcal{A}^k(M),$$

and it is easy to check that $i(X)\omega$ is indeed a smooth $(k-1)$-form. When $k = 0$, we set $i(X)\omega = 0$. Observe that $i(X)\omega$ is also given by

$$(i(X)\omega)_p = i(X_p)\omega_p, \qquad p \in M,$$

where $i(X_p)$ is the interior product (or insertion operator) defined in Section 3.6 (with $i(X_p)\omega_p$ equal to our right hook, $\omega_p \llcorner X_p$). As a consequence, by Proposition 3.22, the operator $i(X)$ is an anti-derivation of degree -1; that is, we have

$$i(X)(\omega \wedge \eta) = (i(X)\omega) \wedge \eta + (-1)^r \omega \wedge (i(X)\eta),$$

for all $\omega \in \mathcal{A}^r(M)$ and all $\eta \in \mathcal{A}^s(M)$.

Remark. Other authors, including Marsden [77], use a left hook instead of a right hook, and denote $i(X)\omega$ as $X \lrcorner \omega$.

4.4   Lie Derivatives

We just saw in Section 4.3 that for any function $f \in C^\infty(M)$ and every vector field $X \in \mathfrak{X}(M)$, the Lie derivative $X[f]$ (or $X(f)$) of f w.r.t. X is defined so that

$$X[f]_p = df_p(X_p).$$

recall that for any manifold M, given any two vector fields $X, Y \in \mathfrak{X}(M)$, the *Lie derivative of X with respect to Y* is given by

$$(L_X Y)_p = \lim_{t \to 0} \frac{(\Phi_t^* Y)_p - Y_p}{t} = \frac{d}{dt} (\Phi_t^* Y)_p \Big|_{t=0},$$

where Φ_t is the local one-parameter group associated with X (Φ is the global flow associated with X; see Warner [109], Chapters 1 and 2, or Gallier and Quaintance [47]), and Φ_t^* is the pull-back of the diffeomorphism Φ_t given by

$$(\Phi_t^* Y)_p = d(\Phi_t^{-1})_{\Phi_t(p)}(Y_{\Phi_t(p)}).$$

Furthermore, to calculate $L_X Y$ recall that

$$L_X Y = [X, Y].$$

Proposition 4.18. *The following identity holds:*

$$X_p[f] = \lim_{t \to 0} \frac{(\Phi_t^* f)(p) - f(p)}{t} = \frac{d}{dt} (\Phi_t^* f)(p) \Big|_{t=0},$$

with $\Phi_t^* f = f \circ \Phi_t$ *(as usual for functions)*.

Proof. Recall that if Φ is the flow of X, then for every $p \in M$, the map $t \mapsto \Phi_t(p)$ is an integral curve of X through p; that is

$$\dot{\Phi}_t(p) = X(\Phi_t(p)), \qquad \Phi_0(p) = p,$$

in some open set containing p. In particular, $\dot{\Phi}_0(p) = X_p$. Then we have

$$
\begin{aligned}
\lim_{t \to 0} \frac{(\Phi_t^* f)(p) - f(p)}{t} &= \lim_{t \to 0} \frac{f(\Phi_t(p)) - f(\Phi_0(p))}{t} \\
&= \frac{d}{dt}(f \circ \Phi_t(p))\Big|_{t=0} \\
&= df_p(\dot{\Phi}_0(p)) = df_p(X_p) = X_p[f].
\end{aligned}
$$

$\square$

We would like to define the Lie derivative of differential forms (and tensor fields). This can be done algebraically or in terms of flows; the two approaches are equivalent, but it seems more natural to give a definition using flows.

Definition 4.14. Let M be a smooth manifold. For every vector field $X \in \mathfrak{X}(M)$, for every k-form $\omega \in \mathcal{A}^k(M)$, the *Lie derivative of $\omega$the with respect to X*, denoted $L_X \omega$, is given by

$$(L_X \omega)_p = \lim_{t \to 0} \frac{(\Phi_t^* \omega)_p - \omega_p}{t} = \frac{d}{dt}(\Phi_t^* \omega)_p \Big|_{t=0},$$

where $\Phi_t^* \omega$ is the pull-back of ω along Φ_t (see Definition 4.8).

Obviously, $L_X \colon \mathcal{A}^k(M) \to \mathcal{A}^k(M)$ is a linear map, but it has many other interesting properties.

We can also define the Lie derivative on tensor fields.

Definition 4.15. The *Lie derivative* on tensor fields is the map $L_X \colon \Gamma(M, T^{r,s}(M)) \to \Gamma(M, T^{r,s}(M))$, obtained by requiring that for any tensor field

$$\alpha = X_1 \otimes \cdots \otimes X_r \otimes \omega_1 \otimes \cdots \otimes \omega_s,$$

where $X_i \in \mathfrak{X}(M)$ and $\omega_j \in \mathcal{A}^1(M)$,

$$\Phi_t^* \alpha = \Phi_t^* X_1 \otimes \cdots \otimes \Phi_t^* X_r \otimes \Phi_t^* \omega_1 \otimes \cdots \otimes \Phi_t^* \omega_s,$$

where $\Phi_t^* X_i$ is the pull-back of the vector field, X_i, and $\Phi_t^* \omega_j$ is the pull-back of one-form ω_j, and then setting

$$(L_X\alpha)_p = \lim_{t \longrightarrow 0} \frac{(\Phi_t^*\alpha)_p - \alpha_p}{t} = \frac{d}{dt}(\Phi_t^*\alpha)_p\bigg|_{t=0}.$$

So, as long we can define the "right" notion of pull-back, the formula giving the Lie derivative of a function, a vector field, a differential form, and more generally a tensor field, is the same.

The Lie derivative of tensors is used in most areas of mechanics, for example in elasticity (the rate of strain tensor) and in fluid dynamics.

We now state, mostly without proofs, a number of properties of Lie derivatives. Most of these proofs are fairly straightforward computations, often tedious, and can be found in most texts, including Warner [109], Morita [82], and Gallot, Hullin, and Lafontaine [48].

Proposition 4.19. *Let M be a smooth manifold. For every vector field $X \in \mathfrak{X}(M)$, the following properties hold:*

(1) For all $\omega \in \mathcal{A}^r(M)$ and all $\eta \in \mathcal{A}^s(M)$,

$$L_X(\omega \wedge \eta) = (L_X\omega) \wedge \eta + \omega \wedge (L_X\eta);$$

that is, L_X is a derivation.
(2) For all $\omega \in \mathcal{A}^k(M)$, for all $Y_1, \ldots, Y_k \in \mathfrak{X}(M)$,

$$L_X(\omega(Y_1, \ldots, Y_k)) = (L_X\omega)(Y_1, \ldots, Y_k)$$

$$+ \sum_{i=1}^{k} \omega(Y_1, \ldots, Y_{i-1}, L_X Y_i, Y_{i+1}, \ldots, Y_k).$$

(3) The Lie derivative commutes with d:

$$L_X \circ d = d \circ L_X.$$

Proof. We only prove (2). First we claim that if $\varphi \colon M \to M$ is a diffeomorphism, then for every $\omega \in \mathcal{A}^k(M)$, for all $X_1, \ldots, X_k \in \mathfrak{X}(M)$,

$$(\varphi^*\omega)(X_1, \ldots, X_k) = \varphi^*(\omega((\varphi^{-1})^*X_1, \ldots, (\varphi^{-1})^*X_k)), \qquad (*)$$

where $(\varphi^{-1})^*X_i$ is the pull-back of the vector field X_i (also equal to the push-forward φ_*X_i of X_i). Recall that

$$((\varphi^{-1})^*Y)_p = d\varphi_{\varphi^{-1}(p)}(Y_{\varphi^{-1}(p)}),$$

for any vector field Y. Then by Definition 4.8, for every $p \in M$, we have

$$(\varphi^*\omega(X_1, \ldots, X_k))(p)$$

$$= \omega_{\varphi(p)}(d\varphi_p(X_1(p)), \ldots, d\varphi_p(X_k(p)))$$

$$= \omega_{\varphi(p)}\big(d\varphi_{\varphi^{-1}(\varphi(p))}(X_1(\varphi^{-1}(\varphi(p)))), \ldots, d\varphi_{\varphi^{-1}(\varphi(p))}(X_k(\varphi^{-1}(\varphi(p))))\big)$$

$$= \omega_{\varphi(p)}(((\varphi^{-1})^*X_1)_{\varphi(p)}, \ldots, ((\varphi^{-1})^*X_k)_{\varphi(p)})$$

$$= (\omega((\varphi^{-1})^*X_1, \ldots, (\varphi^{-1})^*X_k)) \circ \varphi)(p)$$

$$= \varphi^*(\omega((\varphi^{-1})^*X_1, \ldots, (\varphi^{-1})^*X_k))(p),$$

since for any function $g \in C^\infty(M)$, we have $\varphi^*g = g \circ \varphi$.

We know that

$$X_p[f] = \lim_{t \to 0} \frac{(\Phi_t^*f)(p) - f(p)}{t}$$

and for any vector field Y,

$$[X, Y]_p = (L_X Y)_p = \lim_{t \to 0} \frac{(\Phi_t^*Y)_p - Y_p}{t}.$$

Since the one-parameter group associated with $-X$ is Φ_{-t}, (this follows from $\Phi_{-t} \circ \Phi_t = \mathrm{id}$), we have

$$\lim_{t \to 0} \frac{(\Phi_{-t}^*Y)_p - Y_p}{t} = -[X, Y]_p.$$

Now, using $\Phi_t^{-1} = \Phi_{-t}$ and $(*)$, we have

$$(L_X\omega)(Y_1, \ldots, Y_k) = \lim_{t \to 0} \frac{(\Phi_t^*\omega)(Y_1, \ldots, Y_k) - \omega(Y_1, \ldots, Y_k)}{t}$$

$$= \lim_{t \to 0} \frac{\Phi_t^*(\omega(\Phi_{-t}^*Y_1, \ldots, \Phi_{-t}^*Y_k)) - \omega(Y_1, \ldots, Y_k)}{t}$$

$$= \lim_{t \to 0} \frac{\Phi_t^*(\omega(\Phi_{-t}^*Y_1, \ldots, \Phi_{-t}^*Y_k)) - \Phi_t^*(\omega(Y_1, \ldots, Y_k))}{t}$$

$$+ \lim_{t \to 0} \frac{\Phi_t^*(\omega(Y_1, \ldots, Y_k)) - \omega(Y_1, \ldots, Y_k)}{t}.$$

Call the first term A and the second term B. Then as

$$X_p[f] = \lim_{t \to 0} \frac{(\Phi_t^*f)(p) - f(p)}{t},$$

we have

$$B = X[\omega(Y_1, \ldots, Y_k)] = L_X(\omega(Y_1, \ldots, Y_k)).$$

As to A, we have

$$
\begin{aligned}
A = {} & \lim_{t \to 0} \frac{\Phi_t^*(\omega(\Phi_{-t}^* Y_1, \ldots, \Phi_{-t}^* Y_k)) - \Phi_t^*(\omega(Y_1, \ldots, Y_k))}{t} \\[2mm]
= {} & \lim_{t \to 0} \Phi_t^* \left(\frac{\omega(\Phi_{-t}^* Y_1, \ldots, \Phi_{-t}^* Y_k) - \omega(Y_1, \ldots, Y_k)}{t} \right) \\[2mm]
= {} & \lim_{t \to 0} \Phi_t^* \left(\frac{\omega(\Phi_{-t}^* Y_1, \ldots, \Phi_{-t}^* Y_k) - \omega(Y_1, \Phi_{-t}^* Y_2, \ldots, \Phi_{-t}^* Y_k)}{t} \right) \\[2mm]
& + \lim_{t \to 0} \Phi_t^* \left(\frac{\omega(Y_1, \Phi_{-t}^* Y_2, \ldots, \Phi_{-t}^* Y_k) - \omega(Y_1, Y_2, \Phi_{-t}^* Y_3, \ldots, \Phi_{-t}^* Y_k)}{t} \right) \\[2mm]
& + \cdots + \lim_{t \to 0} \Phi_t^* \left(\frac{\omega(Y_1, \ldots, Y_{k-1}, \Phi_{-t}^* Y_k) - \omega(Y_1, \ldots, Y_k)}{t} \right) \\[2mm]
= {} & \lim_{t \to 0} \Phi_t^* \left(\frac{\omega(\Phi_{-t}^* Y_1 - Y_1, \Phi_{-t}^* Y_2, \ldots, \Phi_{-t}^* Y_k)}{t} \right) \\[2mm]
& + \lim_{t \to 0} \Phi_t^* \left(\frac{\omega(Y_1, \Phi_{-t}^* Y_2 - Y_2, \ldots, \Phi_{-t}^* Y_k)}{t} \right) \\[2mm]
& + \cdots + \lim_{t \to 0} \Phi_t^* \left(\frac{\omega(Y_1, \ldots, Y_{k-1}, \Phi_{-t}^* Y_k - Y_k)}{t} \right) \\[2mm]
= {} & \lim_{t \to 0} \Phi_t^* \left(\omega \left(\frac{\Phi_{-t}^* Y_1 - Y_1}{t}, \ldots, \Phi_{-t}^* Y_k \right) \right) \\[2mm]
& + \lim_{t \to 0} \Phi_t^* \left(\omega \left(Y_1, \frac{\Phi_{-t}^* Y_2 - Y_2}{t}, \ldots, \Phi_{-t}^* Y_k \right) \right) \\[2mm]
& + \cdots + \lim_{t \to 0} \Phi_t^* \left(\omega \left(Y_1, \ldots, Y_{k-1}, \frac{\Phi_{-t}^* Y_k - Y_k}{t} \right) \right) \\[2mm]
= {} & \sum_{i=1}^{k} \omega(Y_1, \ldots, -[X, Y_i], \ldots, Y_k),
\end{aligned}
$$

since $\lim_{t \to 0} \Phi_t^* = \mathrm{id}$. When we add up A and B, we get

$$A + B = L_X(\omega(Y_1, \ldots, Y_k)) - \sum_{i=1}^{k} \omega(Y_1, \ldots, [X, Y_i], \ldots, Y_k)$$

$$= (L_X \omega)(Y_1, \ldots, Y_k),$$

which finishes the proof. $\qquad\qquad\qquad\qquad\qquad\qquad\qquad\qquad\qquad\qquad\qquad\qquad \square$

Part (2) of Proposition 4.19 shows that the Lie derivative of a differential form can be defined in terms of the Lie derivatives of functions and vector fields:

$$(L_X\omega)(Y_1,\ldots,Y_k) = L_X(\omega(Y_1,\ldots,Y_k)) - \sum_{i=1}^{k}\omega(Y_1,\ldots,Y_{i-1},L_XY_i,Y_{i+1},\ldots,Y_k)$$

$$= X[\omega(Y_1,\ldots,Y_k)] - \sum_{i=1}^{k}\omega(Y_1,\ldots,Y_{i-1},[X,Y_i],Y_{i+1},\ldots,Y_k).$$

However, to best calculate $L_X\omega$, we use *Cartan's formula*. Recall the definition of $i(X)\colon \mathcal{A}^k(M) \to \mathcal{A}^{k-1}(M)$ given in Definition 4.13.

Proposition 4.20 (Cartan's Formula). *Let M be a smooth manifold. For every vector field $X \in \mathfrak{X}(M)$, for every $\omega \in \mathcal{A}^k(M)$, we have*

$$L_X\omega = i(X)d\omega + d(i(X)\omega),$$

that is, $L_X = i(X) \circ d + d \circ i(X)$.

Proof. If $k = 0$, then $L_X f = X[f] = df(X)$ for a function f, and on the other hand, $i(X)f = 0$ and $i(X)df = df(X)$, so the equation holds. If $k \geq 1$, then by Proposition 4.16, we have

$$(i(X)d\omega)(X_1,\ldots,X_k) = d\omega(X,X_1,\ldots,X_k)$$

$$= X[\omega(X_1,\ldots,X_k)] + \sum_{i=1}^{k}(-1)^i X_i[\omega(X,X_1,\ldots,\widehat{X_i},\ldots,X_k)]$$

$$+ \sum_{j=1}^{k}(-1)^j\omega([X,X_j],X_1,\ldots,\widehat{X_j},\ldots,X_k)$$

$$+ \sum_{i<j}(-1)^{i+j}\omega([X_i,X_j],X,X_1,\ldots,\widehat{X_i},\ldots,\widehat{X_j},\ldots,X_k).$$

On the other hand, again by Proposition 4.16, we have

$$d(i(X)\omega)(X_1,\ldots,X_k) = \sum_{i=1}^{k}(-1)^{i-1}X_i[\omega(X,X_1,\ldots,\widehat{X_i},\ldots,X_k)]$$

$$+ \sum_{i<j}(-1)^{i+j}\omega(X,[X_i,X_j],X_1,\ldots,\widehat{X_i},\ldots,\widehat{X_j},\ldots,X_k).$$

Adding up these two equations, and using the fact that ω is alternating, we get

$$(i(X)d\omega + di(X))\omega(X_1, \ldots, X_k) = X[\omega(X_1, \ldots, X_k)]$$

$$+ \sum_{i=1}^{k} (-1)^i \omega([X, X_i], X_1, \ldots, \widehat{X_i}, \ldots, X_k)$$

$$= X[\omega(X_1, \ldots, X_k)] - \sum_{i=1}^{k} \omega(X_1, \ldots, [X, X_i], \ldots, X_k) = (L_X\omega)(X_1, \ldots, X_k),$$

as claimed.   □

Here is an example which demonstrates the usefulness of Cartan's formula. Consider S^1 embedded in $\mathbb{R}^2$ via the parameterization $\psi : (0, 2\pi) \to \mathbb{R}^2$, where $\psi(t) = (\cos t, \sin t)$. Since $\psi'(t) = (-\sin t, \cos t) = (-y, x)$, the vector field $X = -y \, \partial/\partial x + x \, \partial/\partial y$ is tangent to S^1. Consider $\omega = -y \, dx + x \, dy$ as the restriction of the one form in $\mathbb{R}^2$ to S^1. We want to calculate $L_X\omega$ on S^1 using Cartan's formula. This means we must first compute

$$d\omega = d(-y) \wedge dx + d(x) \wedge dy = -dy \wedge dx + dx \wedge dy = 2 \, dx \wedge dy.$$

Next we compute $i(X)d\omega$ as follows: by definition of $i(X)$,

$$i(X)dx = dx(X) = dx \left(-y \frac{\partial}{\partial x} + x \frac{\partial}{\partial y} \right) = -y,$$

$$i(X)dy = dy(X) = dy \left(-y \frac{\partial}{\partial x} + x \frac{\partial}{\partial y} \right) = x.$$

Then the anti-derivation property of $i(X)$ implies that

$$i(X)d\omega = 2i(X)(dx \wedge dy) = 2 \left(i(X)dx \wedge dy - dx \wedge i(X)dy \right) = 2(-y \, dy - x \, dx).$$
$$(*)$$

To complete Cartan's formula, we must calculate $d(i(X)\omega)$. Since $i(X)$ is $C^\infty(S^1)$ linear in X, we have

$$i(X)\omega = -y \, i(X)dx + x \, i(X)dy = y^2 + x^2,$$

and hence

$$d(i(X)\omega) = d(y^2 + x^2) = 2y \, dy + 2x \, dx. \qquad (**)$$

Cartan's formula combines Lines $(*)$ and $(**)$ to give

$$L_X\omega = i(X)d\omega + d(i(X)\omega) = -2y \, dy - 2x \, dx + 2y \, dy + 2x \, dx = 0.$$

The following proposition states more useful identities, some of which can be proved using Cartan's formula.

Proposition 4.21. *Let M be a smooth manifold. For all vector fields $X, Y \in \mathfrak{X}(M)$, for all $\omega \in \mathcal{A}^k(M)$, we have*

(1) $L_X i(Y) - i(Y)L_X = i([X, Y])$.
(2) $L_X L_Y \omega - L_Y L_X \omega = L_{[X,Y]}\omega$.
(3) $L_X i(X)\omega = i(X)L_X\omega$.
(4) $L_{fX}\omega = fL_X\omega + df \wedge i(X)\omega$, *for all* $f \in C^\infty(M)$.
(5) *For any diffeomorphism* $\varphi\colon M \to N$, *for all* $Z \in \mathfrak{X}(N)$ *and all* $\beta \in \mathcal{A}^k(N)$,

$$\varphi^* L_Z \beta = L_{\varphi^* Z} \varphi^* \beta.$$

Finally here is a proposition about the Lie derivative of tensor fields. Obviously, tensor product and contraction of tensor fields are defined pointwise; that is

$$(\alpha \otimes \beta)_p = \alpha_p \otimes \beta_p$$

$$(c_{i,j}\alpha)_p = c_{i,j}\alpha_p,$$

for all $p \in M$, where $c_{i,j}$ is the contraction operator of Definition 2.14.

Proposition 4.22. *Let M be a smooth manifold. For every vector field $X \in \mathfrak{X}(M)$, the Lie derivative $L_X\colon \Gamma(M, T^{\bullet,\bullet}(M)) \to \Gamma(M, T^{\bullet,\bullet}(M))$ is the unique linear local operator satisfying the following properties:*

(1) $L_X f = X[f] = df(X)$, *for all* $f \in C^\infty(M)$.
(2) $L_X Y = [X, Y]$, *for all* $Y \in \mathfrak{X}(M)$.
(3) $L_X(\alpha \otimes \beta) = (L_X\alpha) \otimes \beta + \alpha \otimes (L_X\beta)$, *for all tensor fields* $\alpha \in \Gamma(M, T^{r_1, s_1}(M))$ *and* $\beta \in \Gamma(M, T^{r_2, s_2}(M))$; *that is, L_X is a derivation.*
(4) *For all tensor fields* $\alpha \in \Gamma(M, T^{r,s}(M))$, *with* $r, s > 0$, *for every contraction operator* $c_{i,j}$,

$$L_X(c_{i,j}(\alpha)) = c_{i,j}(L_X\alpha).$$

The proof of Proposition 4.22 can be found in Gallot, Hullin, and Lafontaine [48] (Chapter 1). The following proposition is also useful:

Proposition 4.23. *For every $(0, q)$-tensor $S \in \Gamma(M, (T^*)^{\otimes q}(M))$, we have*

$$(L_X S)(X_1, \ldots, X_q) = X[S(X_1, \ldots, X_q)] - \sum_{i=1}^{q} S(X_1, \ldots, [X, X_i], \ldots, X_q),$$

for all $X_1, \ldots, X_q, X \in \mathfrak{X}(M)$.

There are situations in differential geometry where it is convenient to deal with differential forms taking values in a vector space. This happens when we consider connections and the curvature form on vector bundles and principal bundles (see Chapter 10) and when we study Lie groups, where differential forms valued in a Lie algebra occur naturally.

4.5   Vector-Valued Differential Forms

This section contains background material for Chapter 10, especially for Section 10.2. Let us go back for a moment to differential forms defined on some open subset of $\mathbb{R}^n$. In Section 4.1, a differential form is defined as a smooth map $\omega: U \to \bigwedge^p(\mathbb{R}^n)^*$, and since we have a canonical isomorphism

$$\mu: \bigwedge^p(\mathbb{R}^n)^* \cong \mathrm{Alt}^p(\mathbb{R}^n; \mathbb{R}),$$

such differential forms are real-valued. Now let F be any normed vector space, possibly infinite dimensional. Then $\mathrm{Alt}^p(\mathbb{R}^n; F)$ is also a normed vector space, and by Proposition 3.33, we have a canonical isomorphism

$$\mu_F: \left(\bigwedge^p(\mathbb{R}^n)^*\right) \otimes F \longrightarrow \mathrm{Alt}^p(\mathbb{R}^n; F)$$

defined on generators by

$$\mu_F((v_1^* \wedge \cdots \wedge v_p^*) \otimes f)(u_1, \ldots, u_p) = (\det(v_j^*(u_i)))f,$$

with $v_1^*, \ldots, v_p^* \in (\mathbb{R}^n)^*$, $u_1, \ldots, u_p \in \mathbb{R}^n$, and $f \in F$. Then it is natural to define differential forms with values in F as smooth maps $\omega: U \to \mathrm{Alt}^p(\mathbb{R}^n; F)$. Actually, we can even replace $\mathbb{R}^n$ with any normed vector space, even infinite dimensional as in Cartan [21], but we do not need such generality for our purposes.

Definition 4.16. Let F by any normed vector space. Given any open subset U of $\mathbb{R}^n$, a smooth *differential p-form on U with values in F*, for short a *p-form on U*, is any smooth function $\omega: U \to \mathrm{Alt}^p(\mathbb{R}^n; F)$. The vector space of all p-forms on U is denoted $\mathcal{A}^p(U; F)$. The vector space $\mathcal{A}^*(U; F) = \bigoplus_{p \geq 0} \mathcal{A}^p(U; F)$ is the set of *differential forms on U with values in F*.

Observe that $\mathcal{A}^0(U; F) = C^\infty(U, F)$, the vector space of smooth functions on U with values in F, and $\mathcal{A}^1(U; F) = C^\infty(U, \mathrm{Hom}(\mathbb{R}^n, F))$, the set of smooth functions from U to the set of linear maps from $\mathbb{R}^n$ to F. Also, $\mathcal{A}^p(U; F) = (0)$ for $p > n$.

Of course we would like to have a "good" notion of exterior differential, and we would like as many properties of "ordinary" differential forms as possible to remain valid. As we will see in our somewhat sketchy presentation, these goals can be achieved, except for some properties of the exterior product.

Using the isomorphism

$$\mu_F : \left(\bigwedge^p (\mathbb{R}^n)^* \right) \otimes F \longrightarrow \text{Alt}^p(\mathbb{R}^n; F)$$

and Proposition 3.34, we obtain a convenient expression for differential forms in $\mathcal{A}^*(U; F)$.

Proposition 4.24. *If $(e_1, \ldots, e_n)$ is any basis of $\mathbb{R}^n$ and $(e_1^*, \ldots, e_n^*)$ is its dual basis, then every differential p-form $\omega \in \mathcal{A}^p(U; F)$ can be written uniquely as*

$$\omega(x) = \sum_I e_{i_1}^* \wedge \cdots \wedge e_{i_p}^* \otimes f_I(x) = \sum_I e_I^* \otimes f_I(x) \qquad x \in U, \qquad (*2)$$

where each $f_I : U \to F$ is a smooth function on U.

As explained in Section 3.9, to express the above formula directly on alternating multilinear maps, define the product $\cdot : \text{Alt}^p(\mathbb{R}^n; \mathbb{R}) \times F \to \text{Alt}^p(\mathbb{R}^n; F)$ as follows.

Definition 4.17. For all $\omega \in \text{Alt}^p(\mathbb{R}^n; \mathbb{R})$ and all $f \in F$, define $\omega \cdot f$ by

$$(\omega \cdot f)(u_1, \ldots, u_p) = \omega(u_1, \ldots, u_p)f,$$

for all $u_1, \ldots, u_p \in \mathbb{R}^n$.

Then it is immediately verified that for every $\omega \in \left(\bigwedge^p (\mathbb{R}^n)^* \right) \otimes F$ of the form

$$\omega = u_1^* \wedge \cdots \wedge u_p^* \otimes f,$$

we have

$$\mu_F(u_1^* \wedge \cdots \wedge u_p^* \otimes f) = \mu_F(u_1^* \wedge \cdots \wedge u_p^*) \cdot f.$$

By Proposition 3.36, the above property can be restated as the fact that for any basis $(e_1, \ldots, e_n)$ of $\mathbb{R}^n$, every differential p-form $\omega \in \mathcal{A}^p(U; F)$ can be written uniquely as

$$\omega(x) = \sum_I e_{i_1}^* \wedge \cdots \wedge e_{i_p}^* \cdot f_I(x), \qquad x \in U,$$

where each $f_I : U \to F$ is a smooth function on U.

In order to define a multiplication on differential forms, we use a bilinear form $\Phi : F \times G \to H$; see Section 3.9.

Definition 4.18. For every pair (p, q), we define the multiplication

$$\wedge_\Phi : \left(\left(\overset{p}{\bigwedge}(\mathbb{R}^n)^*\right) \otimes F\right) \times \left(\left(\overset{q}{\bigwedge}(\mathbb{R}^n)^*\right) \otimes G\right) \longrightarrow \left(\overset{p+q}{\bigwedge}(\mathbb{R}^n)^*\right) \otimes H$$

by

$$(\alpha \otimes f) \wedge_\Phi (\beta \otimes g) = (\alpha \wedge \beta) \otimes \Phi(f, g).$$

We can also define a multiplication $\wedge_\Phi$ directly on alternating multilinear maps as follows: For $f \in \mathrm{Alt}^p(\mathbb{R}^n; F)$ and $g \in \mathrm{Alt}^q(\mathbb{R}^n; G)$,

$$(f \wedge_\Phi g)(u_1, \ldots, u_{p+q})$$

$$= \sum_{\sigma \in \mathrm{shuffle}(p,q)} \mathrm{sgn}(\sigma) \, \Phi(f(u_{\sigma(1)}, \ldots, u_{\sigma(p)}), g(u_{\sigma(p+1)}, \ldots, u_{\sigma(p+q)})),$$

where $\mathrm{shuffle}(p, q)$ consists of all (m, n)-"shuffles;" that is, permutations σ of $\{1, \ldots p + q\}$ such that $\sigma(1) < \cdots < \sigma(p)$ and $\sigma(p + 1) < \cdots < \sigma(p + q)$.

Then, we obtain a multiplication

$$\wedge_\Phi : \mathcal{A}^p(U; F) \times \mathcal{A}^q(U; G) \to \mathcal{A}^{p+q}(U; H),$$

defined so that for any differential forms $\omega \in \mathcal{A}^p(U; F)$ and $\eta \in \mathcal{A}^q(U; G)$,

$$(\omega \wedge_\Phi \eta)_x = \omega_x \wedge_\Phi \eta_x, \qquad x \in U.$$

In general, not much can be said about $\wedge_\Phi$, unless Φ has some additional properties. In particular, $\wedge_\Phi$ is generally not associative, and there is no analog of Proposition 4.1. For simplicity of notation, we write $\wedge$ for $\wedge_\Phi$.

Using Φ, we can also define a multiplication

$$\cdot : \mathcal{A}^p(U; F) \times \mathcal{A}^0(U; G) \to \mathcal{A}^p(U; H).$$

Definition 4.19. The multiplication $\cdot : \mathcal{A}^p(U; F) \times \mathcal{A}^0(U; G) \to \mathcal{A}^p(U; H)$ is given by

$$(\omega \cdot f)_x(u_1, \ldots, u_p) = \Phi(\omega_x(u_1, \ldots, u_p), f(x)),$$

for all $x \in U$, all $f \in \mathcal{A}^0(U; G) = C^\infty(U, G)$, and all $u_1, \ldots, u_p \in \mathbb{R}^n$.

This multiplication will be used in the case where $F = \mathbb{R}$ and $G = H$ to obtain a normal form for differential forms.

Generalizing d is no problem. Observe that since a differential p-form is a smooth map $\omega : U \to \mathrm{Alt}^p(\mathbb{R}^n; F)$, its derivative is a map

$$\omega' : U \to \mathrm{Hom}(\mathbb{R}^n, \mathrm{Alt}^p(\mathbb{R}^n; F))$$

such that ω'_x is a linear map from $\mathbb{R}^n$ to $\text{Alt}^p(\mathbb{R}^n; F)$ for every $x \in U$. We can view ω'_x as a multilinear map $\omega'_x \colon (\mathbb{R}^n)^{p+1} \to F$ which is alternating in its last p arguments. As in Section 4.1, the exterior derivative $(d\omega)_x$ is obtained by making ω'_x into an alternating map in all of its $p + 1$ arguments.

Definition 4.20. For every $p \geq 0$, the *exterior differential* $d \colon \mathcal{A}^p(U; F) \to \mathcal{A}^{p+1}(U; F)$ is given by

$$(d\omega)_x(u_1, \ldots, u_{p+1}) = \sum_{i=1}^{p+1} (-1)^{i-1} \omega'_x(u_i)(u_1, \ldots, \widehat{u_i}, \ldots, u_{p+1}),$$

for all $\omega \in \mathcal{A}^p(U; F)$ and all $u_1, \ldots, u_{p+1} \in \mathbb{R}^n$, where the hat over the argument u_i means that it should be omitted.

Note that d depends on the vector space F, so to be very precise we should denote d as d_F. To keep the notation simple, it is customary to drop the subscript F.

For any smooth function $f \in \mathcal{A}^0(U; F) = C^\infty(U, F)$, we get

$$df_x(u) = f'_x(u).$$

Therefore, for smooth functions, *the exterior differential df coincides with the usual derivative f'.*

The important observation following Definition 4.3 also applies here. If $x_i \colon U \to \mathbb{R}$ is the restriction of pr_i to U, then x'_i is the constant map given by

$$x'_i(x) = pr_i, \qquad x \in U.$$

It follows that $dx_i = x'_i$ is the constant function with value $pr_i = e^*_i$. As a consequence, every p-form ω can be uniquely written as

$$\omega_x = \sum_I dx_{i_1} \wedge \cdots \wedge dx_{i_p} \otimes f_I(x), \tag{$*_3$}$$

where each $f_I \colon U \to F$ is a smooth function on U. Using the multiplication $\cdot$ induced by the scalar multiplication in F ($\Phi(\lambda, f) = \lambda f$, with $\lambda \in \mathbb{R}$ and $f \in F$), we see that every p-form ω can be uniquely written as

$$\omega = \sum_I dx_{i_1} \wedge \cdots \wedge dx_{i_p} \cdot f_I. \tag{$*_4$}$$

As for real-valued functions, for any $f \in \mathcal{A}^0(U; F) = C^\infty(U, F)$, we have

$$df_x(u) = \sum_{i=1}^n u_i \frac{\partial f}{\partial x_i}(x) = \sum_{i=1}^n e^*_i(u) \frac{\partial f}{\partial x_i}(x),$$

and so,

$$df = \sum_{i=1}^{n} dx_i \cdot \frac{\partial f}{\partial x_i}. \tag{$*_5$}$$

In general, Proposition 4.3 fails, unless F is finite dimensional (see below). However for any arbitrary F, a weak form of Proposition 4.3 can be salvaged. Again, let $\Phi \colon F \times G \to H$ be a bilinear form, let $\cdot \colon \mathcal{A}^p(U; F) \times \mathcal{A}^0(U; G) \to \mathcal{A}^p(U; H)$ be as defined before Definition 4.20, and let $\wedge_\Phi$ be the wedge product associated with Φ. The following fact is proved in Cartan [21] (Section 2.4):

Proposition 4.25. *For all* $\omega \in \mathcal{A}^p(U; F)$ *and all* $f \in \mathcal{A}^0(U; G)$*, we have*

$$d(\omega \cdot f) = (d\omega) \cdot f + \omega \wedge_\Phi df.$$

Fortunately, $d \circ d$ still vanishes, but this requires a completely different proof, since we can't rely on Proposition 4.2 (see Cartan [21], Section 2.5). Similarly, Proposition 4.2 holds, but a different proof is needed.

Proposition 4.26. *The composition* $\mathcal{A}^p(U; F) \xrightarrow{d} \mathcal{A}^{p+1}(U; F) \xrightarrow{d} \mathcal{A}^{p+2}(U; F)$ *is identically zero for every* $p \geq 0$*; that is*

$$d \circ d = 0,$$

which is an abbreviation for $d^{p+1} \circ d^p = 0$.

To generalize Proposition 4.2, we use Proposition 4.25 with the product $\cdot$ and the wedge product $\wedge_\Phi$ induced by the bilinear form Φ given by scalar multiplication in F; that is, $\Phi(\lambda, f) = \lambda f$, for all $\lambda \in \mathbb{R}$ and all $f \in F$.

Proposition 4.27. *For every* p *form* $\omega \in \mathcal{A}^p(U; F)$ *with* $\omega = dx_{i_1} \wedge \cdots \wedge dx_{i_p} \cdot f$, *we have*

$$d\omega = dx_{i_1} \wedge \cdots \wedge dx_{i_p} \wedge_F df,$$

where $\wedge$ *is the usual wedge product on real-valued forms and* $\wedge_F$ *is the wedge product associated with scalar multiplication in* F.

More explicitly, for a p form $\omega = dx_{i_1} \wedge \cdots \wedge dx_{i_p} \cdot f$, for every $x \in U$, for all $u_1, \ldots, u_{p+1} \in \mathbb{R}^n$, we have

$$(d\omega_x)(u_1, \ldots, u_{p+1}) = \sum_{i=1}^{p+1} (-1)^{i-1} (dx_{i_1} \wedge \cdots \wedge dx_{i_p})_x (u_1, \ldots, \widehat{u_i}, \ldots, u_{p+1}) df_x(u_i).$$

If we use the fact that

$$df = \sum_{i=1}^{n} dx_i \cdot \frac{\partial f}{\partial x_i},$$

we see easily that

$$d\omega = \sum_{j=1}^{n} dx_{i_1} \wedge \cdots \wedge dx_{i_p} \wedge dx_j \cdot \frac{\partial f}{\partial x_j}, \qquad (*6)$$

the direct generalization of the real-valued case, except that the "coefficients" are functions with values in F.

The pull-back of forms in $\mathcal{A}^*(V, F)$ is defined as before. Luckily, Proposition 4.9 holds (see Cartan [21], Section 2.8).

Proposition 4.28. *Let $U \subseteq \mathbb{R}^n$ and $V \subseteq \mathbb{R}^m$ be two open sets and let $\varphi \colon U \to V$ be a smooth map. Then*

(i) $\varphi^*(\omega \wedge \eta) = \varphi^*\omega \wedge \varphi^*\eta$, for all $\omega \in \mathcal{A}^p(V; F)$ and all $\eta \in \mathcal{A}^q(V; F)$.

(ii) $\varphi^*(f) = f \circ \varphi$, for all $f \in \mathcal{A}^0(V; F)$.

(iii) $d\varphi^*(\omega) = \varphi^*(d\omega)$, for all $\omega \in \mathcal{A}^p(V; F)$; that is, the following diagram commutes for all $p \geq 0$.

$$
\begin{array}{ccc}
\mathcal{A}^p(V; F) & \xrightarrow{\;\varphi^*\;} & \mathcal{A}^p(U; F) \\
{\scriptstyle d}\Big\downarrow & & \Big\downarrow{\scriptstyle d} \\
\mathcal{A}^{p+1}(V; F) & \xrightarrow{\;\varphi^*\;} & \mathcal{A}^{p+1}(U; F)
\end{array}
$$

Let us now consider the special case where F has finite dimension m. Pick any basis $(f_1, \ldots, f_m)$ of F. Then as every differential p-form $\omega \in \mathcal{A}^p(U; F)$ can be written uniquely as

$$\omega(x) = \sum_I e_{i_1}^* \wedge \cdots \wedge e_{i_p}^* \cdot f_I(x), \qquad x \in U,$$

where each $f_I \colon U \to F$ is a smooth function on U, by expressing the f_I over the basis $(f_1, \ldots, f_m)$, we see that ω can be written uniquely as

$$\omega = \sum_{i=1}^{m} \omega_i \cdot f_i, \qquad (*7)$$

where $\omega_1, \ldots, \omega_m$ are smooth real-valued differential forms in $\mathcal{A}^p(U; \mathbb{R})$, and we view f_i as the constant map with value f_i from U to F. Then as

$$\omega'_x(u) = \sum_{i=1}^{m} (\omega'_i)_x(u) f_i,$$

for all $u \in \mathbb{R}^n$, we see that

$$d\omega = \sum_{i=1}^{m} d\omega_i \cdot f_i. \tag{$*_8$}$$

Actually, because $d\omega$ is defined independently of bases, the f_i do not need to be linearly independent; any choice of vectors and forms such that

$$\omega = \sum_{i=1}^{k} \omega_i \cdot f_i$$

will do.

Given a bilinear map $\Phi \colon F \times G \to H$, a simple calculation shows that for all $\omega \in \mathcal{A}^p(U; F)$ and all $\eta \in \mathcal{A}^p(U; G)$, we have

$$\omega \wedge_\Phi \eta = \sum_{i=1}^{m} \sum_{j=1}^{m'} \omega_i \wedge \eta_j \cdot \Phi(f_i, g_j),$$

with $\omega = \sum_{i=1}^{m} \omega_i \cdot f_i$ and $\eta = \sum_{j=1}^{m'} \eta_j \cdot g_j$, where $(f_1, \ldots, f_m)$ is a basis of F and $(g_1, \ldots, g_{m'})$ is a basis of G. From this and Proposition 4.25, it follows that Proposition 4.3 holds for finite-dimensional spaces.

Proposition 4.29. *If F, G, H are finite dimensional and $\Phi \colon F \times G \to H$ is a bilinear map, then or all $\omega \in \mathcal{A}^p(U; F)$ and all $\eta \in \mathcal{A}^q(U; G)$,*

$$d(\omega \wedge_\Phi \eta) = d\omega \wedge_\Phi \eta + (-1)^p \omega \wedge_\Phi d\eta.$$

On the negative side, in general, Proposition 4.1 still fails.

A special case of interest is the case where $F = G = H = \mathfrak{g}$ is a Lie algebra, and $\Phi(a, b) = [a, b]$ is the Lie bracket of $\mathfrak{g}$. In this case, using a basis $(f_1, \ldots, f_r)$ of $\mathfrak{g}$, if we write $\omega = \sum_i \alpha_i f_i$ and $\eta = \sum_j \beta_j f_j$, we have

$$\omega \wedge_\Phi \eta = [\omega, \eta] = \sum_{i,j} \alpha_i \wedge \beta_j [f_i, f_j],$$

where for simplicity of notation we dropped the subscript Φ on $[\omega, \eta]$ and the multiplication sign $\cdot$.

Let us figure out what $[\omega, \omega]$ is for a one-form $\omega \in \mathcal{A}^1(U, \mathfrak{g})$. By definition,

$$[\omega, \omega] = \sum_{i,j} \omega_i \wedge \omega_j [f_i, f_j],$$

so

$$[\omega, \omega](u, v) = \sum_{i,j} (\omega_i \wedge \omega_j)(u, v)[f_i, f_j]$$

$$= \sum_{i,j} (\omega_i(u)\omega_j(v) - \omega_i(v)\omega_j(u))[f_i, f_j]$$

$$= \sum_{i,j} \omega_i(u)\omega_j(v)[f_i, f_j] - \sum_{i,j} \omega_i(v)\omega_j(u)[f_i, f_j]$$

$$= \left[\sum_i \omega_i(u)f_i, \sum_j \omega_j(v)f_j\right] - \left[\sum_i \omega_i(v)f_i, \sum_j \omega_j(u)f_j\right]$$

$$= [\omega(u), \omega(v)] - [\omega(v), \omega(u)]$$

$$= 2[\omega(u), \omega(v)].$$

Therefore,

$$[\omega, \omega](u, v) = 2[\omega(u), \omega(v)].$$

Note that in general, $[\omega, \omega] \neq 0$, because ω is vector valued. Of course, for real-valued forms, $[\omega, \omega] = 0$. Using the Jacobi identity of the Lie algebra, we easily find that

$$[[\omega, \omega], \omega] = 0.$$

The generalization of vector-valued differential forms to manifolds is no problem, except that some results involving the wedge product fail for the same reason that they fail in the case of forms on open subsets of $\mathbb{R}^n$.

Definition 4.21. Given a smooth manifold M of dimension n and a vector space F, the set $\mathcal{A}^k(M; F)$ of *differential k-forms on M with values in F* is the set of maps $p \mapsto \omega_p$ with

$\omega_p \in \left(\bigwedge^k T_p^*M\right) \otimes F \cong \mathrm{Alt}^k(T_pM; F)$, which vary smoothly in $p \in M$. This means that the map

$$p \mapsto \omega_p(X_1(p), \ldots, X_k(p))$$

is smooth for all vector fields $X_1, \ldots, X_k \in \mathfrak{X}(M)$.

It can be shown (see Proposition 9.12) that

$$\mathcal{A}^k(M; F) \cong \mathcal{A}^k(M) \otimes_{C^\infty(M)} C^\infty(M; F) \cong \mathrm{Alt}^k_{C^\infty(M)}(\mathfrak{X}(M); C^\infty(M; F)),$$

which reduces to Proposition 4.15 when $F = \mathbb{R}$.

The reader may want to carry out the verification that the theory generalizes to manifolds on her/his own. The following result will be used in the next section.

Proposition 4.30. *If* $\omega \in \mathcal{A}^1(M; F)$ *is a vector valued one-form, then for any two vector fields* X, Y *on* M, *we have*

$$d\omega(X, Y) = X(\omega(Y)) - Y(\omega(X)) - \omega([X, Y]).$$

In the next section we consider some properties of differential forms on Lie groups.

4.6   Differential Forms on Lie Groups and Maurer-Cartan Forms

Given a Lie group G, it is well known that the set of left-invariant vector fields on G is isomorphic to the Lie algebra $\mathfrak{g} = T_1 G$ of G (where 1 denotes the identity element of G); see Warner [109] (Chapter 4) or Gallier and Quaintance [47]. Recall that a vector field X on G is left-invariant iff

$$d(L_a)_b(X_b) = X_{L_a b} = X_{ab},$$

for all $a, b \in G$. In particular, for $b = 1$, we get

$$X_a = d(L_a)_1(X_1).$$

which shows that X is completely determined by its value at 1. The map $X \mapsto X(1)$ is an isomorphism between left-invariant vector fields on G and $\mathfrak{g}$.

The above suggests looking at left-invariant differential forms on G. We will see that the set of left-invariant one-forms on G is isomorphic to $\mathfrak{g}^*$, the dual of $\mathfrak{g}$ as a vector space.

Definition 4.22. Given a Lie group G, a differential form $\omega \in \mathcal{A}^k(G)$ is *left-invariant* iff

$$L_a^* \omega = \omega, \qquad \text{for all } a \in G,$$

where $L_a^* \omega$ is the pull-back of ω by L_a (left multiplication by a). The left-invariant one-forms $\omega \in \mathcal{A}^1(G)$ are also called *Maurer-Cartan forms*.

Here is a simple example of a left-invariant one-form on S^1. Let $g = (\cos t, \sin t) \in S^1$. Then $L_g \colon S^1 \to S^1$ is given by

$$L_g(u, v) = (\cos t\, u - \sin t\, v, \sin t\, u + \cos t\, v) = (x, y).$$

Let $\omega = -y\, dx + x\, dy$. Then

$$
\begin{aligned}
L_g^*\omega &= (-\sin t\, u - \cos t\, v)d(\cos t\, u - \sin t\, v) + (\cos t\, u - \sin t\, v)d(\sin t\, u + \cos t\, v) \\
&= (-\sin t\, u - \cos t\, v)(\cos t\, du - \sin t\, dv) + (\cos t\, u - \sin t\, v)(\sin t\, du + \cos t\, dv) \\
&= -(\sin^2 t + \cos^2 t)v\, du + (\sin^2 t + \cos^2 t)u\, dv \\
&= -v\, du + u\, dv,
\end{aligned}
$$

which (by setting $u \to x$ and $v \to y$) shows that $L_g^*\omega = \omega$.

For a one-form $\omega \in \mathcal{A}^1(G)$ left-invariance means that

$$(L_a^*\omega)_g(u) = \omega_{L_a g}(d(L_a)_g u) = \omega_{ag}(d(L_a)_g u) = \omega_g(u),$$

for all $a, g \in G$ and all $u \in T_g G$. For $a = g^{-1}$, we get

$$\omega_g(u) = \omega_1(d(L_{g^{-1}})_g u) = \omega_1(d(L_g^{-1})_g u),$$

which shows that ω_g is completely determined by ω_1 (the value of ω_g at $g = 1$).

Proposition 4.31. *The map $\omega \mapsto \omega_1$ is an isomorphism between the set of left-invariant one-forms on G and $\mathfrak{g}^*$.*

Proof. First, for any linear form $\alpha \in \mathfrak{g}^*$, the one-form α^L given by

$$\alpha_g^L(u) = \alpha(d(L_g^{-1})_g u), \qquad u \in T_g(G)$$

is left-invariant, because

$$
\begin{aligned}
(L_h^*\alpha^L)_g(u) &= \alpha_{hg}^L(d(L_h)_g(u)) \\
&= \alpha(d(L_{hg}^{-1})_{hg}(d(L_h)_g(u))) \\
&= \alpha(d(L_{hg}^{-1} \circ L_h)_g(u)) \\
&= \alpha(d(L_g^{-1})_g(u)) = \alpha_g^L(u).
\end{aligned}
$$

Secondly, we saw that for every one-form $\omega \in \mathcal{A}^1(G)$,

$$\omega_g(u) = \omega_1(d(L_g^{-1})_g u),$$

so $\omega_1 \in \mathfrak{g}^*$ is the unique element such that $\omega = \omega_1^L$, which shows that the map $\alpha \mapsto \alpha^L$ is an isomorphism whose inverse is the map $\omega \mapsto \omega_1$. $\square$

Now, since every left-invariant vector field is of the form $X = u^L$ for some unique $u \in \mathfrak{g}$, where u^L is the vector field given by $u^L(a) = d(L_a)_1 u$, and since the left-invariance of ω implies that

$$\omega_{ag}(d(L_a)_g u) = \omega_g(u),$$

for $g = 1$, we get $\omega_a(d(L_a)_1 u) = \omega_1(u)$; that is

$$\omega_a(X) = \omega_1(u), \qquad a \in G,$$

which shows that $\omega(X)$ is a constant function on G. It follows that for every vector field Y (not necessarily left-invariant),

$$Y[\omega(X)] = 0.$$

Recall that by Corollary 4.17, we have

$$d\omega(X, Y) = X[\omega(Y)] - Y[\omega(X)] - \omega([X, Y]).$$

Consequently, for all left-invariant vector fields X, Y on G, for every left-invariant one-form ω, we have

$$d\omega(X, Y) = -\omega([X, Y]).$$

If we identify the set of left-invariant vector fields on G with $\mathfrak{g}$ and the set of left-invariant one-forms on G with $\mathfrak{g}^*$, we have

$$d\omega(X, Y) = -\omega([X, Y]), \qquad \omega \in \mathfrak{g}^*, \; X, Y \in \mathfrak{g}.$$

We summarize these facts in the following proposition.

Proposition 4.32. *Let G be any Lie group.*

(1) The set of left-invariant one-forms on G is isomorphic to $\mathfrak{g}^*$, the dual of the Lie algebra $\mathfrak{g}$ of G, via the isomorphism $\omega \mapsto \omega_1$.

(2) For every left-invariant one form ω and every left-invariant vector field X, the value of the function $\omega(X)$ is constant and equal to $\omega_1(X_1)$.

(3) If we identify the set of left-invariant vector fields on G with $\mathfrak{g}$ and the set of left-invariant one-forms on G with $\mathfrak{g}^*$, then

$$d\omega(X, Y) = -\omega([X, Y]), \qquad \omega \in \mathfrak{g}^*, \; X, Y \in \mathfrak{g}.$$

Pick any basis $X_1, \ldots, X_r$ of the Lie algebra $\mathfrak{g}$, and let $\omega_1, \ldots, \omega_r$ be the dual basis of $\mathfrak{g}^*$. There are some constants c_{ij}^k such that

$$[X_i, X_j] = \sum_{k=1}^{r} c_{ij}^k X_k.$$

The constants c_{ij}^k are called the *structure constants* of the Lie algebra $\mathfrak{g}$. Observe that $c_{ji}^k = -c_{ij}^k$.

As $\omega_i([X_p, X_q]) = c_{pq}^i$ and $d\omega_i(X, Y) = -\omega_i([X, Y])$, we have

$$d\omega_i(X, Y) = -c_{pq}^i.$$

Proposition 4.33. *The following equations known as the Maurer-Cartan equations hold:*

$$d\omega_i = -\frac{1}{2} \sum_{j,k} c_{jk}^i \omega_j \wedge \omega_k.$$

Proof. Since

$$\sum_{j,k} c_{jk}^i \omega_j \wedge \omega_k(X_p, X_q) = \sum_{j,k} c_{jk}^i (\omega_j(X_p)\omega_k(X_q) - \omega_j(X_q)\omega_k(X_p))$$

$$= \sum_{j,k} c_{jk}^i \omega_j(X_p)\omega_k(X_q) - \sum_{j,k} c_{jk}^i \omega_j(X_q)\omega_k(X_p)$$

$$= \sum_{j,k} c_{jk}^i \omega_j(X_p)\omega_k(X_q) + \sum_{j,k} c_{kj}^i \omega_j(X_q)\omega_k(X_p)$$

$$= c_{pq}^i + c_{pq}^i = 2c_{pq}^i,$$

we get the equations

$$d\omega_i = -\frac{1}{2} \sum_{j,k} c_{jk}^i \omega_j \wedge \omega_k.$$

$\square$

These equations can be neatly described if we use differential forms valued in $\mathfrak{g}$. Let ω_{MC} be the one-form given by

$$(\omega_{MC})_g(u) = d(L_g^{-1})_g u, \qquad g \in G, \; u \in T_g G.$$

What ω_{MC} does is to "bring back" a vector $v \in T_g G$ to $\mathfrak{g} = T_1 G$. The same computation that showed that α^L is left-invariant if $\alpha \in \mathfrak{g}^*$ shows that ω_{MC} is left-invariant, and obviously $(\omega_{MC})_1 = \text{id}$.

Definition 4.23. Given any Lie group G, the *Maurer-Cartan form on G* is the $\mathfrak{g}$-valued differential 1-form $\omega_{MC} \in \mathcal{A}^1(G, \mathfrak{g})$ given by

$$(\omega_{MC})_g = d(L_g^{-1})_g, \qquad g \in G.$$

The same argument that we used to prove Property (2) of Proposition 4.32 shows that for every left-invariant one-form $\omega \in \mathcal{A}^1(G, \mathfrak{g})$ and every left-invariant vector field $X \in \mathfrak{X}(G)$, the value of the function $\omega(X)$ is constant and equal to $\omega_1(X_1)$. In particular, this holds for the Maurer-Cartan form ω_{MC}. As in Section 4.5, the Lie bracket on $\mathfrak{g}$ induces a multiplication

$$[-, -] \colon \mathcal{A}^p(G; \mathfrak{g}) \times \mathcal{A}^q(G; \mathfrak{g}) \to \mathcal{A}^{p+q}(G; \mathfrak{g})$$

given by

$$[\omega, \eta] = \sum_{ij} \alpha_i \wedge \beta_j \cdot [f_i, f_j],$$

where $(f_1, \ldots, f_r)$ is a basis of $\mathfrak{g}$ and where $\omega = \sum_i \alpha_i \cdot f_i$ and $\eta = \sum_j \beta_j \cdot f_j$. Using the same proof, we obtain the equation

$$[\omega, \omega](X, Y) = 2[\omega(X), \omega(Y)].$$

Recall that for every $g \in G$, conjugation by g is the map given by $a \mapsto gag^{-1}$; that is, $a \mapsto (L_g \circ R_{g^{-1}})a$, and the adjoint map $\mathrm{Ad}(g) \colon \mathfrak{g} \to \mathfrak{g}$ associated with g is the derivative of $\mathbf{Ad}_g = L_g \circ R_{g^{-1}}$ at 1; that is, we have

$$\mathrm{Ad}(g)(u) = d(\mathbf{Ad}_g)_1(u), \qquad u \in \mathfrak{g}.$$

Furthermore, it is obvious that L_g and R_h commute.

Proposition 4.34. *Given any Lie group G, for all $g \in G$, the Maurer-Cartan form ω_{MC} has the following properties:*

(1) $(\omega_{MC})_1 = id_{\mathfrak{g}}$.
(2) For all $g \in G$,

$$R_g^* \omega_{MC} = \mathrm{Ad}(g^{-1}) \circ \omega_{MC}.$$

(3) The 2-form $d\omega_{MC} \in \mathcal{A}^2(G, \mathfrak{g})$ satisfies the Maurer-Cartan equation

$$d\omega_{MC} = -\frac{1}{2}[\omega_{MC}, \omega_{MC}].$$

Proof. Property (1) is obvious.

(2) For simplicity of notation, if we write $\omega = \omega_{MC}$, then

$$
\begin{aligned}
(R_g^* \omega)_h &= \omega_{hg} \circ d(R_g)_h \\
&= d(L_{hg}^{-1})_{hg} \circ d(R_g)_h \\
&= d(L_{hg}^{-1} \circ R_g)_h \\
&= d((L_h \circ L_g)^{-1} \circ R_g)_h \\
&= d(L_g^{-1} \circ L_h^{-1} \circ R_g)_h \\
&= d(L_g^{-1} \circ R_g \circ L_h^{-1})_h \\
&= d(L_{g^{-1}} \circ R_g)_1 \circ d(L_h^{-1})_h \\
&= \mathrm{Ad}(g^{-1}) \circ \omega_h,
\end{aligned}
$$

as claimed.

(3) We can easily express ω_{MC} in terms of a basis of $\mathfrak{g}$. If $X_1, \ldots, X_r$ is a basis of $\mathfrak{g}$ and $\omega_1, \ldots, \omega_r$ is the dual basis, then by Proposition 4.32 (2) and Part (1) of Proposition 4.34, we have $\omega_{MC}(X_i) = (\omega_{MC})_1(X_i) = X_i$, for $i = 1, \ldots, r$, so ω_{MC} is given by

$$
\omega_{MC} = \omega_1 \cdot X_1 + \cdots + \omega_r \cdot X_r, \tag{$*_9$}
$$

under the usual identification of left-invariant vector fields (resp. left-invariant one forms) with elements of $\mathfrak{g}$ (resp. elements of $\mathfrak{g}^*$). Then we have

$$
d\omega_{MC} = d\omega_1 \cdot X_1 + \cdots + d\omega_r \cdot X_r. \tag{$*_{10}$}
$$

We will use the Maurer-Cartan equations

$$
d\omega_i = -\frac{1}{2} \sum_{j,k} c_{jk}^i \omega_j \wedge \omega_k
$$

to obtain the desired equation. Using the fact that the c_{jk}^i are skew-symmetric in j, k, for all $u, v \in \mathfrak{g}$, we have

$$
\begin{aligned}
[\omega_{MC}, \omega_{MC}](u, v) &= \left[\sum_j \omega_j(u) \cdot X_j, \sum_k \omega_j(v) \cdot X_k \right] \\
&= \sum_{i,j,k} \omega_j(u)\omega_k(v) c_{jk}^i \cdot X_i
\end{aligned}
$$

$$= \sum_{i,j,k} c^i_{jk}(\omega_j \wedge \omega_k)(u, v) \cdot X_i$$

$$= -2 \sum_i d\omega_i(u, v) \cdot X_i$$

$$= -2d\omega_{\mathrm{MC}}(u, v),$$

namely

$$d\omega_{\mathrm{MC}} = -\frac{1}{2}[\omega_{\mathrm{MC}}, \omega_{\mathrm{MC}}],$$

as claimed.   □

In the case of a matrix group $G \subseteq \mathrm{GL}(n, \mathbb{R})$, it is easy to see that the Maurer-Cartan form is given explicitly by

$$\omega_{\mathrm{MC}}(v) = g^{-1}v, \qquad v \in T_g G, \ g \in G.$$

Since $T_g G$ is isomorphic to $g\mathfrak{g}$, we have

$$\omega_{\mathrm{MC}}(gv) = v, \quad v \in \mathfrak{g}.$$

The above expression suggests that, with some abuse of notation, ω_{MC} may be denoted by $g^{-1}dg$, where $g = (g_{ij})$ and where dg is an abbreviation for the $n \times n$ matrix (dg_{ij}). Thus, ω_{MC} is a kind of logarithmic derivative of the identity. For $n = 2$, if we write

$$g = \begin{pmatrix} \alpha & \beta \\ \gamma & \delta \end{pmatrix},$$

we get

$$\omega_{\mathrm{MC}} = \frac{1}{\alpha\delta - \beta\gamma} \begin{pmatrix} \delta d\alpha - \beta d\gamma & \delta d\beta - \beta d\delta \\ -\gamma d\alpha + \alpha d\gamma & -\gamma d\beta + \alpha d\delta \end{pmatrix}.$$

Remarks.

(1) The quantity $d\omega_{\mathrm{MC}} + \frac{1}{2}[\omega_{\mathrm{MC}}, \omega_{\mathrm{MC}}]$ is the *curvature* of the *connection* ω_{MC} on G. The Maurer-Cartan equation says that the curvature of the Maurer-Cartan connection is zero. We also say that ω_{MC} is a *flat* connection.
(2) As $d\omega_{\mathrm{MC}} = -\frac{1}{2}[\omega_{\mathrm{MC}}, \omega_{\mathrm{MC}}]$, we get

$$d[\omega_{\mathrm{MC}}, \omega_{\mathrm{MC}}] = 0,$$

which yields

$$[[\omega_{MC}, \omega_{MC}], \omega_{MC}] = 0.$$

It is easy to show that the above expresses the Jacobi identity in $\mathfrak{g}$.

(3) As in the case of real-valued one-forms, for every left-invariant one-form $\omega \in \mathcal{A}^1(G, \mathfrak{g})$, we have

$$\omega_g(u) = \omega_1(d(L_g^{-1})_g u) = \omega_1((\omega_{MC})_g u),$$

for all $g \in G$ and all $u \in T_g G$, and where $\omega_1 \colon \mathfrak{g} \to \mathfrak{g}$ is a linear map. Consequently, there is a bijection between the set of left-invariant one-forms in $\mathcal{A}^1(G, \mathfrak{g})$ and $\mathrm{Hom}(\mathfrak{g}, \mathfrak{g})$.

(4) The Maurer-Cartan form can be used to define the *Darboux derivative* of a map $f \colon M \to G$, where M is a manifold and G is a Lie group. The Darboux derivative of f is the $\mathfrak{g}$-valued one-form $\omega_f \in \mathcal{A}^1(M, \mathfrak{g})$ on M given by

$$\omega_f = f^* \omega_{MC}.$$

Then it can be shown that when M is connected, if f_1 and f_2 have the same Darboux derivative $\omega_{f_1} = \omega_{f_2}$, then $f_2 = L_g \circ f_1$, for some $g \in G$. Elie Cartan also characterized which $\mathfrak{g}$-valued one-forms on M are Darboux derivatives ($d\omega + \frac{1}{2}[\omega, \omega] = 0$ must hold). For more on Darboux derivatives, see Sharpe [100] (Chapter 3) and Malliavin [76] (Chapter III).

4.7 Problems

Problem 4.1. Recall that $d \colon \mathcal{A}^p(U) \to \mathcal{A}^{p+1}(U)$ is given by

$$(d\omega)_x(u_1, \ldots, u_{p+1}) = \sum_{i=1}^{p+1} (-1)^{i-1} \omega_x'(u_i)(u_1, \ldots, \widehat{u_i}, \ldots, u_{p+1}),$$

for all $\omega \in \mathcal{A}^p(U)$, all $x \in U$, and all $u_1, \ldots, u_{p+1} \in \mathbb{R}^n$. Show that $(d\omega)_x$ is alternating in its $p + 1$ arguments.

Problem 4.2. Given the 1-form

$$\omega_{(x,y)} = \frac{-y}{x^2 + y^2} \, dx + \frac{x}{x^2 + y^2} \, dy,$$

on $U = \mathbb{R}^2 - \{0\}$, we have $d\omega = 0$. Show that there is no smooth function f on U such that $df = \omega$.

Hint. See Madsen and Tornehave [75], Chapter 1.

Problem 4.3. Let $U \subseteq \mathbb{R}^n$, $V \subseteq \mathbb{R}^m$, and $W \subseteq \mathbb{R}^p$ be three open subsets. Let $\varphi \colon U \to V$, $\psi \colon V \to W$, and $\mathrm{id} \colon U \to U$ be smooth maps. Show that

$$\mathrm{id}^* = \mathrm{id},$$

$$(\psi \circ \varphi)^* = \varphi^* \circ \psi^*.$$

Problem 4.4. Complete the proof of Proposition 4.12.

Problem 4.5. Complete the proof details for Proposition 4.16.
Hint. See Morita [82] (Chapter 2).

Problem 4.6. Let M be a smooth manifold. For every vector field $X \in \mathfrak{X}(M)$, for all $k \geq 1$, let $i(X) \colon \mathcal{A}^k(M) \to \mathcal{A}^{k-1}(M)$ the linear insertion operator defined in Definition 4.13. Show that

$$i(fX)\omega = fi(X)\omega, \qquad i(X)(f\omega) = fi(X)\omega, \qquad f \in C^\infty(M), \ \omega \in \mathcal{A}^k(M),$$

Also prove that

$$i(X)(\omega \wedge \eta) = (i(X)\omega) \wedge \eta + (-1)^r \omega \wedge (i(X)\eta),$$

for all $\omega \in \mathcal{A}^r(M)$ and all $\eta \in \mathcal{A}^s(M)$.

Problem 4.7. Let M be a smooth manifold. For every vector field $X \in \mathfrak{X}(M)$, prove the following:

(1) For all $\omega \in \mathcal{A}^r(M)$ and all $\eta \in \mathcal{A}^s(M)$,

$$L_X(\omega \wedge \eta) = (L_X\omega) \wedge \eta + \omega \wedge (L_X\eta);$$

(2) The Lie derivative commutes with d:

$$L_X \circ d = d \circ L_X.$$

Hint. See Warner [109], Morita [82], and Gallot, Hullin, and Lafontaine [48].

Problem 4.8. Prove Proposition 4.21.

Problem 4.9. Prove Proposition 4.22.
Hint. See Gallot, Hullin, and Lafontaine [48] (Chapter 1).

Problem 4.10. Show that the composition $\mathcal{A}^p(U; F) \xrightarrow{d} \mathcal{A}^{p+1}(U; F) \xrightarrow{d} \mathcal{A}^{p+2}(U; F)$ is identically zero for every $p \geq 0$; that is

$$d \circ d = 0.$$

Hint. See Cartan [21], Section 2.5.

Problem 4.11.

(i) Prove Proposition 4.25.
   Hint. See Cartan [21], Section 2.4.
(ii) Use Proposition 4.25 with the product $\cdot$ and the wedge product $\wedge_\Phi$ induced by the bilinear form Φ given by scalar multiplication in F to prove Proposition 4.27.

Problem 4.12. Prove Proposition 4.28.
Hint. See Cartan [21], Section 2.8.

Problem 4.13. Let F and G be finite-dimensional vector spaces with $(f_1, \ldots, f_m)$ a basis of F and $(g_1, \ldots, g_{m'})$ a basis of G. Show that

$$\omega \wedge_\Phi \eta = \sum_{i=1}^{m} \sum_{j=1}^{m'} \omega_i \wedge \eta_j \cdot \Phi(f_i, g_j),$$

where $\omega = \sum_{i=1}^{m} \omega_i \cdot f_i$ and $\eta = \sum_{j=1}^{m'} \eta_j \cdot g_j$.

Problem 4.14. Use Problem 4.13 to prove Proposition 4.29.

Chapter 5
Distributions and the Frobenius Theorem

Given any smooth manifold M (of dimension n), for any smooth vector field X on M, it is known that for every point $p \in M$, there is a unique maximal integral curve through p; see Warner [109] (Chapter 1, Theorem 1.48) or Gallier and Quaintance [47]. Furthermore, any two distinct integral curves do not intersect each other, and the union of all the integral curves is M itself. A nonvanishing vector field X can be viewed as the smooth assignment of a one-dimensional vector space to every point of M, namely $p \mapsto \mathbb{R} X_p \subseteq T_p M$, where $\mathbb{R} X_p$ denotes the line spanned by X_p. Thus, it is natural to consider the more general situation where we fix some integer r, with $1 \leq r \leq n$, and we have an assignment $p \mapsto D(p) \subseteq T_p M$, where $D(p)$ is some r-dimensional subspace of $T_p M$ such that $D(p)$ "varies smoothly" with $p \in M$. Is there a notion of integral manifold for such assignments? Do they always exist?

It is indeed possible to generalize the notion of integral curve and to define integral manifolds, but unlike the situation for vector fields ($r = 1$), not every assignment D as above possess an integral manifold. However, there is a necessary and sufficient condition for the existence of integral manifolds given by the Frobenius theorem.

This theorem has several equivalent formulations. First we will present a formulation in terms of vector fields. Then we show that there are advantages in reformulating the notion of involutivity in terms of differential ideals, and we state a differential form version of the Frobenius theorem. The above versions of the Frobenius theorem are "local." We will briefly discuss the notion of foliation and state a global version of the Frobenius theorem.

Since Frobenius' theorem is a standard result of differential geometry, we will omit most proofs, and instead refer the reader to the literature. A complete treatment of Frobenius' theorem can be found in Warner [109], Morita [82], and Lee [73].

The original version of this chapter was revised. The correction to this chapter is available at https://doi.org/10.1007/978-3-030-46047-1_12

© Springer Nature Switzerland AG 2020, corrected publication 2020
J. Gallier, J. Quaintance, *Differential Geometry and Lie Groups*, Geometry and Computing 13, https://doi.org/10.1007/978-3-030-46047-1_5

5.1   Tangential Distributions and Involutive Distributions

Our first task is to define precisely what we mean by a smooth assignment $p \mapsto D(p) \subseteq T_pM$, where $D(p)$ is an r-dimensional subspace. Recall the definition of an *immersed submanifold* given in Warner [109], Chapter 1, Definition 1.2, namely a pair (M, ψ) where $\psi \colon M \to N$ is a smooth injective immersion (which means that $d\psi_p$ is injective for all $p \in M$).

Definition 5.1. Let M be a smooth manifold of dimension n. For any integer r, with $1 \leq r \leq n$, an r-dimensional *tangential distribution* (for short, a *distribution*) is a map $D \colon M \to TM$, such that

(a) $D(p) \subseteq T_pM$ is an r-dimensional subspace for all $p \in M$.
(b) For every $p \in M$, there is some open subset U with $p \in U$, and r smooth vector fields $X_1, \ldots, X_r$ defined on U, such that $(X_1(q), \ldots, X_r(q))$ is a basis of $D(q)$ for all $q \in U$. We say that D is *locally spanned* by $X_1, \ldots, X_r$.

An immersed submanifold N of M is an *integral manifold* of D iff $D(p) = T_pN$ for all $p \in N$. We say that D is *completely integrable* iff there exists an integral manifold of D through every point of M.

We also write D_p for $D(p)$.

Remarks.

(1) An r-dimensional distribution D is just a smooth subbundle of TM.
(2) An integral manifold is only an immersed submanifold, not necessarily an embedded submanifold.
(3) Some authors (such as Lee) reserve the locution "completely integrable" to a seemingly strong condition (see Lee [73], Chapter 19, page 500). This condition is in fact equivalent to "our" definition (which seems the most commonly adopted).
(4) Morita [82] uses a stronger notion of integral manifold. Namely, an integral manifold is actually an embedded manifold. Most of the results including Frobenius theorem still hold, but maximal integral manifolds are immersed but not embedded manifolds, and this is why most authors prefer to use the weaker definition (immersed manifolds).

Here is an example of a distribution which does not have any integral manifolds. This is the two-dimensional distribution in $\mathbb{R}^3$ spanned by the vector fields

$$X = \frac{\partial}{\partial x} + y\frac{\partial}{\partial z}, \qquad Y = \frac{\partial}{\partial y}.$$

To show why this distribution is not integrable, we will need an involutivity condition. Here is the definition.

Definition 5.2. Let M be a smooth manifold of dimension n and let D be an r-dimensional distribution on M. For any smooth vector field X, we say that X *belongs to D* (or *lies in D*) iff $X_p \in D_p$ for all $p \in M$. We say that D is *involutive* iff for any two smooth vector fields X, Y on M, if X and Y belong to D, then $[X, Y]$ also belongs to D.

Proposition 5.1. *Let M be a smooth manifold of dimension n. If an r-dimensional distribution D is completely integrable, then D is involutive.*

Proof. A proof can be found in in Warner [109] (Chapter 1), and Lee [73] (Proposition 19.3). Another proof is given in Morita [82] (Section 2.3), but beware that Morita defines an integral manifold to be an embedded manifold. $\square$

In the example before Definition 5.1, we have

$$[X, Y] = -\frac{\partial}{\partial z},$$

so this distribution is not involutive. Therefore, by Proposition 5.1, this distribution is not completely integrable.

5.2 Frobenius Theorem

Frobenius' theorem asserts that the converse of Proposition 5.1 holds. Although we do not intend to prove it in full, we would like to explain the main idea of the proof of Frobenius' theorem. It turns out that the involutivity condition of two vector fields is equivalent to the commutativity of their corresponding flows, and this is the crucial fact used in the proof.

Definition 5.3. Given a manifold, M, we say that two vector fields X and Y are *mutually commutative* iff $[X, Y] = 0$.

For example, on $\mathbb{R}^2$, the vector fields $\frac{\partial}{\partial x}$ and $\frac{\partial}{\partial y}$ are commutative since $\frac{\partial^2 f}{\partial x \partial y} = \frac{\partial^2 f}{\partial y \partial x}$. On the other hand, the vector fields $\frac{\partial}{\partial x}$ and $x\frac{\partial}{\partial y}$ are not commutative since

$$\left[\frac{\partial}{\partial x}, x\frac{\partial}{\partial y}\right] f = \frac{\partial}{\partial x}\left(x\frac{\partial f}{\partial y}\right) - x\frac{\partial}{\partial y}\left(\frac{\partial f}{\partial x}\right)$$

$$= \frac{\partial f}{\partial y} + x\frac{\partial^2 f}{\partial x \partial y} - x\frac{\partial^2 f}{\partial y \partial x}$$

$$= \frac{\partial f}{\partial y},$$

which in turn implies $\left[\frac{\partial}{\partial x}, x\frac{\partial}{\partial y}\right] = \frac{\partial}{\partial y}$.

Recall that we denote by Φ^X the (global) flow of the vector field X. For every $p \in M$, the map $t \mapsto \Phi^X(t, p) = \gamma_p(t)$ is the maximal integral curve through p. We also write $\Phi_t(p)$ for $\Phi^X(t, p)$ (dropping X). Recall that the map $p \mapsto \Phi_t(p)$ is a diffeomorphism on its domain (an open subset of M). For the next proposition, given two vector fields X and Y, we write Φ for the flow associated with X and Ψ for the flow associated with Y.

Proposition 5.2. *Given a manifold M, for any two smooth vector fields X and Y, the following conditions are equivalent:*

(1) X and Y are mutually commutative (i.e., $[X, Y] = 0$).
(2) Y is invariant under Φ_t; that is, $(\Phi_t)_* Y = Y$, whenever the left-hand side is defined.
(3) X is invariant under Ψ_s; that is, $(\Psi_s)_* X = X$, whenever the left-hand side is defined.
(4) The maps Φ_t and Ψ_t are mutually commutative. This means that

$$\Phi_t \circ \Psi_s = \Psi_s \circ \Phi_t,$$

for all s, t such that both sides are defined.
(5) $\mathcal{L}_X Y = [X, Y] = 0$.
(6) $\mathcal{L}_Y X = [Y, X] = 0$.

(In (5) $\mathcal{L}_X Y$ is the Lie derivative and similarly in (6).)

Proof. A proof can be found in Lee [73] (Chapter 18, Proposition 18.5) and in Morita [82] (Chapter 2, Proposition 2.18). For example, to prove the implication $(2) \implies (4)$, we observe that if φ is a diffeomorphism on some open subset U of M, then the integral curves of $\varphi_* Y$ through a point $p \in M$ are of the form $\varphi \circ \gamma$, where γ is the integral curve of Y through $\varphi^{-1}(p)$. Consequently, the local one-parameter group generated by $\varphi_* Y$ is $\{\varphi \circ \Psi_s \circ \varphi^{-1}\}$. If we apply this to $\varphi = \Phi_t$, as $(\Phi_t)_* Y = Y$, we get $\Phi_t \circ \Psi_s \circ \Phi_t^{-1} = \Psi_s$, and hence $\Phi_t \circ \Psi_s = \Psi_s \circ \Phi_t$. $\square$

In order to state our first version of the Frobenius theorem we make the following definition.

Definition 5.4. Let M be a smooth manifold of dimension n. Given any smooth r-dimensional distribution D on M, a chart (U, φ) is *flat for D* iff

$$\varphi(U) \cong U' \times U'' \subseteq \mathbb{R}^r \times \mathbb{R}^{n-r},$$

where U' and U'' are connected open subsets such that for every $p \in U$, the distribution D is spanned by the vector fields

$$\frac{\partial}{\partial x_1}, \dots, \frac{\partial}{\partial x_r}.$$

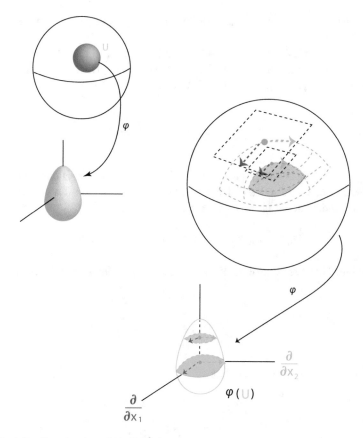

Fig. 5.1 A flat chart for the solid ball B^3. Each slice in $\varphi(U)$ is parallel to the xy-plane and turns into a cap shape inside of B^3.

If (U, φ) is flat for D, then each slice of (U, φ)

$$S_c = \{q \in U \mid x_{r+1} = c_{r+1}, \ldots, x_n = c_n\}$$

is an integral manifold of D, where $x_i = pr_i \circ \varphi$ is the ith-coordinate function on U and $c = (c_{r+1}, \ldots, c_n) \in \mathbb{R}^{n-r}$ is a fixed vector, as illustrated in Figure 5.1.

Theorem 5.3 (Frobenius). *Let M be a smooth manifold of dimension n. A smooth r-dimensional distribution D on M is completely integrable iff it is involutive. Furthermore, for every $p \in U$, there is a flat chart (U, φ) for D with $p \in U$ so that every slice of (U, φ) is an integral manifold of D.*

Proof. A proof of Theorem 5.3 can be found in Warner [109] (Chapter 1, Theorem 1.60), Lee [73] (Chapter 19, Theorem 19.10), and Morita [82] (Chapter 2, Theorem 2.17). Since we already have Proposition 5.1, it is only necessary to prove that if a distribution is involutive, then it is completely integrable. Here is a sketch of the proof, following Morita.

Pick any $p \in M$. As D is a smooth distribution, we can find some chart (U, φ) with $p \in U$, and some vector fields $Y_1, \ldots, Y_r$ so that $Y_1(q), \ldots, Y_r(q)$ are linearly independent and span D_q for all $q \in U$. Locally, we can write

$$Y_i = \sum_{j=1}^{n} a_{ij} \frac{\partial}{\partial x_j}, \qquad i = 1, \ldots, r. \tag{$\dagger$}$$

Since $Y_1, \ldots, Y_r$ are linearly independent, the $r \times n$ matrix (a_{ij}) has rank r, so by renumbering the coordinates if necessary, we may assume that the first r columns are linearly independent in which case the $r \times r$ matrix

$$A(q) = (a_{ij}(q)), \qquad 1 \leq i, j \leq r. \qquad q \in U$$

is invertible. Then the inverse matrix $B(q) = A^{-1}(q)$ defines $r \times r$ functions $b_{ij}(q)$, and let

$$X_i = \sum_{j=1}^{r} b_{ij} Y_j, \qquad j = 1, \ldots, r. \tag{$\dagger\dagger$}$$

Now in matrix form Line ($\dagger$) becomes

$$\begin{pmatrix} Y_1 \\ \vdots \\ Y_r \end{pmatrix} = (A \ R) \begin{pmatrix} \frac{\partial}{\partial x_1} \\ \vdots \\ \frac{\partial}{\partial x_n} \end{pmatrix},$$

for some $r \times (n - r)$ matrix R, and Line ($\dagger\dagger$) becomes

$$\begin{pmatrix} X_1 \\ \vdots \\ X_r \end{pmatrix} = B \begin{pmatrix} Y_1 \\ \vdots \\ Y_r \end{pmatrix},$$

so we get

$$\begin{pmatrix} X_1 \\ \vdots \\ X_r \end{pmatrix} = (I \ BR) \begin{pmatrix} \frac{\partial}{\partial x_1} \\ \vdots \\ \frac{\partial}{\partial x_n} \end{pmatrix},$$

that is,

$$X_i = \frac{\partial}{\partial x_i} + \sum_{j=r+1}^{n} c_{ij} \frac{\partial}{\partial x_j}, \qquad i = 1, \ldots, r, \tag{$*$}$$

where the c_{ij} are functions defined on U. Obviously, $X_1, \ldots, X_r$ are linearly independent and they span D_q for all $q \in U$. Since D is involutive, there are some functions f_k defined on U so that

$$[X_i, X_j] = \sum_{k=1}^{r} f_k X_k.$$

On the other hand, by $(*)$, each $[X_i, X_j]$ is a linear combination of $\frac{\partial}{\partial x_{r+1}}, \ldots, \frac{\partial}{\partial x_n}$. Using $(*)$, we obtain

$$[X_i, X_j] = \sum_{k=1}^{r} f_k X_k = \sum_{k=1}^{r} f_k \frac{\partial}{\partial x_k} + \sum_{k=1}^{r} \sum_{j=r+1}^{n} f_k c_{kj} \frac{\partial}{\partial x_j},$$

and since this is supposed to be a linear combination of $\frac{\partial}{\partial x_{r+1}}, \ldots, \frac{\partial}{\partial x_n}$, we must have $f_k = 0$ for $k = 1, \ldots, r$, which shows that

$$[X_i, X_j] = 0, \qquad 1 \le i, j \le r;$$

that is, the vector fields $X_1, \ldots, X_r$ are mutually commutative.

Let Φ_t^i be the local one-parameter group associated with X_i. By Proposition 5.2 (4), the Φ_t^i commute; that is,

$$\Phi_t^i \circ \Phi_s^j = \Phi_s^j \circ \Phi_t^i \qquad 1 \le i, j \le r,$$

whenever both sides are defined. We can pick a sufficiently small open subset V in $\mathbb{R}^r$ containing the origin and define the map $\Phi \colon V \to U$ by

$$\Phi(t_1, \ldots, t_r) = \Phi_{t_1}^1 \circ \cdots \circ \Phi_{t_r}^r(p).$$

Clearly, Φ is smooth, and using the fact that each X_i is invariant under each Φ_s^j for $j \ne i$, and

$$d\Phi_p^i \left(\frac{\partial}{\partial t_i} \right) = X_i(p),$$

we get

$$d\Phi_p \left(\frac{\partial}{\partial t_i} \right) = X_i(p).$$

As $X_1, \ldots, X_r$ are linearly independent, we deduce that $d\Phi_p \colon T_0\mathbb{R}^r \to T_pM$ is an injection, and thus we may assume by shrinking V if necessary that our map $\Phi \colon V \to M$ is an embedding. But then, $N = \Phi(V)$ is a an immersed submanifold of M, and it only remains to prove that N is an integral manifold of D through p.

Obviously, $T_pN = D_p$, so we just have to prove that $T_qN = D_q$ for all $q \in N$. Now for every $q \in N$, we can write

$$q = \Phi(t_1, \ldots, t_r) = \Phi^1_{t_1} \circ \cdots \circ \Phi^r_{t_r}(p),$$

for some $(t_1, \ldots, t_r) \in V$. Since the Φ^i_t commute for any i, with $1 \le i \le r$, we can write

$$q = \Phi^i_{t_i} \circ \Phi^1_{t_1} \circ \cdots \circ \Phi^{i-1}_{t_{i-1}} \circ \Phi^{i+1}_{t_{i+1}} \circ \cdots \circ \Phi^r_{t_r}(p).$$

If we fix all the t_j but t_i and vary t_i by a small amount, we obtain a curve in N through q, and this is an orbit of Φ^i_t. Therefore, this curve is an integral curve of X_i through q whose velocity vector at q is equal to $X_i(q)$, and so $X_i(q) \in T_qN$. Since the above reasoning holds for all i, we get $T_qN = D_q$, as claimed. Therefore, N is an integral manifold of D through p. $\square$

To best understand how the proof of Theorem 5.3 constructs the integral manifold N, we provide the following example found in Chapter 19 of Lee [73]. Let $D \subset T\mathbb{R}^3$ be the distribution

$$Y_1 := V = x\frac{\partial}{\partial x} + \frac{\partial}{\partial y} + x(y+1)\frac{\partial}{\partial z}$$

$$Y_2 := W = \frac{\partial}{\partial x} + y\frac{\partial}{\partial z}.$$

Given $f \in C^\infty(\mathbb{R}^3)$, observe that

$$
\begin{aligned}
[V, W](f) &= V(W(f)) - W(V(f)) \\
&= \left(x\frac{\partial}{\partial x} + \frac{\partial}{\partial y} + x(y+1)\frac{\partial}{\partial z}\right)\left(\frac{\partial f}{\partial x} + y\frac{\partial f}{\partial z}\right) \\
&\quad - \left(\frac{\partial}{\partial x} + y\frac{\partial}{\partial z}\right)\left(x\frac{\partial f}{\partial x} + \frac{\partial f}{\partial y} + x(y+1)\frac{\partial f}{\partial z}\right) \\
&= \frac{\partial f}{\partial z} - \frac{\partial f}{\partial x} - (y+1)\frac{\partial f}{\partial z} \\
&= -\frac{\partial f}{\partial x} - y\frac{\partial f}{\partial z} = -W(f).
\end{aligned}
$$

Thus D is involutive and Theorem 5.3 is applicable. Our goal is to find a flat chart around the origin.

In order to construct this chart, we note that $\frac{\partial}{\partial z}$ is not in the span of V and W since if $\frac{\partial}{\partial z} = aV + bW$, then

$$\frac{\partial}{\partial z} = (ax + b)\frac{\partial}{\partial x} + a\frac{\partial}{\partial y} + (a(x + 1) + by)\frac{\partial}{\partial z},$$

which in turn implies $a = 0 = b$, a contradiction. This means we may rewrite a basis for D in terms of Line $(*)$ and find that

$$X_1 := X = W = \frac{\partial}{\partial x} + y\frac{\partial}{\partial z}$$

$$X_2 := Y = V - xW = \frac{\partial}{\partial y} + x\frac{\partial}{\partial z}.$$

Alternatively we may obtain X_1, X_2 from the matrix form of Line $(\dagger)$,

$$\begin{pmatrix} Y_1 \\ Y_2 \end{pmatrix} = \begin{pmatrix} x & 1 & x(y + 1) \\ 1 & 0 & y \end{pmatrix} \begin{pmatrix} \frac{\partial}{\partial x} \\ \frac{\partial}{\partial y} \\ \frac{\partial}{\partial z} \end{pmatrix},$$

(with $A = \begin{pmatrix} x & 1 \\ 1 & 0 \end{pmatrix}$), and the matrix form of Line $(\dagger\dagger)$, namely

$$\begin{pmatrix} X_1 \\ X_2 \end{pmatrix} = \begin{pmatrix} 0 & 1 \\ 1 & -x \end{pmatrix} \begin{pmatrix} Y_1 \\ Y_2 \end{pmatrix}.$$

The flow of X is

$$\alpha_u(x, y, z) := \Phi_u^1(x, y, z) = (x + u, y, z + uy),$$

while the flow of Y is

$$\beta_v(x, y, z) := \Phi_v^2(x, y, z) = (x, y + v, z + vx).$$

For a fixed point on the z-axis near the origin, say $(0, 0, w)$, we define $\Phi : \mathbb{R}^3 \to \mathbb{R}^3$ as a composition of the flows, namely

$$\Phi(u, v)(0, 0, w) = \alpha_u \circ \beta_v(0, 0, w) = \alpha_u(0, v, w) = (u, v, w + uv).$$

In other words $\Phi(u, v)(0, 0, w)$ provides the parameterization of $\mathbb{R}^3$ given by

$$x = u, \qquad y = v, \qquad z = w + uv,$$

and thus the flat chart is given by

$$\Phi^{-1}(x, y, z) = (u, v, z - xy).$$

By the paragraph immediately preceding Theorem 5.3, we conclude that the N, the integral manifolds of D, are given by the level sets of $w(x, y, z) = z - xy$.

In preparation for a global version of Frobenius theorem in terms of foliations, we state the following proposition proved in Lee [73] (Chapter 19, Proposition 19.12):

Proposition 5.4. *Let M be a smooth manifold of dimension n and let D be an involutive r-dimensional distribution on M. For every flat chart (U, φ) for D, for every integral manifold N of D, the set $N \cap U$ is a countable disjoint union of open parallel r-dimensional slices of U, each of which is open in N and embedded in M.*

We now describe an alternative method for describing involutivity in terms of differential forms.

5.3   Differential Ideals and Frobenius Theorem

First, we give a smoothness criterion for distributions in terms of one-forms.

Proposition 5.5. *Let M be a smooth manifold of dimension n and let D be an assignment $p \mapsto D_p \subseteq T_pM$ of some r-dimensional subspace of T_pM, for all $p \in M$. Then D is a smooth distribution iff for every $p \in U$, there is some open subset U with $p \in U$, and some linearly independent one-forms $\omega_1, \ldots, \omega_{n-r}$ defined on U, so that*

$$D_q = \{u \in T_qM \mid (\omega_1)_q(u) = \cdots = (\omega_{n-r})_q(u) = 0\}, \qquad \textit{for all } q \in U.$$

Proof. Proposition 5.5 is proved in Lee [73] (Chapter 19, Lemma 19.5). The idea is to either extend a set of linearly independent differential one-forms to a coframe and then consider the dual frame, or to extend some linearly independent vector fields to a frame and then take the dual basis.   □

Proposition 5.5 suggests the following definitions.

Definition 5.5. Let M be a smooth manifold of dimension n and let D be an r-dimensional distribution on M.

1. Some linearly independent one-forms $\omega_1, \ldots, \omega_{n-r}$ defined on some open subset $U \subseteq M$ are called *local defining one-forms* for D if

$$D_q = \{u \in T_qM \mid (\omega_1)_q(u) = \cdots = (\omega_{n-r})_q(u) = 0\}, \qquad \textit{for all } q \in U.$$

2. We say that a k-form $\omega \in \mathcal{A}^k(M)$ *annihilates D* iff

$$\omega_q(X_1(q), \ldots, X_k(q)) = 0,$$

for all $q \in M$ and for all vector fields $X_1, \ldots, X_k$ belonging to D. We write

$$\mathfrak{I}^k(D) = \{\omega \in \mathcal{A}^k(M) \mid \omega_q(X_1(q), \ldots, X_k(q)) = 0\},$$

for all $q \in M$ and for all vector fields $X_1, \ldots, X_k$ belonging to D, and we let

$$\mathfrak{I}(D) = \bigoplus_{k=1}^{n} \mathfrak{I}^k(D).$$

Thus, $\mathfrak{I}(D)$ is the collection of differential forms that "vanish on D." In the classical terminology, a system of local defining one-forms as above is called a *system of Pfaffian equations*.

It turns out that $\mathfrak{I}(D)$ is not only a vector space, but also an ideal of $\mathcal{A}^\bullet(M)$.

Recall that a subspace $\mathfrak{I}$ of $\mathcal{A}^\bullet(M)$ is an *ideal* iff for every $\omega \in \mathfrak{I}$, we have $\theta \wedge \omega \in \mathfrak{I}$ for every $\theta \in \mathcal{A}^\bullet(M)$.

Proposition 5.6. *Let M be a smooth n-dimensional manifold and D be an r-dimensional distribution. If $\mathfrak{I}(D)$ is the space of forms annihilating D, then the following hold:*

(a) $\mathfrak{I}(D)$ is an ideal in $\mathcal{A}^\bullet(M)$.

(b) $\mathfrak{I}(D)$ is locally generated by $n - r$ linearly independent one-forms, which means for every $p \in U$, there is some open subset $U \subseteq M$ with $p \in U$ and a set of linearly independent one-forms $\omega_1, \ldots, \omega_{n-r}$ defined on U, so that

  (i) If $\omega \in \mathfrak{I}^k(D)$, then $\omega \restriction U$ belongs to the ideal in $\mathcal{A}^\bullet(U)$ generated by $\omega_1, \ldots, \omega_{n-r}$; that is,

$$\omega = \sum_{i=1}^{n-r} \theta_i \wedge \omega_i, \qquad on \ U,$$

  for some $(k-1)$-forms $\theta_i \in \mathcal{A}^{k-1}(U)$.

  (ii) If $\omega \in \mathcal{A}^k(M)$ and if there is an open cover by subsets U (as above) such that for every U in the cover, $\omega \restriction U$ belongs to the ideal generated by $\omega_1, \ldots, \omega_{n-r}$, then $\omega \in \mathfrak{I}(D)$.

(c) If $\mathfrak{I} \subseteq \mathcal{A}^\bullet(M)$ is an ideal locally generated by $n - r$ linearly independent one-forms, then there exists a unique smooth r-dimensional distribution D for which $\mathfrak{I} = \mathfrak{I}(D)$.

Proof. Proposition 5.6 is proved in Warner (Chapter 2, Proposition 2.28); see also Morita [82] (Chapter 2, Lemma 2.19) and Lee [73] (Chapter 19, pages 498–500).
□

In order to characterize involutive distributions, we need the notion of a differential ideal.

Definition 5.6. Let M be a smooth manifold of dimension n. An ideal $\mathfrak{I} \subseteq \mathcal{A}^{\bullet}(M)$ is a *differential ideal* iff it is closed under exterior differentiation; that is,

$$d\omega \in \mathfrak{I} \quad \text{whenever} \quad \omega \in \mathfrak{I},$$

which we also express by $d\mathfrak{I} \subseteq \mathfrak{I}$.

Here is the differential ideal criterion for involutivity.

Proposition 5.7. *Let M be a smooth manifold of dimension n. A smooth r-dimensional distribution D is involutive iff the ideal $\mathfrak{I}(D)$ is a differential ideal.*

Proof. Proposition 5.7 is proved in Warner [109] (Chapter 2, Proposition 2.30), Morita [82] (Chapter 2, Proposition 2.20), and Lee [73] (Chapter 19, Proposition 19.19).

Assume D is involutive. Let $\omega \in \mathcal{A}^k(M)$ be any k form on M and let $X_0, \ldots, X_k$ be $k + 1$ smooth vector fields lying in D. Then by Proposition 4.16 and the fact that D is involutive, we deduce that $d\omega(X_0, \ldots, X_k) = 0$. Hence, $d\omega \in \mathfrak{I}(D)$, which means that $\mathfrak{I}(D)$ is a differential ideal.

For the converse, assume $\mathfrak{I}(D)$ is a differential ideal. We know that for any one-form ω,

$$d\omega(X, Y) = X(\omega(Y)) - Y(\omega(X)) - \omega([X, Y]),$$

for any vector fields X, Y. Now, if $\omega_1, \ldots, \omega_{n-r}$ are linearly independent one-forms that define D locally on U, using a bump function, we can extend $\omega_1, \ldots, \omega_{n-r}$ to M, and then using the above equation, for any vector fields X, Y belonging to D, we get

$$\omega_i([X, Y]) = X(\omega_i(Y)) - Y(\omega_i(X)) - d\omega_i(X, Y),$$

and since $\omega_i(X) = \omega_i(Y) = d\omega_i(X, Y) = 0$ (because $\mathfrak{I}(D)$ is a differential ideal and $\omega_i \in \mathfrak{I}(D)$), we get $\omega_i([X, Y]) = 0$ for $i = 1, \ldots, n - r$, which means that $[X, Y]$ belongs to D.
□

Using Proposition 5.6, we can give a more concrete criterion.

Proposition 5.8. *A distribution D is involutive iff for every local defining one-forms $\omega_1, \ldots, \omega_{n-r}$ for D (on some open subset U), there are some one-forms $\omega_{ij} \in \mathcal{A}^1(U)$ so that*

$$d\omega_i = \sum_{j=1}^{n-r} \omega_{ij} \wedge \omega_j \qquad (i = 1, \ldots, n-r).$$

The above conditions are often called the *integrability conditions*.

Definition 5.7. Let M be a smooth manifold of dimension n. Given any ideal $\mathfrak{I} \subseteq \mathcal{A}^\bullet(M)$, an immersed manifold $N = (M, \psi)$ of M (where $\psi \colon N \to M$) is an *integral manifold of* $\mathfrak{I}$ iff

$$\psi^*\omega = 0, \qquad \text{for all } \omega \in \mathfrak{I}.$$

A connected integral manifold of the ideal $\mathfrak{I}$ is *maximal* iff its image is not a proper subset of the image of any other connected integral manifold of $\mathfrak{I}$.

Finally, here is the differential form version of the Frobenius theorem.

Theorem 5.9 (Frobenius Theorem, Differential Ideal Version). *Let M be a smooth manifold of dimension n. If $\mathfrak{I} \subseteq \mathcal{A}^\bullet(M)$ is a differential ideal locally generated by $n - r$ linearly independent one-forms, then for every $p \in M$, there exists a unique maximal, connected, integral manifold of $\mathfrak{I}$ through p, and this integral manifold has dimension r.*

Proof. Theorem 5.9 is proved in Warner [109]. This theorem follows immediately from Theorem 1.64 in Warner [109]. $\square$

Another version of the Frobenius theorem goes as follows; see Morita [82] (Chapter 2, Theorem 2.21).

Theorem 5.10 (Frobenius Theorem, Integrability Conditions Version). *Let M be a smooth manifold of dimension n. An r-dimensional distribution D on M is completely integrable iff for every local defining one-forms $\omega_1, \ldots, \omega_{n-r}$ for D (on some open subset, U), there are some one-forms $\omega_{ij} \in \mathcal{A}^1(U)$ so that we have the integrability conditions*

$$d\omega_i = \sum_{j=1}^{n-r} \omega_{ij} \wedge \omega_j \qquad (i = 1, \ldots, n-r).$$

There are applications of Frobenius theorem (in its various forms) to systems of partial differential equations, but we will not deal with this subject. The reader is advised to consult Lee [73], Chapter 19, and the references there.

5.4   A Glimpse at Foliations and a Global Version of Frobenius Theorem

All the maximal integral manifolds of an r-dimensional involutive distribution on a manifold M yield a decomposition of M with some nice properties, those of a foliation.

Definition 5.8. Let M be a smooth manifold of dimension n. A family $\mathcal{F} = \{\mathcal{F}_\alpha\}_\alpha$ of subsets of M is a *k-dimensional foliation* iff it is a family of pairwise disjoint, connected, immersed k-dimensional submanifolds of M called the *leaves* of the foliation, whose union is M, and such that for every $p \in M$, there is a chart (U, φ) with $p \in U$ called a *flat chart* for the foliation, and the following property holds:

$$\varphi(U) \cong U' \times U'' \subseteq \mathbb{R}^r \times \mathbb{R}^{n-r},$$

where U' and U'' are some connected open subsets, and for every leaf $\mathcal{F}_\alpha$ of the foliation, if $\mathcal{F}_\alpha \cap U \neq \emptyset$, then $\mathcal{F}_\alpha \cap U$ is a countable union of k-dimensional slices given by

$$x_{r+1} = c_{r+1}, \ldots, x_n = c_n,$$

for some constants $c_{r+1}, \ldots, c_n \in \mathbb{R}$.

The structure of a foliation can be very complicated. For instance, the leaves can be dense in M. For example, there are spirals on a torus that form the leaves of a foliation (see Lee [73], Example 19.9). Foliations are in one-to-one correspondence with involutive distributions.

Proposition 5.11. *Let M be a smooth manifold of dimension n. For any foliation $\mathcal{F}$ on M, the family of tangent spaces to the leaves of $\mathcal{F}$ forms an involutive distribution on M.*

The converse to the above proposition may be viewed as a global version of Frobenius theorem.

Theorem 5.12. *Let M be a smooth manifold of dimension n. For every r-dimensional smooth, involutive distribution D on M, the family of all maximal, connected, integral manifolds of D forms a foliation of M.*

Proof. The proof of Theorem 5.12 can be found in Lee [73] (Theorem 19.21).   □

5.5   Problems

Problem 5.1. Prove Proposition 5.1.
Hint. See Warner [109], Chapter 1, and Lee [73], Proposition 19.3.

Problem 5.2. Prove Proposition 5.2.

Hint. See Lee [73], Chapter 18, Proposition 18.5, and Morita [82], Chapter 2, Proposition 2.18.

Problem 5.3. Prove Proposition 5.4.

Hint. See [73], Chapter 19, Proposition 19.12.

Problem 5.4. Prove Proposition 5.5.

Hint. See Lee [73], Chapter 19, Lemma 19.5.

Problem 5.5. Prove Proposition 5.6.

Hint. See Warner, Chapter 2, Proposition 2.28; see also Morita [82], Chapter 2, Lemma 2.19, and Lee [73], Chapter 19, pages 498–500.

Chapter 6
Integration on Manifolds

The purpose of this chapter is to generalize the theory of integration known for functions defined on open subsets of $\mathbb{R}^n$ to manifolds. As a first step, we explain how differential forms defined on an open subset of $\mathbb{R}^n$ are integrated. Then, if M is a smooth manifold of dimension n, and if ω is an n-form on M (with compact support), the integral $\int_M \omega$ is defined by patching together the integrals defined on small-enough open subsets covering M using a partition of unity. If (U, φ) is a chart such that the support of ω is contained in U, then the pullback $(\varphi^{-1})^*\omega$ of ω is an n-form on $\mathbb{R}^n$, so we know how to compute its integral $\int_{\varphi(U)} (\varphi^{-1})^*\omega$. To ensure that these integrals have a consistent value on overlapping charts, we need for M to be orientable. Actually, there is a more general notion of integration on a manifold that uses densities instead differential forms, but we do not need such generality.

In Section 6.1 we define the notion of *orientation* of a manifold. First we define an orientation of a vector space. Then we define an oriented smooth manifold as a manifold that has an atlas consisting of charts such that the transition maps all have positive Jacobian determinants. Technically, a more convenient criterion for orientability is the existence of a differential n-form (where n is the dimension of the manifold) which is nowhere-vanishing. Such an n-form is called a *volume form*. We prove that a smooth manifold (Hausdorff and second-countable) is orientable if and only if it possesses a volume form. We also define orientable diffeomorphisms.

In Section 6.2 we consider the special case of Riemannian manifolds. An orientable Riemannian manifold has a special volume form expressible in a chart in terms of the square root of the determinant of the matrix expressing the metric. Lie groups are always orientable and possess a left-invariant volume form.

In Section 6.3 we explain how to integrate differential forms with compact support defined on an open subset U of $\mathbb{R}^n$. Since a differential n-form on U can be expressed as

$$\omega_x = f(x)dx_1 \wedge \cdots \wedge dx_n,$$

© Springer Nature Switzerland AG 2020
J. Gallier, J. Quaintance, *Differential Geometry and Lie Groups*, Geometry and Computing 13, https://doi.org/10.1007/978-3-030-46047-1_6

where $f \colon U \to \mathbb{R}$ is a smooth function with compact support contained in U, we can define $\int_U \omega$ as $\int_U f(x)dx_1 \cdots dx_n$, the standard Riemann integral of f. We also give a formula for the change of variable induced by a diffeomorphism $\varphi \colon U \to V$.

In Section 6.4, we promote the definition of the integral of a differential form defined on an open subset of $\mathbb{R}^n$ to smooth oriented manifolds. For any n-form ω with compact support on a smooth n-dimensional oriented manifold M, the integral $\int_M \omega$ is computed by patching together the integrals defined on small-enough open subsets covering M using a partition of unity. The orientability of M is needed to ensure that the above integrals have a consistent value on overlapping charts

In preparation for discussing Stokes' theorem, we need to define *manifolds with boundaries*, which is the object of Section 6.6. The idea is to allow a class of manifolds that can be covered with open subsets homeomorphic to open subset of the half space $\mathbb{H}^n = \{(x_1, \ldots, x_n) \in \mathbb{R}^n \mid x_n \geq 0\}$.

In Section 6.7 we define a class of manifolds with boundaries called regular domains, and we prove *Stokes' theorem*, which roughly speaking is stated as

$$\int_{\partial N} \omega = \int_N d\omega,$$

where N is an oriented domain with smooth boundary ∂N. We also present the classical versions of Stokes' theorem in $\mathbb{R}^3$ as well as the divergence theorem. We also mention the class of *manifolds with corners*, which is more general than the class of manifolds with boundaries, for which a version of Stokes' theorem holds.

In Section 6.8 we define the integral of a smooth function f with compact support defined on an orientable Riemannian manifold M. For this we use the canonical volume form Vol_M induced by the Riemannian metric, and let $\int_M f = \int_M f \, \mathrm{Vol}_M$.

Since a Lie group G is orientable, we can pick a left-invariant volume form ω and define the integral of a function f with compact support as $\int_G f = \int_G f \omega$. Such an integral is left-invariant. Roughly speaking this means that the integral does not change if the variable t in $f(t)$ is replaced by st. In general it is not right-invariant (the integral is right-invariant if the integral does not change when the variable t in $f(t)$ is replaced by ts). The failure of right-invariance of the integral is measured by the *modular function* of the group. Technically, $\int_G f \omega = \Delta(g) \int_G f(tg)\omega$. Lie groups for which $\Delta \equiv 1$ are particularly nice, and are called *unimodular*. Compact Lie groups are unimodular, and so are semisimple Lie groups.

6.1 Orientation of Manifolds

Although the notion of orientation of a manifold is quite intuitive it is technically rather subtle. We restrict our discussion to smooth manifolds (the notion of orientation can also be defined for topological manifolds, but more work is involved).

Intuitively, a manifold M is orientable if it is possible to give a consistent orientation to its tangent space T_pM at every point $p \in M$. So, if we go around a closed curve starting at $p \in M$, when we come back to p, the orientation of $T_p M$ should be the same as when we started. For example, if we travel on a Möbius strip (a manifold with boundary) dragging a coin with us, we will come back to our point of departure with the coin flipped. Try it; see Figure 6.3 for an illustration.

To be rigorous, we have to say what it means to orient $T_p M$ (a vector space) and what consistency of orientation means. We begin by quickly reviewing the notion of orientation of a vector space. Let E be a vector space of dimension n. If $u_1, \ldots, u_n$ and $v_1, \ldots, v_n$ are two bases of E, a basic and crucial fact of linear algebra says that there is a unique linear map g mapping each u_i to the corresponding v_i (i.e., $g(u_i) = v_i$, $i = 1, \ldots, n$). Then look at the determinant $\det(g)$ of this map. We know that $\det(g) = \det(P)$, where P is the matrix whose j-th column consists of the coordinates of v_j over the basis $u_1, \ldots, u_n$. Either $\det(g)$ is negative, or it is positive. This leads to the following definition.

Definition 6.1. Let E be a vector space of dimension n with bases $u_1, \ldots, u_n$ and $v_1, \ldots, v_n$. Let g be the unique linear map such that $g(u_i) = v_i$, $i = 1, \ldots, n$. We say $u_1, \ldots, u_n$ and $v_1, \ldots, v_n$ have the *same orientation* iff $\det(g)$ is positive.

Definition 6.1 defines an equivalence relation on bases where two bases are equivalent iff they have the same orientation.

Definition 6.2. Let E be a vector space of dimension n. An *orientation* of E is the choice of one of the two equivalence classes, which amounts to picking some basis as an orientation frame.

For $E = \mathbb{R}$, an orientation is given by e_1 or $-e_1$. Such an orientation is visualized as either right or left translation from the origin. For $E = \mathbb{R}^2$, an orientation is given by (e_1, e_2) or (e_2, e_1), i.e., either counterclockwise or clockwise rotation about the origin. For $E = \mathbb{R}^3$, the orientation is represented by (e_1, e_2, e_3) or (e_2, e_1, e_3), namely the right-handed or left-handed orientation of the i, j, k axis system. See Figure 6.1.

Definition 6.2 is perfectly fine, but it turns out that it is more convenient, in the long term, to use a definition of orientation in terms of differential forms and the exterior algebra $\bigwedge^n E^*$. This approach is especially useful when defining the notion of integration on a manifold. We observe that two bases $u_1, \ldots, u_n$ and $v_1, \ldots, v_n$ have the same orientation iff

$$\omega(u_1, \ldots, u_n) \quad \text{and} \quad \omega(v_1, \ldots, v_n) \quad \text{have the same sign for all } \omega \in \bigwedge^n E^* - \{0\}$$

(where 0 denotes the zero n-form). As $\bigwedge^n E^*$ is one dimensional, picking an orientation of E is equivalent to picking a generator (a one-element basis) ω of $\bigwedge^n E^*$, and to say that $u_1, \ldots, u_n$ has positive orientation iff $\omega(u_1, \ldots, u_n) > 0$.

Definition 6.3. Let E be a vector space of dimension n. Given an orientation (say, given by $\omega \in \bigwedge^n E^*$) of E, a linear map $f : E \to E$ is *orientation preserving*

Fig. 6.1 The two orientations of $\mathbb{R}$, $\mathbb{R}^2$, and $\mathbb{R}^3$.

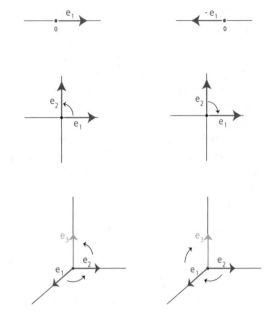

iff $\omega(f(u_1), \ldots, f(u_n)) > 0$ whenever $\omega(u_1, \ldots, u_n) > 0$ (or equivalently, iff $\det(f) > 0$).

To define the orientation of an n-dimensional manifold M we use charts. Given any $p \in M$, for any chart (U, φ) at p, the tangent map $d\varphi^{-1}_{\varphi(p)} \colon \mathbb{R}^n \to T_p M$ makes sense. If $(e_1, \ldots, e_n)$ is the standard basis of $\mathbb{R}^n$, as it gives an orientation to $\mathbb{R}^n$, we can orient $T_p M$ by giving it the orientation induced by the basis $d\varphi^{-1}_{\varphi(p)}(e_1), \ldots, d\varphi^{-1}_{\varphi(p)}(e_n)$. The consistency of orientations of the $T_p M$'s is given by the overlapping of charts. See Figure 6.2.

We require that the Jacobian determinants of all $\varphi_j \circ \varphi_i^{-1}$ have the same sign whenever (U_i, φ_i) and (U_j, φ_j) are any two overlapping charts. Thus, we are led to the definition below. All definitions and results stated in the rest of this section apply to manifolds with or without boundary.

Definition 6.4. Given a smooth manifold M of dimension n, an *orientation atlas* of M is any atlas so that the transition maps $\varphi_i^j = \varphi_j \circ \varphi_i^{-1}$ (from $\varphi_i(U_i \cap U_j)$ to $\varphi_j(U_i \cap U_j)$) all have a positive Jacobian determinant for every point in $\varphi_i(U_i \cap U_j)$. A manifold is *orientable* iff it has some orientation atlas.

We should mention that not every manifold is orientable. The open Mobius strip, i.e., the Mobius strip with circle boundary removed, is not orientable, as demonstrated in Figure 6.3.

Definition 6.4 can be hard to check in practice and there is an equivalent criterion is terms of n-forms which is often more convenient. The idea is that a manifold of dimension n is orientable iff there is a map $p \mapsto \omega_p$, assigning to every point $p \in M$ a nonzero n-form $\omega_p \in \bigwedge^n T_p^* M$, so that this map is smooth.

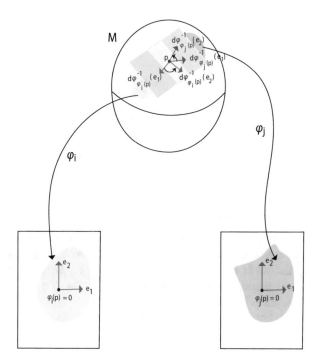

Fig. 6.2 The sphere S^2 with consistent orientation on two overlapping charts.

Fig. 6.3 The Mobius strip
does not have a consistent
orientation. The frame
starting at 1 is reversed when
traveling around the loop to
$1'$.

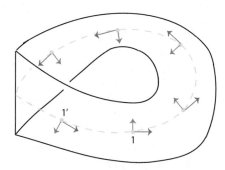

Definition 6.5. If M is an n-dimensional manifold, recall that a smooth section $\omega \in \Gamma(M, \bigwedge^n T^*M)$ is called a (smooth) *n-form*. An *n*-form ω is a *nowhere-vanishing n-form on M* or *volume form on M* iff ω_p is a nonzero form for every $p \in M$. This is equivalent to saying that $\omega_p(u_1, \ldots, u_n) \neq 0$, for all $p \in M$ and all bases $u_1, \ldots, u_n$, of T_pM.

The determinant function $(u_1, \ldots, u_n) \mapsto \det(u_1, \ldots, u_n)$ where the u_i are expressed over the canonical basis $(e_1, \ldots, e_n)$ of $\mathbb{R}^n$, is a volume form on $\mathbb{R}^n$. We will denote this volume form by $\omega_{\mathbb{R}^n} = dx_1 \wedge \cdots \wedge dx_n$. Observe the justification for the term volume form: the quantity $\det(u_1, \ldots, u_n)$ is indeed the (signed) volume of the parallelepiped

$$\{\lambda_1 u_1 + \cdots + \lambda_n u_n \mid 0 \le \lambda_i \le 1,\ 1 \le i \le n\}.$$

A volume form on the sphere $S^n \subseteq \mathbb{R}^{n+1}$ is obtained as follows:

$$\omega_{S^n}(u_1, \ldots u_n) = \det(p, u_1, \ldots u_n),$$

where $p \in S^n$ and $u_1, \ldots u_n \in T_p S^n$. As the u_i are orthogonal to p, this is indeed a volume form.

Observe that if f is a smooth function on M and ω is any n-form, then $f\omega$ is also an n-form.

More interesting is the following proposition.

Proposition 6.1. *(a) If $h: M \to N$ is a local diffeomorphism of manifolds, where $\dim M = \dim N = n$, and $\omega \in \mathcal{A}^n(N)$ is a volume form on N, then $h^*\omega$ is a volume form on M. (b) Assume M has a volume form ω. For every n-form $\eta \in \mathcal{A}^n(M)$, there is a unique smooth function $f \in C^\infty(M)$ so that $\eta = f\omega$. If η is a volume form, then $f(p) \ne 0$ for all $p \in M$.*

Proof.

(a) By definition,

$$h^*\omega_p(u_1, \ldots, u_n) = \omega_{h(p)}(dh_p(u_1), \ldots, dh_p(u_n)),$$

for all $p \in M$ and all $u_1, \ldots, u_n \in T_p M$. As h is a local diffeomorphism, $d_p h$ is a bijection for every p. Thus, if $u_1, \ldots, u_n$ is a basis, then so is $dh_p(u_1), \ldots, dh_p(u_n)$, and as ω is nonzero at every point for every basis, $h^*\omega_p(u_1, \ldots, u_n) \ne 0$.

(b) Pick any $p \in M$ and let (U, φ) be any chart at p. As φ is a diffeomorphism, by (a), we see that $\varphi^{-1*}\omega$ is a volume form on $\varphi(U)$. But then, it is easy to see that $\varphi^{-1*}\eta = g\varphi^{-1*}\omega$, for some unique smooth function g on $\varphi(U)$, and so $\eta = f_U\omega$, for some unique smooth function f_U on U. For any two overlapping charts (U_i, φ_i) and (U_j, φ_j), for every $p \in U_i \cap U_j$, for every basis $u_1, \ldots, u_n$ of $T_p M$, we have

$$\eta_p(u_1, \ldots, u_n) = f_i(p)\omega_p(u_1, \ldots, u_n) = f_j(p)\omega_p(u_1, \ldots, u_n),$$

and as $\omega_p(u_1, \ldots, u_n) \ne 0$, we deduce that f_i and f_j agree on $U_i \cap U_j$. But then the f_i's patch on the overlaps of the cover $\{U_i\}$ of M, and so there is a smooth function f defined on the whole of M and such that $f \restriction U_i = f_i$. As the f_i's are unique, so is f. If η is a volume form, then η_p does not vanish for all $p \in M$, and since ω_p is also a volume form, ω_p does not vanish for all $p \in M$, so $f(p) \ne 0$ for all $p \in M$. □

Remark. If h_1 and h_2 are smooth maps of manifolds, it is easy to prove that

$$(h_2 \circ h_1)^* = h_1^* \circ h_2^*,$$

and that for any smooth map $h \colon M \to N$,

$$h^*(f\omega) = (f \circ h)h^*\omega,$$

where f is any smooth function on N and ω is any n-form on N.

The connection between Definition 6.4 and volume forms is given by the following important theorem whose proof contains a wonderful use of partitions of unity.

Theorem 6.2. *A smooth manifold (Hausdorff and second-countable) is orientable iff it possesses a volume form.*

Proof. First assume that a volume form ω exists on M, and say $n = \dim M$. For any atlas $\{(U_i, \varphi_i)\}_i$ of M, by Proposition 6.1, each n-form $\varphi_i^{-1*}\omega$ is a volume form on $\varphi_i(U_i) \subseteq \mathbb{R}^n$, and

$$\varphi_i^{-1*}\omega = f_i \omega_{\mathbb{R}^n},$$

for some smooth function f_i never zero on $\varphi_i(U_i)$, where $\omega_{\mathbb{R}^n}$ is the volume form on $\mathbb{R}^n$. By composing φ_i with an orientation-reversing linear map if necessary, we may assume that for this new atlas, $f_i > 0$ on $\varphi_i(U_i)$. We claim that the family $(U_i, \varphi_i)_i$ is an orientation atlas. This is because, on any (nonempty) overlap $U_i \cap U_j$, as $\omega = \varphi_j^*(f_j \omega_{\mathbb{R}^n})$ and $(\varphi_j \circ \varphi_i^{-1})^* = (\varphi_i^{-1})^* \circ \varphi_j^*$, we have

$$(\varphi_j \circ \varphi_i^{-1})^*(f_j \omega_{\mathbb{R}^n}) = (\varphi_i^{-1})^* \circ \varphi_j^*(f_j \omega_{\mathbb{R}^n}) = (\varphi_i^{-1})^*\omega = f_i \omega_{\mathbb{R}^n},$$

and by the definition of pull-backs, we see that for every $x \in \varphi_i(U_i \cap U_j)$, if we let $y = \varphi_j \circ \varphi_i^{-1}(x)$, then

$$f_i(x)(\omega_{\mathbb{R}^n})_x(e_1, \ldots, e_n) = (\varphi_j \circ \varphi_i^{-1})_x^*(f_j \omega_{\mathbb{R}^n})(e_1, \ldots, e_n)$$

$$= f_j(y)(\omega_{\mathbb{R}^n})_y(d(\varphi_j \circ \varphi_i^{-1})_x(e_1), \ldots, d(\varphi_j \circ \varphi_i^{-1})_x(e_n))$$

$$= f_j(y)J((\varphi_j \circ \varphi_i^{-1})_x)(\omega_{\mathbb{R}^n})_y(e_1, \ldots, e_n),$$

where $e_1, \ldots, e_n$ is the standard basis of $\mathbb{R}^n$ and $J((\varphi_j \circ \varphi_i^{-1})_x)$ is the Jacobian determinant of $\varphi_j \circ \varphi_i^{-1}$ at x. As both $f_j(y) > 0$ and $f_i(x) > 0$, we have $J((\varphi_j \circ \varphi_i^{-1})_x) > 0$, as desired.

Conversely, assume that $J((\varphi_j \circ \varphi_i^{-1})_x) > 0$, for all $x \in \varphi_i(U_i \cap U_j)$, whenever $U_i \cap U_j \neq \emptyset$. We need to make a volume form on M. For each U_i, let

$$\omega_i = \varphi_i^* \omega_{\mathbb{R}^n},$$

where $\omega_{\mathbb{R}^n}$ is the volume form on $\mathbb{R}^n$. As φ_i is a diffeomorphism, by Proposition 6.1, we see that ω_i is a volume form on U_i. Then if we apply Theorem 1.11 from Chapter 1 of Warner [109], we can find a partition of unity $\{f_i\}$ subordinate to the cover $\{U_i\}$, with the same index set. Let,

$$\omega = \sum_i f_i \omega_i.$$

We claim that ω is a volume form on M.

It is clear that ω is an n-form on M. Now since every $p \in M$ belongs to some U_i, check that on $\varphi_i(U_i)$, we have

$$
\begin{aligned}
\varphi_i^{-1*} \omega &= \sum_{j \in \text{finite set}} \varphi_i^{-1*}(f_j \omega_j) \\
&= \sum_{j \in \text{finite set}} \varphi_i^{-1*}(f_j \varphi_j^* \omega_{\mathbb{R}^n}) \\
&= \sum_{j \in \text{finite set}} (f_j \circ \varphi_i^{-1})(\varphi_i^{-1*} \circ \varphi_j^*) \omega_{\mathbb{R}^n} \\
&= \sum_{j \in \text{finite set}} (f_j \circ \varphi_i^{-1})(\varphi_j \circ \varphi_i^{-1})^* \omega_{\mathbb{R}^n} \\
&= \left(\sum_{j \in \text{finite set}} (f_j \circ \varphi_i^{-1}) J(\varphi_j \circ \varphi_i^{-1}) \right) \omega_{\mathbb{R}^n},
\end{aligned}
$$

and this sum is strictly positive because the Jacobian determinants are positive, and as $\sum_j f_j = 1$ and $f_j \geq 0$, some term must be strictly positive. Therefore, $\varphi_i^{-1*} \omega$ is a volume form on $\varphi_i(U_i)$, so $\varphi_i^* \varphi_i^{-1*} \omega = \omega$ is a volume form on U_i. As this holds for all U_i, we conclude that ω is a volume form on M.   □

Since we showed there is a volume form on the sphere S^n, by Theorem 6.2, the sphere S^n is orientable. It can be shown that the projective spaces $\mathbb{RP}^n$ are non-orientable iff n is even, and thus orientable iff n is odd. In particular, $\mathbb{RP}^2$ is not orientable. Also, even though M may not be orientable, its tangent bundle $T(M)$ is always orientable! (Prove it). It is also easy to show that if $f \colon \mathbb{R}^{n+1} \to \mathbb{R}$ is a smooth submersion, then $M = f^{-1}(0)$ is a smooth orientable manifold. Another nice fact is that every Lie group is orientable.

By Proposition 6.1 (b), given any two volume forms ω_1 and ω_2 on a manifold M, there is a function $f \colon M \to \mathbb{R}$ never 0 on M, such that $\omega_2 = f \omega_1$. This fact suggests the following definition.

Definition 6.6. Given an orientable manifold M, two volume forms ω_1 and ω_2 on M are *equivalent* iff $\omega_2 = f\omega_1$ for some smooth function $f: M \to \mathbb{R}$, such that $f(p) > 0$ for all $p \in M$. An *orientation of* M is the choice of some equivalence class of volume forms on M, and an *oriented manifold* is a manifold together with a choice of orientation. If M is a manifold oriented by the volume form ω, for every $p \in M$, a basis $(b_1, \ldots, b_n)$ of $T_p M$ is *positively oriented* iff $\omega_p(b_1, \ldots, b_n) > 0$, else it is *negatively oriented* (where $n = \dim(M)$).

If M is an orientable manifold, for any two volume forms ω_1 and ω_2 on M, as $\omega_2 = f\omega_1$ for some function f on M which is never zero, f has a constant sign on every connected component of M. Consequently, a connected orientable manifold has two orientations.

We will also need the notion of orientation-preserving diffeomorphism.

Definition 6.7. Let $h: M \to N$ be a diffeomorphism of oriented manifolds M and N, of dimension n, and say the orientation on M is given by the volume form ω_1 while the orientation on N is given by the volume form ω_2. We say that h is *orientation preserving* iff $h^*\omega_2$ determines the same orientation of M as ω_1.

Using Definition 6.7, we can define the notion of a positive atlas.

Definition 6.8. If M is a manifold oriented by the volume form ω, an atlas for M is *positive* iff for every chart (U, φ), the diffeomorphism $\varphi: U \to \varphi(U)$ is orientation preserving, where U has the orientation induced by M and $\varphi(U) \subseteq \mathbb{R}^n$ has the orientation induced by the standard orientation on $\mathbb{R}^n$ (with $\dim(M) = n$).

The proof of Theorem 6.2 shows the following.

Proposition 6.3. *If a manifold M has an orientation atlas, then there is a uniquely determined orientation on M such that this atlas is positive.*

6.2   Volume Forms on Riemannian Manifolds and Lie Groups

Recall that a smooth manifold M is a *Riemannian manifold* iff the vector bundle TM has a Euclidean metric. This means that there is a family $(\langle -, - \rangle_p)_{p \in M}$ of inner products on the tangent spaces $T_p M$, such that $\langle -, - \rangle_p$ depends smoothly on p, which can be expressed by saying that the maps

$$x \mapsto \langle d\varphi_x^{-1}(e_i), d\varphi_x^{-1}(e_j) \rangle_{\varphi^{-1}(x)}, \qquad x \in \varphi(U), \ 1 \leq i, j \leq n$$

are smooth, for every chart (U, φ) of M, where $(e_1, \ldots, e_n)$ is the canonical basis of $\mathbb{R}^n$. We let

$$g_{ij}(x) = \langle d\varphi_x^{-1}(e_i), d\varphi_x^{-1}(e_j) \rangle_{\varphi^{-1}(x)},$$

and we say that the $n \times n$ matrix $(g_{ij}(x))$ is the *local expression of the Riemannian metric on M at x* in the coordinate patch (U, φ).

If a Riemannian manifold M is orientable, then there is a volume form on M with some special properties.

Proposition 6.4. *Let M be a Riemannian manifold with $\dim(M) = n$. If M is orientable, then there is a uniquely determined volume form Vol_M on M with the following property: For every $p \in M$, for every positively oriented orthonormal basis $(b_1, \ldots, b_n)$ of T_pM, we have*

$$(\mathrm{Vol}_M)_p(b_1, \ldots, b_n) = 1.$$

Furthermore, if the above equation holds, then in every orientation preserving local chart (U, φ), we have

$$((\varphi^{-1})^* \mathrm{Vol}_M)_q = \sqrt{\det(g_{ij}(q))} \, dx_1 \wedge \cdots \wedge dx_n, \qquad q \in \varphi(U).$$

Proof. Say the orientation of M is given by $\omega \in \mathcal{A}^n(M)$. For any two positively oriented orthonormal bases $(b_1, \ldots, b_n)$ and $(b'_1, \ldots, b'_n)$ in T_pM, by expressing the second basis over the first, there is an orthogonal matrix $C = (c_{ij})$ so that

$$b'_i = \sum_{j=1}^{n} c_{ij} b_j.$$

We have

$$\omega_p(b'_1, \ldots, b'_n) = \det(C)\omega_p(b_1, \ldots, b_n),$$

and as these bases are positively oriented, we conclude that $\det(C) = 1$ (as C is orthogonal, $\det(C) = \pm 1$). As a consequence, we have a well-defined function $\rho \colon M \to \mathbb{R}$ with $\rho(p) > 0$ for all $p \in M$, such that

$$\rho(p) = \omega_p(b_1, \ldots, b_n)$$

for every positively oriented orthonormal basis $(b_1, \ldots, b_n)$ of T_pM. If we can show that ρ is smooth, then $(\mathrm{Vol}_M)_p = \frac{1}{\rho(p)}\omega_p$ is the required volume form.

Let (U, φ) be a positively oriented chart and consider the vector fields X_j on $\varphi(U)$ given by

$$X_j(q) = d\varphi_q^{-1}(e_j), \qquad q \in \varphi(U), \ 1 \le j \le n.$$

Then $(X_1(q), \ldots, X_n(q))$ is a positively oriented basis of $T_{\varphi^{-1}(q)}$. If we apply Gram-Schmidt orthogonalization, we get an upper triangular matrix $A(q) = (a_{ij}(q))$ of smooth functions on $\varphi(U)$ with $a_{ii}(q) > 0$, such that

$$b_i(q) = \sum_{j=1}^{n} a_{ij}(q) X_j(q), \qquad 1 \le i \le n,$$

and $(b_1(q), \ldots, b_n(q))$ is a positively oriented orthonormal basis of $T_{\varphi^{-1}(q)}$. We have

$$\begin{aligned}
\rho(\varphi^{-1}(q)) &= \omega_{\varphi^{-1}(q)}(b_1(q), \ldots, b_n(q)) \\
&= \det(A(q)) \omega_{\varphi^{-1}(q)}(X_1(q), \ldots, X_n(q)) \\
&= \det(A(q))(\varphi^{-1})^* \omega_q(e_1, \ldots, e_n),
\end{aligned}$$

which shows that ρ is smooth.

If we repeat the end of the proof with $\omega = \mathrm{Vol}_M$, then $\rho \equiv 1$ on M, and the above formula yields

$$((\varphi^{-1})^* \mathrm{Vol}_M)_q = (\det(A(q)))^{-1} dx_1 \wedge \cdots \wedge dx_n.$$

If we compute $\langle b_i(q), b_k(q) \rangle_{\varphi^{-1}(q)}$, we get

$$\delta_{ik} = \langle b_i(q), b_k(q) \rangle_{\varphi^{-1}(q)} = \sum_{j=1}^{n} \sum_{l=1}^{n} a_{ij}(q) g_{jl}(q) a_{kl}(q),$$

and so $I = A(q) G(q) A(q)^\top$, where $G(q) = (g_{jl}(q))$. Thus, $(\det(A(q)))^2 \det(G(q)) = 1$, and since $\det(A(q)) = \prod_i a_{ii}(q) > 0$, we conclude that

$$(\det(A(q)))^{-1} = \sqrt{\det(g_{ij}(q))},$$

which proves the second formula. $\square$

We saw in Section 6.1 that a volume form ω_{S^n} on the sphere $S^n \subseteq \mathbb{R}^{n+1}$ is given by

$$(\omega_{S_n})_p(u_1, \ldots u_n) = \det(p, u_1, \ldots u_n),$$

where $p \in S^n$ and $u_1, \ldots u_n \in T_p S^n$. To be more precise, we consider the n-form $\tilde{\omega}_{\mathbb{R}^n} \in \mathcal{A}^n(\mathbb{R}^{n+1})$ given by the above formula. As

$$(\tilde{\omega}_{\mathbb{R}^n})_p(e_1, \ldots, \widehat{e_i}, \ldots, e_{n+1}) = \det(p, e_1, \ldots, \widehat{e_i}, \ldots, e_{n+1}) = (-1)^{i-1} p_i,$$

where $p = (p_1, \ldots, p_{n+1})$, we have

$$(\tilde{\omega}_{\mathbb{R}^n})_p = \sum_{i=1}^{n+1} (-1)^{i-1} p_i \, dx_1 \wedge \cdots \wedge \widehat{dx_i} \wedge \cdots \wedge dx_{n+1}. \qquad (*)$$

Let $i: S^n \to \mathbb{R}^{n+1}$ be the inclusion map. For every $p \in S^n$ and every basis $(u_1, \ldots, u_n)$ of $T_p S^n$, the $(n+1)$-tuple $(p, u_1, \ldots, u_n)$ is a basis of $\mathbb{R}^{n+1}$, and so $(\tilde{\omega}_{\mathbb{R}^n})_p \neq 0$. Hence, $\tilde{\omega}_{\mathbb{R}^n} \upharpoonright S^n = i^* \tilde{\omega}_{\mathbb{R}^n}$ is a volume form on S^n. If we give S^n the Riemannian structure induced by $\mathbb{R}^{n+1}$, then the discussion above shows that

$$\text{Vol}_{S^n} = \tilde{\omega}_{\mathbb{R}^n} \upharpoonright S^n.$$

To obtain another representation for Vol_{S^n}, let $r: \mathbb{R}^{n+1} - \{0\} \to S^n$ be the map given by

$$r(x) = \frac{x}{\|x\|},$$

and set

$$\omega = r^* \text{Vol}_{S^n},$$

a closed n-form on $\mathbb{R}^{n+1} - \{0\}$. Clearly,

$$\omega \upharpoonright S^n = \text{Vol}_{S^n}.$$

Furthermore

$$\omega_x(u_1, \ldots, u_n) = (\tilde{\omega}_{\mathbb{R}^n})_{r(x)}(dr_x(u_1), \ldots, dr_x(u_n))$$
$$= \|x\|^{-1} \det(x, dr_x(u_1), \ldots, dr_x(u_n)).$$

We leave it as an exercise to prove that ω is given by

$$\omega_x = \frac{1}{\|x\|^n} \sum_{i=1}^{n+1} (-1)^{i-1} x_i \, dx_1 \wedge \cdots \wedge \widehat{dx_i} \wedge \cdots \wedge dx_{n+1}.$$

The procedure used to construct Vol_{S^n} can be generalized to any n-dimensional orientable manifold embedded in $\mathbb{R}^m$. Let U be an open subset of $\mathbb{R}^n$ and $\psi: U \to M \subseteq \mathbb{R}^m$ be an orientation-preserving parametrization. Assume that $x_1, x_2, \ldots, x_m$ are the coordinates of $\mathbb{R}^m$ (the ambient coordinates of M) and that $u_1, u_2, \ldots, u_n$ are the coordinates of U (the local coordinates of M). Let $x = \psi(u)$ be a point in M. Edwards [41] (Theorem 5.6) shows that

$$\text{Vol}_M = \sum_{\substack{(i_1, i_2, \ldots, i_n) \\ 1 \leq i_1 < i_2 < \cdots < i_n \leq m}} n_{i_1, i_2, \ldots, i_n} dx_{i_1} \wedge dx_{i_2} \wedge \cdots \wedge dx_{i_n}, \qquad (**)$$

where

$$n_{i_1, i_2, \cdots, i_n}(x) = \frac{1}{D} \frac{\partial(\psi_{i_1}, \psi_{i_2}, \ldots, \psi_{i_n})}{\partial(u_1, u_2, \ldots, u_n)}, \qquad D = [\det\left(J^\top(\psi)(u) J(\psi)(u)\right)]^{\frac{1}{2}}$$

and $\frac{\partial(\psi_{i_1}, \psi_{i_2}, \ldots, \psi_{i_n})}{\partial(u_1, u_2, \ldots, u_n)}$ is the determinant of the $n \times n$ matrix obtained by selecting rows i_1 through i_n of $d\psi_u$.

If M is a smooth orientable manifold of dimension $m - 1$, Edwards's formula for Vol_M reduces to

$$\mathrm{Vol}_M = \sum_{i=1}^{m} (-1)^{i-1} n_i \, dx_1 \wedge \cdots \wedge dx_{i-1} \wedge dx_{i+1} \wedge \cdots \wedge dx_m, \qquad (***)$$

where $n_i = n_i(x)$ is the ith component of the unit normal vector $N(x)$ on M given by

$$n_i(x) = \frac{(-1)^{i-1}}{D} \frac{\partial(\psi_1, \ldots, \hat{\psi}_i, \ldots, \psi_m)}{\partial(u_1, u_2, \ldots, u_{m-1})}.$$

In particular, if $M = S^n$ embedded in $\mathbb{R}^{n+1}$, for $p \in S^n$, $N(p) = (p_1, p_2, \ldots, p_{n+1})$ and $(***)$ becomes $(*)$.

For a particular example of $(**)$, let $M = S^2$ and $\psi : U \to S^2$ where

$$x = \sin\theta \cos\varphi, \qquad y = \sin\theta \sin\varphi, \qquad z = \cos\theta.$$

and $U = \{(\theta, \varphi) : 0 < \theta < \pi, \ 0 < \varphi < 2\pi\} \subset \mathbb{R}^2$. See Figure 4.1. Clearly

$$J(\psi)(\theta, \varphi) = \begin{pmatrix} \cos\theta \cos\varphi & -\sin\theta \sin\varphi \\ \cos\theta \sin\varphi & \sin\theta \cos\varphi \\ -\sin\theta & 0 \end{pmatrix},$$

which in turn implies

$$D = [\det\left(J^\top(\psi)(\theta, \varphi) J(\psi)(\theta, \varphi)\right)]^{\frac{1}{2}} = \left[\det\begin{pmatrix} 1 & 0 \\ 0 & \sin^2\theta \end{pmatrix}\right]^{\frac{1}{2}} = \sin\theta.$$

Then

$$\mathrm{Vol}_{S^2} = n_{1,2} \, dx \wedge dy + n_{1,3} \, dx \wedge dz + n_{2,3} \, dy \wedge dz,$$

where

$$n_{1,2} = \frac{1}{\sin\theta} \frac{\partial(x,y)}{\partial(\theta,\varphi)} = \frac{1}{\sin\theta}\det\begin{pmatrix}\cos\theta\cos\varphi & -\sin\theta\sin\varphi \\ \cos\theta\sin\varphi & \sin\theta\cos\varphi\end{pmatrix} = \frac{\cos\theta\sin\theta}{\sin\theta}$$

$$= \cos\theta = z$$

$$n_{1,3} = \frac{1}{\sin\theta} \frac{\partial(x,z)}{\partial(\theta,\varphi)} = \frac{1}{\sin\theta}\det\begin{pmatrix}\cos\theta\cos\varphi & -\sin\theta\sin\varphi \\ -\sin\theta & 0\end{pmatrix} = \frac{-\sin^2\theta\sin\varphi}{\sin\theta}$$

$$= -\sin\theta\sin\varphi = -y$$

$$n_{2,3} = \frac{1}{\sin\theta} \frac{\partial(y,z)}{\partial(\theta,\varphi)} = \frac{1}{\sin\theta}\det\begin{pmatrix}\cos\theta\sin\varphi & \sin\theta\cos\varphi \\ -\sin\theta & 0\end{pmatrix} = \frac{\sin^2\theta\cos\varphi}{\sin\theta}$$

$$= \sin\theta\cos\varphi = x.$$

Thus

$$\mathrm{Vol}_{S^2} = n_{1,2}\, dx \wedge dy + n_{1,3}\, dx \wedge dz + n_{2,3}\, dy \wedge dz$$
$$= z\, dx \wedge dy - y\, dx \wedge dz + x\, dy \wedge dz,$$

which agrees with $(*)$ when $n = 2$.

We mention that the orientation of S^n provides a way of orienting projective spaces of *odd* dimension. We know that there is a map $\pi\colon S^n \to \mathbb{RP}^n$ such that $\pi^{-1}([p])$ consists of two antipodal points for every $[p] \in \mathbb{RP}^n$. It can be shown that there is a volume form on $\mathbb{RP}^n$ iff n is odd, in which case

$$\pi^*(\mathrm{Vol}_{\mathbb{RP}^n}) = \mathrm{Vol}_{S^n}.$$

Thus, $\mathbb{RP}^n$ is orientable iff n is odd.

We end this section with an important result regarding orientability of Lie groups. Let G be a Lie group of dimension n. For any basis $(\omega_1, \ldots, \omega_n)$ of the dual $\mathfrak{g}^*$ of the Lie algebra $\mathfrak{g}$ of G, we have the left-invariant one-forms defined by the ω_i, also denoted ω_i, and obviously $(\omega_1, \ldots, \omega_n)$ is a frame for T^*G. Therefore, $\omega = \omega_1 \wedge \cdots \wedge \omega_n$ is an n-form on G that is never zero; that is, a volume form. Since pull-back commutes with $\wedge$, the n-form ω is left-invariant. We summarize this as follows.

Proposition 6.5. *Every Lie group G possesses a left-invariant volume form. Therefore, every Lie group is orientable.*

6.3   Integration in $\mathbb{R}^n$

As we said in Section 4.1, one of the *raison d'être* for differential forms is that they are the objects that can be integrated on manifolds. We will be integrating differential forms that are at least continuous (in most cases, smooth) and with compact support. In the case of forms ω on $\mathbb{R}^n$, this means that the closure of the set $\{x \in \mathbb{R}^n \mid \omega_x \neq 0\}$ is compact. Similarly, for a form $\omega \in \mathcal{A}^*(M)$ where M is a manifold, the support $\text{supp}_M(\omega)$ of ω is the closure of the set $\{p \in M \mid \omega_p \neq 0\}$. We let $\mathcal{A}_c^*(M)$ denote the set of differential forms with compact support on M. If M is a smooth manifold of dimension n, our ultimate goal is to define a linear operator

$$\int_M : \mathcal{A}_c^n(M) \longrightarrow \mathbb{R}$$

which generalizes in a natural way the usual integral on $\mathbb{R}^n$.

In this section we assume that $M = \mathbb{R}^n$ or $M = U$ for some open subset U of $\mathbb{R}^n$. Now every n-form (with compact support) on $\mathbb{R}^n$ is given by

$$\omega_x = f(x)\, dx_1 \wedge \cdots \wedge dx_n,$$

where f is a smooth function with compact support. Thus, we set

$$\int_{\mathbb{R}^n} \omega = \int_{\mathbb{R}^n} f(x) dx_1 \cdots dx_n,$$

where the expression on the right-hand side is the usual Riemann integral of f on $\mathbb{R}^n$. For the reader who would like to review the definition of the Riemann integral, we suggest Sections 23.1 to 23.3 of [105] and Sections 4.1 to 4.3 of [41]. Actually we will need to integrate smooth forms $\omega \in \mathcal{A}_c^n(U)$ with compact support defined on some open subset $U \subseteq \mathbb{R}^n$ (with $\text{supp}(\omega) \subseteq U$). However, this is not a problem since we still have

$$\omega_x = f(x)\, dx_1 \wedge \cdots \wedge dx_n,$$

where $f : U \to \mathbb{R}$ is a smooth function with compact support contained in U, and f can be smoothly extended to $\mathbb{R}^n$ by setting it to 0 on $\mathbb{R}^n - \text{supp}(f)$. We write $\int_U \omega$ for this integral and make the following definition.

Definition 6.9. Let U be an open subset of $\mathbb{R}^n$ and let $\mathcal{A}_c^n(U)$ denote the set of smooth n-forms with compact support contained in U. In other words $\omega \in \mathcal{A}_c^n(U)$ if and only if $\omega_x = f(x)\, dx_1 \wedge \cdots \wedge dx_n$ for some smooth function $f : U \to \mathbb{R}$ and the closure of $\{x \in \mathbb{R}^n \mid \omega_x \neq 0\}$ is a compact set of $\mathbb{R}^n$ contained in U. For $\omega \in \mathcal{A}_c^n(U)$, the expression $\int_U \omega$ is defined as

$$\int_U \omega = \int_U f(x) dx_1 \cdots dx_n, \qquad (*)$$

where the right side of (*) is interpreted as the Riemann integral.

In Definition 6.9, the n-form must be represented as $dx_1 \wedge \cdots \wedge dx_n$. This is not a problem since Proposition 4.1 says that we may switch order within the wedge product by adjusting the functional coefficient with the appropriate negative signs. For example, if $\omega_x = f(x)\, dx_1 \wedge dx_3 \wedge dx_2$, Definition 6.9 implies that

$$\int_U \omega = \int_U f(x)\, dx_1 \wedge dx_3 \wedge dx_2 = -\int_U f(x)\, dx_1\, dx_2\, dx_3.$$

For this reason, $\int_U \omega$ is often called a "signed" integral.

It is crucial for the generalization of the integral to manifolds to see what the change of variable formula looks like in terms of differential forms.

Proposition 6.6. *Let $\varphi \colon U \to V$ be a diffeomorphism between two open subsets of $\mathbb{R}^n$. If the Jacobian determinant $J(\varphi)(x)$ has a constant sign $\delta = \pm 1$ on U, then for every $\omega \in \mathcal{A}_c^n(V)$, we have*

$$\int_U \varphi^*\omega = \delta \int_V \omega.$$

Proof. We know that ω can be written as

$$\omega_x = f(x)\, dx_1 \wedge \cdots \wedge dx_n, \qquad x \in V,$$

where $f \colon V \to \mathbb{R}$ has compact support. From the example after Proposition 4.9 we have

$$(\varphi^*\omega)_y = f(\varphi(y)) J(\varphi)_y\, dy_1 \wedge \cdots \wedge dy_n$$
$$= \delta f(\varphi(y)) |J(\varphi)_y|\, dy_1 \wedge \cdots \wedge dy_n.$$

On the other hand, the change of variable formula (using φ) is

$$\int_{\varphi(U)} f(x)\, dx_1 \cdots dx_n = \int_U f(\varphi(y))\, |J(\varphi)_y|\, dy_1 \cdots dy_n,$$

so the formula follows. $\square$

We will promote the integral on open subsets of $\mathbb{R}^n$ to manifolds using partitions of unity.

6.4 Integration on Manifolds

Definition 6.10. Let M be an oriented manifold of dimension n. We say ω is a smooth n-form on M with compact support if the closure of $\{p \in M \mid \omega_p \neq 0\}$

is compact in M. We denote $\{p \in M \mid \omega_p \neq 0\}$ by $\mathrm{supp}(\omega)$. The set of smooth n-forms on M with compact support is denoted $\mathcal{A}_c^n(M)$ while $\mathcal{A}_c^*(M)$ is the set of all smooth differential forms on M with compact support.

Intuitively, for any n-form $\omega \in \mathcal{A}_c^n(M)$ on a smooth n-dimensional oriented manifold M, the integral $\int_M \omega$ is computed by patching together the integrals defined on small-enough open subsets covering M using a partition of unity. If (U, φ) is a chart such that $\mathrm{supp}(\omega) \subseteq U$, then the form $(\varphi^{-1})^*\omega$ is an n-form on $\mathbb{R}^n$, and the integral $\int_{\varphi(U)} (\varphi^{-1})^*\omega$ makes sense. The orientability of M is needed to ensure that the above integrals have a consistent value on overlapping charts.

Proposition 6.7. *Let M be a smooth oriented manifold of dimension n. There exists a unique linear operator*

$$\int_M : \mathcal{A}_c^n(M) \longrightarrow \mathbb{R}$$

with the following property: For any $\omega \in \mathcal{A}_c^n(M)$, if $\mathrm{supp}(\omega) \subseteq U$, where (U, φ) is a positively oriented chart, then

$$\int_M \omega = \int_{\varphi(U)} (\varphi^{-1})^*\omega. \tag{\dagger}$$

Proof. First, assume that $\mathrm{supp}(\omega) \subseteq U$, where (U, φ) is a positively oriented chart. Then we wish to set

$$\int_M \omega = \int_{\varphi(U)} (\varphi^{-1})^*\omega.$$

However, we need to prove that the above expression does not depend on the choice of the chart. Let (V, ψ) be another chart such that $\mathrm{supp}(\omega) \subseteq V$, so that $\mathrm{supp}(\omega) \subseteq U \cap V$. The map $\theta = \psi \circ \varphi^{-1}$ is a diffeomorphism from $W = \varphi(U \cap V)$ to $W' = \psi(U \cap V)$, and by hypothesis, its Jacobian determinant is positive on W. Since

$$\mathrm{supp}_{\varphi(U)}((\varphi^{-1})^*\omega) \subseteq W, \qquad \mathrm{supp}_{\psi(V)}((\psi^{-1})^*\omega) \subseteq W',$$

and $\theta^* \circ (\psi^{-1})^*\omega = (\varphi^{-1})^* \circ \psi^* \circ (\psi^{-1})^*\omega = (\varphi^{-1})^*\omega$, Proposition 6.6 yields

$$\int_{W'} (\psi^{-1})^*\omega = \int_W \theta^*((\psi^{-1})^*\omega) = \int_W (\varphi^{-1})^*\omega,$$

as claimed.

In the general case, using a partition of unity, for every open cover of M by positively oriented charts (U_i, φ_i), we have a partition of unity $(\rho_i)_{i \in I}$ subordinate to this cover. Recall that

$$\text{supp}(\rho_i) \subseteq U_i, \qquad i \in I.$$

Thus, $\rho_i \omega$ is an n-form whose support is a subset of U_i. Furthermore, as $\sum_i \rho_i = 1$,

$$\omega = \sum_i \rho_i \omega.$$

Define

$$I(\omega) = \sum_i \int_{U_i} \rho_i \omega,$$

where each term in the sum is defined by

$$\int_{U_i} \rho_i \omega = \int_{\varphi_i(U_i)} (\varphi_i^{-1})^* \rho_i \omega,$$

where (U_i, φ_i) is the chart associated with $i \in I$.

It remains to show that $I(\omega)$ does not depend on the choice of open cover and on the choice of partition of unity. Let (V_j, ψ_j) be another open cover by positively oriented charts, and let $(\theta_j)_{j \in J}$ be a partition of unity subordinate to the open cover (V_j). Note that

$$\int_{U_i} \rho_i \theta_j \omega = \int_{V_j} \rho_i \theta_j \omega,$$

since $\text{supp}(\rho_i \theta_j \omega) \subseteq U_i \cap V_j$, and as $\sum_i \rho_i = 1$ and $\sum_j \theta_j = 1$, we have

$$\sum_i \int_{U_i} \rho_i \omega = \sum_{i,j} \int_{U_i} \rho_i \theta_j \omega = \sum_{i,j} \int_{V_j} \rho_i \theta_j \omega = \sum_j \int_{V_j} \theta_j \omega,$$

proving that $I(\omega)$ is indeed independent of the open cover and of the partition of unity. The uniqueness assertion is easily proved using a partition of unity. $\quad\square$

Since the integral at (†) is well defined we are able to make the following definition.

Definition 6.11. Let M be a smooth oriented manifold of dimension n. For $\omega \in \mathcal{A}_c^n(M)$, if (U, φ) is a positively oriented chart, and $\text{supp}(\omega) \subseteq U$, we define $\int_M \omega$ by

$$\int_M \omega = \int_{\varphi(U)} (\varphi^{-1})^* \omega.$$

Given an embedded manifold M in $\mathbb{R}^n$, Definition 6.11 shows that integration of a form over a manifold reduces, after a change of variables, to an appropriate Riemann integral over the parameter space. We will demonstrate the meaning of this sentence by explicitly calculating $\int_{S_2} \text{Vol}_{S_2}$. In Section 6.2 we described a parametrization of S^2 by $\psi : U \to S^2$ where

$$x = \sin\theta\cos\varphi, \qquad y = \sin\theta\sin\varphi, \qquad z = \cos\theta,$$

and $U = \{(\theta,\varphi) : 0 < \theta < \pi,\ 0 < \varphi < 2\pi\} \subset \mathbb{R}^2$. See Figure 4.1. We then showed that

$$\text{Vol}_{S_2} = z\,dx \wedge dy - y\,dx \wedge dz + x\,dy \wedge dz.$$

To calculate $\int_{S_2} \text{Vol}_{S_2}$, we first use (†) of Section 4.3 to calculate

$$
\begin{aligned}
\psi^*(z\,dx \wedge dy) &= \cos\theta(d(\sin\theta\cos\varphi) \wedge d(\sin\theta\sin\varphi)) \\
&= \cos\theta\,((\cos\theta\cos\varphi\,d\theta - \sin\theta\sin\varphi\,d\varphi) \\
&\quad \wedge(\cos\theta\sin\varphi\,d\theta + \sin\theta\cos\varphi\,d\varphi)) \\
&= \cos\theta(\cos^2\varphi\cos\theta\sin\theta + \sin^2\varphi\sin\theta\cos\theta)d\theta \wedge d\varphi \\
&= \cos^2\theta\sin\theta\,d\theta \wedge d\varphi \\
\psi^*(-y\,dx \wedge dz) &= -\sin\theta\sin\varphi\,(d(\sin\theta\cos\varphi) \wedge d(\cos\theta)) \\
&= -\sin\theta\sin\varphi\,((\cos\theta\cos\varphi\,d\theta - \sin\theta\sin\varphi\,d\varphi) \wedge -\sin\theta\,d\theta) \\
&= \sin^3\theta\sin^2\varphi\,d\theta \wedge d\varphi \\
\psi^*(x\,dy \wedge dz) &= \sin\theta\cos\varphi\,(d(\sin\theta\sin\varphi) \wedge d(\cos\theta)) \\
&= \sin\theta\cos\varphi\,((\cos\theta\sin\varphi\,d\theta + \sin\theta\cos\varphi\,d\varphi) \wedge -\sin\theta\,d\theta) \\
&= \sin^3\theta\cos^2\varphi\,d\theta \wedge d\varphi.
\end{aligned}
$$

Then

$$
\begin{aligned}
\varphi^*(\text{Vol}_{S^2}) &= (\cos^2\theta\sin\theta + \sin^3\theta\sin^2\varphi + \sin^3\theta\cos^2\varphi)d\theta \wedge d\varphi \\
&= (\cos^2\theta\sin\theta + \sin^3\theta)d\theta \wedge d\varphi = \sin\theta(\cos^2\theta + \sin^2\theta)d\theta \wedge d\varphi \\
&= \sin\theta\,d\theta \wedge d\varphi,
\end{aligned}
$$

and Line (†) implies that

$$\int_{S^2}(\text{Vol}_{S^2}) = \int_0^{2\pi}\int_0^{\pi} \varphi^*(\text{Vol}_{S^2}) = \int_0^{2\pi}\int_0^{\pi} \sin\theta\,d\theta\,d\varphi = 2\pi\,[-\cos\theta]_0^{\pi} = 4\pi.$$

Observe that 4π is indeed the surface area of S^2, a result we should have expected since we were integrating the volume form.

The integral of Definition 6.11 has the following properties:

Proposition 6.8. *Let M be an oriented manifold of dimension n. The following properties hold:*

(1) If M is connected, then for every n-form $\omega \in \mathcal{A}_c^n(M)$, the sign of $\int_M \omega$ changes when the orientation of M is reversed.

(2) For every n-form $\omega \in \mathcal{A}_c^n(M)$, if $\mathrm{supp}(\omega) \subseteq W$ for some open subset W of M, then

$$\int_M \omega = \int_W \omega,$$

where W is given the orientation induced by M.

(3) If $\varphi \colon M \to N$ is an orientation-preserving diffeomorphism, then for every $\omega \in \mathcal{A}_c^n(N)$, we have

$$\int_N \omega = \int_M \varphi^* \omega.$$

Proof. Use a partition of unity to reduce to the case where $\mathrm{supp}(\omega)$ is contained in the domain of a chart, and then use Proposition 6.6 and (†) from Proposition 6.7.

$\square$

It is also possible to define integration on non-orientable manifolds using densities. The next section will not be used anywhere else in this book and can be omitted.

6.5 Densities ⊛

Definition 6.12. Given a vector space V of dimension $n \geq 1$, a *density* on V is a function $\mu \colon V^n \to \mathbb{R}$ such that for every linear map $f \colon V \to V$, we have

$$\mu(f(v_1), \ldots, f(v_n)) = |\det(f)| \mu(v_1, \ldots, v_n)$$

for all $v_1, \ldots, v_n \in V$.

If $(v_1, \ldots, v_n)$ are linearly dependent, then for any basis $(e_1, \ldots, e_n)$ of V there is a unique linear map f such that $f(e_i) = v_i$ for $i = 1, \ldots, n$, and since $(v_1, \ldots, v_n)$ are linearly dependent, f is singular so $\det(f) = 0$, which implies that

$$\mu(v_1, \ldots, v_n) = |\det(f)| \mu(e_1, \ldots, e_n) = 0$$

for any linearly dependent vectors $v_1, \ldots, v_n \in V$.

In view of this fact, a density is sometimes defined as a function $\mu : \bigwedge^n V \to \mathbb{R}$ such that for every automorphism $f \in \mathbf{GL}(V)$,

$$\mu(f(v_1) \wedge \cdots \wedge f(v_n)) = |\det(f)|\mu(v_1 \wedge \cdots \wedge v_n) \qquad (\dagger\dagger)$$

for all $v_1 \wedge \cdots \wedge v_n \in V$ (with $\mu(0) = 0$). For any nonzero $v_1 \wedge \cdots \wedge v_n, w_1 \wedge \cdots \wedge w_n \in \bigwedge^n V$, because

$$w_1 \wedge \cdots \wedge w_n = \det(P)v_1 \wedge \cdots \wedge v_n,$$

where P is the matrix whose jth column consists of the coefficients of w_j over the basis $(v_1, \ldots, v_n)$, it is not hard to show that Condition $(\dagger\dagger)$ is equivalent to

$$\mu(cw) = |c|\mu(w), \quad w \in \bigwedge^n V, \ c \in \mathbb{R}.$$

Densities are not multilinear, but it is not hard to show that for any fixed n, they form a vector space of dimension 1 which is spanned by the absolute value $|\omega|$ of any nonzero n-form $\omega \in \bigwedge^n V^*$. Let $\mathrm{den}(V)$ be the set of all densities on V. We have the following proposition from Lee [73] (Chapter 14, Proposition 14.26).

Proposition 6.9. *Let V be any vector space of dimension $n \geq 1$. The following properties hold:*

(a) *The set $\mathrm{den}(V)$ is a vector space.*
(b) *For any two densities $\mu_1, \mu_2 \in \mathrm{den}(V)$ and for any basis $(e_1, \ldots, e_n)$ of V, if $\mu_1(e_1, \ldots, e_n) = \mu_2(e_1, \ldots, e_n)$, then $\mu_1 = \mu_2$.*
(c) *For any n-form $\omega \in \bigwedge^n V^*$, the function $|\omega|$ given by*

$$|\omega|(v_1, \ldots, v_n) = |\omega(v_1, \ldots, v_n)|$$

   is a density.
(d) *The vector space $\mathrm{den}(V)$ is a one-dimensional space spanned by $|\omega|$ for any nonzero $\omega \in \bigwedge^n V^*$.*

Proof.

(a) That $\mathrm{den}(V)$ is a vector space is immediate from the definition.
(b) Pick any n vectors $(v_1, \ldots, v_n) \in V^n$. Since $(e_1, \ldots, e_n)$ is a basis of V, there is a unique linear map $f : V \to V$ such that $f(e_i) = v_i$ for $i = 1, \ldots, n$, and since by hypothesis $\mu_1(e_1, \ldots, e_n) = \mu_2(e_1, \ldots, e_n)$, we have

$$\begin{aligned}
\mu_1(v_1, \ldots, v_n) &= \mu_1(f(e_1), \ldots, f(e_n)) \\
&= |\det(f)|\mu_1(e_1, \ldots, e_n) \\
&= |\det(f)|\mu_2(e_1, \ldots, e_n)
\end{aligned}$$

$$= \mu_2(f(e_1), \ldots, f(e_n))$$
$$= \mu_2(v_1, \ldots, v_n),$$

which proves that $\mu_1 = \mu_2$.

(c) If $\omega \in \bigwedge^n V^*$, then

$$|\omega|(f(v_1), \ldots, f(v_n)) = |\omega(f(v_1), \ldots, f(v_n))|$$
$$= |\det(f)\omega(v_1, \ldots, v_n)|$$
$$= |\det(f)| \, |\omega|(v_1, \ldots, v_n),$$

which shows that $|\omega|$ is a density.

(d) Let $(e_1, \ldots, e_n)$ be any basis of V, and let $\omega \in \bigwedge^n V^*$ be any nonzero n-form. For any density μ, we need to show that $\mu = c|\omega|$ for some $c \in \mathbb{R}$. Let

$$a = |\omega|(e_1, \ldots, e_n) = |\omega(e_1, \ldots, e_n)|$$
$$b = \mu(e_1, \ldots, e_n).$$

Since $\omega \neq 0$ and $(e_1, \ldots, e_n)$ is a basis, $\omega(e_1, \ldots, e_n) \neq 0$ so $a \neq 0$, and by Condition (c) $(b/a)|\omega|$ is a density. Since

$$(b/a)|\omega|(e_1, \ldots, e_n) = b = \mu(e_1, \ldots, e_n),$$

by Condition (b) $\mu = (b/a)|\omega|$, as desired.   □

If we denote the vector space of densities on V by $\mathrm{den}(V)$, then given a manifold M, we can form the density bundle $\mathrm{den}(M)$ whose underlying set is the disjoint union of the vector spaces $\mathrm{den}(T_pM)$ for all $p \in M$. This set can be made into a smooth bundle, and a density on M is a smooth section of the density bundle. The main property of densities is that every smooth manifold admits a *global* smooth (positive) density, without any orientability assumptions. Then it is possible to carry out the theory of integration on manifolds using densities instead of volume forms, as we did in this section. This development can be found in Lee [73] (Chapter 14), but we have no need for this extra generality.

It turns out that orientations can be defined as certain functions satisfying a variant of the condition used in Definition 6.12, and this definition clarifies the relationship between volume forms and densities. The sign function is defined such that for any $\lambda \in \mathbb{R}$,

$$\mathrm{sign}(\lambda) = \begin{cases} +1 & \text{if } \lambda > 0 \\ -1 & \text{if } \lambda < 0 \\ 0 & \text{if } \lambda = 0. \end{cases}$$

Definition 6.13. Given a vector space V of dimension $n \geq 1$, an *orientation* on V is a function $o: V^n \to \mathbb{R}$ such that for every linear map $f: V \to V$, we have

$$o(f(v_1), \ldots, f(v_n)) = \text{sign}(\det(f))o(v_1, \ldots, v_n)$$

for all $v_1, \ldots, v_n \in V$.

If $(v_1, \ldots, v_n)$ are linearly dependent, then for any basis $(e_1, \ldots, e_n)$ of V there is a unique linear map f such that $f(e_i) = v_i$ for $i = 1, \ldots, n$, and since $(v_1, \ldots, v_n)$ are linearly dependent, f is singular so $\det(f) = 0$, which implies that

$$o(v_1, \ldots, v_n) = \text{sign}(\det(f))o(e_1, \ldots, e_n) = 0$$

for any linearly dependent vectors $v_1, \ldots, v_n \in V$.

For any two bases $(u_1, \ldots, u_n)$ and $(v_1, \ldots, v_n)$, there is a unique linear map f such that $f(u_i) = v_i$ for $i = 1, \ldots, n$, and $o(u_1, \ldots, u_n) = o(v_1, \ldots, v_n)$ iff $\det(f) > 0$, which is indeed the condition for $(u_1, \ldots, u_n)$ and $(v_1, \ldots, v_n)$ to have the same orientation. There are exactly two orientations o such that $|o(u_1, \ldots, u_n)| = 1$.

Let $\text{Or}(V)$ be the set of all orientations on V. We have the following proposition.

Proposition 6.10. *Let V be any vector space of dimension $n \geq 1$. The following properties hold:*

(a) The set $\text{Or}(V)$ is a vector space.
(b) For any two orientations $o_1, o_2 \in \text{Or}(V)$ and for any basis $(e_1, \ldots, e_n)$ of V, if $o_1(e_1, \ldots, e_n) = o_2(e_1, \ldots, e_n)$, then $o_1 = o_2$.
(c) For any nonzero n-form $\omega \in \bigwedge^n V^*$, the function $o(\omega)$ given by $o(\omega)(v_1, \ldots, v_n) = 0$ if $(v_1, \ldots, v_n)$ are linearly dependent and

$$o(\omega)(v_1, \ldots, v_n) = \frac{\omega(v_1, \ldots, v_n)}{|\omega(v_1, \ldots, v_n)|},$$

if $(v_1, \ldots, v_n)$ are linearly independent, is an orientation.
(d) The vector space $\text{Or}(V)$ is a one-dimensional space, and it is spanned by $o(\omega)$ for any nonzero $\omega \in \bigwedge^n V^*$.

Proof.

(a) That $\text{Or}(V)$ is a vector space is immediate from the definition.
(b) Pick any n vectors $(v_1, \ldots, v_n) \in V^n$. Since $(e_1, \ldots, e_n)$ is a basis of V, there is a unique linear map $f: V \to V$ such that $f(e_i) = v_i$ for $i = 1, \ldots, n$, and since by hypothesis $o_1(e_1, \ldots, e_n) = o_2(e_1, \ldots, e_n)$, we have

$$o_1(v_1, \ldots, v_n) = o_1(f(e_1), \ldots, f(e_n))$$

$$= \text{sign}(\det(f))o_1(e_1, \ldots, e_n)$$
$$= \text{sign}(\det(f))o_2(e_1, \ldots, e_n)$$
$$= o_2(f(e_1), \ldots, f(e_n))$$
$$= o_2(v_1, \ldots, v_n),$$

which proves that $o_1 = o_2$.

(c) Let $\omega \in \bigwedge^n V^*$ be any nonzero form. If $(v_1, \ldots, v_n)$ are linearly independent, then we know that

$$\omega(f(v_1), \ldots, f(v_n)) = \det(f)\,\omega(v_1, \ldots, v_n)$$
$$|\omega|(f(v_1), \ldots, f(v_n)) = |\det(f)|\,|\omega|(v_1, \ldots, v_n).$$

We know that $\det(f) = 0$ iff f is singular, but then $(f(v_1), \ldots, f(v_n))$ are linearly dependent so

$$o(\omega)(f(v_1), \ldots, f(v_n)) = 0 = \text{sign}(\det(f))o(\omega)(v_1, \ldots, v_n).$$

If $\det(f) \neq 0$, then

$$o(\omega)(f(v_1), \ldots, f(v_n)) = \frac{\omega(f(v_1), \ldots, f(v_n))}{|\omega(f(v_1), \ldots, f(v_n))|}$$
$$= \frac{\det(f)}{|\det(f)|} \frac{\omega(f(v_1), \ldots, f(v_n))}{|\omega(f(v_1), \ldots, f(v_n))|}$$
$$= \text{sign}(\det(f))o(\omega)(v_1, \ldots, v_n),$$

which shows that $o(\omega)$ is an orientation.

(d) Let $(e_1, \ldots, e_n)$ be any basis of V, and let $\omega \in \bigwedge^n V^*$ be any nonzero n-form. For any orientation o, we need to show that $o = co(\omega)$ for some $c \in \mathbb{R}$. Let

$$a = o(\omega)(e_1, \ldots, e_n) = \frac{\omega(e_1, \ldots, ve_n)}{|\omega(e_1, \ldots, e_n)|}$$
$$b = o(e_1, \ldots, e_n).$$

Since $\omega \neq 0$ and $(e_1, \ldots, e_n)$ is a basis, $\omega(e_1, \ldots, e_n) \neq 0$ so $a \neq 0$, and by Condition (c) $(b/a)o(\omega)$ is an orientation. Since

$$(b/a)o(\omega)(e_1, \ldots, e_n) = b = o(e_1, \ldots, e_n),$$

by Condition (b) $o = (b/a)o(\omega)$, as desired. $\square$

Part (c) of Proposition 6.10 implies that for every nonzero n-form $\omega \in \bigwedge^n V^*$, there exists some density $|\omega|$ and some orientation $o(\omega)$ such that

$$o(\omega)|\omega| = \omega.$$

This shows that orientations are just normalized volume forms that take exactly two values c and $-c$ on linearly independent vectors (with $c > 0$), whereas densities are absolute values of volume forms. We have the following results showing the relationship between the spaces $\bigwedge^n V^*$, $\mathrm{Or}(V)$, and $\mathrm{den}(V)$.

Proposition 6.11. *Let V be any vector space of dimension $n \geq 1$. For any nonzero n-form $\omega \in \bigwedge^n V^*$, the bilinear map $\Phi \colon \mathrm{Or}(V) \times \mathrm{den}(V) \to \bigwedge^n V^*$ given by*

$$\Phi(\alpha o(\omega), \beta|\omega|) = \alpha\beta\omega, \quad \alpha, \beta \in \mathbb{R}$$

induces an isomorphism $\mathrm{Or}(V) \otimes \mathrm{den}(V) \cong \bigwedge^n V^*$.

Proof. The spaces $\bigwedge^n V^*$, $\mathrm{Or}(V)$, and $\mathrm{den}(V)$ are all one dimensional, and if $\omega \neq 0$, then ω is a basis of $\bigwedge^n V^*$ and Propositions 6.9 and 6.10 show that $o(\omega)$ is a basis of $\mathrm{Or}(V)$ and $|\omega|$ is a basis of $\mathrm{den}(V)$, so the map Φ defines a bilinear map from $\mathrm{Or}(V) \times \mathrm{den}(V)$ to $\bigwedge^n V^*$. Therefore, by the universal mapping property, we obtain a linear map
$\Phi_\otimes \colon \mathrm{Or}(V) \otimes \mathrm{den}(V) \to \bigwedge^n V^*$. Since $\omega \neq 0$, we have

$$o(\omega)|\omega| = \omega,$$

which shows that Φ is surjective, and thus $\Phi_\otimes$ is surjective. Since all the spaces involved are one dimensional, $\mathrm{Or}(V) \otimes \mathrm{den}(V)$ is also one dimensional, so $\Phi_\otimes$ is bijective. $\qquad\square$

Given a manifold M, we can form the orientation bundle $\mathrm{Or}(M)$ whose underlying set is the disjoint union of the vector spaces $\mathrm{Or}(T_p M)$ for all $p \in M$. This set can be made into a smooth bundle, and an orientation of M is a smooth global section of the orientation bundle. Then it can be shown that there is a bundle isomorphism

$$\mathrm{Or}(M) \otimes \mathrm{den}(M) \cong \bigwedge^n T^* M.$$

and since $\mathrm{den}(M)$ always has global sections, we see that there is a global volume form iff $\mathrm{Or}(M)$ has a global section iff M is orientable.

The theory or integration developed so far deals with domains that are not general enough. Indeed, for many applications, we need to integrate over domains with boundaries.

6.6   Manifolds with Boundary

Up to now we have defined manifolds locally diffeomorphic to an open subset of $\mathbb{R}^m$. This excludes many natural spaces such as a closed disk, whose boundary is a circle, a closed ball $\overline{B(1)}$, whose boundary is the sphere S^{m-1}, a compact cylinder $S^1 \times [0, 1]$, whose boundary consist of two circles, a Möbius strip, *etc*. These spaces fail to be manifolds because they have a boundary; that is, neighborhoods of points on their boundaries are not diffeomorphic to open sets in $\mathbb{R}^m$. Perhaps the simplest example is the (closed) upper half space

$$\mathbb{H}^m = \{(x_1, \ldots, x_m) \in \mathbb{R}^m \mid x_m \geq 0\}.$$

Under the natural embedding $\mathbb{R}^{m-1} \cong \mathbb{R}^{m-1} \times \{0\} \hookrightarrow \mathbb{R}^m$, the subset $\partial \mathbb{H}^m$ of $\mathbb{H}^m$ defined by

$$\partial \mathbb{H}^m = \{x \in \mathbb{H}^m \mid x_m = 0\}$$

is isomorphic to $\mathbb{R}^{m-1}$, and is called the *boundary of* $\mathbb{H}^m$ When $m = 0$ we have $\mathbb{H}^0 = \emptyset$ and $\partial \mathbb{H}^0 = \emptyset$. We also define the *interior* of $\mathbb{H}^m$ as

$$\mathrm{Int}(\mathbb{H}^m) = \mathbb{H}^m - \partial \mathbb{H}^m.$$

Now if U and V are open subsets of $\mathbb{H}^m$, where $\mathbb{H}^m \subseteq \mathbb{R}^m$ has the subset topology, and if $f : U \to V$ is a continuous function, we need to explain what we mean by f being smooth.

Definition 6.14. Let U and V be open subsets of $\mathbb{H}^m$. We say that $f : U \to V$ as above is *smooth* if it has an extension $\widetilde{f} : \widetilde{U} \to \widetilde{V}$, where $\widetilde{U}$ and $\widetilde{V}$ are open subsets of $\mathbb{R}^m$ with $U \subseteq \widetilde{U}$ and $V \subseteq \widetilde{V}$, and with $\widetilde{f}$ a smooth function. We say that f is a (smooth) *diffeomorphism* iff f^{-1} exists and if both f and f^{-1} are smooth, as just defined.

To define a *manifold with boundary*, we replace everywhere $\mathbb{R}$ by $\mathbb{H}$ in the definition of a chart and in the definition of an atlas (see Tu [105], Chapter 6, §22, or Gallier and Quaintance [47]). So, for instance, given a topological space M, a *chart* is now pair (U, φ), where U is an open subset of M and $\varphi : U \to \Omega$ is a homeomorphism onto an open subset $\Omega = \varphi(U)$ of $\mathbb{H}^{n_\varphi}$ (for some $n_\varphi \geq 1$), *etc*. Thus, we obtain the following.

Definition 6.15. Given some integer $n \geq 1$ and given some k such that k is either an integer $k \geq 1$ or $k = \infty$, a C^k-*manifold of dimension n with boundary* consists of a topological space M together with an equivalence class $\overline{\mathcal{A}}$ of C^k n-atlases on M (where the charts are now defined in terms of open subsets of $\mathbb{H}^n$). Any atlas $\mathcal{A}$ in the equivalence class $\overline{\mathcal{A}}$ is called a *differentiable structure of class C^k (and*

Fig. 6.4 A two-dimensional
manifold with red boundary.

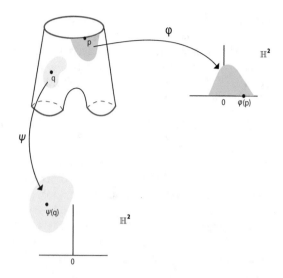

dimension n) on M. We say that *M* is *modeled on* $\mathbb{H}^n$ (Figure 6.4). When $k = \infty$,
we say that *M* is a *smooth manifold with boundary*.

It remains to define what is the boundary of a manifold with boundary.

Definition 6.16. Let *M* be a manifold with boundary as defined by Definition 6.15.
The *boundary* ∂M of *M* is the set of all points $p \in M$, such that there is some chart
$(U_\alpha, \varphi_\alpha)$, with $p \in U_\alpha$ and $\varphi_\alpha(p) \in \partial\mathbb{H}^n$. We also let $\text{Int}(M) = M - \partial M$ and call
it the *interior* of *M*.

 Do not confuse the boundary ∂M and the interior $\text{Int}(M)$ of a manifold
with boundary embedded in $\mathbb{R}^N$ with the topological notions of boundary
and interior of *M* as a topological space. In general, they are different. For
example, if *M* is the subset $[0, 1) \cup \{2\}$ of the real line, then its manifold
boundary is $\partial M = \{0\}$, and its topological boundary is $\text{Bd}(M) = \{0, 1, 2\}$.

Note that manifolds are also manifolds with boundary: their boundary is just
empty. We shall still reserve the word "manifold" for these, but for emphasis, we
will sometimes call them "boundaryless."

The definition of tangent spaces, tangent maps, *etc.* is easily extended to
manifolds with boundary. The reader should note that if *M* is a manifold with
boundary of dimension *n*, the tangent space $T_p M$ is defined for all $p \in M$ and
has dimension *n*, *even* for boundary points $p \in \partial M$. The only notion that requires
more care is that of a submanifold. For more on this, see Hirsch [56], Chapter 1,
Section 4. One should also beware that the product of two manifolds with boundary
is generally **not** a manifold with boundary (consider the product $[0, 1] \times [0, 1]$ of two
line segments). There is a generalization of the notion of a manifold with boundary

called *manifold with corners*, and such manifolds are closed under products (see Hirsch [56], Chapter 1, Section 4, Exercise 12).

If M is a manifold with boundary, we see that $\text{Int}(M)$ is a manifold without boundary. What about ∂M? Interestingly, the boundary ∂M of a manifold with boundary M of dimension n is a manifold of dimension $n - 1$. For this we need the following proposition.

Proposition 6.12. *If M is a manifold with boundary of dimension n, for any $p \in \partial M$ on the boundary on M, for any chart (U, φ) with $p \in M$, we have $\varphi(p) \in \partial \mathbb{H}^n$.*

Proof. Since $p \in \partial M$, by definition, there is some chart (V, ψ) with $p \in V$ and $\psi(p) \in \partial \mathbb{H}^n$. Let (U, φ) be any other chart, with $p \in M$, and assume that $q = \varphi(p) \in \text{Int}(\mathbb{H}^n)$. The transition map $\psi \circ \varphi^{-1} \colon \varphi(U \cap V) \to \psi(U \cap V)$ is a diffeomorphism, and $q = \varphi(p) \in \text{Int}(\mathbb{H}^n)$. By the inverse function theorem, there is some open $W \subseteq \varphi(U \cap V) \cap \text{Int}(\mathbb{H}^n) \subseteq \mathbb{R}^n$, with $q \in W$, so that $\psi \circ \varphi^{-1}$ maps W homeomorphically onto some subset Ω open in $\text{Int}(\mathbb{H}^n)$, with $\psi(p) \in \Omega$, contradicting the hypothesis, $\psi(p) \in \partial \mathbb{H}^n$. $\square$

Using Proposition 6.12, we immediately derive the fact that ∂M is a manifold of dimension $n - 1$. We obtain charts on ∂M by considering the charts $(U \cap \partial M, L \circ \varphi)$, where (U, φ) is a chart on M such that $U \cap \partial M = \varphi^{-1}(\partial \mathbb{H}^n) \neq \emptyset$ and $L \colon \partial \mathbb{H}^n \to \mathbb{R}^{n-1}$ is the natural isomorphism (see Hirsch [56], Chapter 1, Section 4).

6.7   Integration on Regular Domains and Stokes' Theorem

Given a manifold M, we define a class of subsets with boundaries that can be integrated on, and for which Stokes' theorem holds. In Warner [109] (Chapter 4), such subsets are called *regular domains*, and in Madsen and Tornehave [75] (Chapter 10), they are called *domains with smooth boundary*.

Definition 6.17. Let M be a smooth manifold of dimension n. A subset $N \subseteq M$ is called a *domain with smooth boundary* (or *codimension zero submanifold with boundary*) iff for every $p \in M$, there is a chart (U, φ) with $p \in U$ such that

$$\varphi(U \cap N) = \varphi(U) \cap \mathbb{H}^n, \qquad (*)$$

where $\mathbb{H}^n$ is the closed upper-half space

$$\mathbb{H}^n = \{(x_1, \ldots, x_n) \in \mathbb{R}^n \mid x_n \geq 0\}.$$

Note that $(*)$ is automatically satisfied when p is an interior or an exterior point of N, since we can pick a chart such that $\varphi(U)$ is contained in an open half space of $\mathbb{R}^n$ defined by either $x_n > 0$ or $x_n < 0$. If p is a boundary point of N, then $\varphi(p)$ has its last coordinate equal to 0; see Figure 6.5.

Fig. 6.5 The subset N, the peach region of the torus M, is a domain with smooth boundary.

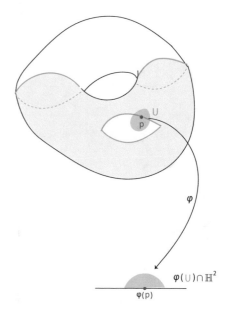

φ

$\varphi(U) \cap \mathbb{H}^2$

$\varphi(p)$

If M is orientable, then any orientation of M induces an orientation of ∂N, the boundary of N. This follows from the following proposition:

Proposition 6.13. *Let $\varphi: \mathbb{H}^n \to \mathbb{H}^n$ be a diffeomorphism with everywhere positive Jacobian determinant. Then φ induces a diffeomorphism $\Phi: \partial\mathbb{H}^n \to \partial\mathbb{H}^n$, which viewed as a diffeomorphism of $\mathbb{R}^{n-1}$, also has everywhere positive Jacobian determinant.*

Proof. By the inverse function theorem, every interior point of $\mathbb{H}^n$ is the image of an interior point, so φ maps the boundary to itself. If $\varphi = (\varphi_1, \ldots, \varphi_n)$, then

$$\Phi = (\varphi_1(x_1, \ldots, x_{n-1}, 0), \ldots, \varphi_{n-1}(x_1, \ldots, x_{n-1}, 0)),$$

since $\varphi_n(x_1, \ldots, x_{n-1}, 0) = 0$. It follows that $\frac{\partial\varphi_n}{\partial x_i}(x_1, \ldots, x_{n-1}, 0) = 0$ for $i = 1, \ldots, n-1$, and as φ maps $\mathbb{H}^n$ to itself,

$$\frac{\partial\varphi_n}{\partial x_n}(x_1, \ldots, x_{n-1}, 0) > 0.$$

Now the Jacobian matrix of φ at $q = \varphi(p) \in \partial\mathbb{H}^n$ is of the form

$$J(\varphi)(q) = \begin{pmatrix} & & & * \\ & d\Phi_q & & \vdots \\ & & & * \\ 0 & \cdots & 0 & \frac{\partial\varphi_n}{\partial x_n}(q) \end{pmatrix}$$

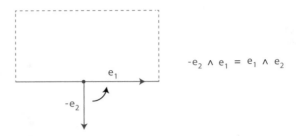

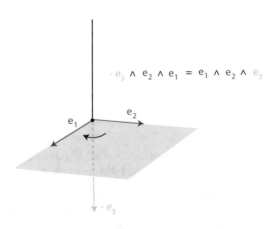

Fig. 6.6 The boundary orientations of $\partial \mathbb{H}^2$ and $\partial \mathbb{H}^3$.

and since $\frac{\partial \varphi_n}{\partial x_n}(q) > 0$ and by hypothesis $\det(J(\varphi)_q) > 0$, we have $\det(J(\Phi)_q) > 0$, as claimed. $\square$

In order to make Stokes' formula sign free, if $\mathbb{R}^n$ has the orientation given by $dx_1 \wedge \cdots \wedge dx_n$, then $\partial \mathbb{H}^n$ is given the orientation given by $(-1)^n dx_1 \wedge \cdots \wedge dx_{n-1}$ if $n \geq 2$, and -1 for $n = 1$. In particular $\partial \mathbb{H}_2$ is oriented by e_1 while $\partial \mathbb{H}_3$ is oriented by $-e_1 \wedge e_2 = e_2 \wedge e_1$. See Figure 6.6.

Definition 6.18. Given any domain with smooth boundary $N \subseteq M$, a tangent vector $w \in T_p M$ at a boundary point $p \in \partial N$ is *outward directed* iff there is a chart (U, φ) with $p \in U$, $\varphi(U \cap N) = \varphi(U) \cap \mathbb{H}^n$, and $d\varphi_p(w)$ has a negative n^{th} coordinate $pr_n(d\varphi_p(w))$; see Figure 6.7.

Let (V, ψ) be another chart with $p \in V$. The transition map

$$\theta = \psi \circ \varphi^{-1} : \varphi(U \cap V) \to \psi(U \cap V)$$

induces a map

Fig. 6.7 An example of an outward directed tangent vector to N. Notice this red tangent vector points away from N.

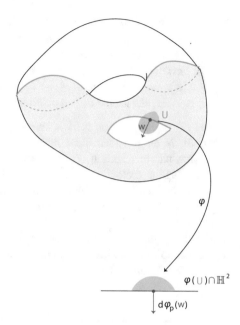

$$\varphi(U \cap V) \cap \mathbb{H}^n \longrightarrow \psi(U \cap V) \cap \mathbb{H}^n$$

which restricts to a diffeomorphism

$$\Theta \colon \varphi(U \cap V) \cap \partial\mathbb{H}^n \to \psi(U \cap V) \cap \partial\mathbb{H}^n.$$

The proof of Proposition 6.13 shows that the Jacobian matrix of $d\theta_q$ at $q = \varphi(p) \in \partial\mathbb{H}^n$ is of the form

$$J(\theta)(q) = \begin{pmatrix} & & & * \\ & J(\Theta)_q & & \vdots \\ & & & * \\ 0 & \cdots & 0 & \frac{\partial\theta_n}{\partial x_n}(q) \end{pmatrix}$$

with $\theta = (\theta_1, \ldots, \theta_n)$, and that $\frac{\partial\theta_n}{\partial x_n}(q) > 0$. As $d\psi_p = d(\psi \circ \varphi^{-1})_q \circ d\varphi_p$, we see that for any $w \in T_pM$ with $pr_n(d\varphi_p(w)) < 0$, since $pr_n(d\psi_p(w)) = \frac{\partial\theta_n}{\partial x_n}(q) \, pr_n(d\varphi_p(w))$, we also have $pr_n(d\psi_p(w)) < 0$ (recall that $\theta = \psi \circ \varphi^{-1}$). Therefore, the negativity condition of Definition 6.18 does not depend on the chart at p. The following proposition is then easy to show.

Proposition 6.14. *Let $N \subseteq M$ be a domain with smooth boundary, where M is a smooth manifold of dimension n.*

(1) The boundary ∂N of N is a smooth manifold of dimension $n - 1$.

(2) Assume M is oriented. If $n \geq 2$, there is an induced orientation on ∂N
determined as follows: For every $p \in \partial N$, if $v_1 \in T_p M$ is an outward directed
tangent vector, then a basis $(v_2, \ldots, v_n)$ for $T_p \partial N$ is positively oriented iff
the basis $(v_1, v_2, \ldots, v_n)$ for $T_p M$ is positively oriented. When $n = 1$, every
$p \in \partial N$ has the orientation $+1$ iff for every outward directed tangent vector
$v_1 \in T_p M$, the vector v_1 is a positively oriented basis of $T_p M$.

Part (2) of Proposition 6.14 is summarized as "outward pointing vector first."
When M is an n-dimensional embedded manifold in $\mathbb{R}^m$ with an orientation
preserving parametrization $\psi : U \to \mathbb{R}^m$, for any point $p = \psi(q) \in \partial N$, let
v_1 be a tangent vector pointing away from N. This means $d\psi_q(-e_n) = v_1$. To
complete the basis of $T_p M$ in a manner consistent with the positive orientation of
U given by $dx_1 \wedge \cdots \wedge dx_n$, we choose an ordered basis $(v_2, \cdots, v_n)$ of $T_p \partial N$
such that $d\psi_q((-1)^n e_1) = v_2$ and $d\psi_q(e_i) = v_{i+1}$ whenever $2 \leq i \leq n - 1$.
Intuitively, ψ maps the positive orientation of U to a positive orientation of $T_p M$
with the condition that the first vector in the orientation frame of $T_p M$ points away
from N. See Figure 6.8.

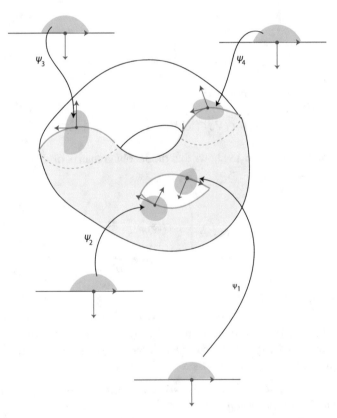

Fig. 6.8 The orientation of $T_p \partial N$ consistent with the positive orientation of $\mathbb{R}^2$.

Another way to describe the induced orientation of ∂N is through the insertion operator; see Definition 4.13. Let ω be a volume form on M, let $p \in \partial N$, and let $v_1 \in T_p M$ be an outward directed tangent vector. The volume form on ∂N is given by $i_{v_1}\omega$ where

$$i_{v_1}\omega(v_2, \cdots, v_n) = \omega(v_1, v_2, \cdots, v_n).$$

If M is oriented, then for every n-form $\omega \in \mathcal{A}_c^n(M)$, the integral $\int_N \omega$ is well defined. More precisely, Proposition 6.7 can be generalized to domains with a smooth boundary. This can be shown in various ways. The most natural way to proceed is to prove an extension of Proposition 6.6 using a slight generalization of the change of variable formula.

Proposition 6.15. *Let $\varphi: U \to V$ be a diffeomorphism between two open subsets of $\mathbb{R}^n$, and assume that φ maps $U \cap \mathbb{H}^n$ to $V \cap \mathbb{H}^n$. Then for every smooth function $f: V \to \mathbb{R}$ with compact support,*

$$\int_{V \cap \mathbb{H}^n} f(x)dx_1 \cdots dx_n = \int_{U \cap \mathbb{H}^n} f(\varphi(y)) |J(\varphi)_y| dy_1 \cdots dy_n.$$

One alternative way to define $\int_N \omega$ involves covering N with special kinds of open subsets arising from regular simplices (see Warner [109], Chapter 4).

Remark. Another alternative way to proceed is to apply techniques of measure theory. In Madsen and Tornehave [75] it is argued that integration theory goes through for continuous n-forms with compact support. If σ is a volume form on M, then for every continuous function with compact support f, the map

$$f \mapsto I_\sigma(f) = \int_M f\sigma$$

is a linear positive operator[1] (which means that $I(f) \geq 0$ for $f \geq 0$). By Riesz' representation theorem (see Rudin [92], Chapter 2), I_σ determines a positive Borel measure μ_σ which satisfies

$$\int_M f d\mu_\sigma = \int_M f\sigma$$

for all continuous functions f with compact support. Since any C^1 n-form ω can be written uniquely as $\omega = f\sigma$ for some C^1 function f, we can set

$$\int_N \omega = \int_M 1_N f\sigma,$$

---

[1] In fact, a Radon measure.

where 1_N is the function with value 1 on N and 0 outside N.

We now have all the ingredient to state and prove Stokes's formula. Our proof is based on the proof found in Section 23.5 of Tu [105]. Alternative proofs can be found in many places (for example, Warner [109] (Chapter 4), Bott and Tu [12] (Chapter 1), and Madsen and Tornehave [75] (Chapter 10).

Theorem 6.16 (Stokes' Theorem). *Let $N \subseteq M$ be a domain with smooth boundary, where M is a smooth oriented manifold of dimension n, give ∂N the orientation induced by M, and let $i : \partial N \to M$ be the inclusion map. For every differential form with compact support $\omega \in \mathcal{A}_c^{n-1}(M)$ with $N \cap \mathrm{supp}(\omega)$ compact, we have*

$$\int_{\partial N} i^*\omega = \int_N d\omega.$$

In particular, if $N = M$ is a smooth oriented manifold with boundary, then

$$\int_{\partial M} i^*\omega = \int_M d\omega, \qquad\qquad (\ast\ast\ast)$$

and if M is a smooth oriented manifold without boundary, then

$$\int_M d\omega = 0.$$

Of course, $i^*\omega$ is the restriction of ω to ∂N, and for simplicity of notation $i^*\omega$ is usually written as ω, and Stokes' formula is written as

$$\int_{\partial N} \omega = \int_N d\omega.$$

Proof Based on Tu [105]. We select a covering $\{(U_i, \varphi_i)\}_{i \in I}$ of M and we restrict to those charts (U_i, φ_i) such that $\varphi_i(U_i \cap N) = \varphi_i(U_i) \cap \mathbb{H}^n$ is diffeomorphic to either $\mathbb{R}^n$ or $\mathbb{H}^n$ via an orientation preserving diffeomorphism. Note that each U_i has a nonempty intersection with N. Let $(\rho_i)_{i \in I}$ be a partition of unity subordinate to this cover. An adaptation of the proof of Proposition 6.7 shows that $\rho_i \omega$ is an $(n-1)$-form on M with compact support in U_i.

Assume that Stokes' theorem is true for $\mathbb{R}^n$ and $\mathbb{H}^n$. Then Stokes' theorem will hold for all U_i which are diffeomorphic to either $\mathbb{R}^n$ or $\mathbb{H}^n$. Observe that the paragraph preceding Proposition 6.14 implies that $\partial N \cap U_i = \partial U_i$. Since $\sum_i \rho_i = 1$, we have

$$\int_{\partial N} \omega = \int_{\partial N} \sum_i \rho_i \omega$$

$$= \sum_i \int_{\partial N} \rho_i \omega, \qquad \text{since } \sum_i \rho_i \omega \text{ is finite}$$

$$= \sum_i \int_{\partial U_i} \rho_i \omega, \qquad \text{since } \operatorname{supp}(\rho_i \omega) \subseteq U_i$$

$$= \sum_i \int_{U_i} d(\rho_i \omega), \qquad \text{by assumption that Stokes' is true for } U_i$$

$$= \sum_i \int_N d(\rho \omega), \qquad \text{since } \operatorname{supp}(d(\rho_i \omega)) \subseteq U_i \cap N$$

$$= \int_N d\left(\sum_i \rho_i \omega\right) = \int_N d\omega.$$

Thus it remains to prove Stokes' theorem for $\mathbb{R}^n$ and $\mathbb{H}^n$. Since ω is now assumed to be an $(n-1)$-form on $\mathbb{R}^n$ or $\mathbb{H}^n$ with compact support,

$$\omega = \sum_{i=1}^n f_i \, dx_1 \wedge \cdots \wedge \widehat{dx_i} \wedge \cdots \wedge dx_n,$$

where each f_i is a smooth function with compact support in $\mathbb{R}^n$ or $\mathbb{H}^n$. By using the $\mathbb{R}$-linearity of the exterior derivative and the integral operator, we may assume that ω has only one term, namely

$$\omega = f \, dx_1 \wedge \cdots \wedge \widehat{dx_i} \wedge \cdots \wedge dx_n,$$

and

$$d\omega = \sum_{j=1}^n \frac{\partial f}{\partial x_j} dx_j \wedge dx_1 \wedge \cdots \wedge \widehat{dx_i} \wedge \cdots \wedge dx_n$$

$$= (-1)^{i-1} \frac{\partial f}{\partial x_i} dx_1 \wedge \cdots \wedge dx_i \wedge \cdots \wedge dx_n.$$

where f is smooth function on $\mathbb{R}^n$ such that $\operatorname{supp}(f)$ is contained in the interior of the n-cube $[-a, a]^n$ for some fixed $a > 0$.

To verify Stokes' theorem for $\mathbb{R}^n$, we evaluate $\int_{\mathbb{R}^n} d\omega$ as an iterated integral via Fubini's theorem. (See Edwards [41], Theorem 4.1.) In particular, we find that

$$\int_{\mathbb{R}^n} d\omega = \int_{\mathbb{R}^n} (-1)^{i-1} \frac{\partial f}{\partial x_i} dx_1 \cdots dx_i \cdots dx_n$$

$$= (-1)^{i-1} \int_{\mathbb{R}^{n-1}} \left(\int_{-\infty}^{\infty} \frac{\partial f}{\partial x_i} dx_i \right) dx_1 \cdots \widehat{dx_i} \cdots dx_n$$

$$= (-1)^{i-1} \int_{\mathbb{R}^{n-1}} \left(\int_{-a}^{a} \frac{\partial f}{\partial x_i} dx_i \right) dx_1 \cdots \widehat{dx_i} \cdots dx_n$$

$$= (-1)^{i-1} \int_{\mathbb{R}^{n-1}} 0 \, dx_1 \cdots \widehat{dx_i} \cdots dx_n \qquad \text{since supp}(f) \subset [-a,a]^n$$

$$= 0 = \int_{\emptyset} \omega = \int_{\partial \mathbb{R}^n} \omega.$$

The verification of Stokes' theorem for $\mathbb{H}^n$ involves the analysis of two cases. For the first case assume $i \neq n$. Since $\partial \mathbb{H}^n$ is given by $x_n = 0$, then $dx_n \equiv 0$ on $\partial \mathbb{H}^n$. An application of Fubini's theorem shows that

$$\int_{\mathbb{H}^n} d\omega = \int_{\mathbb{H}^n} (-1)^{i-1} \frac{\partial f}{\partial x_i} dx_1 \cdots dx_i \cdots dx_n$$

$$= (-1)^{i-1} \int_{\mathbb{H}^{n-1}} \left(\int_{-\infty}^{\infty} \frac{\partial f}{\partial x_i} dx_i \right) dx_1 \cdots \widehat{dx_i} \cdots dx_n$$

$$= (-1)^{i-1} \int_{\mathbb{H}^{n-1}} \left(\int_{-a}^{a} \frac{\partial f}{\partial x_i} dx_i \right) dx_1 \cdots \widehat{dx_i} \cdots dx_n$$

$$= (-1)^{i-1} \int_{\mathbb{H}^{n-1}} 0 \, dx_1 \cdots \widehat{dx_i} \cdots dx_n \qquad \text{since supp}(f) \subset [-a,a]^n$$

$$= 0 = \int_{\partial \mathbb{H}^n} f \, dx_1 \cdots \widehat{dx_i} \cdots dx_n, \qquad \text{since } dx_n \equiv 0 \text{ on } \partial \mathbb{H}^n.$$

It remains to analyze the case $i = n$. Fubini's theorem implies

$$\int_{\mathbb{H}^n} d\omega = \int_{\mathbb{H}^n} (-1)^{n-1} \frac{\partial f}{\partial x_n} dx_1 \cdots dx_n$$

$$= (-1)^{n-1} \int_{\mathbb{R}^{n-1}} \left(\int_0^{\infty} \frac{\partial f}{\partial x_n} dx_n \right) dx_1, \cdots dx_{n-1}$$

$$= (-1)^{n-1} \int_{\mathbb{R}^{n-1}} \left(\int_0^{a} \frac{\partial f}{\partial x_n} dx_n \right) dx_1 \cdots dx_{n-1}$$

$$= (-1)^n \int_{\mathbb{R}^{n-1}} f(x^1, \cdots, x^{n-1}, 0) \, dx_1 \cdots dx_{n-1}, \quad \text{since supp}(f) \subset [-a,a]^n$$

$$= \int_{\partial \mathbb{H}^n} \omega,$$

where the last equality follows from the fact that $(-1)^n \mathbb{R}^{n-1}$ is the induced boundary orientation of $\partial \mathbb{H}^n$. □

Stokes' theorem, as presented in Theorem 6.16, unifies the integral theorems of vector calculus since the classical integral theorems of vector calculus are particular examples of (∗∗∗) when M is an n-dimensional manifold embedded in $\mathbb{R}^3$. If $n = 3$, $\omega \in \mathcal{A}_c^2(M)$, and (∗∗∗) becomes the Divergence theorem. Given a smooth function $F \colon \mathbb{R}^3 \to \mathbb{R}^3$, recall that the *divergence* of F is the smooth real-valued function $\mathrm{div}\, F \colon \mathbb{R}^3 \to \mathbb{R}$ where

$$\mathrm{div}\, F = \frac{\partial F_1}{\partial x_1} + \frac{\partial F_2}{\partial x_2} + \frac{\partial F_3}{\partial x_3},$$

and (x_1, x_2, x_3) are the standard coordinates of $\mathbb{R}^3$ (often represented as (x, y, z)). The Divergence theorem is as follows:

Proposition 6.17 (Divergence Theorem). *Let $F \colon \mathbb{R}^3 \to \mathbb{R}^3$ be a smooth vector field defined on a neighborhood of M, a compact oriented smooth 3-dimensional manifold with boundary. Then*

$$\int_M \mathrm{div}\, F \,\mathrm{Vol}_M = \int_{\partial M} F \cdot N \,\mathrm{Vol}_{\partial M}, \tag{1}$$

where $N(x) = (n_1(x), n_2(x), n_3(x))$ is the unit outer normal vector field on ∂M,

$$\mathrm{Vol}_{\partial M} = n_1 \, dx_2 \wedge dx_3 - n_2 \, dx_1 \wedge dx_3 + n_3 \, dx_1 \wedge dx_2$$

and ∂M is positively oriented as the boundary of $M \subseteq \mathbb{R}^3$.

In calculus books (1) is often written as

$$\int\int\int \mathrm{div}\, F \, dx \, dy \, dz = \int F \cdot N \, dS, \tag{2}$$

where dS is the surface area differential. In particular if ∂M is parametrized by $\varphi(x, y) = (x, y, f(x, y))$,

$$\int F \cdot N \, dS = \int\int F \cdot \left(-\frac{\partial f}{\partial x}, -\frac{\partial f}{\partial y}, 1 \right) dx \, dy. \tag{3}$$

The verification of (3) is an application of Equation (∗∗) from Section 6.2. In particular

$$J(\varphi)(x, y) = \begin{pmatrix} 1 & 0 \\ 0 & 1 \\ \frac{\partial f}{\partial x} & \frac{\partial f}{\partial y} \end{pmatrix},$$

which in turn implies

$$D = \det\left[J(\varphi)^{\top}(x, y) J(\varphi)(x, y)\right]^{\frac{1}{2}} = \sqrt{1 + \left(\frac{\partial f}{\partial x}\right)^2 + \left(\frac{\partial f}{\partial y}\right)^2}.$$

Hence

$$n_{1,2} = \frac{\det\begin{pmatrix} 1 & 0 \\ 0 & 1 \end{pmatrix}}{D} = \frac{1}{\sqrt{1 + \left(\frac{\partial f}{\partial x}\right)^2 + \left(\frac{\partial f}{\partial y}\right)^2}}$$

$$n_{1,3} = \frac{\det\begin{pmatrix} 1 & 0 \\ \frac{\partial f}{\partial x} & \frac{\partial f}{\partial y} \end{pmatrix}}{D} = \frac{\frac{\partial f}{\partial y}}{\sqrt{1 + \left(\frac{\partial f}{\partial x}\right)^2 + \left(\frac{\partial f}{\partial y}\right)^2}}$$

$$n_{2,3} = \frac{\det\begin{pmatrix} 0 & 1 \\ \frac{\partial f}{\partial x} & \frac{\partial f}{\partial y} \end{pmatrix}}{D} = \frac{-\frac{\partial f}{\partial x}}{\sqrt{1 + \left(\frac{\partial f}{\partial x}\right)^2 + \left(\frac{\partial f}{\partial y}\right)^2}}$$

and

$$dS = n_{1,2}\, dx \wedge dy + n_{1,3}\, dx \wedge dz + n_{2,3}\, dy \wedge dz$$

$$= \frac{dx \wedge dy}{\sqrt{1 + \left(\frac{\partial f}{\partial x}\right)^2 + \left(\frac{\partial f}{\partial y}\right)^2}} + \frac{\frac{\partial f}{\partial y}\, dx \wedge dz}{\sqrt{1 + \left(\frac{\partial f}{\partial x}\right)^2 + \left(\frac{\partial f}{\partial y}\right)^2}} + \frac{-\frac{\partial f}{\partial x}\, dy \wedge dz}{\sqrt{1 + \left(\frac{\partial f}{\partial x}\right)^2 + \left(\frac{\partial f}{\partial y}\right)^2}}.$$

Since $z = f(x, y)$,

$$\varphi^*(dS) = \frac{dx \wedge dy}{\sqrt{1 + \left(\frac{\partial f}{\partial x}\right)^2 + \left(\frac{\partial f}{\partial y}\right)^2}} + \frac{\frac{\partial f}{\partial y}\, dx \wedge (\frac{\partial f}{\partial x}\, dx + \frac{\partial f}{\partial y}\, dy)}{\sqrt{1 + \left(\frac{\partial f}{\partial x}\right)^2 + \left(\frac{\partial f}{\partial y}\right)^2}} + \frac{-\frac{\partial f}{\partial x}\, dy \wedge (\frac{\partial f}{\partial x}\, dx + \frac{\partial f}{\partial y}\, dy)}{\sqrt{1 + \left(\frac{\partial f}{\partial x}\right)^2 + \left(\frac{\partial f}{\partial y}\right)^2}}$$

$$= \sqrt{1 + \left(\frac{\partial f}{\partial x}\right)^2 + \left(\frac{\partial f}{\partial y}\right)^2}\, dx \wedge dy.$$

Furthermore,

$$N = \frac{\frac{\partial \varphi}{\partial x} \times \frac{\partial \varphi}{\partial y}}{\left\| \frac{\partial \varphi}{\partial x} \times \frac{\partial \varphi}{\partial y} \right\|} = \frac{1}{\sqrt{1 + \left(\frac{\partial f}{\partial x}\right)^2 + \left(\frac{\partial f}{\partial y}\right)^2}} \left(-\frac{\partial f}{\partial x}, -\frac{\partial f}{\partial y}, 1 \right).$$

Substituting the expressions for N and $\varphi^*(dS)$ into $\int F \cdot N \, dS$ gives the right side of (3).

If $n = 2$, $\omega \in \mathcal{A}_c^1(M)$, and $(***)$ becomes the classical Stokes' theorem. Given a smooth function $F: \mathbb{R}^3 \to \mathbb{R}^3$, recall that the *curl* of F is the smooth function $\mathrm{curl}\, F: \mathbb{R}^3 \to \mathbb{R}^3$

$$\mathrm{curl}\, F = \left(\frac{\partial F_3}{\partial x_2} - \frac{\partial F_2}{\partial x_3}, \frac{\partial F_1}{\partial x_3} - \frac{\partial F_3}{\partial x_1}, \frac{\partial F_2}{\partial x_1} - \frac{\partial F_1}{\partial x_2} \right).$$

The classical Stokes' theorem is as follows:

Proposition 6.18. *Let M be an oriented compact 2-dimensional manifold with boundary locally parametrized in $\mathbb{R}^3$ by the orientation-preserving local diffeomorphism $\psi : U \to \mathbb{R}^3$ such that $\psi(u, v) = (x_1, x_2, x_3) \in M$. Define*

$$N = \frac{\frac{\partial \psi}{\partial u} \times \frac{\partial \psi}{\partial v}}{\left\| \frac{\partial \psi}{\partial u} \times \frac{\partial \psi}{\partial v} \right\|}$$

to be the smooth outward unit normal vector field on M. Let n be the outward directed tangent vector field on ∂M. Let $T = N \times n$. Given $F: \mathbb{R}^3 \to \mathbb{R}^3$, a smooth vector field defined on an open subset of $\mathbb{R}^3$ containing M,

$$\int_M \mathrm{curl}\, F \cdot N \, \mathrm{Vol}_M = \int_{\partial M} F \cdot T \, \mathrm{Vol}_{\partial M}, \tag{4}$$

where Vol_M is defined as in $\mathrm{Vol}_{\partial M}$ of Proposition 6.17 and $\mathrm{Vol}_{\partial M} = ds$, the line integral form.

If M is parametrized by $\varphi(x, y) = (x, y, f(x, y))$, we have shown that the left side of (4) may be written as

$$\int_M \mathrm{curl}\, F \cdot N \, \mathrm{Vol}_M = \int \mathrm{curl}\, F \cdot N \, dS = \int \int \mathrm{curl}\, F \cdot \left(-\frac{\partial f}{\partial x}, -\frac{\partial f}{\partial y}, 1 \right) dx \, dy.$$

Many calculus books represent the right side of (4) as

$$\int_{\partial M} F \cdot T \, ds = \int F \cdot d\mathbf{r}, \tag{5}$$

where $d\mathbf{r} = (dx, dy, dz)$. Once again the verification of (5) is an application of Equation $(**)$ from Section 6.2. Let $\psi(x) = (x, y(x), z(x))$ be a parameterization of ∂M. Then $J(\psi)(x) = (1, y_x, z_x)^\top$, where $y_x = \frac{dy}{dx}$ and $z_x = \frac{dz}{dx}$. Then

$$
D = \det\left[J(\psi)^\top(x) J(\psi)(x) \right]^{\frac{1}{2}} = \sqrt{1 + y_x^2 + z_x^2},
$$

$$
ds = \frac{dx + y_x\, dy + z_x\, dz}{\sqrt{1 + y_x^2 + z_x^2}},
$$

and

$$
\psi^* ds = \sqrt{1 + y_x^2 + z_x^2}\, dx.
$$

Furthermore

$$
T = \frac{J(\psi)(x)}{\sqrt{1 + y_x^2 + z_x^2}} = \frac{(1, y_x, z_x)^\top}{\sqrt{1 + y_x^2 + z_x^2}}.
$$

Substituting the expressions for T and $\psi^* ds$ into the left side of (5) gives

$$
\int_{\partial M} F \cdot T\, ds = \int F \cdot \left(1, \frac{dy}{dx}, \frac{dz}{dx}\right) dx = \int F \cdot (dx, dy, dz) = \int F \cdot d\mathbf{r}.
$$

Thus the classical form of Stokes' theorem often appears as

$$
\int\int \operatorname{curl} F \cdot \left(-\frac{\partial f}{\partial x}, -\frac{\partial f}{\partial y}, 1\right) dx\, dy = \int F \cdot \left(1, \frac{dy}{dx}, \frac{dz}{dx}\right) dx = \int F \cdot d\mathbf{r},
$$

where M is parametrized via $\varphi(x, y) = (x, y, f(x, y))$.

The orientation frame (n, T, N) given in Proposition 6.18 provides the standard orientation of $\mathbb{R}^3$ given by (e_1, e_2, e_3) and is visualized as follows. Pick a preferred side of the surface. This choice is represented by N. At each boundary point, draw the outward pointing tangent vector n which is locally perpendicular (in the tangent plane) to the boundary curve. To determine T, pretend you are a bug on the side of the surface selected by N. You must walk along the boundary curve in the direction that keeps the boundary of the surface your *right*. Then $T = N \times n$ and (n, T, N) is oriented via the right-hand rule in the same manner as (e_1, e_2, e_3); see Figure 6.9.

For those readers who wish to learn more about the connections between the classical integration theorems of vector calculus and Stokes' theorem, we refer them to Edwards [41] (Chapter 5, Section 7).

The version of Stokes' theorem that we have presented applies to domains with smooth boundaries, but there are many situations where it is necessary to deal

Fig. 6.9 The orientation frame (n, T, N) for the bell shaped surface M. Notice the bug must walk along the boundary in a counter clockwise direction.

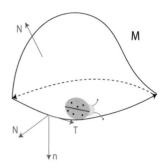

with domains with singularities, for example corners (as a cube, a tetrahedron, *etc.*). Manifolds with corners form a nice class of manifolds that allow such a generalization of Stokes' theorem.

To model corners, we adapt the idea that we used when we defined charts of manifolds with boundaries but instead of using the closed half space $\mathbb{H}^m$, we use the closed convex cone

$$\overline{\mathbb{R}^m_+} = \{(x_1, \ldots, x_m) \in \mathbb{R}^m \mid x_1 \geq 0, \ldots, x_m \geq 0\}.$$

The *boundary* $\partial\overline{\mathbb{R}^m_+}$ of $\overline{\mathbb{R}^m_+}$ is the space

$$\partial\overline{\mathbb{R}^m_+} = \{(x_1, \ldots, x_m) \in \mathbb{R}^m \mid x_1 \geq 0, \ldots, x_m \geq 0, \ x_i = 0 \text{ for some } i$$

which can also be written as

$$\partial\overline{\mathbb{R}^m_+} = H_1 \cup \cdots \cup H_m,$$

with

$$H_i = \{(x_1, \ldots, x_m) \in \overline{\mathbb{R}^m_+} \mid x_i = 0\}.$$

The set of *corner points* of $\overline{\mathbb{R}^m_+}$ is the subset

$$\{(x_1, \ldots, x_m) \in \overline{\mathbb{R}^m_+} \mid \exists i \exists j \, (i \neq j), \ x_i = 0 \text{ and } x_j = 0\}.$$

Equivalently, the set of corner points is the union of all intersections $H_{i_1} \cap \cdots \cap H_{i_k}$ for all finite subsets $\{i_1, \ldots, i_k\}$ of $\{1, \ldots, m\}$ with $k \geq 2$. See Figure 6.10.

Definition 6.19. Given a topological space M, *a chart with corners* is a pair (U, φ) where U is some open subset of M and φ is a homeomorphism of U onto some open subset of $\overline{\mathbb{R}^m_+}$ (with the subspace topology of $\mathbb{R}^m$). Compatible charts, atlases, and equivalent atlases are defined as usual, and a *smooth manifold with corners*

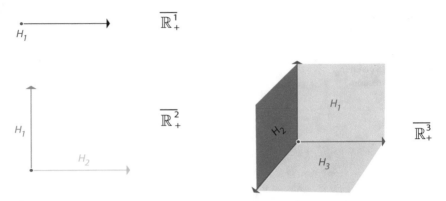

Fig. 6.10 The closed convex cones $\overline{\mathbb{R}}_+^1$, $\overline{\mathbb{R}}_+^2$, and $\overline{\mathbb{R}}_+^3$. Corner points are in red.

is a topological space together with an equivalence class of atlases of charts with corners.

A point $p \in M$ is a *corner point* if there is a chart (U, φ) with $p \in U$ such that $\varphi(p)$ is a corner point of $\overline{\mathbb{R}}_+^m$.

It is not hard to show that the definition of corner point does not depend on the chart (U, φ) with $p \in U$. See Figure 6.11.

Now, in general, the boundary of a smooth manifold with corners is not a smooth manifold with corners. For example, $\partial \overline{\mathbb{R}}_+^m$ is not a smooth manifold with corners, but it is the union of smooth manifolds with corners, since $\partial \overline{\mathbb{R}}_+^m = H_1 \cup \cdots \cup H_m$, and each H_i is a smooth manifold with corners. We can use this fact to define $\int_{\partial M} \omega$ where ω is an $(n-1)$-form whose support in contained in the domain of a chart with corners (U, φ) by setting

$$\int_{\partial M} \omega = \sum_{i=1}^{m} \int_{H_i} (\varphi^{-1})^* \omega,$$

where each H_i is given a suitable orientation. Then it is not hard to prove a version of Stokes' theorem for manifolds with corners. For a detailed exposition, see Lee [73], Chapter 14. An even more general class of manifolds with singularities (in $\mathbb{R}^N$) for which Stokes' theorem is valid is discussed in Lang [70] (Chapter XVII, §3).

6.8   Integration on Riemannian Manifolds and Lie Groups

We saw in Section 6.2 that every orientable Riemannian manifold has a uniquely defined volume form Vol_M (see Proposition 6.4).

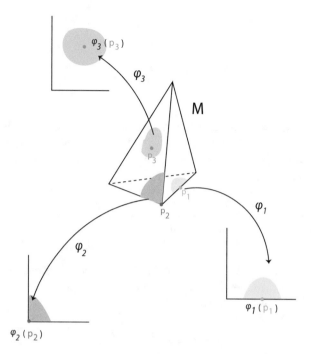

Fig. 6.11 The three types of charts on M, a manifold with corners. Note that p_2 is a corner point of M.

Definition 6.20. Given any smooth real-valued function f with compact support on M, we define the *integral of f over M* by

$$\int_M f = \int_M f \, \mathrm{Vol}_M .$$

Actually it is possible to define the integral $\int_M f$ using densities even if M is not orientable, but we do not need this extra generality. If M is compact, then $\int_M 1_M = \int_M \mathrm{Vol}_M$ is the *volume* of M (where 1_M is the constant function with value 1).

If M and N are Riemannian manifolds, then we have the following version of Proposition 6.8 (3).

Proposition 6.19. *If M and N are oriented Riemannian manifolds and if $\varphi \colon M \to N$ is an orientation preserving diffeomorphism, then for every function $f \in C^\infty(N)$ with compact support, we have*

$$\int_N f \, \mathrm{Vol}_N = \int_M f \circ \varphi \, |\det(d\varphi)| \, \mathrm{Vol}_M ,$$

where $f \circ \varphi \, |\det(d\varphi)|$ denotes the function $p \mapsto f(\varphi(p))|\det(d\varphi_p)|$, with $d\varphi_p \colon T_p M \to T_{\varphi(p)} N$. In particular, if φ is an orientation preserving isometry

(see Definition 6 in Chapter 3 of O'Neill [84], or Gallier and Quaintance [47]),
then

$$\int_N f \, \text{Vol}_N = \int_M f \circ \varphi \, \text{Vol}_M.$$

We often denote $\int_M f \, \text{Vol}_M$ by $\int_M f(t)dt$.

If $f: M \to \mathbb{C}$ is a smooth complex valued-function then we can write $f = u + iv$ for two real-valued functions $u: M \to \mathbb{R}$ and $v: M \to \mathbb{R}$ with $u(p) = \Re(f(p))$ and $v(p) = \text{Im}(f(p))$ for all $p \in M$. Then, if f has compact support so do u and v, and we define $\int_M f \, \text{Vol}_M$ by

$$\int_M f \, \text{Vol}_M = \int_M u \, \text{Vol}_M + i \int_M v \, \text{Vol}_M.$$

Remark. A volume form on an orientable Riemannian manifold (M, g) yields a linear form $v_g: C_0(M) \to \mathbb{R}$ with domain the vector space $C_0(M)$ of continuous real-valued functions on M with compact support, given by

$$v_g(f) = \int_M f \, \text{Vol}_M.$$

This linear form turns out to a *Radon measure* (see Sakai [93] (Chapter II, Section 5)). This Radon measure can be used to define an analog of the Lebesgue integral on the Riemannian manifold M, and to define measurable sets and measurable functions on the manifold M (see Sakai [93] (Chapter II, Section 5)). Given a diffeomorphism $\Phi: M \to N$ between two Riemannian manifolds (M, g) and (N, h), the Radon measure v_h on N can be pulled back to a Radon measure $\Phi^* v_h$ on M given by

$$(\Phi^* v_h)(f) = v_h(f \circ \Phi^{-1}).$$

If $\tilde{g}$ denotes the canonical Riemannian metric on $T_p M$ given by g_p, then we can establish a relationship between $\exp_p^* v_g$ and $v_{\tilde{g}}$ that involves the Ricci curvature at p, where $\exp_p: T_p M \to M$ is the exponential map at p. Actually, in general $\exp_p$ is only defined in an open ball $B_r(0)$ in $T_p M$ centered at the origin, and it is a diffeomorphism on this open ball. The following result can be shown; see Sakai [93] (Chapter 5, Section Lemma 5.4). We use polar coordinates in $T_p M$, which means that every nonzero $x \in T_p M$ is expressed at $x = tu$ with $t = \|x\|$ and $u = x/\|x\|$.

Proposition 6.20. *Given a Riemannian manifold (M, g), for every $p \in M$, in normal coordinates, near p, we have*

$$\exp_p^* v_g = \theta \, v_{\tilde{g}},$$

where θ is the function given by $\theta(t, u) = t^{n-1} \sqrt{\det(g_{ij}(\exp_p tu))}$. Furthermore, we have

$$\theta(t, u) = t^{n-1} - \frac{1}{6}\mathrm{Ric}_p(u, u)\, t^{n+1} + o(t^{n+2}).$$

Recall from Proposition 6.4 that if (M, g) is an oriented Riemannian manifold, then the volume form Vol_M satisfies the following property: in every orientation preserving local chart (U, φ), we have

$$((\varphi^{-1})^*\mathrm{Vol}_M)_q = \sqrt{\det(g_{ij}(q))}\, dx_1 \wedge \cdots \wedge dx_n, \qquad q \in \varphi(U).$$

In particular, if $\varphi = \exp_p^{-1}$ is a chart specified by the inverse of the exponential map on some small enough open subset containing p, then near p we obtain

$$(\exp_p^* \mathrm{Vol}_M)_{tu} = \left(1 - \frac{1}{6}\mathrm{Ric}_p(u, u)\, t^2 + o(t^3)\right) dx_1 \wedge \cdots \wedge dx_n.$$

Thus we can think of the Ricci curvature as a measure of the deviation of the volume form on M (at tu) from the Euclidean volume form.

If G is a Lie group, we know from Section 6.2 that G is always orientable and that G possesses left-invariant volume forms. Since $\dim(\bigwedge^n \mathfrak{g}^*) = 1$ if $\dim(G) = n$, and since every left-invariant volume form is determined by its value at the identity, the space of left-invariant volume forms on G has dimension 1. If we pick some left-invariant volume form ω defining the orientation of G, then every other left-invariant volume form is proportional to ω.

Definition 6.21. Let G be a Lie group and ω be a left-invariant volume form. Given any smooth real-valued function f with compact support on G, we define the *integral of f over G (w.r.t. ω)* by

$$\int_G f = \int_G f\omega.$$

This integral depends on ω, but since ω is defined up to some positive constant, so is the integral. When G is compact, we usually pick ω so that

$$\int_G \omega = 1.$$

If $f : G \to \mathbb{C}$ is a smooth complex valued-function then we can write $f = u + iv$ for two real-valued functions $u : G \to \mathbb{R}$ and $v : G \to \mathbb{R}$ as before and we define

$$\int_G f\omega = \int_G u\,\omega + i \int_G v\,\omega.$$

For every $g \in G$, as ω is left-invariant, $L_g^* \omega = \omega$, so L_g^* is an orientation-preserving diffeomorphism, and by Proposition 6.8 (3),

$$\int_G f\omega = \int_G L_g^*(f\omega),$$

so using Proposition 4.12, we get

$$\int_G f = \int_G f\omega = \int_G L_g^*(f\omega) = \int_G L_g^* f\, L_g^* \omega = \int_G L_g^* f\, \omega = \int_G (f \circ L_g)\omega = \int_G f \circ L_g.$$

Thus we proved the following proposition.

Proposition 6.21. *Given any left-invariant volume form ω on a Lie group G, for any smooth function f with compact support, we have*

$$\int_G f = \int_G f \circ L_g,$$

a property called left-invariance.

It is then natural to ask when our integral is right-invariant; that is, when

$$\int_G f = \int_G f \circ R_g.$$

Observe that $R_g^* \omega$ is left-invariant, since

$$L_h^* R_g^* \omega = R_g^* L_h^* \omega = R_g^* \omega.$$

It follows that $R_g^* \omega$ is some constant multiple of ω, and so there is a function $\overline{\Delta} : G \to \mathbb{R}$ such that

$$R_g^* \omega = \overline{\Delta}(g)\omega.$$

One can check that $\overline{\Delta}$ is smooth, and we let

$$\Delta(g) = |\overline{\Delta}(g)|.$$

Since

$$\overline{\Delta}(gh)\omega = R_{gh}^* \omega = \left(R_h \circ R_g\right)^* \omega = R_g^* \circ R_h^* \omega = \overline{\Delta}(g)\overline{\Delta}(h)\omega,$$

we deduce that

$$\Delta(gh) = \Delta(g)\Delta(h),$$

so Δ is a homomorphism of G into $\mathbb{R}_+$.

Definition 6.22. The function Δ defined above is called the *modular function of* G.

Proposition 6.22. *Given any left-invariant volume form ω on a Lie group G, for any smooth function f with compact support, we have*

$$\int_G f\omega = \Delta(g) \int_G (f \circ R_g)\omega.$$

Proof. By Proposition 6.8 (3), for a fixed $g \in G$, as R_g^* is an orientation-preserving diffeomorphism,

$$\int_G f\omega = \int_G R_g^*(f\omega) = \int_G R_g^* f \, R_g^* \omega = \int_G (f \circ R_g)\Delta(g)\omega,$$

or equivalently,

$$\int_G f\omega = \Delta(g) \int_G (f \circ R_g)\omega,$$

which is the desired formula.   □

Consequently, our integral is right-invariant iff $\Delta \equiv 1$ on G. Thus, our integral is not always right-invariant. When it is, *i.e.*, when $\Delta \equiv 1$ on G, we say that *G is unimodular*.

Proposition 6.23. *Any compact Lie group G is unimodular.*

Proof. In this case,

$$1 = \int_G \omega = \int_G 1_G\omega = \int_G \Delta(g)\omega = \Delta(g) \int_G \omega = \Delta(g),$$

for all $g \in G$.   □

Therefore, for a *compact Lie group G*, our integral is both left- and right-invariant. We say that our integral is *bi-invariant*.

As a matter of notation, the integral $\int_G f = \int_G f\omega$ is often written $\int_G f(g)dg$. Then left-invariance can be expressed as

$$\int_G f(g)dg = \int_G f(hg)dg,$$

and right-invariance as

$$\int_G f(g)dg = \int_G f(gh)dg,$$

for all $h \in G$.

If ω is left-invariant, then it can be shown (see Dieudonné [35], Chapter XIV, Section 3) that

$$\int_G f(g^{-1})\Delta(g^{-1})dg = \int_G f(g)dg.$$

Consequently, if G is unimodular, then

$$\int_G f(g^{-1})dg = \int_G f(g)dg.$$

Proposition 6.24. *If ω_l is any left-invariant volume form on G and if ω_r is any right-invariant volume form on G, then*

$$\omega_r(g) = c\Delta(g^{-1})\omega_l(g),$$

for some constant $c \neq 0$.

Proof. Indeed, define the form ω by $\omega(g) = \overline{\Delta}(g^{-1})\omega_l(g)$. By Proposition 4.12 we have

$$(R_h^*\omega)_g = \overline{\Delta}((gh)^{-1})(R_h^*\omega_l)_g$$
$$= \overline{\Delta}(h^{-1})\overline{\Delta}(g^{-1})\overline{\Delta}(h)(\omega_l)_g$$
$$= \overline{\Delta}(g^{-1})(\omega_l)_g,$$

which shows that ω is right-invariant, and thus $\omega_r(g) = c\Delta(g^{-1})\omega_l(g)$, as claimed (since $\overline{\Delta}(g^{-1}) = \pm\Delta(g^{-1})$).   □

Actually, the following property holds.

Proposition 6.25. *For any Lie group G, for any $g \in G$, we have*

$$\Delta(g) = |\det(\mathrm{Ad}(g^{-1}))|.$$

Proof. For this recall that $\mathrm{Ad}(g) = d(L_g \circ R_{g^{-1}})_1$. For any left-invariant n-form $\omega \in \bigwedge^n \mathfrak{g}^*$, we claim that

$$(R_g^*\omega)_h = \det(\mathrm{Ad}(g^{-1}))\omega_h,$$

which shows that $\Delta(g) = |\det(\mathrm{Ad}(g^{-1}))|$. Indeed, for all $v_1, \ldots, v_n \in T_hG$, we have

$(R_g^* \omega)_h(v_1, \ldots, v_n)$

$$= \omega_{hg}(d(R_g)_h(v_1), \ldots, d(R_g)_h(v_n))$$

$$= \omega_{hg}(d(L_g \circ L_{g^{-1}} \circ R_g \circ L_h \circ L_{h^{-1}})_h(v_1), \ldots, d(L_g \circ L_{g^{-1}} \circ R_g \circ L_h \circ L_{h^{-1}})_h(v_n))$$

$$= \omega_{hg}(d(L_h \circ L_g \circ L_{g^{-1}} \circ R_g \circ L_{h^{-1}})_h(v_1), \ldots, d(L_h \circ L_g \circ L_{g^{-1}} \circ R_g \circ L_{h^{-1}})_h(v_n))$$

$$= \omega_{hg}(d(L_{hg} \circ L_{g^{-1}} \circ R_g \circ L_{h^{-1}})_h(v_1), \ldots, d(L_{hg} \circ L_{g^{-1}} \circ R_g \circ L_{h^{-1}})_h(v_n))$$

$$= \omega_{hg}\big(d(L_{hg})_1(\mathrm{Ad}(g^{-1})(d(L_{h^{-1}})_h(v_1))), \ldots, d(L_{hg})_1(\mathrm{Ad}(g^{-1})(d(L_{h^{-1}})_h(v_n)))\big)$$

$$= (L_{hg}^* \omega)_1\big(\mathrm{Ad}(g^{-1})(d(L_{h^{-1}})_h(v_1)), \ldots, \mathrm{Ad}(g^{-1})(d(L_{h^{-1}})_h(v_n))\big)$$

$$= \omega_1\big(\mathrm{Ad}(g^{-1})(d(L_{h^{-1}})_h(v_1)), \ldots, \mathrm{Ad}(g^{-1})(d(L_{h^{-1}})_h(v_n))\big)$$

$$= \det(\mathrm{Ad}(g^{-1}))\, \omega_1\big(d(L_{h^{-1}})_h(v_1), \ldots, d(L_{h^{-1}})_h(v_n)\big)$$

$$= \det(\mathrm{Ad}(g^{-1}))\, (L_{h^{-1}}^* \omega)_h(v_1, \ldots, v_n)$$

$$= \det(\mathrm{Ad}(g^{-1}))\, \omega_h(v_1, \ldots, v_n),$$

where we used the left-invariance of ω twice. $\qquad\square$

In general, if G is not unimodular, then $\omega_l \neq \omega_r$. A simple example provided by Vinroot [108] is the group G of direct affine transformations of the real line, which can be viewed as the group of matrices of the form

$$g = \begin{pmatrix} x & y \\ 0 & 1 \end{pmatrix}, \qquad x, y, \in \mathbb{R}, \ x > 0.$$

Let $A = \begin{pmatrix} a & b \\ 0 & 1 \end{pmatrix} \in G$ and define $T : G \to G$ as

$$T(g) = Ag = \begin{pmatrix} a & b \\ 0 & 1 \end{pmatrix}\begin{pmatrix} x & y \\ 0 & 1 \end{pmatrix} = \begin{pmatrix} ax & ay + b \\ 0 & 1 \end{pmatrix}.$$

Since G is homeomorphic to $\mathbb{R}^+ \times \mathbb{R}$, $T(g)$ is also represented by $T(x, y) = (ax, ay + b)$. Then the Jacobian matrix of T is given by

$$J(T)(x, y) = \begin{pmatrix} a & 0 \\ 0 & a \end{pmatrix},$$

which implies that $\det(J(T)(x, y)) = a^2$. Let $F : G \to \mathbb{R}^+$ be a smooth function on G with compact support. Furthermore assume that $\Theta(x, y) = F(x, y)x^{-2}$ is also smooth on G with compact support. Since $\Theta \circ T(x, y) = \Theta(ax, ay + b) = F(ax, ay + b)(ax)^{-2}$, Proposition 6.19 implies that

$$\int_G F(x, y)x^{-2}dx\,dy = \int_G \Theta(x, y) \circ T \mid \det(J(T)(x, y)) \mid dx\,dy$$

$$= \int_G F(ax, ay + b)(ax)^{-2}a^2\,dx\,dy = \int_G F \circ T x^{-2}dx\,dy.$$

In summary we have shown for $g = \begin{pmatrix} x & y \\ 0 & 1 \end{pmatrix}$, we have

$$\int_G F(Ag)x^{-2}\,dx\,dy = \int_G F(g)x^{-2}dx\,dy$$

which implies that the left-invariant volume form on G is

$$\omega_l = \frac{dx\,dy}{x^2}.$$

To define a right-invariant volume form on G, define $S \colon G \to G$ as

$$S(g) = gA = \begin{pmatrix} x & y \\ 0 & 1 \end{pmatrix}\begin{pmatrix} a & b \\ 0 & 1 \end{pmatrix} = \begin{pmatrix} ax & bx + y \\ 0 & 1 \end{pmatrix},$$

which is represented by $S(x, y) = (ax, bx + y)$. Then the Jacobian matrix of S is

$$J(S)(x, y) = \begin{pmatrix} a & 0 \\ b & 1 \end{pmatrix},$$

and $\det(J(S)(x, y)) = a$. Using $F(x, y)$ as above and $\Theta(x, y) = F(x, y)x^{-1}$, we find that

$$\int_G F(x, y)x^{-1}dx\,dy = \int_G \Theta(x, y) \circ S \mid \det(J(S)(x, y)) \mid dx\,dy$$

$$= \int_G F(ax, bx + y)(ax)^{-1}a\,dx\,dy = \int_G F \circ S x^{-1}dx\,dy,$$

which implies that

$$\omega_r = \frac{dx\,dy}{x}.$$

Note that $\Delta(g) = |x^{-1}|$.

Observe that $\Delta(A) = |a^{-1}|$, $F \circ R_A = F(ax, bx + y)$, and that

$$\frac{1}{|a|}\int_G F \circ R_A\,\omega_l = \frac{1}{|a|}\int_G F(ax, bx + y)\frac{dx\,dy}{x^2}$$

$$= \frac{1}{|a|} \int_G F(u, v) \, \frac{\frac{du}{a}(dv - \frac{b}{a}\, du)}{\left(\frac{u}{a}\right)^2}, \qquad u = ax, \quad v = bx + y$$

$$= \int_G F(u, v) \, \frac{du\, dv}{u^2} = \int_G F(u, v) \, w_l,$$

which is a special case of Proposition 6.22.

Remark. By the Riesz' representation theorem, ω defines a positive measure μ_ω which satisfies

$$\int_G f \, d\mu_\omega = \int_G f \omega.$$

Using what we have shown, this measure is left-invariant. Such measures are called *left Haar measures*, and similarly we have *right Haar measures*.

It can be shown that every two left Haar measures on a Lie group are proportional (see Knapp [66], Chapter VIII). Given a left Haar measure μ, the function Δ such that

$$\mu(R_g h) = \Delta(g)\mu(h)$$

for all $g, h \in G$ is the *modular function* of G. However, beware that some authors, including Knapp, use $\Delta(g^{-1})$ instead of $\Delta(g)$. As above, we have

$$\Delta(g) = |\det(\mathrm{Ad}(g^{-1}))|.$$

Beware that authors who use $\Delta(g^{-1})$ instead of $\Delta(g)$ give a formula where $\mathrm{Ad}(g)$ appears instead of $\mathrm{Ad}(g^{-1})$. Again, G is *unimodular* iff $\Delta \equiv 1$.

It can be shown that compact, semisimple, reductive, and nilpotent Lie groups are unimodular (for instance, see Knapp [66], Chapter VIII). On such groups, left Haar measures are also right Haar measures (and vice versa). In this case, we can speak of *Haar measures* on G. For more details on Haar measures on locally compact groups and Lie groups, we refer the reader to Folland [42] (Chapter 2), Helgason [55] (Chapter 1), and Dieudonné [35] (Chapter XIV).

6.9   Problems

Problem 6.1. Prove the following: if $f : \mathbb{R}^{n+1} \to \mathbb{R}$ is a smooth submersion, then $M = f^{-1}(0)$ is a smooth orientable manifold.

Problem 6.2. Let $r : \mathbb{R}^{n+1} - \{0\} \to S^n$ be the map given by

$$r(x) = \frac{x}{\|x\|},$$

and set

$$\omega = r^* \mathrm{Vol}_{S^n}.$$

a closed n-form on $\mathbb{R}^{n+1} - \{0\}$. Clearly,

$$\omega \restriction S^n = \mathrm{Vol}_{S^n}.$$

Show that ω is given by

$$\omega_x = \frac{1}{\|x\|^n} \sum_{i=1}^{n+1} (-1)^{i-1} x_i \, dx_1 \wedge \cdots \wedge \widehat{dx_i} \wedge \cdots \wedge dx_{n+1}.$$

Problem 6.3. Recall that $\pi : S^n \to \mathbb{RP}^n$ is the map such that $\pi^{-1}([p])$ consists of two antipodal points for every $[p] \in \mathbb{RP}^n$. Show there is a volume form on $\mathbb{RP}^n$ iff n is odd, given by

$$\pi^*(\mathrm{Vol}_{\mathbb{RP}^n}) = \mathrm{Vol}_{S^n}.$$

Problem 6.4. Complete the proof of Proposition 6.7 by using a partition of unity argument to show the uniqueness of the linear operator

$$\int_M : \mathcal{A}_c^n(M) \longrightarrow \mathbb{R}$$

which satisfies the following property: For any $\omega \in \mathcal{A}_c^n(M)$, if $\mathrm{supp}(\omega) \subseteq U$, where (U, φ) is a positively oriented chart, then

$$\int_M \omega = \int_{\varphi(U)} (\varphi^{-1})^* \omega.$$

Problem 6.5. Complete the proof sketch details of Proposition 6.8.

Problem 6.6. Recall that a density may be defined as a function $\mu : \bigwedge^n V \to \mathbb{R}$ such that for every automorphism $f \in \mathbf{GL}(V)$,

$$\mu(f(v_1) \wedge \cdots \wedge f(v_n)) = |\det(f)| \mu(v_1 \wedge \cdots \wedge v_n) \qquad (\dagger\dagger)$$

for all $v_1 \wedge \cdots \wedge v_n \in V$ (with $\mu(0) = 0$). Show that Condition $(\dagger\dagger)$ is equivalent to

$$\mu(cw) = |c| \mu(w), \quad w \in \bigwedge^n V, \, c \in \mathbb{R}.$$

Problem 6.7. Prove Proposition 6.14.

Problem 6.8. Prove Proposition 6.15.

Problem 6.9. Prove Proposition 6.19.

Problem 6.10. Prove Proposition 6.20.
Hint. See Sakai [93], Chapter 5, Section Lemma 5.4.

Problem 6.11. Let G be a Lie group with a left-invariant volume form ω. Show that

$$\int_G f(g^{-1})\Delta(g^{-1})dg = \int_G f(g)dg.$$

Hint. See Dieudonné [35], Chapter XIV, Section 3.

Chapter 7
Spherical Harmonics and Linear Representations of Lie Groups

This chapter and the next focus on topics that are somewhat different from the more geometric and algebraic topics discussed in the previous chapters. Indeed, the focus of this chapter is on the types of functions that can be defined on a manifold, the sphere S^n in particular, and this involves some analysis. A main theme of this chapter is to generalize Fourier analysis on the circle to higher dimensional spheres. One of our goals is to understand the structure of the space $L^2(S^n)$ of real-valued square integrable functions on the sphere S^n, and its complex analog $L^2_{\mathbb{C}}(S^n)$. Both are Hilbert spaces if we equip them with the inner product

$$\langle f, g \rangle_{S^n} = \int_{S^n} f(t)g(t)\, dt = \int_{S^n} fg \, \mathrm{Vol}_{S^n},$$

and in the complex case with the Hermitian inner product

$$\langle f, g \rangle_{S^n} = \int_{S^n} f(t)\overline{g(t)}\, dt = \int_{S^n} f\overline{g} \, \mathrm{Vol}_{S^n}.$$

This means that if we define the L^2-norm associated with the above inner product as $\|f\| = \sqrt{\langle f, f \rangle}$, then $L^2(S^n)$ and $L^2_{\mathbb{C}}(S^n)$ are complete normed vector spaces (see Section 7.1 for a review of Hilbert spaces). It turns out that each of $L^2(S^n)$ and $L^2_{\mathbb{C}}(S^n)$ contains a countable family of very nice finite-dimensional subspaces $\mathcal{H}_k(S^n)$ (and $\mathcal{H}^{\mathbb{C}}_k(S^n)$), where $\mathcal{H}_k(S^n)$ is the space of (real) *spherical harmonics* on S^n, that is, the restrictions of the harmonic homogeneous polynomials of degree k (in $n+1$ real variables) to S^n (and similarly for $\mathcal{H}^{\mathbb{C}}_k(S^n)$); these polynomials satisfy the Laplace equation

$$\Delta P = 0,$$

© Springer Nature Switzerland AG 2020
J. Gallier, J. Quaintance, *Differential Geometry and Lie Groups*, Geometry and Computing 13, https://doi.org/10.1007/978-3-030-46047-1_7

where the operator Δ is the (Euclidean) *Laplacian*,

$$\Delta = \frac{\partial^2}{\partial x_1^2} + \cdots + \frac{\partial^2}{\partial x_{n+1}^2}.$$

Remarkably, each space $\mathcal{H}_k(S^n)$ (resp. $\mathcal{H}_k^{\mathbb{C}}(S^n)$) is the eigenspace of the Laplace-Beltrami operator Δ_{S^n} on S^n, a generalization to Riemannian manifolds of the standard Laplacian (in fact, $\mathcal{H}_k(S^n)$ is the eigenspace for the eigenvalue $-k(n+k-1)$). As a consequence, the spaces $\mathcal{H}_k(S^n)$ (resp. $\mathcal{H}_k^{\mathbb{C}}(S^n)$) are pairwise orthogonal. Furthermore (and this is where analysis comes in), the set of all finite linear combinations of elements in $\bigcup_{k=0}^{\infty} \mathcal{H}_k(S^n)$ (resp. $\bigcup_{k=0}^{\infty} \mathcal{H}_k^{\mathbb{C}}(S^n)$) is dense in $L^2(S^n)$ (resp. dense in $L_{\mathbb{C}}^2(S^n)$). These two facts imply the following fundamental result about the structure of the spaces $L^2(S^n)$ and $L_{\mathbb{C}}^2(S^n)$.

The family of spaces $\mathcal{H}_k(S^n)$ (resp. $\mathcal{H}_k^{\mathbb{C}}(S^n)$) yields a Hilbert space direct sum decomposition

$$L^2(S^n) = \bigoplus_{k=0}^{\infty} \mathcal{H}_k(S^n) \qquad (\text{resp.} \quad L_{\mathbb{C}}^2(S^n) = \bigoplus_{k=0}^{\infty} \mathcal{H}_k^{\mathbb{C}}(S^n)),$$

which means that the summands are closed, pairwise orthogonal, and that every $f \in L^2(S^n)$ (resp. $f \in L_{\mathbb{C}}^2(S^n)$) is the sum of a converging series

$$f = \sum_{k=0}^{\infty} f_k$$

in the L^2-norm, where the $f_k \in \mathcal{H}_k(S^n)$ (resp. $f_k \in \mathcal{H}_k^{\mathbb{C}}(S^n)$) are uniquely determined functions. Furthermore, given any orthonormal basis $(Y_k^1, \ldots, Y_k^{a_{k,n+1}})$ of $\mathcal{H}_k(S^n)$, we have

$$f_k = \sum_{m_k=1}^{a_{k,n+1}} c_{k,m_k} Y_k^{m_k}, \qquad \text{with} \quad c_{k,m_k} = \langle f, Y_k^{m_k} \rangle_{S^n}.$$

The coefficients c_{k,m_k} are "generalized" *Fourier coefficients* with respect to the Hilbert basis $\{Y_k^{m_k} \mid 1 \le m_k \le a_{k,n+1}, \ k \ge 0\}$; see Theorems 7.18 and 7.19.

In Section 7.2 we begin by reviewing the simple case $n = 1$, where S^1 is a circle, which corresponds to standard Fourier analysis. In this case, there is a simple expression in polar coordinates for the Laplacian Δ_{S_1} on the circle, and we are led to the equation

$$\Delta_{S^1} g = -k^2 g.$$

We find that $\mathcal{H}_0(S^1) = \mathbb{R}$, and $\mathcal{H}_k(S^1)$ is the two-dimensional space spanned by $\cos k\theta$ and $\sin k\theta$ for $k \geq 1$. We also determine explicitly the harmonic polynomials in two variables.

In Section 7.3 we consider the sphere S^2. This time we need to find the Laplacian Δ_{S^2} on the sphere. This is an old story, and we use the formula in terms of spherical coordinates. Then we need to solve the equation

$$\Delta_{S^2} g = -k(k-1)g.$$

This is a classical problem that was solved in the early 1780s by the separation of variables method. After some labor, we are led to the general Legendre equation. The solutions are the *associated Legendre functions* $P_k^m(t)$, which are defined in terms of the *Legendre polynomials*. The upshot is that the functions

$$\cos m\varphi \, P_k^m(\cos\theta), \qquad \sin m\varphi \, P_k^m(\cos\theta)$$

are eigenfunctions of the Laplacian Δ_{S^2} on the sphere for the eigenvalue $-k(k+1)$. For k fixed, as $0 \leq m \leq k$, we get $2k+1$ linearly independent functions, so $\mathcal{H}_k(S^2)$ has dimension $2k+1$. These functions are the *spherical harmonics*, but they are usually expressed in a different notation ($y_l^m(\theta, \varphi)$ with $-l \leq m \leq l$). Expressed in Cartesian coordinates, these are the homogeneous harmonic polynomials.

In order to generalize the above cases to $n \geq 3$, we need to define the Laplace-Beltrami operator on a manifold, which is done in Section 7.4. We also find a formula relating the Laplacian on $\mathbb{R}^{n+1}$ to the Laplacian Δ_{S^n} on S^n. The Hilbert sum decomposition of $L^2(S^n)$ is accomplished in Section 7.5.

In Section 7.6 we describe the zonal spherical functions Z_k^τ on S^n and show that they essentially come from certain polynomials generalizing the Legendre polynomials known as the *Gegenbauer polynomials*. For any fixed point τ on S^n and any constant $c \in \mathbb{C}$, the zonal spherical function Z_k^τ is the unique homogeneous harmonic polynomial of degree k such that $Z_k^\tau(\tau) = c$, and Z_k^τ is invariant under any rotation fixing τ.

An interesting property of the zonal spherical functions is a formula for obtaining the kth spherical harmonic component of a function $f \in L_\mathbb{C}^2(S^n)$; see Proposition 7.27. Another important property of the zonal spherical functions Z_k^τ is that they generate $\mathcal{H}_k^\mathbb{C}(S^n)$. A closer look at the Gegenbauer polynomials is taken in Section 7.7.

In Section 7.8 we prove the Funk-Hecke formula. This formula basically allows one to perform a sort of convolution of a "kernel function" with a spherical function in a convenient way. The Funk-Hecke formula was used in a ground-breaking paper by Basri and Jacobs [9] to compute the reflectance function r from the lighting function ℓ as a pseudo-convolution $K \star \ell$ (over S^2) with the *Lambertian kernel* K.

The final Sections 7.9 and 7.11 are devoted to more advanced material which is presented without proofs.

The purpose of Section 7.9 is to generalize the results about the structure of the space of functions $L^2_{\mathbb{C}}(S^n)$ defined on the sphere S^n, especially the results of Sections 7.5 and 7.6 (such as Theorem 7.19, except part (3)), to homogeneous spaces G/K where G is a compact Lie group and K is a closed subgroup of G.

The first step is to consider the Hilbert space $L^2_{\mathbb{C}}(G)$ where G is a compact Lie group and to find a Hilbert sum decomposition of this space. The key to this generalization is the notion of (unitary) linear representation of the group G.

The result that we are alluding to is a famous theorem known as the *Peter–Weyl theorem* about unitary representations of compact Lie groups (Herman Klauss Hugo Weyl, 1885–1955).

The Peter–Weyl theorem can be generalized to any representation $V \colon G \to \mathrm{Aut}(E)$ of G into a separable Hilbert space E, and we obtain a Hilbert sum decomposition of E in terms of subspaces E_ρ of E.

The next step is to consider the subspace $L^2_{\mathbb{C}}(G/K)$ of $L^2_{\mathbb{C}}(G)$ consisting of the functions that are right-invariant under the action of K. These can be viewed as functions on the homogeneous space G/K. Again we obtain a Hilbert sum decomposition. It is also interesting to consider the subspace $L^2_{\mathbb{C}}(K\backslash G/K)$ of functions in $L^2_{\mathbb{C}}(G)$ consisting of the functions that are both left- and right-invariant under the action of K. The functions in $L^2_{\mathbb{C}}(K\backslash G/K)$ can be viewed as functions on the homogeneous space G/K that are invariant under the left action of K.

Convolution makes the space $L^2_{\mathbb{C}}(G)$ into a noncommutative algebra. Remarkably, it is possible to characterize when $L^2_{\mathbb{C}}(K\backslash G/K)$ is commutative (under convolution) in terms of a simple criterion about the irreducible representations of G. In this situation, (G, K) is a called a *Gelfand pair*.

When (G, K) is a Gelfand pair, it is possible to define a well-behaved notion of *Fourier transform* on $L^2_{\mathbb{C}}(K\backslash G/K)$. Gelfand pairs and the Fourier transform are briefly considered in Section 7.11.

7.1 Hilbert Spaces and Hilbert Sums

The material in this chapter assumes that the reader has some familiarity with the concepts of a Hilbert space and a Hilbert basis. We present this section to review these important concepts. Many of the proofs are omitted and are found in traditional sources such as Bourbaki [15], Dixmier [36], Lang [68, 69], and Rudin [92]. The special case of separable Hilbert spaces is treated very nicely in Deitmar [29].

We begin our review by recalling the definition of a Hermitian space. To do this we need to define the notion of a Hermitian form.

Definition 7.1. Given two vector spaces E and F over $\mathbb{C}$, a function $f \colon E \to F$ is *semilinear* if

$$f(u + v) = f(u) + f(v)$$

$$f(\lambda u) = \overline{\lambda} u,$$

for all $u, v \in E$ and $\lambda \in \mathbb{C}$.

Definition 7.2. Given a complex vector space E, a function $\varphi \colon E \times E \to \mathbb{C}$ is a *sesquilinear* form if it is linear in its first argument and semilinear in its second argument, which means that

$$\varphi(u_1 + u_2, v) = \varphi(u_1, v) + \varphi(u_2, v)$$

$$\varphi(u, v_1 + v_2) = \varphi(u, v_1) + \varphi(u, v_2)$$

$$\varphi(\lambda u, v) = \lambda \varphi(u, v)$$

$$\varphi(u, \lambda v) = \overline{\lambda} \varphi(u, v),$$

for all $u, v, u_1, u_2, v_1, v_2 \in E$ and $\lambda \in \mathbb{C}$. A function $\varphi \colon E \times E \to \mathbb{C}$ is a *Hermitian form* if it is sesquilinear and if

$$\varphi(u, v) = \overline{\varphi(v, u)},$$

for all $u, v \in E$.

Definition 7.3. Given a complex vector space E, a Hermitian form $\varphi \colon E \times E \to \mathbb{C}$ is *positive definite* if $\varphi(u, u) > 0$ for all $u \neq 0$. A pair $\langle E, \varphi \rangle$ where E is a complex vector space and φ is a Hermitian form on E is called a *Hermitian (or unitary)* space if φ is positive definite.

The standard example of a Hermitian form on $\mathbb{C}^n$ is the map φ defined such that

$$\varphi((x_1, \ldots, x_n), (y_1, \ldots, y_n)) = x_1 \overline{y_1} + x_2 \overline{y_2} + \cdots + x_n \overline{y_n}.$$

This map is also positive definite and makes $\mathbb{C}^n$ into a Hermitian space.

Given a Hermitian space $\langle E, \varphi \rangle$, we can readily show that the function $\| \ \| \colon E \to \mathbb{R}$ defined such that $\langle u, u \rangle = \|u\| = \varphi(u, u)$ is a norm on E. Thus, E is a normed vector space. If E is also complete, then it is a very interesting space.

Recall that completeness has to do with the convergence of Cauchy sequences. A normed vector space $\langle E, \| \ \| \rangle$ is automatically a metric space under the metric d defined such that $d(u, v) = \|v - u\|$. This leads us to the following definition.

Definition 7.4. Given a metric space E with metric d, a sequence $(a_n)_{n \geq 1}$ of elements $a_n \in E$ is a *Cauchy sequence* iff for every $\epsilon > 0$, there is some $N \geq 1$ such that

$$d(a_m, a_n) < \epsilon \quad \text{for all} \quad m, n \geq N.$$

We say that E is *complete* iff every Cauchy sequence converges to a limit (which is unique, since a metric space is Hausdorff).

Every finite-dimensional vector space over $\mathbb{R}$ or $\mathbb{C}$ is complete. One can show by induction that given any basis $(e_1, \ldots, e_n)$ of E, the linear map $h \colon \mathbb{C}^n \to E$ defined such that

$$h((z_1, \ldots, z_n)) = z_1 e_1 + \cdots + z_n e_n$$

is a homeomorphism (using the *sup*-norm on $\mathbb{C}^n$). One can also use the fact that any two norms on a finite-dimensional vector space over $\mathbb{R}$ or $\mathbb{C}$ are equivalent (see Lang [69], Dixmier [36], or Schwartz [96]).

However, if E has infinite dimension, it may not be complete. When a Hermitian space is complete, a number of properties that hold for finite-dimensional Hermitian spaces also hold for infinite-dimensional spaces. For example, any closed subspace has an orthogonal complement, and in particular, a finite-dimensional subspace has an orthogonal complement. Hermitian spaces that are also complete play an important role in analysis. Since they were first studied by Hilbert, they are called Hilbert spaces.

Definition 7.5. A (complex) Hermitian space $\langle E, \varphi \rangle$ which is a complete normed vector space under the norm $\| \ \|$ induced by φ is called a *Hilbert space*. A real Euclidean space $\langle E, \varphi \rangle$ which is complete under the norm $\| \ \|$ induced by φ is called a *real Hilbert space*.

All the results in this section hold for complex Hilbert spaces as well as for real Hilbert spaces. We state all results for the complex case only, since they also apply to the real case, and since the proofs in the complex case need a little more care.

Example 7.1. The space l^2 of all countably infinite sequences $x = (x_i)_{i \in \mathbb{N}}$ of complex numbers such that $\sum_{i=0}^{\infty} |x_i|^2 < \infty$ is a Hilbert space. It will be shown later that the map $\varphi \colon l^2 \times l^2 \to \mathbb{C}$ defined such that

$$\varphi\left((x_i)_{i \in \mathbb{N}}, (y_i)_{i \in \mathbb{N}}\right) = \sum_{i=0}^{\infty} x_i \overline{y_i}$$

is well defined, and that l^2 is a Hilbert space under φ. In fact, we will prove a more general result (Proposition 7.3).

Example 7.2. The set $C^{\infty}[a, b]$ of smooth functions $f \colon [a, b] \to \mathbb{C}$ is a Hermitian space under the Hermitian form

$$\langle f, g \rangle = \int_a^b f(x)\overline{g(x)}dx,$$

but it is not a Hilbert space because it is not complete (see Section 6.8 for the definition of the integral of a complex-valued function). It is possible to construct its completion $L^2([a, b])$, which turns out to be the space of Lebesgue square-integrable functions on $[a, b]$.

One of the most important facts about finite-dimensional Hermitian (and Euclidean) spaces is that they have orthonormal bases. This implies that, up to isomorphism, every finite-dimensional Hermitian space is isomorphic to $\mathbb{C}^n$ (for some $n \in \mathbb{N}$) and that the inner product is given by

$$\langle (x_1, \ldots, x_n), (y_1, \ldots, y_n) \rangle = \sum_{i=1}^{n} x_i \overline{y_i}.$$

Furthermore, every subspace W has an orthogonal complement $W^\perp$, and the inner product induces a natural duality between E and E^*, where E^* is the space of linear forms on E.

When E is a Hilbert space, E may be infinite dimensional, often of uncountable dimension. Thus, we can't expect that E always have an orthonormal basis. However, if we modify the notion of basis so that a "Hilbert basis" is an orthogonal family that is also dense in E, i.e., every $v \in E$ is the limit of a sequence of finite combinations of vectors from the Hilbert basis, then we can recover most of the "nice" properties of finite-dimensional Hermitian spaces. For instance, if $(u_k)_{k \in K}$ is a Hilbert basis, for every $v \in E$, we can define the Fourier coefficients $c_k = \langle v, u_k \rangle / \|u_k\|$, and then, v is the "sum" of its Fourier series $\sum_{k \in K} c_k u_k$. However, the cardinality of the index set K can be very large, and it is necessary to define what it means for a family of vectors indexed by K to be summable. It turns out that every Hilbert space is isomorphic to a space of the form $l^2(K)$, where $l^2(K)$ is a generalization of the space of Example 7.1 (see Theorem 7.7, usually called the Riesz-Fischer theorem).

Definition 7.6. Given a Hilbert space E, a family $(u_k)_{k \in K}$ of nonnull vectors is an *orthogonal family* iff the u_k are pairwise orthogonal, i.e., $\langle u_i, u_j \rangle = 0$ for all $i \neq j$ $(i, j \in K)$, and an *orthonormal family* iff $\langle u_i, u_j \rangle = \delta_{i,j}$, for all $i, j \in K$. A *total orthogonal family (or system)* or *Hilbert basis* is an orthogonal family that is dense in E. This means that for every $v \in E$, for every $\epsilon > 0$, there is some finite subset $I \subseteq K$ and some family $(\lambda_i)_{i \in I}$ of complex numbers, such that

$$\left\| v - \sum_{i \in I} \lambda_i u_i \right\| < \epsilon.$$

Given an orthogonal family $(u_k)_{k \in K}$, for every $v \in E$, for every $k \in K$, the scalar $c_k = \langle v, u_k \rangle / \|u_k\|^2$ is called the *k-th Fourier coefficient of v over* $(u_k)_{k \in K}$.

Remark. The terminology Hilbert basis is misleading, because a Hilbert basis $(u_k)_{k \in K}$ is not necessarily a basis in the algebraic sense. Indeed, in general, $(u_k)_{k \in K}$ does not span E. Intuitively, it takes linear combinations of the u_k's with infinitely many nonnull coefficients to span E. Technically, this is achieved in terms of limits.

In order to avoid the confusion between bases in the algebraic sense and Hilbert bases, some authors refer to algebraic bases as *Hamel bases* and to total orthogonal families (or Hilbert bases) as *Schauder bases*.

Definition 7.7. Given an orthogonal family $(u_k)_{k \in K}$ of a Hilbert space E, for any finite subset I of K, and for any family $(\lambda_i)_{i \in I}$ of complex numbers, we call sums of the form $\sum_{i \in I} \lambda_i u_i$ *partial sums of Fourier series*, and if these partial sums converge to a limit denoted as $\sum_{k \in K} c_k u_k$, we call $\sum_{k \in K} c_k u_k$ a *Fourier series*.

However, we have to make sense of such sums! Indeed, when K is unordered or uncountable, the notion of limit or sum has not been defined. This can be done as follows (for more details, see Dixmier [36]).

Definition 7.8. Given a normed vector space E (say, a Hilbert space), for any nonempty index set K, we say that a family $(u_k)_{k \in K}$ of vectors in E is *summable with sum* $v \in E$ iff for every $\epsilon > 0$, there is some finite subset I of K, such that,

$$\left\| v - \sum_{j \in J} u_j \right\| < \epsilon$$

for every finite subset J with $I \subseteq J \subseteq K$. We say that the family $(u_k)_{k \in K}$ is *summable* iff there is some $v \in E$ such that $(u_k)_{k \in K}$ is summable with sum v. A family $(u_k)_{k \in K}$ is a *Cauchy family* iff for every $\epsilon > 0$, there is a finite subset I of K, such that,

$$\left\| \sum_{j \in J} u_j \right\| < \epsilon$$

for every finite subset J of K with $I \cap J = \emptyset$.

If $(u_k)_{k \in K}$ is summable with sum v, we usually denote v as $\sum_{k \in K} u_k$.

Remark. The notion of summability implies that the sum of a family $(u_k)_{k \in K}$ is independent of any order on K. In this sense, it is a kind of "commutative summability." More precisely, it is easy to show that for every bijection $\varphi \colon K \to K$ (intuitively, a reordering of K), the family $(u_k)_{k \in K}$ is summable iff the family $(u_l)_{l \in \varphi(K)}$ is summable, and if so, they have the same sum.

To state some important properties of Fourier coefficients the following technical proposition, whose proof is found in Bourbaki [15], will be needed.

Proposition 7.1. *Let E be a complete normed vector space (say, a Hilbert space).*

(1) For any nonempty index set K, a family $(u_k)_{k \in K}$ is summable iff it is a Cauchy family.

(2) Given a family $(r_k)_{k \in K}$ of nonnegative reals $r_k \geq 0$, if there is some real number $B > 0$ such that $\sum_{i \in I} r_i < B$ for every finite subset I of K, then $(r_k)_{k \in K}$ is

summable and $\sum_{k \in K} r_k = r$, where r is least upper bound of the set of finite
sums $\sum_{i \in I} r_i$ ($I \subseteq K$).

The following proposition gives some of the main properties of Fourier coefficients. Among other things, at most countably many of the Fourier coefficient may be nonnull, and the partial sums of a Fourier series converge. Given an orthogonal family $(u_k)_{k \in K}$, we let $U_k = \mathbb{C} u_k$.

Proposition 7.2. *Let E be a Hilbert space, $(u_k)_{k \in K}$ an orthogonal family in E, and V the closure of the subspace generated by $(u_k)_{k \in K}$. The following properties hold:*

(1) For every $v \in E$, for every finite subset $I \subseteq K$, we have

$$\sum_{i \in I} |c_i|^2 \leq \|v\|^2,$$

where the $c_k = \langle v, u_k \rangle / \|u_k\|^2$ are the Fourier coefficients of v.

(2) For every vector $v \in E$, if $(c_k)_{k \in K}$ are the Fourier coefficients of v, the following conditions are equivalent:

(2a) $v \in V$
(2b) The family $(c_k u_k)_{k \in K}$ is summable and $v = \sum_{k \in K} c_k u_k$.
(2c) The family $(|c_k|^2)_{k \in K}$ is summable and $\|v\|^2 = \sum_{k \in K} |c_k|^2$;

(3) The family $(|c_k|^2)_{k \in K}$ is summable, and we have the Bessel inequality:

$$\sum_{k \in K} |c_k|^2 \leq \|v\|^2.$$

As a consequence, at most countably many of the c_k may be nonzero. The family $(c_k u_k)_{k \in K}$ forms a Cauchy family, and thus, the Fourier series $\sum_{k \in K} c_k u_k$ converges in E to some vector $u = \sum_{k \in K} c_k u_k$.

See Figure 7.1.

Proof.

(1) Let

$$u_I = \sum_{i \in I} c_i u_i$$

for any finite subset I of K. We claim that $v - u_I$ is orthogonal to u_i for every $i \in I$. Indeed,

$$\langle v - u_I, u_i \rangle = \left\langle v - \sum_{j \in I} c_j u_j, u_i \right\rangle$$

Fig. 7.1 A schematic illustration of Proposition 7.2. Figure (i.) illustrates Condition (2b), while Figure (ii.) illustrates Condition (3). Note E is the purple oval and V is the magenta oval. In both cases, take a vector of E, form the Fourier coefficients c_k, then form the Fourier series $\sum_{k \in K} c_k u_k$. Condition (2b) ensures v equals its Fourier series since $v \in V$. However, if $v \notin V$, the Fourier series does not equal v. Eventually, we will discover that $V = E$, which implies that Fourier series converges to its vector v.

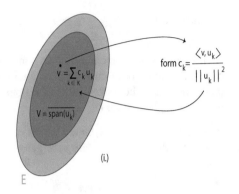

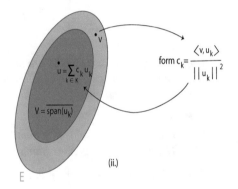

$$= \langle v, u_i \rangle - \sum_{j \in I} c_j \langle u_j, u_i \rangle$$

$$= \langle v, u_i \rangle - c_i \|u_i\|^2$$

$$= \langle v, u_i \rangle - \langle v, u_i \rangle = 0,$$

since $\langle u_j, u_i \rangle = 0$ for all $i \neq j$ and $c_i = \langle v, u_i \rangle / \|u_i\|^2$. As a consequence, we have

$$\|v\|^2 = \left\| v - \sum_{i \in I} c_i u_i + \sum_{i \in I} c_i u_i \right\|^2$$

$$= \left\| v - \sum_{i \in I} c_i u_i \right\|^2 + \left\| \sum_{i \in I} c_i u_i \right\|^2$$

$$= \left\| v - \sum_{i \in I} c_i u_i \right\|^2 + \sum_{i \in I} |c_i|^2,$$

since the u_i are pairwise orthogonal; that is,

$$\|v\|^2 = \left\|v - \sum_{i \in I} c_i u_i \right\|^2 + \sum_{i \in I} |c_i|^2,$$

which in turn implies

$$\sum_{i \in I} |c_i|^2 \le \|v\|^2,$$

as claimed.

(2) We prove the chain of implications $(a) \Rightarrow (b) \Rightarrow (c) \Rightarrow (a)$.

$(a) \Rightarrow (b)$ If $v \in V$, since V is the closure of the subspace spanned by $(u_k)_{k \in K}$, for every $\epsilon > 0$, there is some finite subset I of K and some family $(\lambda_i)_{i \in I}$ of complex numbers, such that

$$\left\|v - \sum_{i \in I} \lambda_i u_i \right\| < \epsilon.$$

Now for every finite subset J of K such that $I \subseteq J$, we have

$$\left\|v - \sum_{i \in I} \lambda_i u_i \right\|^2 = \left\|v - \sum_{j \in J} c_j u_j + \sum_{j \in J} c_j u_j - \sum_{i \in I} \lambda_i u_i \right\|^2$$

$$= \left\|v - \sum_{j \in J} c_j u_j \right\|^2 + \left\|\sum_{j \in J} c_j u_j - \sum_{i \in I} \lambda_i u_i \right\|^2,$$

since $I \subseteq J$ and the u_j (with $j \in J$) are orthogonal to $v - \sum_{j \in J} c_j u_j$ by the argument in (1), which shows that

$$\left\|v - \sum_{j \in J} c_j u_j \right\| \le \left\|v - \sum_{i \in I} \lambda_i u_i \right\| < \epsilon,$$

and thus, that the family $(c_k u_k)_{k \in K}$ is summable with sum v, so that

$$v = \sum_{k \in K} c_k u_k.$$

$(b) \Rightarrow (c)$ If $v = \sum_{k \in K} c_k u_k$, then for every $\epsilon > 0$, there some finite subset I of K, such that

$$\left\|v - \sum_{j \in J} c_j u_j \right\| < \sqrt{\epsilon},$$

for every finite subset J of K such that $I \subseteq J$, and since we proved in (1) that

$$\|v\|^2 = \left\| v - \sum_{j \in J} c_j u_j \right\|^2 + \sum_{j \in J} |c_j|^2,$$

we get

$$\|v\|^2 - \sum_{j \in J} |c_j|^2 < \epsilon,$$

which proves that $(|c_k|^2)_{k \in K}$ is summable with sum $\|v\|^2$.

$(c) \Rightarrow (a)$ Finally, if $(|c_k|^2)_{k \in K}$ is summable with sum $\|v\|^2$, for every $\epsilon > 0$, there is some finite subset I of K such that

$$\|v\|^2 - \sum_{j \in J} |c_j|^2 < \epsilon^2$$

for every finite subset J of K such that $I \subseteq J$, and again, using the fact that

$$\|v\|^2 = \left\| v - \sum_{j \in J} c_j u_j \right\|^2 + \sum_{j \in J} |c_j|^2,$$

we get

$$\left\| v - \sum_{j \in J} c_j u_j \right\| < \epsilon,$$

which proves that $(c_k u_k)_{k \in K}$ is summable with sum $\sum_{k \in K} c_k u_k = v$, and $v \in V$.

(3) Since Part (1) implies $\sum_{i \in I} |c_i|^2 \leq \|v\|^2$ for every finite subset I of K, by Proposition 7.1 (2), the family $(|c_k|^2)_{k \in K}$ is summable. The Bessel inequality

$$\sum_{k \in K} |c_k|^2 \leq \|v\|^2$$

is an obvious consequence of the inequality $\sum_{i \in I} |c_i|^2 \leq \|v\|^2$ (for every finite $I \subseteq K$). Now, for every natural number $n \geq 1$, if K_n is the subset of K consisting of all c_k such that $|c_k| \geq 1/n$, (i.e., $n|c_k| \geq 1$ whenever $c_k \in K_n$), the number of elements in each K_n is finite since

$$\sum_{k \in K_n} |nc_k|^2 \leq n^2 \sum_{k \in K} |c_k|^2 \leq n^2 \|v\|^2$$

converges. Hence, at most a countable number of the c_k may be nonzero.

Since $(|c_k|^2)_{k \in K}$ is summable with sum c, Proposition 7.1 (1) shows that for every $\epsilon > 0$, there is some finite subset I of K such that

$$\sum_{j \in J} |c_j|^2 < \epsilon^2$$

for every finite subset J of K such that $I \cap J = \emptyset$. Since

$$\left\| \sum_{j \in J} c_j u_j \right\|^2 = \sum_{j \in J} |c_j|^2,$$

we get

$$\left\| \sum_{j \in J} c_j u_j \right\| < \epsilon.$$

This proves that $(c_k u_k)_{k \in K}$ is a Cauchy family, which, by Proposition 7.1 (1), implies that $(c_k u_k)_{k \in K}$ is summable, since E is complete. Thus, the Fourier series $\sum_{k \in K} c_k u_k$ is summable, with its sum denoted $u \in V$.   □

Proposition 7.2 suggests looking at the space of sequences $(z_k)_{k \in K}$ (where $z_k \in \mathbb{C}$) such that $(|z_k|^2)_{k \in K}$ is summable. Indeed, such spaces are Hilbert spaces, and it turns out that every Hilbert space is isomorphic to one of those. Such spaces are the infinite-dimensional version of the spaces $\mathbb{C}^n$ under the usual Euclidean norm.

Definition 7.9. Given any nonempty index set K, the space $l^2(K)$ is the set of all sequences $(z_k)_{k \in K}$, where $z_k \in \mathbb{C}$, such that $(|z_k|^2)_{k \in K}$ is summable, i.e., $\sum_{k \in K} |z_k|^2 < \infty$.

Remarks.

(1) When K is a finite set of cardinality n, $l^2(K)$ is isomorphic to $\mathbb{C}^n$.
(2) When $K = \mathbb{N}$, the space $l^2(\mathbb{N})$ corresponds to the space l^2 of Example 7.1. In that example we claimed that l^2 was a Hermitian space, and in fact, a Hilbert space. We now state this fact for any index set K. For a proof of Proposition 7.3 we refer the reader to Schwartz [96]).

Proposition 7.3. *Given any nonempty index set K, the space $l^2(K)$ is a Hilbert space under the Hermitian product*

$$\langle (x_k)_{k \in K}, (y_k)_{k \in K} \rangle = \sum_{k \in K} x_k \overline{y_k}.$$

The subspace consisting of sequences $(z_k)_{k \in K}$ such that $z_k = 0$, except perhaps for finitely many k, is a dense subspace of $l^2(K)$.

We just need two more propositions before being able to prove that every Hilbert space is isomorphic to some $l^2(K)$.

Proposition 7.4. *Let E be a Hilbert space, and $(u_k)_{k \in K}$ an orthogonal family in E. The following properties hold:*

(1) *For every family* $(\lambda_k)_{k\in K} \in l^2(K)$, *the family* $(\lambda_k u_k)_{k\in K}$ *is summable. Furthermore,* $v = \sum_{k\in K} \lambda_k u_k$ *is the only vector such that* $c_k = \lambda_k$ *for all* $k \in K$, *where the* c_k *are the Fourier coefficients of* v.

(2) *For any two families* $(\lambda_k)_{k\in K} \in l^2(K)$ *and* $(\mu_k)_{k\in K} \in l^2(K)$, *if* $v = \sum_{k\in K} \lambda_k u_k$ *and* $w = \sum_{k\in K} \mu_k u_k$, *we have the following equation, also called Parseval identity:*

$$\langle v, w \rangle = \sum_{k\in K} \lambda_k \overline{\mu_k}.$$

Proof.

(1) The fact that $(\lambda_k)_{k\in K} \in l^2(K)$ means that $(|\lambda_k|^2)_{k\in K}$ is summable. The proof given in Proposition 7.2 (3) applies to the family $(|\lambda_k|^2)_{k\in K}$ (instead of $(|c_k|^2)_{k\in K}$), and yields the fact that $(\lambda_k u_k)_{k\in K}$ is summable. Letting $v = \sum_{k\in K} \lambda_k u_k$, recall that $c_k = \langle v, u_k \rangle / \|u_k\|^2$. Pick some $k \in K$. Since $\langle -, - \rangle$ is continuous, for every $\epsilon > 0$, there is some $\eta > 0$ such that

$$| \langle v, u_k \rangle - \langle w, u_k \rangle | = | \langle v - w, u_k \rangle | < \epsilon \|u_k\|^2$$

whenever

$$\|v - w\| < \eta.$$

However, since for every $\eta > 0$, there is some finite subset I of K such that

$$\left\| v - \sum_{j\in J} \lambda_j u_j \right\| < \eta$$

for every finite subset J of K such that $I \subseteq J$, we can pick $J = I \cup \{k\}$, and letting $w = \sum_{j\in J} \lambda_j u_j$, we get

$$\left| \langle v, u_k \rangle - \left\langle \sum_{j\in J} \lambda_j u_j, u_k \right\rangle \right| < \epsilon \|u_k\|^2.$$

However,

$$\langle v, u_k \rangle = c_k \|u_k\|^2 \quad \text{and} \quad \left\langle \sum_{j\in J} \lambda_j u_j, u_k \right\rangle = \lambda_k \|u_k\|^2,$$

and thus, the above proves that $|c_k - \lambda_k| < \epsilon$ for every $\epsilon > 0$, and thus, that $c_k = \lambda_k$.

(2) Since $\langle -, - \rangle$ is continuous, for every $\epsilon > 0$, there are some $\eta_1 > 0$ and $\eta_2 > 0$, such that

$$|\langle x, y \rangle| < \epsilon$$

whenever $\|x\| < \eta_1$ and $\|y\| < \eta_2$. Since $v = \sum_{k \in K} \lambda_k u_k$ and $w = \sum_{k \in K} \mu_k u_k$, there is some finite subset I_1 of K such that

$$\left\| v - \sum_{i \in I} \lambda_i u_i \right\| < \eta_1$$

for every finite subset I of K such that $I_1 \subseteq I$, and there is some finite subset I_2 of K such that

$$\left\| w - \sum_{i \in I} \mu_i u_i \right\| < \eta_2$$

for every finite subset I of K such that $I_2 \subseteq I$. Letting $I = I_1 \cup I_2$, we get

$$\left| \left\langle v - \sum_{i \in I} \lambda_i u_i, \, w - \sum_{i \in I} \mu_i u_i \right\rangle \right| < \epsilon.$$

Furthermore,

$$\langle v, w \rangle = \left\langle v - \sum_{i \in I} \lambda_i u_i + \sum_{i \in I} \lambda_i u_i, \, w - \sum_{i \in I} \mu_i u_i + \sum_{i \in I} \mu_i u_i \right\rangle$$

$$= \left\langle v - \sum_{i \in I} \lambda_i u_i, \, w - \sum_{i \in I} \mu_i u_i \right\rangle + \sum_{i \in I} \lambda_i \overline{\mu_i},$$

since the u_i are orthogonal to $v - \sum_{i \in I} \lambda_i u_i$ and $w - \sum_{i \in I} \mu_i u_i$ for all $i \in I$. This proves that for every $\epsilon > 0$, there is some finite subset I of K such that

$$\left| \langle v, w \rangle - \sum_{i \in I} \lambda_i \overline{\mu_i} \right| < \epsilon.$$

We already know from Proposition 7.3 that $(\lambda_k \overline{\mu_k})_{k \in K}$ is summable, and since $\epsilon > 0$ is arbitrary, we get

$$\langle v, w \rangle = \sum_{k \in K} \lambda_k \overline{\mu_k}.$$

$\square$

The next proposition states properties characterizing Hilbert bases (total orthogonal families).

Fig. 7.2 A schematic
illustration of Proposition 7.5.
Since $(u_k)_{k \in K}$ is a Hilbert
basis, $V = E$. Then given a
vector of E, if we form the
Fourier coefficients c_k, then
form the Fourier series
$\sum_{k \in K} c_k u_k$, we are ensured
that v is equal to its Fourier
series.

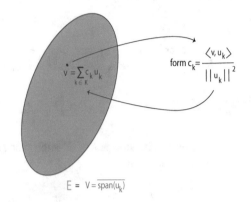

Proposition 7.5. *Let E be a Hilbert space, and let $(u_k)_{k \in K}$ be an orthogonal family in E. The following properties are equivalent:*

(1) *The family $(u_k)_{k \in K}$ is a total orthogonal family.*
(2) *For every vector $v \in E$, if $(c_k)_{k \in K}$ are the Fourier coefficients of v, then the family $(c_k u_k)_{k \in K}$ is summable and $v = \sum_{k \in K} c_k u_k$.*
(3) *For every vector $v \in E$, we have the Parseval identity:*

$$\|v\|^2 = \sum_{k \in K} |c_k|^2.$$

(4) *For every vector $u \in E$, if $\langle u, u_k \rangle = 0$ for all $k \in K$, then $u = 0$.*

See Figure 7.2.

Proof. The equivalence of (1), (2), and (3) is an immediate consequence of Proposition 7.2 and Proposition 7.4. It remains to show that (1) and (4) are equivalent.

(1) $\Rightarrow$ (4) If $(u_k)_{k \in K}$ is a total orthogonal family and $\langle u, u_k \rangle = 0$ for all $k \in K$, since $u = \sum_{k \in K} c_k u_k$ where $c_k = \langle u, u_k \rangle / \|u_k\|^2$, we have $c_k = 0$ for all $k \in K$, and $u = 0$.

(4) $\Rightarrow$ (1) Conversely, assume that the closure V, where V is the subspace generated by $(u_k)_{k \in K}$, is different from E. Then we have $E = V \oplus V^\perp$, where $V^\perp$ is the orthogonal complement of V, and $V^\perp$ is nontrivial since $V \neq E$. As a consequence, there is some nonnull vector $u \in V^\perp$. But then, u is orthogonal to every vector in V, and in particular,

$$\langle u, u_k \rangle = 0$$

for all $k \in K$, which, by assumption, implies that $u = 0$, contradicting the fact that $u \neq 0$.   □

At last, we can prove that every Hilbert space is isomorphic to some Hilbert space $l^2(K)$ for some suitable K.

First, we need the fact that every Hilbert space has some Hilbert basis. This proof uses *Zorn's Lemma* (see Rudin [92]).

Proposition 7.6. *Let E be a Hilbert space. Given any orthogonal family $(u_k)_{k \in K}$ in E, there is a total orthogonal family $(u_l)_{l \in L}$ containing $(u_k)_{k \in K}$.*

All Hilbert bases for a Hilbert space E have index sets K of the same cardinality. For a proof, see Bourbaki [15].

Definition 7.10. A Hilbert space E is *separable* if its Hilbert bases are countable.

Theorem 7.7 (Riesz-Fischer). *For every Hilbert space E, there is some nonempty set K such that E is isomorphic to the Hilbert space $l^2(K)$. More specifically, for any Hilbert basis $(u_k)_{k \in K}$ of E, the maps $f : l^2(K) \to E$ and $g : E \to l^2(K)$ defined such that*

$$f\left((\lambda_k)_{k \in K}\right) = \sum_{k \in K} \lambda_k u_k \quad and \quad g(u) = \left(\langle u, u_k \rangle / \|u_k\|^2\right)_{k \in K} = (c_k)_{k \in K},$$

are bijective linear isometries such that $g \circ f = \mathrm{id}$ and $f \circ g = \mathrm{id}$.

Proof. By Proposition 7.4 (1), the map f is well defined, and it clearly linear. By Proposition 7.2 (3), the map g is well defined, and it is also clearly linear. By Proposition 7.2 (2b), we have

$$f(g(u)) = u = \sum_{k \in K} c_k u_k,$$

and by Proposition 7.4 (1), we have

$$g(f\left((\lambda_k)_{k \in K}\right)) = (\lambda_k)_{k \in K},$$

and thus $g \circ f = \mathrm{id}$ and $f \circ g = \mathrm{id}$. By Proposition 7.4 (2), the linear map g is an isometry. Therefore, f is a linear bijection and an isometry between $l^2(K)$ and E, with inverse g.   □

Remark. The surjectivity of the map $g : E \to l^2(K)$ is known as the *Riesz-Fischer* theorem.

Having done all this hard work, we sketch how these results apply to Fourier series. Again, we refer the readers to Rudin [92] or Lang [68, 69] for a comprehensive exposition.

Let $\mathcal{C}(T)$ denote the set of all periodic continuous functions $f : [-\pi, \pi] \to \mathbb{C}$ with period 2π. There is a Hilbert space $L^2(T)$ containing $\mathcal{C}(T)$ and such that $\mathcal{C}(T)$ is dense in $L^2(T)$, whose inner product is given by

$$\langle f, g \rangle = \int_{-\pi}^{\pi} f(x)\overline{g(x)}dx.$$

The Hilbert space $L^2(T)$ is the space of *Lebesgue square-integrable periodic functions* (of period 2π).

It turns out that the family $(e^{ikx})_{k \in \mathbb{Z}}$ is a total orthogonal family in $L^2(T)$, because it is already dense in $\mathcal{C}(T)$ (for instance, see Rudin [92]). Then the Riesz-Fischer theorem says that for every family $(c_k)_{k \in \mathbb{Z}}$ of complex numbers such that

$$\sum_{k \in \mathbb{Z}} |c_k|^2 < \infty,$$

there is a unique function $f \in L^2(T)$ such that f is equal to its Fourier series

$$f(x) = \sum_{k \in \mathbb{Z}} c_k e^{ikx},$$

where the Fourier coefficients c_k of f are given by the formula

$$c_k = \frac{1}{2\pi} \int_{-\pi}^{\pi} f(t)e^{-ikt}dt.$$

The Parseval theorem says that

$$\sum_{k=-\infty}^{+\infty} c_k \overline{d_k} = \frac{1}{2\pi} \int_{-\pi}^{\pi} f(t)\overline{g(t)}dt$$

for all $f, g \in L^2(T)$, where c_k and d_k are the Fourier coefficients of f and g.

Thus, there is an isomorphism between the two Hilbert spaces $L^2(T)$ and $l^2(\mathbb{Z})$, which is the deep reason why the Fourier coefficients "work." Theorem 7.7 implies that the Fourier series $\sum_{k \in \mathbb{Z}} c_k e^{ikx}$ of a function $f \in L^2(T)$ converges to f in the L^2-sense, i.e., in the mean-square sense. This does not necessarily imply that the Fourier series converges to f pointwise! This is a subtle issue, and for more on this subject, the reader is referred to Lang [68, 69] or Schwartz [97, 98].

An alternative Hilbert basis for $L^2(T)$ is given by $\{\cos kx, \sin kx\}_{k=0}^{\infty}$. This particular Hilbert basis will play an important role in representing the spherical harmonics on S^1 as seen in the next section.

7.2 Spherical Harmonics on the Circle

For the remainder of this chapter we discuss spherical harmonics and take a glimpse at the linear representation of Lie groups. Spherical harmonics on the sphere S^2 have interesting applications in computer graphics and computer vision, so this material is important not only for theoretical reasons but also for practical reasons.

Joseph Fourier (1768–1830) invented Fourier series in order to solve the heat equation [43]. Using Fourier series, every square-integrable periodic function f (of period 2π) can be expressed uniquely as the sum of a power series of the form

$$f(\theta) = a_0 + \sum_{k=1}^{\infty} (a_k \cos k\theta + b_k \cos k\theta),$$

where the *Fourier coefficients* a_k, b_k of f are given by the formulae

$$a_0 = \frac{1}{2\pi} \int_{-\pi}^{\pi} f(\theta) \, d\theta, \quad a_k = \frac{1}{\pi} \int_{-\pi}^{\pi} f(\theta) \cos k\theta \, d\theta, \quad b_k = \frac{1}{\pi} \int_{-\pi}^{\pi} f(\theta) \sin k\theta \, d\theta,$$

for $k \geq 1$. The reader will find the above formulae in Fourier's famous book [43] in Chapter III, Section 233, page 256, essentially using the notation that we use today.

This remarkable discovery has many theoretical and practical applications in physics, signal processing, engineering, *etc.* We can describe Fourier series in a more conceptual manner if we introduce the following inner product on square-integrable functions:

$$\langle f, g \rangle = \int_{-\pi}^{\pi} f(\theta) g(\theta) \, d\theta,$$

which we will also denote by

$$\langle f, g \rangle = \int_{S^1} f(\theta) g(\theta) \, d\theta,$$

where S^1 denotes the unit circle. After all, periodic functions of (period 2π) can be viewed as functions on the circle. With this inner product, the space $L^2(S^1)$ is a complete normed vector space, that is, a Hilbert space. Furthermore, if we define the subspaces $\mathcal{H}_k(S^1)$ of $L^2(S^1)$ so that $\mathcal{H}_0(S^1)$ $(= \mathbb{R})$ is the set of constant functions and $\mathcal{H}_k(S^1)$ is the two-dimensional space spanned by the functions $\cos k\theta$ and $\sin k\theta$, then it turns out that we have a Hilbert sum decomposition

$$L^2(S^1) = \bigoplus_{k=0}^{\infty} \mathcal{H}_k(S^1)$$

into pairwise orthogonal subspaces, where $\bigcup_{k=0}^{\infty} \mathcal{H}_k(S^1)$ is dense in $L^2(S^1)$. The functions $\cos k\theta$ and $\sin k\theta$ are also orthogonal in $\mathcal{H}_k(S^1)$.

Now it turns out that the spaces $\mathcal{H}_k(S^1)$ arise naturally when we look for homogeneous solutions of the Laplace equation $\Delta f = 0$ in $\mathbb{R}^2$ (Pierre-Simon

Laplace, 1749–1827). Roughly speaking, a homogeneous function in $\mathbb{R}^2$ is a function that can be expressed in polar coordinates (r, θ) as

$$f(r, \theta) = r^k g(\theta).$$

Recall that the Laplacian on $\mathbb{R}^2$ expressed in Cartesian coordinates (x, y) is given by

$$\Delta f = \frac{\partial^2 f}{\partial x^2} + \frac{\partial^2 f}{\partial y^2},$$

where $f \colon \mathbb{R}^2 \to \mathbb{R}$ is a function which is at least of class C^2. In polar coordinates (r, θ), where $(x, y) = (r \cos \theta, r \sin \theta)$ and $r > 0$, since

$$\frac{\partial f}{\partial r} = \cos \theta \frac{\partial f}{\partial x} + \sin \theta \frac{\partial f}{\partial y},$$

$$\frac{\partial^2 f}{\partial r^2} = \cos^2 \theta \frac{\partial^2 f}{\partial x^2} + \sin^2 \theta \frac{\partial^2 f}{\partial y^2} + 2 \sin \theta \cos \theta \frac{\partial^2 f}{\partial x \partial y},$$

and

$$\frac{\partial^2 f}{\partial \theta^2} = -r \left(\cos \theta \frac{\partial f}{\partial x} + \sin \theta \frac{\partial f}{\partial y} \right) + r^2 \left(\sin^2 \theta \frac{\partial^2 f}{\partial x^2} - 2 \sin \theta \cos \theta \frac{\partial^2 f}{\partial x \partial y} + \cos^2 \theta \frac{\partial^2 f}{\partial y^2} \right)$$

$$= -r \frac{\partial f}{\partial r} + r^2 \left(\sin^2 \theta \frac{\partial^2 f}{\partial x^2} - 2 \sin \theta \cos \theta \frac{\partial^2 f}{\partial x \partial y} + \cos^2 \theta \frac{\partial^2 f}{\partial y^2} \right),$$

we find that

$$\frac{\partial^2 f}{\partial r^2} + \frac{1}{r^2} \frac{\partial^2 f}{\partial \theta^2} = \frac{\partial^2 f}{\partial x^2} + \frac{\partial^2 f}{\partial y^2} - \frac{1}{r} \frac{\partial f}{\partial r},$$

which implies that the Laplacian (in polar coordinates) is given by

$$\Delta f = \frac{1}{r} \frac{\partial}{\partial r} \left(r \frac{\partial f}{\partial r} \right) + \frac{1}{r^2} \frac{\partial^2 f}{\partial \theta^2}.$$

If we restrict f to the unit circle S^1, then the Laplacian on S^1 is given by

$$\Delta_{s^1} f = \frac{\partial^2 f}{\partial \theta^2}.$$

It turns out that *the space $\mathcal{H}_k(S^1)$ is the eigenspace of Δ_{S^1} for the eigenvalue $-k^2$*.

To show this, we consider another question, namely *what are the harmonic functions on* $\mathbb{R}^2$; that is, the functions f that are solutions of the *Laplace equation*

$$\Delta f = 0.$$

Our ancestors had the idea that the above equation can be solved by *separation of variables*. This means that we write $f(r, \theta) = F(r)g(\theta)$, where $F(r)$ and $g(\theta)$ are independent functions. To make things easier, let us assume that $F(r) = r^k$ for some integer $k \geq 0$, which means that we assume that f is a *homogeneous function of degree k*. Recall that a function $f : \mathbb{R}^2 \to \mathbb{R}$ is *homogeneous of degree k* iff

$$f(tx, ty) = t^k f(x, y) \qquad \text{for all } t > 0.$$

Now, using the Laplacian in polar coordinates, we get

$$
\begin{aligned}
\Delta f &= \frac{1}{r}\frac{\partial}{\partial r}\left(r\frac{\partial(r^k g(\theta))}{\partial r}\right) + \frac{1}{r^2}\frac{\partial^2(r^k g(\theta))}{\partial\theta^2}\\
&= \frac{1}{r}\frac{\partial}{\partial r}\left(kr^k g\right) + r^{k-2}\frac{\partial^2 g}{\partial\theta^2}\\
&= r^{k-2}k^2 g + r^{k-2}\frac{\partial^2 g}{\partial\theta^2}\\
&= r^{k-2}(k^2 g + \Delta_{S^1} g).
\end{aligned}
$$

Thus, we deduce that

$$\Delta f = 0 \quad \text{iff} \quad \Delta_{S^1} g = -k^2 g;$$

that is, g is an eigenfunction of Δ_{S^1} for the eigenvalue $-k^2$. But the above equation is equivalent to the second-order differential equation

$$\frac{d^2 g}{d\theta^2} + k^2 g = 0,$$

whose general solution is given by

$$g(\theta) = a_n \cos k\theta + b_n \sin k\theta.$$

In summary, we showed the following facts.

Proposition 7.8. *The integers* $0, -1, -4, -9, \ldots, -k^2, \ldots$ *are eigenvalues of* Δ_{S^1}, *and the functions* $\cos k\theta$ *and* $\sin k\theta$ *are eigenfunctions for the eigenvalue* $-k^2$, *with* $k \geq 0$.

It looks like the dimension of the eigenspace corresponding to the eigenvalue $-k^2$ is 1 when $k = 0$, and 2 when $k \geq 1$.

It can indeed be shown that Δ_{S^1} has no other eigenvalues and that the dimensions claimed for the eigenspaces are correct. Observe that if we go back to our homogeneous harmonic functions $f(r, \theta) = r^k g(\theta)$, we see that this space is spanned by the functions

$$u_k = r^k \cos k\theta, \qquad v_k = r^k \sin k\theta.$$

Now, $(x + iy)^k = r^k(\cos k\theta + i \sin k\theta)$, and since $\Re(x + iy)^k = u_k$ and $\mathrm{Im}\,(x + iy)^k = v_k$ are homogeneous polynomials, we see that u_k and v_k are homogeneous polynomials called *harmonic polynomials*. For example, here is a list of a basis for the harmonic polynomials (in two variables) of degree $k = 0, 1, 2, 3, 4$, listed as $\tilde{u}_k = \cos k\theta$, $\tilde{v}_k = \sin k\theta$:

$k = 0$	1
$k = 1$	$x,\ y$
$k = 2$	$x^2 - y^2,\ 2xy$
$k = 3$	$x^3 - 3xy^2,\ 3x^2y - y^3$
$k = 4$	$x^4 - 6x^2y^2 + y^4,\ x^3y - xy^3.$

To derive these formulas, we simply expand $(x + iy)^k$ via the binomial theorem and take $\tilde{u}_k$ as the real part, and $\tilde{v}_k$ as the imaginary part.

Therefore, the eigenfunctions of the Laplacian on S^1 are the restrictions of the harmonic polynomials on $\mathbb{R}^2$ to S^1, and we have a Hilbert sum decomposition $L^2(S^1) = \bigoplus_{k=0}^{\infty} \mathcal{H}_k(S^1)$. It turns out that this phenomenon generalizes to the sphere $S^n \subseteq \mathbb{R}^{n+1}$ for all $n \geq 1$.

Let us take a look at next case $n = 2$.

7.3   Spherical Harmonics on the 2-Sphere

The material of section is very classical and can be found in many places, for example Andrews, Askey, and Roy [1] (Chapter 9), Sansone [94] (Chapter III), Hochstadt [58] (Chapter 6), and Lebedev [72] (Chapter). We recommend the exposition in Lebedev [72] because we find it particularly clear and uncluttered. We have also borrowed heavily from some lecture notes by Hermann Gluck for a course he offered in 1997–1998.

Our goal is to find the homogeneous solutions of the Laplace equation $\Delta f = 0$ in $\mathbb{R}^3$, and to show that they correspond to spaces $\mathcal{H}_k(S^2)$ of eigenfunctions of the Laplacian Δ_{S^2} on the 2-sphere

$$S^2 = \{(x, y, z) \in \mathbb{R}^3 \mid x^2 + y^2 + z^2 = 1\}.$$

Then the spaces $\mathcal{H}_k(S^2)$ will give us a Hilbert sum decomposition of the Hilbert space $L^2(S^2)$ of square-integrable functions on S^2. This is the generalization of Fourier series to the 2-sphere and the functions in the spaces $\mathcal{H}_k(S^2)$ are called *spherical harmonics*.

The Laplacian in $\mathbb{R}^3$ is of course given by

$$\Delta f = \frac{\partial^2 f}{\partial x^2} + \frac{\partial^2 f}{\partial y^2} + \frac{\partial^2 f}{\partial z^2}.$$

If we use spherical coordinates

$$x = r \sin \theta \cos \varphi$$

$$y = r \sin \theta \sin \varphi$$

$$z = r \cos \theta,$$

in $\mathbb{R}^3$, where $0 \leq \theta < \pi, 0 \leq \varphi < 2\pi$ and $r > 0$ (recall that φ is the so-called *azimuthal angle* in the xy-plane originating at the x-axis and θ is the so-called *polar angle* from the z-axis, angle defined in the plane obtained by rotating the xz-plane around the z-axis by the angle φ), then since

$$\frac{\partial f}{\partial r} = \sin \theta \cos \varphi \frac{\partial f}{\partial x} + \sin \theta \sin \varphi \frac{\partial f}{\partial y} + \cos \theta \frac{\partial f}{\partial z},$$

$$\frac{\partial^2 f}{\partial r^2} = \sin^2 \theta \cos^2 \varphi \frac{\partial^2 f}{\partial x^2} + \sin^2 \theta \sin^2 \varphi \frac{\partial^2 f}{\partial y^2} + \cos^2 \theta \frac{\partial^2 f}{\partial z^2}$$

$$+ 2 \sin^2 \theta \sin \varphi \cos \varphi \frac{\partial^2 f}{\partial x \partial y} + 2 \sin \theta \cos \theta \cos \varphi \frac{\partial^2 f}{\partial x \partial z}$$

$$+ 2 \cos \theta \sin \theta \sin \varphi \frac{\partial^2 f}{\partial y \partial z},$$

$$\frac{\partial f}{\partial \theta} = r \cos \theta \cos \varphi \frac{\partial f}{\partial x} + r \cos \theta \sin \varphi \frac{\partial f}{\partial y} - r \sin \theta \frac{\partial f}{\partial z},$$

$$\frac{\partial^2 f}{\partial \theta^2} = -r\frac{\partial f}{\partial r} + r^2 \cos^2 \theta \cos^2 \varphi \frac{\partial^2 f}{\partial x^2} + r^2 \cos^2 \theta \sin^2 \varphi \frac{\partial^2 f}{\partial y^2} + r^2 \sin^2 \theta \frac{\partial^2 f}{\partial z^2}$$

$$+ 2r^2 \cos^2 \theta \cos \varphi \sin \varphi \frac{\partial^2 f}{\partial x \partial y} - 2r^2 \cos \theta \sin \theta \cos \varphi \frac{\partial^2 f}{\partial x \partial z}$$

$$- 2r^2 \cos \theta \sin \theta \sin \varphi \frac{\partial^2 f}{\partial y \partial z},$$

$$\frac{\partial f}{\partial \varphi} = -r \sin \theta \sin \varphi \frac{\partial f}{\partial x} + r \sin \theta \cos \varphi \frac{\partial f}{\partial y},$$

and

$$\frac{\partial^2 f}{\partial \varphi^2} = -r \sin \theta \cos \varphi \frac{\partial f}{\partial x} - r \sin \theta \sin \varphi \frac{\partial f}{\partial y} + r^2 \sin^2 \theta \sin^2 \varphi \frac{\partial^2 f}{\partial x^2}$$

$$+ r^2 \sin^2 \theta \cos^2 \varphi \frac{\partial^2 f}{\partial y^2} - 2r^2 \sin^2 \theta \cos \varphi \sin \varphi \frac{\partial^2 f}{\partial x \partial y},$$

we discover that

$$\frac{\partial^2 f}{\partial r^2} + \frac{1}{r^2}\frac{\partial^2 f}{\partial \theta^2} + \frac{1}{r^2 \sin^2 \theta}\frac{\partial^2 f}{\partial \varphi^2} + \frac{2}{r}\frac{\partial f}{\partial r} + \frac{\cos \theta}{r^2 \sin \theta}\frac{\partial f}{\partial \theta} = \frac{\partial^2 f}{\partial x^2} + \frac{\partial^2 f}{\partial y^2} + \frac{\partial^2 f}{\partial z^2},$$

which implies that the Laplacian in spherical coordinates is given by

$$\Delta f = \frac{1}{r^2}\frac{\partial}{\partial r}\left(r^2 \frac{\partial f}{\partial r}\right) + \frac{1}{r^2}\Delta_{S^2} f,$$

where

$$\Delta_{S^2} f = \frac{1}{\sin \theta}\frac{\partial}{\partial \theta}\left(\sin \theta \frac{\partial f}{\partial \theta}\right) + \frac{1}{\sin^2 \theta}\frac{\partial^2 f}{\partial \varphi^2},$$

is the Laplacian on the sphere S^2. Let us look for homogeneous harmonic functions $f(r, \theta, \varphi) = r^k g(\theta, \varphi)$ on $\mathbb{R}^3$; that is, solutions of the Laplace equation

$$\Delta f = 0.$$

We get

$$\Delta f = \frac{1}{r^2}\frac{\partial}{\partial r}\left(r^2 \frac{\partial (r^k g)}{\partial r}\right) + \frac{1}{r^2}\Delta_{S^2}(r^k g)$$

$$= \frac{1}{r^2} \frac{\partial}{\partial r} \left(kr^{k+1}g \right) + r^{k-2} \Delta_{S^2} g$$

$$= r^{k-2} k(k+1)g + r^{k-2} \Delta_{S^2} g$$

$$= r^{k-2} (k(k+1)g + \Delta_{S^2} g).$$

Therefore,

$$\Delta f = 0 \quad \text{iff} \quad \Delta_{S^2} g = -k(k+1)g;$$

that is, g is an eigenfunction of Δ_{S^2} for the eigenvalue $-k(k+1)$.

We can look for solutions of the above equation using the separation of variables method. If we let $g(\theta, \varphi) = \Theta(\theta)\Phi(\varphi)$, then we get the equation

$$\frac{\Phi}{\sin\theta} \frac{\partial}{\partial\theta} \left(\sin\theta \frac{\partial\Theta}{\partial\theta} \right) + \frac{\Theta}{\sin^2\theta} \frac{\partial^2\Phi}{\partial\varphi^2} = -k(k+1)\Theta\Phi;$$

that is, dividing by $\Theta\Phi$ and multiplying by $\sin^2\theta$,

$$\frac{\sin\theta}{\Theta} \frac{\partial}{\partial\theta} \left(\sin\theta \frac{\partial\Theta}{\partial\theta} \right) + k(k+1)\sin^2\theta = -\frac{1}{\Phi} \frac{\partial^2\Phi}{\partial\varphi^2}.$$

Since Θ and Φ are independent functions, the above is possible only if both sides are equal to a constant, say μ. This leads to two equations

$$\frac{\partial^2\Phi}{\partial\varphi^2} + \mu\Phi = 0$$

$$\frac{\sin\theta}{\Theta} \frac{\partial}{\partial\theta} \left(\sin\theta \frac{\partial\Theta}{\partial\theta} \right) + k(k+1)\sin^2\theta - \mu = 0.$$

However, we want Φ to be periodic in φ since we are considering functions on the sphere, so μ must be of the form $\mu = m^2$ for some nonnegative integer m. Then we know that the space of solutions of the equation

$$\frac{\partial^2\Phi}{\partial\varphi^2} + m^2\Phi = 0$$

is two dimensional and is spanned by the two functions

$$\Phi(\varphi) = \cos m\varphi, \qquad \Phi(\varphi) = \sin m\varphi.$$

We still have to solve the equation

$$\sin\theta \frac{\partial}{\partial\theta}\left(\sin\theta \frac{\partial\Theta}{\partial\theta}\right) + (k(k+1)\sin^2\theta - m^2)\Theta = 0,$$

which is equivalent to

$$\sin^2\theta\,\Theta'' + \sin\theta\cos\theta\,\Theta' + (k(k+1)\sin^2\theta - m^2)\Theta = 0.$$

a variant of Legendre's equation. For this, we use the change of variable $t = \cos\theta$, and we consider the function u given by $u(\cos\theta) = \Theta(\theta)$ (recall that $0 \le \theta < \pi$), so we get the second-order differential equation

$$(1-t^2)u'' - 2tu' + \left(k(k+1) - \frac{m^2}{1-t^2}\right)u = 0$$

sometimes called the *general Legendre equation* (Adrien-Marie Legendre, 1752–1833). The trick to solve this equation is to make the substitution

$$u(t) = (1-t^2)^{\frac{m}{2}}v(t);$$

see Lebedev [72], Chapter 7, Section 7.12. Then we get

$$(1-t^2)v'' - 2(m+1)tv' + (k(k+1) - m(m+1))v = 0.$$

When $m = 0$, we get the *Legendre equation*:

$$(1-t^2)v'' - 2tv' + k(k+1)v = 0;$$

see Lebedev [72], Chapter 7, Section 7.3.

This equation has two fundamental solutions $P_k(t)$ and $Q_k(t)$ called the *Legendre functions of the first and second kinds*. The $P_k(t)$ are actually polynomials and the $Q_k(t)$ are given by power series that diverge for $t = 1$, so we only keep the *Legendre polynomials* $P_k(t)$. The Legendre polynomials can be defined in various ways. One definition is in terms of *Rodrigues' formula*:

$$P_n(t) = \frac{1}{2^n n!}\frac{d^n}{dt^n}(t^2-1)^n;$$

see Lebedev [72], Chapter 4, Section 4.2. In this version of the Legendre polynomials they are normalized so that $P_n(1) = 1$. There is also the following recurrence relation:

$$P_0 = 1$$

$$P_1 = t$$

$$(n+1)P_{n+1} = (2n+1)tP_n - nP_{n-1} \qquad n \geq 1;$$

see Lebedev [72], Chapter 4, Section 4.3. For example, the first six Legendre polynomials are

$$1$$

$$t$$

$$\frac{1}{2}(3t^2 - 1)$$

$$\frac{1}{2}(5t^3 - 3t)$$

$$\frac{1}{8}(35t^4 - 30t^2 + 3)$$

$$\frac{1}{8}(63t^5 - 70t^3 + 15t).$$

Let us now return to our differential equation

$$(1 - t^2)v'' - 2(m+1)tv' + (k(k+1) - m(m+1))v = 0. \qquad (*)$$

Observe that if we differentiate with respect to t, we get the equation

$$(1 - t^2)v''' - 2(m+2)tv'' + (k(k+1) - (m+1)(m+2))v' = 0.$$

This shows that if v is a solution of our Equation $(*)$ for given k and m, then v' is a solution of the same equation for k and $m+1$. Thus, if $P_k(t)$ solves $(*)$ for given k and $m = 0$, then $P_k'(t)$ solves $(*)$ for the same k and $m = 1$, $P_k''(t)$ solves $(*)$ for the same k and $m = 2$, and in general $d^m/dt^m(P_k(t))$ solves $(*)$ for k and m. Therefore, our original equation

$$(1 - t^2)u'' - 2tu' + \left(k(k+1) - \frac{m^2}{1 - t^2} \right) u = 0 \qquad (†)$$

has the solution

$$u(t) = (1 - t^2)^{\frac{m}{2}} \frac{d^m}{dt^m}(P_k(t)) := P_m^k(t).$$

The function $u(t)$ is traditionally denoted $P_k^m(t)$ and called an *associated Legendre function*; see Lebedev [72], Chapter 7, Section 7.12. The index k is often called the *band index*. Obviously, $P_k^m(t) \equiv 0$ if $m > k$ and $P_k^0(t) = P_k(t)$, the Legendre polynomial of degree k. An associated Legendre function is not a polynomial in general, and because of the factor $(1 - t^2)^{\frac{m}{2}}$, it is only defined on the closed interval $[-1, 1]$.

 Certain authors add the factor $(-1)^m$ in front of the expression for the associated Legendre function $P_k^m(t)$, as in Lebedev [72], Chapter 7, Section 7.12, see also Footnote 29 on Page 193. This seems to be a common practice in the quantum mechanics literature where it is called the *Condon Shortley phase factor*.

The associated Legendre functions satisfy various recurrence relations that allow us to compute them. For example, for fixed $m \geq 0$, we have (see Lebedev [72], Chapter 7, Section 7.12) the recurrence

$$(k - m + 1)P_{k+1}^m(t) = (2k + 1)t P_k^m(t) - (k + m) P_{k-1}^m(t), \qquad k \geq 1,$$

and for fixed $k \geq 2$, we have

$$P_k^{m+2}(t) = \frac{2(m + 1)t}{(t^2 - 1)^{\frac{1}{2}}} P_k^{m+1}(t) + (k - m)(k + m + 1) P_k^m(t), \qquad 0 \leq m \leq k - 2,$$

which can also be used to compute P_k^m starting from

$$P_k^0(t) = P_k(t)$$

$$P_k^1(t) = \frac{kt}{(t^2 - 1)^{\frac{1}{2}}} P_k(t) - \frac{k}{(t^2 - 1)^{\frac{1}{2}}} P_{k-1}(t).$$

Observe that the recurrence relation for m fixed yields the following equation for $k = m$ (as $P_{m-1}^m = 0$):

$$P_{m+1}^m(t) = (2m + 1)t P_m^m(t).$$

It also easy to see that

$$P_m^m(t) = \frac{(2m)!}{2^m m!} (1 - t^2)^{\frac{m}{2}}.$$

Observe that

$$\frac{(2m)!}{2^m m!} = (2m - 1)(2m - 3) \cdots 5 \cdot 3 \cdot 1,$$

an expression that is sometimes denoted $(2m - 1)!!$ and called the *double factorial*.

 Beware that some papers in computer graphics adopt the definition of associated Legendre functions with the scale factor $(-1)^m$ added, so this factor is present in these papers, for example Green [51].

The equation above allows us to "lift" P_m^m to the higher band $m+1$. The computer graphics community (see Green [51]) uses the following three rules to compute $P_k^m(t)$ where $0 \leq m \leq k$:

(1) Compute

$$P_m^m(t) = \frac{(2m)!}{2^m m!} (1 - t^2)^{\frac{m}{2}}.$$

If $m = k$, stop. Otherwise do Step 2 once.

(2) Compute $P_{m+1}^m(t) = (2m + 1)t\, P_m^m(t)$. If $k = m + 1$, stop. Otherwise, iterate Step 3.

(3) Starting from $i = m + 1$, compute

$$(i - m + 1)P_{i+1}^m(t) = (2i + 1)t\, P_i^m(t) - (i + m)P_{i-1}^m(t)$$

until $i + 1 = k$.

If we recall that Equation (†) was obtained from the equation

$$\sin^2 \theta\, \Theta'' + \sin \theta \cos \theta\, \Theta' + (k(k + 1) \sin^2 \theta - m^2)\Theta = 0$$

using the substitution $u(\cos \theta) = \Theta(\theta)$, we see that

$$\Theta(\theta) = P_k^m(\cos \theta)$$

is a solution of the above equation. Putting everything together, as $f(r, \theta, \varphi) = r^k \Theta(\theta)\Phi(\varphi)$, we proved the following facts.

Proposition 7.9. *The homogeneous functions*

$$f(r, \theta, \varphi) = r^k \cos m\varphi\, P_k^m(\cos \theta), \qquad f(r, \theta, \varphi) = r^k \sin m\varphi\, P_k^m(\cos \theta)$$

are solutions of the Laplacian Δ in $\mathbb{R}^3$, and the functions

$$\cos m\varphi\, P_k^m(\cos \theta), \qquad \sin m\varphi\, P_k^m(\cos \theta)$$

are eigenfunctions of the Laplacian Δ_{S^2} on the sphere for the eigenvalue $-k(k+1)$.

For k fixed, as $0 \leq m \leq k$, we get $2k + 1$ linearly independent functions.

The notation for the above functions varies quite a bit, essentially because of the choice of normalization factors used in various fields (such as physics, seismology, geodesy, spectral analysis, magnetics, quantum mechanics, *etc.*). We will adopt the notation y_l^m, where l is a nonnegative integer but m is allowed to be negative, with $-l \leq m \leq l$. Thus, we set

$$y_l^m(\theta, \varphi) = \begin{cases} N_l^0 P_l(\cos\theta) & \text{if } m = 0 \\ \sqrt{2}N_l^m \cos m\varphi \, P_l^m(\cos\theta) & \text{if } m > 0 \\ \sqrt{2}N_l^m \sin(-m\varphi) \, P_l^{-m}(\cos\theta) & \text{if } m < 0 \end{cases}$$

for $l = 0, 1, 2, \ldots$, and where the N_l^m are scaling factors. In physics and computer graphics, N_l^m is chosen to be

$$N_l^m = \sqrt{\frac{(2l+1)(l-|m|)!}{4\pi(l+|m|)!}}.$$

Definition 7.11. The functions y_l^m are called the *real spherical harmonics of degree l and order m*. The index l is called the *band index*.

The functions, y_l^m, have some very nice properties, but to explain these we need to recall the Hilbert space structure of the space $L^2(S^2)$ of square-integrable functions on the sphere. Recall that we have an inner product on $L^2(S^2)$ given by

$$\langle f, g \rangle = \int_{S^2} fg \, \mathrm{Vol}_{S^2} = \int_0^{2\pi} \int_0^{\pi} f(\theta, \varphi)g(\theta, \varphi) \sin\theta d\theta d\varphi,$$

where $f, g \in L^2(S^2)$ and where Vol_{S^2} is the volume form on S^2 (induced by the metric on $\mathbb{R}^3$). With this inner product, $L^2(S^2)$ is a complete normed vector space using the norm $\|f\| = \sqrt{\langle f, f \rangle}$ associated with this inner product; that is, $L^2(S^2)$ is a *Hilbert space*. Now, it can be shown that the Laplacian Δ_{S^2} on the sphere is a self-adjoint linear operator with respect to this inner product. As the functions $y_{l_1}^{m_1}$ and $y_{l_2}^{m_2}$ with $l_1 \neq l_2$ are eigenfunctions corresponding to distinct eigenvalues $(-l_1(l_1+1)$ and $-l_2(l_2+1))$, they are orthogonal; that is,

$$\langle y_{l_1}^{m_1}, y_{l_2}^{m_2} \rangle = 0, \qquad \text{if } \quad l_1 \neq l_2.$$

It is also not hard to show that for a fixed l,

$$\langle y_l^{m_1}, y_l^{m_2} \rangle = \delta_{m_1, m_2};$$

that is, the functions y_l^m with $-l \leq m \leq l$ form an orthonormal system, and we denote by $\mathcal{H}_l(S^2)$ the $(2l+1)$-dimensional space spanned by these functions.

It turns out that the functions y_l^m form a basis of the eigenspace E_l of Δ_{S^2} associated with the eigenvalue $-l(l+1)$, so that $E_l = \mathcal{H}_l(S^2)$, and that Δ_{S^2} has no other eigenvalues. More is true. It turns out that $L^2(S^2)$ is the orthogonal Hilbert sum of the eigenspaces $\mathcal{H}_l(S^2)$. This means that the $\mathcal{H}_l(S^2)$ are

(1) mutually orthogonal
(2) closed, and

(3) the space $L^2(S^2)$ is the Hilbert sum $\bigoplus_{l=0}^{\infty} \mathcal{H}_l(S^2)$, which means that for every function $f \in L^2(S^2)$, there is a unique sequence of spherical harmonics $f_j \in \mathcal{H}_l(S^2)$ so that

$$f = \sum_{l=0}^{\infty} f_l;$$

that is, the sequence $\sum_{j=0}^{l} f_j$ converges to f (in the norm on $L^2(S^2)$). Observe that each f_l is a unique linear combination $f_l = \sum_{m_l} a_{m_l l}\, y_l^{m_l}$.

Therefore, (3) gives us a *Fourier decomposition on the sphere* generalizing the familiar Fourier decomposition on the circle. Furthermore, the *Fourier coefficients* $a_{m_l l}$ can be computed using the fact that the y_l^m form an orthonormal Hilbert basis:

$$a_{m_l l} = \langle f, y_l^{m_l} \rangle.$$

We also have the corresponding homogeneous harmonic functions $H_l^m(r, \theta, \varphi)$ on $\mathbb{R}^3$ given by

$$H_l^m(r, \theta, \varphi) = r^l y_l^m(\theta, \varphi).$$

If one starts computing explicitly the H_l^m for small values of l and m, one finds that it is always possible to express these functions in terms of the Cartesian coordinates x, y, z as *homogeneous polynomials*! This remarkable fact holds in general: The eigenfunctions of the Laplacian Δ_{S^2}, and thus the spherical harmonics, are the restrictions of homogeneous harmonic polynomials in $\mathbb{R}^3$. Here is a list of bases of the homogeneous harmonic polynomials of degree k in three variables up to $k = 4$ (thanks to Herman Gluck).

$k = 0$	1
$k = 1$	x, y, z
$k = 2$	$x^2 - y^2, \; x^2 - z^2, \; xy, \; xz, \; yz$
$k = 3$	$x^3 - 3xy^2, \; 3x^2y - y^3, \; x^3 - 3xz^2, \; 3x^2z - z^3,$
	$y^3 - 3yz^2, \; 3y^2z - z^3, \; xyz$
$k = 4$	$x^4 - 6x^2y^2 + y^4, \; x^4 - 6x^2z^2 + z^4, \; y^4 - 6y^2z^2 + z^4,$
	$x^3y - xy^3, \; x^3z - xz^3, \; y^3z - yz^3,$
	$3x^2yz - yz^3, \; 3xy^2z - xz^3, \; 3xyz^2 - x^3y.$

Subsequent sections will be devoted to a proof of the important facts stated earlier.

7.4   The Laplace-Beltrami Operator

In order to define rigorously the Laplacian on the sphere $S^n \subseteq \mathbb{R}^{n+1}$ and establish its relationship with the Laplacian on $\mathbb{R}^{n+1}$, we need the definition of the Laplacian on a Riemannian manifold (M, g), the *Laplace-Beltrami operator* (Eugenio Beltrami, 1835–1900). A more general definition of the the Laplace-Beltrami operator as an operator on differential forms is given in Section 8.3. In this chapter we only need the definition of the Laplacian on functions.

Recall that a Riemannian metric g on a manifold M is a smooth family of inner products $g = (g_p)$, where g_p is an inner product on the tangent space $T_p M$ for every $p \in M$. The inner product g_p on $T_p M$ establishes a canonical duality between $T_p M$ and $T_p^* M$, namely, we have the isomorphism $\flat \colon T_p M \to T_p^* M$ defined such that for every $u \in T_p M$, the linear form $u^\flat \in T_p^* M$ is given by

$$u^\flat(v) = g_p(u, v), \qquad v \in T_p M.$$

The inverse isomorphism $\sharp \colon T_p^* M \to T_p M$ is defined such that for every $\omega \in T_p^* M$, the vector $\omega^\sharp$ is the unique vector in $T_p M$ so that

$$g_p(\omega^\sharp, v) = \omega(v), \qquad v \in T_p M.$$

The isomorphisms $\flat$ and $\sharp$ induce isomorphisms between vector fields $X \in \mathfrak{X}(M)$ and one-forms $\omega \in \mathcal{A}^1(M)$. In particular, for every smooth function $f \in C^\infty(M)$, the vector field corresponding to the one-form df is the *gradient* grad f of f. The gradient of f is uniquely determined by the condition

$$g_p((\text{grad } f)_p, v) = df_p(v), \qquad v \in T_p M, \ p \in M.$$

Definition 7.12. Let (M, g) be a Riemannian manifold. If ∇_X is the covariant derivative associated with the Levi-Civita connection induced by the metric g, then the *divergence* of a vector field $X \in \mathfrak{X}(M)$ is the function div $X \colon M \to \mathbb{R}$ defined so that

$$(\text{div } X)(p) = \text{tr}(Y(p) \mapsto (\nabla_Y X)_p);$$

namely, for every p, $(\text{div } X)(p)$ is the trace of the linear map $Y(p) \mapsto (\nabla_Y X)_p$. Then the *Laplace-Beltrami operator*, for short, *Laplacian*, is the linear operator $\Delta \colon C^\infty(M) \to C^\infty(M)$ given by

$$\Delta f = \operatorname{div} \operatorname{grad} f.$$

Remark. The definition just given differs from the definition given in Section 8.3 by a negative sign. We adopted this sign convention to conform with most of the literature on spherical harmonics (where the negative sign is omitted). A consequence of this choice is that the eigenvalues of the Laplacian are negative.

For more details on the Laplace-Beltrami operator, we refer the reader to Chapter 8 or to Gallot, Hulin, and Lafontaine [48] (Chapter 4) or O'Neill [84] (Chapter 3), Postnikov [88] (Chapter 13), Helgason [55] (Chapter 2), or Warner [109] (Chapters 4 and 6).

All this being rather abstract, it is useful to know how grad f, div X, and Δf are expressed in a chart. If (U, φ) is a chart of M, with $p \in M$, and if as usual

$$\left(\left(\frac{\partial}{\partial x_1}\right)_p, \ldots, \left(\frac{\partial}{\partial x_n}\right)_p\right)$$

denotes the basis of $T_p M$ induced by φ, the local expression of the metric g at p is given by the $n \times n$ matrix $(g_{ij})_p$, with

$$(g_{ij})_p = g_p\left(\left(\frac{\partial}{\partial x_i}\right)_p, \left(\frac{\partial}{\partial x_j}\right)_p\right).$$

The matrix $(g_{ij})_p$ is symmetric, positive definite, and its inverse is denoted $(g^{ij})_p$. We also let $|g|_p = \det(g_{ij})_p$. For simplicity of notation we often omit the subscript p. Then it can be shown that for every function $f \in C^\infty(M)$, in local coordinates given by the chart (U, φ), we have

$$\operatorname{grad} f = \sum_{ij} g^{ij} \frac{\partial f}{\partial x_j} \frac{\partial}{\partial x_i},$$

where as usual

$$\frac{\partial f}{\partial x_j}(p) = \left(\frac{\partial}{\partial x_j}\right)_p f = \frac{\partial(f \circ \varphi^{-1})}{\partial u_j}(\varphi(p)),$$

and $(u_1, \ldots, u_n)$ are the coordinate functions in $\mathbb{R}^n$. There are formulae for div X and Δf involving the Christoffel symbols. Let

$$X = \sum_{i=1}^{n} X_i \frac{\partial}{\partial x_i},$$

be a vector field expressed over a chart (U, φ). Recall that the Christoffel symbol Γ^k_{ij} is defined as

$$\Gamma^k_{ij} = \frac{1}{2} \sum_{l=1}^{n} g^{kl} \left(\partial_i g_{jl} + \partial_j g_{il} - \partial_l g_{ij} \right), \qquad (*)$$

where $\partial_k g_{ij} = \frac{\partial}{\partial x_k}(g_{ij})$. Then

$$\operatorname{div} X = \sum_{i=1}^{n} \left[\frac{\partial X_i}{\partial x_i} + \sum_{j=1}^{n} \Gamma^i_{ij} X_j \right],$$

and

$$\Delta f = \sum_{i,j} g^{ij} \left[\frac{\partial^2 f}{\partial x_i \partial x_j} - \sum_{k=1}^{n} \Gamma^k_{ij} \frac{\partial f}{\partial x_k} \right],$$

whenever $f \in C^\infty(M)$; see pages 86 and 87 of O'Neill [84].

We take a moment to use O'Neill formula to re-derive the expression for the Laplacian on $\mathbb{R}^2$ in terms of polar coordinates (r, θ), where $x = r \cos\theta$ and $y = r \sin\theta$. Note that

$$\frac{\partial}{\partial x_1} = \frac{\partial}{\partial r} = (\cos\theta, \sin\theta)$$

$$\frac{\partial}{\partial x_2} = \frac{\partial}{\partial \theta} = (-r \sin\theta, r \cos\theta),$$

which in turn gives

$$g_{ij} = \begin{pmatrix} 1 & 0 \\ 0 & r^2 \end{pmatrix} \qquad g^{ij} = \begin{pmatrix} 1 & 0 \\ 0 & r^{-2} \end{pmatrix}.$$

Some computations show that the only nonzero Christoffel symbols are

$$\Gamma^2_{12} = \Gamma^2_{21} = \frac{1}{r} \qquad \Gamma^1_{22} = -r;$$

see Gallier and Quaintance [47]. Hence

$$\Delta f = \sum_{i,j=1}^{2} g^{ij} \left[\frac{\partial^2 f}{\partial x_i \partial x_j} - \sum_{k=1}^{2} \Gamma^k_{ij} \frac{\partial f}{\partial x_k} \right]$$

$$= g^{11}\left[\frac{\partial^2 f}{\partial x_1^2} - \sum_{k=1}^{2}\Gamma_{11}^{k}\frac{\partial f}{\partial x_k}\right] + g^{22}\left[\frac{\partial^2 f}{\partial x_2^2} - \sum_{k=1}^{2}\Gamma_{22}^{k}\frac{\partial f}{\partial x_k}\right]$$

$$= \frac{\partial^2 f}{\partial r^2} + \frac{1}{r^2}\left[\frac{\partial^2 f}{\partial \theta^2} - \Gamma_{22}^{1}\frac{\partial f}{\partial r}\right]$$

$$= \frac{\partial^2 f}{\partial r^2} + \frac{1}{r^2}\left[\frac{\partial^2 f}{\partial \theta^2} + r\frac{\partial f}{\partial r}\right]$$

$$= \frac{1}{r^2}\frac{\partial^2 f}{\partial \theta^2} + \frac{\partial^2 f}{\partial r^2} + \frac{1}{r}\frac{\partial f}{\partial r}$$

$$= \frac{1}{r^2}\frac{\partial^2 f}{\partial \theta^2} + \frac{1}{r}\frac{\partial}{\partial r}\left(r\frac{\partial f}{\partial r}\right).$$

O'Neill's formula may also be used to re-derive the expression for the Laplacian on $\mathbb{R}^3$ in terms of spherical coordinates (r, θ, φ) where

$$x = r\sin\theta\cos\varphi$$
$$y = r\sin\theta\sin\varphi$$
$$z = r\cos\theta.$$

We have

$$\frac{\partial}{\partial x_1} = \frac{\partial}{\partial r} = \sin\theta\cos\varphi\frac{\partial}{\partial x} + \sin\theta\sin\varphi\frac{\partial}{\partial y} + \cos\theta\frac{\partial}{\partial z} = \widehat{r}$$

$$\frac{\partial}{\partial x_2} = \frac{\partial}{\partial \theta} = r\left(\cos\theta\cos\varphi\frac{\partial}{\partial x} + \cos\theta\sin\varphi\frac{\partial}{\partial y} - \sin\theta\frac{\partial}{\partial z}\right) = r\widehat{\theta}$$

$$\frac{\partial}{\partial x_3} = \frac{\partial}{\partial \varphi} = r\left(-\sin\theta\sin\varphi\frac{\partial}{\partial x} + \sin\theta\cos\varphi\frac{\partial}{\partial y}\right) = r\widehat{\varphi}.$$

Observe that $\widehat{r}, \widehat{\theta}$ and $\widehat{\varphi}$ are pairwise orthogonal. Therefore, the matrix (g_{ij}) is given by

$$(g_{ij}) = \begin{pmatrix} 1 & 0 & 0 \\ 0 & r^2 & 0 \\ 0 & 0 & r^2\sin^2\theta \end{pmatrix}$$

and $|g| = r^4\sin^2\theta$. The inverse of (g_{ij}) is

$$(g^{ij}) = \begin{pmatrix} 1 & 0 & 0 \\ 0 & r^{-2} & 0 \\ 0 & 0 & r^{-2}\sin^{-2}\theta \end{pmatrix}.$$

By using Line (∗), it is not hard to show that $\Gamma_{ij}^k = 0$ except for

$$\Gamma_{22}^1 = -\frac{1}{2}g^{11}\partial_1 g_{22} = -\frac{1}{2}\frac{\partial}{\partial r}r^2 = -r$$

$$\Gamma_{33}^1 = -\frac{1}{2}g^{11}\partial_1 g_{33} = -\frac{1}{2}\frac{\partial}{\partial r}r^2 \sin^2\theta = -r\sin^2\theta$$

$$\Gamma_{12}^2 = \Gamma_{21}^2 = \frac{1}{2}g^{22}\partial_1 g_{22} = \frac{1}{2r^2}\frac{\partial}{\partial r}r^2 = \frac{1}{r}$$

$$\Gamma_{33}^2 = -\frac{1}{2}g^{22}\partial_2 g_{33} = -\frac{1}{2r^2}\frac{\partial}{\partial\theta}r^2\sin^2\theta = -\sin\theta\cos\theta$$

$$\Gamma_{13}^3 = \Gamma_{31}^3 = \frac{1}{2}g^{33}\partial_1 g_{33} = \frac{1}{2r^2\sin^2\theta}\frac{\partial}{\partial r}r^2\sin^2\theta = \frac{1}{r}$$

$$\Gamma_{23}^3 = \Gamma_{32}^2 = \frac{1}{2}g^{33}\partial_2 g_{33} = \frac{1}{2r^2\sin^2\theta}\frac{\partial}{\partial\theta}r^2\sin^2\theta = \cot\theta.$$

Then

$$\Delta f = \sum_{i,j=1}^3 g^{ij}\left[\frac{\partial^2 f}{\partial x_i\partial x_j} - \sum_{k=1}^3 \Gamma_{ij}^k\frac{\partial f}{\partial x_k}\right]$$

$$= g^{11}\left[\frac{\partial^2 f}{\partial x_1^2} - \sum_{k=1}^3 \Gamma_{11}^k\frac{\partial f}{\partial x_k}\right] + g^{22}\left[\frac{\partial^2 f}{\partial x_2^2} - \sum_{k=1}^3 \Gamma_{22}^k\frac{\partial f}{\partial x_k}\right]$$

$$+ g^{33}\left[\frac{\partial^2 f}{\partial x_3^2} - \sum_{k=1}^3 \Gamma_{33}^k\frac{\partial f}{\partial x_k}\right]$$

$$= \frac{\partial^2 f}{\partial r^2} + \frac{1}{r^2}\left[\frac{\partial^2 f}{\partial\theta^2} - \Gamma_{22}^1\frac{\partial f}{\partial r}\right] + \frac{1}{r^2\sin^2\theta}\left[\frac{\partial^2 f}{\partial\varphi^2} - \left[\Gamma_{33}^1\frac{\partial f}{\partial r} + \Gamma_{33}^2\frac{\partial f}{\partial\theta}\right]\right]$$

$$= \frac{\partial^2 f}{\partial r^2} + \frac{1}{r^2}\left[\frac{\partial^2 f}{\partial\theta^2} + r\frac{\partial f}{\partial r}\right] + \frac{1}{r^2\sin^2\theta}\left[\frac{\partial^2 f}{\partial\varphi^2} + r\sin^2\theta\frac{\partial f}{\partial r} + \sin\theta\cos\theta\frac{\partial f}{\partial\theta}\right]$$

$$= \frac{\partial^2 f}{\partial r^2} + \frac{2}{r}\frac{\partial f}{\partial r} + \frac{1}{r^2}\left[\frac{\partial^2 f}{\partial\theta^2} + \frac{\cos\theta}{\sin\theta}\frac{\partial f}{\partial\theta}\right] + \frac{1}{r^2\sin^2\theta}\frac{\partial^2 f}{\partial\varphi^2}$$

$$= \frac{1}{r^2}\frac{\partial}{\partial r}\left(r^2\frac{\partial f}{\partial r}\right) + \frac{1}{r^2\sin\theta}\frac{\partial}{\partial\theta}\left(\sin\theta\frac{\partial f}{\partial\theta}\right) + \frac{1}{r^2\sin^2\theta}\frac{\partial^2 f}{\partial\varphi^2}.$$

O'Neill's formulae for the divergence and the Laplacian can be tedious to calculate since they involve knowing the Christoffel symbols. Fortunately there are other formulas for the divergence and the Laplacian which only involve (g_{ij}) and (g^{ij}) and hence will be more convenient for our purposes: For every vector field $X \in \mathfrak{X}(M)$ expressed in local coordinates as

$$X = \sum_{i=1}^{n} X_i \frac{\partial}{\partial x_i},$$

we have

$$\operatorname{div} X = \frac{1}{\sqrt{|g|}} \sum_{i=1}^{n} \frac{\partial}{\partial x_i} \left(\sqrt{|g|}\, X_i \right), \tag{$\dagger$}$$

and for every function $f \in C^{\infty}(M)$, the Laplacian Δf is given by

$$\Delta f = \frac{1}{\sqrt{|g|}} \sum_{i,j} \frac{\partial}{\partial x_i} \left(\sqrt{|g|}\, g^{ij} \frac{\partial f}{\partial x_j} \right). \tag{$**$}$$

A detailed proof of Equation ($\dagger$) is given in Helgason [55] (Chapter II, Lemma 2.5). This formula is also stated in Postnikov [88] (Chapter 13, Section 6) and O'Neill [84] (Chapter 7, Exercise 5).

One should check that for $M = \mathbb{R}^n$ with its standard coordinates, the Laplacian is given by the familiar formula

$$\Delta f = \frac{\partial^2 f}{\partial x_1^2} + \cdots + \frac{\partial^2 f}{\partial x_n^2}.$$

By using Equation ($**$), we quickly rediscover the Laplacian in spherical coordinates, namely

$$
\begin{aligned}
\Delta f &= \frac{1}{r^2 \sin\theta} \sum_{i=1}^{3} \sum_{j=1}^{3} \frac{\partial}{\partial x_i} \left(r^2 \sin\theta\, g^{ij} \frac{\partial f}{\partial x_j} \right) \\
&= \frac{1}{r^2 \sin\theta} \left[\frac{\partial}{\partial r} \left(r^2 \sin\theta \frac{\partial f}{\partial r} \right) + \frac{\partial}{\partial \theta} \left(r^2 \sin\theta\, r^{-2} \frac{\partial f}{\partial \theta} \right) \right.\\
&\qquad\qquad \left. + \frac{\partial}{\partial \varphi} \left(r^2 \sin\theta\, r^{-2} \sin^{-2}\theta \frac{\partial f}{\partial \varphi} \right) \right] \\
&= \frac{1}{r^2} \frac{\partial}{\partial r} \left(r^2 \frac{\partial f}{\partial r} \right) + \frac{1}{r^2 \sin\theta} \frac{\partial}{\partial \theta} \left(\sin\theta \frac{\partial f}{\partial \theta} \right) + \frac{1}{r^2 \sin^2\theta} \frac{\partial^2 f}{\partial \varphi^2}.
\end{aligned}
$$

Since (θ, φ) are coordinates on the sphere S^2 via

$$x = \sin\theta \cos\varphi$$
$$y = \sin\theta \sin\varphi$$
$$z = \cos\theta,$$

we see that in these coordinates, the metric $(\widetilde{g}_{ij})$ on S^2 is given by the matrix

$$(\widetilde{g}_{ij}) = \begin{pmatrix} 1 & 0 \\ 0 & \sin^2 \theta \end{pmatrix},$$

that $|\widetilde{g}| = \sin^2 \theta$, and that the inverse of $(\widetilde{g}_{ij})$ is

$$(\widetilde{g}^{ij}) = \begin{pmatrix} 1 & 0 \\ 0 & \sin^{-2} \theta \end{pmatrix}.$$

It follows immediately that

$$\Delta_{S^2} f = \frac{1}{\sin \theta} \frac{\partial}{\partial \theta} \left(\sin \theta \frac{\partial f}{\partial \theta} \right) + \frac{1}{\sin^2 \theta} \frac{\partial^2 f}{\partial \varphi^2},$$

so we have verified that

$$\Delta f = \frac{1}{r^2} \frac{\partial}{\partial r} \left(r^2 \frac{\partial f}{\partial r} \right) + \frac{1}{r^2} \Delta_{S^2} f.$$

Let us now generalize the above formula to the Laplacian Δ on $\mathbb{R}^{n+1}$, and the Laplacian Δ_{S^n} on S^n, where

$$S^n = \{(x_1, \ldots, x_{n+1}) \in \mathbb{R}^{n+1} \mid x_1^2 + \cdots + x_{n+1}^2 = 1\}.$$

Following Morimoto [81] (Chapter 2, Section 2), let us use "polar coordinates." The map from $\mathbb{R}_+ \times S^n$ to $\mathbb{R}^{n+1} - \{0\}$ given by

$$(r, \sigma) \mapsto r\sigma$$

is clearly a diffeomorphism. Thus, for any system of coordinates $(u_1, \ldots, u_n)$ on S^n, the tuple $(u_1, \ldots, u_n, r)$ is a system of coordinates on $\mathbb{R}^{n+1} - \{0\}$ called *polar coordinates*. Let us establish the relationship between the Laplacian Δ, on $\mathbb{R}^{n+1} - \{0\}$ in polar coordinates and the Laplacian Δ_{S^n} on S^n in local coordinates $(u_1, \ldots, u_n)$.

Proposition 7.10. *If Δ is the Laplacian on $\mathbb{R}^{n+1} - \{0\}$ in polar coordinates $(u_1, \ldots, u_n, r)$ and Δ_{S^n} is the Laplacian on the sphere S^n in local coordinates $(u_1, \ldots, u_n)$, then*

$$\Delta f = \frac{1}{r^n} \frac{\partial}{\partial r} \left(r^n \frac{\partial f}{\partial r} \right) + \frac{1}{r^2} \Delta_{S^n} f.$$

Proof. Let us compute the $(n+1) \times (n+1)$ matrix $G = (g_{ij})$ expressing the metric on $\mathbb{R}^{n+1}$ in polar coordinates and the $n \times n$ matrix $\widetilde{G} = (\widetilde{g}_{ij})$ expressing the metric on S^n. Recall that if $\sigma \in S^n$, then $\sigma \cdot \sigma = 1$, and so

$$\frac{\partial \sigma}{\partial u_i} \cdot \sigma = 0,$$

as

$$\frac{\partial \sigma}{\partial u_i} \cdot \sigma = \frac{1}{2} \frac{\partial (\sigma \cdot \sigma)}{\partial u_i} = 0.$$

If $x = r\sigma$ with $\sigma \in S^n$, we have

$$\frac{\partial x}{\partial u_i} = r \frac{\partial \sigma}{\partial u_i}, \qquad 1 \le i \le n,$$

and

$$\frac{\partial x}{\partial r} = \sigma.$$

It follows that

$$g_{ij} = \frac{\partial x}{\partial u_i} \cdot \frac{\partial x}{\partial u_j} = r^2 \frac{\partial \sigma}{\partial u_i} \cdot \frac{\partial \sigma}{\partial u_j} = r^2 \widetilde{g}_{ij}$$

$$g_{in+1} = \frac{\partial x}{\partial u_i} \cdot \frac{\partial x}{\partial r} = r \frac{\partial \sigma}{\partial u_i} \cdot \sigma = 0$$

$$g_{n+1n+1} = \frac{\partial x}{\partial r} \cdot \frac{\partial x}{\partial r} = \sigma \cdot \sigma = 1.$$

Consequently, we get

$$G = \begin{pmatrix} r^2 \widetilde{G} & 0 \\ 0 & 1 \end{pmatrix},$$

$|g| = r^{2n} |\widetilde{g}|$, and

$$G^{-1} = \begin{pmatrix} r^{-2} \widetilde{G}^{-1} & 0 \\ 0 & 1 \end{pmatrix}.$$

Using the above equations and

$$\Delta f = \frac{1}{\sqrt{|g|}} \sum_{i,j} \frac{\partial}{\partial x_i} \left(\sqrt{|g|} \, g^{ij} \frac{\partial f}{\partial x_j} \right),$$

we get

$$\Delta f = \frac{1}{r^n \sqrt{|\tilde{g}|}} \sum_{i,j=1}^{n} \frac{\partial}{\partial x_i} \left(r^n \sqrt{|\tilde{g}|} \frac{1}{r^2} \tilde{g}^{ij} \frac{\partial f}{\partial x_j} \right) + \frac{1}{r^n \sqrt{|\tilde{g}|}} \frac{\partial}{\partial r} \left(r^n \sqrt{|\tilde{g}|} \frac{\partial f}{\partial r} \right)$$

$$= \frac{1}{r^2 \sqrt{|\tilde{g}|}} \sum_{i,j=1}^{n} \frac{\partial}{\partial x_i} \left(\sqrt{|\tilde{g}|} \tilde{g}^{ij} \frac{\partial f}{\partial x_j} \right) + \frac{1}{r^n} \frac{\partial}{\partial r} \left(r^n \frac{\partial f}{\partial r} \right)$$

$$= \frac{1}{r^2} \Delta_{S^n} f + \frac{1}{r^n} \frac{\partial}{\partial r} \left(r^n \frac{\partial f}{\partial r} \right),$$

as claimed.   □

It is also possible to express Δ_{S^n} in terms of $\Delta_{S^{n-1}}$. If $e_{n+1} = (0, \ldots, 0, 1) \in \mathbb{R}^{n+1}$, then we can view S^{n-1} as the intersection of S^n with the hyperplane $x_{n+1} = 0$; that is, as the set

$$S^{n-1} = \{\sigma \in S^n \mid \sigma \cdot e_{n+1} = 0\}.$$

If $(u_1, \ldots, u_{n-1})$ are local coordinates on S^{n-1}, then $(u_1, \ldots, u_{n-1}, \theta)$ are local coordinates on S^n, by setting

$$\sigma = \sin\theta \, \tilde{\sigma} + \cos\theta \, e_{n+1},$$

with $\tilde{\sigma} \in S^{n-1}$ and $0 \leq \theta < \pi$.

Proposition 7.11. *We have*

$$\Delta_{S^n} f = \frac{1}{\sin^{n-1}\theta} \frac{\partial}{\partial\theta} \left(\sin^{n-1}\theta \frac{\partial f}{\partial\theta} \right) + \frac{1}{\sin^2\theta} \Delta_{S^{n-1}} f.$$

Proof. Note that $\tilde{\sigma} \cdot \tilde{\sigma} = 1$, which in turn implies

$$\frac{\partial\tilde{\sigma}}{\partial u_i} \cdot \tilde{\sigma} = 0.$$

Furthermore, $\tilde{\sigma} \cdot e_{n+1} = 0$, and hence

$$\frac{\partial\tilde{\sigma}}{\partial u_i} \cdot e_{n+1} = 0.$$

By using these local coordinate systems, we find the relationship between Δ_{S^n} and $\Delta_{S^{n-1}}$ as follows: First observe that

$$\frac{\partial\sigma}{\partial u_i} = \sin\theta \frac{\partial\tilde{\sigma}}{\partial u_i} + 0 \, e_{n+1} \qquad \frac{\partial\sigma}{\partial\theta} = \cos\theta \, \tilde{\sigma} - \sin\theta \, e_{n+1}.$$

If $\widetilde{G} = (\widetilde{g}_{ij})$ represents the metric on S^n and $\widehat{G} = (\widehat{g}_{ij})$ is the restriction of this metric to S^{n-1} as defined above then for $1 \leq i, j \leq n-1$, we have

$$\widetilde{g}_{ij} = \frac{\partial \sigma}{\partial u_i} \cdot \frac{\partial \sigma}{\partial u_j} = \sin^2 \theta \frac{\partial \widetilde{\sigma}}{\partial u_i} \cdot \frac{\partial \widetilde{\sigma}}{\partial u_j} = \sin^2 \theta \, \widehat{g}_{ij}$$

$$\widetilde{g}_{in} = \frac{\partial \sigma}{\partial u_i} \cdot \frac{\partial \sigma}{\partial \theta} = \left(\sin \theta \frac{\partial \widetilde{\sigma}}{\partial u_i} + 0 \, e_{n+1} \right) \cdot (\cos \theta \, \widetilde{\sigma} - \sin \theta \, e_{n+1}) = 0$$

$$\widetilde{g}_{nn} = \frac{\partial \sigma}{\partial \theta} \cdot \frac{\partial \sigma}{\partial \theta} = (\cos \theta \, \widetilde{\sigma} - \sin \theta \, e_{n+1}) \cdot (\cos \theta \, \widetilde{\sigma} - \sin \theta \, e_{n+1}) = \cos^2 \theta + \sin^2 \theta = 1.$$

These calculations imply that

$$\widetilde{G} = \begin{pmatrix} \sin^2 \theta \, \widehat{G} & 0 \\ 0 & 1 \end{pmatrix},$$

$|\widetilde{g}| = \sin^{2n-2} \theta |\widehat{g}|$, and that

$$\widetilde{G}^{-1} = \begin{pmatrix} \sin^{-2} \theta \, \widehat{G}^{-1} & 0 \\ 0 & 1 \end{pmatrix}.$$

Hence

$$\Delta_{S^n} f = \frac{1}{\sin^{n-1} \theta \sqrt{|\widehat{g}|}} \sum_{i,j=1}^{n-1} \frac{\partial}{\partial u_i} \left(\sin^{n-1} \theta \sqrt{|\widehat{g}|} \frac{1}{\sin^2 \theta} \widehat{g}^{ij} \frac{\partial f}{\partial u_j} \right)$$

$$+ \frac{1}{\sin^{n-1} \theta \sqrt{|\widehat{g}|}} \frac{\partial}{\partial \theta} \left(\sin^{n-1} \theta \sqrt{|\widehat{g}|} \frac{\partial f}{\partial \theta} \right)$$

$$= \frac{1}{\sin^{n-1} \theta} \frac{\partial}{\partial \theta} \left(\sin^{n-1} \theta \frac{\partial f}{\partial \theta} \right) + \frac{1}{\sin^2 \theta \sqrt{|\widehat{g}|}} \sum_{i,j=1}^{n-1} \frac{\partial}{\partial u_i} \left(\sqrt{|\widehat{g}|} \widehat{g}^{ij} \frac{\partial f}{\partial u_j} \right)$$

$$= \frac{1}{\sin^{n-1} \theta} \frac{\partial}{\partial \theta} \left(\sin^{n-1} \theta \frac{\partial f}{\partial \theta} \right) + \frac{1}{\sin^2 \theta} \Delta_{S^{n-1}} f,$$

as claimed.   □

A fundamental property of the divergence is known as *Green's formula*. There are actually two Green's formulae, but we will only need the version for an orientable manifold without boundary given in Proposition 8.15. Recall that Green's formula states that if M is a compact, orientable, Riemannian manifold without boundary, then, for every smooth vector field $X \in \mathfrak{X}(M)$, we have

$$\int_M (\operatorname{div} X) \operatorname{Vol}_M = 0,$$

where Vol_M is the volume form on M induced by the metric.

Definition 7.13. If M is a compact, orientable Riemannian manifold, then for any two smooth functions $f, h \in C^\infty(M)$, we define $\langle f, h \rangle_M$ by

$$\langle f, h \rangle_M = \int_M f h \operatorname{Vol}_M.$$

Then, it is not hard to show that $\langle -, - \rangle_M$ is an inner product on $C^\infty(M)$.

An important property of the Laplacian on a compact, orientable Riemannian manifold is that it is a self-adjoint operator. This fact is proved in the more general case of an inner product on differential forms in Proposition 8.8, but it is instructive to give another proof in the special case of functions using Green's formula.

First we need the following two properties: For any two functions $f, h \in C^\infty(M)$, and any vector field $X \in \mathfrak{X}(M)$, we have:

$$\operatorname{div}(f X) = f \operatorname{div} X + X(f) = f \operatorname{div} X + g(\operatorname{grad} f, X)$$

$$\operatorname{grad} f\ (h) = g(\operatorname{grad} f, \operatorname{grad} h) = \operatorname{grad} h\ (f).$$

Using the above identities, we obtain the following important result.

Proposition 7.12. *Let M be a compact, orientable, Riemannian manifold without boundary. The Laplacian on M is self-adjoint; that is, for any two functions $f, h \in C^\infty(M)$, we have*

$$\langle \Delta f, h \rangle_M = \langle f, \Delta h \rangle_M,$$

or equivalently

$$\int_M f \Delta h \operatorname{Vol}_M = \int_M h \Delta f \operatorname{Vol}_M.$$

Proof. By the two identities before Proposition 7.12,

$$f \Delta h = f \operatorname{div} \operatorname{grad} h = \operatorname{div}(f \operatorname{grad} h) - g(\operatorname{grad} f, \operatorname{grad} h)$$

and

$$h \Delta f = h \operatorname{div} \operatorname{grad} f = \operatorname{div}(h \operatorname{grad} f) - g(\operatorname{grad} h, \operatorname{grad} f),$$

so we get

$$f \Delta h - h \Delta f = \operatorname{div}(f \operatorname{grad} h - h \operatorname{grad} f).$$

By Green's formula,

$$\int_M (f\Delta h - h\Delta f)\text{Vol}_M = \int_M \text{div}(f\,\text{grad}\,h - h\,\text{grad}\,f)\text{Vol}_M = 0,$$

which proves that Δ is self-adjoint. $\square$

The importance of Proposition 7.12 lies in the fact that as $\langle -, - \rangle_M$ is an inner product on $\mathcal{C}^\infty(M)$, the eigenspaces of Δ for distinct eigenvalues are pairwise orthogonal. We will make heavy use of this property in the next section on harmonic polynomials.

7.5 Harmonic Polynomials, Spherical Harmonics, and $L^2(S^n)$

Harmonic homogeneous polynomials and their restrictions to S^n, where

$$S^n = \{(x_1, \ldots, x_{n+1}) \in \mathbb{R}^{n+1} \mid x_1^2 + \cdots + x_{n+1}^2 = 1\},$$

turn out to play a crucial role in understanding the structure of the eigenspaces of the Laplacian on S^n (with $n \geq 1$). The results in this section appear in one form or another in Stein and Weiss [102] (Chapter 4), Morimoto [81] (Chapter 2), Helgason [55] (Introduction, Section 3), Dieudonné [31] (Chapter 7), Axler, Bourdon, and Ramey [7] (Chapter 5), and Vilenkin [107] (Chapter IX). Some of these sources assume a fair amount of mathematical background, and consequently uninitiated readers will probably find the exposition rather condensed, especially Helgason. We tried hard to make our presentation more "user-friendly."

Recall that a homogeneous polynomial P of degree k in n variables $x_1, \ldots, x_n$ is an expression of the form

$$P = \sum_{\substack{\alpha_1 + \cdots + \alpha_n = k \\ (\alpha_1, \ldots, \alpha_n) \in \mathbb{N}^k}} a_{(\alpha_1, \ldots, \alpha_n)}\, x_1^{\alpha_1} \cdots x_n^{\alpha_n},$$

where the coefficients $a_{(\alpha_1, \ldots, \alpha_n)}$ are either real or complex numbers. We view such a homogeneous polynomial as a function $P \colon \mathbb{R}^n \to \mathbb{C}$, or as a function $P \colon \mathbb{R}^n \to \mathbb{R}$ when the coefficients are all real. The Laplacian ΔP of P is defined by

$$\Delta P = \sum_{\substack{\alpha_1 + \cdots + \alpha_n = k \\ (\alpha_1, \ldots, \alpha_n) \in \mathbb{N}^k}} a_{(\alpha_1, \ldots, \alpha_n)} \left(\frac{\partial^2}{\partial x_1^2} + \cdots + \frac{\partial^2}{\partial x_n^2} \right)(x_1^{\alpha_1} \cdots x_n^{\alpha_n}).$$

Definition 7.14. Let $\mathcal{P}_k(n+1)$ (resp. $\mathcal{P}_k^{\mathbb{C}}(n+1)$) denote the space of homogeneous polynomials of degree k in $n+1$ variables with real coefficients (resp. complex coefficients), and let $\mathcal{P}_k(S^n)$ (resp. $\mathcal{P}_k^{\mathbb{C}}(S^n)$) denote the restrictions of homogeneous polynomials in $\mathcal{P}_k(n+1)$ to S^n (resp. the restrictions of homogeneous polynomials in $\mathcal{P}_k^{\mathbb{C}}(n+1)$ to S^n). Let $\mathcal{H}_k(n+1)$ (resp. $\mathcal{H}_k^{\mathbb{C}}(n+1)$) denote the space of *(real) harmonic polynomials* (resp. *complex harmonic polynomials*), with

$$\mathcal{H}_k(n+1) = \{P \in \mathcal{P}_k(n+1) \mid \Delta P = 0\}$$

and

$$\mathcal{H}_k^{\mathbb{C}}(n+1) = \{P \in \mathcal{P}_k^{\mathbb{C}}(n+1) \mid \Delta P = 0\}.$$

Harmonic polynomials are sometimes called *solid harmonics*. Finally, let $\mathcal{H}_k(S^n)$ (resp. $\mathcal{H}_k^{\mathbb{C}}(S^n)$) denote the space of *(real) spherical harmonics* (resp. *complex spherical harmonics*) be the set of restrictions of harmonic polynomials in $\mathcal{H}_k(n+1)$ to S^n (resp. restrictions of harmonic polynomials in $\mathcal{H}_k^{\mathbb{C}}(n+1)$ to S^n).

Definition 7.15. A function $f: \mathbb{R}^n \to \mathbb{R}$ (resp. $f: \mathbb{R}^n \to \mathbb{C}$) is *homogeneous of degree k* iff

$$f(tx) = t^k f(x), \qquad \text{for all } x \in \mathbb{R}^n \text{ and } t > 0.$$

The restriction map $\rho: \mathcal{H}_k(n+1) \to \mathcal{H}_k(S^n)$ is a surjective linear map. In fact, it is a bijection. Indeed, if $P \in \mathcal{H}_k(n+1)$, observe that

$$P(x) = \|x\|^k \, P\left(\frac{x}{\|x\|}\right), \qquad \text{with} \quad \frac{x}{\|x\|} \in S^n,$$

for all $x \neq 0$. Consequently, if $P \upharpoonright S^n = Q \upharpoonright S^n$, that is $P(\sigma) = Q(\sigma)$ for all $\sigma \in S^n$, then

$$P(x) = \|x\|^k \, P\left(\frac{x}{\|x\|}\right) = \|x\|^k \, Q\left(\frac{x}{\|x\|}\right) = Q(x)$$

for all $x \neq 0$, which implies $P = Q$ (as P and Q are polynomials). Therefore, we have a linear isomorphism between $\mathcal{H}_k(n+1)$ and $\mathcal{H}_k(S^n)$ (and between $\mathcal{H}_k^{\mathbb{C}}(n+1)$ and $\mathcal{H}_k^{\mathbb{C}}(S^n)$).

It will be convenient to introduce some notation to deal with homogeneous polynomials. Given $n \geq 1$ variables $x_1, \ldots, x_n$, and any n-tuple of nonnegative integers $\alpha = (\alpha_1, \ldots, \alpha_n)$, let $|\alpha| = \alpha_1 + \cdots + \alpha_n$, let $x^\alpha = x_1^{\alpha_1} \cdots x_n^{\alpha_n}$, and let $\alpha! = \alpha_1! \cdots \alpha_n!$. Then every homogeneous polynomial P of degree k in the variables $x_1, \ldots, x_n$ can be written uniquely as

$$P = \sum_{|\alpha|=k} c_\alpha x^\alpha,$$

with $c_\alpha \in \mathbb{R}$ or $c_\alpha \in \mathbb{C}$. It is well known that $\mathcal{P}_k(n)$ is a (real) vector space of dimension

$$d_k = \binom{n+k-1}{k}$$

and $\mathcal{P}_k^\mathbb{C}(n)$ is a complex vector space of the same dimension d_k. For example, $\mathcal{P}_2(3)$ is a vector space of dimension 6 with basis $\{x_1 x_2, x_1 x_3, x_2 x_3, x_1^2, x_2^2, x_3^2\}$.

We can define an Hermitian inner product on $\mathcal{P}_k^\mathbb{C}(n)$ whose restriction to $\mathcal{P}_k(n)$ is an inner product by viewing a homogeneous polynomial as a differential operator as follows.

Definition 7.16. For every $P = \sum_{|\alpha|=k} c_\alpha x^\alpha \in \mathcal{P}_k^\mathbb{C}(n)$, let

$$\partial(P) = \sum_{|\alpha|=k} c_\alpha \frac{\partial^k}{\partial x_1^{\alpha_1} \cdots \partial x_n^{\alpha_n}}.$$

Then for any two polynomials $P, Q \in \mathcal{P}_k^\mathbb{C}(n)$, let

$$\langle P, Q \rangle = \partial(P)\overline{Q}.$$

Observe that $\langle x^\alpha, x^\beta \rangle = 0$ unless $\alpha = \beta$, in which case we have $\langle x^\alpha, x^\alpha \rangle = \alpha!$. For example, in $\mathcal{P}_2(3)$, if $x^\alpha = x_1^2$ and $x^\beta = x_1 x_2$, then

$$\langle x_1^2, x_1 x_2 \rangle = \frac{\partial^2}{dx_1^2} x_1 x_2 = 0,$$

while

$$\langle x_1^2, x_1^2 \rangle = \frac{\partial^2}{dx_1^2} x_1^2 = 2!.$$

Then a simple computation shows that

$$\left\langle \sum_{|\alpha|=k} a_\alpha x^\alpha, \sum_{|\alpha|=k} b_\alpha x^\alpha \right\rangle = \sum_{|\alpha|=k} \alpha! \, a_\alpha \overline{b}_\alpha.$$

Therefore, $\langle P, Q \rangle$ is indeed an inner product. Also observe that

$$\partial(x_1^2 + \cdots + x_n^2) = \frac{\partial^2}{\partial x_1^2} + \cdots + \frac{\partial^2}{\partial x_n^2} = \Delta.$$

Another useful property of our inner product is this: For $P \in \mathcal{P}_k^{\mathbb{C}}(n)$, $Q \in \mathcal{P}_j^{\mathbb{C}}(n)$, and $R \in \mathcal{P}_{k-j}^{\mathbb{C}}(n)$,

$$\langle P, QR \rangle = \langle \partial(Q)P, R \rangle.$$

Indeed:

$$\begin{aligned}
\langle P, QR \rangle &= \overline{\langle QR, P \rangle} \\
&= \overline{\partial(QR)\overline{P}} \\
&= \overline{\partial(Q)(\partial(R)\overline{P})} \\
&= \overline{\partial(R)(\partial(Q)\overline{P})} \\
&= \overline{\langle R, \partial(Q)P \rangle} \\
&= \langle \partial(Q)P, R \rangle.
\end{aligned}$$

In particular,

$$\langle (x_1^2 + \cdots + x_n^2)P, Q \rangle = \langle P, \partial(x_1^2 + \cdots + x_n^2)Q \rangle = \langle P, \Delta Q \rangle.$$

Let us write $\|x\|^2$ for $x_1^2 + \cdots + x_n^2$. Using our inner product, we can prove the following important theorem.

Theorem 7.13. *The map* $\Delta \colon \mathcal{P}_k(n) \to \mathcal{P}_{k-2}(n)$ *is surjective for all* $n, k \geq 2$ *(and similarly for* $\Delta \colon \mathcal{P}_k^{\mathbb{C}}(n) \to \mathcal{P}_{k-2}^{\mathbb{C}}(n)$*). Furthermore, we have the following orthogonal direct sum decompositions:*

$$\mathcal{P}_k(n) = \mathcal{H}_k(n) \oplus \|x\|^2 \, \mathcal{H}_{k-2}(n) \oplus \cdots \oplus \|x\|^{2j} \, \mathcal{H}_{k-2j}(n) \oplus \cdots \oplus \|x\|^{2[k/2]} \, \mathcal{H}_{[k/2]}(n)$$

and

$$\mathcal{P}_k^{\mathbb{C}}(n) = \mathcal{H}_k^{\mathbb{C}}(n) \oplus \|x\|^2 \, \mathcal{H}_{k-2}^{\mathbb{C}}(n) \oplus \cdots \oplus \|x\|^{2j} \, \mathcal{H}_{k-2j}^{\mathbb{C}}(n) \oplus \cdots \oplus \|x\|^{2[k/2]} \, \mathcal{H}_{[k/2]}^{\mathbb{C}}(n),$$

with the understanding that only the first term occurs on the right-hand side when $k < 2$.

Proof. If the map $\Delta \colon \mathcal{P}_k^{\mathbb{C}}(n) \to \mathcal{P}_{k-2}^{\mathbb{C}}(n)$ is not surjective, then some nonzero polynomial $Q \in \mathcal{P}_{k-2}^{\mathbb{C}}(n)$ is orthogonal to the image of Δ, i.e., $= \langle Q, \Delta P \rangle$. Since $P = \|x\|^2 \, Q \in \mathcal{P}_k^{\mathbb{C}}(n)$, and $0 = \langle Q, \Delta P \rangle$, a fact established earlier shows that

$$0 = \langle Q, \Delta P \rangle = \langle \|x\|^2 \, Q, P \rangle = \langle P, P \rangle,$$

which implies that $P = \|x\|^2 Q = 0$, and thus $Q = 0$, a contradiction. The same proof is valid in the real case.

We claim that we have an orthogonal direct sum decomposition

$$\mathcal{P}_k^{\mathbb{C}}(n) = \mathcal{H}_k^{\mathbb{C}}(n) \oplus \|x\|^2 \mathcal{P}_{k-2}^{\mathbb{C}}(n),$$

and similarly in the real case, with the understanding that the second term is missing if $k < 2$.

If $k = 0, 1$, then $\mathcal{P}_k^{\mathbb{C}}(n) = \mathcal{H}_k^{\mathbb{C}}(n)$, so this case is trivial. Assume $k \geq 2$. Since $\operatorname{Ker} \Delta = \mathcal{H}_k^{\mathbb{C}}(n)$ and Δ is surjective, $\dim(\mathcal{P}_k^{\mathbb{C}}(n)) = \dim(\mathcal{H}_k^{\mathbb{C}}(n)) + \dim(\mathcal{P}_{k-2}^{\mathbb{C}}(n))$, so it is sufficient to prove that $\mathcal{H}_k^{\mathbb{C}}(n)$ is orthogonal to $\|x\|^2 \mathcal{P}_{k-2}^{\mathbb{C}}(n)$. Now, if $H \in \mathcal{H}_k^{\mathbb{C}}(n)$ and $P = \|x\|^2 Q \in \|x\|^2 \mathcal{P}_{k-2}^{\mathbb{C}}(n)$, we have

$$\langle \|x\|^2 Q, H \rangle = \langle Q, \Delta H \rangle = 0,$$

so $\mathcal{H}_k^{\mathbb{C}}(n)$ and $\|x\|^2 \mathcal{P}_{k-2}^{\mathbb{C}}(n)$ are indeed orthogonal. Using induction, we immediately get the orthogonal direct sum decomposition

$$\mathcal{P}_k^{\mathbb{C}}(n) = \mathcal{H}_k^{\mathbb{C}}(n) \oplus \|x\|^2 \mathcal{H}_{k-2}^{\mathbb{C}}(n) \oplus \cdots \oplus \|x\|^{2j} \mathcal{H}_{k-2j}^{\mathbb{C}}(n) \oplus \cdots \oplus \|x\|^{2[k/2]} \mathcal{H}_{[k/2]}^{\mathbb{C}}(n)$$

and the corresponding real version.   □

Remark. Theorem 7.13 also holds for $n = 1$.

Theorem 7.13 has some important corollaries. Since every polynomial in $n + 1$ variables is the sum of homogeneous polynomials, we get

Corollary 7.14. *The restriction to S^n of every polynomial (resp. complex polynomial) in $n + 1 \geq 2$ variables is a sum of restrictions to S^n of harmonic polynomials (resp. complex harmonic polynomials).*

We can also derive a formula for the dimension of $\mathcal{H}_k(n)$ (and $\mathcal{H}_k^{\mathbb{C}}(n)$).

Corollary 7.15. *The dimension $a_{k,n}$ of the space of harmonic polynomials $\mathcal{H}_k(n)$ is given by the formula*

$$a_{k,n} = \binom{n + k - 1}{k} - \binom{n + k - 3}{k - 2}$$

if $n, k \geq 2$, with $a_{0,n} = 1$ and $a_{1,n} = n$, and similarly for $\mathcal{H}_k^{\mathbb{C}}(n)$. As $\mathcal{H}_k(n + 1)$ is isomorphic to $\mathcal{H}_k(S^n)$ (and $\mathcal{H}_k^{\mathbb{C}}(n + 1)$ is isomorphic to $\mathcal{H}_k^{\mathbb{C}}(S^n)$) we have

$$\dim(\mathcal{H}_k^{\mathbb{C}}(S^n)) = \dim(\mathcal{H}_k(S^n)) = a_{k,n+1} = \binom{n + k}{k} - \binom{n + k - 2}{k - 2}.$$

Proof. The cases $k = 0$ and $k = 1$ are trivial, since in this case $\mathcal{H}_k(n) = \mathcal{P}_k(n)$. For $k \geq 2$, the result follows from the direct sum decomposition

$$\mathcal{P}_k(n) = \mathcal{H}_k(n) \oplus \|x\|^2 \mathcal{P}_{k-2}(n)$$

proved earlier. The proof is identical in the complex case.   □

Observe that when $n = 2$, we get $a_{k,2} = 2$ for $k \geq 1$, and when $n = 3$, we get $a_{k,3} = 2k + 1$ for all $k \geq 0$, which we already knew from Section 7.3. The formula even applies for $n = 1$ and yields $a_{k,1} = 0$ for $k \geq 2$.

Remark. It is easy to show that

$$a_{k,n+1} = \binom{n+k-1}{n-1} + \binom{n+k-2}{n-1}$$

for $k \geq 2$; see Morimoto [81] (Chapter 2, Theorem 2.4) or Dieudonné [31] (Chapter 7, Formula 99), where a different proof technique is used.

Definition 7.17. Let $L^2(S^n)$ be the space of (real) square-integrable functions on the sphere S^n. We have an inner product on $L^2(S^n)$ given by

$$\langle f, g \rangle_{S^n} = \int_{S^n} fg \, \text{Vol}_{S^n},$$

where $f, g \in L^2(S^n)$ and where Vol_{S^n} is the volume form on S^n (induced by the metric on $\mathbb{R}^{n+1}$). With this inner product, $L^2(S^n)$ is a complete normed vector space using the norm $\|f\| = \|f\|_2 = \sqrt{\langle f, f \rangle}$ associated with this inner product; that is, $L^2(S^n)$ is a *Hilbert space*. In the case of complex-valued functions, we use the Hermitian inner product

$$\langle f, g \rangle_{S^n} = \int_{S^n} f \, \overline{g} \, \text{Vol}_{S^n},$$

and we get the complex Hilbert space $L^2_{\mathbb{C}}(S^n)$ (see Section 6.8 for the definition of the integral of a complex-valued function).

We also denote by $C(S^n)$ the space of continuous (real) functions on S^n with the L^∞ norm; that is,

$$\|f\|_\infty = \sup\{|f(x)|\}_{x \in S^n},$$

and by $C_{\mathbb{C}}(S^n)$ the space of continuous complex-valued functions on S^n also with the L^∞ norm. Recall that $C(S^n)$ is dense in $L^2(S^n)$ (and $C_{\mathbb{C}}(S^n)$ is dense in $L^2_{\mathbb{C}}(S^n)$); see Rudin [92] (Chapter 3). The following proposition shows why the spherical harmonics play an important role.

Proposition 7.16. *The set of all finite linear combinations of elements in $\bigcup_{k=0}^\infty \mathcal{H}_k(S^n)$ (resp. $\bigcup_{k=0}^\infty \mathcal{H}_k^{\mathbb{C}}(S^n)$) is*

(i) dense in $C(S^n)$ (resp. in $C_{\mathbb{C}}(S^n)$) with respect to the L^∞-norm;

(ii) *dense in $L^2(S^n)$ (resp. dense in $L^2_{\mathbb{C}}(S^n)$).*

Proof.

(i) As S^n is compact, by the Stone-Weierstrass approximation theorem (Lang [68], Chapter III, Corollary 1.3), if g is continuous on S^n, then it can be approximated uniformly by polynomials P_j restricted to S^n. By Corollary 7.14, the restriction of each P_j to S^n is a linear combination of elements in $\bigcup_{k=0}^{\infty} \mathcal{H}_k(S^n)$.

(ii) We use the fact that $C(S^n)$ is dense in $L^2(S^n)$. Given $f \in L^2(S^n)$, for every $\epsilon > 0$, we can choose a continuous function g so that $\|f - g\|_2 < \epsilon/2$. By (i), we can find a linear combination h of elements in $\bigcup_{k=0}^{\infty} \mathcal{H}_k(S^n)$ so that $\|g - h\|_{\infty} < \epsilon/(2\sqrt{\mathrm{vol}(S^n)})$, where $\mathrm{vol}(S^n)$ is the volume of S^n (really, area). Thus we get

$$\|f-h\|_2 \leq \|f-g\|_2 + \|g-h\|_2 < \epsilon/2 + \sqrt{\mathrm{vol}(S^n)}\,\|g-h\|_{\infty} < \epsilon/2 + \epsilon/2 = \epsilon,$$

which proves (ii). The proof in the complex case is identical. $\square$

We need one more proposition before showing that the spaces $\mathcal{H}_k(S^n)$ constitute an orthogonal Hilbert space decomposition of $L^2(S^n)$.

Proposition 7.17. *For every harmonic polynomial $P \in \mathcal{H}_k(n + 1)$ (resp. $P \in \mathcal{H}^{\mathbb{C}}_k(n + 1)$), the restriction $H \in \mathcal{H}_k(S^n)$ (resp. $H \in \mathcal{H}^{\mathbb{C}}_k(S^n)$) of P to S^n is an eigenfunction of Δ_{S^n} for the eigenvalue $-k(n + k - 1)$.*

Proof. We have

$$P(r\sigma) = r^k H(\sigma), \qquad r > 0, \ \sigma \in S^n,$$

and by Proposition 7.10, for any $f \in C^{\infty}(\mathbb{R}^{n+1})$, we have

$$\Delta f = \frac{1}{r^n} \frac{\partial}{\partial r}\left(r^n \frac{\partial f}{\partial r}\right) + \frac{1}{r^2}\Delta_{S^n} f.$$

Consequently,

$$\Delta P = \Delta(r^k H) = \frac{1}{r^n} \frac{\partial}{\partial r}\left(r^n \frac{\partial(r^k H)}{\partial r}\right) + \frac{1}{r^2}\Delta_{S^n}(r^k H)$$

$$= \frac{1}{r^n} \frac{\partial}{\partial r}\left(kr^{n+k-1}H\right) + r^{k-2}\Delta_{S^n} H$$

$$= \frac{1}{r^n}k(n + k - 1)r^{n+k-2}H + r^{k-2}\Delta_{S^n} H$$

$$= r^{k-2}(k(n + k - 1)H + \Delta_{S^n} H).$$

Thus,

$$\Delta P = 0 \quad \text{iff} \quad \Delta_{S^n} H = -k(n + k - 1)H,$$

as claimed.   □

From Proposition 7.17, we deduce that the space $\mathcal{H}_k(S^n)$ is a subspace of the eigenspace E_k of Δ_{S^n} associated with the eigenvalue $-k(n+k-1)$ (and similarly for $\mathcal{H}_k^{\mathbb{C}}(S^n)$). Remarkably, $E_k = \mathcal{H}_k(S^n)$, but it will take more work to prove this.

What we can deduce immediately is that $\mathcal{H}_k(S^n)$ and $\mathcal{H}_l(S^n)$ are pairwise orthogonal whenever $k \neq l$. This is because, by Proposition 7.12, the Laplacian is self-adjoint, and thus any two eigenspaces E_k and E_l are pairwise orthogonal whenever $k \neq l$, and as $\mathcal{H}_k(S^n) \subseteq E_k$ and $\mathcal{H}_l(S^n) \subseteq E_l$, our claim is indeed true. Furthermore, by Proposition 7.15, each $\mathcal{H}_k(S^n)$ is finite dimensional, and thus closed. Finally, we know from Proposition 7.16 that $\bigcup_{k=0}^{\infty} \mathcal{H}_k(S^n)$ is dense in $L^2(S^n)$. But then we can apply a standard result from Hilbert space theory (for example, see Lang [68], Chapter V, Proposition 1.9) to deduce the following important result.

Theorem 7.18. *The family of spaces $\mathcal{H}_k(S^n)$ (resp. $\mathcal{H}_k^{\mathbb{C}}(S^n)$) yields a Hilbert space direct sum decomposition*

$$L^2(S^n) = \bigoplus_{k=0}^{\infty} \mathcal{H}_k(S^n) \qquad (resp. \quad L_{\mathbb{C}}^2(S^n) = \bigoplus_{k=0}^{\infty} \mathcal{H}_k^{\mathbb{C}}(S^n)),$$

which means that the summands are closed, pairwise orthogonal, and that every $f \in L^2(S^n)$ (resp. $f \in L_{\mathbb{C}}^2(S^n)$) is the sum of a converging series

$$f = \sum_{k=0}^{\infty} f_k$$

in the L^2-norm, where the $f_k \in \mathcal{H}_k(S^n)$ (resp. $f_k \in \mathcal{H}_k^{\mathbb{C}}(S^n)$) are uniquely determined functions. Furthermore, given any orthonormal basis $(Y_k^1, \ldots, Y_k^{a_{k,n+1}})$ of $\mathcal{H}_k(S^n)$, we have

$$f_k = \sum_{m_k=1}^{a_{k,n+1}} c_{k,m_k} Y_k^{m_k}, \qquad with \quad c_{k,m_k} = \langle f, Y_k^{m_k} \rangle_{S^n}.$$

The coefficients c_{k,m_k} are "generalized" *Fourier coefficients* with respect to the Hilbert basis $\{Y_k^{m_k} \mid 1 \leq m_k \leq a_{k,n+1}, \, k \geq 0\}$. We can finally prove the main theorem of this section.

Theorem 7.19.

(1) The eigenspaces (resp. complex eigenspaces) of the Laplacian Δ_{S^n} on S^n are the spaces of spherical harmonics

$$E_k = \mathcal{H}_k(S^n) \qquad (resp. \quad E_k = \mathcal{H}_k^{\mathbb{C}}(S^n)),$$

and E_k corresponds to the eigenvalue $-k(n+k-1)$.
(2) We have the Hilbert space direct sum decompositions

$$L^2(S^n) = \bigoplus_{k=0}^{\infty} E_k \qquad (resp. \quad L_{\mathbb{C}}^2(S^n) = \bigoplus_{k=0}^{\infty} E_k).$$

(3) The complex polynomials of the form $(c_1 x_1 + \cdots + c_{n+1} x_{n+1})^k$, with $c_1^2 + \cdots + c_{n+1}^2 = 0$, span the space $\mathcal{H}_k^{\mathbb{C}}(n+1) \cong \mathcal{H}_k^{\mathbb{C}}(S^n)$, for $k \geq 1$.

Proof. We follow essentially the proof in Helgason [55] (Introduction, Theorem 3.1). In (1) and (2) we only deal with the real case, the proof in the complex case being identical.

(1) We already know that the integers $-k(n+k-1)$ are eigenvalues of Δ_{S^n} and that $\mathcal{H}_k(S^n) \subseteq E_k$. We will prove that Δ_{S^n} has no other eigenvalues and no other eigenvectors using the Hilbert basis $\{Y_k^{m_k} \mid 1 \leq m_k \leq a_{k,n+1}, k \geq 0\}$ given by Theorem 7.18. Let λ be any eigenvalue of Δ_{S^n} and let $f \in L^2(S^n)$ be any eigenfunction associated with λ so that

$$\Delta f = \Delta_{S^n} f = \lambda f.$$

We have a unique series expansion

$$f = \sum_{k=0}^{\infty} \sum_{m_k=1}^{a_{k,n+1}} c_{k,m_k} Y_k^{m_k},$$

with $c_{k,m_k} = \langle f, Y_k^{m_k} \rangle_{S^n}$. Now, as Δ_{S^n} is self-adjoint and $\Delta_{S_n} Y_k^{m_k} = -k(n+k-1)Y_k^{m_k}$, the Fourier coefficients d_{k,m_k} of Δf are given by

$$d_{k,m_k} = \langle \Delta_{S_n} f, Y_k^{m_k} \rangle_{S^n} = \langle f, \Delta_{S^n} Y_k^{m_k} \rangle_{S^n} = -k(n+k-1)\langle f, Y_k^{m_k} \rangle_{S^n}$$
$$= -k(n+k-1)c_{k,m_k}.$$

On the other hand, as $\Delta f = \lambda f$, the Fourier coefficients of Δf are given by

$$d_{k,m_k} = \lambda c_{k,m_k}.$$

By uniqueness of the Fourier expansion, we must have

$$\lambda c_{k,m_k} = -k(n+k-1)c_{k,m_k} \qquad \text{for all } k \geq 0.$$

Since $f \neq 0$, there some k such that $c_{k,m_k} \neq 0$, and we must have

$$\lambda = -k(n+k-1)$$

for any such k. However, the function $k \mapsto -k(n+k-1)$ reaches its maximum for $k = -\frac{n-1}{2}$, and as $n \geq 1$, it is strictly decreasing for $k \geq 0$, which implies that k is unique and that

$$c_{j,m_j} = 0 \qquad \text{for all } j \neq k.$$

Therefore $f \in \mathcal{H}_k(S^n)$, and the eigenvalues of Δ_{S^n} are exactly the integers $-k(n+k-1)$, so $E_k = \mathcal{H}_k(S^n)$ as claimed.

Since we just proved that $E_k = \mathcal{H}_k(S^n)$,

(2) follows immediately from the Hilbert decomposition given by Theorem 7.18.

(3) If $H = (c_1 x_1 + \cdots + c_{n+1} x_{n+1})^k$, with $c_1^2 + \cdots + c_{n+1}^2 = 0$, then for $k \leq 1$ it is obvious that $\Delta H = 0$, and for $k \geq 2$ we have

$$\Delta H = k(k-1)(c_1^2 + \cdots + c_{n+1}^2)(c_1 x_1 + \cdots + c_{n+1} x_{n+1})^{k-2} = 0,$$

so $H \in \mathcal{H}_k^{\mathbb{C}}(n+1)$. A simple computation shows that for every $Q \in \mathcal{P}_k^{\mathbb{C}}(n+1)$, if $c = (c_1, \ldots, c_{n+1})$, then we have

$$\partial(Q)(c_1 x_1 + \cdots + c_{n+1} x_{n+1})^m = m(m-1) \cdots (m-k+1)Q(c)(c_1 x_1 + \cdots$$

$$+ c_{n+1} x_{n+1})^{m-k},$$

for all $m \geq k \geq 1$.

Assume that $\mathcal{H}_k^{\mathbb{C}}(n+1)$ is not spanned by the complex polynomials of the form $(c_1 x_1 + \cdots + c_{n+1} x_{n+1})^k$, with $c_1^2 + \cdots + c_{n+1}^2 = 0$, for $k \geq 1$. Then some $Q \in \mathcal{H}_k^{\mathbb{C}}(n+1)$ is orthogonal to all polynomials of the form $H = (c_1 x_1 + \cdots + c_{n+1} x_{n+1})^k$, with $c_1^2 + \cdots + c_{n+1}^2 = 0$. Recall that

$$\langle P, \partial(Q)H \rangle = \langle QP, H \rangle$$

and apply this equation to $P = Q(c)$, H and Q. Since

$$\partial(Q)H = \partial(Q)(c_1 x_1 + \cdots + c_{n+1} x_{n+1})^k = k!Q(c),$$

and as Q is orthogonal to H, we get

$$k!\langle Q(c), Q(c)\rangle = \langle Q(c), k!Q(c)\rangle = \langle Q(c), \partial(Q)H\rangle$$
$$= \langle Q\,Q(c), H\rangle = Q(c)\langle Q, H\rangle = 0,$$

which implies $Q(c) = 0$. Consequently, $Q(x_1, \ldots, x_{n+1})$ vanishes on the complex algebraic variety

$$\{(x_1, \ldots, x_{n+1}) \in \mathbb{C}^{n+1} \mid x_1^2 + \cdots + x_{n+1}^2 = 0\}.$$

By the Hilbert *Nullstellensatz*, some power Q^m belongs to the ideal $(x_1^2 + \cdots + x_{n+1}^2)$ generated by $x_1^2 + \cdots + x_{n+1}^2$. Now, if $n \geq 2$, it is well known that the polynomial $x_1^2 + \cdots + x_{n+1}^2$ is irreducible so the ideal $(x_1^2 + \cdots + x_{n+1}^2)$ is a prime ideal, and thus Q is divisible by $x_1^2 + \cdots + x_{n+1}^2$. However, we know from the proof of Theorem 7.13 that we have an orthogonal direct sum

$$\mathcal{P}_k^{\mathbb{C}}(n+1) = \mathcal{H}_k^{\mathbb{C}}(n+1) \oplus \|x\|^2 \mathcal{P}_{k-2}^{\mathbb{C}}(n+1).$$

Since $Q \in \mathcal{H}_k^{\mathbb{C}}(n+1)$ and Q is divisible by $x_1^2 + \cdots + x_{n+1}^2$, we must have $Q = 0$. Therefore, if $n \geq 2$, we proved (3). However, when $n = 1$, we know from Section 7.2 that the complex harmonic homogeneous polynomials in two variables $P(x, y)$ are spanned by the real and imaginary parts U_k, V_k of the polynomial $(x + iy)^k = U_k + iV_k$. Since $(x - iy)^k = U_k - iV_k$ we see that

$$U_k = \frac{1}{2}\left((x + iy)^k + (x - iy)^k\right), \qquad V_k = \frac{1}{2i}\left((x + iy)^k - (x - iy)^k\right),$$

and as $1 + i^2 = 1 + (-i)^2 = 0$, the space $\mathcal{H}_k^{\mathbb{C}}(\mathbb{R}^2)$ is spanned by $(x + iy)^k$ and $(x - iy)^k$ (for $k \geq 1$), so (3) holds for $n = 1$ as well. $\quad\square$

As an illustration of Part (3) of Theorem 7.19, the polynomials $(x_1 + i\cos\theta x_2 + i\sin\theta x_3)^k$ are harmonic. Of course, the real and imaginary part of a complex harmonic polynomial $(c_1 x_1 + \cdots + c_{n+1} x_{n+1})^k$ are real harmonic polynomials.

7.6 Zonal Spherical Functions and Gegenbauer Polynomials

In this section we describe the zonal spherical functions Z_k^τ on S^n and show that they essentially come from certain polynomials generalizing the Legendre polynomials known as the *Gegenbauer polynomials*. An interesting property of the zonal spherical functions is a formula for obtaining the kth spherical harmonic component of a function $f \in L_{\mathbb{C}}^2(S^n)$; see Proposition 7.27. Another important property of the zonal spherical functions Z_k^τ is that they generate $\mathcal{H}_k^{\mathbb{C}}(S^n)$.

Most proofs will be omitted. We refer the reader to Stein and Weiss [102] (Chapter 4) and Morimoto [81] (Chapter 2) for a complete exposition with proofs.

In order to define zonal spherical functions we will need the following proposition.

Proposition 7.20. *If P is any (complex) polynomial in n variables such that*

$$P(R(x)) = P(x) \qquad \text{for all rotations } R \in \mathbf{SO}(n), \text{ and all } x \in \mathbb{R}^n,$$

then P is of the form

$$P(x) = \sum_{j=0}^{m} c_j (x_1^2 + \cdots + x_n^2)^j,$$

for some $c_0, \ldots, c_m \in \mathbb{C}$.

Proof. Write P as the sum of its homogeneous pieces $P = \sum_{l=0}^{k} Q_l$, where Q_l is homogeneous of degree l. For every $\epsilon > 0$ and every rotation R, we have

$$\sum_{l=0}^{k} \epsilon^l Q_l(x) = P(\epsilon x) = P(R(\epsilon x)) = P(\epsilon R(x)) = \sum_{l=0}^{k} \epsilon^l Q_l(R(x)),$$

which implies that

$$Q_l(R(x)) = Q_l(x), \qquad l = 0, \ldots, k.$$

If we let $F_l(x) = \|x\|^{-l} Q_l(x)$, then F_l is a homogeneous function of degree 0 since

$$F_l(tx) = \|tx\|^{-l} Q_l(tx) = t^{-l} \|x\| \, t^l Q_l(x) = F_l(x).$$

Furthermore, F_l is invariant under all rotations since

$$F_l(R(x)) = \|R(x)\|^{-l} Q_l(R(x)) = \|x\|^{-l} Q_l(x) = F_l(x).$$

This is only possible if F_l is a constant function, thus $F_l(x) = a_l$ for all $x \in \mathbb{R}^n$. But then, $Q_l(x) = a_l \|x\|^l$. Since Q_l is a polynomial, l must be even whenever $a_l \neq 0$. It follows that

$$P(x) = \sum_{j=0}^{m} c_j \|x\|^{2j}$$

with $c_j = a_{2j}$ for $j = 0, \ldots, m$, and where m is the largest integer $\leq k/2$. $\qquad\square$

Proposition 7.20 implies that if a polynomial function on the sphere S^n, in particular a spherical harmonic, is invariant under all rotations, then it is a constant.

If we relax this condition to invariance under all rotations leaving some given point $\tau \in S^n$ invariant, then we obtain zonal harmonics.

The following theorem from Morimoto [81] (Chapter 2, Theorem 2.24) gives the relationship between zonal harmonics and the Gegenbauer polynomials:

Theorem 7.21. *Fix any $\tau \in S^n$. For every constant $c \in \mathbb{C}$, there is a unique homogeneous harmonic polynomial $Z_k^\tau \in \mathcal{H}_k^{\mathbb{C}}(n+1)$ satisfying the following conditions:*

(1) $Z_k^\tau(\tau) = c$;
(2) For every rotation $R \in \mathbf{SO}(n+1)$, if $R\tau = \tau$, then $Z_k^\tau(R(x)) = Z_k^\tau(x)$ for all $x \in \mathbb{R}^{n+1}$.

Furthermore, we have

$$Z_k^\tau(x) = c \, \|x\|^k \, P_{k,n}\left(\frac{x}{\|x\|} \cdot \tau\right),$$

for some polynomial $P_{k,n}(t)$ of degree k.

Remark. The proof given in Morimoto [81] is essentially the same as the proof of Theorem 2.12 in Stein and Weiss [102] (Chapter 4), but Morimoto makes an implicit use of Proposition 7.20 above. Also, Morimoto states Theorem 7.21 only for $c = 1$, but the proof goes through for any $c \in \mathbb{C}$, including $c = 0$, and we will need this extra generality in the proof of the Funk-Hecke formula.

Proof. Let $e_{n+1} = (0, \ldots, 0, 1) \in \mathbb{R}^{n+1}$, and for any $\tau \in S^n$, let R_τ be some rotation such that $R_\tau(e_{n+1}) = \tau$. Assume $Z \in \mathcal{H}_k^{\mathbb{C}}(n+1)$ satisfies Conditions (1) and (2), and let Z' be given by $Z'(x) = Z(R_\tau(x))$. As $R_\tau(e_{n+1}) = \tau$, we have $Z'(e_{n+1}) = Z(\tau) = c$. Furthermore, for any rotation S such that $S(e_{n+1}) = e_{n+1}$, observe that

$$R_\tau \circ S \circ R_\tau^{-1}(\tau) = R_\tau \circ S(e_{n+1}) = R_\tau(e_{n+1}) = \tau,$$

and so, as Z satisfies property (2) for the rotation $R_\tau \circ S \circ R_\tau^{-1}$, we get

$$Z'(S(x)) = Z(R_\tau \circ S(x)) = Z(R_\tau \circ S \circ R_\tau^{-1} \circ R_\tau(x)) = Z(R_\tau(x)) = Z'(x),$$

which proves that Z' is a harmonic polynomial satisfying Properties (1) and (2) with respect to e_{n+1}. Therefore, we may assume that $\tau = e_{n+1}$.
Write

$$Z(x) = \sum_{j=0}^{k} x_{n+1}^{k-j} P_j(x_1, \ldots, x_n),$$

where $P_j(x_1, \ldots, x_n)$ is a homogeneous polynomial of degree j. Since Z is invariant under every rotation R fixing e_{n+1}, and since the monomials x_{n+1}^{k-j} are clearly invariant under such a rotation, we deduce that every $P_j(x_1, \ldots, x_n)$ is invariant under all rotations of $\mathbb{R}^n$ (clearly, there is a one-to-one correspondence between the rotations of $\mathbb{R}^{n+1}$ fixing e_{n+1} and the rotations of $\mathbb{R}^n$). By Proposition 7.20, we conclude that

$$P_j(x_1, \ldots, x_n) = c_j(x_1^2 + \cdots + x_n^2)^{\frac{j}{2}},$$

which implies that $P_j = 0$ if j is odd. Thus we can write

$$Z(x) = \sum_{i=0}^{[k/2]} c_i x_{n+1}^{k-2i}(x_1^2 + \cdots + x_n^2)^i, \tag{$\dagger$}$$

where $[k/2]$ is the greatest integer m such that $2m \leq k$. If $k < 2$, then $Z(x) = c_0$, so $c_0 = c$ and Z is uniquely determined. If $k \geq 2$, we know that Z is a harmonic polynomial so we assert that $\Delta Z = 0$. For $i \leq j \leq n$,

$$\frac{\partial}{\partial x_j}(x_1^2 + \cdots + x_j^2 + \cdots x_n^2)^i = 2ix_j(x_1^2 + \cdots + x_n^2)^{i-1},$$

and

$$\frac{\partial^2}{\partial x_j^2}(x_1^2 + \cdots + x_j^2 + \cdots + x_n^2)^i = 2i(x_1^2 + \cdots x_n^2)^{i-1} + 4x_j^2 i(i-1)(x_1^2 + \cdots + x_n^2)^{i-2}$$

$$= 2i(x_1^2 + \cdots x_n^2)^{i-2}[x_1^2 + \cdots + x_n^2 + 2(i-1)x_j^2].$$

Since $\Delta(x_1^2 + \cdots + x_n^2)^i = \sum_{j=1}^n \frac{\partial^2}{\partial x_j^2}(x_1^2 + \cdots + x_j^2 + \cdots + x_n^2)^i$, we find that

$$\Delta(x_1^2 + \cdots + x_n^2)^i = 2i(x_1^2 + \cdots + x_n^2)^{i-2}\sum_{j=1}^n[x_1^2 + \cdots + x_n^2 + 2(i-1)x_j^2]$$

$$= 2i(x_1^2 + \cdots + x_n^2)^{i-2}\left[n(x_1^2 + \cdots + x_n^2) + 2(i-1)\sum_{j=1}^n x_j^2\right]$$

$$= 2i(x_1^2 + \cdots + x_n^2)^{i-2}[n(x_1^2 + \cdots + x_n^2) + 2(i-1)(x_1^2 + \cdots + x_n^2)]$$

$$= 2i(n + 2i - 2)(x_1^2 + \cdots + x_n^2)^{i-1}.$$

Thus

$$\Delta x_{n+1}^{k-2i}(x_1^2 + \cdots + x_n^2)^i = (k-2i)(k-2i-1)x_{n+1}^{k-2i-2}(x_1^2 + \cdots + x_n^2)^i$$
$$+ x_{n+1}^{k-2i}\Delta(x_1^2 + \cdots + x_n^2)^i$$
$$= (k-2i)(k-2i-1)x_{n+1}^{k-2i-2}(x_1^2 + \cdots + x_n^2)^i$$
$$+ 2i(n+2i-2)x_{n+1}^{k-2i}(x_1^2 + \cdots + x_n^2)^{i-1},$$

and so we get

$$\Delta Z = \sum_{i=0}^{[k/2]-1}((k-2i)(k-2i-1)c_i + 2(i+1)(n+2i)c_{i+1})\,x_{n+1}^{k-2i-2}(x_1^2 + \cdots + x_n^2)^i.$$

Then $\Delta Z = 0$ yields the relations

$$2(i+1)(n+2i)c_{i+1} = -(k-2i)(k-2i-1)c_i, \qquad i = 0, \ldots, [k/2]-1, \quad (\dagger\dagger)$$

which shows that Z is uniquely determined up to the constant c_0. Since we are requiring $Z(e_{n+1}) = c$, we get $c_0 = c$, and Z is uniquely determined. Now on S^n we have $x_1^2 + \cdots + x_{n+1}^2 = 1$, so if we let $t = x_{n+1}$, for $c_0 = 1$, we get a polynomial in one variable

$$P_{k,n}(t) = \sum_{i=0}^{[k/2]}c_i t^{k-2i}(1-t^2)^i. \qquad (*)$$

Thus we proved that when $Z(e_{n+1}) = c$, we have

$$Z(x) = c\,\|x\|^k\,P_{k,n}\left(\frac{x_{n+1}}{\|x\|}\right) = c\,\|x\|^k\,P_{k,n}\left(\frac{x}{\|x\|} \cdot e_{n+1}\right).$$

When $Z(\tau) = c$, we write $Z = Z' \circ R_\tau^{-1}$ with $Z' = Z \circ R_\tau$ and where R_τ is a rotation such that $R_\tau(e_{n+1}) = \tau$. Then, as $Z'(e_{n+1}) = c$, using the formula above for Z', we have

$$Z(x) = Z'(R_\tau^{-1}(x)) = c\,\left\|R_\tau^{-1}(x)\right\|^k\,P_{k,n}\left(\frac{R_\tau^{-1}(x)}{\left\|R_\tau^{-1}(x)\right\|} \cdot e_{n+1}\right)$$

$$= c\,\|x\|^k\,P_{k,n}\left(\frac{x}{\|x\|} \cdot R_\tau(e_{n+1})\right)$$

$$= c\,\|x\|^k\,P_{k,n}\left(\frac{x}{\|x\|} \cdot \tau\right),$$

since R_τ is an isometry. $\square$

To best understand the proof of Theorem 7.21, we let $n = 2, k = 3$, and construct $Z(x) \in \mathcal{H}_3^{\mathbb{C}}(3)$ such that Z satisfies Conditions (1) and (2) with $\tau = e_3$. Line (†) implies that

$$Z(x) = c_0 x_3^3 + c_1 x_3 (x_1^2 + x_2^2).$$

The conditions of Line (††) show that $c_1 = -3/2 c_0$. Hence

$$Z(x) = c x_3^3 - \frac{3}{2} c x_3 (x_1^2 + x_2^2),$$

where we let $c = c_0$. We want to rewrite $Z(x)$ via $P_{3,2}(t)$, where $P_{3,2}(t)$ is given by Line (∗) as

$$P_{3,2}(t) = t^3 - \frac{3}{2} t (1 - t^2).$$

Then a simple verification shows that

$$Z(x) = c \, \|x\|^3 \, P_{3,2} \left(\frac{x_3}{\|x\|} \right).$$

Definition 7.18. The function, Z_k^{τ}, is called a *zonal function* and its restriction to S^n is a *zonal spherical function*. The polynomial $P_{k,n}(t)$ is called the *Gegenbauer polynomial* of degree k and dimension $n + 1$ or *ultraspherical polynomial*. By definition, $P_{k,n}(1) = 1$.

The proof of Theorem 7.21 shows that for k even, say $k = 2m$, the polynomial $P_{2m,n}$ is of the form

$$P_{2m,n}(t) = \sum_{j=0}^{m} c_{m-j} t^{2j} (1 - t^2)^{m-j},$$

and for k odd, say $k = 2m + 1$, the polynomial $P_{2m+1,n}$ is of the form

$$P_{2m+1,n}(t) = \sum_{j=0}^{m} c_{m-j} t^{2j+1} (1 - t^2)^{m-j}.$$

Consequently, $P_{k,n}(-t) = (-1)^k P_{k,n}(t)$, for all $k \geq 0$. The proof also shows that the "natural basis" for these polynomials consists of the polynomials, $t^i (1 - t^2)^{\frac{k-i}{2}}$, with $k - i$ even. Indeed, with this basis, there are simple recurrence equations for computing the coefficients of $P_{k,n}(t)$.

Remark. Morimoto [81] calls the polynomials $P_{k,n}(t)$ "Legendre polynomials." For $n = 2$, they are indeed the Legendre polynomials. Stein and Weiss denote our (and Morimoto's) $P_{k,n}(t)$ by $P_k^{\frac{n-1}{2}}(t)$ (up to a constant factor), and Dieudonné [31] (Chapter 7) by $G_{k,n+1}(t)$.

When $n = 2$, using the notation of Section 7.3, the zonal spherical functions on S^2 are the spherical harmonics y_l^0 for which $m = 0$; that is (up to a constant factor),

$$y_l^0(\theta, \varphi) = \sqrt{\frac{(2l + 1)}{4\pi}}\, P_l(\cos\theta),$$

where P_l is the Legendre polynomial of degree l. For example, for $l = 2$, $P_l(t) = \frac{1}{2}(3t^2 - 1)$.

Zonal spherical functions have many important properties. One such property is associated with the reproducing kernel of $\mathcal{H}_k^{\mathbb{C}}(S^n)$.

Definition 7.19. Let $\mathcal{H}_k^{\mathbb{C}}(S^n)$ be the space of spherical harmonics. Let $a_{k,n+1}$ be the dimension of $\mathcal{H}_k^{\mathbb{C}}(S^n)$ where

$$a_{k,n+1} = \binom{n+k}{k} - \binom{n+k-2}{k-2},$$

if $n \geq 1$ and $k \geq 2$, with $a_{0,n+1} = 1$ and $a_{1,n+1} = n + 1$. Let $(Y_k^1, \ldots, Y_k^{a_{k,n+1}})$ be any orthonormal basis of $\mathcal{H}_k^{\mathbb{C}}(S^n)$, and define $F_k(\sigma, \tau)$ by

$$F_k(\sigma, \tau) = \sum_{i=1}^{a_{k,n+1}} Y_k^i(\sigma)\overline{Y_k^i(\tau)}, \qquad \sigma, \tau \in S^n.$$

The function $F_k(\sigma, \tau)$ is the *reproducing kernel* of $\mathcal{H}_k^{\mathbb{C}}(S^n)$.

The following proposition is easy to prove (see Morimoto [81], Chapter 2, Lemma 1.19 and Lemma 2.20).

Proposition 7.22. *The function F_k is independent of the choice of orthonormal basis. Furthermore, for every orthogonal transformation $R \in \mathbf{O}(n + 1)$, we have*

$$F_k(R\sigma, R\tau) = F_k(\sigma, \tau), \qquad \sigma, \tau \in S^n.$$

Clearly, F_k is a symmetric function. Since we can pick an orthonormal basis of real orthogonal functions for $\mathcal{H}_k^{\mathbb{C}}(S^n)$ (pick a basis of $\mathcal{H}_k(S^n)$), Proposition 7.22 shows that F_k is a real-valued function.

The function F_k satisfies the following property which justifies its name as the reproducing kernel for $\mathcal{H}_k^{\mathbb{C}}(S^n)$:

Remark. In the proofs below, integration is performed with respect to the repeated variable.

Proposition 7.23. *For every spherical harmonic* $H \in \mathcal{H}_j^{\mathbb{C}}(S^n)$, *we have*

$$\int_{S^n} H(\tau) F_k(\sigma, \tau) \, \mathrm{Vol}_{S^n} = \delta_{jk} H(\sigma), \qquad \sigma, \tau \in S^n,$$

for all $j, k \geq 0$.

Proof. When $j \neq k$, since $\mathcal{H}_k^{\mathbb{C}}(S^n)$ and $\mathcal{H}_j^{\mathbb{C}}(S^n)$ are orthogonal and since $F_k(\sigma, \tau) = \sum_{i=1}^{a_{k,n+1}} Y_k^i(\sigma) \overline{Y_k^i(\tau)}$, it is clear that the integral in Proposition 7.23 vanishes. When $j = k$, we have

$$\int_{S^n} H(\tau) F_k(\sigma, \tau) \, \mathrm{Vol}_{S^n} = \int_{S^n} H(\tau) \sum_{i=1}^{a_{k,n+1}} Y_k^i(\sigma) \overline{Y_k^i(\tau)} \, \mathrm{Vol}_{S^n}$$

$$= \sum_{i=1}^{a_{k,n+1}} Y_k^i(\sigma) \int_{S^n} H(\tau) \overline{Y_k^i(\tau)} \, \mathrm{Vol}_{S^n}$$

$$= \sum_{i=1}^{a_{k,n+1}} Y_k^i(\sigma) \langle H, Y_k^i \rangle$$

$$= H(\sigma),$$

since $(Y_k^1, \ldots, Y_k^{a_{k,n+1}})$ is an orthonormal basis.   □

Remark. In Stein and Weiss [102] (Chapter 4), the function $F_k(\sigma, \tau)$ is denoted by $Z_\sigma^{(k)}(\tau)$ and it is called the *zonal harmonic of degree k with pole σ*.

Before we investigate the relationship between $F_k(\sigma, \tau)$ and $Z_k^\tau(\sigma)$, we need two technical propositions. Both are proven in Morimoto [81]. The first, Morimoto [81] (Chapter 2, Lemma 2.21), is needed to prove the second, Morimoto [81] (Chapter 2, Lemma 2.23).

Proposition 7.24. *For all* $\sigma, \tau, \sigma', \tau' \in S^n$, *with* $n \geq 1$, *the following two conditions are equivalent:*

(i) *There is some orthogonal transformation* $R \in \mathbf{O}(n + 1)$ *such that* $R(\sigma) = \sigma'$ *and* $R(\tau) = \tau'$.

(ii) $\sigma \cdot \tau = \sigma' \cdot \tau'$.

Propositions 7.22 and 7.24 immediately yield the following.

Proposition 7.25. *For all* $\sigma, \tau, \sigma', \tau' \in S^n$, *if* $\sigma \cdot \tau = \sigma' \cdot \tau'$, *then* $F_k(\sigma, \tau) = F_k(\sigma', \tau')$. *Consequently, there is some function* $\varphi \colon \mathbb{R} \to \mathbb{R}$ *such that* $F_k(\sigma, \tau) = \varphi(\sigma \cdot \tau)$.

We claim that the $\varphi(\sigma \cdot \tau)$ of Proposition 7.25 is a zonal spherical function $Z_k^\tau(\sigma)$.

To see why this is true, define $Z(r^k\sigma) := r^k F_k(\sigma, \tau)$ for a fixed τ. By the definition of $F_k(\sigma, \tau)$, it is clear that Z is a homogeneous harmonic polynomial. The value $F_k(\tau, \tau)$ does not depend of τ, because by transitivity of the action of $SO(n+1)$ on S^n, for any other $\sigma \in S^n$, there is some rotation R so that $R\tau = \sigma$, and by Proposition 7.22, we have $F_k(\sigma, \sigma) = F_k(R\tau, R\tau) = F_k(\tau, \tau)$.

To compute $F_k(\tau, \tau)$, since

$$F_k(\tau, \tau) = \sum_{i=1}^{a_{k,n+1}} \left\| Y_k^i(\tau) \right\|^2,$$

and since $(Y_k^1, \ldots, Y_k^{a_{k,n+1}})$ is an orthonormal basis of $\mathcal{H}_k^{\mathbb{C}}(S^n)$, observe that

$$a_{k,n+1} = \sum_{i=1}^{a_{k,n+1}} \langle Y_k^i, Y_k^i \rangle$$

$$= \sum_{i=1}^{a_{k,n+1}} \int_{S^n} \left\| Y_k^i(\tau) \right\|^2 \, \mathrm{Vol}_{S^n}$$

$$= \int_{S^n} \left(\sum_{i=1}^{a_{k,n+1}} \left\| Y_k^i(\tau) \right\|^2 \right) \mathrm{Vol}_{S^n}$$

$$= \int_{S^n} F_k(\tau, \tau) \, \mathrm{Vol}_{S^n} = F_k(\tau, \tau) \, \mathrm{vol}(S^n).$$

Therefore,

$$F_k(\tau, \tau) = \frac{a_{k,n+1}}{\mathrm{vol}(S^n)}.$$

Beware that Morimoto [81] uses the normalized measure on S^n, so the factor involving $\mathrm{vol}(S^n)$ does not appear.

Remark. The volume of the n-sphere is given by

$$\mathrm{vol}(S^{2d}) = \frac{2^{d+1}\pi^d}{1 \cdot 3 \cdots (2d-1)} \quad \text{if} \quad d \geq 1 \quad \text{and} \quad \mathrm{vol}(S^{2d+1}) = \frac{2\pi^{d+1}}{d!} \quad \text{if} \quad d \geq 0.$$

These formulae will be proved in Section 7.8 just after the proof of Theorem 7.36.

Now, if $R\tau = \tau$, Proposition 7.22 shows that

$$Z(R(r^k\sigma)) = Z(r^k R(\sigma)) = r^k F_k(R\sigma, \tau) = r^k F_k(R\sigma, R\tau) = r^k F_k(\sigma, \tau) = Z(r^k\sigma).$$

Therefore, the function Z satisfies Conditions (1) and (2) of Theorem 7.21 with $c = \frac{a_{k,n+1}}{\text{vol}(S^n)}$, and by uniqueness, we conclude that Z is the zonal function Z_k^τ whose restriction to S^n is the zonal spherical function

$$F_k(\sigma, \tau) = \frac{a_{k,n+1}}{\text{vol}(S^n)} \, P_{k,n}(\sigma \cdot \tau).$$

Consequently, we have obtained the so-called *addition formula*:

Proposition 7.26 (Addition Formula). *If $(Y_k^1, \ldots, Y_k^{a_{k,n+1}})$ is any orthonormal basis of*
$\mathcal{H}_k^{\mathbb{C}}(S^n)$, *then*

$$P_{k,n}(\sigma \cdot \tau) = \frac{\text{vol}(S^n)}{a_{k,n+1}} \sum_{i=1}^{a_{k,n+1}} Y_k^i(\sigma) \overline{Y_k^i(\tau)}.$$

Again, beware that Morimoto [81] does not have the factor $\text{vol}(S^n)$.

For $n = 1$, we can write $\sigma = (\cos\theta, \sin\theta)$ and $\tau = (\cos\varphi, \sin\varphi)$, and it is easy to see that the addition formula reduces to

$$P_{k,1}(\cos(\theta - \varphi)) = \cos k\theta \cos k\varphi + \sin k\theta \sin k\varphi = \cos k(\theta - \varphi),$$

the standard addition formula for trigonometric functions.

Proposition 7.26 implies that $P_{k,n}(t)$ has real coefficients. Furthermore Proposition 7.23 is reformulated as

$$\frac{a_{k,n+1}}{\text{vol}(S^n)} \int_{S^n} P_{k,n}(\sigma \cdot \tau) H(\tau) \, \text{Vol}_{S^n} = \delta_{jk} H(\sigma), \tag{rk}$$

showing that the Gengenbauer polynomials are reproducing kernels. A neat application of this formula is a formula for obtaining the kth spherical harmonic component of a function $f \in L_{\mathbb{C}}^2(S^n)$.

Proposition 7.27. *For every function $f \in L_{\mathbb{C}}^2(S^n)$, if $f = \sum_{k=0}^{\infty} f_k$ is the unique decomposition of f over the Hilbert sum $\bigoplus_{k=0}^{\infty} \mathcal{H}_k^{\mathbb{C}}(S^k)$, then f_k is given by*

$$f_k(\sigma) = \frac{a_{k,n+1}}{\text{vol}(S^n)} \int_{S^n} f(\tau) P_{k,n}(\sigma \cdot \tau) \, \text{Vol}_{S^n},$$

for all $\sigma \in S^n$.

Proof. If we recall that $\mathcal{H}_k^{\mathbb{C}}(S^k)$ and $\mathcal{H}_j^{\mathbb{C}}(S^k)$ are orthogonal for all $j \neq k$, using the Formula (rk), we have

$$\frac{a_{k,n+1}}{\text{vol}(S^n)} \int_{S^n} f(\tau) P_{k,n}(\sigma \cdot \tau) \, \text{Vol}_{S^n} = \frac{a_{k,n+1}}{\text{vol}(S^n)} \int_{S^n} \sum_{j=0}^{\infty} f_j(\tau) P_{k,n}(\sigma \cdot \tau) \, \text{Vol}_{S^n}$$

$$= \frac{a_{k,n+1}}{\text{vol}(S^n)} \sum_{j=0}^{\infty} \int_{S^n} f_j(\tau) P_{k,n}(\sigma \cdot \tau) \, \text{Vol}_{S^n}$$

$$= \frac{a_{k,n+1}}{\text{vol}(S^n)} \int_{S^n} f_k(\tau) P_{k,n}(\sigma \cdot \tau) \, \text{Vol}_{S^n}$$

$$= f_k(\sigma),$$

as claimed. $\square$

Another important property of the zonal spherical functions Z_k^{τ} is that they generate $\mathcal{H}_k^{\mathbb{C}}(S^n)$. In order to prove this fact, we use the following proposition.

Proposition 7.28. *If $H_1, \ldots, H_m \in \mathcal{H}_k^{\mathbb{C}}(S^n)$ are linearly independent, then there are m points $\sigma_1, \ldots, \sigma_m$ on S^n so that the $m \times m$ matrix $(H_j(\sigma_i))$ is invertible.*

Proof. We proceed by induction on m. The case $m = 1$ is trivial. For the induction step, we may assume that we found m points $\sigma_1, \ldots, \sigma_m$ on S^n so that the $m \times m$ matrix $(H_j(\sigma_i))$ is invertible. Consider the function

$$\sigma \mapsto \begin{vmatrix} H_1(\sigma) & \cdots & H_m(\sigma) & H_{m+1}(\sigma) \\ H_1(\sigma_1) & \cdots & H_m(\sigma_1) & H_{m+1}(\sigma_1) \\ \vdots & \ddots & \vdots & \vdots \\ H_1(\sigma_m) & \cdots & H_m(\sigma_m) & H_{m+1}(\sigma_m) \end{vmatrix}.$$

Since $H_1, \ldots, H_{m+1}$ are linearly independent, the above function does not vanish for all σ, since otherwise, by expanding this determinant with respect to the first row, we would get a linear dependence among the H_j's where the coefficient of H_{m+1} is nonzero. Therefore, we can find σ_{m+1} so that the $(m+1) \times (m+1)$ matrix $(H_j(\sigma_i))$ is invertible. $\square$

Definition 7.20. We say that $a_{k,n+1}$ points, $\sigma_1, \ldots, \sigma_{a_{k,n+1}}$ on S^n form a *fundamental system* iff the $a_{k,n+1} \times a_{k,n+1}$ matrix $(P_{n,k}(\sigma_i \cdot \sigma_j))$ is invertible.

Theorem 7.29. *The following properties hold:*

(i) There is a fundamental system $\sigma_1, \ldots, \sigma_{a_{k,n+1}}$ for every $k \geq 1$.

(ii) Every spherical harmonic $H \in \mathcal{H}_k^{\mathbb{C}}(S^n)$ can be written as

$$H(\sigma) = \sum_{j=1}^{a_{k,n+1}} c_j \, P_{k,n}(\sigma_j \cdot \sigma),$$

for some unique $c_j \in \mathbb{C}$.

Proof.

(i) By the addition formula,

$$P_{k,n}(\sigma_i \cdot \sigma_j) = \frac{\mathrm{vol}(S^n)}{a_{k,n+1}} \sum_{l=1}^{a_{k,n+1}} Y_k^l(\sigma_i)\overline{Y_k^l(\sigma_j)}$$

for any orthonormal basis $(Y_k^1, \ldots, Y_k^{a_{k,n+1}})$. It follows that the matrix $(P_{k,n}(\sigma_i \cdot \sigma_j))$ can be written as

$$(P_{k,n}(\sigma_i \cdot \sigma_j)) = \frac{\mathrm{vol}(S^n)}{a_{k,n+1}}\, YY^*,$$

where $Y = (Y_k^l(\sigma_i))$, and by Proposition 7.28, we can find $\sigma_1, \ldots, \sigma_{a_{k,n+1}} \in S^n$ so that Y and thus also Y^* are invertible, and so $(P_{n,k}(\sigma_i \cdot \sigma_j))$ is invertible.

(ii) Again, by the addition formula,

$$P_{k,n}(\sigma \cdot \sigma_j) = \frac{\mathrm{vol}(S^n)}{a_{k,n+1}} \sum_{i=1}^{a_{k,n+1}} Y_k^i(\sigma)\overline{Y_k^i(\sigma_j)}.$$

However, as $(Y_k^1, \ldots, Y_k^{a_{k,n+1}})$ is an orthonormal basis, Part (i) proved that the matrix Y^* is invertible, so the $Y_k^i(\sigma)$ can be expressed uniquely in terms of the $P_{k,n}(\sigma \cdot \sigma_j)$, as claimed. $\square$

Statement (ii) of Theorem 7.29 shows that the set of $P_{k,n}(\sigma \cdot \tau) = \frac{\mathrm{vol}(S^n)}{a_{k,n+1}} F_k(\sigma, \tau)$ do indeed generate $\mathcal{H}_k^{\mathbb{C}}(S^n)$.

We end this section with a neat geometric characterization of the zonal spherical functions as given in Stein and Weiss [102]. For this, we need to define the notion of a parallel on S^n. A *parallel of S^n orthogonal to a point* $\tau \in S^n$ is the intersection of S^n with any (affine) hyperplane orthogonal to the line through the center of S^n and τ. See Figure 7.3 Clearly, any rotation R leaving τ fixed leaves every parallel orthogonal to τ globally invariant, and for any two points σ_1 and σ_2, on such a parallel, there is a rotation leaving τ fixed that maps σ_1 to σ_2. Consequently, the zonal function Z_k^τ defined by τ is constant on the parallels orthogonal to τ. In fact, this property characterizes zonal harmonics, up to a constant.

The theorem below is proved in Stein and Weiss [102] (Chapter 4, Theorem 2.12). The proof uses Proposition 7.20 and it is very similar to the proof of Theorem 7.21. To save space, it is omitted.

Theorem 7.30. *Fix any point* $\tau \in S^n$. *A spherical harmonic* $Y \in \mathcal{H}_k^{\mathbb{C}}(S^n)$ *is constant on parallels orthogonal to* τ *iff* $Y = cZ_k^\tau$ *for some constant* $c \in \mathbb{C}$.

In the next section we show how the Gegenbauer polynomials can actually be computed.

Fig. 7.3 The purple planes are parallels of S^2 orthogonal to the red point τ. Any rotation around the red axis maps each parallel to itself.

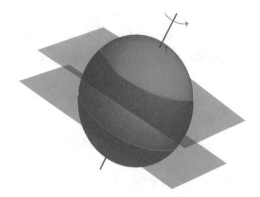

7.7 More on the Gegenbauer Polynomials

The Gegenbauer polynomials are characterized by a formula generalizing the Rodrigues formula defining the Legendre polynomials (see Section 7.3). The expression

$$\left(k + \frac{n-2}{2}\right)\left(k - 1 + \frac{n-2}{2}\right)\cdots\left(1 + \frac{n-2}{2}\right)$$

can be expressed in terms of the Γ function as

$$\frac{\Gamma\left(k + \frac{n}{2}\right)}{\Gamma\left(\frac{n}{2}\right)}.$$

Recall that the Γ function is a generalization of factorial that satisfies the equation

$$\Gamma(z + 1) = z\Gamma(z).$$

For $z = x + iy$ with $x > 0$, $\Gamma(z)$ is given by

$$\Gamma(z) = \int_0^\infty t^{z-1} e^{-t}\, dt,$$

where the integral converges absolutely. If n is an integer $n \geq 0$, then $\Gamma(n+1) = n!$.

It is proved in Morimoto [81] (Chapter 2, Theorem 2.35) that

Proposition 7.31. *The Gegenbauer polynomial $P_{k,n}$ is given by Rodrigues' formula:*

$$P_{k,n}(t) = \frac{(-1)^k}{2^k} \frac{\Gamma\left(\frac{n}{2}\right)}{\Gamma\left(k + \frac{n}{2}\right)} \frac{1}{(1 - t^2)^{\frac{n-2}{2}}} \frac{d^k}{dt^k} (1 - t^2)^{k + \frac{n-2}{2}},$$

with $n \geq 2$.

The Gegenbauer polynomials satisfy the following orthogonality properties with respect to the kernel $(1 - t^2)^{\frac{n-2}{2}}$ (see Morimoto [81], Chapter 2, Theorem 2.34):

Proposition 7.32. *The Gegenbauer polynomial $P_{k,n}$ have the following properties:*

$$\int_{-1}^{-1} (P_{k,n}(t))^2 (1 - t^2)^{\frac{n-2}{2}} \, dt = \frac{\mathrm{vol}(S^n)}{a_{k,n+1}\mathrm{vol}(S^{n-1})}$$

$$\int_{-1}^{-1} P_{k,n}(t) P_{l,n}(t)(1 - t^2)^{\frac{n-2}{2}} \, dt = 0, \qquad k \neq l.$$

The Gegenbauer polynomials satisfy a second-order differential equation generalizing the Legendre equation from Section 7.3.

Proposition 7.33. *The Gegenbauer polynomial $P_{k,n}$ are solutions of the differential equation*

$$(1 - t^2)P_{k,n}''(t) - nt\, P_{k,n}'(t) + k(k + n - 1)P_{k,n}(t) = 0.$$

Proof. If we let $\tau = e_{n+1}$, then the function H given by $H(\sigma) = P_{k,n}(\sigma \cdot \tau) = P_{k,n}(\cos \theta)$ belongs to $\mathcal{H}_k^{\mathbb{C}}(S^n)$, so

$$\Delta_{S^n} H = -k(k + n - 1)H.$$

Recall from Section 7.4 that

$$\Delta_{S^n} f = \frac{1}{\sin^{n-1}\theta} \frac{\partial}{\partial \theta}\left(\sin^{n-1}\theta \, \frac{\partial f}{\partial \theta}\right) + \frac{1}{\sin^2\theta}\Delta_{S^{n-1}} f,$$

in the local coordinates where

$$\sigma = \sin\theta \, \widetilde{\sigma} + \cos\theta \, e_{n+1},$$

with $\widetilde{\sigma} \in S^{n-1}$ and $0 \leq \theta < \pi$. If we make the change of variable $t = \cos\theta$, then it is easy to see that the above formula becomes

$$\Delta_{S^n} f = (1 - t^2)\frac{\partial^2 f}{\partial t^2} - nt\frac{\partial f}{\partial t} + \frac{1}{1 - t^2}\Delta_{S^{n-1}} f$$

(see Morimoto [81], Chapter 2, Theorem 2.9). But H being zonal, it only depends on θ, that is on t, so $\Delta_{S^{n-1}} H = 0$, and thus

$$-k(k + n - 1)P_{k,n}(t) = \Delta_{S^n} P_{k,n}(t) = (1 - t^2)\frac{\partial^2 P_{k,n}}{\partial t^2} - nt\frac{\partial P_{k,n}}{\partial t},$$

which yields our equation.   □

Note that for $n = 2$, the differential equation of Proposition 7.33 is the Legendre equation from Section 7.3.

The Gegenbauer polynomials also appear as coefficients in some simple generating functions. The following proposition is proved in Morimoto [81] (Chapter 2, Theorem 2.53 and Theorem 2.55):

Proposition 7.34. *For all r and t such that $-1 < r < 1$ and $-1 \le t \le 1$, for all $n \ge 1$, we have the following generating formula:*

$$\sum_{k=0}^{\infty} a_{k,n+1} \, r^k P_{k,n}(t) = \frac{1 - r^2}{(1 - 2rt + r^2)^{\frac{n+1}{2}}}.$$

Furthermore, for all r and t such that $0 \le r < 1$ and $-1 \le t \le 1$, if $n = 1$, then

$$\sum_{k=1}^{\infty} \frac{r^k}{k} \, P_{k,1}(t) = -\frac{1}{2} \log(1 - 2rt + r^2),$$

and if $n \ge 2$, then

$$\sum_{k=0}^{\infty} \frac{n-1}{2k+n-1} \, a_{k,n+1} \, r^k P_{k,n}(t) = \frac{1}{(1 - 2rt + r^2)^{\frac{n-1}{2}}}.$$

In Stein and Weiss [102] (Chapter 4, Section 2), the polynomials $P_k^\lambda(t)$, where $\lambda > 0$, are defined using the following generating formula:

$$\sum_{k=0}^{\infty} r^k P_k^\lambda(t) = \frac{1}{(1 - 2rt + r^2)^\lambda}.$$

Each polynomial $P_k^\lambda(t)$ has degree k and is called an *ultraspherical polynomial of degree k associated with λ.* In view of Proposition 7.34, we see that

$$P_k^{\frac{n-1}{2}}(t) = \frac{n-1}{2k+n-1} \, a_{k,n+1} \, P_{k,n}(t),$$

as we mentioned earlier. There is also an integral formula for the Gegenbauer polynomials known as *Laplace representation*; see Morimoto [81] (Chapter 2, Theorem 2.52).

7.8   The Funk–Hecke Formula

The Funk–Hecke formula (also known as Hecke–Funk formula) basically allows one to perform a sort of convolution of a "kernel function" with a spherical function in a convenient way. Given a measurable function K on $[-1, 1]$ such that the integral

$$\int_{-1}^{1} |K(t)|(1 - t^2)^{\frac{n-2}{2}} \, dt < \infty,$$

(which means the integral makes sense), given a function $f \in L_{\mathbb{C}}^2(S^n)$, we can view the expression

$$K \star f(\sigma) = \int_{S^n} K(\sigma \cdot \tau) f(\tau) \, \mathrm{Vol}_{S^n}$$

as a sort of *convolution* of K and f.

Actually, the use of the term convolution is really unfortunate because in a "true" convolution $f * g$, either the argument of f or the argument of g should be multiplied by the inverse of the variable of integration, which means that the integration should really be taking place over the group $\mathbf{SO}(n + 1)$. We will come back to this point later. For the time being, let us call the expression $K \star f$ defined above a *pseudo-convolution*. Now, if f is expressed in terms of spherical harmonics as

$$f = \sum_{k=0}^{\infty} \sum_{m_k=1}^{a_{k,n+1}} c_{k,m_k} Y_k^{m_k},$$

then the Funk–Hecke formula states that

$$K \star Y_k^{m_k}(\sigma) = \alpha_k Y_k^{m_k}(\sigma),$$

for some fixed constant α_k, and so

$$K \star f = \sum_{k=0}^{\infty} \sum_{m_k=1}^{a_{k,n+1}} \alpha_k c_{k,m_k} Y_k^{m_k}.$$

Thus, if the constants α_k are known, then it is "cheap" to compute the pseudo-convolution $K \star f$.

This method was used in a ground-breaking paper by Basri and Jacobs [9] to compute the reflectance function r from the lighting function ℓ as a pseudo-convolution $K \star \ell$ (over S^2) with the *Lambertian kernel* K given by

$$K(\sigma \cdot \tau) = \max(\sigma \cdot \tau, 0).$$

Below, we give a proof of the Funk–Hecke formula due to Morimoto [81] (Chapter 2, Theorem 2.39); see also Andrews, Askey, and Roy [1] (Chapter 9). This formula was first published by Funk in 1916 and then by Hecke in 1918. But before we get to the Funk–Hecke formula, we need the following auxiliary proposition.

Proposition 7.35. *Let $\sigma \in S^n$ be given by the local coordinates on S^n where*

$$\sigma = \sqrt{1 - t^2}\, \widetilde{\sigma} + t\, e_{n+1},$$

with $\widetilde{\sigma} \in S^{n-1}$ and $-1 \leq t \leq 1$. The volume form on S^n is given by

$$\mathrm{Vol}_{S^n} = (1 - t^2)^{\frac{n-2}{2}}\, \mathrm{Vol}_{S^{n-1}}\, dt.$$

Proof. We need to compute the determinant of the $n \times n$ matrix $g = (g_{ij})$ expressing the Riemannian metric on S^n in this local coordinate system. Say the local coordinates on S^{n-1} are $t_1, \ldots, t_{n-1}$. Given $\sigma = \sqrt{1 - t^2}\, \widetilde{\sigma} + t\, e_{n+1}$, we compute

$$\frac{\partial \sigma}{\partial t_i} = \sqrt{1 - t^2}\, \frac{\partial \widetilde{\sigma}}{\partial t_i}$$

$$\frac{\partial \sigma}{\partial t} = -\frac{t}{\sqrt{1 - t^2}}\, \widetilde{\sigma} + e_{n+1},$$

and then using the fact that $\widetilde{\sigma}$ and e_{n+1} are orthogonal unit vectors,

$$g_{ij} = \frac{\partial \sigma}{\partial t_i} \cdot \frac{\partial \sigma}{\partial t_j} = (1 - t^2) \frac{\partial \widetilde{\sigma}}{\partial t_i} \cdot \frac{\partial \widetilde{\sigma}}{\partial t_i} \qquad 1 \leq i, j \leq n - 1$$

$$g_{in} = g_{ni} = \frac{\partial \sigma}{\partial t_i} \cdot \frac{\partial \sigma}{\partial t} = 0 \qquad\qquad\quad 1 \leq i \leq n - 1$$

$$g_{nn} = \frac{\partial \sigma}{\partial t} \cdot \frac{\partial \sigma}{\partial t} = \frac{t^2}{1 - t^2} + 1 = \frac{1}{1 - t^2}.$$

If we let $\widetilde{g}$ be the $(n - 1) \times (n - 1)$ matrix given by

$$\widetilde{g}_{ij} = \frac{\partial \widetilde{\sigma}}{\partial t_i} \cdot \frac{\partial \widetilde{\sigma}}{\partial t_j},$$

then g is the matrix

$$g = \begin{pmatrix} (1 - t^2)\widetilde{g} & 0 \\ 0 & \frac{1}{1 - t^2} \end{pmatrix},$$

and since $\widetilde{g}$ is an $(n-1)\times(n-1)$ matrix,

$$\sqrt{\det(g)} = (1-t^2)^{\frac{n-2}{2}}\sqrt{\det(\widetilde{g})},$$

as Proposition 6.4 implies $\mathrm{Vol}_{S^{n-1}} = \sqrt{\det(\widetilde{g})}dt_1 \wedge \cdots \wedge dt_{n-1}$, it follows that

$$\mathrm{Vol}_{S^n} = (1-t^2)^{\frac{n-2}{2}}\,\mathrm{Vol}_{S^{n-1}}\,dt,$$

as claimed. $\square$

Theorem 7.36 (Funk–Hecke Formula). *Given any measurable function K on $[-1,1]$ such that the integral*

$$\int_{-1}^{1} |K(t)|(1-t^2)^{\frac{n-2}{2}}\,dt$$

makes sense, for every function $H \in \mathcal{H}_k^{\mathbb{C}}(S^n)$, we have

$$\int_{S^n} K(\sigma\cdot\xi)H(\xi)\,\mathrm{Vol}_{S^n} = \left(\mathrm{vol}(S^{n-1})\int_{-1}^{1} K(t)P_{k,n}(t)(1-t^2)^{\frac{n-2}{2}}\,dt\right)H(\sigma).$$

Observe that when $n = 2$, the term $(1-t^2)^{\frac{n-2}{2}}$ is missing and we are simply requiring that $\int_{-1}^{1}|K(t)|\,dt$ makes sense.

Proof. We first prove the formula in the case where $H(\xi) = P_{k,n}(\xi\cdot\tau)$ for some fixed $\tau \in S^n$, and then use the fact that the $P_{k,n}$'s are reproducing kernels (Formula (rk)).

For any fixed $\tau \in S^n$ and every $\sigma \in S^n$, define F by

$$F(\sigma,\tau) = \int_{S^n} K(\sigma\cdot\xi)H(\xi)\,\mathrm{Vol}_{S^n} = \int_{S^n} K(\sigma\cdot\xi)P_{k,n}(\xi\cdot\tau)\,\mathrm{Vol}_{S^n}.$$

Since the volume form on the sphere is invariant under orientation-preserving isometries, for every $R \in \mathbf{SO}(n+1)$, we have

$$F(R\sigma, R\tau) = F(\sigma,\tau),$$

which means that $F(\sigma,\tau)$ is a function of $\sigma\cdot\tau$. On the other hand, for σ fixed, it is not hard to see that as a function in τ, the function $F(\sigma,-)$ is a spherical harmonic. This is because the function given by $H(\xi) = P_{k,n}(\xi\cdot\tau)$ may be viewed as a function of τ, namely $H(\tau) = P_{k,n}(\xi\cdot\tau)$. Furthermore $H \in \mathcal{H}_k^{\mathbb{C}}(S^n)$, and H satisfies the equation

$$\Delta_{S^n}H(\tau) = -k(k+n-1)H(\tau),$$

with respect to the τ coordinates. This implies

$$\Delta_{S^n} F(\sigma, -) = -k(k + n - 1)F(\sigma, -),$$

since

$$-k(k + n - 1)F(\sigma, \tau) = -k(k + n - 1) \int_{S^n} K(\sigma \cdot \xi) P_{k,n}(\xi \cdot \tau) \, \mathrm{Vol}_{S^n}$$

$$= \int_{S^n} K(\sigma \cdot \xi)(-k(k + n - 1)H(\tau)) \, \mathrm{Vol}_{S^n}$$

$$= \int_{S^n} K(\sigma \cdot \xi) \Delta_{S^n} H(\tau) \, \mathrm{Vol}_{S^n}$$

$$= \Delta_{S_n} \int_{S^n} K(\sigma \cdot \xi) H(\xi) \, \mathrm{Vol}_{S^n}$$

$$= \Delta_{S_n} \int_{S^n} K(\sigma \cdot \xi) F(\sigma, \tau).$$

Thus $F(\sigma, -) \in \mathcal{H}_k^{\mathbb{C}}(S^n)$. Now for every rotation R that fixes σ,

$$F(\sigma, \tau) = F(R\sigma, R\tau) = F(\sigma, R\tau),$$

which means that $F(\sigma, -)$ satisfies Condition (2) of Theorem 7.21. By Theorem 7.21, we get

$$F(\sigma, \tau) = F(\sigma, \sigma) P_{k,n}(\sigma \cdot \tau),$$

since

$$F(\sigma, \sigma) = cP_{k,n}(\sigma \cdot \sigma) = cP_{k,n}(1) = c.$$

We now want to explicitly compute $F(\sigma, \sigma) = c$. In order to do so, we apply Proposition 7.35 and find that for $\sigma = e_{n+1}$,

$$F(\sigma, \sigma) = \int_{S^n} K(\sigma \cdot \xi) P_{k,n}(\xi \cdot \sigma) \, \mathrm{Vol}_{S^n}$$

$$= \int_{S^n} K(e_{n+1} \cdot \xi) P_{k,n}(\xi \cdot e_{n+1}) \, \mathrm{Vol}_{S^n}$$

$$= \mathrm{vol}(S^{n-1}) \int_{-1}^{1} K(t) P_{k,n}(t)(1 - t^2)^{\frac{n-2}{2}} \, dt,$$

and thus,

$$F(\sigma, \tau) = \left(\text{vol}(S^{n-1}) \int_{-1}^{1} K(t) P_{k,n}(t)(1 - t^2)^{\frac{n-2}{2}} \, dt \right) P_{k,n}(\sigma \cdot \tau),$$

which is the Funk–Hecke formula when $H(\sigma) = P_{k,n}(\sigma \cdot \tau)$.

Let us now consider any function $H \in \mathcal{H}_k^{\mathbb{C}}(S^n)$. Recall that by the reproducing kernel property (rk), we have

$$\frac{a_{k,n+1}}{\text{vol}(S^n)} \int_{S^n} P_{k,n}(\xi \cdot \tau) H(\tau) \, \text{Vol}_{S^n} = H(\xi).$$

Then we can compute $\int_{S^n} K(\sigma \cdot \xi) H(\xi) \, \text{Vol}_{S^n}$ using Fubini's Theorem and the Funk–Hecke formula in the special case where $H(\sigma) = P_{k,n}(\sigma \cdot \tau)$, as follows:

$$\int_{S^n} K(\sigma \cdot \xi) H(\xi) \, \text{Vol}_{S^n}$$

$$= \int_{S^n} K(\sigma \cdot \xi) \left(\frac{a_{k,n+1}}{\text{vol}(S^n)} \int_{S^n} P_{k,n}(\xi \cdot \tau) H(\tau) \, \text{Vol}_{S^n} \right) \text{Vol}_{S^n}$$

$$= \frac{a_{k,n+1}}{\text{vol}(S^n)} \int_{S^n} H(\tau) \left(\int_{S^n} K(\sigma \cdot \xi) P_{k,n}(\xi \cdot \tau) \, \text{Vol}_{S^n} \right) \text{Vol}_{S^n}$$

$$= \frac{a_{k,n+1}}{\text{vol}(S^n)} \int_{S^n} H(\tau) \left(\left(\text{vol}(S^{n-1}) \int_{-1}^{1} K(t) P_{k,n}(t)(1 - t^2)^{\frac{n-2}{2}} \, dt \right) P_{k,n}(\sigma \cdot \tau) \right) \text{Vol}_{S^n}$$

$$= \left(\text{vol}(S^{n-1}) \int_{-1}^{1} K(t) P_{k,n}(t)(1 - t^2)^{\frac{n-2}{2}} \, dt \right) \left(\frac{a_{k,n+1}}{\text{vol}(S^n)} \int_{S^n} P_{k,n}(\sigma \cdot \tau) H(\tau) \, \text{Vol}_{S^n} \right)$$

$$= \left(\text{vol}(S^{n-1}) \int_{-1}^{1} K(t) P_{k,n}(t)(1 - t^2)^{\frac{n-2}{2}} \, dt \right) H(\sigma),$$

which proves the Funk–Hecke formula in general. $\square$

Remark. The formula

$$\text{Vol}_{S^n} = (1 - t^2)^{\frac{n-2}{2}} \, \text{Vol}_{S^{n-1}} \, dt.$$

can be recursively integrated to obtain a closed form for $\text{vol}(S^n)$. We follow Morimoto [81] and let $t = \sqrt{u}$. Then $dt = \frac{1}{2} u^{-\frac{1}{2}}$ and the integral of the previous line becomes

$$\text{vol}(S^n) = \text{vol}(S^{n-1}) \int_{-1}^{1} (1 - t^2)^{\frac{n-2}{2}} \, dt$$

$$= 2\text{vol}(S^{n-1}) \int_{0}^{1} (1 - t^2)^{\frac{n-2}{2}} \, dt$$

$$= \text{vol}(S^{n-1}) \int_0^1 (1-u)^{\frac{n-2}{2}} u^{-\frac{1}{2}}\, du$$

$$= \text{vol}(S^{n-1}) B\left(\frac{n}{2}, \frac{1}{2}\right),$$

where the last equality made use of the beta function formula

$$B(x, y) = \int_0^1 t^{x-1}(1-t)^{y-1}\, dt, \qquad \text{Re}\, x > 0, \qquad \text{Re}\, y > 0.$$

Since

$$B(x, y) = \frac{\Gamma(x)\Gamma(y)}{\Gamma(x+y)},$$

(see Theorem 1.1.4 of Andrews, Askey, and Roy [1]), our calculations imply that

$$\text{vol}(S^n) = \frac{\Gamma(\frac{1}{2})\Gamma(\frac{n}{2})}{\Gamma(\frac{n+1}{2})}\text{vol}(S^{n-1}) = \frac{\sqrt{\pi}\,\Gamma(\frac{n}{2})}{\Gamma(\frac{n+1}{2})}\text{vol}(S^{n-1}),$$

where the last equality used $\Gamma\left(\frac{1}{2}\right) = \sqrt{\pi}$. We now recursively apply this formula $n - 1$ times to obtain

$$\text{vol}(S^n) = \frac{(\sqrt{\pi})^n \Gamma(\frac{1}{2})}{\Gamma(\frac{n+1}{2})}\text{vol}(S^0) = \frac{2\pi^{\frac{n+1}{2}}}{\Gamma(\frac{n+1}{2})},$$

since $\text{vol}(S^0) = 0$.

It is now a matter of evaluating $\Gamma\left(\frac{n+1}{2}\right)$. If n is odd, say $n = 2d + 1$,

$$\text{vol}(S^{2d+1}) = \frac{2\pi^{\frac{2d+2}{2}}}{\Gamma(d+1)} = \frac{2\pi^{d+1}}{d!}.$$

If n is even, say $n = 2d$, by using the formula $\Gamma(x+1) = x\Gamma(x)$, we find that

$$\Gamma\left(\frac{2d+1}{2}\right) = \Gamma\left(\frac{2d-1}{2} + 1\right)$$

$$= \left(\frac{2d-1}{2}\right)\cdots\left(\frac{3}{2}\right)\left(\frac{1}{2}\right)\Gamma\left(\frac{1}{2}\right)$$

$$= \frac{(2d-1)\cdots 3 \cdot 1\sqrt{\pi}}{2^d}.$$

Then

$$\text{vol}(S^{2d}) = \frac{2\pi^{\frac{2d+1}{2}}}{\Gamma(\frac{2d+1}{2})} = \frac{2^{d+1}\pi^d}{(2d-1)\cdots 3\cdot 1}.$$

The Funk–Hecke formula can be used to derive an "addition theorem" for the ultraspherical polynomials (Gegenbauer polynomials). We omit this topic and we refer the interested reader to Andrews, Askey, and Roy [1] (Chapter 9, Section 9.8).

Remark. Oddly, in their computation of $K \star \ell$, Basri and Jacobs [9] first expand K in terms of spherical harmonics as

$$K = \sum_{n=0}^{\infty} k_n Y_n^0,$$

and then use the Funk–Hecke formula to compute $K \star Y_n^m$. They get (see page 222)

$$K \star Y_n^m = \alpha_n Y_n^m, \quad \text{with} \quad \alpha_n = \sqrt{\frac{4\pi}{2n+1}}\, k_n,$$

for some constant k_n given on page 230 of their paper (see below). However, there is no need to expand K, as the Funk–Hecke formula yields directly

$$K \star Y_n^m(\sigma) = \int_{S^2} K(\sigma \cdot \xi) Y_n^m(\xi)\, \text{Vol}_{S^n} = \left(\text{vol}(S^1) \int_{-1}^{1} K(t) P_n(t)\, dt\right) Y_n^m(\sigma),$$

where $P_n(t)$ is the standard Legendre polynomial of degree n, since we are in the case of S^2. By the definition of K ($K(t) = \max(t, 0)$) and since $\text{vol}(S^1) = 2\pi$, we get

$$K \star Y_n^m = \left(2\pi \int_0^1 t P_n(t)\, dt\right) Y_n^m,$$

which is equivalent to Basri and Jacobs' formula (14), since their α_n on page 222 is given by

$$\alpha_n = \sqrt{\frac{4\pi}{2n+1}}\, k_n,$$

but from page 230,

$$k_n = \sqrt{(2n+1)\pi} \int_0^1 t P_n(t)\, dt.$$

What remains to be done is to compute $\int_0^1 t P_n(t) \, dt$, which is done by using the Rodrigues Formula and integrating by parts (see Appendix A of Basri and Jacobs [9]).

In the next section we show how spherical harmonics fit into the broader framework of linear representations of (Lie) groups.

7.9   Linear Representations of Compact Lie Groups: A Glimpse ⊛

The purpose of this section and the next is to generalize the results about the structure of the space of functions $L_{\mathbb{C}}^2(S^n)$ defined on the sphere S^n, especially the results of Sections 7.5 and 7.6 (such as Theorem 7.19, except Part (3)), to homogeneous spaces G/K where G is a compact Lie group and K is a closed subgroup of G.

The first step is to consider the Hilbert space $L_{\mathbb{C}}^2(G)$ where G is a compact Lie group and to find a Hilbert sum decomposition of this space. The key to this generalization is the notion of (unitary) linear representation of the group G. The space $L_{\mathbb{C}}^2(S^n)$ is replaced by $L_{\mathbb{C}}^2(G)$, and each subspace $\mathcal{H}_k^{\mathbb{C}}(S^n)$ involved in the Hilbert sum

$$L_{\mathbb{C}}^2(S^n) = \bigoplus_{k=0}^{\infty} \mathcal{H}_k^{\mathbb{C}}(S^n)$$

is replaced by a subspace $\mathfrak{a}_\rho$ of $L_{\mathbb{C}}^2(G)$ isomorphic to a finite-dimensional algebra of $n_\rho \times n_\rho$ matrices. More precisely, there is a basis of $\mathfrak{a}_\rho$ consisting of n_ρ^2 functions $m_{ij}^{(\rho)}$ (from G to $\mathbb{C}$) and if for every $g \in G$ we form the matrix

$$M_\rho(g) = \frac{1}{n_\rho} \begin{pmatrix} m_{11}^{(\rho)}(g) & \ldots & m_{1n_\rho}^{(\rho)}(g) \\ \vdots & \ddots & \vdots \\ m_{n_\rho 1}^{(\rho)}(g) & \ldots & m_{n_\rho n_\rho}^{(\rho)}(g) \end{pmatrix}, \qquad (*)$$

then the matrix $M_\rho(g)$ is unitary and $M_\rho(g_1 g_2) = M_\rho(g_1) M_\rho(g_2)$ for all $g_1, g_2 \in G$. This means that the map $g \mapsto M_\rho(g)$ is a *unitary representation* of G in the vector space $\mathbb{C}^{n_\rho}$. Furthermore, this representation is irreducible. Thus, the set of indices ρ is the set of equivalence classes of irreducible unitary representations of G.

The result that we are sketching is a famous theorem known as the *Peter–Weyl theorem* about unitary representations of compact Lie groups (Herman, Klauss, Hugo Weyl, 1885–1955).

The Peter–Weyl theorem can be generalized to any representation $V \colon G \to \mathrm{Aut}(E)$ of G into a separable Hilbert space E (see Definition 7.10), and we obtain a

Hilbert sum decomposition of E in terms of subspaces E_ρ of E. The corresponding subrepresentations are not irreducible but each nontrivial E_ρ splits into a Hilbert sum whose subspaces correspond to irreducible representations.

The next step is to consider the subspace $L^2_{\mathbb{C}}(G/K)$ of $L^2_{\mathbb{C}}(G)$ consisting of the functions that are right-invariant under the action of K. These can be viewed as functions on the homogeneous space G/K. Again, we obtain a Hilbert sum decomposition

$$L^2_{\mathbb{C}}(G/K) = \bigoplus_\rho L_\rho = L^2_{\mathbb{C}}(G/K) \cap \mathfrak{a}_\rho.$$

It is also interesting to consider the subspace $L^2_{\mathbb{C}}(K\backslash G/K)$ of functions in $L^2_{\mathbb{C}}(G)$ consisting of the functions that are both left- and right-invariant under the action of K. The functions in $L^2_{\mathbb{C}}(K\backslash G/K)$ can be viewed as functions on the homogeneous space G/K that are invariant under the left action of K.

Convolution makes the space $L^2_{\mathbb{C}}(G)$ into a noncommutative algebra. Remarkably, it is possible to characterize when $L^2_{\mathbb{C}}(K\backslash G/K)$ is commutative (under convolution) in terms of a simple criterion about the irreducible representations of G. In this situation, (G, K) is a called a *Gelfand pair*.

When (G, K) is a Gelfand pair, it is possible to define a well-behaved notion of *Fourier transform* on $L^2_{\mathbb{C}}(K\backslash G/K)$. Gelfand pairs and the Fourier transform are briefly considered in Section 7.11.

First we review the notion of a linear representation of a group. A good and easy-going introduction to representations of Lie groups can be found in Hall [53]. We begin with finite-dimensional representations.

Definition 7.21. Given a Lie group G and a vector space V of dimension n, a *linear representation* of G of *dimension* (or *degree*) n is a group homomorphism $U: G \to \mathbf{GL}(V)$ such that the map $g \mapsto U(g)(u)$ is continuous for every $u \in V$, where $\mathbf{GL}(V)$ denotes the group of invertible linear maps from V to itself. The space V, called the *representation space*, may be a real or a complex vector space. If V has a Hermitian (resp. Euclidean) inner product $\langle -, - \rangle$, we say that $U: G \to \mathbf{GL}(V)$ is a *unitary representation* iff

$$\langle U(g)(u), U(g)(v) \rangle = \langle u, v \rangle, \qquad \text{for all } g \in G \text{ and all } u, v \in V.$$

Thus, a linear representation of G is a map $U: G \to \mathbf{GL}(V)$ satisfying the properties:

$$U(gh) = U(g)U(h)$$
$$U(g^{-1}) = U(g)^{-1}$$
$$U(1) = I.$$

For simplicity of language, we usually abbreviate *linear representation* as *representation*. The representation space V is also called a G-*module*, since the representation $U : G \to \mathbf{GL}(V)$ is equivalent to the left action $\cdot : G \times V \to V$, with $g \cdot v = U(g)(v)$. The representation such that $U(g) = I$ for all $g \in G$ is called the *trivial representation*.

As an example, we describe a class of representations of $\mathbf{SL}(2, \mathbb{C})$, the group of complex matrices with determinant $+1$,

$$\begin{pmatrix} a & b \\ c & d \end{pmatrix}, \qquad ad - bc = 1.$$

Recall that $\mathcal{P}_k^{\mathbb{C}}(2)$ denotes the vector space of complex homogeneous polynomials of degree k in two variables (z_1, z_2). For every matrix $A \in \mathbf{SL}(2, \mathbb{C})$, with

$$A = \begin{pmatrix} a & b \\ c & d \end{pmatrix},$$

for every homogeneous polynomial $Q \in \mathcal{P}_k^{\mathbb{C}}(2)$, we define $U_k(A)(Q(z_1, z_2))$ by

$$U_k(A)(Q(z_1, z_2)) = Q(dz_1 - bz_2, -cz_1 + az_2).$$

If we think of the homogeneous polynomial $Q(z_1, z_2)$ as a function $Q\begin{pmatrix} z_1 \\ z_2 \end{pmatrix}$ of the vector $\begin{pmatrix} z_1 \\ z_2 \end{pmatrix}$, then

$$U_k(A)\left(Q\begin{pmatrix} z_1 \\ z_2 \end{pmatrix}\right) = QA^{-1}\begin{pmatrix} z_1 \\ z_2 \end{pmatrix} = Q\begin{pmatrix} d & -b \\ -c & a \end{pmatrix}\begin{pmatrix} z_1 \\ z_2 \end{pmatrix}.$$

The expression above makes it clear that

$$U_k(AB) = U_k(A)U_k(B)$$

for any two matrices $A, B \in \mathbf{SL}(2, \mathbb{C})$, so U_k is indeed a representation of $\mathbf{SL}(2, \mathbb{C})$ into $\mathcal{P}_k^{\mathbb{C}}(2)$.

One might wonder why we considered $\mathbf{SL}(2, \mathbb{C})$ rather than $\mathbf{SL}(2, \mathbb{R})$. This is because it can be shown that $\mathbf{SL}(2, \mathbb{R})$ has *no* nontrivial unitary (finite-dimensional) representations! For more on representations of $\mathbf{SL}(2, \mathbb{R})$, see Dieudonné [31] (Chapter 14).

Given any basis $(e_1, \dots, e_n)$ of V, each $U(g)$ is represented by an $n \times n$ matrix $U(g) = (U_{ij}(g))$. We may think of the scalar functions $g \mapsto U_{ij}(g)$ as *special functions* on G. As explained in Dieudonné [31] (see also Vilenkin [107]), essentially all special functions (Legendre polynomials, ultraspherical polynomials, Bessel functions, *etc.*) arise in this way by choosing some suitable G and V.

There is a natural and useful notion of equivalence of representations:

Definition 7.22. Given any two representations $U_1 : G \to \mathbf{GL}(V_1)$ and $U_2 : G \to \mathbf{GL}(V_2)$, a *G-map* (or *morphism of representations*) $\varphi : U_1 \to U_2$ is a linear map $\varphi : V_1 \to V_2$ so that the following diagram commutes for every $g \in G$:

$$
\begin{array}{ccc}
V_1 & \xrightarrow{\ U_1(g)\ } & V_1 \\
{\scriptstyle \varphi}\downarrow & & \downarrow{\scriptstyle \varphi} \\
V_2 & \xrightarrow{\ U_2(g)\ } & V_2,
\end{array}
$$

i.e.

$$
\varphi(U_1(g)(v)) = U_2(g)(\varphi(v)), \qquad v \in V_1.
$$

The space of all G-maps between two representations as above is denoted $\mathrm{Hom}_G(U_1, U_2)$. Two representations $U_1 : G \to \mathbf{GL}(V_1)$ and $U_2 : G \to \mathbf{GL}(V_2)$ are *equivalent* iff $\varphi : V_1 \to V_2$ is an invertible linear map (which implies that $\dim V_1 = \dim V_2$). In terms of matrices, the representations $U_1 : G \to \mathbf{GL}(V_1)$ and $U_2 : G \to \mathbf{GL}(V_2)$ are equivalent iff there is some invertible $n \times n$ matrix, P, so that

$$
U_2(g) = P U_1(g) P^{-1}, \qquad g \in G.
$$

If $W \subseteq V$ is a subspace of V, then in some cases, a representation $U : G \to \mathbf{GL}(V)$ yields a representation $U : G \to \mathbf{GL}(W)$. This is interesting because under certain conditions on G (*e.g.*, G compact) every representation may be decomposed into a "sum" of the so-called irreducible representations (defined below), and thus the study of all representations of G boils down to the study of irreducible representations of G; for instance, see Knapp [66] (Chapter 4, Corollary 4.7), or Bröcker and tom Dieck [18] (Chapter 2, Proposition 1.9).

Definition 7.23. Let $U : G \to \mathbf{GL}(V)$ be a representation of G. If $W \subseteq V$ is a subspace of V, then we say that W is *invariant* (or *stable*) under U iff $U(g)(w) \in W$, for all $g \in G$ and all $w \in W$. If W is invariant under U, then we have a homomorphism, $U : G \to \mathbf{GL}(W)$, called a *subrepresentation* of G. A representation $U : G \to \mathbf{GL}(V)$ with $V \neq (0)$ is *irreducible* iff it only has the two subrepresentations $U : G \to \mathbf{GL}(W)$ corresponding to $W = (0)$ or $W = V$.

It can be shown that the representations U_k of $\mathbf{SL}(2, \mathbb{C})$ defined earlier are irreducible, and that every representation of $\mathbf{SL}(2, \mathbb{C})$ is equivalent to one of the U_k's (see Bröcker and tom Dieck [18], Chapter 2, Section 5). The representations U_k are also representations of $\mathbf{SU}(2)$. Again, they are irreducible representations of $\mathbf{SU}(2)$, and they constitute all of them (up to equivalence). The reader should consult Hall [53] for more examples of representations of Lie groups.

An easy but crucial lemma about irreducible representations is "Schur's Lemma."

Lemma 7.37 (Schur's Lemma). *Let $U_1 \colon G \to \mathbf{GL}(V)$ and $U_2 \colon G \to \mathbf{GL}(W)$ be any two real or complex representations of a group G. If U_1 and U_2 are irreducible, then the following properties hold:*

(i) Every G-map $\varphi \colon U_1 \to U_2$ is either the zero map or an isomorphism.

(ii) If U_1 is a complex representation, then every G-map $\varphi \colon U_1 \to U_1$ is of the form $\varphi = \lambda \mathrm{id}$, for some $\lambda \in \mathbb{C}$.

Proof.

(i) Observe that the kernel $\mathrm{Ker}\ \varphi \subseteq V$ of φ is invariant under U_1. Indeed, for every $v \in \mathrm{Ker}\ \varphi$ and every $g \in G$, we have

$$\varphi(U_1(g)(v)) = U_2(g)(\varphi(v)) = U_2(g)(0) = 0,$$

so $U_1(g)(v) \in \mathrm{Ker}\ \varphi$. Thus, $U_1 \colon G \to \mathbf{GL}(\mathrm{Ker}\ \varphi)$ is a subrepresentation of U_1, and as U_1 is irreducible, either $\mathrm{Ker}\ \varphi = (0)$ or $\mathrm{Ker}\ \varphi = V$. In the second case, $\varphi = 0$. If $\mathrm{Ker}\ \varphi = (0)$, then φ is injective. However, $\varphi(V) \subseteq W$ is invariant under U_2, since for every $v \in V$ and every $g \in G$,

$$U_2(g)(\varphi(v)) = \varphi(U_1(g)(v)) \in \varphi(V),$$

and as $\varphi(V) \neq (0)$ (as $V \neq (0)$ since U_1 is irreducible) and U_2 is irreducible, we must have $\varphi(V) = W$; that is, φ is an isomorphism.

(ii) Since V is a complex vector space, the linear map φ has some eigenvalue $\lambda \in \mathbb{C}$. Let $E_\lambda \subseteq V$ be the eigenspace associated with λ. The subspace E_λ is invariant under U_1, since for every $u \in E_\lambda$ and every $g \in G$, we have

$$\varphi(U_1(g)(u)) = U_1(g)(\varphi(u)) = U_1(g)(\lambda u) = \lambda U_1(g)(u),$$

so $U_1 \colon G \to \mathbf{GL}(E_\lambda)$ is a subrepresentation of U_1, and as U_1 is irreducible and $E_\lambda \neq (0)$, we must have $E_\lambda = V$. $\square$

An interesting corollary of Schur's Lemma is the following fact:

Proposition 7.38. *Every complex irreducible representation $U \colon G \to \mathbf{GL}(V)$ of a commutative group G is one dimensional.*

Proof. Since G is abelian, we claim that for every $g \in G$, the map $\tau_g \colon V \to V$ given by $\tau_g(v) = U(g)(v)$ for all $v \in V$ is a G-map. This amounts to checking that the following diagram commutes

$$
\begin{array}{ccc}
V & \xrightarrow{\ U(g_1)\ } & V \\
{\scriptstyle \tau_g}\downarrow & & \downarrow{\scriptstyle \tau_g} \\
V & \xrightarrow{\ U(g_1)\ } & V
\end{array}
$$

for all $g, g_1 \in G$. This is equivalent to checking that

$$\tau_g(U(g_1)(v)) = U(g)(U(g_1)(v)) = U(gg_1)(v)$$
$$= U(g_1)(\tau_g(v)) = U(g_1)(U(g)(v)) = U(g_1g)(v)$$

for all $v \in V$, that is, $U(gg_1)(v) = U(g_1g)(v)$, which holds since G is commutative (so $gg_1 = g_1g$).

By Schur's lemma (Lemma 7.37 (ii)), $\tau_g = \lambda_g \mathrm{id}$ for some $\lambda_g \in \mathbb{C}$. It follows that any subspace of V is invariant. If the representation is irreducible, we must have $\dim(V) = 1$ since otherwise V would contain a one-dimensional invariant subspace, contradicting the assumption that U is irreducible.   □

Let us now restrict our attention to compact Lie groups. If G is a compact Lie group, then it is known that it has a left- and right-invariant volume form ω_G, so we can define the integral of a (real or complex) continuous function f defined on G by

$$\int_G f = \int_G f \, \omega_G,$$

also denoted $\int_G f \, d\mu_G$ or simply $\int_G f(t) \, dt$, with ω_G normalized so that $\int_G \omega_G = 1$. (See Section 6.8, or Knapp [66], Chapter 8, or Warner [109], Chapters 4 and 6.) Because G is compact, the *Haar measure* μ_G induced by ω_G is both left- and right-invariant (G is a *unimodular group*), and our integral has the following invariance properties:

$$\int_G f(t) \, dt = \int_G f(st) \, dt = \int_G f(tu) \, dt = \int_G f(t^{-1}) \, dt,$$

for all $s, u \in G$ (see Section 6.8).

Since G is a compact Lie group, we can use an "averaging trick" to show that every (finite-dimensional) representation is equivalent to a unitary representation; see Bröcker and tom Dieck [18] (Chapter 2, Theorem 1.7) or Knapp [66] (Chapter 4, Proposition 4.6).

If we define the Hermitian inner product

$$\langle f, g \rangle = \int_G f \, \overline{g} \, \omega_G,$$

then with this inner product the space of square-integrable functions $L^2_{\mathbb{C}}(G)$ is a *Hilbert space* (in fact, a separable Hilbert space).

Definition 7.24. The *convolution* $f * g$ of two functions $f, g \in L^2_{\mathbb{C}}(G)$ is given by

$$(f * g)(x) = \int_G f(xt^{-1})g(t)dt = \int_G f(t)g(t^{-1}x)dt.$$

In general, $f * g \neq g * f$, unless G is commutative. With the convolution product, $L_{\mathbb{C}}^2(G)$ becomes an associative algebra (noncommutative in general).

This leads us to consider unitary representations of G into the infinite-dimensional vector space $L_{\mathbb{C}}^2(G)$, and more generally into a Hilbert space E.

Given a Hilbert space E, the definition of a unitary representation $U: G \to \text{Aut}(E)$ is the same as in Definition 7.21, except that $\mathbf{GL}(E)$ is replaced by the group of automorphisms (unitary operators) $\text{Aut}(E)$ of the Hilbert space E, and

$$\langle U(g)(u), U(g)(v) \rangle = \langle u, v \rangle$$

with respect to the inner product on E. Also, in the definition of an irreducible representation $U: G \to \text{Aut}(E)$, we require that the only *closed* subrepresentations $U: G \to \text{Aut}(W)$ of the representation $U: G \to \text{Aut}(E)$ correspond to $W = (0)$ or $W = E$. Here, a subrepresentation $U: G \to \text{Aut}(W)$ is closed if W is closed in E.

The *Peter–Weyl theorem* gives a decomposition of $L_{\mathbb{C}}^2(G)$ as a Hilbert sum of spaces that correspond to all the irreducible unitary representations of G. We present a version of the Peter–Weyl theorem found in Dieudonné [31] (Chapters 3–8) and Dieudonné [32] (Chapter XXI, Sections 1–4), which contains complete proofs. Other versions can be found in Bröcker and tom Dieck [18] (Chapter 3), Knapp [66] (Chapter 4), or Duistermaat and Kolk [38] (Chapter 4). A good preparation for these fairly advanced books is Deitmar [29].

Theorem 7.39 (Peter–Weyl (1927)). *Given a compact Lie group G, there is a decomposition of the associative algebra (under convolution $*$) $L_{\mathbb{C}}^2(G)$ as a Hilbert sum*

$$L_{\mathbb{C}}^2(G) = \bigoplus_{\rho \in R(G)} \mathfrak{a}_\rho$$

of countably many two-sided ideals $\mathfrak{a}_\rho$, where each $\mathfrak{a}_\rho$ is isomorphic to a finite-dimensional algebra of $n_\rho \times n_\rho$ complex matrices, where the set of indices $R(G)$ corresponds to the set of equivalence classes of irreducible representations of G. More precisely, for each $\rho \in R(G)$, there is a basis of $\mathfrak{a}_\rho$ consisting of n_ρ^2 pairwise orthogonal continuous functions $m_{ij}^{(\rho)}: G \to \mathbb{C}$, that is

$$\langle m_{ij}^{(\rho)}, m_{hk}^{(\rho')} \rangle = 0$$

unless $\rho = \rho'$, $i = h$, and $j = k$, and satisfying the properties

$$m_{ij}^{(\rho)} * m_{hk}^{(\rho)} = \delta_{jh} m_{ik}^{(\rho)} \qquad m_{ij}^{(\rho)}(e) = \delta_{ij} n_\rho$$

$$\langle m_{ij}^{(\rho)}, m_{ij}^{(\rho)} \rangle = n_\rho \qquad m_{ji}^{(\rho)}(g) = \overline{m_{ij}^{(\rho)}(g^{-1})},$$

and if for any $g \in G$ we form the $n_\rho \times n_\rho$ matrix $M_\rho(g)$ given by

$$M_\rho(g) = \frac{1}{n_\rho} \begin{pmatrix} m_{11}^{(\rho)}(g) & \cdots & m_{1n_\rho}^{(\rho)}(g) \\ \vdots & \ddots & \vdots \\ m_{n_\rho 1}^{(\rho)}(g) & \cdots & m_{n_\rho n_\rho}^{(\rho)}(g) \end{pmatrix},$$

then the matrix $M_\rho(g)$ is unitary, $M_\rho(g_1 g_2) = M_\rho(g_1) M_\rho(g_2)$, and the map $g \mapsto M_\rho(g)$ is an **irreducible unitary representation** *of G in the vector space $\mathbb{C}^{n_\rho}$ (M_ρ is a group homomorphism $M_\rho \colon G \to \mathbf{GL}(\mathbb{C}^{n_\rho})$). Furthermore, every irreducible unitary representation of G is equivalent to some M_ρ. The function u_ρ given by*

$$u_\rho(g) = \sum_{j=1}^{n_\rho} m_{jj}^{(\rho)}(g) = n_\rho \mathrm{tr}(M_\rho(g))$$

is the unit of the algebra $\mathfrak{a}_\rho$, and the orthogonal projection of $L^2_{\mathbb{C}}(G)$ onto $\mathfrak{a}_\rho$ is the map

$$f \mapsto u_\rho * f = f * u_\rho;$$

that is, convolution with u_ρ.

The Peter–Weyl theorem implies that *all irreducible unitary representations of a compact Lie group are finite dimensional.* The constant functions on G form a one-dimensional ideal $\mathfrak{a}_{\rho_0}$ called the *trivial* ideal, corresponding to the trivial representation ρ_0 (such that $M_{\rho_0}(g) = 1$ for all $g \in G$). The fact that the $m_{ij}^{(\rho)}$ form an orthogonal system implies that

$$\int_G m_{ij}^{(\rho)}(g)\, dg = 0 \quad \text{for all } \rho \neq \rho_0.$$

Theorem 7.39 implies that the countable family of functions

$$\left(\frac{1}{\sqrt{n_\rho}} m_{ij}^{(\rho)} \right)_{\rho \in R(G),\, 1 \le i, j \le n_\rho}$$

is a *Hilbert basis* of $L^2_{\mathbb{C}}(G)$.

Remark. We will often refer to the decomposition of the Hilbert space $L^2_{\mathbb{C}}(G)$ in terms of the ideals $\mathfrak{a}_\rho$ as the *master decomposition* of $L^2_{\mathbb{C}}(G)$.

A complete proof of Theorem 7.39 is given in Dieudonné [32], Chapter XXI, Section 2, but see also Sections 3 and 4.

Remark. The Peter–Weyl theorem actually holds for any compact topological metrizable group, not just for a compact Lie group.

Definition 7.25. The function $\chi_\rho = \frac{1}{n_\rho} u_\rho = \mathrm{tr}(M_\rho)$ is the *character* of G associated with the representation M_ρ.

The functions χ_ρ satisfy the following properties:

$$\chi_\rho(e) = n_\rho$$

$$\chi_\rho(sts^{-1}) = \chi_\rho(t) \quad \text{for all } s, t \in G$$

$$\chi_\rho(s^{-1}) = \overline{\chi_\rho(s)} \quad \text{for all } s \in G$$

$$\chi_\rho * \chi_{\rho'} = 0 \quad \text{if } \rho \neq \rho'$$

$$\chi_\rho * \chi_\rho = \frac{1}{n_\rho} \chi_\rho.$$

Furthermore, the characters form an orthonormal Hilbert basis of the Hilbert subspace of $L^2_{\mathbb{C}}(G)$ consisting of the *central functions*, namely those functions $f \in L^2_{\mathbb{C}}(G)$ such that for every $s \in G$,

$$f(sts^{-1}) = f(t) \quad \text{almost everywhere.}$$

So, we have

$$\int_G \chi_\rho(t)\overline{\chi_{\rho'}(t)}\, dt = 0 \quad \text{if } \rho \neq \rho', \qquad \int_G |\chi_\rho(t)|^2 dt = 1,$$

and

$$\int_g \chi_\rho(g)\, dg = 0 \quad \text{for all } \rho \neq \rho_0.$$

If G (compact) is commutative, then by Proposition 7.38 all representations M_ρ are one dimensional. Then each character $s \mapsto \chi_\rho(s)$ is a continuous homomorphism of G into $\mathbf{U}(1)$, the group of unit complex numbers. For the torus group $S_1 = \mathbf{T} = \mathbb{R}/\mathbb{Z}$, the characters are the homomorphisms $\theta \mapsto e^{k2\pi i\theta}$, with $k \in \mathbb{N}$. This is the special case of Fourier analysis on the circle.

An important corollary of the Peter–Weyl theorem is that every compact Lie group is isomorphic to a matrix group.

Theorem 7.40. *For every compact Lie group G, there is some integer $N \geq 1$ and an isomorphism of G onto a closed subgroup of $\mathbf{U}(N)$.*

The proof of Theorem 7.40 can be found in Dieudonné [32], Chapter XXI, Theorem 21.13.1) or Knapp [66] (Chapter 4, Corollary 4.22).

There is more to the Peter–Weyl theorem: It gives a description of all unitary representations of G into a separable Hilbert space.

7.10   Consequences of the Peter–Weyl Theorem

Recall that a Hilbert space is separable if it has a countable total orthogonal family, also called a Hilbert basis; see Definition 7.10.

If $f : G \to E$ is function from a compact Lie group G to a Hilbert space E and if for all $z \in E$ the function $s \mapsto \langle f(s), z \rangle$ is integrable and the function $s \mapsto \| f(s) \|$ is integrable, then it can be shown that the map

$$z \mapsto \int_G \langle f(s), z \rangle ds \quad \text{for all } z \in E$$

is a bounded linear functional on $L^2_{\mathbb{C}}(G)$ (using the dominated convergence theorem). By the Riesz representation theorem for Hilbert spaces, there is a unique $y \in E$ such that

$$\langle y, z \rangle = \int_G \langle f(s), z \rangle ds \quad \text{for all } z \in E;$$

see Dieudonné [35] (Chapter XIII, Proposition 13.10.4).

Definition 7.26. If $f : G \to E$ is a function from a compact Lie group G to a Hilbert space E, under the conditions on f stated above, the unique vector $y \in E$ such that

$$\langle y, z \rangle = \int_G \langle f(s), z \rangle ds \quad \text{for all } z \in E$$

is denoted by

$$\int_G f(s) \, ds$$

and is called the *weak integral* (for short, *integral*) of f.

Theorem 7.41. *Given a compact Lie group G, if $V : G \to \mathrm{Aut}(E)$ is a unitary representation of G in a separable Hilbert space E, using the notation of Theorem 7.39, for every $\rho \in R(G)$, for every $x \in E$ the map*

$$x \mapsto V_{u_\rho}(x) = \int_G u_\rho(s)(V(s)(x)) \, ds$$

is an orthogonal projection of E onto a closed subspace E_ρ, where the expression on the right-hand side is the weak integral of the function $s \mapsto u_\rho(s)(V(s)(x))$. Furthermore, E is the Hilbert sum

$$E = \bigoplus_{\rho \in R(G)} E_\rho$$

of those E_ρ such that $E_\rho \neq (0)$. Each such E_ρ is invariant under V, but the subrepresentation of V in E_ρ is not necessarily irreducible. However, each E_ρ is a (finite or countable) Hilbert sum of closed subspaces invariant under V, and the subrepresentations of V corresponding to these subspaces of E_ρ are all equivalent to $M_{\overline{\rho}}$, where M_ρ is defined as in Theorem 7.39, and $M_{\overline{\rho}}$ is the representation of G given by $M_{\overline{\rho}}(g) = \overline{M_\rho(g)}$ for all $g \in G$. These representations are all irreducible. As a consequence, every irreducible unitary representation of G is equivalent to some representation of the form M_ρ. For any closed subspace F of E, if F is invariant under V, then F is the Hilbert sum of the orthogonal spaces $F \cap E_\rho$ for those $\rho \in R(G)$ for which $F \cap E_\rho$ is not reduced to 0, and each nontrivial subspace $F \cap E_\rho$ is itself the Hilbert sum of closed subspaces invariant under V, and such that the corresponding subrepresentations are all irreducible and equivalent to $M_{\overline{\rho}}$.

If $E_\rho \neq (0)$, we say that the irreducible representation $M_{\overline{\rho}}$ is *contained* in the representation V.

Definition 7.27. If E_ρ is finite dimensional, then $\dim(E_\rho) = d_\rho n_\rho$ for some positive integer d_ρ. The integer d_ρ is called the *multiplicity* of $M_{\overline{\rho}}$ in V.

An interesting special case of Theorem 7.41 is the case of the so-called regular representation of G in $L^2_{\mathbb{C}}(G)$ itself, that is, $E = L^2_{\mathbb{C}}(G)$.

Definition 7.28. The (left) *regular representation* $\mathbf{R}$ (or λ) of G in $L^2_{\mathbb{C}}(G)$ is defined by

$$(\mathbf{R}_s(f))(t) = \lambda_s(f)(t) = f(s^{-1}t), \qquad f \in L^2_{\mathbb{C}}(G),\ s, t \in G.$$

We have

$$(\mathbf{R}_{s_1}(\mathbf{R}_{s_2}(f)))(t) = (\mathbf{R}_{s_2}(f))(s_1^{-1}t) = f(s_2^{-1}s_1^{-1}t) = f((s_1 s_2)^{-1}t) = (\mathbf{R}_{s_1 s_2}(f))(t),$$

which shows that

$$(\mathbf{R}_{s_1} \circ \mathbf{R}_{s_2})(f) = \mathbf{R}_{s_1 s_2}(f),$$

namely, $\mathbf{R}$ is a representation of G in $L^2_{\mathbb{C}}(G)$. Observe that if we had defined $\mathbf{R}_s(f)$ as $f(st)$ instead of $f(s^{-1}t)$, then would get $(\mathbf{R}_{s_1} \circ \mathbf{R}_{s_2})(f) = \mathbf{R}_{s_2 s_1}(f)$, so this version of $\mathbf{R}_s(f)$ would not be a representation since the above composition is

$\mathbf{R}_{s_2 s_1}(f)$ rather than $\mathbf{R}_{s_1 s_2}(f)$. This is the reason for using s^{-1} instead of s in the definition of $\mathbf{R}_s(f) = f(s^{-1}t)$.

Theorem 7.41 implies that we also get a Hilbert sum $L^2_{\mathbb{C}}(G) = \bigoplus_{\rho \in R(G)} E_\rho$, and it turns out that $E_\rho = \mathfrak{a}_\rho$, where $\mathfrak{a}_\rho$ is the ideal occurring in the master decomposition of $L^2_{\mathbb{C}}(G)$, so again we get the Hilbert sum

$$L^2_{\mathbb{C}}(G) = \bigoplus_{\rho \in R(G)} \mathfrak{a}_\rho$$

of the master decomposition. This time, the $\mathfrak{a}_\rho$ generally *do not correspond* to irreducible subrepresentations of $\mathbf{R}$. However, $\mathfrak{a}_\rho$ splits into $d_\rho = n_\rho$ minimal left ideals $\mathfrak{b}_j^{(\rho)}$, where $\mathfrak{b}_j^{(\rho)}$ is spanned by the jth column of M_ρ, that is,

$$\mathfrak{a}_\rho = \bigoplus_{j=1}^{n_\rho} \mathfrak{b}_j^{(\rho)} \quad \text{and} \quad \mathfrak{b}_j^{(\rho)} = \bigoplus_{k=1}^{n_\rho} \mathbb{C} m_{kj}^{(\rho)},$$

and all the subrepresentations $\mathbf{R}: G \to \mathbf{GL}(\mathfrak{b}_j^{(\rho)})$ of G in $\mathfrak{b}_j^{(\rho)}$ are equivalent to $M_{\overline{\rho}}$, and thus are irreducible (see Dieudonné [31], Chapter 3).

Finally, assume that besides the compact Lie group G, we also have a closed subgroup K of G. Then we know that $M = G/K$ is a manifold called a *homogeneous space*, and G acts on M on the left. For example, if $G = \mathbf{SO}(n+1)$ and $K = \mathbf{SO}(n)$, then $S^n = \mathbf{SO}(n+1)/\mathbf{SO}(n)$ (see Warner [109], Chapter 3, or Gallier and Quaintance [47]).

Definition 7.29. The subspace of $L^2_{\mathbb{C}}(G)$ consisting of the functions $f \in L^2_{\mathbb{C}}(G)$ that are right-invariant under the action of K, that is, such that

$$f(su) = f(s) \qquad \text{for all } s \in G \text{ and all } u \in K,$$

forms a closed subspace of $L^2_{\mathbb{C}}(G)$ denoted by $L^2_{\mathbb{C}}(G/K)$.

Since a function as above is constant on every left coset sK ($s \in G$), such a function can be viewed as a function on the homogeneous space G/K. For example, if $G = \mathbf{SO}(n+1)$ and $K = \mathbf{SO}(n)$, then $L^2_{\mathbb{C}}(G/K) = L^2_{\mathbb{C}}(S^n)$.

It turns out that $L^2_{\mathbb{C}}(G/K)$ is invariant under the regular representation $\mathbf{R}$ of G in $L^2_{\mathbb{C}}(G)$, so we get a subrepresentation (of the regular representation) of G in $L^2_{\mathbb{C}}(G/K)$.

The corollary of the Peter–Weyl theorem (Theorem 7.41) gives us a Hilbert sum decomposition of $L^2_{\mathbb{C}}(G/K)$ of the form

$$L^2_{\mathbb{C}}(G/K) = \bigoplus_\rho L_\rho = L^2_{\mathbb{C}}(G/K) \cap \mathfrak{a}_\rho,$$

for the same ρ's as before. However, these subrepresentations of $\mathbf{R}$ in L_ρ are not necessarily irreducible. What happens is that there is some d_ρ with $0 \le d_\rho \le n_\rho$, so that if $d_\rho \ge 1$, then L_ρ is the direct sum of the subspace spanned by the first d_ρ columns of M_ρ. The number d_ρ can be characterized as follows.

If we consider the restriction of the representation $M_\rho \colon G \to \mathbf{GL}(\mathbb{C}^{n_\rho})$ to K, then this representation is generally not irreducible, so $\mathbb{C}^{n_\rho}$ splits into subspaces $F_{\sigma_1}, \dots, F_{\sigma_r}$ such that the restriction of the subrepresentation M_ρ to F_{σ_i} is an irreducible representation of K. Then d_ρ is the multiplicity of the trivial representation σ_0 of K if it occurs. For this reason, d_ρ is also denoted $(\rho : \sigma_0)$ (see Dieudonné [31], Chapter 6 and Dieudonné [33], Chapter XXII, Sections 4–5).

Definition 7.30. The subspace of $L^2_{\mathbb{C}}(G)$ consisting of the functions $f \in L^2_{\mathbb{C}}(G)$ that are left-invariant under the action of K, that is, such that

$$f(ts) = f(s) \qquad \text{for all } s \in G \text{ and all } t \in K,$$

is a closed subspace of $L^2_{\mathbb{C}}(G)$ denoted $L^2_{\mathbb{C}}(K \backslash G)$.

We get a Hilbert sum decomposition of $L^2_{\mathbb{C}}(K \backslash G)$ of the form

$$L^2_{\mathbb{C}}(K \backslash G) = \bigoplus_\rho L'_\rho = \bigoplus_\rho L^2_{\mathbb{C}}(K \backslash G) \cap \mathfrak{a}_\rho,$$

and for the same d_ρ as before, L'_ρ is the direct sum of the subspace spanned by the first d_ρ rows of M_ρ.

Finally, we consider the following algebra.

Definition 7.31. The space $L^2_{\mathbb{C}}(K \backslash G / K)$ is defined by

$$L^2_{\mathbb{C}}(K \backslash G / K) = L^2_{\mathbb{C}}(G/K) \cap L^2_{\mathbb{C}}(K \backslash G)$$

$$= \{ f \in L^2_{\mathbb{C}}(G) \mid f(tsu) = f(s) \} \qquad \text{for all } s \in G \text{ and all } t, u \in K.$$

Functions in $L^2_{\mathbb{C}}(K \backslash G / K)$ can be viewed as functions on the homogeneous space G/K that are invariant under the left action of K. These functions are constant on the double cosets KsK ($s \in G$).

In the case where $G = \mathbf{SO}(3)$ and $K = \mathbf{SO}(2)$, these are the functions on S^2 that are invariant under the action of $\mathbf{SO}(2)$ (more precisely, a subgroup of $\mathbf{SO}(3)$ leaving invariant some chosen element of S^2). The functions in $L^2_{\mathbb{C}}(K \backslash G / K)$ are reminiscent of zonal spherical functions, and indeed these functions are often called *spherical functions*, as in Helgason [55] (Chapter 4).

From our previous discussion, we see that we have a Hilbert sum decomposition

$$L^2_{\mathbb{C}}(K \backslash G / K) = \bigoplus_\rho L_\rho \cap L'_\rho$$

and each $L_\rho \cap L'_\rho$ for which $d_\rho \geq 1$ is a matrix algebra of dimension d_ρ^2 having as a basis the functions $m_{ij}^{(\rho)}$ for $1 \leq i, j \leq d_\rho$. As a consequence, *the algebra $L_{\mathbb{C}}^2(K\backslash G/K)$ is commutative iff $d_\rho \leq 1$ for all ρ.*

7.11 Gelfand Pairs, Spherical Functions, and Fourier Transform ⊛

In this section we investigate briefly what happens when the algebra $L_{\mathbb{C}}^2(K\backslash G/K)$ is commutative. In this case, the space $L_{\mathbb{C}}^2(K\backslash G/K)$ is a Hilbert sum of one-dimensional subspaces spanned by the functions $\omega_\rho = (1/n_\rho)m_{11}^{(\rho)}$, which are called *zonal spherical harmonics.*

It is also the case that $L_{\mathbb{C}}^2(G/K)$ is a Hilbert sum of n_ρ-dimensional subspaces L_ρ, where L_ρ is spanned by the left translates of ω_ρ. Finally, it is possible to define a well-behaved notion of Fourier transform on $L_{\mathbb{C}}^2(K\backslash G/K)$, in the sense that the Fourier transform on $L_{\mathbb{C}}^2(K\backslash G/K)$ satisfies the fundamental relation

$$\mathcal{F}(f * g) = \mathcal{F}(f)\mathcal{F}(g).$$

Observe that in order for this equation to hold, convolution has to be commutative. This is why the Fourier transform is defined on $L_{\mathbb{C}}^2(K\backslash G/K)$, where convolution is commutative, rather than the whole of $L_{\mathbb{C}}^2(G/K)$.

Definition 7.32. Given a compact Lie group G and a closed subgroup K, if the algebra $L_{\mathbb{C}}^2(K\backslash G/K)$ is commutative (for the convolution product), we say that (G, K) is a *Gelfand pair*; see Dieudonné [31], Chapter 8 and Dieudonné [33], Chapter XXII, Sections 6–7.

In this case, the L_ρ in the Hilbert sum decomposition of $L_{\mathbb{C}}^2(G/K)$ are nontrivial of dimension n_ρ iff $(\rho : \sigma_0) = d_\rho = 1$, and the subrepresentation $\mathbf{U}$ (of the regular representation $\mathbf{R}$) of G into L_ρ is irreducible and equivalent to $M_{\overline{\rho}}$. The space L_ρ is generated by the functions $m_{11}^{(\rho)}, \ldots, m_{n_\rho 1}^{(\rho)}$, but the function

$$\omega_\rho(s) = \frac{1}{n_\rho} m_{11}^{(\rho)}(s)$$

plays a special role.

Definition 7.33. Given a compact Lie group G and a closed subgroup K, if (G, K) is a Gelfand pair, then function $\omega_\rho = \frac{1}{n_\rho} m_{11}^{(\rho)}$ is called a *zonal spherical function*, for short a *spherical function*. The set of zonal spherical functions on G/K is denoted $S(G/K)$.

Because G is compact, $S(G/K)$ it is a countable set in bijection with the set of equivalence classes of representations $\rho \in R(G)$ such that $(\rho : \sigma_0) = 1$.

Spherical functions defined in Definition 7.33 are generalizations of the zonal functions on S^n of Definition 7.18. They have some interesting properties, some of which are listed below. In particular, they are a key ingredient in generalizing the notion of Fourier transform on the homogeneous space G/K.

First, ω_ρ is a continuous function, even a smooth function since G is a Lie group. The function ω_ρ is such that $\omega_\rho(e) = 1$ (where e is the identity element of the group, G), and

$$\omega_\rho(ust) = \omega_\rho(s) \qquad \text{for all } s \in G \text{ and all } u, t \in K.$$

In addition, ω_ρ is of positive type. A function $f : G \to \mathbb{C}$ is of *positive type* iff

$$\sum_{j,k=1}^{n} f(s_j^{-1} s_k) z_j \bar{z}_k \geq 0,$$

for every finite set $\{s_1, \ldots, s_n\}$ of elements of G and every finite tuple $(z_1, \ldots, z_n) \in \mathbb{C}^n$.

When $L_\mathbb{C}^2(K \backslash G/K)$ is commutative, it is the Hilbert sum of all the 1-dimensional subspaces $\mathbb{C}\omega_\rho$ for all $\rho \in R(G)$ such that $d_\rho = 1$. The orthogonal projection of $L_\mathbb{C}^2(K \backslash G/K)$ onto $\mathbb{C}\omega_\rho$ is given by

$$g \mapsto g * \omega_\rho \quad g \in L_\mathbb{C}^2(K \backslash G/K).$$

Since $\mathbb{C}\omega_\rho$ is an ideal in the algebra $L_\mathbb{C}^2(K \backslash G/K)$, there is some homomorphism $\xi_\rho : L_\mathbb{C}^2(K \backslash G/K) \to \mathbb{C}$ such that

$$g * \omega_\rho = \xi_\rho(g)\omega_\rho \quad g \in L_\mathbb{C}^2(K \backslash G/K).$$

To be more precise, ξ_ρ has the property

$$\xi_\rho(g_1 * g_2) = \xi_\rho(g_1)\xi_\rho(g_2) \quad \text{for all } g_1, g_2 \in L_\mathbb{C}^2(K \backslash G/K).$$

In other words, ξ_ρ is a character of the algebra $L_\mathbb{C}^2(K \backslash G/K)$ (see below for the definition of characters).

Because the subrepresentation $\mathbf{R}$ of G into L_ρ is irreducible (if (G/K) is a Gelfand pair, all nontrivial L_ρ are one dimensional), the function ω_ρ generates L_ρ under left translation. This means the following: If we recall that for any function f on G,

$$\lambda_s(f)(t) = f(s^{-1}t), \qquad s, t \in G,$$

then L_ρ is generated by the functions $\lambda_s(\omega_\rho)$, as s varies in G.

It can be shown that a (non-identically zero) function ω in the set $C_{\mathbb{C}}(K\backslash G/K)$ of continuous complex-valued functions in $L^2_{\mathbb{C}}(K\backslash G/K)$ belongs to $S(G/K)$ iff the functional equation

$$\int_K \omega(xsy)\,ds = \omega(x)\omega(y) \tag{$*$}$$

holds for all $s \in K$ and all $x, y \in G$.

The space $S(G/K)$ is also in bijection with the characters of the algebra $L^2_{\mathbb{C}}(K\backslash G/K)$.

Definition 7.34. If (G, K) is a Gelfand pair a *character* of the commutative algebra of $L^2_{\mathbb{C}}(K\backslash G/K)$ is a non-identically zero linear map $\xi \colon L^2_{\mathbb{C}}(K\backslash G/K) \to \mathbb{C}$ such that $\xi(f * g) = \xi(f)\xi(g)$ for all $f, g \in L^2_{\mathbb{C}}(K\backslash G/K)$. Let $\mathbb{X}_0$ denote the set of characters of $L^2_{\mathbb{C}}(K\backslash G/K)$.

Then it can be shown that for every character $\xi \in \mathbb{X}_0$, there is a unique spherical function $\omega \in S(G/K)$ such that

$$\xi(f) = \int_G f(s)\omega(s)\,ds.$$

It follows that there is a bijection between $S(G/K)$ and $\mathbb{X}_0$. All this is explained in Dieudonné [31] (Chapters 8 and 9) and Dieudonné [33] (Chapter XXII, Sections 6–9).

It is remarkable that fairly general criteria (due to Gelfand) for a pair (G, K) to be a Gelfand pair exist. This is certainly the case if G is commutative and $K = (e)$; this situation corresponds to commutative harmonic analysis. If G is a semisimple compact connected Lie group and if $\sigma \colon G \to G$ is an involutive automorphism of G (that is, $\sigma^2 = \mathrm{id}$), if K is the subgroup of fixed points of σ

$$K = \{s \in G \mid \sigma(s) = s\},$$

then it can be shown that (G, K) is a Gelfand pair. Involutive automorphism as above was determined explicitly by E. Cartan.

It turns out that $G = \mathbf{SO}(n + 1)$ and $K = \mathbf{SO}(n)$ form a Gelfand pair corresponding to the above situation (see Dieudonné [31], Chapters 7–8 and Dieudonné [34], Chapter XXIII, Section 38). In this particular case, $\rho = k$ is any nonnegative integer and $L_\rho = E_k$, the eigenspace of the Laplacian on S^n corresponding to the eigenvalue $-k(n + k - 1)$; all this was shown in Section 7.5. Therefore, the regular representation of $\mathbf{SO}(n+1)$ into $E_k = \mathcal{H}^{\mathbb{C}}_k(S^n)$ is irreducible. This can be proved more directly; for example, see Helgason [55] (Introduction, Theorem 3.1) or Bröcker and tom Dieck [18] (Chapter 2, Proposition 5.10).

The zonal spherical harmonics ω_k can be expressed in terms of the *ultraspherical polynomials* (also called *Gegenbauer polynomials*) $P_k^{(n-1)/2}$ (up to a constant factor); this was discussed in Sections 7.6 and 7.7. The reader should also consult Stein and Weiss [102] (Chapter 4), Morimoto [81] (Chapter 2), and Dieudonné [31] (Chapter 7). For $n = 2$, $P_k^{\frac{1}{2}}$ is just the ordinary Legendre polynomial (up to a constant factor).

Returning to arbitrary Gelfand pairs (G compact), the Fourier transform is defined as follows. For any function $f \in L_{\mathbb{C}}^2(K \backslash G / K)$, the Fourier transform $\mathcal{F}(f)$ is a function defined *on the space* $S(G/K)$.

Definition 7.35. If (G, K) is a Gelfand pair (with G a compact group), the *Fourier transform* $\mathcal{F}(f)$ of a function $f \in L_{\mathbb{C}}^2(K \backslash G / K)$ is the function $\mathcal{F}(f) \colon S(G/K) \to \mathbb{C}$ given by

$$\mathcal{F}(f)(\omega) = \int_G f(s)\omega(s^{-1})\, ds \quad \omega \in S(G/K).$$

More explicitly, because $\omega_\rho = \frac{1}{n_\rho} m_{11}^{(\rho)}$ and $m_{11}^{(\rho)}(s^{-1}) = \overline{m_{11}^{(\rho)}(s)}$, the Fourier transform $\mathcal{F}(f)$ is the countable family

$$\rho \mapsto \frac{1}{n_\rho}\left\langle f, m_{11}^{(\rho)} \right\rangle = \int_G f(s)\omega_\rho(s^{-1})\, ds$$

for all $\rho \in R(G)$ such that $(\rho : \sigma_0) = 1$.

This Fourier transform is often called the *spherical Fourier transform* or *spherical transform*, as in Helgason [55] (Chapter 4). It appears that it was first introduced by Harish-Chandra around 1957.

The Fourier transform on $L_{\mathbb{C}}^2(K \backslash G / K)$ satisfies the fundamental relation

$$\mathcal{F}(f * g) = \mathcal{F}(g * f) = \mathcal{F}(f)\mathcal{F}(g).$$

Observe that in order for this equation to hold, convolution has to be commutative. This is why the Fourier transform is defined on $L_{\mathbb{C}}^2(K \backslash G / K)$ rather than the whole of $L_{\mathbb{C}}^2(G/K)$. For a Gelfand pair, convolution on $L_{\mathbb{C}}^2(K \backslash G / K)$ is commutative.

The notion of Gelfand pair and of the Fourier transform can be generalized to locally compact unimodular groups that are not necessarily compact, but we will not discuss this here. Let us just say that when G is a commutative locally compact group and $K = (e)$, then Equation (∗) implies that

$$\omega(xy) = \omega(x)\omega(y),$$

which means that the functions ω are characters of G, so $S(G/K)$ is the Pontrjagin *dual* group $\widehat{G}$ of G, which is the group of characters of G (continuous homomorphisms of G into the group $U(1)$). In this case, the Fourier transform $\mathcal{F}(f)$ is defined for every function $f \in L_{\mathbb{C}}^1(G)$ as a function on the characters of G. This is the case of commutative harmonic analysis, as discussed in Folland [42] and Deitmar [29].

For more on Gelfand pairs, curious readers may consult Dieudonné [31] (Chapters 8 and 9) and Dieudonné [33] (Chapter XXII, Sections 6–9). Another approach to spherical functions (not using Gelfand pairs) is discussed in Helgason [55] (Chapter 4). Helgason [54] contains a short section on Gelfand pairs (chapter III, Section 12).

The material in this section belongs to the overlapping areas of *representation theory* and *noncommutative harmonic analysis*. These are deep and vast areas. Besides the references cited earlier, for noncommutative harmonic analysis, the reader may consult Knapp [65], Folland [42], Taylor [103], or Varadarajan [106], but they may find the pace rather rapid. Another great survey on both topics is Kirillov [64], although it is not geared for the beginner. In a different direction, namely Fourier analysis on *finite* groups, Audrey Terras's book [104] contains some fascinating material.

7.12 Problems

Problem 7.1. Let E be a complex vector space of dimension n.

(i) Show that given any basis $(e_1, \ldots, e_n)$ of E, the linear map $h \colon \mathbb{C}^n \to E$ defined such that

$$h((z_1, \ldots, z_n)) = z_1 e_1 + \cdots + z_n e_n$$

is a homeomorphism (using the *sup*-norm on $\mathbb{C}^n$).

(ii) Use Part (i.) and the fact that any two norms on a finite-dimensional vector space over $\mathbb{R}$ or $\mathbb{C}$ are equivalent to prove that E is complete.

Problem 7.2. Let E be a normed vector space. Let K be a nonempty index set. Show that for every bijection $\varphi \colon K \to K$ (intuitively, a reordering of K), the family $(u_k)_{k \in K}$ is summable iff the family $(u_l)_{l \in \varphi(K)}$ is summable, and if so, they have the same sum.

Problem 7.3. Prove Proposition 7.1.

Problem 7.4. Let Θ be a function of the independent variable θ. Take second order differential equation

$$\sin^2 \theta \, \Theta'' + \sin \theta \cos \theta \, \Theta' + (k(k+1) \sin^2 \theta - m^2)\Theta = 0,$$

and use the change of variable $t = \cos \theta$ to obtain

$$(1 - t^2)u'' - 2tu' + \left(k(k+1) - \frac{m^2}{1 - t^2} \right) u = 0.$$

Then make the substitution

$$u(t) = (1 - t^2)^{\frac{m}{2}} v(t);$$

to obtain

$$(1 - t^2)v'' - 2(m + 1)tv' + (k(k + 1) - m(m + 1))v = 0.$$

Hint. See Lebedev [72], Chapter 7, Section 7.12.

Problem 7.5. Recall that the Legendre polynomial $P_n(t)$ is defined as

$$P_n(t) = \frac{1}{2^n n!} \frac{d^n}{dt^n} (t^2 - 1)^n.$$

(i) Show that $P_k(t)$ is a solution to the second order differential equation

$$(1 - t^2)v'' - 2tv' + k(k + 1)v = 0.$$

(ii) Show that the Legendre polynomials satisfy the following recurrence relation:

$$P_0 = 1$$
$$P_1 = t$$
$$(n + 1)P_{n+1} = (2n + 1)tP_n - nP_{n-1} \qquad n \geq 1;$$

Hint. See Lebedev [72], Chapter 4, Section 4.3.

Problem 7.6. Recall that the associated Legendre function $P_m^k(t)$ is defined by

$$P_m^k(t) = (1 - t^2)^{\frac{m}{2}} \frac{d^m}{dt^m} (P_k(t)),$$

where $P_k(t)$ is the Legendre polynomial of order k.

(i) For fixed $m \geq 0$, prove the recurrence relation

$$(k - m + 1)P_{k+1}^m(t) = (2k + 1)tP_k^m(t) - (k + m)P_{k-1}^m(t), \qquad k \geq 1.$$

(ii) Fore fixed $k \geq 2$, prove the recurrence relation

$$P_k^{m+2}(t) = \frac{2(m + 1)t}{(t^2 - 1)^{\frac{1}{2}}} P_k^{m+1}(t) + (k-m)(k+m+1)P_k^m(t), \qquad 0 \leq m \leq k-2.$$

Hint. See Lebedev [72], Chapter 7, Section 7.12.

Problem 7.7. Let M be a n-dimensional Riemannian manifold with chart (U, φ). If for $p \in M$

$$\left(\left(\frac{\partial}{\partial x_1} \right)_p, \ldots, \left(\frac{\partial}{\partial x_n} \right)_p \right)$$

denotes the basis of $T_p M$ induced by φ, the local expression of the metric g at p is given by the $n \times n$ matrix $(g_{ij})_p$, with

$$(g_{ij})_p = g_p \left(\left(\frac{\partial}{\partial x_i} \right)_p, \left(\frac{\partial}{\partial x_j} \right)_p \right).$$

Its inverse is denoted $(g^{ij})_p$. We also let $|g|_p = \det(g_{ij})_p$. Show that for every function $f \in C^\infty(M)$, in local coordinates given by the chart (U, φ), we have

$$\operatorname{grad} f = \sum_{ij} g^{ij} \frac{\partial f}{\partial x_j} \frac{\partial}{\partial x_i},$$

where as usual

$$\frac{\partial f}{\partial x_j}(p) = \left(\frac{\partial}{\partial x_j} \right)_p f = \frac{\partial (f \circ \varphi^{-1})}{\partial u_j}(\varphi(p)),$$

and $(u_1, \ldots, u_n)$ are the coordinate functions in $\mathbb{R}^n$.

Problem 7.8. Let M be a Riemannian manifold of dimension n with chart (U, φ). Let

$$X = \sum_{i=1}^n X_i \frac{\partial}{\partial x_i},$$

be a vector field expressed over this chart. Recall that the Christoffel symbol Γ_{ij}^k is defined as

$$\Gamma_{ij}^k = \frac{1}{2} \sum_{l=1}^n g^{kl} \left(\partial_i g_{jl} + \partial_j g_{il} - \partial_l g_{ij} \right), \tag{$*$}$$

where $\partial_k g_{ij} = \frac{\partial}{\partial x_k}(g_{ij})$. Show that

$$\operatorname{div} X = \sum_{i=1}^n \left[\frac{\partial X_i}{\partial x_i} + \sum_{j=1}^n \Gamma_{ij}^i X_j \right],$$

and that

$$\Delta f = \sum_{i,j} g^{ij} \left[\frac{\partial^2 f}{\partial x_i \partial x_j} - \sum_{k=1}^{n} \Gamma_{ij}^k \frac{\partial f}{\partial x_k} \right],$$

whenever $f \in C^\infty(M)$.
Hint. See Pages 86 and 87 of O'Neill [84].

Problem 7.9. Let M be a Riemannian manifold of dimension n with chart (U, φ). For every vector field $X \in \mathfrak{X}(M)$ expressed in local coordinates as

$$X = \sum_{i=1}^{n} X_i \frac{\partial}{\partial x_i},$$

show that

$$\operatorname{div} X = \frac{1}{\sqrt{|g|}} \sum_{i=1}^{n} \frac{\partial}{\partial x_i} \left(\sqrt{|g|}\, X_i \right),$$

and for every function $f \in C^\infty(M)$, show that

$$\Delta f = \frac{1}{\sqrt{|g|}} \sum_{i,j} \frac{\partial}{\partial x_i} \left(\sqrt{|g|}\, g^{ij} \frac{\partial f}{\partial x_j} \right).$$

Hint. See Helgason [55], Chapter II, Lemma 2.5, Postnikov [88], Chapter 13, Section 6, and O'Neill [84], Chapter 7, Exercise 5.

Problem 7.10. Let M be a Riemannian manifold with metric g. For any two functions $f, h \in C^\infty(M)$, and any vector field $X \in \mathfrak{X}(M)$, show that

$$\operatorname{div}(fX) = f \operatorname{div} X + X(f) = f \operatorname{div} X + g(\operatorname{grad} f, X)$$

$$\operatorname{grad} f\,(h) = g(\operatorname{grad} f, \operatorname{grad} h) = \operatorname{grad} h\,(f).$$

Problem 7.11. Recall that $a_{k,n}$ denotes the dimension of $\mathcal{H}_k(n)$. Show that

$$a_{k,n+1} = \binom{n+k-1}{n-1} + \binom{n+k-2}{n-1}$$

for $k \geq 2$.
Hint. See Morimoto [81], Chapter 2, Theorem 2.4, or Dieudonné [31], Chapter 7, Formula 99.

Problem 7.12. Let $C(S^n)$ the space of continuous (real) functions on S^n. Show that $C(S^n)$ is dense in $L^2(S^n)$, where $L^2(S^n)$ is the space of (real) square-integrable functions on the sphere S^n. with norm given by

$$\langle f, f \rangle_{S^n} = \int_{S^n} f^2 \, \mathrm{Vol}_{S^n}.$$

Problem 7.13. Prove Theorem 7.18.
Hint. See Lang [68], Chapter V, Proposition 1.9.

Problem 7.14. Prove Proposition 7.22.
Hint. See Morimoto [81], Chapter 2, Lemma 1.19 and Lemma 2.20.

Problem 7.15. Prove Propositions 7.23 and 7.25.
Hint. See Morimoto [81] Chapter 2, Lemma 2.21 and Morimoto [81] Chapter 2, Lemma 2.23.

Problem 7.16. Prove Theorem 7.30.
Hint. See Stein and Weiss [102], Chapter 4, Theorem 2.12.

Problem 7.17. Show that the Gegenbauer polynomial $P_{k,n}$ is given by Rodrigues' formula:

$$P_{k,n}(t) = \frac{(-1)^k}{2^k} \frac{\Gamma\left(\frac{n}{2}\right)}{\Gamma\left(k+\frac{n}{2}\right)} \frac{1}{(1-t^2)^{\frac{n-2}{2}}} \frac{d^k}{dt^k}(1-t^2)^{k+\frac{n-2}{2}},$$

with $n \geq 2$.
Hint. See Morimoto [81], Chapter 2, Theorem 2.35.

Problem 7.18. Prove Proposition 7.32.
Hint. See Morimoto [81], Chapter 2, Theorem 2.34.

Problem 7.19. Prove Proposition 7.34.
Hint. See Morimoto [81], Chapter 2, Theorem 2.53 and Theorem 2.55.

Chapter 8
Operators on Riemannian Manifolds: Hodge Laplacian, Laplace-Beltrami Laplacian, the Bochner Laplacian, and Weitzenböck Formulae

The Laplacian is a very important operator because it shows up in many of the equations used in physics to describe natural phenomena such as heat diffusion or wave propagation. Therefore, it is highly desirable to generalize the Laplacian to functions defined on a manifold. Furthermore, in the late 1930s, Georges de Rham (inspired by Élie Cartan) realized that it was fruitful to define a version of the Laplacian operating on differential forms, because of a fundamental and almost miraculous relationship between harmonics forms (those in the kernel of the Laplacian) and the de Rham cohomology groups on a (compact, orientable) smooth manifold. Indeed, as we will see in Section 8.6, for every cohomology group $H_{\mathrm{DR}}^k(M)$, every cohomology class $[\omega] \in H_{\mathrm{DR}}^k(M)$ is represented by a *unique harmonic k-form* ω. The connection between analysis and topology lies deep and has many important consequences. For example, *Poincaré duality* follows as an "easy" consequence of the Hodge theorem.

Technically, the Hodge Laplacian can be defined on differential forms using the Hodge $*$ operator (Section 3.5). On functions, there is an alternate and equivalent definition of the Laplacian using only the covariant derivative and obtained by generalizing the notions of gradient and divergence to functions on manifolds.

Another version of the Laplacian on k-forms can be defined in terms of a generalization of the Levi-Civita connection $\nabla \colon \mathfrak{X}(M) \times \mathfrak{X}(M) \to \mathfrak{X}(M)$ to k-forms viewed as a linear map

$$\nabla \colon \mathcal{A}^k(M) \to \mathrm{Hom}_{C^\infty(M)}(\mathfrak{X}(M), \mathcal{A}^k(M)),$$

and in terms of a certain adjoint ∇^* of ∇, a linear map

$$\nabla^* \colon \mathrm{Hom}_{C^\infty(M)}(\mathfrak{X}(M), \mathcal{A}^k(M)) \to \mathcal{A}^k(M).$$

For this, we will define an inner product $(-, -)$ on k-forms and an inner product $((-, -))$ on $\mathrm{Hom}_{C^\infty(M)}(\mathfrak{X}(M), \mathcal{A}^k(M))$ and define ∇^* so that

© Springer Nature Switzerland AG 2020
J. Gallier, J. Quaintance, *Differential Geometry and Lie Groups*, Geometry and Computing 13, https://doi.org/10.1007/978-3-030-46047-1_8

$$(\nabla^* A, \omega) = ((A, \nabla\omega))$$

for all $A \in \mathrm{Hom}_{C^\infty(M)}(\mathfrak{X}(M), \mathcal{A}^k(M))$ and all $\omega \in \mathcal{A}^k(M)$.

We obtain the *Bochner Laplacian* (or *connection Laplacian*) $\nabla^*\nabla$. Then it is natural to wonder how the Hodge Laplacian Δ differs from the connection Laplacian $\nabla^*\nabla$?

Remarkably, there is a formula known as *Weitzenböck's formula* (or *Bochner's formula*) of the form

$$\Delta = \nabla^*\nabla + C(R_\nabla),$$

where $C(R_\nabla)$ is a contraction of a version of the curvature tensor on differential forms (a fairly complicated term). In the case of one-forms,

$$\Delta = \nabla^*\nabla + \mathrm{Ric},$$

where Ric is a suitable version of the Ricci curvature operating on one-forms.

Weitzenböck-type formulae are at the root of the so-called "Bochner technique," which consists in exploiting curvature information to deduce topological information. For example, if the Ricci curvature on a compact orientable Riemannian manifold is strictly positive, then $H^1_{\mathrm{DR}}(M) = (0)$, a theorem due to Bochner.

8.1   The Gradient and Hessian Operators on Riemannian Manifolds

In preparation for defining the (Hodge) Laplacian, we define the gradient of a function on a Riemannian manifold, as well as the Hessian, which plays an important role in optimization theory. Unlike the situation where M is a vector space (M is flat), the Riemannian metric on M is critically involved in the definition of the gradient and of the Hessian.

If $(M, \langle -, - \rangle)$ is a Riemannian manifold of dimension n, then for every $p \in M$, the inner product $\langle -, - \rangle_p$ on T_pM yields a canonical isomorphism $\flat \colon T_pM \to T_p^*M$, as explained in Sections 2.2 and 9.7. Namely, for any $u \in T_pM$, $u^\flat = \flat(u)$ is the linear form in T_p^*M defined by

$$u^\flat(v) = \langle u, v \rangle_p, \qquad v \in T_pM.$$

Recall that the inverse of the map $\flat$ is the map $\sharp \colon T_p^*M \to T_pM$. As a consequence, for every smooth function $f \in C^\infty(M)$, we get smooth vector field $\mathrm{grad}\, f = (df)^\sharp$ defined so that

$$(\mathrm{grad}\, f)_p = (df_p)^\sharp.$$

Definition 8.1. For every smooth function f over a Riemannian manifold $(M, \langle -, - \rangle)$, the vector field grad f defined by

$$\langle (\operatorname{grad} f)_p, u \rangle_p = df_p(u), \quad \text{for all} \quad u \in T_p M, \text{ and all } p \in M,$$

is the *gradient* of the function f.

Definition 8.2. Let $(M, \langle -, - \rangle)$ be a Riemannian manifold. For any vector field $X \in \mathfrak{X}(M)$, the one-form $X^\flat \in \mathcal{A}^1(M)$ is given by

$$(X^\flat)_p = (X_p)^\flat.$$

The one-form $X^\flat$ is uniquely defined by the equation

$$(X^\flat)_p(v) = \langle X_p, v \rangle_p, \quad \text{for all } p \in M \text{ and all } v \in T_p M.$$

In view of this equation, the one-form $X^\flat$ is an insertion operator in the sense discussed in Section 3.6 just after Proposition 3.22, so it is also denoted by $i_X g$, where $g = \langle -, - \rangle$ is the Riemannian metric on M.

In the special case $X = \operatorname{grad} f$, we have

$$(\operatorname{grad} f)^\flat_p(v) = \langle (\operatorname{grad} f)_p, v \rangle = df_p(v),$$

and since $dd = 0$, we deduce that

$$d(\operatorname{grad} f)^\flat = 0.$$

Therefore, for an arbitrary vector field X, the 2-form $dX^\flat$ measures the extent to which X is a gradient field.

If (U, φ) is a chart of M, with $p \in M$, and if

$$\left(\left(\frac{\partial}{\partial x_1} \right)_p, \ldots, \left(\frac{\partial}{\partial x_n} \right)_p \right)$$

denotes the basis of $T_p M$ induced by φ, the local expression of the metric g at p is given by the $n \times n$ matrix $(g_{ij})_p$, with

$$(g_{ij})_p = g_p \left(\left(\frac{\partial}{\partial x_i} \right)_p, \left(\frac{\partial}{\partial x_j} \right)_p \right).$$

The inverse is denoted by $(g^{ij})_p$. We often omit the subscript p and observe that for every function $f \in C^\infty(M)$,

$$\operatorname{grad} f = \sum_{ij} g^{ij} \frac{\partial f}{\partial x_j} \frac{\partial}{\partial x_i}.$$

It is instructive to look at the following special case of the preceding formula. Let $f \in C^\infty(M)$, where M is a two-dimensional manifold. For each $p \in M$, let $\{\frac{\partial}{\partial x_1}, \frac{\partial}{\partial x_2}\}$ be the basis for the tangent space $T_p(M)$. Let $v = a\frac{\partial}{\partial x_1} + b\frac{\partial}{\partial x_2} \in T_p(M)$. Then

$$\operatorname{grad} f = \left(g^{11} \frac{\partial f}{\partial x_1} + g^{12} \frac{\partial f}{\partial x_2} \right) \frac{\partial}{\partial x_1} + \left(g^{21} \frac{\partial f}{\partial x_1} + g^{22} \frac{\partial f}{\partial x_2} \right) \frac{\partial}{\partial x_2}.$$

Since $g_{12} = g_{21}$, $g^{12} = g^{21}$, and

$$\begin{pmatrix} g^{11} & g^{12} \\ g^{21} & g^{22} \end{pmatrix} \begin{pmatrix} g_{11} & g_{12} \\ g_{21} & g_{22} \end{pmatrix} = \begin{pmatrix} g_{11} & g_{12} \\ g_{21} & g_{22} \end{pmatrix} \begin{pmatrix} g^{11} & g^{12} \\ g^{21} & g^{22} \end{pmatrix} = \begin{pmatrix} 1 & 0 \\ 0 & 1 \end{pmatrix},$$

we discover that

$$
\begin{aligned}
\langle \operatorname{grad} f, v\rangle &= \left\langle \left(g^{11} \frac{\partial f}{\partial x_1} + g^{12} \frac{\partial f}{\partial x_2} \right) \frac{\partial}{\partial x_1} + \left(g^{21} \frac{\partial f}{\partial x_1} + g^{22} \frac{\partial f}{\partial x_2} \right) \frac{\partial}{\partial x_2}, \ a\frac{\partial}{\partial x_1} + b\frac{\partial}{\partial x_2} \right\rangle \\
&= a\left(g^{11} \frac{\partial f}{\partial x_1} + g^{12} \frac{\partial f}{\partial x_2} \right) g_{11} + b\left(g^{11} \frac{\partial f}{\partial x_1} + g^{12} \frac{\partial f}{\partial x_2} \right) g_{12} \\
&\quad + a\left(g^{21} \frac{\partial f}{\partial x_1} + g^{22} \frac{\partial f}{\partial x_2} \right) g_{12} + b\left(g^{21} \frac{\partial f}{\partial x_1} + g^{22} \frac{\partial f}{\partial x_2} \right) g_{22} \\
&= \left[a(g^{11}g_{11} + g^{21}g_{21}) + b(g^{11}g_{12} + g^{21}g_{22}) \right] \frac{\partial f}{\partial x_1} \\
&\quad + \left[a(g^{12}g_{11} + g^{22}g_{21}) + b(g^{12}g_{21} + g^{22}g_{22}) \right] \frac{\partial f}{\partial x_2} \\
&= a\frac{\partial f}{\partial x_1} + b\frac{\partial f}{\partial x_2} = \left(\frac{\partial f}{\partial x_1}, \frac{\partial f}{\partial x_2} \right) \begin{pmatrix} a \\ b \end{pmatrix} = df_p(v).
\end{aligned}
$$

We now define the Hessian of a function. For this we assume that ∇ is the Levi-Civita connection.

Definition 8.3. The *Hessian* $\operatorname{Hess}(f)$ (or $\nabla^2(f)$) of a function $f \in C^\infty(M)$ is the $(0, 2)$-tensor defined by

$$\operatorname{Hess}(f)(X, Y) = X(Y(f)) - (\nabla_X Y)(f) = X(df(Y)) - df(\nabla_X Y),$$

for all vector fields $X, Y \in \mathfrak{X}(M)$.

Remark. The Hessian of f is defined in various ways throughout the literature. For our purposes, Definition 8.3 is sufficient, but for completeness sake, we point out two alternative formulations of $\mathrm{Hess}(f)(X, Y)$.

The first reformulation utilizes the covariant derivative of a one-form. Let $X \in \mathfrak{X}(M)$ and $\theta \in \mathcal{A}^1(M)$. The covariant derivative $\nabla_X \theta$ of any one-form may be defined as

$$(\nabla_X \theta)(Y) := X(\theta(Y)) - \theta(\nabla_X Y).$$

Thus the Hessian of f may be written as

$$\mathrm{Hess}(f)(X, Y) = (\nabla_X df)(Y).$$

The Hessian of f also appears in the literature as

$$\mathrm{Hess}(f)(X, Y) = (\nabla df)(X, Y) = (\nabla_X df)(Y),$$

which means that the $(0, 2)$-tensor $\mathrm{Hess}(f)$ is given by

$$\mathrm{Hess}(f) = \nabla df.$$

Since by definition $\nabla_X f = df(X)$, we can also write $\mathrm{Hess}(f) = \nabla \nabla f$, but we find this expression confusing.

Proposition 8.1. *The Hessian is given by the equation*

$$\mathrm{Hess}(f)(X, Y) = \langle \nabla_X(\mathrm{grad}\, f), Y \rangle, \quad X, Y \in \mathfrak{X}(M).$$

Proof. We have

$$
\begin{aligned}
X(Y(f)) &= X(df(Y)) \\
&= X(\langle \mathrm{grad}\, f, Y \rangle) \\
&= \langle \nabla_X(\mathrm{grad}\, f), Y \rangle + \langle \mathrm{grad}\, f, \nabla_X Y \rangle \\
&= \langle \nabla_X(\mathrm{grad}\, f), Y \rangle + (\nabla_X Y)(f)
\end{aligned}
$$

which yields

$$\langle \nabla_X(\mathrm{grad}\, f), Y \rangle = X(Y(f)) - (\nabla_X Y)(f) = \mathrm{Hess}(f)(X, Y),$$

as claimed. $\qquad\qquad\square$

The Hessian can also be defined in terms of Lie derivatives; this is the approach followed by Petersen [86] (Chapter 2, Section 1.3). This approach utilizes the observation that the Levi-Civita connection can be defined in terms of the Lie derivative of the Riemannian metric g on M by the equation

$$2g(\nabla_X Y, Z) = (L_Y g)(X, Z) + (d(i_Y g))(X, Z), \quad X, Y, Z \in \mathfrak{X}(M).$$

Proposition 8.2. *The Hessian of f is given by*

$$\mathrm{Hess}(f) = \frac{1}{2} L_{\mathrm{grad}\, f}\, g.$$

Proof. To prove the above equation, we use the fact that $d(i_{\mathrm{grad}\, f}\, g) = d(\mathrm{grad}\, f)^\flat = 0$ and Proposition 8.1. We have

$$\begin{aligned}
2\mathrm{Hess}(f)(X, Y) &= 2g(\nabla_X(\mathrm{grad}\, f), Y), \qquad\qquad \text{by Proposition 8.1}\\
&= (L_{\mathrm{grad}\, f}\, g)(X, Y) + d(i_{\mathrm{grad}\, f}\, g)(X, Y)\\
&= (L_{\mathrm{grad}\, f}\, g)(X, Y),
\end{aligned}$$

as claimed.   □

Since ∇ is torsion-free, we get

$$\mathrm{Hess}(f)(X, Y) = X(Y(f)) - (\nabla_X Y)(f) = Y(X(f)) - (\nabla_Y X)(f) = \mathrm{Hess}(f)(Y, X),$$

which means that the Hessian is a *symmetric* $(0, 2)$-tensor.

Since the Hessian is a symmetric bilinear form, it is determined by the quadratic form $X \mapsto \mathrm{Hess}(f)(X, X)$, and it can be recovered by polarization from this quadratic form. There is also a way to compute $\mathrm{Hess}(f)(X, X)$ using geodesics. When geodesics are easily computable, this is usually the simplest way to compute the Hessian.

Proposition 8.3. *Given any $p \in M$ and any tangent vector $X \in T_p M$, if γ is a geodesic such that $\gamma(0) = p$ and $\gamma'(0) = X$, then at p, we have*

$$\mathrm{Hess}(f)_p(X, X) = \frac{d^2}{dt^2}\, f(\gamma(t))\Big|_{t=0}.$$

Proof. To prove the above formula, following Jost [62], we have

$$\begin{aligned}
X(X(f))(p) &= \gamma'\langle(\mathrm{grad}\, f)_p, \gamma'\rangle\\
&= \gamma'\left(\frac{d}{dt}\, f(\gamma(t))\Big|_{t=0}\right)
\end{aligned}$$

$$= \frac{d^2}{dt^2} f(\gamma(t)) \Big|_{t=0}.$$

Furthermore, since γ is a geodesic, $\nabla_{\gamma'}\gamma' = 0$, so we get

$$\mathrm{Hess}(f)_p(X, X) = X(X(f))(p) - (\nabla_X X)(f)(p) = X(X(f))(p),$$

which proves our claim.   □

Proposition 8.4. *In local coordinates with respect to a chart, if we write*

$$df = \sum_{i=1}^{n} \frac{\partial f}{\partial x_i} dx_i,$$

then

$$\mathrm{Hess}\, f = \sum_{i,j=1}^{n} \left(\frac{\partial^2 f}{\partial x_i \partial x_j} - \sum_{k=1}^{n} \frac{\partial f}{\partial x_k} \Gamma_{ij}^k \right) dx_i \otimes dx_j,$$

where the Γ_{ij}^k are the Christoffel symbols of the connection in the chart, namely

$$\Gamma_{ij}^k = \frac{1}{2} \sum_{l=1}^{n} g^{kl} \left(\partial_i g_{jl} + \partial_j g_{il} - \partial_l g_{ij} \right),$$ (∗)

with $\partial_k g_{ij} = \frac{\partial}{\partial x_k}(g_{ij})$.

The formula of Proposition 8.4 is shown in O'Neill [84]. If (g_{ij}) is the standard Euclidean metric, the Christoffel symbols vanish and O'Neill's formula becomes

$$\mathrm{Hess}\, f = \sum_{i,j=1}^{n} \frac{\partial^2 f}{\partial x_i \partial x_j} dx_i \otimes dx_j.$$

For another example of the preceding formula, take $f \in C^\infty(\mathbb{R}^2)$ and let us compute Hess f in terms of polar coordinates (r, θ), where $x = r\cos\theta$, and $y = r\sin\theta$. Note that

$$\frac{\partial}{\partial x_1} = \frac{\partial}{\partial r} = (\cos\theta, \sin\theta)$$

$$\frac{\partial}{\partial x_2} = \frac{\partial}{\partial \theta} = (-r\sin\theta, r\cos\theta),$$

which in turn gives

$$g_{ij} = \begin{pmatrix} 1 & 0 \\ 0 & r^2 \end{pmatrix} \qquad g^{ij} = \begin{pmatrix} 1 & 0 \\ 0 & r^{-2} \end{pmatrix}.$$

A computation shows that the only nonzero Christoffel symbols were

$$\Gamma^2_{12} = \Gamma^2_{21} = \frac{1}{r} \qquad \Gamma^1_{22} = -r.$$

Hence

$$
\begin{aligned}
\text{Hess } f &= \sum_{i,j=1}^{2} \left(\frac{\partial^2 f}{\partial x_i \partial x_j} - \sum_{k=1}^{2} \frac{\partial f}{\partial x_k} \Gamma^k_{ij} \right) dx_i \otimes dx_j \\
&= \frac{\partial^2 f}{\partial r^2} \, dr \otimes dr + \left(\frac{\partial^2 f}{\partial r \partial \theta} - \frac{\partial f}{\partial \theta} \Gamma^2_{12} \right) dr \otimes d\theta \\
&\quad + \left(\frac{\partial^2 f}{\partial \theta \partial r} - \frac{\partial f}{\partial \theta} \Gamma^2_{21} \right) d\theta \otimes dr + \left(\frac{\partial^2 f}{\partial^2 \theta} - \frac{\partial f}{\partial r} \Gamma^1_{22} \right) d\theta \otimes d\theta \\
&= \frac{\partial^2 f}{\partial r^2} \, dr \otimes dr + \left(\frac{\partial^2 f}{\partial r \partial \theta} - \frac{1}{r} \frac{\partial f}{\partial \theta} \right) dr \otimes d\theta \\
&\quad + \left(\frac{\partial^2 f}{\partial r \partial \theta} - \frac{1}{r} \frac{\partial f}{\partial \theta} \right) d\theta \otimes dr + \left(\frac{\partial^2 f}{\partial^2 \theta} + r \frac{\partial f}{\partial r} \right) d\theta \otimes d\theta.
\end{aligned}
$$

If we write $X = x_1 \frac{\partial}{\partial r} + x_2 \frac{\partial}{\partial \theta}$ and $Y = y_1 \frac{\partial}{\partial r} + y_2 \frac{\partial}{\partial \theta}$, then

$$
\text{Hess } f(X, Y) = \begin{pmatrix} x_1 & x_2 \end{pmatrix} \begin{pmatrix} \dfrac{\partial^2 f}{\partial r^2} & \dfrac{\partial^2 f}{\partial r \partial \theta} - \dfrac{1}{r} \dfrac{\partial f}{\partial \theta} \\ \dfrac{\partial^2 f}{\partial r \partial \theta} - \dfrac{1}{r} \dfrac{\partial f}{\partial \theta} & \dfrac{\partial^2 f}{\partial^2 \theta} + r \dfrac{\partial f}{\partial r} \end{pmatrix} \begin{pmatrix} y_1 \\ y_2 \end{pmatrix}.
$$

Definition 8.4. A function $f \in C^\infty(M)$ is *convex* (resp. *strictly convex*) iff its Hessian $\text{Hess}(f)$ is positive semi-definite (resp. positive definite).

The computation of the gradient of a function defined either on the Stiefel manifold or on the Grassmannian manifold is instructive. Let us first consider the Stiefel manifold $S(k, n)$. Recall that $S(k, n)$ is the set of all orthonormal k-frames, where an orthonormal k-frame is a k-tuples of orthonormal vectors $(u_1, \ldots, u_k)$ with $u_i \in \mathbb{R}^n$. Then $\mathbf{SO}(n)$ acts transitively on $S(k, n)$ via the action $\cdot \colon \mathbf{SO}(n) \times S(k, n) \to S(k, n)$

$$R \cdot (u_1, \ldots, u_k) = (Ru_1, \ldots, Ru_k).$$

and that the stabilizer of this action is

$$H = \left\{ \begin{pmatrix} I & 0 \\ 0 & R \end{pmatrix} \,\middle|\, R \in \mathbf{SO}(n-k) \right\}.$$

It follows (see Warner [109], Chapter 3, and Gallier and Quaintance [47]) that $S(k,n) \cong G/H$, with $G = \mathbf{SO}(n)$ and $H \cong \mathbf{SO}(n-k)$. Observe that the points of $G/H \cong S(k,n)$ are the cosets QH, with $Q \in \mathbf{SO}(n)$. If we write $Q = [Y \; Y_\perp]$, where Y consists of the first k columns of Q and $Y_\perp$ consists of the last $n-k$ columns of Q, it is clear that $[Q]$ is uniquely determined by Y. We also found that $\mathfrak{g}/\mathfrak{h} \cong \mathfrak{m}$ where

$$\mathfrak{m} = \left\{ \begin{pmatrix} T & -A^\top \\ A & 0 \end{pmatrix} \,\middle|\, T \in \mathfrak{so}(k), \; A \in \mathrm{M}_{n-k,k}(\mathbb{R}) \right\}.$$

The inner product on $\mathfrak{m}$ is given by

$$\langle X, Y \rangle = -\frac{1}{2}\mathrm{tr}(XY) = \frac{1}{2}\mathrm{tr}(X^\top Y), \qquad X, Y \in \mathfrak{m}.$$

The vector space $\mathfrak{m}$ is the tangent space $T_o S(k,n)$ to $S(k,n)$ at $o = [H]$, the coset of the point corresponding to H. For any other point $[Q] \in G/H \cong S(k,n)$, the tangent space $T_{[Q]}S(k,n)$ is given by

$$T_{[Q]}S(k,n) = \left\{ Q \begin{pmatrix} S & -A^\top \\ A & 0 \end{pmatrix} \,\middle|\, S \in \mathfrak{so}(k), \; A \in \mathrm{M}_{n-k,k}(\mathbb{R}) \right\}.$$

For every $n \times k$ matrix $Y \in S(k,n)$, this observation implies that tangent vectors to $S(k,n)$ at Y are of the form

$$X = YS + Y_\perp A,$$

where S is any $k \times k$ skew-symmetric matrix, A is any $(n-k) \times k$ matrix, and $[Y \; Y_\perp]$ is an orthogonal matrix. Given any differentiable function $F \colon S(k,n) \to \mathbb{R}$, if we let F_Y be the $n \times k$ matrix of partial derivatives

$$F_Y = \left(\frac{\partial F}{\partial Y_{ij}} \right),$$

we then have

$$dF_Y(X) = \mathrm{tr}(F_Y^\top X).$$

The gradient $\mathrm{grad}(F)_Y$ of F at Y is the uniquely defined tangent vector to $S(k,n)$ at Y such that

$$\langle \mathrm{grad}(F)_Y, X \rangle = dF_Y(X) = \mathrm{tr}(F_Y^\top X), \quad \text{for all} \quad X \in T_Y S(k,n).$$

For short, if write $Z = \text{grad}(F)_Y$, then it can be shown that Z must satisfy the equation

$$\text{tr}(F_Y^\top X) = \text{tr}\left(Z^\top\left(I - \frac{1}{2}YY^\top\right)X\right),$$

and since Z is of the form $Z = YS + Y_\perp A$, and since

$$Y^\top Y = I_{k\times k}, \qquad Y_\perp^\top Y = 0, \qquad Y_\perp^\top Y_\perp = I_{(n-k)\times(n-k)},$$

we get

$$\text{tr}(F_Y^\top X) = \text{tr}\left((S^\top Y^\top + A^\top Y_\perp^\top)\left(I - \frac{1}{2}YY^\top\right)X\right)$$

$$= \text{tr}\left(\left(\frac{1}{2}S^\top Y^\top + A^\top Y_\perp^\top\right)X\right)$$

for all $X \in T_Y S(k, n)$. The above equation implies that we must find $Z = YS + Y_\perp A$ such that

$$F_Y^\top = \frac{1}{2}S^\top Y^\top + A^\top Y_\perp^\top,$$

which is equivalent to

$$F_Y = \frac{1}{2}YS + Y_\perp A.$$

From the above equation, we deduce that

$$Y_\perp^\top F_Y = A$$

$$Y^\top F_Y = \frac{1}{2}S.$$

Since S is skew-symmetric, we get

$$F_Y^\top Y = -\frac{1}{2}S,$$

so

$$S = Y^\top F_Y - F_Y^\top Y,$$

and thus,

$$Z = YS + Y_\perp A$$
$$= Y(Y^\top F_Y - F_Y^\top Y) + Y_\perp Y_\perp^\top F_Y$$
$$= (YY^\top + Y_\perp Y_\perp^\top)F_Y - YF_Y^\top Y$$
$$= F_Y - YF_Y^\top Y.$$

Therefore, we proved that the gradient of F at Y is given by

$$\mathrm{grad}(F)_Y = F_Y - YF_Y^\top Y.$$

Let us now turn to the Grassmannian $G(k, n)$. Recall that $G(k, n)$ is the set of all linear k-dimensional subspaces of $\mathbb{R}^n$, where the k-dimensional subspace U of $\mathbb{R}$ is spanned by k linearly independent vectors $u_1, \ldots, u_k$ in $\mathbb{R}^n$; write $U = \mathrm{span}(u_1, \ldots, u_k)$. It can be shown that the action $\cdot : \mathbf{SO}(n) \times G(k, n) \to G(k, n)$

$$R \cdot U = \mathrm{span}(Ru_1, \ldots, Ru_k).$$

is well defined, transitive, and has the property that stabilizer of U is the set of matrices in $\mathbf{SO}(n)$ with the form

$$R = \begin{pmatrix} S & 0 \\ 0 & T \end{pmatrix},$$

where $S \in \mathbf{O}(k)$, $T \in \mathbf{O}(n - k)$ and $\det(S)\det(T) = 1$. We denote this group by $S(\mathbf{O}(k) \times \mathbf{O}(n - k))$. Since $\mathbf{SO}(n)$ is a connected, compact semisimple Lie group whenever $n \geq 3$, this implies that

$$G(k, n) \cong \mathbf{SO}(n)/S(\mathbf{O}(k) \times \mathbf{O}(n - k))$$

is a naturally reductive homogeneous manifold whenever $n \geq 3$. It can be shown that $\mathfrak{g}/\mathfrak{h} \cong \mathfrak{m}$ where

$$\mathfrak{m} = \left\{ \begin{pmatrix} 0 & -A^\top \\ A & 0 \end{pmatrix} \,\middle|\, A \in \mathbf{M}_{n-k,k}(\mathbb{R}) \right\};$$

see Gallier and Quaintance [47]. For any point $[Q] \in G(k, n)$ with $Q \in \mathbf{SO}(n)$, if we write $Q = [Y \; Y_\perp]$, where Y denotes the first k columns of Q and $Y_\perp$ denotes the last $n - k$ columns of Q, the tangent vectors $X \in T_{[Q]}G(k, n)$ are of the form

$$X = [Y \; Y_\perp] \begin{pmatrix} 0 & -A^\top \\ A & 0 \end{pmatrix} = [Y_\perp A \; -YA^\top], \quad A \in \mathbf{M}_{n-k,k}(\mathbb{R}).$$

This implies that the tangent vectors to $G(k, n)$ at Y are of the form

$$X = Y_\perp A,$$

where A is any $(n - k) \times k$ matrix. We would like to compute the gradient at Y of a function $F : G(k, n) \to \mathbb{R}$. Again, if write $Z = \text{grad}(F)_Y$, then Z must satisfy the equation

$$\text{tr}(F_Y^\top X) = \langle Z, X \rangle = \text{tr}(Z^\top X), \quad \text{for all} \quad X \in T_{[Y]}G(k, n).$$

Since Z is of the form $Z = Y_\perp A$, we get

$$\text{tr}(F_Y^\top X) = \text{tr}(A^\top Y_\perp^\top X), \quad \text{for all} \quad X \in T_{[Y]}G(k, n),$$

which implies that

$$F_Y^\top = A^\top Y_\perp^\top;$$

that is,

$$F_Y = Y_\perp A.$$

The above yields

$$A = Y_\perp^\top F_Y,$$

so we have

$$Z = Y_\perp Y_\perp^\top F_Y = (I - YY^\top)F_Y.$$

Therefore, the gradient of F at Y is given by

$$\text{grad}(F)_Y = F_Y - YY^\top F_Y.$$

Since the geodesics in the Stiefel manifold and in the Grassmannian can be determined explicitly (see Gallier and Quaintance [47]), we can find the Hessian of a function using the formula

$$\text{Hess}(f)_p(X, X) = \frac{d^2}{dt^2} f(\gamma(t)) \Big|_{t=0}.$$

Let us do this for a function F defined on the Grassmannian, the computation on the Stiefel manifold being more complicated; see Edelman, Arias, and Smith [40] for details.

For any two tangent vectors $X_1, X_2 \in T_Y G(k, n)$ to $G(k, n)$ at Y, define $F_{YY}(X_1, X_2)$ by

$$F_{YY}(X_1, X_2) = \sum_{ij,kl} (F_{YY})_{ij,kl} (X_1)_{ij} (X_2)_{kl},$$

with

$$(F_{YY})_{ij,kl} = \frac{\partial^2 F}{\partial Y_{ij} \partial Y_{kl}}.$$

By using Proposition 8.3, Edelman, Arias, and Smith [40] find that a somewhat lengthy computation yields

$$\mathrm{Hess}(F)_Y(X_1, X_2) = F_{YY}(X_1, X_2) - \mathrm{tr}(X_1^\top X_2 Y^\top F_Y),$$

where

$$F_Y = \left(\frac{\partial F}{\partial Y_{ij}} \right),$$

as above, when we found a formula for the gradient of F at Y.

8.2 The Hodge $*$ Operator on Riemannian Manifolds

Let M be an n-dimensional Riemann manifold. By Section 2.1 the inner product $\langle -, - \rangle_p$ on $T_p M$ induces an inner product on $T_p^* M$ defined as follows.

Definition 8.5. For any Riemannian manifold M, the inner product $\langle -, - \rangle_p$ on $T_p M$ induces an inner product on $T_p^* M$ given by

$$\langle w_1, w_2 \rangle := \langle w_1^\sharp, w_2^\sharp \rangle, \qquad w_1, w_2 \in T_p^* M.$$

This inner product on $T_p^* M$ defines an inner product on $\bigwedge^k T_p^* M$, with

$$\langle u_1 \wedge \cdots \wedge u_k, v_1 \wedge \cdots \wedge v_k \rangle_\wedge = \det(\langle u_i, v_j \rangle),$$

for all $u_i, v_i \in T_p^* M$, and extending $\langle -, - \rangle$ by bilinearity.

Therefore, for any two k-forms $\omega, \eta \in \mathcal{A}^k(M)$, we get the smooth function $\langle \omega, \eta \rangle$ given by

$$\langle \omega, \eta \rangle(p) = \langle \omega_p, \eta_p \rangle_p.$$

Furthermore, if M is oriented, then we can apply the results of Section 3.5, so the vector bundle $T^* M$ is oriented (by giving $T_p^* M$ the orientation induced by the orientation of $T_p M$, for every $p \in M$), and for every $p \in M$, we get a Hodge $*$-operator

$$*: \bigwedge^k T_p^* M \rightarrow \bigwedge^{n-k} T_p^* M.$$

Then given any k-form $\omega \in \mathcal{A}^k(M)$, we can define $*\omega$ by

$$(*\omega)_p = *(\omega_p), \qquad p \in M.$$

We have to check that $*\omega$ is indeed a smooth form in $\mathcal{A}^{n-k}(M)$, but this is not hard to do in local coordinates (for help, see Morita [82], Chapter 4, Section 1). Therefore, if M is a Riemannian oriented manifold of dimension n, we have Hodge $*$-operators

$$*: \mathcal{A}^k(M) \rightarrow \mathcal{A}^{n-k}(M).$$

Observe that $*1$ is just the volume form Vol_M induced by the metric. Indeed, we know from Section 2.2 that in local coordinates $x_1, \ldots, x_n$ near p, the metric on $T_p^* M$ is given by the inverse (g^{ij}) of the metric (g_{ij}) on $T_p M$, and by the results of Section 3.5 (Proposition 3.17),

$$*(1) = \frac{1}{\sqrt{\det(g^{ij})}} dx_1 \wedge \cdots \wedge dx_n$$

$$= \sqrt{\det(g_{ij})} \, dx_1 \wedge \cdots \wedge dx_n = \mathrm{Vol}_M.$$

Proposition 3.16 yields the following:

Proposition 8.5. *If M is a Riemannian oriented manifold of dimension n, then we have the following properties:*

(i) $*(f\omega + g\eta) = f * \omega + g * \eta$, *for all* $\omega, \eta \in \mathcal{A}^k(M)$ *and all* $f, g \in C^\infty(M)$.
(ii) $** = (-\mathrm{id})^{k(n-k)}$.
(iii) $\omega \wedge *\eta = \eta \wedge *\omega = \langle \omega, \eta \rangle \mathrm{Vol}_M$, *for all* $\omega, \eta \in \mathcal{A}^k(M)$.
(iv) $*(\omega \wedge *\eta) = *(\eta \wedge *\omega) = \langle \omega, \eta \rangle$, *for all* $\omega, \eta \in \mathcal{A}^k(M)$.
(v) $\langle *\omega, *\eta \rangle = \langle \omega, \eta \rangle$, *for all* $\omega, \eta \in \mathcal{A}^k(M)$.

Recall that exterior differentiation d is a map $d: \mathcal{A}^k(M) \rightarrow \mathcal{A}^{k+1}(M)$. Using the Hodge $*$-operator, we can define an operator $\delta: \mathcal{A}^k(M) \rightarrow \mathcal{A}^{k-1}(M)$ that will turn out to be adjoint to d with respect to an inner product on $\mathcal{A}^\bullet(M)$.

Definition 8.6. Let M be an oriented Riemannian manifold of dimension n. For any k, with $1 \leq k \leq n$, let

$$\delta = (-1)^{n(k+1)+1} * d * .$$

Clearly, $\delta \colon \mathcal{A}^k(M) \to \mathcal{A}^{k-1}(M)$, and $\delta = 0$ on $\mathcal{A}^0(M) = C^\infty(M)$.

Here is an example of Definition 8.6. Let $M = \mathbb{R}^3$ and $\omega = x\, dx \wedge dy$. Since $\{dx, dy, dz\}$ is an orthonormal basis of $T_p^*\mathbb{R}^3$, we apply Proposition 8.5 (i) and the calculations of Section 3.5 to discover that

$$*x\, dx \wedge dy = x * dx \wedge dy = x\, dz.$$

Then

$$d(x\, dz) = d(x) \wedge dz = dx \wedge dz,$$

and

$$* dx \wedge dz = -dy.$$

Since $n = 3$ and $k = 2$, these calculations imply that

$$\delta\, x\, dx \wedge dy = (-1)^{3(3)+1}(-dy) = -dy.$$

By using the definition of δ, the fact that $d \circ d = 0$, and Proposition 8.5 (ii), it is easy to prove the following proposition.

Proposition 8.6. *Let M be an oriented Riemannian manifold of dimension n. Let d the exterior derivative as defined in Definition 4.9. Let δ be as defined in Definition 8.6. Then*

$$*\delta = (-1)^k d*, \quad \delta* = (-1)^{k+1} * d, \quad \delta \circ \delta = 0.$$

8.3 The Hodge Laplacian and the Hodge Divergence Operators on Riemannian Manifolds

Using d and δ, we can generalize the Laplacian to an operator on differential forms.

Definition 8.7. Let M be an oriented Riemannian manifold of dimension n. The *Hodge Laplacian* or *Laplace-Beltrami operator*, for short *Laplacian*, is the operator $\Delta \colon \mathcal{A}^k(M) \to \mathcal{A}^k(M)$ defined by

$$\Delta = d\delta + \delta d.$$

A form, $\omega \in \mathcal{A}^k(M)$ such that $\Delta\omega = 0$, is a *harmonic form*. In particular, a function $f \in \mathcal{A}^0(M) = C^\infty(M)$ such that $\Delta f = 0$ is called a *harmonic function*.

To demonstrate the Hodge Laplacian, we let $M = \mathbb{R}^3$ and calculate $\Delta \omega$, where

$$\omega = f_{12}\, dx \wedge dy + f_{13}\, dx \wedge dz + f_{23}\, dy \wedge dz.$$

We first determine $d\delta\omega$. Since $n = 3$ and $k = 2$, $\delta = *d*$. Since dx, dy, dz is an orthonormal basis, we use the calculations of Section 3.5 and Proposition 8.5 (i) to determine $\delta\omega$. Note that

$$*\omega = f_{12} * dx \wedge dy + f_{13} * dx \wedge dz + f_{23} * dy \wedge dz = f_{12}\, dz - f_{13}\, dy + f_{23}\, dx.$$

Then

$$d(f_{12}\, dz - f_{13}\, dy + f_{23}\, dx) = \left(-\frac{\partial f_{13}}{\partial x} - \frac{\partial f_{23}}{\partial y}\right) dx \wedge dy + \left(\frac{\partial f_{12}}{\partial x} - \frac{\partial f_{23}}{\partial z}\right) dx \wedge dz$$
$$+ \left(\frac{\partial f_{12}}{\partial y} + \frac{\partial f_{13}}{\partial z}\right) dy \wedge dz,$$

and

$$\delta\omega = *d(f_{12}\, dz - f_{13}\, dy + f_{23}\, dx)$$
$$= \left(-\frac{\partial f_{13}}{\partial x} - \frac{\partial f_{23}}{\partial y}\right) dz - \left(\frac{\partial f_{12}}{\partial x} - \frac{\partial f_{23}}{\partial z}\right) dy + \left(\frac{\partial f_{12}}{\partial y} + \frac{\partial f_{13}}{\partial z}\right) dx.$$

Thus

$$d\delta\omega = \left(-\frac{\partial^2 f_{13}}{\partial x^2} - \frac{\partial^2 f_{23}}{\partial x \partial y} - \frac{\partial^2 f_{12}}{\partial y \partial z} - \frac{\partial^2 f_{13}}{\partial z^2}\right) dx \wedge dz$$
$$+ \left(-\frac{\partial^2 f_{13}}{\partial x \partial y} - \frac{\partial^2 f_{23}}{\partial y^2} + \frac{\partial^2 f_{12}}{\partial x \partial z} - \frac{\partial^2 f_{23}}{\partial z^2}\right) dy \wedge dz$$
$$+ \left(-\frac{\partial^2 f_{12}}{\partial x^2} + \frac{\partial^2 f_{23}}{\partial x \partial z} - \frac{\partial^2 f_{12}}{\partial y^2} - \frac{\partial^2 f_{13}}{\partial y \partial z}\right) dx \wedge dy.$$

It remains to compute $\delta d\omega$. Observe that

$$d\omega = \left(\frac{\partial f_{12}}{\partial z} - \frac{\partial f_{13}}{\partial y} + \frac{\partial f_{23}}{\partial x}\right) dx \wedge dy \wedge dz.$$

Since $d\omega$ is a three form, $\delta = (-1) * d*$. Once again we go through a three step process to calculate δ. First

$$*d\omega = \frac{\partial f_{12}}{\partial z} - \frac{\partial f_{13}}{\partial y} + \frac{\partial f_{23}}{\partial x}.$$

Next

$$d * d\omega = \left(\frac{\partial^2 f_{12}}{\partial x \partial z} - \frac{\partial^2 f_{13}}{\partial x \partial y} + \frac{\partial^2 f_{23}}{\partial x^2}\right) dx + \left(\frac{\partial^2 f_{12}}{\partial y \partial z} - \frac{\partial^2 f_{13}}{\partial y^2} + \frac{\partial^2 f_{23}}{\partial x \partial y}\right) dy$$

$$+ \left(\frac{\partial^2 f_{12}}{\partial z^2} - \frac{\partial^2 f_{13}}{\partial y \partial z} + \frac{\partial^2 f_{23}}{\partial x \partial z}\right) dz.$$

Lastly

$$\delta d\omega = (-1) * d * d\omega = -\left(\frac{\partial^2 f_{12}}{\partial x \partial z} - \frac{\partial^2 f_{13}}{\partial x \partial y} + \frac{\partial^2 f_{23}}{\partial x^2}\right) dy \wedge dz$$

$$+ \left(\frac{\partial^2 f_{12}}{\partial y \partial z} - \frac{\partial^2 f_{13}}{\partial y^2} + \frac{\partial^2 f_{23}}{\partial x \partial y}\right) dx \wedge dz - \left(\frac{\partial^2 f_{12}}{\partial z^2} - \frac{\partial^2 f_{13}}{\partial y \partial z} + \frac{\partial^2 f_{23}}{\partial x \partial z}\right) dx \wedge dy.$$

Finally we discover that

$$\Delta\omega = d\delta\omega + \delta d\omega$$

$$= \left(-\frac{\partial^2 f_{13}}{\partial x^2} - \frac{\partial^2 f_{23}}{\partial x \partial y} - \frac{\partial^2 f_{12}}{\partial y \partial z} - \frac{\partial^2 f_{13}}{\partial z^2}\right) dx \wedge dz$$

$$+ \left(-\frac{\partial^2 f_{13}}{\partial x \partial y} - \frac{\partial^2 f_{23}}{\partial y^2} + \frac{\partial^2 f_{12}}{\partial x \partial z} - \frac{\partial^2 f_{23}}{\partial z^2}\right) dy \wedge dz$$

$$+ \left(-\frac{\partial^2 f_{12}}{\partial x^2} + \frac{\partial^2 f_{23}}{\partial x \partial z} - \frac{\partial^2 f_{12}}{\partial y^2} - \frac{\partial^2 f_{13}}{\partial y \partial z}\right) dx \wedge dy$$

$$- \left(\frac{\partial^2 f_{12}}{\partial x \partial z} - \frac{\partial^2 f_{13}}{\partial x \partial y} + \frac{\partial^2 f_{23}}{\partial x^2}\right) dy \wedge dz$$

$$+ \left(\frac{\partial^2 f_{12}}{\partial y \partial z} - \frac{\partial^2 f_{13}}{\partial y^2} + \frac{\partial^2 f_{23}}{\partial x \partial y}\right) dx \wedge dz$$

$$- \left(\frac{\partial^2 f_{12}}{\partial z^2} - \frac{\partial^2 f_{13}}{\partial y \partial z} + \frac{\partial^2 f_{23}}{\partial x \partial z}\right) dx \wedge dy$$

$$= \left(-\frac{\partial^2 f_{12}}{\partial x^2} - \frac{\partial^2 f_{12}}{\partial y^2} - \frac{\partial^2 f_{12}}{\partial z^2}\right) dx \wedge dy$$

$$+ \left(-\frac{\partial^2 f_{13}}{\partial x^2} - \frac{\partial^2 f_{13}}{\partial y^2} - \frac{\partial^2 f_{13}}{\partial z^2}\right) dx \wedge dz$$

$$+ \left(-\frac{\partial^2 f_{23}}{\partial x^2} - \frac{\partial^2 f_{23}}{\partial y^2} - \frac{\partial^2 f_{23}}{\partial z^2}\right) dy \wedge dz.$$

Notice that the coefficients of the two-form $\Delta\omega$ are given by the negative of the harmonic operator on functions as defined in Section 7.5. In fact, if $M = \mathbb{R}^n$ with the Euclidean metric and f is a smooth function, a laborious computation yields

$$\Delta f = -\sum_{i=1}^{n} \frac{\partial^2 f}{\partial x_i^2};$$

that is, the usual Laplacian *with a negative sign* in front. (The computation can be found in Morita [82], Example 4.12, or Jost [62], Chapter 2, Section 2.1).

By using Proposition 8.6, it is easy to see that Δ commutes with $*$; that is,

$$\Delta* = *\Delta.$$

We have

$$
\begin{aligned}
\Delta* &= (d\delta + \delta d)* = d\delta * + \delta d* \\
&= (-1)^{k+1}d * d + (-1)^k \delta * \delta \\
&= (-1)^{k+1}(-1)^{k+1} * \delta d + (-1)^k \delta * \delta, \quad \text{since } * \text{ acts on a } k+1 \text{ form} \\
&= *\delta d + (-1)^k(-1)^k d\delta *, \quad \text{since } * \text{ acts on a } k-1 \text{ form} \\
&= *(\delta d + d\delta) = *\Delta.
\end{aligned}
$$

Definition 8.8. Let M be an oriented Riemannian manifold of dimension n. Given any vector field $X \in \mathfrak{X}(M)$, its *Hodge divergence* div X is defined by

$$\text{div } X = \delta X^\flat.$$

Now for a function $f \in C^\infty(M)$, we have $\delta f = 0$, so $\Delta f = \delta df$. However,

$$\text{div}(\text{grad } f) = \delta(\text{grad } f)^\flat = \delta((df)^\sharp)^\flat) = \delta df,$$

so

$$\Delta f = \text{div grad } f,$$

as in the case of $\mathbb{R}^n$.

Remark. Since the definition of δ involves two occurrences of the Hodge $*$-operator, δ also makes sense on non-orientable manifolds by using a local definition. Therefore, the Laplacian Δ and the divergence also make sense on non-orientable manifolds.

In the rest of this section we assume that M is orientable.

The relationship between δ and d can be made clearer by introducing an inner product on forms with compact support. Recall that $\mathcal{A}_c^k(M)$ denotes the space of k-forms with compact support (an infinite-dimensional vector space). Let $k \geq 1$.

Definition 8.9. For any two k-forms with compact support $\omega, \eta \in \mathcal{A}_c^k(M)$, set

$$(\omega, \eta) = \int_M \langle \omega, \eta \rangle \, \mathrm{Vol}_M = \int_M \langle \omega, \eta \rangle * (1).$$

If $k = 0$, then $\omega, \eta \in C^\infty(M)$ and we define

$$(\omega, \eta) = \int_M \omega \, \eta \, \mathrm{Vol}_M.$$

Using Proposition 8.5 (iii), we have

$$(\omega, \eta) = \int_M \langle \omega, \eta \rangle \, \mathrm{Vol}_M = \int_M \omega \wedge *\eta = \int_M \eta \wedge *\omega,$$

so it is easy to check that $(-, -)$ is indeed an inner product on k-forms with compact support. We can extend this inner product to forms with compact support in $\mathcal{A}_c^\bullet(M) = \bigoplus_{k=0}^n \mathcal{A}_c^k(M)$ by making $\mathcal{A}_c^h(M)$ and $\mathcal{A}_c^k(M)$ orthogonal if $h \neq k$.

Proposition 8.7. *If M is an n-dimensional orientable Riemannian manifold, then δ is (formally) adjoint to d; that is,*

$$(d\omega, \eta) = (\omega, \delta\eta),$$

for all $\omega \in \mathcal{A}_c^{k-1}(M)$ and $\eta \in \mathcal{A}_c^k(M)$ with compact support.

Proof. By linearity and orthogonality of the $\mathcal{A}_c^k(M)$, the proof reduces to the case where $\omega \in \mathcal{A}_c^{k-1}(M)$ and $\eta \in \mathcal{A}_c^k(M)$ (both with compact support). By definition of δ and the fact that

$$** = (-\mathrm{id})^{(k-1)(n-k+1)}$$

for $*\colon \mathcal{A}^{k-1}(M) \to \mathcal{A}^{n-k+1}(M)$, we have

$$*\delta = (-1)^k d*,$$

and since

$$d(\omega \wedge *\eta) = d\omega \wedge *\eta + (-1)^{k-1} \omega \wedge d * \eta$$
$$= d\omega \wedge *\eta - \omega \wedge *\delta\eta$$

we get

$$\int_M d(\omega \wedge *\eta) = \int_M d\omega \wedge *\eta - \int_M \omega \wedge *\delta\eta$$
$$= (d\omega, \eta) - (\omega, \delta\eta).$$

However, by Stokes' theorem (Theorem 6.16),

$$\int_M d(\omega \wedge *\eta) = 0,$$

so $(d\omega, \eta) - (\omega, \delta\eta) = 0$; that is, $(d\omega, \eta) = (\omega, \delta\eta)$, as claimed.   □

Corollary 8.8. *If M is an n-dimensional orientable Riemannian manifold, then the Laplacian Δ is self-adjoint; that is,*

$$(\Delta\omega, \eta) = (\omega, \Delta\eta),$$

for all k-forms ω, η with compact support.

Proof. Using Proposition 8.7 several times we have

$$(\Delta\omega, \eta) = (d\delta\omega + \delta d\omega, \eta)$$
$$= (d\delta\omega, \eta) + (\delta d\omega, \eta)$$
$$= (\delta\omega, \delta\eta) + (d\omega, d\eta)$$
$$= (\omega, d\delta\eta) + (\omega, \delta d\eta)$$
$$= (\omega, d\delta\eta + \delta d\eta) = (\omega, \Delta\eta).$$

□

We also obtain the following useful fact:

Proposition 8.9. *If M is an n-dimensional orientable Riemannian manifold, then for every k-form ω with compact support, $\Delta\omega = 0$ iff $d\omega = 0$ and $\delta\omega = 0$.*

Proof. Since $\Delta = d\delta + \delta d$, it is obvious that if $d\omega = 0$ and $\delta\omega = 0$, then $\Delta\omega = 0$. Conversely,

$$(\Delta\omega, \omega) = ((d\delta + \delta d)\omega, \omega) = (d\delta\omega, \omega) + (\delta d\omega, \omega) = (\delta\omega, \delta\omega) + (d\omega, d\omega).$$

Thus, if $\Delta\omega = 0$, then $(\delta\omega, \delta\omega) = (d\omega, d\omega) = 0$, which implies $d\omega = 0$ and $\delta\omega = 0$.   □

As a consequence of Proposition 8.9, if M is a connected, orientable, compact Riemannian manifold, then every harmonic function on M is a constant. Indeed, if M is compact, then f is a 0-form of compact support, and if $\Delta f = 0$, then $df = 0$. Since f is connected, f is a constant function.

8.4   The Hodge and Laplace–Beltrami Laplacians of Functions

For practical reasons we need a formula for the Hodge Laplacian of a function
$f \in C^\infty(M)$, in local coordinates. If (U, φ) is a chart near p, as usual, let

$$\frac{\partial f}{\partial x_j}(p) = \frac{\partial(f \circ \varphi^{-1})}{\partial u_j}(\varphi(p)),$$

where $(u_1, \ldots, u_n)$ are the coordinate functions in $\mathbb{R}^n$. Write $|g| = \det(g_{ij})$, where
(g_{ij}) is the symmetric, positive definite matrix giving the metric in the chart (U, φ).

Proposition 8.10. *If M is an n-dimensional orientable Riemannian manifold, then
for every local chart (U, φ), for every function $f \in C^\infty(M)$, we have*

$$\Delta f = -\frac{1}{\sqrt{|g|}} \sum_{i,j} \frac{\partial}{\partial x_i}\left(\sqrt{|g|}\, g^{ij}\, \frac{\partial f}{\partial x_j}\right).$$

Proof. We follow Jost [62], Chapter 2, Section 1. Pick any function $h \in C^\infty(M)$
with compact support. We have

$$\int_M (\Delta f)h * (1) = (\Delta f, h)$$

$$= (\delta df, h)$$

$$= (df, dh)$$

$$= \int_M \langle df, dh \rangle * (1)$$

$$= \int_M \sum_{ij} g^{ij} \frac{\partial f}{\partial x_i} \frac{\partial h}{\partial x_j} \sqrt{|g|}\, dx_1 \cdots dx_n$$

$$= -\int_M \sum_{ij} \frac{1}{\sqrt{|g|}} \frac{\partial}{\partial x_j}\left(\sqrt{|g|}\, g^{ij} \frac{\partial f}{\partial x_i}\right) h \sqrt{|g|}\, dx_1 \cdots dx_n$$

$$= -\int_M \sum_{ij} \frac{1}{\sqrt{|g|}} \frac{\partial}{\partial x_j}\left(\sqrt{|g|}\, g^{ij} \frac{\partial f}{\partial x_i}\right) h * (1),$$

where we have used integration by parts in the second to last line. Since the above
equation holds for all h, we get our result.   □

It turns out that in a Riemannian manifold, the divergence of a vector field and
the Laplacian of a function can be given by a definition that uses the covariant

derivative instead of the Hodge $*$-operator. We did this in Section 7.4. A comparison of Proposition 8.10 with Line $(**)$ of Section 7.4 shows that the definition of the Hodge Laplacian of a function differs by a sign factor with the definition of the Laplacian provided by Definition 7.12. We reconcile the difference between these two definitions by defining the notion of *connection divergence* and *connection Laplacian* via the negation of the quantity described in Definition 7.12.

Definition 8.10. Let M be a Riemannian manifold. If ∇ is the Levi-Civita connection induced by the Riemannian metric, then the *connection divergence* (for short *divergence*) of a vector field $X \in \mathfrak{X}(M)$ is the function $\operatorname{div}_C X \colon M \to \mathbb{R}$ defined so that

$$(\operatorname{div}_C X)(p) = \operatorname{tr}(Y(p) \mapsto (-\nabla_Y X)_p);$$

namely, for every p, $(\operatorname{div}_C X)(p)$ is the trace of the linear map $Y(p) \mapsto (-\nabla_Y X)_p$. The *connection Laplacian* of $f \in C^\infty M$ is defined as

$$\Delta_C f = \operatorname{div}_C \operatorname{grad} f.$$

The connection divergence and the connection Laplacian make sense even if M is non-orientable. This is also true for the Hodge divergence and the Hodge Laplacian. Because of the sign change provided by Definition 8.10, the Hodge Laplacian Δf agrees with the connection Laplacian $\Delta_C f$. Thus, we will not distinguish between the two notions of Laplacian on a function.

Since the connection Laplacian and the Hodge Laplacian (for functions) agree, we should expect that the two variants of the divergence operator also agree. This is indeed the case but a proof is not so easily found in the literature. We are aware of two proofs: one is found in Petersen [86] (Chapter 7, Proposition 32) for compact orientable manifolds, and the other in Rosenberg [90] for orientable manifolds, closer to the proof of Proposition 8.10. We present the second proof because it applies to a more general situation and yields an explicit formula.

Proposition 8.11. *If M is an n-dimensional orientable Riemannian manifold, then for every local chart (U, φ), for every vector field $X \in \mathfrak{X}(M)$, we have*

$$\operatorname{div} X = -\sum_{i=1}^{n} \frac{1}{\sqrt{|g|}} \frac{\partial}{\partial x_i} \left(\sqrt{|g|} X_i \right).$$

Proof. (Following Rosenberg [90].) Let (U, φ) be a chart for M. Within this chart, any $X \in \mathfrak{X}(M)$ is expressed as $X = \sum_{i=1}^{n} X_i \frac{\partial}{\partial x_i}$. Take $f \in C^\infty(M)$ with compact support and compute

$$(X, \operatorname{grad} f) = \int_M \langle X, \operatorname{grad} f \rangle * (1)$$

$$= \int_M df(X) * (1)$$

$$= \int_M \sum_{i=1}^n X_i \frac{\partial f}{\partial x_i} \sqrt{|g|}\, dx_1 \cdots dx_n$$

$$= -\int_M \sum_{i=1}^n \frac{1}{\sqrt{|g|}} \frac{\partial}{\partial x_i} \left(\sqrt{|g|}X_i\right) f \sqrt{|g|}\, dx_1 \cdots dx_n,$$

where the last equality follows from integration by parts. We claim $(X, \operatorname{grad} f) = (\operatorname{div} X, f)$ since

$$(\operatorname{div} X, f) = (\delta X^\flat, f)$$

$$= (X^\flat, df), \qquad \text{by Proposition 8.7}$$

$$= ((X^\flat)^\sharp, (df)^\sharp), \qquad \text{definition of inner product on one forms}$$

$$= (X, (df)^\sharp)$$

$$= (X, \operatorname{grad} f), \qquad \text{by the remark preceding Definition 8.1.}$$

Thus we have shown

$$(\operatorname{div} X, f) = -\int_M \sum_{i=1}^n \frac{1}{\sqrt{|g|}} \frac{\partial}{\partial x_i} \left(\sqrt{|g|}X_i\right) f * (1) = \left\langle -\sum_{i=1}^n \frac{1}{\sqrt{|g|}} \frac{\partial}{\partial x_i} \left(\sqrt{|g|}X_i\right), f \right\rangle$$

for all $f \in C^\infty(M)$ with compact support, and this concludes the proof. $\square$

By comparing the expression for div X provided by Proposition 8.11 with the expression of $\operatorname{div}_C X$ given by Line (†) of Section 7.4, we have the following proposition.

Proposition 8.12. *If M is an orientable Riemannian manifold, then for every vector field $X \in \mathfrak{X}(M)$, the connection divergence is given by*

$$\operatorname{div}_C X = \delta X^\flat = \operatorname{div} X.$$

Consequently, for the Laplacian, we have

$$\Delta f = \delta df = \operatorname{div} \operatorname{grad} f.$$

Proposition 8.12 shows there is no need to distinguish between the Hodge divergence and the connection divergence. Thus we will use the notation div X to simply denote the divergence of a vector field over $T(M)$.

Our next result shows relationship between div X and the Lie derivative of the volume form.

8.5   Divergence and Lie Derivative of the Volume Form

Proposition 8.13. *Let* M *be an* n*-dimensional Riemannian manifold. For any vector field* $X \in \mathfrak{X}(M)$, *we have*

$$L_X \operatorname{Vol}_M = -(\operatorname{div} X)\operatorname{Vol}_M,$$

where $\operatorname{div} X$ *is the connection divergence of* X.

Proof. (Following O'Neill [84] (Chapter 7, Lemma 21).) Let $X_1, X_2, \ldots X_n$ be an orthonormal frame on M such that $\operatorname{Vol}_M(X_1, \ldots, X_n) = 1$. Then $L_X(\operatorname{Vol}_M(X_1, \ldots, X_n)) = L_X(1) = X(1) = 0$, and Proposition 4.19 (2) implies

$$(L_X \operatorname{Vol}_M)(X_1, \ldots, X_n) = -\sum_{i=1}^n \operatorname{Vol}_M(X_1, \ldots, L_X X_i, \ldots X_n).$$

Fix i and set $L_X X_i = [X, X_i] = \sum_{j=1}^n f_{ij} X_j$. Since Vol_M is multilinear and skew-symmetric, we find that

$$\operatorname{Vol}_M(X_1, \ldots, L_X X_i, \ldots X_n) = \operatorname{Vol}_M(X_1, \ldots, \sum_{j=1}^n f_{ij} X_j, \ldots X_n)$$

$$= \sum_{j=1}^n \operatorname{Vol}_M(X_1, \ldots, f_{ij} X_j, \ldots X_n)$$

$$= f_{ii} \operatorname{Vol}_M(X_1, \ldots X_i, \ldots X_n) = f_{ii}.$$

By varying i we discover that

$$(L_X \operatorname{Vol}_M)(X_1, \ldots, X_n) = -\sum_{i=1}^n \operatorname{Vol}_M(X_1, \ldots, L_X X_i, \ldots X_n) = -\sum_{i=1}^n f_{ii}.$$

On the other hand, since $(\operatorname{div} X)(p) = \operatorname{tr}(Y(p) \mapsto (-\nabla_Y X)_p)$, $X_1, \ldots, X_n$ is an orthonormal frame, and ∇ is the Levi-Civita connection (which is torsion free), the equation before Definition 2.2 implies that

$$-\operatorname{div} X = \sum_{i=1}^n \langle \nabla_{X_i} X, X_i \rangle$$

$$= -\sum_{i=1}^n \langle [X, X_i], X_i \rangle + \sum_{i=1}^n \langle \nabla_X X_i, X_i \rangle, \quad \text{since } \nabla_{X_i} X - \nabla_X X_i = [X_i, X]$$

$$= -\sum_{i=1}^{n} \langle [X, X_i], X_i \rangle, \qquad \text{since } 0 = \nabla_X \langle X_i, X_i \rangle = 2 \langle \nabla_X X_i, X_i \rangle$$

$$= -\sum_{i=1}^{n} \langle \sum_{j=1}^{n} f_{ij} X_j, X_i \rangle = -\sum_{i=1}^{n} f_{ii} \langle X_i, X_i \rangle = -\sum_{i=1}^{n} f_{ii}.$$

Thus we have shown

$$-\mathrm{div}\, X = (L_X \mathrm{Vol}_M)(X_1, \ldots, X_n),$$

which is equivalent to the statement found in the proposition.   □

Proposition 8.13 is interesting in its own right since it is used in the proof of Green's theorem. But before stating and proving Green's theorem, we reformulate Proposition 8.13 through the application of Cartan's formula.

Proposition 8.14. *The following formula holds:*

$$(\mathrm{div}\, X)\mathrm{Vol}_M = -d(i(X)\mathrm{Vol}_M).$$

Proof. By Cartan's formula (Proposition 4.20), $L_X = i(X) \circ d + d \circ i(X)$; as $d\mathrm{Vol}_M = 0$ (since Vol_M is a top form), Proposition 8.13 implies

$$(\mathrm{div}\, X)\mathrm{Vol}_M = -d(i(X)\mathrm{Vol}_M).$$

□

The above formulae also holds for a local volume form (*i.e.*, for a volume form on a local chart).

Proposition 8.15 (Green's Formula). *If M is an orientable and compact Riemannian manifold without boundary, then for every vector field $X \in \mathfrak{X}(M)$, we have*

$$\int_M (\mathrm{div}\, X)\, \mathrm{Vol}_M = 0.$$

Proof. Proofs of Proposition 8.15 can be found in Gallot, Hulin, and Lafontaine [48] (Chapter 4, Proposition 4.9) and Helgason [55] (Chapter 2, Section 2.4). Since Proposition 8.13 implies that

$$(\mathrm{div}\, X)\mathrm{Vol}_M = -d(i(X)\mathrm{Vol}_M),$$

we have

$$\int_M (\mathrm{div}\, X)\, \mathrm{Vol}_M = -\int_M d(i(X)\mathrm{Vol}_M) = -\int_{\partial M} i(X)\mathrm{Vol}_M = 0$$

where the last equality follows by Stokes' theorem, since $\partial M = 0$. $\qquad\square$

We end this section by discussing an alternative definition for the operator $\delta\colon \mathcal{A}^1(M) \to \mathcal{A}^0(M)$ in terms of the covariant derivative (see Gallot, Hulin, and Lafontaine [48], Chapter 4). For any one-form $\omega \in \mathcal{A}^1(M)$, and any $X, Y \in \mathfrak{X}(M)$, define

$$(\nabla_X \omega)(Y) := X(\omega(Y)) - \omega(\nabla_X Y).$$

It turns out that

$$\delta\omega = -\mathrm{tr}\, \nabla\omega,$$

where the trace should be interpreted as the trace of the $\mathbb{R}$-bilinear map $X, Y \mapsto (\nabla_X \omega)(Y)$, as in Chapter 2 (see Proposition 2.3). This means that in any chart (U, φ),

$$\delta\omega = -\sum_{i=1}^{n}(\nabla_{E_i}\omega)(E_i),$$

for any orthonormal frame field $(E_1, \ldots, E_n)$ over U. By applying this trace definition of $\delta\omega$, it can be shown that

$$\delta(f\,df) = f\Delta f - \langle \mathrm{grad}\, f, \mathrm{grad}\, f\rangle.$$

Proposition 8.16. *For any orientable, compact manifold M, we have*

$$(\Delta f, f) = \int_M f\Delta f \,\mathrm{Vol}_M = \int_M \langle \mathrm{grad}\, f, \mathrm{grad}\, f\rangle \mathrm{Vol}_M.$$

Proof. Proposition 8.12 implies that

$$\delta(f\,df) = \delta((f\,df)^\sharp)^\flat) = \mathrm{div}(f\,df)^\sharp,$$

and since Green's formula implies that

$$\int_M \delta(f\,df)\,\mathrm{Vol}_M = \int_M \mathrm{div}(f\,df)^\sharp \,\mathrm{Vol}_M = 0,$$

we conclude that

$$(\Delta f, f) = \int_M f \Delta f \, \mathrm{Vol}_M = \int_M \langle \mathrm{grad}\, f, \mathrm{grad}\, f \rangle \mathrm{Vol}_M,$$

for any orientable, compact manifold M.   □

There is a generalization of the formula expressing $\delta\omega$ over an orthonormal frame $E_1, \ldots, E_n$ for a one-form ω that applies to any differential form. In fact, there are formulae expressing both d and δ over an orthonormal frame and its coframe, and these are often handy in proofs. The formula for $\delta\omega$ will be used in the proof of Theorem 8.26.

Recall that for every vector field $X \in \mathfrak{X}(M)$, the interior product $i(X) \colon \mathcal{A}^{k+1}(M) \to \mathcal{A}^k(M)$ is defined by

$$(i(X)\omega)(Y_1, \ldots, Y_k) = \omega(X, Y_1, \ldots, Y_k),$$

for all $Y_1, \ldots, Y_k \in \mathfrak{X}(M)$.

Proposition 8.17. *Let M be a compact, orientable Riemannian manifold. For every $p \in M$, for every local chart (U, φ) with $p \in M$, if $(E_1, \ldots, E_n)$ is an orthonormal frame over U and $(\theta_1, \ldots, \theta_n)$ is its dual coframe, then for every k-form $\omega \in \mathcal{A}^k(M)$, we have:*

$$d\omega = \sum_{i=1}^n \theta_i \wedge \nabla_{E_i}\omega$$

$$\delta\omega = -\sum_{i=1}^n i(E_i)\nabla_{E_i}\omega.$$

A proof of Proposition 8.17 can be found in Petersen [86] (Chapter 7, Proposition 37) or Jost [62] (Chapter 3, Lemma 3.3.4). When ω is a one-form, $\delta\omega_p$ is just a number, and indeed

$$\delta\omega = -\sum_{i=1}^n i(E_i)\nabla_{E_i}\omega = -\sum_{i=1}^n (\nabla_{E_i}\omega)(E_i),$$

as stated earlier.

8.6   Harmonic Forms, the Hodge Theorem, and Poincaré Duality

Let us now assume that M is orientable and compact.

Definition 8.11. Let M be an orientable and compact Riemannian manifold of dimension n. For every k, with $0 \leq k \leq n$, let

$$\mathbb{H}^k(M) = \{\omega \in \mathcal{A}^k(M) \mid \Delta\omega = 0\},$$

the space of *harmonic k-forms*.

The following proposition is left as an easy exercise:

Proposition 8.18. *Let M be an orientable and compact Riemannian manifold of dimension n. The Laplacian commutes with the Hodge $*$-operator, and we have a linear map*

$$*: \mathbb{H}^k(M) \to \mathbb{H}^{n-k}(M).$$

One of the deepest and most important theorems about manifolds is the *Hodge decomposition theorem*, which we now state.

Theorem 8.19 (Hodge Decomposition Theorem). *Let M be an orientable and compact Riemannian manifold of dimension n. For every k, with $0 \leq k \leq n$, the space $\mathbb{H}^k(M)$ is finite dimensional, and we have the following orthogonal direct sum decomposition of the space of k-forms:*

$$\mathcal{A}^k(M) = \mathbb{H}^k(M) \oplus d(\mathcal{A}^{k-1}(M)) \oplus \delta(\mathcal{A}^{k+1}(M)).$$

The proof of Theorem 8.19 involves a lot of analysis and it is long and complicated. A complete proof can be found in Warner [109] (Chapter 6). Other treatments of Hodge theory can be found in Morita [82] (Chapter 4) and Jost [62] (Chapter 2).

The Hodge decomposition theorem has a number of important corollaries, one of which is *Hodge theorem*:

Theorem 8.20 (Hodge Theorem). *Let M be an orientable and compact Riemannian manifold of dimension n. For every k, with $0 \leq k \leq n$, there is an isomorphism between $\mathbb{H}^k(M)$ and the de Rham cohomology vector space $H_{DR}^k(M)$:*

$$H_{DR}^k(M) \cong \mathbb{H}^k(M).$$

Proof. Since by Proposition 8.9, every harmonic form $\omega \in \mathbb{H}^k(M)$ is closed, we get a linear map from $\mathbb{H}^k(M)$ to $H_{DR}^k(M)$ by assigning its cohomology class $[\omega]$ to ω. This map is injective. Indeed, if $[\omega] = 0$ for some $\omega \in \mathbb{H}^k(M)$, then $\omega = d\eta$ for some $\eta \in \mathcal{A}^{k-1}(M)$ so

$$(\omega, \omega) = (d\eta, \omega) = (\eta, \delta\omega).$$

But, as $\omega \in \mathbb{H}^k(M)$ we have $\delta\omega = 0$ by Proposition 8.9, so $(\omega, \omega) = 0$; that is, $\omega = 0$.

Our map is also surjective. This is the hard part of Hodge theorem. By the Hodge decomposition theorem, for every closed form $\omega \in \mathcal{A}^k(M)$, we can write

$$\omega = \omega_H + d\eta + \delta\theta,$$

with $\omega_H \in \mathbb{H}^k(M)$, $\eta \in \mathcal{A}^{k-1}(M)$, and $\theta \in \mathcal{A}^{k+1}(M)$. Since ω is closed and $\omega_H \in \mathbb{H}^k(M)$, we have $d\omega = 0$ and $d\omega_H = 0$, thus

$$d\delta\theta = 0$$

and so

$$0 = (d\delta\theta, \theta) = (\delta\theta, \delta\theta);$$

that is, $\delta\theta = 0$. Therefore, $\omega = \omega_H + d\eta$, which implies $[\omega] = [\omega_H]$ with $\omega_H \in \mathbb{H}^k(M)$, proving the surjectivity of our map.   □

The Hodge theorem also implies the *Poincaré duality theorem*. If M is a compact, orientable, n-dimensional smooth manifold, for each k, with $0 \leq k \leq n$, we define a bilinear map

$$((-, -)) \colon H_{\mathrm{DR}}^k(M) \times H_{\mathrm{DR}}^{n-k}(M) \longrightarrow \mathbb{R}$$

by setting

$$(([\omega], [\eta])) = \int_M \omega \wedge \eta.$$

We need to check that this definition does not depend on the choice of closed forms in the cohomology classes $[\omega]$ and $[\eta]$. However, if $\omega + d\alpha$ is another representative in $[\omega]$ and $\eta + d\beta$ is another representative in $[\eta]$, as $d\omega = d\eta = 0$, we have

$$d(\alpha \wedge \eta + (-1)^k \omega \wedge \beta + \alpha \wedge d\beta) = d\alpha \wedge \eta + \omega \wedge d\beta + d\alpha \wedge d\beta,$$

so by Stokes' theorem,

$$\int_M (\omega + d\alpha) \wedge (\eta + d\beta) = \int_M \omega \wedge \eta + \int_M d(\alpha \wedge \eta + (-1)^k \omega \wedge \beta + \alpha \wedge d\beta)$$

$$= \int_M \omega \wedge \eta.$$

Theorem 8.21 (Poincaré Duality). *If M is a compact, orientable, smooth manifold of dimension n, then the bilinear map*

$$((-,-)) \colon H_{\mathrm{DR}}^k(M) \times H_{\mathrm{DR}}^{n-k}(M) \longrightarrow \mathbb{R}$$

defined above is a nondegenerate pairing, and hence yields an isomorphism

$$H_{\mathrm{DR}}^k(M) \cong (H_{\mathrm{DR}}^{n-k}(M))^*.$$

Proof. Pick any Riemannian metric on M. It is enough to show that for every nonzero cohomology class $[\omega] \in H_{\mathrm{DR}}^k(M)$, there is some $[\eta] \in H_{\mathrm{DR}}^{n-k}(M)$ such that

$$(([\omega], [\eta])) = \int_M \omega \wedge \eta \neq 0.$$

By the Hodge theorem, we may assume that ω is a nonzero harmonic form. By Proposition 8.18, $\eta = *\omega$ is also harmonic and $\eta \in \mathbb{H}^{n-k}(M)$. Then, we get

$$(\omega, \omega) = \int_M \omega \wedge *\omega = (([\omega], [\eta])),$$

and indeed, $(([\omega], [\eta])) \neq 0$, since $\omega \neq 0$. $\qquad\qquad\qquad\qquad\qquad\qquad \square$

8.7 The Bochner Laplacian, Weitzenböck Formula, and the Bochner Technique

Let M be a compact orientable Riemannian manifold.[1] The goal of this section is to define another notion of Laplacian on k-forms in terms of a generalization of the Levi-Civita connection $\nabla \colon \mathfrak{X}(M) \times \mathfrak{X}(M) \to \mathfrak{X}(M)$ to k-forms viewed as a linear map

$$\nabla \colon \mathcal{A}^k(M) \to \mathrm{Hom}_{C^\infty(M)}(\mathfrak{X}(M), \mathcal{A}^k(M)),$$

and in terms of a certain adjoint ∇^* of ∇, a linear map

$$\nabla^* \colon \mathrm{Hom}_{C^\infty(M)}(\mathfrak{X}(M), \mathcal{A}^k(M)) \to \mathcal{A}^k(M).$$

Since we already have an inner product $(-,-)$ on k-forms as explained in Section 8.3, we will define an inner product $((-,-))$ on $\mathrm{Hom}_{C^\infty(M)}(\mathfrak{X}(M), \mathcal{A}^k(M))$ and define ∇^* so that

---

[1] The Bochner Laplacian makes sense for noncompact manifolds as long as we consider forms with compact support, but we have no need for this more general setting.

$$(\nabla^* A, \omega) = ((A, \nabla \omega))$$

for all $A \in \operatorname{Hom}_{C^\infty(M)}(\mathfrak{X}(M), \mathcal{A}^k(M))$ and all $\omega \in \mathcal{A}^k(M)$.

Our exposition is heavily inspired by Petersen [86] (Chapter 7, Section 3.2), but Petersen deals with the more general case of a vector bundle and we restrict our attention to the simpler case of a Riemannian manifold.

The definition of the inner product $((-, -))$ on $\operatorname{Hom}_{C^\infty(M)}(\mathfrak{X}(M), \mathcal{A}^k(M))$ is accomplished in four steps.

1. First, we define the connection $\nabla \colon \mathcal{A}^k(M) \to \operatorname{Hom}_{C^\infty(M)}(\mathfrak{X}(M), \mathcal{A}^k(M))$ on k-forms. We define the covariant derivative $\nabla_X \omega$ of any k-form $\omega \in \mathcal{A}^k(M)$ as the k-form given by

$$(\nabla_X \omega)(Y_1, \ldots, Y_k) = X(\omega(Y_1, \ldots, Y_k)) - \sum_{j=1}^{k} \omega(Y_1, \ldots, \nabla_X Y_j, \ldots, Y_k);$$

$$(\dagger)$$

see Proposition 9.13 for a justification. We can view ∇ as a linear map

$$\nabla \colon \mathcal{A}^k(M) \to \operatorname{Hom}_{C^\infty(M)}(\mathfrak{X}(M), \mathcal{A}^k(M)),$$

where $\nabla \omega$ is the $C^\infty(M)$-linear map $X \mapsto \nabla_X \omega$.

2. The second step is to define the adjoint of a linear map in $\operatorname{Hom}_{C^\infty(M)}(\mathfrak{X}(M), \mathcal{A}^k(M))$. We use two inner products, one on differential forms and one on vector fields.

The inner product $\langle -, - \rangle_p$ on $T_p M$ (with $p \in M$) induces an inner product on differential forms, namely

$$(\omega, \eta) = \int_M \langle \omega, \eta \rangle \operatorname{Vol}_M = \int_M \langle \omega, \eta \rangle * (1),$$

as we explained in Section 8.3.

We also obtain an inner product on vector fields if, for any two vector field $X, Y \in \mathfrak{X}(M)$, we define $(X, Y)_{\mathfrak{X}}$ by

$$(X, Y)_{\mathfrak{X}} = \int_M \langle X, Y \rangle \operatorname{Vol}_M,$$

where $\langle X, Y \rangle$ is the function defined pointwise by

$$\langle X, Y \rangle(p) = \langle X(p), Y(p) \rangle_p.$$

Now for any linear map $A \in \operatorname{Hom}_{C^\infty(M)}(\mathfrak{X}(M), \mathcal{A}^k(M))$, let A^* be the adjoint of A defined by

$$(AX, \theta) = (X, A^*\theta)_{\mathfrak{X}},$$

for all vector fields $X \in \mathfrak{X}(M)$ and all k-forms $\theta \in \mathcal{A}^k(M)$. It can be verified that $A^* \in \mathrm{Hom}_{C^\infty(M)}(\mathcal{A}^k(M), \mathfrak{X}(M))$.

3. In the third step, given $A, B \in \mathrm{Hom}_{C^\infty(M)}(\mathfrak{X}(M), \mathcal{A}^k(M))$, the expression $\mathrm{tr}(A^*B)$ is a smooth function on M, and it can be verified that

$$\langle\langle A, B\rangle\rangle = \mathrm{tr}(A^*B)$$

defines a nondegenerate pairing on $\mathrm{Hom}_{C^\infty(M)}(\mathfrak{X}(M), \mathcal{A}^k(M))$. Using this pairing, we obtain the ($\mathbb{R}$-valued) inner product on $\mathrm{Hom}_{C^\infty(M)}(\mathfrak{X}(M), \mathcal{A}^k(M))$ given by

$$((A, B)) = \int_M \mathrm{tr}(A^*B)\,\mathrm{Vol}_M.$$

4. The fourth and final step is to define the (formal) adjoint ∇^* of $\nabla\colon \mathcal{A}^k(M) \to \mathrm{Hom}_{C^\infty(M)}(\mathfrak{X}(M), \mathcal{A}^k(M))$ as the linear map $\nabla^*\colon \mathrm{Hom}_{C^\infty(M)}(\mathfrak{X}(M), \mathcal{A}^k(M)) \to \mathcal{A}^k(M)$ defined implicitly by

$$(\nabla^* A, \omega) = ((A, \nabla\omega));$$

that is,

$$(\nabla^* A, \omega) = \int_M \langle \nabla^* A, \omega\rangle\,\mathrm{Vol}_M = \int_M \langle\langle A, \nabla\omega\rangle\rangle\,\mathrm{Vol}_M$$

$$= \int_M \mathrm{tr}(A^*\nabla\omega)\,\mathrm{Vol}_M = ((A, \nabla\omega)),$$

for all $A \in \mathrm{Hom}_{C^\infty(M)}(\mathfrak{X}(M), \mathcal{A}^k(M))$ and all $\omega \in \mathcal{A}^k(M)$.

 The notation ∇^* for the adjoint of ∇ should not be confused with the dual connection on T^*M of a connection ∇ on TM! In the second interpretation, ∇^* denotes the connection on $\mathcal{A}^*(M)$ induced by the original connection ∇ on TM. The argument type (differential form or vector field) should make it clear which ∇ is intended, but it might have been better to use a notation such as $\nabla^\top$ instead of ∇^*.

What we just did also applies to $\mathcal{A}^*(M) = \bigoplus_{k=0}^n \mathcal{A}^k(M)$ (where $\dim(M) = n$), and so we can view the connection ∇ as a linear map $\nabla\colon \mathcal{A}^*(M) \to \mathrm{Hom}_{C^\infty(M)}(\mathfrak{X}(M), \mathcal{A}^*(M))$, and its adjoint as a linear map $\nabla^*\colon \mathrm{Hom}_{C^\infty(M)}(\mathfrak{X}(M), \mathcal{A}^*(M)) \to \mathcal{A}^*(M)$.

Definition 8.12. Given a compact, orientable Riemannian manifold M, the *Bochner Laplacian* (or *connection Laplacian*) $\nabla^*\nabla$ is defined as the composition of the connection $\nabla\colon \mathcal{A}^*(M) \to \mathrm{Hom}_{C^\infty(M)}(\mathfrak{X}(M), \mathcal{A}^*(M))$ with its adjoint $\nabla^*\colon \mathrm{Hom}_{C^\infty(M)}(\mathfrak{X}(M), \mathcal{A}^*(M)) \to \mathcal{A}^*(M)$, as defined above.

Observe that

$$(\nabla^*\nabla\omega, \omega) = ((\nabla\omega, \nabla\omega)) = \int_M \langle\langle\nabla\omega, \nabla\omega\rangle\rangle \, \mathrm{Vol}_M,$$

for all $\omega \in \mathcal{A}^k(M)$. Consequently, the "harmonic forms" ω with respect to $\nabla^*\nabla$ must satisfy

$$\nabla\omega = 0,$$

but this condition is not equivalent to the harmonicity of ω with respect to the Hodge Laplacian.

Thus, *in general, $\nabla^*\nabla$ and Δ are different operators.* The relationship between the two is given by formulae involving contractions of the curvature tensor and are known as *Weitzenböck formulae.* We will state such a formula in case of one-forms later on. In order to do this, we need to give another definition of the Bochner Laplacian using *second covariant derivatives* of forms.

If $\omega \in \mathcal{A}^1(M)$ is a one-form, then the covariant derivative of ω defines a $(0, 2)$-tensor T given by $T(Y, Z) = (\nabla_Y\omega)(Z)$. Thus, we can define the second covariant derivative $\nabla^2_{X,Y}\omega$ of ω as the covariant derivative of T (see Proposition 9.13); that is,

$$\nabla^2_{X,Y}\omega = (\nabla_X T)(Y, Z) = X(T(Y, Z)) - T(\nabla_X Y, Z) - T(Y, \nabla_X Z).$$

Proposition 8.22. *The following formula holds for any one-form $\omega \in \mathcal{A}^1(M)$:*

$$\nabla^2_{X,Y}\omega = \nabla_X(\nabla_Y\omega) - \nabla_{\nabla_X Y}\omega;$$

Proof. We have

$$(\nabla^2_{X,Y}\omega)(Z) = X((\nabla_Y\omega)(Z)) - (\nabla_{\nabla_X Y}\omega)(Z) - (\nabla_Y\omega)(\nabla_X Z)$$

$$= X((\nabla_Y\omega)(Z)) - (\nabla_Y\omega)(\nabla_X Z) - (\nabla_{\nabla_X Y}\omega)(Z)$$

$$= X(\beta(Z)) - \beta(\nabla_X Z) - (\nabla_{\nabla_X Y}\omega)(Z), \quad \beta \text{ is the one-form } \nabla_Y\omega$$

$$= \nabla_X\beta(Z) - (\nabla_{\nabla_X Y}\omega)(Z), \quad \text{definition of covariant derivative}$$

$$\text{given by } (\dagger)$$

$$= (\nabla_X(\nabla_Y\omega))(Z) - (\nabla_{\nabla_X Y}\omega)(Z).$$

Therefore,

$$\nabla^2_{X,Y}\omega = \nabla_X(\nabla_Y\omega) - \nabla_{\nabla_X Y}\omega,$$

as claimed. $\square$

Note that $\nabla^2_{X,Y}\omega$ is formally the same as the second covariant derivative $\nabla^2_{X,Y}Z$ with ω replacing Z; see Gallier and Quaintance [47].

It is natural to generalize the second covariant derivative $\nabla^2_{X,Y}$ to k-forms as follows.

Definition 8.13. Given any k-form $\omega \in \mathcal{A}^k(M)$, for any two vector fields $X, Y \in \mathfrak{X}(M)$, we define $\nabla^2_{X,Y}\omega$ by

$$\nabla^2_{X,Y}\omega = \nabla_X(\nabla_Y\omega) - \nabla_{\nabla_X Y}\omega.$$

We also need the definition of the trace of $\nabla^2_{X,Y}\omega$.

Definition 8.14. Given any local chart (U, φ) and given any orthonormal frame $(E_1, \ldots, E_n)$ over U, we can define the *trace* $\mathrm{tr}(\nabla^2\omega)$ of $\nabla^2_{X,Y}\omega$ given by

$$\mathrm{tr}(\nabla^2\omega) = \sum_{i=1}^{n} \nabla^2_{E_i,E_i}\omega.$$

It is easily seen that $\mathrm{tr}(\nabla^2\omega)$ does not depend on the choice of local chart and orthonormal frame.

By using this notion of trace, we may calculate the connection Laplacian as follows:

Proposition 8.23. *If M is a compact, orientable, Riemannian manifold, then the connection Laplacian $\nabla^*\nabla$ is given by*

$$\nabla^*\nabla\omega = -\mathrm{tr}(\nabla^2\omega),$$

for all differential forms $\omega \in \mathcal{A}^*(M)$.

The proof of Proposition 8.23, which is quite technical, can be found in Petersen [86] (Chapter 7, Proposition 34).

Given any one-forms $\omega \in \mathcal{A}^1(M)$, it is natural to ask what is the one-form

$$\nabla^2_{X,Y}\omega - \nabla^2_{Y,X}\omega.$$

To answer this question, we need to first recall the definition of the curvature tensor. Given $X, Y, Z \in \mathfrak{X}(M)$, the curvature tensor $R(X, Y)Z$ is the $(1, 3)$-tensor defined by

$$R(X, Y)(Z) = \nabla_{[X,Y]}Z + \nabla_Y\nabla_X Z - \nabla_X\nabla_Y Z.$$

Assuming that ∇ is the Levi-Civita connection, the following result can be shown.

Proposition 8.24. *The following equation holds:*

$$R(X, Y)Z = \nabla^2_{Y,X}Z - \nabla^2_{X,Y}Z.$$

For a proof, see Gallot, Hullin, and Lafontaine [48] or Gallier and Quaintance [47].

We are now in a position to answer the preceding question. The answer is given by the following proposition which plays a crucial role in the proof of a version of Bochner's formula:

Proposition 8.25. *For any vector fields* $X, Y, Z \in \mathfrak{X}(M)$ *and any one-form* $\omega \in \mathcal{A}^1(M)$ *on a Riemannian manifold* M, *we have*

$$((\nabla^2_{X,Y} - \nabla^2_{Y,X})\omega)(Z) = \omega(R(X, Y)Z).$$

Proof. (Adapted from Gallot, Hullin, and Lafontaine [48], Lemma 4.13.) It is proven in Section 9.7 that

$$(\nabla_X \omega)^\sharp = \nabla_X \omega^\sharp.$$

We claim that we also have

$$(\nabla^2_{X,Y}\omega)^\sharp = \nabla^2_{X,Y}\omega^\sharp.$$

This is because

$$
\begin{aligned}
(\nabla^2_{X,Y}\omega)^\sharp &= (\nabla_X(\nabla_Y\omega))^\sharp - (\nabla_{\nabla_X Y}\omega)^\sharp \\
&= \nabla_X(\nabla_Y\omega)^\sharp - \nabla_{\nabla_X Y}\omega^\sharp \\
&= \nabla_X(\nabla_Y\omega^\sharp) - \nabla_{\nabla_X Y}\omega^\sharp \\
&= \nabla^2_{X,Y}\omega^\sharp.
\end{aligned}
$$

Thus, using Proposition 8.24 we deduce that

$$((\nabla^2_{X,Y} - \nabla^2_{Y,X})\omega)^\sharp = (\nabla^2_{X,Y} - \nabla^2_{Y,X})\omega^\sharp = R(Y, X)\omega^\sharp.$$

Consequently,

$$
\begin{aligned}
((\nabla^2_{X,Y} - \nabla^2_{Y,X})\omega)(Z) &= \langle((\nabla^2_{X,Y} - \nabla^2_{Y,X})\omega)^\sharp, Z\rangle \\
&= \langle R(Y, X)\omega^\sharp, Z\rangle \\
&= R(Y, X, \omega^\sharp, Z) \\
&= R(X, Y, Z, \omega^\sharp)
\end{aligned}
$$

$$= \langle R(X, Y)Z, \omega^{\sharp} \rangle$$
$$= \omega(R(X, Y)Z),$$

using properties of the Riemann tensor; see Gallot, Hullin, and Lafontaine [48] or Gallier and Quaintance [47]. □

We are now ready to prove the Weitzenböck formulae for one-forms.

Theorem 8.26 (Weitzenböck–Bochner Formula). *If M is a compact, orientable, Riemannian manifold, then for every one-form $\omega \in \mathcal{A}^1(M)$, we have*

$$\Delta\omega = \nabla^*\nabla\omega + \mathrm{Ric}(\omega),$$

where $\mathrm{Ric}(\omega)$ is the one-form given by

$$\mathrm{Ric}(\omega)(X) = \omega(\mathrm{Ric}^{\sharp}(X)),$$

and where $\mathrm{Ric}^{\sharp}$ is the Ricci curvature viewed as a $(1,1)$-tensor (that is, $\langle \mathrm{Ric}^{\sharp}(u), v \rangle_p = \mathrm{Ric}(u, v)$, for all $u, v \in T_pM$ and all $p \in M$).

Proof. (Adapted from Gallot, Hullin, and Lafontaine [48], Proposition 4.36.) For any $p \in M$, pick any normal local chart (U, φ) with $p \in U$, and pick any orthonormal frame $(E_1, \ldots, E_n)$ over U. Because (U, φ) is a normal chart at p, we have $(\nabla_{E_j} E_j)_p = 0$ for all i, j. Recall from the discussion at the end of Section 8.3 as a special case of Proposition 8.17 that for every one-form ω, we have

$$\delta\omega = -\sum_i \nabla_{E_i}\omega(E_i),$$

where $\delta\omega \in C^{\infty}(M)$. Then $d\delta(w)$ is the one form defined via

$$d(\delta\omega)(X) = -\sum_i d(\nabla_{E_i}\omega(E_i))(X) = -\sum_i \nabla_X \nabla_{E_i}\omega(E_i),$$

since $(\nabla_X f)_p = df_p(X_p)$ for all $X \in \mathfrak{X}(M)$. Also recall Proposition 4.16, which states that

$$d\omega(X, Y) = X(\omega(Y)) - Y(\omega(X)) - \omega([X, Y]).$$

By definition,

$$\nabla_X\omega(Y) = X(\omega(Y)) - \omega(\nabla_X Y)$$
$$\nabla_Y\omega(X) = Y(\omega(X)) - \omega(\nabla_Y X).$$

Hence

$$d\omega(X, Y) = \nabla_X \omega(Y) + \omega(\nabla_X Y) - \nabla_Y \omega(X) - \omega(\nabla_Y X) - \omega([X, Y])$$
$$= \nabla_X \omega(Y) - \nabla_X \omega(Y) + \omega(\nabla_X Y - \nabla_Y X) - \omega([X, Y]).$$

Since we are using the Levi-Civita connection, $\nabla_X Y - \nabla_Y X = [X, Y]$, and the preceding calculation becomes

$$d\omega(X, Y) = \nabla_X \omega(Y) - \nabla_Y \omega(X).$$

Let β be the two-form $d\omega$. Note that $\nabla_{E_i} \beta$ is also a two-form. We use Proposition 8.17 to calculate the one-form $\delta\beta$ as follows:

$$(\delta\beta)(X) = -\sum_i \left(i(E_i)\nabla_{E_i}\beta \right)(X) = -\sum_i \nabla_{E_i}\beta(E_i, X).$$

In other words, we found that

$$(\delta d\omega)(X) = -\sum_i \nabla_{E_i} d\omega(E_i, X) = -\sum_i \nabla_{E_i}\nabla_{E_i}\omega(X) + \sum_i \nabla_{E_i}\nabla_X \omega(E_i),$$

where the last equality is an application of Proposition 4.16. Thus, we get

$$\Delta\omega(X) = -\sum_i \nabla_{E_i}\nabla_{E_i}\omega(X) + \sum_i (\nabla_{E_i}\nabla_X - \nabla_X\nabla_{E_i})\omega(E_i)$$
$$= -\sum_i \nabla^2_{E_i, E_i}\omega(X) + \sum_i (\nabla^2_{E_i, X} - \nabla^2_{X, E_i})\omega(E_i)$$
$$= \nabla^*\nabla\omega(X) + \sum_i \omega(R(E_i, X)E_i)$$
$$= \nabla^*\nabla\omega(X) + \omega(\mathrm{Ric}^\sharp(X)),$$

using the fact that $(\nabla_{E_j} E_j)_p = 0$ for all i, j, and using Proposition 8.25 and Proposition 8.23.   □

For simplicity of notation, we will write $\mathrm{Ric}(u)$ for $\mathrm{Ric}^\sharp(u)$. There should be no confusion, since $\mathrm{Ric}(u, v)$ denotes the Ricci curvature, a $(0, 2)$-tensor. There is another way to express $\mathrm{Ric}(\omega)$ which will be useful in the proof of the next theorem.

Proposition 8.27. *The Weitzenböck formula can be written as*

$$\Delta\omega = \nabla^*\nabla\omega + (\mathrm{Ric}(\omega^\sharp))^\flat.$$

Proof. Observe that

$$\text{Ric}(\omega)(Z) = \omega(\text{Ric}(Z))$$
$$= \langle \omega^\sharp, \text{Ric}(Z) \rangle$$
$$= \langle \text{Ric}(Z), \omega^\sharp \rangle$$
$$= \text{Ric}(Z, \omega^\sharp)$$
$$= \text{Ric}(\omega^\sharp, Z)$$
$$= \langle \text{Ric}(\omega^\sharp), Z \rangle$$
$$= ((\text{Ric}(\omega^\sharp))^\flat(Z),$$

and thus,

$$\text{Ric}(\omega)(Z) = ((\text{Ric}(\omega^\sharp))^\flat(Z).$$

Consequently the Weitzenböck formula can be written as

$$\Delta\omega = \nabla^*\nabla\omega + (\text{Ric}(\omega^\sharp))^\flat.$$

$\square$

The Weitzenböck–Bochner formula implies the following theorem due to Bochner:

Theorem 8.28 (Bochner). *If M is a compact, orientable, connected Riemannian manifold, then the following properties hold:*

(i) *If the Ricci curvature is nonnegative, that is $\text{Ric}(u, u) \geq 0$ for all $p \in M$ and all $u \in T_p M$, and if $\text{Ric}(u, u) > 0$ for some $p \in M$ and all $u \in T_p M$, then $H^1_{\text{DR}} M = (0)$.*

(ii) *If the Ricci curvature is nonnegative, then $\nabla\omega = 0$ for all $\omega \in \mathcal{A}^1(M)$, and $\dim H^1_{\text{DR}} M \leq \dim M$.*

Proof. (After Gallot, Hullin, and Lafontaine [48]; Theorem 4.37.)

(i) Assume $H^1_{\text{DR}} M \neq (0)$. Then by the Hodge theorem, Theorem 8.20, there is some nonzero harmonic one-form ω. The Weitzenböck–Bochner formula implies that

$$(\Delta\omega, \omega) = (\nabla^*\nabla\omega, \omega) + ((\text{Ric}(\omega^\sharp))^\flat, \omega).$$

Since $\Delta\omega = 0$, we get

$$0 = (\nabla^*\nabla\omega, \omega) + ((\text{Ric}(\omega^\sharp))^\flat, \omega)$$

$$= ((\nabla\omega, \nabla\omega)) + \int_M \langle (\mathrm{Ric}(\omega^\sharp))^\flat, \omega \rangle \, \mathrm{Vol}_M$$

$$= ((\nabla\omega, \nabla\omega)) + \int_M \langle \mathrm{Ric}(\omega^\sharp), \omega^\sharp \rangle \, \mathrm{Vol}_M$$

$$= ((\nabla\omega, \nabla\omega)) + \int_M \mathrm{Ric}(\omega^\sharp, \omega^\sharp) \, \mathrm{Vol}_M.$$

However, $((\nabla\omega, \nabla\omega)) \geq 0$, and by the assumption on the Ricci curvature, the integrand is nonnegative and strictly positive at some point, so the integral is strictly positive, a contradiction.

(ii) Again, for any one-form ω, we have

$$(\Delta\omega, \omega) = ((\nabla\omega, \nabla\omega)) + \int_M \mathrm{Ric}(\omega^\sharp, \omega^\sharp) \, \mathrm{Vol}_M,$$

so if the Ricci curvature is nonnegative, $\Delta\omega = 0$ iff $\nabla\omega = 0$. This means that ω is invariant by parallel transport (see Section 10.4), and thus ω is completely determined by its value ω_p at some point $p \in M$, so there is an injection $\mathbb{H}^1(M) \longrightarrow T_p^* M$, which implies that $\dim H_{\mathrm{DR}}^1 M = \dim \mathbb{H}^1(M) \leq \dim M$.

$\square$

There is a version of the Weitzenböck formula for p-forms, but it involves a more complicated curvature term and its proof is also more complicated; see Petersen [86] (Chapter 7). The Bochner technique can also be generalized in various ways, in particular, to *spin manifolds*, but these considerations are beyond the scope of these notes. Let us just say that Weitzenböck formulae involving the Dirac operator play an important role in physics and 4-manifold geometry. We refer the interested reader to Gallot, Hulin, and Lafontaine [48] (Chapter 4), Petersen [86] (Chapter 7), Jost [62] (Chapter 3), and Berger [10] (Section 15.6), for more details on Weitzenböck formulae and the Bochner technique.

8.8 Problems

Problem 8.1. Let M be a Riemannian manifold and let $f \in C^\infty(M)$. In local coordinates with respect to a chart, if we write

$$df = \sum_{i=1}^n \frac{\partial f}{\partial x_i} dx_i,$$

show that

$$\text{Hess } f = \sum_{i,j=1}^{n} \left(\frac{\partial^2 f}{\partial x_i \partial x_j} - \sum_{k=1}^{n} \frac{\partial f}{\partial x_k} \Gamma_{ij}^{k} \right) dx_i \otimes dx_j,$$

where the Γ_{ij}^{k} are the Christoffel symbols of the connection in the chart, namely

$$\Gamma_{ij}^{k} = \frac{1}{2} \sum_{l=1}^{n} g^{kl} \left(\partial_i g_{jl} + \partial_j g_{il} - \partial_l g_{ij} \right), \qquad (*)$$

with $\partial_k g_{ij} = \frac{\partial}{\partial x_k}(g_{ij})$.

Hint. See O'Neill [84].

Problem 8.2. Let $G(k, n)$ be the set of all linear k-dimensional subspaces of $\mathbb{R}^n$. Using the notation of Section 8.1 show that for any two tangent vectors $X_1, X_2 \in T_Y G(k, n)$ to $G(k, n)$ at Y,

$$\text{Hess}(F)_Y(X_1, X_2) = F_{YY}(X_1, X_2) - \text{tr}(X_1^{\top} X_2 Y^{\top} F_Y),$$

where

$$F_Y = \left(\frac{\partial F}{\partial Y_{ij}} \right),$$

and where

$$F_{YY}(X_1, X_2) = \sum_{ij,kl} (F_{YY})_{ij,kl}(X_1)_{ij}(X_2)_{kl},$$

with

$$(F_{YY})_{ij,kl} = \frac{\partial^2 F}{\partial Y_{ij} \partial Y_{kl}}.$$

Hint. Use Proposition 8.3, and Edelman, Arias, and Smith [40].

Problem 8.3. Prove Proposition 8.5.

Hint. See Proposition 3.16.

Problem 8.4. Prove Proposition 8.6.

Problem 8.5. For any one-form $\omega \in \mathcal{A}^1(M)$, and any $X, Y \in \mathfrak{X}(M)$, recall that

$$(\nabla_X \omega)(Y) := X(\omega(Y)) - \omega(\nabla_X Y).$$

It turns out that

$$\delta\omega = -\text{tr}\,\nabla\omega,$$

where the trace should be interpreted as the trace of the $\mathbb{R}$-bilinear map $X, Y \mapsto (\nabla_X\omega)(Y)$, as in Chapter 2 (see Proposition 2.3). By applying this trace definition of $\delta\omega$, show that

$$\delta(f\,df) = f\,\Delta f - \langle \text{grad}\,f, \text{grad}\,f \rangle.$$

Problem 8.6.   Prove Proposition 8.17.
Hint. See Petersen [86], Chapter 7, Proposition 37, or Jost [62], Chapter 3, Lemma 3.3.4.

Problem 8.7.   Prove Proposition 8.18.

Problem 8.8.   Prove Proposition 8.23.
Hint. See Petersen [86], Chapter 7, Proposition 34.

Chapter 9
Bundles, Metrics on Bundles, and Homogeneous Spaces

A transitive action $\cdot\colon G \times X \to X$ of a group G on a set X yields a description of X as a quotient G/G_x, where G_x is the stabilizer of any element, $x \in X$ (see Warner [109], Chapter 3, of Gallier and Quaintance [47]). The points of X are identified with the left cosets gG_x ($g \in G$). If X is a "well-behaved" topological space (say a locally compact Hausdorff space), G is a "well-behaved" topological group (say a locally compact topological group which is countable at infinity), and the action is continuous, then G/G_x is homeomorphic to X (see Bourbaki [17], Chapter IX, Section 5, Proposition 6, or Gallier and Quaintance [47]). In particular these conditions are satisfied if G is a Lie group and X is a manifold. Intuitively, the above theorem says that G can be viewed as a family of "fibers" gG_x, all isomorphic to G_x, these fibers being parametrized by the "base space" X, and varying smoothly when the point corresponding to the coset gG_x moves in X. We have an example of what is called a fiber bundle, in fact, a principal fiber bundle. This view of G as a family of fibers gG_x, with $x \in X$, is a special case of the notion of a fiber bundle.

Intuitively, a fiber bundle over B is a family $E = (E_b)_{b \in B}$ of spaces E_b (fibers) indexed by B and varying smoothly as b moves in B, such that every E_b is diffeomorphic to some prespecified space F. The space E is called the total space, B the base space, and F the fiber. A way to define such a family is to specify a surjective map $\pi\colon E \to B$. We will assume that E, B, and F are smooth manifolds and that π is a smooth map. The theory of bundles can be developed for topological spaces but we do need such generality. The type of bundles that we just described is too general and to develop a useful theory it is necessary to assume that locally, a bundle looks like a product. Technically, this is achieved by assuming that there is some open cover $\mathcal{U} = (U_\alpha)_{\alpha \in I}$ of B and that there is a family $(\varphi_\alpha)_{\alpha \in I}$ of diffeomorphisms

$$\varphi_\alpha\colon \pi^{-1}(U_\alpha) \to U_\alpha \times F.$$

© Springer Nature Switzerland AG 2020
J. Gallier, J. Quaintance, *Differential Geometry and Lie Groups*, Geometry and Computing 13, https://doi.org/10.1007/978-3-030-46047-1_9

Intuitively, above U_α, the open subset $\pi^{-1}(U_\alpha)$ looks like a product. The maps φ_α are called *local trivializations*.

The last important ingredient in the notion of a fiber bundle is the specification of the "twisting" of the bundle; that is, how the fiber $E_b = \pi^{-1}(b)$ gets twisted as b moves in the base space B. Technically, such twisting manifests itself on overlaps $U_\alpha \cap U_\beta \neq \emptyset$. It turns out that we can write

$$\varphi_\alpha \circ \varphi_\beta^{-1}(b, x) = (b, g_{\alpha\beta}(b)(x))$$

for all $b \in U_\alpha \cap U_\beta$ and all $x \in F$. The term $g_{\alpha\beta}(b)$ is a diffeomorphism of F. Then we require that the family of diffeomorphisms $g_{\alpha\beta}(b)$ belongs to a Lie group G, which is expressed by specifying that the maps $g_{\alpha\beta}$, called transitions maps, are maps

$$g_{\alpha\beta} : U_\alpha \cap U_\beta \to G.$$

The purpose of the group G, called the structure group, is to specify the "twisting" of the bundle.

Fiber bundles are defined in Section 9.1. The family of transition maps $g_{\alpha\beta}$ satisfies an important condition on nonempty overlaps $U_\alpha \cap U_\beta \cap U_\gamma$ called the *cocycle condition*:

$$g_{\alpha\beta}(b)g_{\beta\gamma}(b) = g_{\alpha\gamma}(b)$$

(where $g_{\alpha\beta}(b), g_{\beta\gamma}(b), g_{\alpha\gamma}(b) \in G$), for all α, β, γ such that $U_\alpha \cap U_\beta \cap U_\gamma \neq \emptyset$ and all $b \in U_\alpha \cap U_\beta \cap U_\gamma$.

In Section 9.2, we define bundle morphisms, and the notion of equivalence of bundles over the same base, following Hirzebruch [57] and Chern [22]. We show that two bundles (over the same base) are equivalent if and only if they are isomorphic.

In Section 9.3, we describe the construction of a fiber bundle with prescribed fiber F and structure group G from a base manifold, B, an open cover $\mathcal{U} = (U_\alpha)_{\alpha \in I}$ of B, and a family of maps $g_{\alpha\beta} : U_\alpha \cap U_\beta \to G$ satisfying the cocycle condition, called a *cocycle*. This construction is the basic tool for constructing new bundles from old ones. This construction is applied to define the notion of pullback bundle.

Section 9.4 is devoted to a special kind of fiber bundle called *vector bundles*. A vector bundle is a fiber bundle for which the fiber is a finite-dimensional vector space V, and the structure group is a subgroup of the group of linear isomorphisms (**GL**$(n, \mathbb{R})$ or **GL**$(n, \mathbb{C})$, where $n = \dim V$). Typical examples of vector bundles are the tangent bundle TM and the cotangent bundle T^*M of a manifold M. We define maps of vector bundles and equivalence of vector bundles. The construction of a vector bundle in terms of a cocycle also applies to vector bundles. We give a criterion for a vector bundle to be trivial (isomorphic to $B \times V$) in terms of the existence of a frame of global sections.

In Section 9.5 we describe various operations on vector bundles: Whitney sums, tensor products, tensor powers, exterior powers, symmetric powers, dual bundles, and $\mathcal{H}om$ bundles. We also define the complexification of a real vector bundle.

In Section 9.6 we discuss properties of the sections of a vector bundle ξ. We prove that the space of sections $\Gamma(\xi)$ is finitely generated projective $C^\infty(B)$-module. We also prove various useful isomorphisms.

Section 9.7 is devoted to the the covariant derivative of tensor fields, and to the duality between vector fields and differential forms.

In Section 9.8 we explain how to give a vector bundle a Riemannian metric. This is achieved by supplying a smooth family $(\langle -, - \rangle_b)_{b \in B}$ of inner products on each fiber $\pi^{-1}(b)$ above $b \in B$. We describe the notion of reduction of the structure group and define orientable vector bundles.

In Section 9.9 we consider the special case of fiber bundles for which the fiber coincides with the structure group G, which acts on itself by left translations. Such fiber bundles are called *principal bundles*. It turns out that a principal bundle can be defined in terms of a free right action of Lie group on a smooth manifold. When principal bundles are defined in terms of free right actions, the notion of bundle morphism is also defined in terms of equivariant maps.

There are two constructions that allow us to reduce the study of fiber bundles to the study of principal bundles. Given a fiber bundle ξ with fiber F, we can construct a principal bundle $P(\xi)$ obtained by replacing the fiber F by the group G. Conversely, given a principal bundle ξ and an effective action of G on a manifold F, we can construct the fiber bundle $\xi[F]$ obtained by replacing G by F. The *Borel construction* provides a direct construction of $\xi[F]$. The maps

$$\xi \mapsto \xi[F] \quad \text{and} \quad \xi \mapsto P(\xi)$$

induce a bijection between equivalence classes of principal G-bundles and fiber bundles (with structure group G). Furthermore, ξ is a trivial bundle iff $P(\xi)$ is a trivial bundle. The equivalence of fiber bundles and principal bundles (over the same base B, and with the same structure group G) is the key to the classification of fiber bundles, but we do not discuss this deep result.

Section 9.10 is devoted to principal bundles that arise from proper and free actions of a Lie group. When the base space is a homogenous space, which means that it arises from a transitive action of a Lie group, then the total space is a principal bundle. There are many illustrations of this situation involving $\mathbf{SO}(n + 1)$ and $\mathbf{SU}(n + 1)$.

9.1   Fiber Bundles

We begin by carefully stating the definition of a fiber bundle because we believe that it clarifies the notions of vector bundles and principal fiber bundles, the concepts that are our primary concern. The following definition is not the most general, but it is sufficient for our needs.

Definition 9.1. A *fiber bundle with (typical) fiber F* F G *and structure group G* is a tuple $\xi = (E, \pi, B, F, G)$, where E, B, F are smooth manifolds, $\pi: E \to B$ is a smooth surjective map, G is a Lie group of diffeomorphisms of F, and there is some open cover $\mathcal{U} = (U_\alpha)_{\alpha \in I}$ of B and a family $\varphi = (\varphi_\alpha)_{\alpha \in I}$ of diffeomorphisms

$$\varphi_\alpha: \pi^{-1}(U_\alpha) \to U_\alpha \times F.$$

The space B is called the *base space*, E is called the *total space*, F is called the *(typical) fiber*, and each φ_α is called a *(local) trivialization*. The pair $(U_\alpha, \varphi_\alpha)$ is called a *bundle chart*, and the family $\{(U_\alpha, \varphi_\alpha)\}$ is a *trivializing cover*. For each $b \in B$, the space $\pi^{-1}(b)$ is called the *fiber above b*; it is also denoted by E_b, and $\pi^{-1}(U_\alpha)$ is also denoted by $E \upharpoonright U_\alpha$; see Figure 9.1.

The following properties hold:

(a) (Local triviality) The diagram

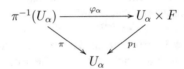

commutes for all $\alpha \in I$, where $p_1: U_\alpha \times F \to U_\alpha$ is the first projection. Equivalently, for all $(b, y) \in U_\alpha \times F$,

Fig. 9.1 The spiky cylinder E is a typical fiber bundle with base B as the purple cylinder and fiber isomorphic to a line segment.

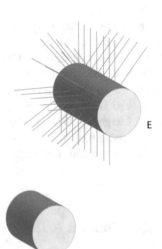

$$\pi \circ \varphi_\alpha^{-1}(b, y) = b.$$

(b) (Fiber diffeomorphism) For every $(U_\alpha, \varphi_\alpha)$ and every $b \in U_\alpha$, because $p_1 \circ \varphi_\alpha = \pi$, by (a) the restriction of φ_α to $E_b = \pi^{-1}(b)$ is a diffeomorphism between E_b and $\{b\} \times F$, so we have the diffeomorphism

$$\varphi_{\alpha,b} \colon E_b \to F$$

given by

$$\varphi_{\alpha,b}(Z) = (p_2 \circ \varphi_\alpha)(Z), \quad \text{for all } Z \in E_b;$$

see Figure 9.2. Furthermore, for all U_α, U_β in $\mathcal{U}$ such that $U_\alpha \cap U_\beta \neq \emptyset$, for every $b \in U_\alpha \cap U_\beta$, there is a relationship between $\varphi_{\alpha,b}$ and $\varphi_{\beta,b}$ which gives the twisting of the bundle.

(c) (Fiber twisting) The diffeomorphism

$$\varphi_{\alpha,b} \circ \varphi_{\beta,b}^{-1} \colon F \to F$$

is an element of the group G.

(d) (Transition maps) The map $g_{\alpha\beta} \colon U_\alpha \cap U_\beta \to G$ defined by

$$g_{\alpha\beta}(b) = \varphi_{\alpha,b} \circ \varphi_{\beta,b}^{-1},$$

is smooth. The maps $g_{\alpha\beta}$ are called the *transition maps* of the fiber bundle.

A fiber bundle $\xi = (E, \pi, B, F, G)$ is also referred to, somewhat loosely, as a *fiber bundle over B* or a *G-bundle*, and it is customary to use the notation

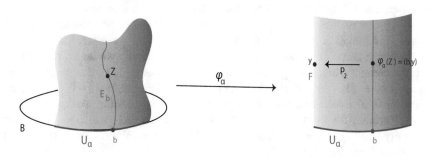

Fig. 9.2 An illustration of $\varphi_{\alpha,b} \colon E_b \to F$ over $B = S^1$.

$$F \longrightarrow E \longrightarrow B,$$

even though it is imprecise (the group G is missing!), and it clashes with the notation for short exact sequences. Observe that the bundle charts $(U_\alpha, \varphi_\alpha)$ are similar to the charts of a manifold.

Actually, Definition 9.1 is too restrictive because it does not allow for the addition of compatible bundle charts, for example when considering a refinement of the cover $\mathcal{U}$. This problem can easily be fixed using a notion of equivalence of trivializing covers analogous to the equivalence of atlases for manifolds (see Remark (2) below). Also observe that (b), (c), and (d) imply that the isomorphism $\varphi_\alpha \circ \varphi_\beta^{-1} : (U_\alpha \cap U_\beta) \times F \to (U_\alpha \cap U_\beta) \times F$ is related to the smooth map $g_{\alpha\beta} : U_\alpha \cap U_\beta \to G$ by the identity

$$\varphi_\alpha \circ \varphi_\beta^{-1}(b, x) = (b, g_{\alpha\beta}(b)(x)), \qquad (*)$$

for all $b \in U_\alpha \cap U_\beta$ and all $x \in F$.

We interpret $g_{\alpha\beta}(b)(x)$ as the action of the group element $g_{\alpha\beta}(b)$ on x; see Figure 9.4.

Intuitively, a fiber bundle over B is a family $E = (E_b)_{b \in B}$ of spaces E_b (fibers) indexed by B and varying smoothly as b moves in B, such that every E_b is diffeomorphic to F. The bundle $E = B \times F$, where π is the first projection, is called the *trivial bundle* (over B). The trivial bundle $B \times F$ is often denoted ϵ^F. The local triviality Condition (a) says that *locally*, that is over every subset U_α from some open cover of the base space B, the bundle $\xi \restriction U_\alpha$ is trivial. Note that if G is the trivial one-element group, then the fiber bundle is trivial. In fact, the purpose of the group G is to specify the "twisting" of the bundle; that is, how the fiber E_b gets twisted as b moves in the base space B.

A Möbius strip is an example of a nontrivial fiber bundle where the base space B is the circle S^1, the fiber space F is the closed interval $[-1, 1]$, and the structural group is $G = \{1, -1\}$, where -1 is the reflection of the interval $[-1, 1]$ about its midpoint 0. The total space E is the strip obtained by rotating the line segment $[-1, 1]$ around the circle, keeping its midpoint in contact with the circle, and gradually twisting the line segment so that after a full revolution, the segment has been tilted by π. See Figure 9.3.

Note that $U_1 = \{-\pi < x < \frac{\pi}{2}\}$, $U_2 = \{0 < x < \frac{3\pi}{2}\}$, while $U_1 \cap U_2 = V \cup W$ where $V = \{0 < x < \frac{\pi}{2}\}$ and $W = \{-\pi < x < -\frac{\pi}{2}\}$. The transition map is $\varphi_1 \circ \varphi_2^{-1}(b, x) = (b, g_{12}(b)x)$ where $g_{12}(b) = 1$ if $b \in V$ and $g_{12}(b) = -1$ if $b \in W$.

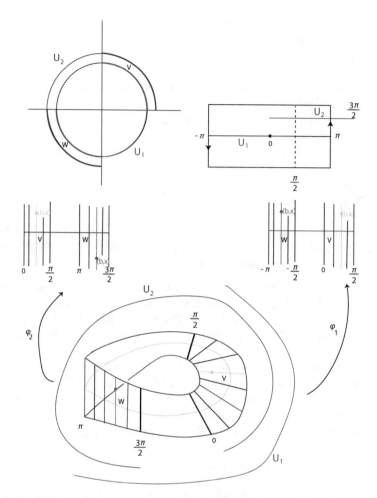

Fig. 9.3 The Möbius strip as a line bundle over the unit circle.

A Klein bottle is also a fiber bundle for which both the base space and the fiber are the circle, S^1, while $G = \{-1, 1\}$. Again, the reader should work out the details for this example.

Other examples of fiber bundles are:

(1) $\mathbf{SO}(n+1)$, an $\mathbf{SO}(n)$-bundle over the sphere S^n with fiber $\mathbf{SO}(n)$. (for $n \geq 0$).
(2) $\mathbf{SU}(n+1)$, an $\mathbf{SU}(n)$-bundle over the sphere S^{2n+1} with fiber $\mathbf{SU}(n)$ (for $n \geq 0$).
(3) $\mathbf{SL}(2, \mathbb{R})$, an $\mathbf{SO}(2)$-bundle over the upper-half space H, with fiber $\mathbf{SO}(2)$.
(4) $\mathbf{GL}(n, \mathbb{R})$, an $\mathbf{O}(n)$-bundle over the space $\mathbf{SPD}(n)$ of symmetric, positive definite matrices, with fiber $\mathbf{O}(n)$.

(5) $\mathbf{GL}^+(n, \mathbb{R})$, an $\mathbf{SO}(n)$-bundle over the space, $\mathbf{SPD}(n)$ of symmetric, positive definite matrices, with fiber $\mathbf{SO}(n)$.

(6) $\mathbf{SO}(n+1)$, an $\mathbf{O}(n)$-bundle over the real projective space $\mathbb{RP}^n$, with fiber $\mathbf{O}(n)$ (for $n \geq 0$).

(7) $\mathbf{SU}(n + 1)$, an $\mathbf{U}(n)$-bundle over the complex projective space $\mathbb{CP}^n$, with fiber $\mathbf{U}(n)$ (for $n \geq 0$).

(8) $\mathbf{O}(n)$, an $\mathbf{O}(k) \times \mathbf{O}(n - k)$-bundle over the Grassmannian $G(k, n)$, with fiber $\mathbf{O}(k) \times \mathbf{O}(n - k)$.

(9) $\mathbf{SO}(n)$, an $S(\mathbf{O}(k) \times \mathbf{O}(n - k))$-bundle over the Grassmannian $G(k, n)$, with fiber $S(\mathbf{O}(k) \times \mathbf{O}(n - k))$.

(10) $\mathbf{SO}(n)$, an $\mathbf{SO}(n - k)$-bundle over the Stiefel manifold $S(k, n)$, with $1 \leq k \leq n - 1$.

(11) The Lorentz group, $\mathbf{SO}_0(n, 1)$, is an $\mathbf{SO}(n)$-bundle over the space $\mathcal{H}_n^+(1)$ consisting of one sheet of the hyperbolic paraboloid $\mathcal{H}_n(1)$, with fiber $\mathbf{SO}(n)$ (see Gallier and Quaintance [47]).

Observe that in all the examples above, $F = G$; that is, the typical fiber is identical to the group G. Special bundles of this kind are called *principal fiber bundles*.

The above definition is slightly different (but equivalent) to the definition given in Bott and Tu [12], page 47–48. Definition 9.1 is closer to the one given in Hirzebruch [57].

Bott and Tu and Hirzebruch assume that G acts effectively (or faithfully) on the left on the fiber F. This means that there is a smooth action $\cdot \colon G \times F \to F$. If G is a Lie group and if F is a manifold, an action $\varphi \colon G \times F \to F$ is *smooth* if φ is smooth. Then for every $g \in G$, the map $\varphi_g \colon M \to M$ is a diffeomorphism. Also recall that G *acts effectively* (or *faithfully*) on F iff for every $g \in G$,

$$\text{if} \quad g \cdot x = x \quad \text{for all } x \in F, \quad \text{then} \quad g = 1.$$

Every $g \in G$ induces a diffeomorphism $\varphi_g \colon F \to F$, defined by

$$\varphi_g(x) = g \cdot x, \quad \text{for all } x \in F.$$

The fact that G acts effectively on F means that the map $g \mapsto \varphi_g$ is injective. This justifies viewing G as a group of diffeomorphisms of F, and from now on we will denote $\varphi_g(x)$ by $g(x)$.

We observed that Definition 9.1 is too restrictive because it does not allow for the addition of compatible bundle charts. We can fix this problem as follows:

Definition 9.2. Let $\xi = (E, \pi, B, F, G)$ be fiber bundle defined as in Definition 9.1. Given a trivializing cover $\{(U_\alpha, \varphi_\alpha)\}$, for any open U of B and any diffeomorphism

$$\varphi \colon \pi^{-1}(U) \to U \times F,$$

we say that (U, φ) is *compatible with the trivializing cover* $\{(U_\alpha, \varphi_\alpha)\}$ iff whenever $U \cap U_\alpha \neq \emptyset$, there is some smooth map $g_\alpha : U \cap U_\alpha \to G$, so that

$$\varphi \circ \varphi_\alpha^{-1}(b, x) = (b, g_\alpha(b)(x)),$$

for all $b \in U \cap U_\alpha$ and all $x \in F$. Two trivializing covers are *equivalent* iff every bundle chart of one cover is compatible with the other cover. This is equivalent to saying that the union of two trivializing covers is a trivializing cover.

Definition 9.2 implies the following:

Definition 9.3. Using the conventions of Definition 9.1, a fiber bundle is a tuple $(E, \pi, B, F, G, \{(U_\alpha, \varphi_\alpha)\})$, where $\{(U_\alpha, \varphi_\alpha)\}$ is an equivalence class of trivializing covers. As for manifolds, given a trivializing cover $\{(U_\alpha, \varphi_\alpha)\}$, the set of all bundle charts compatible with $\{(U_\alpha, \varphi_\alpha)\}$ is a maximal trivializing cover equivalent to $\{(U_\alpha, \varphi_\alpha)\}$; see Figure 9.4.

A special case of the above occurs when we have a trivializing cover $\{(U_\alpha, \varphi_\alpha)\}$ with $\mathcal{U} = \{U_\alpha\}$ an open cover of B, and another open cover $\mathcal{V} = (V_\beta)_{\beta \in J}$ of B which is a refinement of $\mathcal{U}$. This means that there is a map $\tau : J \to I$, such that $V_\beta \subseteq U_{\tau(\beta)}$ for all $\beta \in J$. Then for every $V_\beta \in \mathcal{V}$, since $V_\beta \subseteq U_{\tau(\beta)}$, the restriction of $\varphi_{\tau(\beta)}$ to V_β is a trivialization

$$\varphi_\beta' : \pi^{-1}(V_\beta) \to V_\beta \times F,$$

and Conditions (b) and (c) are still satisfied, so $(V_\beta, \varphi_\beta')$ is compatible with $\{(U_\alpha, \varphi_\alpha)\}$.

The family of transition functions $(g_{\alpha\beta})$ satisfies the following crucial conditions.

Fig. 9.4 A schematic illustration of the transition between two elements of a trivializing cover.

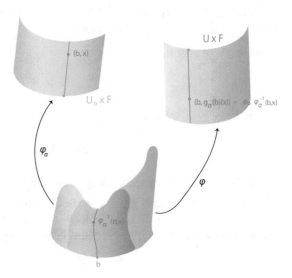

Fig. 9.5 A schematic illustration of the cocycle condition. The three sheets of the bundle actually glue together into a single sheet.

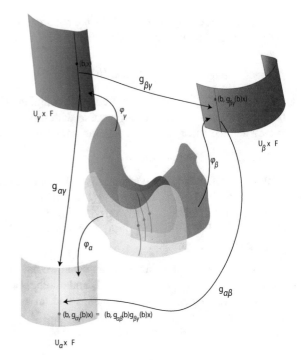

Definition 9.4. Given a fiber bundle $\xi = (E, \pi, B, F, G, \{(U_\alpha, \varphi_\alpha)\})$ with family of transition functions $(g_{\alpha\beta})$, the *cocycle condition* is the set of equations

$$g_{\alpha\beta}(b)g_{\beta\gamma}(b) = g_{\alpha\gamma}(b)$$

(where $g_{\alpha\beta}(b), g_{\beta\gamma}(b), g_{\alpha\gamma}(b) \in G$), for all α, β, γ such that $U_\alpha \cap U_\beta \cap U_\gamma \neq \emptyset$ and all $b \in U_\alpha \cap U_\beta \cap U_\gamma$; see Figure 9.5.

Setting $\alpha = \beta = \gamma$, we get

$$g_{\alpha\alpha} = \text{id},$$

and setting $\gamma = \alpha$, we get

$$g_{\beta\alpha} = g_{\alpha\beta}^{-1}.$$

Again, beware that this means that $g_{\beta\alpha}(b) = g_{\alpha\beta}^{-1}(b)$, where $g_{\alpha\beta}^{-1}(b)$ is the inverse of $g_{\beta\alpha}(b)$ in G. In general, $g_{\alpha\beta}^{-1}$ is **not** the functional inverse of $g_{\beta\alpha}$.

Experience shows that most objects of interest in geometry (vector fields, differential forms, *etc.*) arise as sections of certain bundles. Furthermore, deciding

Fig. 9.6 An illustration of a global section of $E \cong B \times \mathbb{R}$ where B is the unit disk.

whether or not a bundle is trivial often reduces to the existence of a (global) section. Thus, we define the important concept of a section right away.

Definition 9.5. Given a fiber bundle $\xi = (E, \pi, B, F, G)$, a *smooth section* of ξ is a smooth map $s\colon B \to E$, so that $\pi \circ s = \mathrm{id}_B$. Given an open subset U of B, a *(smooth) section of ξ over U* is a smooth map $s\colon U \to E$, so that $\pi \circ s(b) = b$, for all $b \in U$; we say that s is a *local section* of ξ. The set of all sections over U is denoted $\Gamma(U, \xi)$, and $\Gamma(B, \xi)$ (for short, $\Gamma(\xi)$) is the set of *global sections* of ξ; see Figure 9.6.

Here is an observation that proves useful for constructing global sections. Let $s\colon B \to E$ be a global section of a bundle ξ. For every trivialization $\varphi_\alpha\colon \pi^{-1}(U_\alpha) \to U_\alpha \times F$, let $s_\alpha\colon U_\alpha \to E$ and $\sigma_\alpha\colon U_\alpha \to F$ be given by

$$s_\alpha = s \upharpoonright U_\alpha \quad \text{and} \quad \sigma_\alpha = pr_2 \circ \varphi_\alpha \circ s_\alpha,$$

so that

$$s_\alpha(b) = \varphi_\alpha^{-1}(b, \sigma_\alpha(b)).$$

Obviously, $\pi \circ s_\alpha = \mathrm{id}$, so s_α is a local section of ξ, and σ_α is a function $\sigma_\alpha\colon U_\alpha \to F$.

We claim that on overlaps, we have

$$\sigma_\alpha(b) = g_{\alpha\beta}(b)\sigma_\beta(b).$$

See Figure 9.7.

Proof. Indeed, recall that

$$\varphi_\alpha \circ \varphi_\beta^{-1}(b, x) = (b, g_{\alpha\beta}(b)x),$$

for all $b \in U_\alpha \cap U_\beta$ and all $x \in F$, and as $s_\alpha = s \upharpoonright U_\alpha$ and $s_\beta = s \upharpoonright U_\beta$, s_α and s_β agree on $U_\alpha \cap U_\beta$. Consequently, from

$$s_\alpha(b) = \varphi_\alpha^{-1}(b, \sigma_\alpha(b)) \quad \text{and} \quad s_\beta(b) = \varphi_\beta^{-1}(b, \sigma_\beta(b)),$$

Fig. 9.7 An illustration of
the gluing of two local
sections s_α and s_β.

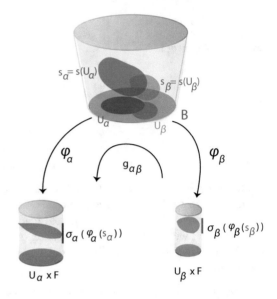

we get

$$\varphi_\alpha^{-1}(b, \sigma_\alpha(b)) = s_\alpha(b) = s_\beta(b) = \varphi_\beta^{-1}(b, \sigma_\beta(b)) = \varphi_\alpha^{-1}(b, g_{\alpha\beta}(b)\sigma_\beta(b)),$$

which implies $\sigma_\alpha(b) = g_{\alpha\beta}(b)\sigma_\beta(b)$, as claimed.   □

Conversely, assume that we have a collection of functions $\sigma_\alpha \colon U_\alpha \;\to\; F$, satisfying

$$\sigma_\alpha(b) = g_{\alpha\beta}(b)\sigma_\beta(b)$$

on overlaps. Let $s_\alpha \colon U_\alpha \to E$ be given by

$$s_\alpha(b) = \varphi_\alpha^{-1}(b, \sigma_\alpha(b)).$$

Each s_α is a local section and we claim that these sections agree on overlaps, so they patch and define a global section s.

Proof. We need to show that

$$s_\alpha(b) = \varphi_\alpha^{-1}(b, \sigma_\alpha(b)) = \varphi_\beta^{-1}(b, \sigma_\beta(b)) = s_\beta(b),$$

for $b \in U_\alpha \cap U_\beta$; that is,

$$(b, \sigma_\alpha(b)) = \varphi_\alpha \circ \varphi_\beta^{-1}(b, \sigma_\beta(b)),$$

and since $\varphi_\alpha \circ \varphi_\beta^{-1}(b, \sigma_\beta(b)) = (b, g_{\alpha\beta}(b)\sigma_\beta(b))$, and by hypothesis, $\sigma_\alpha(b) = g_{\alpha\beta}(b)\sigma_\beta(b)$, our equation $s_\alpha(b) = s_\beta(b)$ is verified. $\square$

9.2   Bundle Morphisms, Equivalent Bundles, and Isomorphic Bundles

Now that we have defined a fiber bundle, it is only natural to analyze mappings between two fiber bundles. The notion of a map between fiber bundles is more subtle than one might think because of the structure group G. Let us begin with the simpler case where $G = \mathrm{Diff}(F)$, the group of *all* smooth diffeomorphisms of F.

Definition 9.6. If $\xi_1 = (E_1, \pi_1, B_1, F, \mathrm{Diff}(F))$ and $\xi_2 = (E_2, \pi_2, B_2, F, \mathrm{Diff}(F))$ are two fiber bundles with **the same typical fiber** F **and the same structure group** $G = \mathrm{Diff}(F)$, a *bundle map (or bundle morphism)* $f : \xi_1 \to \xi_2$ is a pair $f = (f_E, f_B)$ of smooth maps $f_E : E_1 \to E_2$ and $f_B : B_1 \to B_2$, such that

(a) The following diagram commutes:

$$
\begin{array}{ccc}
E_1 & \xrightarrow{\ f_E\ } & E_2 \\
\Big\downarrow{\scriptstyle \pi_1} & & \Big\downarrow{\scriptstyle \pi_2} \\
B_1 & \xrightarrow[\ f_B\ ]{} & B_2
\end{array}
$$

(b) For every $b \in B_1$, the map of fibers

$$
f_E \restriction \pi_1^{-1}(b) : \pi_1^{-1}(b) \to \pi_2^{-1}(f_B(b))
$$

is a diffeomorphism (*preservation of the fiber*).

A bundle map $f : \xi_1 \to \xi_2$ is an *isomorphism* if there is some bundle map $g : \xi_2 \to \xi_1$, called the *inverse of* f, such that

$$
g_E \circ f_E = \mathrm{id} \quad \text{and} \quad f_E \circ g_E = \mathrm{id}.
$$

The bundles ξ_1 and ξ_2 are called *isomorphic*.

Given two fiber bundles $\xi_1 = (E_1, \pi_1, B, F, \mathrm{Diff}(F))$ and $\xi_2 = (E_2, \pi_2, B, F, \mathrm{Diff}(F))$ over the same base space B, a *bundle map (or bundle morphism)* $f : \xi_1 \to \xi_2$ is a pair $f = (f_E, f_B)$, where $f_B = \mathrm{id}$ (the identity map). Such a bundle map is an *isomorphism* if it has an inverse as defined above. In this case, we say that the bundles ξ_1 and ξ_2 over B are *isomorphic*.

Observe that the commutativity of the diagram in Definition 9.6 implies that f_B is actually determined by f_E. Also, when f is an isomorphism, the surjectivity of π_1 and π_2 implies that

$$g_B \circ f_B = \text{id} \quad \text{and} \quad f_B \circ g_B = \text{id}.$$

Thus when $f = (f_E, f_B)$ is an isomorphism, both f_E and f_B are diffeomorphisms.

Remark. Some authors do not require the "preservation" of fibers. However, it is automatic for bundle isomorphisms.

Let us take a closer look at what it means for a bundle map to preserve fibers. When we have a bundle map $f: \xi_1 \rightarrow \xi_2$ as above, for every $b \in B$, for any trivializations $\varphi_\alpha : \pi_1^{-1}(U_\alpha) \rightarrow U_\alpha \times F$ of ξ_1 and $\varphi_\beta' : \pi_2^{-1}(V_\beta) \rightarrow V_\beta \times F$ of ξ_2, with $b \in U_\alpha$ and $f_B(b) \in V_\beta$, we have the map

$$\varphi_\beta' \circ f_E \circ \varphi_\alpha^{-1} : (U_\alpha \cap f_B^{-1}(V_\beta)) \times F \rightarrow V_\beta \times F.$$

Consequently, as φ_α and φ_β' are diffeomorphisms and as f is a diffeomorphism on fibers, we have a map $\rho_{\alpha,\beta} : U_\alpha \cap f_B^{-1}(V_\beta) \rightarrow \text{Diff}(F)$, such that

$$\varphi_\beta' \circ f_E \circ \varphi_\alpha^{-1}(b, x) = (f_B(b), \rho_{\alpha,\beta}(b)(x)),$$

for all $b \in U_\alpha \cap f_B^{-1}(V_\beta)$ and all $x \in F$; see Figure 9.8.

Since we may always pick U_α and V_β so that $f_B(U_\alpha) \subseteq V_\beta$, we may also write ρ_α instead of $\rho_{\alpha,\beta}$, with $\rho_\alpha : U_\alpha \rightarrow G$. Then observe that locally, f_E is given as the composition

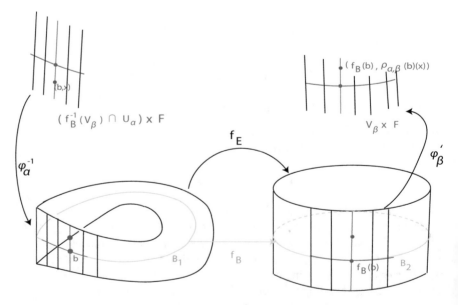

Fig. 9.8 The construction of the map $\varphi_\beta' \circ f_E \circ \varphi_\alpha^{-1}$ between the Möbius strip bundle ξ_1 and the cylinder bundle ξ_2.

$$\pi_1^{-1}(U_\alpha) \xrightarrow{\varphi_\alpha} U_\alpha \times F \xrightarrow{\tilde{f}_\alpha} V_\beta \times F \xrightarrow{\varphi_\beta'^{-1}} \pi_2^{-1}(V_\beta)$$

$$z \longrightarrow (b, x) \longrightarrow (f_B(b), \rho_\alpha(b)(x)) \longrightarrow \varphi_\beta'^{-1}(f_B(b), \rho_\alpha(b)(x)),$$

with $\tilde{f}_\alpha(b, x) = (f_B(b), \rho_\alpha(b)(x))$, that is,

$$f_E(z) = \varphi_\beta'^{-1}(f_B(b), \rho_\alpha(b)(x)), \qquad \text{with } z \in \pi_1^{-1}(U_\alpha) \text{ and } (b, x) = \varphi_\alpha(z).$$

Conversely, if (f_E, f_B) is a pair of smooth maps satisfying the commutative diagram of Definition 9.6 and the above conditions hold locally, then as φ_α, $\varphi_\beta'^{-1}$, and $\rho_\alpha(b)$ are diffeomorphisms on fibers, we see that f_E is a diffeomorphism on fibers.

In the general case where the structure group G is not the whole group of diffeomorphisms $\mathrm{Diff}(F)$, there is no guarantee that $\rho_\alpha(b) \in G$. This is the case if ξ is a vector bundle or a principal bundle, but if ξ is a fiber bundle, following Hirzebruch [57], we use the local conditions above to define the "right notion" of bundle map, namely Definition 9.7. Another advantage of this definition is that two bundles (with the same fiber, structure group, and base) are isomorphic iff they are equivalent (see Proposition 9.1 and Proposition 9.2).

Definition 9.7. Given two fiber bundles $\xi_1 = (E_1, \pi_1, B_1, F, G)$ and $\xi_2 = (E_2, \pi_2, B_2, F, G)$ with the same fiber and the same structure group, a *bundle map* $f: \xi_1 \to \xi_2$ is a pair $f = (f_E, f_B)$ of smooth maps $f_E: E_1 \to E_2$ and $f_B: B_1 \to B_2$, such that:

(a) The diagram

$$\begin{array}{ccc} E_1 & \xrightarrow{f_E} & E_2 \\ \pi_1 \downarrow & & \downarrow \pi_2 \\ B_1 & \xrightarrow{f_B} & B_2 \end{array}$$

commutes.

(b) There is an open cover $\mathcal{U} = (U_\alpha)_{\alpha \in I}$ for B_1, an open cover $\mathcal{V} = (V_\beta)_{\beta \in J}$ for B_2, a family $\varphi = (\varphi_\alpha)_{\alpha \in I}$ of trivializations $\varphi_\alpha: \pi_1^{-1}(U_\alpha) \to U_\alpha \times F$ for ξ_1, a family $\varphi' = (\varphi_\beta')_{\beta \in J}$ of trivializations $\varphi_\beta': \pi_2^{-1}(V_\beta) \to V_\beta \times F$ for ξ_2, such that for every $b \in B$, there are some trivializations $\varphi_\alpha: \pi_1^{-1}(U_\alpha) \to U_\alpha \times F$ and $\varphi_\beta': \pi_2^{-1}(V_\beta) \to V_\beta \times F$, with $f_B(U_\alpha) \subseteq V_\beta$, $b \in U_\alpha$ and some smooth map

$$\rho_\alpha: U_\alpha \to G,$$

such that $\varphi_\beta' \circ f_E \circ \varphi_\alpha^{-1}: U_\alpha \times F \to V_\alpha \times F$ is given by

$$\varphi'_\beta \circ f_E \circ \varphi_\alpha^{-1}(b, x) = (f_B(b), \rho_\alpha(b)(x)),$$

for all $b \in U_\alpha$ and all $x \in F$.

See Figure 9.8. A bundle map is an *isomorphism* if it has an inverse as in Definition 9.6. If the bundles ξ_1 and ξ_2 are over the same base B, then we also require $f_B = \mathrm{id}$.

As we remarked in the discussion before Definition 9.7, Condition (b) ensures that the maps of fibers

$$f_E \restriction \pi_1^{-1}(b) \colon \pi_1^{-1}(b) \to \pi_2^{-1}(f_B(b))$$

are diffeomorphisms. In the special case where ξ_1 and ξ_2 have the same base, $B_1 = B_2 = B$, we require $f_B = \mathrm{id}$, and we can use the same cover (*i.e.*, $\mathcal{U} = \mathcal{V}$), in which case Condition (b) becomes: There is some smooth map $\rho_\alpha \colon U_\alpha \to G$, such that

$$\varphi'_\alpha \circ f \circ \varphi_\alpha^{-1}(b, x) = (b, \rho_\alpha(b)(x)),$$

for all $b \in U_\alpha$ and all $x \in F$.

Definition 9.8. We say that a bundle ξ with base B and structure group G is trivial iff ξ is isomorphic to the product bundle $B \times F$, according to the notion of isomorphism of Definition 9.7.

We can also define the notion of equivalence for fiber bundles over the *same base space* B (see Hirzebruch [57], Section 3.2, Chern [22], Section 5, and Husemoller [60], Chapter 5). We will see shortly that two bundles over the same base are equivalent iff they are isomorphic.

Definition 9.9. Given two fiber bundles $\xi_1 = (E_1, \pi_1, B, F, G)$ and $\xi_2 = (E_2, \pi_2, B, F, G)$ over the same base space B, we say that ξ_1 and ξ_2 are *equivalent* if there is an open cover $\mathcal{U} = (U_\alpha)_{\alpha \in I}$ for B, a family $\varphi = (\varphi_\alpha)_{\alpha \in I}$ of trivializations $\varphi_\alpha \colon \pi_1^{-1}(U_\alpha) \to U_\alpha \times F$ for ξ_1, a family $\varphi' = (\varphi'_\alpha)_{\alpha \in I}$ of trivializations $\varphi'_\alpha \colon \pi_2^{-1}(U_\alpha) \to U_\alpha \times F$ for ξ_2, and a family $(\rho_\alpha)_{\alpha \in I}$ of smooth maps $\rho_\alpha \colon U_\alpha \to G$, such that

$$g'_{\alpha\beta}(b) = \rho_\alpha(b) g_{\alpha\beta}(b) \rho_\beta(b)^{-1}, \qquad \text{for all } b \in U_\alpha \cap U_\beta;$$

see Figure 9.9.

Since the trivializations are bijections, the family $(\rho_\alpha)_{\alpha \in I}$ is unique. The conditions for two fiber bundles to be equivalent are local. Nevertheless, they are strong enough to imply that equivalent bundles over the same base are isomorphic (see Proposition 9.2).

The following proposition shows that isomorphic fiber bundles are equivalent.

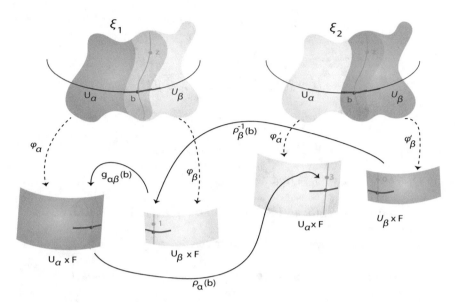

Fig. 9.9 An illustration of the mapping $g'_{\alpha\beta}(b) = \rho_\alpha(b)g_{\alpha\beta}(b)\rho_\beta(b)^{-1}$. Point 0 is $\varphi'_\beta(z') = (b, x)$. Point 1 is $(b, \rho_\beta^{-1}(b)(x))$. Point 2 is $(b, g_{\alpha\beta}(b)\rho_\beta^{-1}(b)(x))$, while Point 3 is $(b, \rho_\alpha(b)g_{\alpha\beta}(b)\rho_\beta^{-1}(b)(x)) = (b, g'_{\alpha\beta}(b)(x))$.

Proposition 9.1. *If two fiber bundles* $\xi_1 = (E_1, \pi_1, B, F, G)$ *and* $\xi_2 = (E_2, \pi_2, B, F, G)$ *over the same base space* B *are isomorphic, then they are equivalent.*

Proof. Let $f\colon \xi_1 \to \xi_2$ be a bundle isomorphism. In a slight abuse of notation, we also let $f\colon E_1 \to E_2$ be the isomorphism between E_1 and E_2. Then by Definition 9.7 we know that for some suitable open cover of the base B, and some trivializing families (φ_α) for ξ_1 and (φ'_α) for ξ_2, there is a family of maps $\rho_\alpha\colon U_\alpha \to G$, so that

$$\varphi'_\alpha \circ f \circ \varphi_\alpha{}^{-1}(b, x) = (b, \rho_\alpha(b)(x)),$$

for all $b \in U_\alpha$ and all $x \in F$. Recall that

$$\varphi_\alpha \circ \varphi_\beta^{-1}(b, x) = (b, g_{\alpha\beta}(b)(x)),$$

for all $b \in U_\alpha \cap U_\beta$ and all $x \in F$. This is equivalent to

$$\varphi_\beta^{-1}(b, x) = \varphi_\alpha^{-1}(b, g_{\alpha\beta}(b)(x)),$$

so it is notationally advantageous to introduce ψ_α such that $\psi_\alpha = \varphi_\alpha^{-1}$. Then we have

$$\psi_\beta(b, x) = \psi_\alpha(b, g_{\alpha\beta}(b)(x)), \qquad\qquad (*)$$

and

$$\varphi'_\alpha \circ f \circ \varphi_\alpha^{-1}(b, x) = (b, \rho_\alpha(b)(x))$$

becomes

$$\psi_\alpha(b, x) = f^{-1} \circ \psi'_\alpha(b, \rho_\alpha(b)(x)). \qquad\qquad (**)$$

By applying (*) and (**) we have

$$\psi_\beta(b, x) = \psi_\alpha(b, g_{\alpha\beta}(b)(x)) = f^{-1} \circ \psi'_\alpha(b, \rho_\alpha(b)(g_{\alpha\beta}(b)(x))).$$

On the other hand, applying (**) then (*) gives

$$\psi_\beta(b, x) = f^{-1} \circ \psi'_\beta(b, \rho_\beta(b)(x)) = f^{-1} \circ \psi'_\alpha(b, g'_{\alpha\beta}(b)(\rho_\beta(b)(x))),$$

from which we deduce

$$\rho_\alpha(b)(g_{\alpha\beta}(b)(x)) = g'_{\alpha\beta}(b)(\rho_\beta(b)(x)),$$

that is

$$g'_{\alpha\beta}(b) = \rho_\alpha(b)g_{\alpha\beta}(b)\rho_\beta(b)^{-1}, \qquad \text{for all } b \in U_\alpha \cap U_\beta,$$

as claimed.   □

Remark. If $\xi_1 = (E_1, \pi_1, B_1, F, G)$ and $\xi_2 = (E_2, \pi_2, B_2, F, G)$ are two bundles over different bases and $f: \xi_1 \to \xi_2$ is a bundle isomorphism, with $f = (f_B, f_E)$, then f_E and f_B are diffeomorphisms, and it is easy to see that we get the conditions

$$g'_{\alpha\beta}(f_B(b)) = \rho_\alpha(b)g_{\alpha\beta}(b)\rho_\beta(b)^{-1}, \qquad \text{for all } b \in U_\alpha \cap U_\beta.$$

The converse of Proposition 9.1 also holds.

Proposition 9.2. *If two fiber bundles* $\xi_1 = (E_1, \pi_1, B, F, G)$ *and* $\xi_2 = (E_2, \pi_2, B, F, G)$ *over the same base space* B *are equivalent, then they are isomorphic.*

Proof. Assume that ξ_1 and ξ_2 are equivalent. Then for some suitable open cover of the base B and some trivializing families (φ_α) for ξ_1 and (φ'_α) for ξ_2, there is a family of maps $\rho_\alpha: U_\alpha \to G$, so that

$$g'_{\alpha\beta}(b) = \rho_\alpha(b)g_{\alpha\beta}(b)\rho_\beta(b)^{-1}, \qquad \text{for all } b \in U_\alpha \cap U_\beta,$$

which can be written as

$$g'_{\alpha\beta}(b)\rho_\beta(b) = \rho_\alpha(b)g_{\alpha\beta}(b).$$

For every U_α, define f_α as the composition

$$\pi_1^{-1}(U_\alpha) \xrightarrow{\varphi_\alpha} U_\alpha \times F \xrightarrow{\tilde{f}_\alpha} U_\alpha \times F \xrightarrow{\varphi'_\alpha{}^{-1}} \pi_2^{-1}(U_\alpha)$$

$$z \longrightarrow (b, x) \longrightarrow (b, \rho_\alpha(b)(x)) \longrightarrow \varphi'_\alpha{}^{-1}(b, \rho_\alpha(b)(x));$$

that is,

$$f_\alpha(z) = \varphi'_\alpha{}^{-1}(b, \rho_\alpha(b)(x)), \qquad \text{with } z \in \pi_1^{-1}(U_\alpha) \text{ and } (b, x) = \varphi_\alpha(z).$$

Since $f_\alpha = \varphi'_\alpha{}^{-1} \circ \tilde{f}_\alpha \circ \varphi_\alpha$, the definition of f_α implies that

$$\varphi'_\alpha \circ f_\alpha \circ \varphi_\alpha{}^{-1}(b, x) = (b, \rho_\alpha(b)(x)),$$

for all $b \in U_\alpha$ and all $x \in F$, and locally f_α is a bundle isomorphism with respect to ρ_α. If we can prove that any two f_α and f_β agree on the overlap $U_\alpha \cap U_\beta$, then the f_α's patch and yield a bundle isomorphism between ξ_1 and ξ_2.

Now, on $U_\alpha \cap U_\beta$,

$$\varphi_\alpha \circ \varphi_\beta^{-1}(b, x) = (b, g_{\alpha\beta}(b)(x))$$

yields

$$\varphi_\beta^{-1}(b, x) = \varphi_\alpha^{-1}(b, g_{\alpha\beta}(b)(x)).$$

We need to show that for every $z \in U_\alpha \cap U_\beta$,

$$f_\alpha(z) = \varphi'_\alpha{}^{-1}(b, \rho_\alpha(b)(x)) = \varphi'_\beta{}^{-1}(b, \rho_\beta(b)(x')) = f_\beta(z),$$

where $\varphi_\alpha(z) = (b, x)$ and $\varphi_\beta(z) = (b, x')$.

From $\varphi_\alpha^1(b, x) = z = \varphi_\beta^{-1}(b, x') = \varphi_\alpha^{-1}(b, g_{\alpha\beta}(b)(x'))$, we get

$$x = g_{\alpha\beta}(b)(x').$$

We also have

$$\varphi'_\beta{}^{-1}(b, \rho_\beta(b)(x')) = \varphi'_\alpha{}^{-1}(b, g'_{\alpha\beta}(b)(\rho_\beta(b)(x'))),$$

and since $g'_{\alpha\beta}(b)\rho_\beta(b) = \rho_\alpha(b)g_{\alpha\beta}(b)$ and $x = g_{\alpha\beta}(b)(x')$, we get

$$\varphi'^{-1}_\beta(b, \rho_\beta(b)(x')) = \varphi'^{-1}_\alpha(b, g'_{\alpha\beta}(b)(\rho_\beta(b))(x')) = \varphi'^{-1}_\alpha(b, \rho_\alpha(b)(g_{\alpha\beta}(b))(x'))$$

$$= \varphi'^{-1}_\alpha(b, \rho_\alpha(b)(x)),$$

as desired. Therefore, the f_α's patch to yield a bundle map f, with respect to the family of maps $\rho_\alpha : U_\alpha \to G$.

The map f is bijective because it is an isomorphism on fibers, but it remains to show that it is a diffeomorphism. This is a local matter, and as the φ_α and φ'_α are diffeomorphisms, it suffices to show that the map $\widetilde{f_\alpha} : U_\alpha \times F \longrightarrow U_\alpha \times F$ given by

$$(b, x) \mapsto (b, \rho_\alpha(b)(x))$$

is a diffeomorphism. For this, observe that in local coordinates, the Jacobian matrix of this map is of the form

$$J = \begin{pmatrix} I & 0 \\ C & J(\rho_\alpha(b)) \end{pmatrix},$$

where I is the identity matrix and $J(\rho_\alpha(b))$ is the Jacobian matrix of $\rho_\alpha(b)$. Since $\rho_\alpha(b)$ is a diffeomorphism, $\det(J) \neq 0$, and by the inverse function theorem, the map $\widetilde{f_\alpha}$ is a diffeomorphism, as desired. $\square$

Remark. If in Proposition 9.2, $\xi_1 = (E_1, \pi_1, B_1, F, G)$ and $\xi_2 = (E_2, \pi_2, B_2, F, G)$ are two bundles over different bases and if we have a diffeomorphism $f_B : B_1 \to B_2$, and the conditions

$$g'_{\alpha\beta}(f_B(b)) = \rho_\alpha(b)g_{\alpha\beta}(b)\rho_\beta(b)^{-1}, \qquad \text{for all } b \in U_\alpha \cap U_\beta$$

hold, then there is a bundle isomorphism (f_B, f_E) between ξ_1 and ξ_2.

It follows from Proposition 9.1 and Proposition 9.2 that two bundles over the same base are equivalent iff they are isomorphic, a very useful fact. Actually, we can use the proof of Proposition 9.2 to show that any bundle morphism $f : \xi_1 \to \xi_2$ between two fiber bundles over the same base B is a bundle isomorphism. Because a bundle morphism f as above is fiber preserving, f is bijective, but it is not obvious that its inverse is smooth.

Proposition 9.3. *Any bundle morphism $f : \xi_1 \to \xi_2$ between two fiber bundles over the same base B is an isomorphism.*

Proof. Since f is bijective this is a local matter, and it is enough to prove that each $\widetilde{f_\alpha} : U_\alpha \times F \longrightarrow U_\alpha \times F$ is a diffeomorphism, since f can be written as

$$f = \varphi'_\alpha{}^{-1} \circ \widetilde{f_\alpha} \circ \varphi_\alpha,$$

with

$$\widetilde{f_\alpha}(b, x) = (b, \rho_\alpha(b)(x)).$$

However, the end of the proof of Proposition 9.2 shows that $\widetilde{f_\alpha}$ is a diffeomorphism.

□

9.3   Bundle Constructions via the Cocycle Condition

Given a fiber bundle $\xi = (E, \pi, B, F, G)$, we observed that the family $g = (g_{\alpha\beta})$ of transition maps $g_{\alpha\beta} \colon U_\alpha \cap U_\beta \to G$ induced by a trivializing family $\varphi = (\varphi_\alpha)_{\alpha \in I}$ relative to the open cover $\mathcal{U} = (U_\alpha)_{\alpha \in I}$ for B satisfies the *cocycle condition*

$$g_{\alpha\beta}(b)g_{\beta\gamma}(b) = g_{\alpha\gamma}(b),$$

for all α, β, γ such that $U_\alpha \cap U_\beta \cap U_\gamma \neq \emptyset$ and all $b \in U_\alpha \cap U_\beta \cap U_\gamma$.

Without altering anything, we may assume that $g_{\alpha\beta}$ is the (unique) function from $\emptyset$ to G, when $U_\alpha \cap U_\beta = \emptyset$. Then we call a family $g = (g_{\alpha\beta})_{(\alpha,\beta) \in I \times I}$ as above a $\mathcal{U}$-*cocycle*, or simply a *cocycle*.

Remarkably, given such a cocycle g relative to $\mathcal{U}$, a fiber bundle ξ_g over B with fiber F and structure group G having g as family of transition functions can be constructed.

In view of Proposition 9.1, we make the following definition.

Definition 9.10. We say that two cocycles $g = (g_{\alpha\beta})_{(\alpha,\beta) \in I \times I}$ and $g' = (g'_{\alpha\beta})_{(\alpha,\beta) \in I \times I}$ are *equivalent* if there is a family $(\rho_\alpha)_{\alpha \in I}$ of smooth maps $\rho_\alpha \colon U_\alpha \to G$, such that

$$g'_{\alpha\beta}(b) = \rho_\alpha(b)g_{\alpha\beta}(b)\rho_\beta(b)^{-1}, \qquad \text{for all } b \in U_\alpha \cap U_\beta.$$

Theorem 9.4. *Given two smooth manifolds B and F, a Lie group G acting effectively on F, an open cover $\mathcal{U} = (U_\alpha)_{\alpha \in I}$ of B, and a cocycle $g = (g_{\alpha\beta})_{(\alpha,\beta) \in I \times I}$, there is a fiber bundle $\xi_g = (E, \pi, B, F, G)$ whose transition maps are the maps in the cocycle g. Furthermore, if g and g' are equivalent cocycles, then ξ_g and $\xi_{g'}$ are isomorphic.*

Proof Sketch. First, we define the space Z as the disjoint sum

$$Z = \coprod_{\alpha \in I} (U_\alpha \times F).$$

We define the relation $\simeq$ on $Z \times Z$ as follows: For all $(b, x) \in U_\beta \times F$ and $(b, y) \in U_\alpha \times F$, if $U_\alpha \cap U_\beta \neq \emptyset$,

$$(b, x) \simeq (b, y) \quad \text{iff} \quad y = g_{\alpha\beta}(b)(x).$$

We let $E = Z/ \simeq$, and we give E the largest topology such that the injections $\eta_\alpha \colon U_\alpha \times F \to Z$ are smooth. The cocycle condition ensures that $\simeq$ is indeed an equivalence relation. We define $\pi \colon E \to B$ by $\pi([b, x]) = b$. If $p \colon Z \to E$ is the the quotient map, observe that the maps $p \circ \eta_\alpha \colon U_\alpha \times F \to E$ are injective, and that

$$\pi \circ p \circ \eta_\alpha(b, x) = b.$$

Thus,

$$p \circ \eta_\alpha \colon U_\alpha \times F \to \pi^{-1}(U_\alpha)$$

is a bijection, and we define the trivializing maps by setting

$$\varphi_\alpha = (p \circ \eta_\alpha)^{-1}.$$

It is easily verified that the corresponding transition functions are the original $g_{\alpha\beta}$. There are some details to check. A complete proof (the only one we could find!) is given in Steenrod [101], Part I, Section 3, Theorem 3.2. The fact that ξ_g and $\xi_{g'}$ are isomorphic when g and g' are equivalent follows from Proposition 9.2 (see Steenrod [101], Part I, Section 2, Lemma 2.10).   □

Remark. (For readers familiar with sheaves.) Hirzebruch defines the sheaf G_∞, where $G_\infty(U) = \Gamma(U, G_\infty)$ is the group of smooth functions $g \colon U \to G$, where U is some open subset of B and G is a Lie group acting effectively (on the left) on the fiber F. The group operation on $\Gamma(U, G_\infty)$ is induced by multiplication in G; that is, given two (smooth) functions $g \colon U \to G$ and $h \colon U \to G$,

$$gh(b) = g(b)h(b), \quad \text{for all } b \in U.$$

 Beware that gh is **not** function composition, unless G itself is a group of functions, which is the case for vector bundles.

Our conditions (b) and (c) are then replaced by the following equivalent condition: For all U_α, U_β in $\mathcal{U}$ such that $U_\alpha \cap U_\beta \neq \emptyset$, there is some $g_{\alpha\beta} \in \Gamma(U_\alpha \cap U_\beta, G_\infty)$ such that

$$\varphi_\alpha \circ \varphi_\beta^{-1}(b, x) = (b, g_{\alpha\beta}(b)(x)),$$

for all $b \in U_\alpha \cap U_\beta$ and all $x \in F$.

The classic source on fiber bundles is Steenrod [101]. The most comprehensive treatment of fiber bundles and vector bundles is probably given in Husemoller [60]. A more extensive list of references is given at the end of Section 9.9.

Remark. (The following paragraph is intended for readers familiar with Čech cohomology.) The cocycle condition makes it possible to view a fiber bundle over B as a member of a certain *(Čech) cohomology set* $\check{H}^1(B, \mathcal{G})$, where $\mathcal{G}$ denotes a certain sheaf of functions from the manifold B into the Lie group G, as explained in Hirzebruch [57], Section 3.2. However, this requires defining a noncommutative version of Čech cohomology (at least, for $\check{H}^1$), and clarifying when two open covers and two trivializations define the same fiber bundle over B, or equivalently, defining when two fiber bundles over B are equivalent. If the bundles under considerations are line bundles (see Definition 9.13), then $\check{H}^1(B, \mathcal{G})$ is actually a group. In this case, $G = \mathbf{GL}(1, \mathbb{R}) \cong \mathbb{R}^*$ in the real case, and $G = \mathbf{GL}(1, \mathbb{C}) \cong \mathbb{C}^*$ in the complex case (where $\mathbb{R}^* = \mathbb{R} - \{0\}$ and $\mathbb{C}^* = \mathbb{C} - \{0\}$), and the sheaf $\mathcal{G}$ is the sheaf of smooth (real-valued or complex-valued) functions vanishing nowhere. The group $\check{H}^1(B, \mathcal{G})$ plays an important role, especially when the bundle is a holomorphic line bundle over a complex manifold. In the latter case, it is called the *Picard group* of B.

Remark. (The following paragraph is intended for readers familiar with Čech cohomology.) Obviously, if we start with a fiber bundle $\xi = (E, \pi, B, F, G)$ whose transition maps are the cocycle $g = (g_{\alpha\beta})$, and form the fiber bundle ξ_g, the bundles ξ and ξ_g are equivalent. This leads to a characterization of the set of equivalence classes of fiber bundles over a base space B as the *cohomology set* $\check{H}^1(B, \mathcal{G})$.

In the present case, the sheaf $\mathcal{G}$ is defined such that $\Gamma(U, \mathcal{G})$ is the group of smooth maps from the open subset U of B to the Lie group G. Since G is not abelian, the coboundary maps have to be interpreted multiplicatively. If we define the sets of cochains $C^k(\mathcal{U}, \mathcal{G})$, so that

$$C^0(\mathcal{U}, \mathcal{G}) = \prod_\alpha \mathcal{G}(U_\alpha), \quad C^1(\mathcal{U}, \mathcal{G}) = \prod_{\alpha < \beta} \mathcal{G}(U_\alpha \cap U_\beta),$$

$$C^2(\mathcal{U}, \mathcal{G}) = \prod_{\alpha < \beta < \gamma} \mathcal{G}(U_\alpha \cap U_\beta \cap U_\gamma),$$

etc., then it is natural to define

$$\delta_0 : C^0(\mathcal{U}, \mathcal{G}) \to C^1(\mathcal{U}, \mathcal{G})$$

by

$$(\delta_0 g)_{\alpha\beta} = g_\alpha^{-1} g_\beta,$$

for any $g = (g_\alpha)$, with $g_\alpha \in \Gamma(U_\alpha, \mathcal{G})$. As to

$$\delta_1 \colon C^1(\mathcal{U}, \mathcal{G}) \to C^2(\mathcal{U}, \mathcal{G}),$$

since the cocycle condition in the usual case is

$$g_{\alpha\beta} + g_{\beta\gamma} = g_{\alpha\gamma},$$

we set

$$(\delta_1 g)_{\alpha\beta\gamma} = g_{\alpha\beta} g_{\beta\gamma} g_{\alpha\gamma}^{-1},$$

for any $g = (g_{\alpha\beta})$, with $g_{\alpha\beta} \in \Gamma(U_\alpha \cap U_\beta, \mathcal{G})$. Note that a cocycle $g = (g_{\alpha\beta})$ is indeed an element of $Z^1(\mathcal{U}, \mathcal{G})$, and the condition for being in the kernel of

$$\delta_1 \colon C^1(\mathcal{U}, \mathcal{G}) \to C^2(\mathcal{U}, \mathcal{G})$$

is the cocycle condition

$$g_{\alpha\beta}(b) g_{\beta\gamma}(b) = g_{\alpha\gamma}(b),$$

for all $b \in U_\alpha \cap U_\beta \cap U_\gamma$. In the commutative case, two cocycles g and g' are equivalent if their difference is a boundary, which can be stated as

$$g'_{\alpha\beta} + \rho_\beta = g_{\alpha\beta} + \rho_\alpha = \rho_\alpha + g_{\alpha\beta},$$

where $\rho_\alpha \in \Gamma(U_\alpha, \mathcal{G})$, for all $\alpha \in I$. In the present case, two cocycles g and g' are equivalent iff there is a family $(\rho_\alpha)_{\alpha \in I}$, with $\rho_\alpha \in \Gamma(U_\alpha, \mathcal{G})$, such that

$$g'_{\alpha\beta}(b) = \rho_\alpha(b) g_{\alpha\beta}(b) \rho_\beta(b)^{-1},$$

for all $b \in U_\alpha \cap U_\beta$. This is the same condition of equivalence defined earlier. Thus, it is easily seen that if $g, h \in Z^1(\mathcal{U}, \mathcal{G})$, then ξ_g and ξ_h are equivalent iff g and h correspond to the same element of the cohomology set $\check{H}^1(\mathcal{U}, \mathcal{G})$.

As usual, $\check{H}^1(B, \mathcal{G})$ is defined as the direct limit of the directed system of sets $\check{H}^1(\mathcal{U}, \mathcal{G})$ over the preordered directed family of open covers. For details, see Hirzebruch [57], Section 3.1. In summary, there is a bijection between the equivalence classes of fiber bundles over B (with fiber F and structure group G) and the cohomology set $\check{H}^1(B, \mathcal{G})$. In the case of line bundles, it turns out that $\check{H}^1(B, \mathcal{G})$ is in fact a group.

As an application of Theorem 9.4, we define the notion of *pullback* (or *induced*) bundle.

Definition 9.11. Let $\xi = (E, \pi, B, F, G)$ is a fiber bundle and assume we have a smooth map $f \colon N \to B$. The *pullback bundle* $f^* \xi = (f^* E, \pi^*, N, F, G)$ is the bundle over N, induced by the bundle map $(f^*, f) \colon f^* \xi \to \xi$

$$
\begin{array}{ccc}
f^*E & \xrightarrow{\ f^*\ } & E \\
{\scriptstyle \pi^*}\downarrow & & \downarrow{\scriptstyle \pi} \\
N & \xrightarrow{\ f\ } & B,
\end{array}
$$

where f^*E is a pullback in the categorical sense. This means that for any other bundle ξ' over N and any bundle map

$$
\begin{array}{ccc}
E' & \xrightarrow{\ f'\ } & E \\
{\scriptstyle \pi'}\downarrow & & \downarrow{\scriptstyle \pi} \\
N & \xrightarrow{\ f\ } & B,
\end{array}
$$

there is a unique bundle map $(\widetilde{f'}, \mathrm{id})\colon \xi' \to f^*\xi$, so that $(f', f) = (f^*, f)\circ(\widetilde{f'}, \mathrm{id})$ as illustrated by the following diagram.

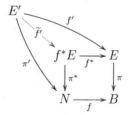

Definition 9.11 implies there is an isomorphism (natural)

$$
\mathrm{Hom}(\xi', \xi) \cong \mathrm{Hom}(\xi,' f^*\xi).
$$

As a consequence, by Proposition 9.3, for any bundle map between ξ' and ξ,

$$
\begin{array}{ccc}
E' & \xrightarrow{\ f'\ } & E \\
{\scriptstyle \pi'}\downarrow & & \downarrow{\scriptstyle \pi} \\
N & \xrightarrow{\ f\ } & B,
\end{array}
$$

there is an isomorphism, $\xi' \cong f^*\xi$.

The bundle $f^*\xi$ can be constructed as follows: Pick any open cover (U_α) of B, then $(f^{-1}(U_\alpha))$ is an open cover of N, and check that if $(g_{\alpha\beta})$ is a cocycle for ξ, then the maps $g_{\alpha\beta} \circ f\colon f^{-1}(U_\alpha) \cap f^{-1}(U_\beta) \to G$ satisfy the cocycle conditions. Then, $f^*\xi$ is the bundle defined by the cocycle $(g_{\alpha\beta} \circ f)$. We leave as an exercise to show that the pullback bundle $f^*\xi$ can be defined explicitly if we set

$$f^*E = \{(n, e) \in N \times E \mid f(n) = \pi(e)\},$$

$\pi^* = pr_1$ and $f^* = pr_2$. For any trivialization $\varphi_\alpha \colon \pi^{-1}(U_\alpha) \to U_\alpha \times F$ of ξ, we have

$$(\pi^*)^{-1}(f^{-1}(U_\alpha)) = \{(n, e) \in N \times E \mid n \in f^{-1}(U_\alpha), e \in \pi^{-1}(f(n))\},$$

and so we have a bijection $\widetilde{\varphi}_\alpha \colon (\pi^*)^{-1}(f^{-1}(U_\alpha)) \to f^{-1}(U_\alpha) \times F$, given by

$$\widetilde{\varphi}_\alpha(n, e) = (n, pr_2(\varphi_\alpha(e))).$$

By giving f^*E the smallest topology that makes each $\widetilde{\varphi}_\alpha$ a diffeomorphism, $\widetilde{\varphi}_\alpha$ is a trivialization of $f^*\xi$ over $f^{-1}(U_\alpha)$, and $f^*\xi$ is a smooth bundle. Note that the fiber of $f^*\xi$ over a point $n \in N$ is isomorphic to the fiber $\pi^{-1}(f(n))$ of ξ over $f(n)$. If $g \colon M \to N$ is another smooth map of manifolds, it is easy to check that

$$(f \circ g)^*\xi = g^*(f^*\xi).$$

Definition 9.12. Given a bundle $\xi = (E, \pi, B, F, G)$ and a submanifold N of B, we define the *restriction of ξ to N* as the bundle $\xi \upharpoonright N = (\pi^{-1}(N), \pi \upharpoonright \pi^{-1}(N), B, F, G)$.

There are two particularly interesting special cases of fiber bundles:

(1) *Vector bundles*, which are fiber bundles for which the typical fiber is a finite-dimensional vector space V, and the structure group is a subgroup of the group of linear isomorphisms ($\mathbf{GL}(n, \mathbb{R})$ or $\mathbf{GL}(n, \mathbb{C})$, where $n = \dim V$).
(2) *Principal fiber bundles*, which are fiber bundles for which the fiber F is equal to the structure group G, with G acting on itself by left translation.

First we discuss vector bundles.

9.4 Vector Bundles

Given a real vector space V, we denote by $\mathbf{GL}(V)$ (or $\mathrm{Aut}(V)$) the group of linear invertible maps from V to V. If V has dimension n, then $\mathbf{GL}(V)$ has dimension n^2. Obviously, $\mathbf{GL}(V)$ is isomorphic to $\mathbf{GL}(n, \mathbb{R})$, so we often write $\mathbf{GL}(n, \mathbb{R})$ instead of $\mathbf{GL}(V)$, but this may be slightly confusing if V is the dual space W^* of some other space W. If V is a complex vector space, we also denote by $\mathbf{GL}(V)$ (or $\mathrm{Aut}(V)$) the group of linear invertible maps from V to V, but this time $\mathbf{GL}(V)$ is isomorphic to $\mathbf{GL}(n, \mathbb{C})$, so we often write $\mathbf{GL}(n, \mathbb{C})$ instead of $\mathbf{GL}(V)$.

Definition 9.13. A *rank n real smooth vector bundle with fiber V* is a tuple $\xi = (E, \pi, B, V)$ such that $(E, \pi, B, V, \mathbf{GL}(V))$ is a smooth fiber bundle, the fiber V is a real vector space of dimension n, and the following conditions hold:

(a) For every $b \in B$, the fiber $\pi^{-1}(b)$ is an n-dimensional (real) vector space.
(b) For every trivialization $\varphi_\alpha \colon \pi^{-1}(U_\alpha) \to U_\alpha \times V$, for every $b \in U_\alpha$, the restriction of φ_α to the fiber $\pi^{-1}(b)$ is a linear isomorphism $\pi^{-1}(b) \longrightarrow V$.

A *rank n complex smooth vector bundle with fiber V* is a tuple, $\xi = (E, \pi, B, V)$, where $(E, \pi, B, V, \mathbf{GL}(V))$ is a smooth fiber bundle such that the fiber V is an n-dimensional complex vector space (viewed as a real smooth manifold), and Conditions (a) and (b) above hold (for complex vector spaces). When $n = 1$, a vector bundle is called a *line bundle*.

Definition 9.14. A *holomorphic vector bundle* is a fiber bundle where E, B are complex manifolds, V is a complex vector space of dimension n, the map π is holomorphic, the φ_α are biholomorphic, and the transition functions $g_{\alpha\beta}$ are holomorphic. When $n = 1$, a holomorphic vector bundle is called a *holomorphic line bundle*.

The trivial vector bundle $E = B \times V$ is often denoted ϵ^V. When $V = \mathbb{R}^k$, we also use the notation ϵ^k. Given a (smooth) manifold M of dimension n, the tangent bundle TM and the cotangent bundle T^*M are rank n vector bundles. Let us compute the transition functions for the tangent bundle TM, where M is a smooth manifold of dimension n.

For every $p \in M$, the tangent space T_pM consists of all equivalence classes of triples (U, φ, x), where (U, φ) is a chart with $p \in U$, $x \in \mathbb{R}^n$, and the equivalence relation on triples is given by

$$(U, \varphi, x) \equiv (V, \psi, y) \quad \text{iff} \quad (\psi \circ \varphi^{-1})'_{\varphi(p)}(x) = y.$$

We have a natural isomorphism $\theta_{U,\varphi,p} \colon \mathbb{R}^n \to T_pM$ between $\mathbb{R}^n$ and T_pM given by

$$\theta_{U,\varphi,p} \colon x \mapsto [(U, \varphi, x)], \qquad x \in \mathbb{R}^n.$$

Observe that for any two overlapping charts (U, φ) and (V, ψ),

$$\theta_{V,\psi,p}^{-1} \circ \theta_{U,\varphi,p} = (\psi \circ \varphi^{-1})'_z$$

for all $p \in U \cap V$, with $z = \varphi(p) = \psi(p)$. We let TM be the disjoint union

$$TM = \bigcup_{p \in M} T_pM,$$

define the projection $\pi \colon TM \to M$ so that $\pi(v) = p$ if $v \in T_pM$, and we give TM the weakest topology that makes the functions $\widetilde{\varphi} \colon \pi^{-1}(U) \to \mathbb{R}^{2n}$ given by

$$\widetilde{\varphi}(v) = (\varphi \circ \pi(v), \theta_{U,\varphi,\pi(v)}^{-1}(v))$$

continuous, where (U, φ) is any chart of M. Each function $\widetilde{\varphi} \colon \pi^{-1}(U) \to \varphi(U) \times \mathbb{R}^n$ is a homeomorphism, and given any two overlapping charts (U, φ) and (V, ψ), since $\theta_{V,\psi,p}^{-1} \circ \theta_{U,\varphi,p} = (\psi \circ \varphi^{-1})'_z$, with $z = \varphi(p) = \psi(p)$, the transition map

$$\widetilde{\psi} \circ \widetilde{\varphi}^{-1} \colon \varphi(U \cap V) \times \mathbb{R}^n \longrightarrow \psi(U \cap V) \times \mathbb{R}^n$$

is given by

$$\widetilde{\psi} \circ \widetilde{\varphi}^{-1}(z, x) = (\psi \circ \varphi^{-1}(z), (\psi \circ \varphi^{-1})'_z(x)), \qquad (z, x) \in \varphi(U \cap V) \times \mathbb{R}^n.$$

It is clear that $\widetilde{\psi} \circ \widetilde{\varphi}^{-1}$ is smooth. Moreover, the bijection

$$\tau_U \colon \pi^{-1}(U) \to U \times \mathbb{R}^n$$

given by

$$\tau_U(v) = (\pi(v), \theta_{U,\varphi,\pi(v)}^{-1}(v))$$

satisfies $pr_1 \circ \tau_U = \pi$ on $\pi^{-1}(U)$ and is a linear isomorphism restricted to fibers, so it is a trivialization for TM. For any two overlapping charts $(U_\alpha, \varphi_\alpha)$ and (U_β, φ_β), the transition function, $g_{\alpha\beta} \colon U_\alpha \cap U_\beta \to \mathbf{GL}(n, \mathbb{R})$ is given by

$$g_{\alpha\beta}(p) = (\varphi_\alpha \circ \varphi_\beta^{-1})'_{\varphi(p)}.$$

See Figure 9.10.

We can also compute trivialization maps for T^*M. This time, T^*M is the disjoint union

$$T^*M = \bigcup_{p \in M} T_p^*M,$$

and $\pi \colon T^*M \to M$ is given by $\pi(\omega) = p$ if $\omega \in T_p^*M$, where T_p^*M is the dual of the tangent space T_pM. For each chart (U, φ), by dualizing the map $\theta_{U,\varphi,p} \colon \mathbb{R}^n \to T_pM$, we obtain an isomorphism $\theta_{U,\varphi,p}^\top \colon T_p^*M \to (\mathbb{R}^n)^*$. Composing $\theta_{U,\varphi,p}^\top$ with the isomorphism $\iota \colon (\mathbb{R}^n)^* \to \mathbb{R}^n$ (induced by the canonical basis $(e_1, \ldots, e_n)$ of $\mathbb{R}^n$ and its dual basis), we get an isomorphism $\theta_{U,\varphi,p}^* = \iota \circ \theta_{U,\varphi,p}^\top \colon T_p^*M \to \mathbb{R}^n$. Then define the bijection

$$\widetilde{\varphi}^* \colon \pi^{-1}(U) \to \varphi(U) \times \mathbb{R}^n \subseteq \mathbb{R}^{2n}$$

by

$$\widetilde{\varphi}^*(\omega) = (\varphi \circ \pi(\omega), \theta_{U,\varphi,\pi(\omega)}^*(\omega)),$$

Fig. 9.10 An illustration of the line bundle TM over the curve M. The diagram in the upper left illustrates U and V, two overlapping charts of M. The middle illustrates the tangent bundle TM, the two trivialization maps, τ_U and τ_V, and the chart maps, $\widetilde{\varphi}$ and $\widetilde{\psi}$.

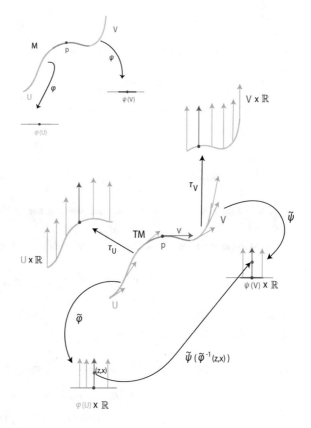

with $\omega \in \pi^{-1}(U)$. We give T^*M the weakest topology that makes the functions $\widetilde{\varphi}^*$ continuous, and then each function $\widetilde{\varphi}^*$ is a homeomorphism. Given any two overlapping charts (U, φ) and (V, ψ), as

$$\theta_{V,\psi,p}^{-1} \circ \theta_{U,\varphi,p} = (\psi \circ \varphi^{-1})'_{\varphi(p)},$$

by dualization we get

$$\theta_{U,\varphi,p}^{\top} \circ (\theta_{V,\psi,p}^{\top})^{-1} = \theta_{U,\varphi,p}^{\top} \circ (\theta_{V,\psi,p}^{-1})^{\top} = ((\psi \circ \varphi^{-1})'_{\varphi(p)})^{\top},$$

then

$$\theta_{V,\psi,p}^{\top} \circ (\theta_{U,\varphi,p}^{\top})^{-1} = (((\psi \circ \varphi^{-1})'_{\varphi(p)})^{\top})^{-1},$$

and so

$$\iota \circ \theta_{V,\psi,p}^{\top} \circ (\theta_{U,\varphi,p}^{\top})^{-1} \circ \iota^{-1} = \iota \circ (((\psi \circ \varphi^{-1})'_{\varphi(p)})^{\top})^{-1} \circ \iota^{-1};$$

that is,

$$\theta^*_{V,\psi,p} \circ (\theta^*_{U,\varphi,p})^{-1} = \iota \circ (((\psi \circ \varphi^{-1})'_{\varphi(p)})^\top)^{-1} \circ \iota^{-1}.$$

Consequently, the transition map

$$\widetilde{\psi}^* \circ (\widetilde{\varphi}^*)^{-1} \colon \varphi(U \cap V) \times \mathbb{R}^n \longrightarrow \psi(U \cap V) \times \mathbb{R}^n$$

is given by

$$\widetilde{\psi}^* \circ (\widetilde{\varphi}^*)^{-1}(z, x) = (\psi \circ \varphi^{-1}(z), \iota \circ (((\psi \circ \varphi^{-1})'_z)^\top)^{-1} \circ \iota^{-1}(x)), \quad (z, x) \in \varphi(U \cap V) \times \mathbb{R}^n.$$

If we view $(\psi \circ \varphi^{-1})'_z$ as a matrix, then we can forget ι and the second component of $\widetilde{\psi}^* \circ (\widetilde{\varphi}^*)^{-1}(z, x)$ is $(((\psi \circ \varphi^{-1})'_z)^\top)^{-1}x$.

We also have trivialization maps $\tau^*_U \colon \pi^{-1}(U) \to U \times (\mathbb{R}^n)^*$ for T^*M, given by

$$\tau^*_U(\omega) = (\pi(\omega), \theta^\top_{U,\varphi,\pi(\omega)}(\omega)),$$

for all $\omega \in \pi^{-1}(U)$. The transition function $g^*_{\alpha\beta} \colon U_\alpha \cap U_\beta \to \mathbf{GL}(n, \mathbb{R})$ is given by

$$\begin{aligned}
g^*_{\alpha\beta}(p)(\eta) &= \theta^\top_{U_\alpha,\varphi_\alpha,\pi(\eta)} \circ (\theta^\top_{U_\beta,\varphi_\beta,\pi(\eta)})^{-1}(\eta) \\
&= ((\theta^{-1}_{U_\alpha,\varphi_\alpha,\pi(\eta)} \circ \theta_{U_\beta,\varphi_\beta,\pi(\eta)})^\top)^{-1}(\eta) \\
&= (((\varphi_\alpha \circ \varphi^{-1}_\beta)'_{\varphi(p)})^\top)^{-1}(\eta),
\end{aligned}$$

with $\eta \in (\mathbb{R}^n)^*$. Also note that $\mathbf{GL}(n, \mathbb{R})$ should really be $\mathbf{GL}((\mathbb{R}^n)^*)$, but $\mathbf{GL}((\mathbb{R}^n)^*)$ is isomorphic to $\mathbf{GL}(n, \mathbb{R})$. We conclude that

$$g^*_{\alpha\beta}(p) = (g_{\alpha\beta}(p)^\top)^{-1}, \qquad \text{for every } p \in M.$$

This is a general property of dual bundles; see Property (f) in Section 9.5.

Maps of vector bundles are maps of fiber bundles such that the isomorphisms between fibers are linear.

Definition 9.15. Given two vector bundles $\xi_1 = (E_1, \pi_1, B_1, V)$ and $\xi_2 = (E_2, \pi_2, B_2, V)$ with the same typical fiber V, a *bundle map (or bundle morphism)* $f \colon \xi_1 \to \xi_2$ is a pair $f = (f_E, f_B)$ of smooth maps $f_E \colon E_1 \to E_2$ and $f_B \colon B_1 \to B_2$, such that:

(a) The following diagram commutes:

$$E_1 \xrightarrow{f_E} E_2$$

$$\pi_1 \downarrow \qquad \downarrow \pi_2$$

$$B_1 \xrightarrow{f_B} B_2$$

(b) For every $b \in B_1$, the map of fibers

$$f_E \upharpoonright \pi_1^{-1}(b) : \pi_1^{-1}(b) \to \pi_2^{-1}(f_B(b))$$

is a bijective linear map.

A bundle map *isomorphism* $f : \xi_1 \to \xi_2$ is defined as in Definition 9.6. Given two vector bundles $\xi_1 = (E_1, \pi_1, B, V)$ and $\xi_2 = (E_2, \pi_2, B, V)$ over the same base space B, we require $f_B = \mathrm{id}$.

Remark. Some authors do not require the preservation of fibers; that is, the map

$$f_E \upharpoonright \pi_1^{-1}(b) : \pi_1^{-1}(b) \to \pi_2^{-1}(f_B(b))$$

is simply a linear map. It is automatically bijective for bundle isomorphisms.

Note that Definition 9.15 does not include Condition (b) of Definition 9.7. However, because the restrictions of the maps φ_α, φ'_β, and f_E to the fibers are linear isomorphisms, it turns out that Condition (b) (of Definition 9.7) does hold.

Indeed, if $f_B(U_\alpha) \subseteq V_\beta$, then

$$\varphi'_\beta \circ f_E \circ \varphi_\alpha^{-1} : U_\alpha \times V \longrightarrow V_\beta \times V$$

is a smooth map of the form

$$\varphi'_\beta \circ f_E \circ \varphi_\alpha^{-1}(b, x) = (f_B(b), \rho_\alpha(b)(x))$$

for all $b \in U_\alpha$ and all $x \in V$, where $\rho_\alpha(b)$ is some linear isomorphism of V. Because $\varphi'_\beta \circ f_E \circ \varphi_\alpha^{-1}$ is smooth, the map $b \mapsto \rho_\alpha(b)$ is smooth, therefore, there is a smooth map $\rho_\alpha : U_\alpha \to \mathbf{GL}(V)$ so that

$$\varphi'_\beta \circ f \circ \varphi_\alpha^{-1}(b, x) = (f_B(b), \rho_\alpha(b)(x)),$$

and a vector bundle map is a fiber bundle map.

Definition 9.9 (equivalence of bundles) also applies to vector bundles (just replace G by $\mathbf{GL}(n, \mathbb{R})$ or $\mathbf{GL}(n, \mathbb{C})$) and defines the notion of equivalence of vector bundles over B. Since vector bundle maps are fiber bundle maps, Propositions 9.1 and 9.2 immediately yield

Proposition 9.5. *Two vector bundles* $\xi_1 = (E_1, \pi_1, B, V)$ *and* $\xi_2 = (E_2, \pi_2, B, V)$ *over the same base space B are equivalent iff they are isomorphic.*

Since a vector bundle map is a fiber bundle map, Proposition 9.3 also yields the useful fact:

Proposition 9.6. *Any vector bundle map* $f: \xi_1 \to \xi_2$ *between two vector bundles over the same base B is an isomorphism.*

Proposition 9.6 is proved in Milnor and Stasheff [78] for continuous vector bundles (see Lemma 2.3), and in Madsen and Tornehave [75] for smooth vector bundles as well as continuous vector bundles (see Lemma 15.10). The definition of a continuous vector bundle is similar to the definition of a smooth vector bundle, except that the manifolds are topological manifolds instead of smooth manifolds, and the maps involved are continuous rather than smooth.

Theorem 9.4 also holds for vector bundles and yields a technique for constructing new vector bundles over some base B.

Theorem 9.7. *Given a smooth manifold B, an n-dimensional (real, resp. complex) vector space V, an open cover* $\mathcal{U} = (U_\alpha)_{\alpha \in I}$ *of B, and a cocycle* $g = (g_{\alpha\beta})_{(\alpha,\beta) \in I \times I}$ *(with* $g_{\alpha\beta}: U_\alpha \cap U_\beta \to \mathbf{GL}(n, \mathbb{R})$*, resp.* $g_{\alpha\beta}: U_\alpha \cap U_\beta \to \mathbf{GL}(n, \mathbb{C})$*), there is a vector bundle* $\xi_g = (E, \pi, B, V)$ *whose transition maps are the maps in the cocycle g. Furthermore, if g and g' are equivalent cocycles, then* ξ_g *and* $\xi_{g'}$ *are equivalent.*

Observe that a cocycle $g = (g_{\alpha\beta})_{(\alpha,\beta) \in I \times I}$ is given by a family of matrices in $\mathbf{GL}(n, \mathbb{R})$ (resp. $\mathbf{GL}(n, \mathbb{C})$).

A vector bundle ξ always has a global section, namely the *zero section*, which assigns the element $0 \in \pi^{-1}(b)$ to every $b \in B$. A global section s is a *nonzero section* iff $s(b) \neq 0$ for all $b \in B$.

It is usually difficult to decide whether a bundle has a nonzero section. This question is related to the nontriviality of the bundle, and there is a useful test for triviality.

Assume ξ is a trivial rank n vector bundle. There is a bundle isomorphism $f: B \times V \to \xi$. For every $b \in B$, we know that $f(b, -)$ is a linear isomorphism, so for any choice of a basis $(e_1, \ldots, e_n)$ of V, we get a basis $(f(b, e_1), \ldots, f(b, e_n))$ of the fiber $\pi^{-1}(b)$. Thus, we have n global sections $s_1(b) = f(b, e_1), \ldots, s_n(b) = f(b, e_n)$ such that $(s_1(b), \ldots, s_n(b))$ forms a basis of the fiber $\pi^{-1}(b)$, for every $b \in B$.

Definition 9.16. Let $\xi = (E, \pi, B, V)$ be a rank n vector bundle. For any open subset $U \subseteq B$, an n-tuple of local sections $(s_1, \ldots, s_n)$ over U is called a *frame over U* iff $(s_1(b), \ldots, s_n(b))$ is a basis of the fiber $\pi^{-1}(b)$, for every $b \in U$. See Figure 9.11. If $U = B$, then the s_i are global sections and $(s_1, \ldots, s_n)$ is called a *frame* (of ξ).

The notion of a frame is due to Élie Cartan who (after Darboux) made extensive use of them under the name of *moving frame* (and the *moving frame method*). Cartan's terminology is intuitively clear: As a point b moves in U, the frame $(s_1(b), \ldots, s_n(b))$ moves from fiber to fiber. Physicists refer to a frame as a choice of *local gauge*.

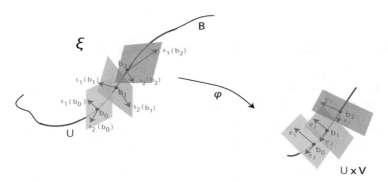

Fig. 9.11 A frame of $\xi = (E, \pi, B, \mathbb{R}^2)$ over U obtained from a local trivialization. For $i \le 0 \le 2$, $s_1(b_i) = \varphi^{-1}(b_i, e_1)$ and $s_2(b_i) = \varphi^{-1}(b_i, e_2)$, where e_1 and e_2 are the standard basis vectors of $\mathbb{R}^2$.

The converse of the property established just before Definition 9.16 is also true.

Proposition 9.8. *A rank n vector bundle ξ is trivial iff it possesses a frame of global sections.*

Proof. (Adapted from Milnor and Stasheff [78], Theorem 2.2.) We only need to prove that if ξ has a frame $(s_1, \ldots, s_n)$, then it is trivial. Pick a basis $(e_1, \ldots, e_n)$ of V, and define the map $f : B \times V \to \xi$ as follows:

$$f(b, v) = \sum_{i=1}^{n} v_i s_i(b),$$

where $v = \sum_{i=1}^{n} v_i e_i$. Clearly, f is bijective on fibers, smooth, and a map of vector bundles. By Proposition 9.6, the bundle map, f, is an isomorphism. $\square$

The above considerations show that if ξ is any rank n vector bundle, not necessarily trivial, then for any local trivialization $\varphi_\alpha : \pi^{-1}(U_\alpha) \to U_\alpha \times V$, there are always frames over U_α. Indeed, for every choice of a basis $(e_1, \ldots, e_n)$ of the typical fiber V, if we set

$$s_i^\alpha(b) = \varphi_\alpha^{-1}(b, e_i), \qquad b \in U_\alpha, \ 1 \le i \le n,$$

then $(s_1^\alpha, \ldots, s_n^\alpha)$ is a frame over U_α. See Figure 9.11.

Definition 9.17. Given any two vector spaces V and W, both of dimension n, we denote by $\mathrm{Iso}(V, W)$ the space of all linear isomorphisms between V and W. The *space of n-frames $F(V)$* is the set of bases of V.

Since every basis $(v_1, \ldots, v_n)$ of V is in one-to-one correspondence with the map from $\mathbb{R}^n$ to V given by $e_i \mapsto v_i$, where $(e_1, \ldots, e_n)$ is the canonical basis of $\mathbb{R}^n$ (so, $e_i = (0, \ldots, 1, \ldots 0)$ with the 1 in the ith slot), we have an isomorphism,

$$F(V) \cong \mathrm{Iso}(\mathbb{R}^n, V).$$

(The choice of a basis in V also yields an isomorphism $\mathrm{Iso}(\mathbb{R}^n, V) \cong \mathbf{GL}(n, \mathbb{R})$, so $F(V) \cong \mathbf{GL}(n, \mathbb{R})$.)

Definition 9.18. For any rank n vector bundle ξ, we can form the *frame bundle* $F(\xi)$, by replacing the fiber $\pi^{-1}(b)$ over any $b \in B$ by $F(\pi^{-1}(b))$.

In fact, $F(\xi)$ can be constructed using Theorem 9.4. Indeed, identifying $F(V)$ with $\mathrm{Iso}(\mathbb{R}^n, V)$, the group $\mathbf{GL}(n, \mathbb{R})$ acts on $F(V)$ effectively on the left *via*

$$A \cdot v = v \circ A^{-1}.$$

(The only reason for using A^{-1} instead of A is that we want a left action.) The resulting bundle has typical fiber $F(V) \cong \mathbf{GL}(n, \mathbb{R})$, and turns out to be a principal bundle. We will take a closer look at principal bundles in Section 9.9.

We conclude this section with an example of a bundle that plays an important role in algebraic geometry, the *canonical line bundle on* $\mathbb{RP}^n$. Let $H_n^{\mathbb{R}} \subseteq \mathbb{RP}^n \times \mathbb{R}^{n+1}$ be the subset

$$H_n^{\mathbb{R}} = \{(L, v) \in \mathbb{RP}^n \times \mathbb{R}^{n+1} \mid v \in L\},$$

where $\mathbb{RP}^n$ is viewed as the set of lines L in $\mathbb{R}^{n+1}$ through 0, or more explicitly,

$$H_n^{\mathbb{R}} = \{((x_0: \cdots : x_n), \lambda(x_0, \ldots, x_n)) \mid (x_0: \cdots : x_n) \in \mathbb{RP}^n, \lambda \in \mathbb{R}\}.$$

Geometrically, $H_n^{\mathbb{R}}$ consists of the set of lines $[(x_0, \ldots, x_n)]$ associated with points $(x_0: \cdots : x_n)$ of $\mathbb{RP}^n$. If we consider the projection $\pi: H_n^{\mathbb{R}} \to \mathbb{RP}^n$ of $H_n^{\mathbb{R}}$ onto $\mathbb{RP}^n$, we see that each fiber is isomorphic to $\mathbb{R}$. We claim that $H_n^{\mathbb{R}}$ is a line bundle. For this, we exhibit trivializations, leaving as an exercise the fact that $H_n^{\mathbb{R}}$ is a manifold of dimension $n + 1$.

Recall the open cover $U_0, \ldots, U_n$ of $\mathbb{RP}^n$, where

$$U_i = \{(x_0: \cdots : x_n) \in \mathbb{RP}^n \mid x_i \neq 0\}.$$

Then the maps $\varphi_i: \pi^{-1}(U_i) \to U_i \times \mathbb{R}$ given by

$$\varphi_i((x_0: \cdots : x_n), \lambda(x_0, \ldots, x_n)) = ((x_0: \cdots : x_n), \lambda x_i)$$

are trivializations. The transition function $g_{ij}: U_i \cap U_j \to \mathbf{GL}(1, \mathbb{R})$ is given by

$$g_{ij}(x_0: \cdots : x_n)(u) = \frac{x_i}{x_j} u,$$

where we identify $\mathbf{GL}(1, \mathbb{R})$ and $\mathbb{R}^* = \mathbb{R} - \{0\}$.

Interestingly, the bundle $H_n^{\mathbb{R}}$ is nontrivial for all $n \geq 1$. For this, by Proposition 9.8 and since $H_n^{\mathbb{R}}$ is a line bundle, it suffices to prove that every global section vanishes at some point. So, let σ be any section of $H_n^{\mathbb{R}}$. Composing the projection, $p \colon S^n \longrightarrow \mathbb{RP}^n$, with σ, we get a smooth function, $s = \sigma \circ p \colon S^n \longrightarrow H_n^{\mathbb{R}}$, and we have

$$s(x) = (p(x), f(x)x),$$

for every $x \in S^n$, where $f \colon S^n \to \mathbb{R}$ is a smooth function. Moreover, f satisfies

$$f(-x) = -f(x),$$

since $s(-x) = (p(-x), -f(-x)x) = (p(x), -f(-x)x) = (p(x), f(x)x) = s(x)$. As S^n is connected and f is continuous, by the intermediate value theorem, there is some x such that $f(x) = 0$, and thus, σ vanishes, as desired.

The reader should look for a geometric representation of $H_1^{\mathbb{R}}$. It turns out that $H_1^{\mathbb{R}}$ is an open Möbius strip; that is, a Möbius strip with its boundary deleted (see Milnor and Stasheff [78], Chapter 2). There is also a complex version of the canonical line bundle on $\mathbb{CP}^n$, with

$$H_n = \{(L, v) \in \mathbb{CP}^n \times \mathbb{C}^{n+1} \mid v \in L\},$$

where $\mathbb{CP}^n$ is viewed as the set of lines L in $\mathbb{C}^{n+1}$ through 0. These bundles are also nontrivial. Furthermore, unlike the real case, the dual bundle H_n^* is not isomorphic to H_n. Indeed, H_n^* turns out to have nonzero global holomorphic sections!

9.5 Operations on Vector Bundles

Because the fibers of a vector bundle are vector spaces all isomorphic to some given space V, we can perform operations on vector bundles that extend familiar operations on vector spaces, such as: direct sum, tensor product, (linear) function space, and dual space. Basically, the same operation is applied on fibers. It is usually more convenient to define operations on vector bundles in terms of operations on cocycles, using Theorem 9.7.

(a) (*Whitney Sum* or *Direct Sum*)
   If $\xi = (E, \pi, B, V)$ is a rank m vector bundle and $\xi' = (E', \pi', B, W)$ is a rank n vector bundle, both over the *same base* B, then their Whitney sum $\xi \oplus \xi'$ is the rank $(m + n)$ vector bundle whose fiber over any $b \in B$ is the direct sum $E_b \oplus E_b'$; that is, the vector bundle with typical fiber $V \oplus W$ (given by Theorem 9.7) specified by the cocycle whose matrices are

$$\begin{pmatrix} g_{\alpha\beta}(b) & 0 \\ 0 & g'_{\alpha\beta}(b) \end{pmatrix}, \qquad b \in U_\alpha \cap U_\beta.$$

(b) (*Tensor Product*)

If $\xi = (E, \pi, B, V)$ is a rank m vector bundle and $\xi' = (E', \pi', B, W)$ is a rank n vector bundle, both over the *same base* B, then their tensor product $\xi \otimes \xi'$ is the rank mn vector bundle whose fiber over any $b \in B$ is the tensor product $E_b \otimes E'_b$; that is, the vector bundle with typical fiber $V \otimes W$ (given by Theorem 9.7) specified by the cocycle whose matrices are

$$g_{\alpha\beta}(b) \otimes g'_{\alpha\beta}(b), \qquad b \in U_\alpha \cap U_\beta.$$

(Here, we identify a matrix with the corresponding linear map.)

(c) (*Tensor Power*)

If $\xi = (E, \pi, B, V)$ is a rank m vector bundle, then for any $k \geq 0$, we can define the tensor power bundle $\xi^{\otimes k}$, whose fiber over any $b \in B$ is the tensor power $E_b^{\otimes k}$, and with typical fiber $V^{\otimes k}$. (When $k = 0$, the fiber is $\mathbb{R}$ or $\mathbb{C}$). The bundle $\xi^{\otimes k}$ is determined by the cocycle

$$g_{\alpha\beta}^{\otimes k}(b), \qquad b \in U_\alpha \cap U_\beta.$$

(d) (*Exterior Power*)

If $\xi = (E, \pi, B, V)$ is a rank m vector bundle, then for any $k \geq 0$, we can define the exterior power bundle $\bigwedge^k \xi$, whose fiber over any $b \in B$ is the exterior power $\bigwedge^k E_b$, and with typical fiber $\bigwedge^k V$. The bundle $\bigwedge^k \xi$ is determined by the cocycle

$$\bigwedge^k g_{\alpha\beta}(b), \qquad b \in U_\alpha \cap U_\beta.$$

Using (a), we also have the *exterior algebra bundle* $\bigwedge \xi = \bigoplus_{k=0}^m \bigwedge^k \xi$. (When $k = 0$, the fiber is $\mathbb{R}$ or $\mathbb{C}$).

(e) (*Symmetric Power*)

If $\xi = (E, \pi, B, V)$ is a rank m vector bundle, then for any $k \geq 0$, we can define the symmetric power bundle $S^k \xi$, whose fiber over any $b \in B$ is the symmetric power $S^k E_b$, and with typical fiber $S^k V$. (When $k = 0$, the fiber is $\mathbb{R}$ or $\mathbb{C}$). The bundle $S^k \xi$ is determined by the cocycle

$$S^k g_{\alpha\beta}(b), \qquad b \in U_\alpha \cap U_\beta.$$

(f) (*Tensor Bundle of Type (r, s)*)

If $\xi = (E, \pi, B, V)$ is a rank m vector bundle, then for any $r, s \geq 0$, we can define the bundle $T^{r,s}\xi$ whose fiber over any $b \in \xi$ is the tensor space $T^{r,s}E_b$, and with typical fiber $T^{r,s}V$. The bundle $T^{r,s}\xi$ is determined by the cocycle

$$g_{\alpha\beta}^{\otimes r}(b) \otimes ((g_{\alpha\beta}(b)^\top)^{-1})^{\otimes s}(b), \qquad b \in U_\alpha \cap U_\beta.$$

(g) (*Dual Bundle*)

If $\xi = (E, \pi, B, V)$ is a rank m vector bundle, then its dual bundle ξ^* is the rank m vector bundle whose fiber over any $b \in B$ is the dual space E_b^*; that is, the vector bundle with typical fiber V^* (given by Theorem 9.7) specified by the cocycle whose matrices are

$$(g_{\alpha\beta}(b)^\top)^{-1}, \qquad b \in U_\alpha \cap U_\beta.$$

The reason for this seemingly complicated formula is this: For any trivialization $\varphi_\alpha \colon \pi^{-1}(U_\alpha) \to U_\alpha \times V$, for any $b \in B$, recall that the restriction $\varphi_{\alpha,b} \colon \pi^{-1}(b) \to V$ of φ_α to $\pi^{-1}(b)$ is a linear isomorphism. By dualization we get a map $\varphi_{\alpha,b}^\top \colon V^* \to (\pi^{-1}(b))^*$, and thus $\varphi_{\alpha,b}^*$ for ξ^* is given by

$$\varphi_{\alpha,b}^* = (\varphi_{\alpha,b}^\top)^{-1} \colon (\pi^{-1}(b))^* \to V^*.$$

As $g_{\alpha\beta}^*(b) = \varphi_{\alpha,b}^* \circ (\varphi_{\beta,b}^*)^{-1}$, we get

$$\begin{aligned}
g_{\alpha\beta}^*(b) &= (\varphi_{\alpha,b}^\top)^{-1} \circ \varphi_{\beta,b}^\top \\
&= ((\varphi_{\beta,b}^\top)^{-1} \circ \varphi_{\alpha,b}^\top)^{-1} \\
&= ((\varphi_{\beta,b}^{-1})^\top \circ \varphi_{\alpha,b}^\top)^{-1} \\
&= ((\varphi_{\alpha,b} \circ \varphi_{\beta,b}^{-1})^\top)^{-1} \\
&= (g_{\alpha\beta}(b)^\top)^{-1},
\end{aligned}$$

as claimed.

(h) (*Hom Bundle*)

If $\xi = (E, \pi, B, V)$ is a rank m vector bundle and $\xi' = (E', \pi', B, W)$ is a rank n vector bundle, both over the same base B, then their *Hom* bundle $Hom(\xi, \xi')$ is the rank mn vector bundle whose fiber over any $b \in B$ is $Hom(E_b, E_b')$; that is, the vector bundle with typical fiber $Hom(V, W)$. The transition functions of this bundle are obtained as follows: For any trivializations $\varphi_\alpha \colon \pi^{-1}(U_\alpha) \to U_\alpha \times V$ and $\varphi_\alpha' \colon (\pi')^{-1}(U_\alpha) \to U_\alpha \times W$, for any $b \in B$, recall that the restrictions $\varphi_{\alpha,b} \colon \pi^{-1}(b) \to V$ and $\varphi_{\alpha,b}' \colon (\pi')^{-1}(b) \to W$ are linear isomorphisms. We have a linear isomorphism $\varphi_{\alpha,b}^{Hom} \colon Hom(\pi^{-1}(b), (\pi')^{-1}(b)) \longrightarrow$

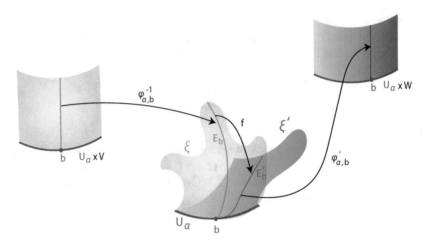

Fig. 9.12 A schematic illustration of $\varphi_{\alpha,b}^{\text{Hom}}$.

$\text{Hom}(V, W)$ given by

$$\varphi_{\alpha,b}^{\text{Hom}}(f) = \varphi_{\alpha,b}' \circ f \circ \varphi_{\alpha,b}^{-1}, \qquad f \in \text{Hom}(\pi^{-1}(b), (\pi')^{-1}(b)).$$

Then, $g_{\alpha\beta}^{\text{Hom}}(b) = \varphi_{\alpha,b}^{\text{Hom}} \circ (\varphi_{\beta,b}^{\text{Hom}})^{-1}$. See Figure 9.12.

As an illustration of (d), consider the exterior power $\bigwedge^r T^*M$, where M is a manifold of dimension n. We have trivialization maps $\tau_U^* \colon \pi^{-1}(U) \to U \times \bigwedge^r(\mathbb{R}^n)^*$ for $\bigwedge^r T^*M$, given by

$$\tau_U^*(\omega) = (\pi(\omega), \bigwedge^r \theta_{U,\varphi,\pi(\omega)}^\top(\omega)),$$

for all $\omega \in \pi^{-1}(U)$. The transition function $g_{\alpha\beta}^{\bigwedge^r} \colon U_\alpha \cap U_\beta \to \mathbf{GL}(n, \mathbb{R})$ is given by

$$g_{\alpha\beta}^{\bigwedge^r}(p)(\omega) = (\bigwedge^r(((\varphi_\alpha \circ \varphi_\beta^{-1})_{\varphi(p)}')^\top)^{-1})(\omega),$$

for all $\omega \in \pi^{-1}(U)$. Consequently,

$$g_{\alpha\beta}^{\bigwedge^r}(p) = \bigwedge^r(g_{\alpha\beta}(p)^\top)^{-1},$$

for every $p \in M$, a special case of (f).

In view of the canonical isomorphism $\text{Hom}(V, W) \cong V^* \otimes W$, it is easy to show the following results.

Proposition 9.9. *The vector bundle* $\mathcal{H}om(\xi, \xi')$, *is isomorphic to* $\xi^* \otimes \xi'$. *Similarly,* ξ^{**} *is isomorphic to* ξ. *We also have the isomorphism*

$$T^{r,s}\xi \cong \xi^{\otimes r} \otimes (\xi^*)^{\otimes s}.$$

Do not confuse the space of bundle morphisms $\mathrm{Hom}(\xi, \xi')$ with the bundle $\mathcal{H}om(\xi, \xi')$. However, observe that $\mathrm{Hom}(\xi, \xi')$ is the set of global sections of $\mathcal{H}om(\xi, \xi')$.

Remark. For rank 1 vector bundles, namely line bundles, it is easy to show that the set of equivalence classes of line bundles over a base B forms a group, where the group operation is $\otimes$, the inverse is $*$ (dual), and the identity element is the trivial bundle. This is the *Picard group* of B.

In general, the dual ξ^* of a bundle is *not* isomorphic to the original bundle ξ. This is because V^* is *not* canonically isomorphic to V, and to get a bundle isomorphism between ξ and ξ^*, we need canonical isomorphisms between the fibers. However, if ξ is real, then (using a partition of unity) ξ can be given a Euclidean metric, and so ξ and ξ^* are isomorphic.

It is *not* true in general that a complex vector bundle is isomorphic to its dual because a Hermitian metric only induces a canonical isomorphism between E^* and $\overline{E}$, where $\overline{E}$ is the conjugate of E, with scalar multiplication in $\overline{E}$ given by $(z, w) \mapsto \overline{w}z$.

Remark. Given a real vector bundle, ξ, the *complexification* $\xi_{\mathbb{C}}$ $\xi_{\mathbb{C}}$of ξ is the complex vector bundle defined by

$$\xi_{\mathbb{C}} = \xi \otimes_{\mathbb{R}} \epsilon_{\mathbb{C}},$$

where $\epsilon_{\mathbb{C}} = B \times \mathbb{C}$ is the trivial complex line bundle. Given a complex vector bundle ξ, by viewing its fiber as a real vector space we obtain the real vector bundle $\xi_{\mathbb{R}}$.

Proposition 9.10. *The following facts hold.*

(1) For every real vector bundle ξ,

$$(\xi_{\mathbb{C}})_{\mathbb{R}} \cong \xi \oplus \xi.$$

(2) For every complex vector bundle ξ,

$$(\xi_{\mathbb{R}})_{\mathbb{C}} \cong \xi \oplus \xi^*.$$

9.6   Properties of Vector Bundle Sections

It can be shown (see Madsen and Tornehave [75], Chapter 15) that for every real smooth vector bundle ξ, there is some integer k such that ξ has a complement η in ϵ^k, where $\epsilon^k = B \times \mathbb{R}^k$ is the trivial rank k vector bundle, so that

$$\xi \oplus \eta = \epsilon^k.$$

This fact can be used to prove an interesting property of the space of global sections $\Gamma(\xi)$.

First, observe that $\Gamma(\xi)$ is not just a real vector space, but also a $C^\infty(B)$-module (see Section 2.11). Indeed, for every smooth function $f : B \to \mathbb{R}$ and every smooth section $s : B \to E$, the map $fs : B \to E$ given by

$$(fs)(b) = f(b)s(b), \qquad b \in B,$$

is a smooth section of ξ.

In general, $\Gamma(\xi)$ is not a free $C^\infty(B)$-module unless ξ is trivial. However, the above remark implies that

$$\Gamma(\xi) \oplus \Gamma(\eta) = \Gamma(\epsilon^k),$$

where $\Gamma(\epsilon^k)$ is a free $C^\infty(B)$-module of dimension $\dim(\xi) + \dim(\eta)$.

This proves that $\Gamma(\xi)$ is a finitely generated $C^\infty(B)$-module which is a summand of a free $C^\infty(B)$-module. Such modules are *projective modules*; see Definition 2.23 in Section 2.11. Therefore, $\Gamma(\xi)$ is a finitely generated projective $C^\infty(B)$-module.

The following isomorphisms can be shown (see Madsen and Tornehave [75], Chapter 16).

Proposition 9.11. *The following isomorphisms hold for vector bundles:*

$$\Gamma(\mathcal{H}om(\xi, \eta)) \cong \operatorname{Hom}_{C^\infty(B)}(\Gamma(\xi), \Gamma(\eta))$$

$$\Gamma(\xi \otimes \eta) \cong \Gamma(\xi) \otimes_{C^\infty(B)} \Gamma(\eta)$$

$$\Gamma(\xi^*) \cong \operatorname{Hom}_{C^\infty(B)}(\Gamma(\xi), C^\infty(B)) = (\Gamma(\xi))^*$$

$$\Gamma(\overset{k}{\bigwedge} \xi) \cong \underset{C^\infty(B)}{\overset{k}{\bigwedge}} (\Gamma(\xi)).$$

Using the operations on vector bundles described in Section 9.5, we can define the set of vector valued differential forms $\mathcal{A}^k(M; F)$ defined in Section 4.5 as the set of smooth sections of the vector bundle $\left(\bigwedge^k T^*M \right) \otimes \epsilon_F$; that is, as

$$\mathcal{A}^k(M; F) = \Gamma\left(\left(\overset{k}{\bigwedge} T^*M\right) \otimes \epsilon_F\right),$$

where ϵ_F is the trivial vector bundle $\epsilon_F = M \times F$.

Proposition 9.12. *We have the following isomorphisms:*

$$\mathcal{A}^k(M; F) \cong \mathcal{A}^k(M) \otimes_{C^\infty(M)} C^\infty(M; F) \cong \mathrm{Alt}^k_{C^\infty(M)}(\mathfrak{X}(M); C^\infty(M; F)).$$

Proof. By Proposition 9.11 and since $\Gamma(\epsilon_F) \cong C^\infty(M; F)$ and $\mathcal{A}^k(M) = \Gamma\left(\bigwedge^k T^*M\right)$, we have

$$\mathcal{A}^k(M; F) = \Gamma\left(\left(\overset{k}{\bigwedge} T^*M\right) \otimes \epsilon_F\right)$$

$$\cong \Gamma\left(\overset{k}{\bigwedge} T^*M\right) \otimes_{C^\infty(M)} \Gamma(\epsilon_F)$$

$$= \mathcal{A}^k(M) \otimes_{C^\infty(M)} C^\infty(M; F)$$

$$\cong \overset{k}{\underset{C^\infty(M)}{\bigwedge}} (\Gamma(TM))^* \otimes_{C^\infty(M)} C^\infty(M; F)$$

$$\cong \mathrm{Hom}_{C^\infty(M)}\left(\overset{k}{\underset{C^\infty(M)}{\bigwedge}} \Gamma(TM), C^\infty(M; F)\right)$$

$$\cong \mathrm{Alt}^k_{C^\infty(M)}(\mathfrak{X}(M); C^\infty(M; F)),$$

with all of the spaces viewed as $C^\infty(M)$-modules, and where we used the fact that $\mathfrak{X}(X) = \Gamma(TM)$ is a projective module, and that Proposition 3.5 is still valid for exterior powers over a commutative ring. Therefore,

$$\mathcal{A}^k(M; F) \cong \mathcal{A}^k(M) \otimes_{C^\infty(M)} C^\infty(M; F) \cong \mathrm{Alt}^k_{C^\infty(M)}(\mathfrak{X}(M); C^\infty(M; F)),$$

which reduces to Proposition 4.15 when $F = \mathbb{R}$.   □

In Section 10.2, we will consider a generalization of the above situation where the trivial vector bundle ϵ_F is replaced by any vector bundle $\xi = (E, \pi, B, V)$, and where $M = B$.

9.7   Duality Between Vector Fields and Differential Forms and Covariant Derivatives of Tensor Fields

Given a manifold M, the covariant derivative ∇_X given by a connection ∇ on TM can be extended to a covariant derivative $\nabla_X^{r,s}$ defined on tensor fields in $\Gamma(M, T^{r,s}(M))$ for all $r, s \geq 0$, where

$$T^{r,s}(M) = T^{\otimes r} M \otimes (T^* M)^{\otimes s}.$$

We already have $\nabla_X^{1,0} = \nabla_X$ and it is natural to set $\nabla_X^{0,0} f = X[f] = df(X)$. Recall that there is an isomorphism between the set of tensor fields $\Gamma(M, T^{r,s}(M))$, and the set of $C^\infty(M)$-multilinear maps

$$\Phi \colon \underbrace{\mathcal{A}^1(M) \times \cdots \times \mathcal{A}^1(M)}_{r} \times \underbrace{\mathfrak{X}(M) \times \cdots \times \mathfrak{X}(M)}_{s} \longrightarrow C^\infty(M),$$

where $\mathcal{A}^1(M)$ and $\mathfrak{X}(M)$ are $C^\infty(M)$-modules.

The next proposition is left as an exercise. For help, see O'Neill [84], Chapter 2, Proposition 13 and Theorem 15.

Proposition 9.13. *For every vector field* $X \in \mathfrak{X}(M)$, *there is a unique family of* $\mathbb{R}$-*linear map* $\nabla^{r,s} \colon \Gamma(M, T^{r,s}(M)) \to \Gamma(M, T^{r,s}(M))$, *with* $r, s \geq 0$, *such that*

(a) $\nabla_X^{0,0} f = df(X)$, *for all* $f \in C^\infty(M)$ *and* $\nabla_X^{1,0} = \nabla_X$, *for all* $X \in \mathfrak{X}(M)$.
(b) $\nabla_X^{r_1+r_2, s_1+s_2}(S \otimes T) = \nabla_X^{r_1,s_1}(S) \otimes T + S \otimes \nabla_X^{r_2,s_2}(T)$, *for all* $S \in \Gamma(M, T^{r_1, s_1}(M))$ *and all* $T \in \Gamma(M, T^{r_2, s_2}(M))$.
(c) $\nabla_X^{r-1,s-1}(c_{ij}(S)) = c_{ij}(\nabla_X^{r,s}(S))$, *for all* $S \in \Gamma(M, T^{r,s}(M))$ *and all contractions,* c_{ij}, *of* $\Gamma(M, T^{r,s}(M))$.

Furthermore,

$$(\nabla_X^{0,1}\theta)(Y) = X[\theta(Y)] - \theta(\nabla_X Y),$$

for all $X, Y \in \mathfrak{X}(M)$ *and all one-forms,* $\theta \in \mathcal{A}^1(M)$, *and for every* $S \in \Gamma(M, T^{r,s}(M))$, *with* $r + s \geq 2$, *the covariant derivative* $\nabla_X^{r,s}(S)$ *is given by*

$$(\nabla_X^{r,s} S)(\theta_1, \ldots, \theta_r, X_1, \ldots, X_s) = X[S(\theta_1, \ldots, \theta_r, X_1, \ldots, X_s)]$$

$$- \sum_{i=1}^{r} S(\theta_1, \ldots, \nabla_X^{0,1}\theta_i, \ldots, \theta_r, X_1, \ldots, X_s)$$

$$- \sum_{j=1}^{s} S(\theta_1, \ldots, \ldots, \theta_r, X_1, \ldots, \nabla_X X_j, \ldots, X_s),$$

for all $X_1, \ldots, X_s \in \mathfrak{X}(M)$ *and all one-forms,* $\theta_1, \ldots, \theta_r \in \mathcal{A}^1(M)$.

In particular, for $S = g$, the Riemannian metric on M (a $(0, 2)$ tensor), we get

$$\nabla_X(g)(Y, Z) = X(g(Y, Z)) - g(\nabla_X Y, Z) - g(Y, \nabla_X Z),$$

for all $X, Y, Z \in \mathfrak{X}(M)$. We will see later on that a connection on M is compatible with a metric g iff $\nabla_X(g) = 0$.

Definition 9.19. The *covariant differential* $\nabla^{r,s} S$ of a tensor $S \in \Gamma(M, T^{r,s}(M))$ is the $(r, s + 1)$-tensor given by

$$(\nabla^{r,s} S)(\theta_1, \ldots, \theta_r, X, X_1, \ldots, X_s) = (\nabla_X^{r,s} S)(\theta_1, \ldots, \theta_r, X_1, \ldots, X_s),$$

for all $X, X_j \in \mathfrak{X}(M)$ and all $\theta_i \in \mathcal{A}^1(M)$.

For simplicity of notation we usually omit the superscripts r and s. In particular, if $r = 1$ and $s = 0$, in which case S is a vector field, the covariant derivative ∇S is defined so that

$$(\nabla S)(X) = \nabla_X S.$$

If $(M, \langle -, - \rangle)$ is a Riemannian manifold, then the inner product $\langle -, - \rangle_p$ on $T_p M$ establishes a canonical duality between $T_p M$ and $T_p^* M$, as explained in Section 2.2. Namely, we have the isomorphism $\flat \colon T_p M \to T_p^* M$, defined such that for every $u \in T_p M$, the linear form $u^\flat \in T_p^* M$ is given by

$$u^\flat(v) = \langle u, v \rangle_p \qquad v \in T_p M.$$

The inverse isomorphism $\sharp \colon T_p^* M \to T_p M$ is defined such that for every $\omega \in T_p^* M$, the vector $\omega^\sharp$ is the unique vector in $T_p M$ so that

$$\langle \omega^\sharp, v \rangle_p = \omega(v), \qquad v \in T_p M.$$

The isomorphisms $\flat$ and $\sharp$ induce isomorphisms between vector fields $X \in \mathfrak{X}(M)$ and one-forms $\omega \in \mathcal{A}^1(M)$: A vector field $X \in \mathfrak{X}(M)$ yields the one-form $X^\flat \in \mathcal{A}^1(M)$ given by

$$(X^\flat)_p = (X_p)^\flat,$$

and a one-form $\omega \in \mathcal{A}^1(M)$ yields the vector field $\omega^\sharp \in \mathfrak{X}(M)$ given by

$$(\omega^\sharp)_p = (\omega_p)^\sharp,$$

so that

$$\omega_p(v) = \langle (\omega_p)^\sharp, v \rangle_p, \qquad v \in T_p M, \ p \in M.$$

In particular, for every smooth function $f \in C^\infty(M)$, the vector field corresponding to the one-form df is the *gradient* grad f, of f. The gradient of f is uniquely determined by the condition

$$\langle (\text{grad } f)_p, v \rangle_p = df_p(v), \qquad v \in T_p M, \; p \in M.$$

Recall from Proposition 9.13 that the covariant derivative $\nabla_X \omega$ of any one-form $\omega \in \mathcal{A}^1(M)$ is the one-form given by

$$(\nabla_X \omega)(Y) = X(\omega(Y)) - \omega(\nabla_X Y).$$

If ∇ is a metric connection, then the vector field $(\nabla_X \omega)^\sharp$ corresponding to $\nabla_X \omega$ is nicely expressed in terms of $\omega^\sharp$. Indeed, we have the following proposition.

Proposition 9.14. *If ∇ is a metric connection on a smooth manifold M, then for every vector field X and every one-form ω we have*

$$(\nabla_X \omega)^\sharp = \nabla_X \omega^\sharp.$$

Proof. We have

$$
\begin{aligned}
(\nabla_X \omega)(Y) &= X(\omega(Y)) - \omega(\nabla_X Y) \\
&= X(\langle \omega^\sharp, Y \rangle) - \langle \omega^\sharp, \nabla_X Y \rangle \\
&= \langle \nabla_X \omega^\sharp, Y \rangle + \langle \omega^\sharp, \nabla_X Y \rangle - \langle \omega^\sharp, \nabla_X Y \rangle \\
&= \langle \nabla_X \omega^\sharp, Y \rangle,
\end{aligned}
$$

where we used the fact that the connection is compatible with the metric in the third line, and so

$$(\nabla_X \omega)^\sharp = \nabla_X \omega^\sharp,$$

as claimed.   □

9.8   Metrics on Vector Bundles, Reduction of Structure Groups, and Orientation

Because the fibers of a vector bundle are vector spaces, the definition of a Riemannian metric on a manifold can be lifted to vector bundles.

Definition 9.20. Given a (real) rank n vector bundle $\xi = (E, \pi, B, V)$, we say that ξ is *Euclidean* iff there is a family $(\langle -, - \rangle_b)_{b \in B}$ of inner products on each

fiber $\pi^{-1}(b)$, such that $\langle -, - \rangle_b$ depends smoothly on b, which means that for every trivializing map
$\varphi_\alpha \colon \pi^{-1}(U_\alpha) \to U_\alpha \times V$, for every frame, $(s_1, \ldots, s_n)$, on U_α, the maps

$$ b \mapsto \langle s_i(b), s_j(b) \rangle_b, \qquad b \in U_\alpha, \ 1 \le i, j \le n $$

are smooth. We say that $\langle -, - \rangle$ is a *Euclidean metric* (or *Riemannian metric*) on ξ. If ξ is a complex rank n vector bundle $\xi = (E, \pi, B, V)$, we say that ξ is *Hermitian* iff there is a family $(\langle -, - \rangle_b)_{b \in B}$ of Hermitian inner products on each fiber $\pi^{-1}(b)$, such that $\langle -, - \rangle_b$ depends smoothly on b. We say that $\langle -, - \rangle$ is a *Hermitian metric* on ξ. For any smooth manifold M, if TM is a Euclidean vector bundle, then we say that M is a *Riemannian manifold*.

Now, given a real (resp. complex) vector bundle ξ, since B is *paracompact* because it is a manifold a Euclidean metric (resp. Hermitian metric) exists on ξ. This is a consequence of the existence of partitions of unity (see Warner [109], Chapter 1, or Gallier and Quaintance [47]).

Theorem 9.15. *Every real (resp. complex) vector bundle admits a Euclidean (resp. Hermitian) metric. In particular, every smooth manifold admits a Riemannian metric.*

Proof. Let (U_α) be a trivializing open cover for ξ and pick any frame $(s_1^\alpha, \ldots, s_n^\alpha)$ over U_α. For every $b \in U_\alpha$, the basis $(s_1^\alpha(b), \ldots, s_n^\alpha(b))$ defines a Euclidean (resp. Hermitian) inner product $\langle -, - \rangle_b$ on the fiber $\pi^{-1}(b)$, by declaring $(s_1^\alpha(b), \ldots, s_n^\alpha(b))$ orthonormal w.r.t. this inner product. (For $x = \sum_{i=1}^n x_i s_i^\alpha(b)$ and $y = \sum_{i=1}^n y_i s_i^\alpha(b)$, let $\langle x, y \rangle_b = \sum_{i=1}^n x_i y_i$, resp. $\langle x, y \rangle_b = \sum_{i=1}^n x_i \overline{y}_i$, in the complex case.) The $\langle -, - \rangle_b$ (with $b \in U_\alpha$) define a metric on $\pi^{-1}(U_\alpha)$, denote it $\langle -, - \rangle_\alpha$. Now, using a partition of unity, glue these inner products using a partition of unity (f_α) subordinate to (U_α), by setting

$$ \langle x, y \rangle = \sum_\alpha f_\alpha \langle x, y \rangle_\alpha. $$

We verify immediately that $\langle -, - \rangle$ is a Euclidean (resp. Hermitian) metric on ξ. $\square$

The existence of metrics on vector bundles allows the so-called reduction of structure group. Recall that the transition maps of a real (resp. complex) vector bundle ξ are functions $g_{\alpha\beta} \colon U_\alpha \cap U_\beta \to \mathbf{GL}(n, \mathbb{R})$ (resp. $\mathbf{GL}(n, \mathbb{C})$). Let $\mathbf{GL}^+(n, \mathbb{R})$ be the subgroup of $\mathbf{GL}(n, \mathbb{R})$ consisting of those matrices of positive determinant (resp. $\mathbf{GL}^+(n, \mathbb{C})$ be the subgroup of $\mathbf{GL}(n, \mathbb{C})$ consisting of those matrices of positive determinant).

Definition 9.21. For every real (resp. complex) vector bundle ξ, if it is possible to find a cocycle $g = (g_{\alpha\beta})$ for ξ with values in a subgroup H of $\mathbf{GL}(n, \mathbb{R})$ (resp. of $\mathbf{GL}(n, \mathbb{C})$), then we say that the *structure group of ξ can be reduced to H*. We say that ξ is *orientable* if its structure group can be reduced to $\mathbf{GL}^+(n, \mathbb{R})$ (resp. $\mathbf{GL}^+(n, \mathbb{C})$).

Proposition 9.16.

(a) *The structure group of a rank n real vector bundle ξ can be reduced to $\mathbf{O}(n)$; it can be reduced to $\mathbf{SO}(n)$ iff ξ is orientable.*

(b) *The structure group of a rank n complex vector bundle ξ can be reduced to $\mathbf{U}(n)$; it can be reduced to $\mathbf{SU}(n)$ iff ξ is orientable.*

Proof. We prove (a), the proof of (b) being similar. Using Theorem 9.15, put a metric on ξ. For every U_α in a trivializing cover for ξ and every $b \in B$, by Gram-Schmidt, orthonormal bases for $\pi^{-1}(b)$ exist. Consider the family of trivializing maps $\widetilde{\varphi}_\alpha \colon \pi^{-1}(U_\alpha) \to U_\alpha \times V$ such that $\widetilde{\varphi}_{\alpha,b} \colon \pi^{-1}(b) \longrightarrow V$ maps orthonormal bases of the fiber to orthonormal bases of V. Then, it is easy to check that the corresponding cocycle takes values in $\mathbf{O}(n)$ and if ξ is orientable, the determinants being positive, these values are actually in $\mathbf{SO}(n)$. $\square$

Remark. If ξ is a Euclidean rank n vector bundle, then by Proposition 9.16, we may assume that ξ is given by some cocycle $(g_{\alpha\beta})$, where $g_{\alpha\beta}(b) \in \mathbf{O}(n)$, for all $b \in U_\alpha \cap U_\beta$. We saw in Section 9.5 (f) that the dual bundle ξ^* is given by the cocycle

$$(g_{\alpha\beta}(b)^\top)^{-1}, \qquad b \in U_\alpha \cap U_\beta.$$

As $g_{\alpha\beta}(b)$ is an orthogonal matrix, $(g_{\alpha\beta}(b)^\top)^{-1} = g_{\alpha\beta}(b)$, and thus, any Euclidean bundle is isomorphic to its dual. As we noted earlier, this is *false* for Hermitian bundles.

Definition 9.22. Let $\xi = (E, \pi, B, V)$ be a rank n vector bundle and assume ξ is orientable. A family of trivializing maps $\varphi_\alpha \colon \pi^{-1}(U_\alpha) \to U_\alpha \times V$ is *oriented* iff for all α, β, the transition function $g_{\alpha\beta}(b)$ has positive determinant for all $b \in U_\alpha \cap U_\beta$. Two oriented families of trivializing maps $\varphi_\alpha \colon \pi^{-1}(U_\alpha) \to U_\alpha \times V$ and $\psi_\beta \colon \pi^{-1}(W_\beta) \to W_\alpha \times V$ are *equivalent* iff for every $b \in U_\alpha \cap W_\beta$, the map $pr_2 \circ \varphi_\alpha \circ \psi_\beta^{-1} \restriction \{b\} \times V \colon V \longrightarrow V$ has positive determinant.

It is easily checked that this is an equivalence relation and that it partitions all the oriented families of trivializations of ξ into two equivalence classes. Either equivalence class is called an *orientation* of ξ.

If M is a manifold and $\xi = TM$, the tangent bundle of M, we know from Section 9.4 that the transition functions of TM are of the form

$$g_{\alpha\beta}(p)(u) = (\varphi_\alpha \circ \varphi_\beta^{-1})'_{\varphi(p)}(u),$$

where each $\varphi_\alpha \colon U_\alpha \to \mathbb{R}^n$ is a chart of M. Consequently, TM is orientable iff the Jacobian of $(\varphi_\alpha \circ \varphi_\beta^{-1})'_{\varphi(p)}$ is positive, for every $p \in M$. This is equivalent to the condition of Definition 6.4 for M to be orientable. Therefore, we have the following result.

Proposition 9.17. *The tangent bundle TM of a manifold M is orientable iff M is orientable.*

 The notion of orientability of a vector bundle $\xi = (E, \pi, B, V)$ is *not* equivalent to the orientability of its total space E. Indeed, if we look at the transition functions of the total space of TM given in Section 9.4, we see that TM, *as a manifold, is always orientable*, even if M is not orientable.

Indeed, the transition functions of the tangent bundle TM are of the form

$$\widetilde{\psi} \circ \widetilde{\varphi}^{-1}(z, x) = ((\psi \circ \varphi^{-1}(z), (\psi \circ \varphi^{-1})'_z(x)), \qquad (z, x) \in \varphi(U \cap V) \times \mathbb{R}^n.$$

Since $(\psi \circ \varphi^{-1})'_z$ is a linear map, its derivative at any point is equal to itself, and it follows that the derivative of $\widetilde{\psi} \circ \widetilde{\varphi}^{-1}$ at (z, x) is given by

$$(\widetilde{\psi} \circ \widetilde{\varphi}^{-1})'_{(z,x)}(u, v) = ((\psi \circ \varphi^{-1})'_z(u), (\psi \circ \varphi^{-1})'_z(v)), \qquad (u, v) \in \mathbb{R}^n \times \mathbb{R}^n.$$

Then the Jacobian matrix of this map is of the form

$$J = \begin{pmatrix} A & 0 \\ 0 & A \end{pmatrix}$$

where A is an $n \times n$ matrix, since $(\psi \circ \varphi^{-1})'_z(u)$ does not involve the variables in v and $(\psi \circ \varphi^{-1})'_z(v)$ does not involve the variables in u. Therefore $\det(J) = \det(A)^2$, which shows that the transition functions have positive Jacobian determinant, and thus that TM is orientable.

Yet, *as a bundle, TM is orientable iff M is orientable.*

On the positive side, we have the following result.

Proposition 9.18. *If $\xi = (E, \pi, B, V)$ is an orientable vector bundle and its base B is an orientable manifold, then E is orientable too.*

Proof. To see this, assume that B is a manifold of dimension m, ξ is a rank n vector bundle with fiber V, let $((U_\alpha, \psi_\alpha))_\alpha$ be an atlas for B, let $\varphi_\alpha : \pi^{-1}(U_\alpha) \to U_\alpha \times V$ be a collection of trivializing maps for ξ, and pick any isomorphism, $\iota : V \to \mathbb{R}^n$. Then, we get maps

$$(\psi_\alpha \times \iota) \circ \varphi_\alpha : \pi^{-1}(U_\alpha) \longrightarrow \mathbb{R}^m \times \mathbb{R}^n.$$

It is clear that these maps form an atlas for E. Check that the corresponding transition maps for E are of the form

$$(x, y) \mapsto (\psi_\beta \circ \psi_\alpha^{-1}(x), g_{\alpha\beta}(\psi_\alpha^{-1}(x))y).$$

Moreover, if B and ξ are orientable, we can check that these transition maps have positive Jacobian. □

The notion of subbundle is defined as follows:

Definition 9.23. Given two vector bundles $\xi = (E, \pi, B, V)$ and $\xi' = (E', \pi', B, V')$ over the same base B, we say that ξ is a *subbundle* of ξ' iff E is a submanifold of E', V is a subspace of V', and for every $b \in B$, the fiber $\pi^{-1}(b)$ is a subspace of the fiber $(\pi')^{-1}(b)$.

If ξ is a subbundle of ξ', we can form the *quotient bundle* ξ'/ξ as the bundle over B whose fiber at $b \in B$ is the quotient space $(\pi')^{-1}(b)/\pi^{-1}(b)$. We leave it as an exercise to define trivializations for ξ'/ξ. In particular, if N is a submanifold of M, then TN is a subbundle of $TM \upharpoonright N$ and the quotient bundle $(TM \upharpoonright N)/TN$ is called the *normal bundle* of N in M.

The fact that every bundle admits a metric allows us to define the notion of *orthogonal complement* of a subbundle. We state the following theorem without proof. The reader is invited to consult Milnor and Stasheff [78] for a proof (Chapter 3).

Proposition 9.19. *Let ξ and η be two vector bundles with ξ a subbundle of η. Then there exists a subbundle $\xi^{\perp}$ of η, such that every fiber of $\xi^{\perp}$ is the orthogonal complement of the fiber of ξ in the fiber of η over every $b \in B$, and*

$$\eta \cong \xi \oplus \xi^{\perp}.$$

In particular, if N is a submanifold of a Riemannian manifold M, then the orthogonal complement of TN in $TM \upharpoonright N$ is isomorphic to the normal bundle $(TM \upharpoonright N)/TN$.

9.9 Principal Fiber Bundles

We now consider principal bundles. Such bundles arise in terms of Lie groups acting on manifolds. Let $L(G)$ be the group of left translations of the group G, that is, the set of all homomorphisms $L_g : G \to G$ given by $L_g(g') = gg'$, for all $g, g' \in G$. The map $g \mapsto L_g$ is an isomorphism between the groups G and $L(G)$ whose inverse is given by $L \mapsto L(1)$ (where $L \in L(G)$).

Definition 9.24. Let G be a Lie group. A *principal fiber bundle*, for short a *principal bundle*, is a fiber bundle $\xi = (E, \pi, B, G, L(G))$ in which the fiber is G and the structure group is $L(G)$, that is, G viewed as its group of left translations (i.e., G acts on itself by multiplication on the left). This means that every transition function $g_{\alpha\beta} : U_\alpha \cap U_\beta \to L(G)$ satisfies

$$g_{\alpha\beta}(b)(h) = (g_{\alpha\beta}(b)(1))h,$$

for all $b \in U_\alpha \cap U_\beta$ and all $h \in G$. A principal G-bundle is denoted $\xi = (E, \pi, B, G)$.

In view of the isomorphism between $L(G)$ and G we allow ourselves the (convenient) abuse of notation

$$g_{\alpha\beta}(b)(h) = g_{\alpha\beta}(b)h,$$

where on the left, $g_{\alpha\beta}(b)$ is viewed as a left translation of G, and on the right as an element of G.

When we want to emphasize that a principal bundle has structure group G, we use the locution *principal G-bundle*.

It turns out that if $\xi = (E, \pi, B, G)$ is a principal bundle, then G acts on the total space E, on the right. For the next proposition, recall that a right action $\cdot : X \times G \rightarrow X$ is *free* iff for every $g \in G$, if $g \neq 1$, then $x \cdot g \neq x$ for all $x \in X$.

Proposition 9.20. *If* $\xi = (E, \pi, B, G)$ *is a principal bundle, then there is a right action of G on E. This action takes each fiber to itself and is free. Moreover, E/G is diffeomorphic to B.*

Proof. We show how to define the right action and leave the rest as an exercise. Let $\{(U_\alpha, \varphi_\alpha)\}$ be some trivializing cover defining ξ. For every $z \in E$, pick some U_α so that $\pi(z) \in U_\alpha$, and let $\varphi_\alpha(z) = (b, h)$, where $b = \pi(z)$ and $h \in G$. For any $g \in G$, we set

$$z \cdot g = \varphi_\alpha^{-1}(b, hg).$$

If we can show that this action does not depend on the choice of U_α, then it is clear that it is a free action. Suppose that we also have $b = \pi(z) \in U_\beta$ and that $\varphi_\beta(z) = (b, h')$. By definition of the transition functions, we have

$$h' = g_{\beta\alpha}(b)h \quad \text{and} \quad \varphi_\beta(z \cdot g) = \varphi_\beta(\varphi_\alpha^{-1}(b, hg)) = (b, g_{\beta\alpha}(b)(hg)).$$

However,

$$g_{\beta\alpha}(b)(hg) = (g_{\beta\alpha}(b)h)g = h'g,$$

hence

$$z \cdot g = \varphi_\beta^{-1}(b, h'g),$$

which proves that our action does not depend on the choice of U_α. $\square$

Observe that the action of Proposition 9.20 is defined by

$$z \cdot g = \varphi_\alpha^{-1}(b, \varphi_{\alpha,b}(z)g), \quad \text{with} \quad b = \pi(z),$$

for all $z \in E$ and all $g \in G$.

It is clear that this action satisfies the following two properties: For every $(U_\alpha, \varphi_\alpha)$,

(1) $\pi(z \cdot g) = \pi(z)$, and
(2) $\varphi_\alpha(z \cdot g) = \varphi_\alpha(z) \cdot g$, for all $z \in E$ and all $g \in G$,

where we define the right action of G on $U_\alpha \times G$ so that $(b, h) \cdot g = (b, hg)$.

Definition 9.25. A trivializing map φ_α satisfying Condition (2) above is *G-equivariant* (or *equivariant*).

The following proposition shows that it is possible to define a principal G-bundle using a suitable right action and equivariant trivializations:

Proposition 9.21. *Let E be a smooth manifold, G be a Lie group, and let $\cdot : E \times G \to E$ be a smooth right action of G on E satisfying the following properties:*

(a) *The right action of G on E is free;*
(b) *The orbit space $B = E/G$ is a smooth manifold under the quotient topology, and the projection $\pi : E \to E/G$ is smooth;*
(c) *There is a family of local trivializations $\{(U_\alpha, \varphi_\alpha)\}$, where $\{U_\alpha\}$ is an open cover for $B = E/G$, and each*

$$\varphi_\alpha : \pi^{-1}(U_\alpha) \to U_\alpha \times G$$

is an equivariant diffeomorphism, which means that

$$\varphi_\alpha(z \cdot g) = \varphi_\alpha(z) \cdot g,$$

for all $z \in \pi^{-1}(U_\alpha)$ and all $g \in G$, where the right action of G on $U_\alpha \times G$ is $(b, h) \cdot g = (b, hg)$.

If $\pi : E \to E/G$ is the quotient map, then $\xi = (E, \pi, E/G, G)$ is a principal G-bundle.

Proof. Since the action of G on E is free, every orbit $b = z \cdot G$ is isomorphic to G, and so every fiber $\pi^{-1}(b)$ is isomorphic to G. Thus, given that we have trivializing maps, we just have to prove that G acts by left translation on itself. Pick any (b, h) in $U_\beta \times G$ and let $z \in \pi^{-1}(U_\beta)$ be the unique element such that $\varphi_\beta(z) = (b, h)$. Then as

$$\varphi_\beta(z \cdot g) = \varphi_\beta(z) \cdot g, \quad \text{for all } g \in G,$$

we have

$$\varphi_\beta(\varphi_\beta^{-1}(b, h) \cdot g) = \varphi_\beta(z \cdot g) = \varphi_\beta(z) \cdot g = (b, h) \cdot g,$$

which implies that

$$\varphi_\beta^{-1}(b, h) \cdot g = \varphi_\beta^{-1}((b, h) \cdot g).$$

Consequently,

$$\varphi_\alpha \circ \varphi_\beta^{-1}(b, h) = \varphi_\alpha \circ \varphi_\beta^{-1}((b, 1) \cdot h) = \varphi_\alpha(\varphi_\beta^{-1}(b, 1) \cdot h) = \varphi_\alpha \circ \varphi_\beta^{-1}(b, 1) \cdot h,$$

and since

$$\varphi_\alpha \circ \varphi_\beta^{-1}(b, h) = (b, g_{\alpha\beta}(b)(h)) \quad \text{and} \quad \varphi_\alpha \circ \varphi_\beta^{-1}(b, 1) = (b, g_{\alpha\beta}(b)(1))$$

we get

$$g_{\alpha\beta}(b)(h) = g_{\alpha\beta}(b)(1)h.$$

The above shows that $g_{\alpha\beta}(b) \colon G \to G$ is the left translation by $g_{\alpha\beta}(b)(1)$, and thus the transition functions $g_{\alpha\beta}(b)$ constitute the group of left translations of G, and ξ is indeed a principal G-bundle. $\square$

Bröcker and tom Dieck [18] (Chapter I, Section 4) and Duistermaat and Kolk [38] (Appendix A) define principal bundles using the conditions of Proposition 9.21. Propositions 9.20 and 9.21 show that this alternate definition is equivalent to ours (Definition 9.24).

It turns out that when we use the definition of a principal bundle in terms of the conditions of Proposition 9.21, it is convenient to define bundle maps in terms of equivariant maps. As we will see shortly, a map of principal bundles is a fiber bundle map.

Definition 9.26. If $\xi_1 = (E_1, \pi_1, B_1, G)$ and $\xi_2 = (E_2, \pi_2, B_2, G)$ are two principal bundles, a *bundle map (or bundle morphism)* $f \colon \xi_1 \to \xi_2$ is a pair, $f = (f_E, f_B)$ of smooth maps $f_E \colon E_1 \to E_2$ and $f_B \colon B_1 \to B_2$, such that:

(a) The following diagram commutes:

$$
\begin{array}{ccc}
E_1 & \xrightarrow{\;f_E\;} & E_2 \\
\pi_1 \downarrow & & \downarrow \pi_2 \\
B_1 & \xrightarrow[\;f_B\;]{} & B_2
\end{array}
$$

(b) The map f_E is *G-equivariant*; that is,

$$f_E(a \cdot g) = f_E(a) \cdot g, \qquad \text{for all } ensurematha \in E_1 \text{ and all } g \in G.$$

A bundle map is an *isomorphism* if it has an inverse as in Definition 9.6. If the bundles ξ_1 and ξ_2 are over the same base B, then we also require $f_B = \text{id}$.

At first glance, it is not obvious that a map of principal bundles satisfies Condition (b) of Definition 9.7. If we define $\widetilde{f}_\alpha : U_\alpha \times G \to V_\beta \times G$ by

$$\widetilde{f}_\alpha = \varphi_\beta' \circ f_E \circ \varphi_\alpha^{-1},$$

then locally f_E is expressed as

$$f_E = \varphi_\beta'^{-1} \circ \widetilde{f}_\alpha \circ \varphi_\alpha.$$

Furthermore, it is trivial that if a map is equivariant and invertible, then its inverse is equivariant. Consequently, since

$$\widetilde{f}_\alpha = \varphi_\beta' \circ f_E \circ \varphi_\alpha^{-1},$$

as φ_α^{-1}, φ_β' and f_E are equivariant, $\widetilde{f}_\alpha$ is also equivariant, and so $\widetilde{f}_\alpha$ is a map of (trivial) principal bundles. Thus, it is enough to prove that for every map of principal bundles

$$\varphi : U_\alpha \times G \to V_\beta \times G,$$

there is some smooth map $\rho_\alpha : U_\alpha \to G$, so that

$$\varphi(b, g) = (f_B(b), \rho_\alpha(b)(g)), \qquad \text{for all } b \in U_\alpha \text{ and all } g \in G.$$

Indeed, we have the following.

Proposition 9.22. *For every map of trivial principal bundles*

$$\varphi : U_\alpha \times G \to V_\beta \times G,$$

there are smooth maps $f_B : U_\alpha \to V_\beta$ and $r_\alpha : U_\alpha \to G$, so that

$$\varphi(b, g) = (f_B(b), r_\alpha(b)g), \qquad \text{for all } b \in U_\alpha \text{ and all } g \in G.$$

In particular, φ is a diffeomorphism on fibers.

Proof. As φ is a map of principal bundles

$$\varphi(b, 1) = (f_B(b), r_\alpha(b)), \qquad \text{for all } b \in U_\alpha,$$

for some smooth maps $f_B : U_\alpha \to V_\beta$ and $r_\alpha : U_\alpha \to G$. Now, using equivariance, we get

$$\varphi(b, g) = \varphi((b, 1)g) = \varphi(b, 1) \cdot g = (f_B(b), r_\alpha(b)) \cdot g = (f_B(b), r_\alpha(b)g),$$

as claimed.   □

Consequently, the map $\rho_\alpha : U_\alpha \to G$ given by

$$\rho_\alpha(b)(g) = r_\alpha(b)g \qquad \text{for all } b \in U_\alpha \text{ and all } g \in G$$

satisfies

$$\varphi(b, g) = (f_B(b), \rho_\alpha(b)(g)), \qquad \text{for all } b \in U_\alpha \text{ and all } g \in G,$$

and a map of principal bundles is indeed a fiber bundle map (as in Definition 9.7). Since a principal bundle map is a fiber bundle map, Proposition 9.3 also yields the useful fact:

Proposition 9.23. *Any map $f : \xi_1 \to \xi_2$ between two principal bundles over the same base B is an isomorphism.*

A natural question is to ask whether a fiber bundle ξ is isomorphic to a trivial bundle. If so, we say that ξ is trivial. (By the way, the triviality of bundles comes up in physics, in particular, field theory.) Generally, this is a very difficult question, but a first step can be made by showing that it reduces to the question of triviality for principal bundles.

Indeed, if $\xi = (E, \pi, B, F, G)$ is a fiber bundle with fiber F, using Theorem 9.4, we can construct a principal fiber bundle $P(\xi)$ using the transition functions $\{g_{\alpha\beta}\}$ of ξ, but using G itself as the fiber (acting on itself by left translation) instead of F.

Definition 9.27. Let $\xi = (E, \pi, B, F, G)$ is a fiber bundle with fiber F, and $P(\xi)$ be the bundle obtained by replacing the fiber of ξ with G (as described in the preceding paragraph). We call $P(\xi)$ the *principal bundle associated with ξ*.

For example, the principal bundle associated with a vector bundle is the *frame bundle*, discussed at the end of Section 9.4.

Then given two fiber bundles ξ and ξ', we see that ξ and ξ' are isomorphic iff $P(\xi)$ and $P(\xi')$ are isomorphic (Steenrod [101], Part I, Section 8, Theorem 8.2). More is true: the fiber bundle ξ is trivial iff the principal fiber bundle $P(\xi)$ is trivial (see Steenrod [101], Part I, Section 8, Corollary 8.4). Moreover, there is a test for the triviality of a principal bundle, the existence of a (global) section.

The following proposition, although easy to prove, is crucial:

Proposition 9.24. *If ξ is a principal bundle, then ξ is trivial iff it possesses some global section.*

Proof. If $f : B \times G \to \xi$ is an isomorphism of principal bundles over the same base B, then for every $g \in G$, the map $b \mapsto f(b, g)$ is a section of ξ.

Conversely, let $s : B \to E$ be a section of ξ. Then, observe that the map $f : B \times G \to \xi$ given by

$$f(b, g) = s(b)g$$

is a map of principal bundles. By Proposition 9.23, it is an isomorphism, so ξ is trivial. $\square$

Generally, in geometry, many objects of interest arise as global sections of some suitable bundle (or sheaf): vector fields, differential forms, tensor fields, *etc.*

Definition 9.28. Given a principal bundle $\xi = (E, \pi, B, G)$ and given a manifold F, if G acts effectively on F from the left, using Theorem 9.4, we can construct a fiber bundle $\xi[F]$ from ξ, with F as typical fiber, and such that $\xi[F]$ has the same transitions functions as ξ. The fiber bundle $\xi[F]$ is called the *fiber bundle induced by ξ*.

In the case of a principal bundle, there is another slightly more direct construction that takes us from principal bundles to fiber bundles (see Duistermaat and Kolk [38], Chapter 2, and Davis and Kirk [28], Chapter 4, Definition 4.6, where it is called the *Borel construction*). This construction is of independent interest, so we describe it briefly (for an application of this construction, see Duistermaat and Kolk [38], Chapter 2).

As ξ is a principal bundle, recall that G acts on E from the right, so we have a right action of G on $E \times F$, *via*

$$(z, f) \cdot g = (z \cdot g, g^{-1} \cdot f).$$

Consequently, we obtain the orbit set $E \times F / \sim$, denoted $E \times_G F$, where $\sim$ is the equivalence relation

$$(z, f) \sim (z', f') \quad \text{iff} \quad (\exists g \in G)(z' = z \cdot g, \ f' = g^{-1} \cdot f).$$

Note that the composed map

$$E \times F \xrightarrow{pr_1} E \xrightarrow{\pi} B$$

factors through $E \times_G F$ as a map $p \colon E \times_G F \to B$ given by

$$p([z, f]) = \pi(pr_1(z, f)),$$

as illustrated in the diagram below

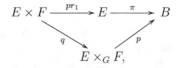

since

$$\pi(pr_1(z, f)) = \pi(z) = \pi(z \cdot g) = \pi(pr_1(z \cdot g, g^{-1} \cdot f)),$$

which means that the definition of p does not depend on the choice of representative in the equivalence class $[(z, f)]$.

The following proposition is not hard to show:

Proposition 9.25. *If $\xi = (E, \pi, B, G)$ is a principal bundle and F is any manifold such that G acts effectively on F from the left, then $\xi[F] = (E \times_G F, p, B, F, G)$ is a fiber bundle with fiber F and structure group G, and $\xi[F]$ and ξ have the same transition functions.*

Proof. Let us verify that the charts of ξ yield charts for $\xi[F]$. For any U_α in an open cover for B, we have a diffeomorphism

$$\varphi_\alpha : \pi^{-1}(U_\alpha) \to U_\alpha \times G.$$

The first step is to show that that there is an isomorphism

$$(U_\alpha \times G) \times_G F \cong U_\alpha \times F,$$

where, as usual, G acts on $U_\alpha \times G$ via $(z, h) \cdot g = (z, hg)$, Two pairs $((b_1, g_1), f_1)$ and $((b_2, g_2), f_2)$ are equivalent iff there is some $g \in G$ such that

$$(b_2, g_2) = (b_1, g_1) \cdot g, \quad f_2 = g^{-1} \cdot f_1,$$

which implies that $(b_2, g_2) = (b_1, g_1 g)$, so $b_1 = b_2$ and $g_2 = g_1 g$. It follows that $g = g_1^{-1} g_2$ and the equivalence class of $((b_1, g_1), f_1)$ consists of all pairs of the form $((b_1, g_2), g_2^{-1} g_1 \cdot f_1)$ for all $g_2 \in G$. The map $\theta : (U_\alpha \times G) \times_G F \to U_\alpha \times F$ given by

$$\theta([((b_1, g_2), g_2^{-1} g_1 \cdot f_1)]) = (b_1, g_1 \cdot f_1)$$

is well defined on the equivalence class $[((b_1, g_1), f_1)]$, and it is clear that it is a bijection since G being a group the map $g_1 \mapsto g_1 \cdot f_1$ is bijective.

We also have an isomorphism

$$p^{-1}(U_\alpha) \cong \pi^{-1}(U_\alpha) \times_G F,$$

and since $\varphi_\alpha : \pi^{-1}(U_\alpha) \to U_\alpha \times G$ induces an isomorphism

$$\pi^{-1}(U_\alpha) \times_G F \xrightarrow{\quad \xi \quad} (U_\alpha \times G) \times_G F,$$

and we have an isomorphism $\theta \colon (U_\alpha \times G) \times_G F \to U_\alpha \times F$, so we have an isomorphism $p^{-1}(U_\alpha) \longrightarrow U_\alpha \times F$ and we get the commutative diagram

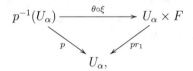

which yields a local trivialization $\theta \circ \xi$ for $\xi[F]$. It is easy to see that the transition functions of $\xi[F]$ are the same as the transition functions of ξ.   □

Now if we start with a fiber bundle ξ with fiber F and structure group G, if we make the associated principal bundle $P(\xi)$, and then the induced fiber bundle $P(\xi)[F]$, what is the relationship between ξ and $P(\xi)[F]$?

The answer is: ξ and $P(\xi)[F]$ are *equivalent* (this is because the transition functions are the same).

Now, if we start with a principal G-bundle ξ, make the fiber bundle $\xi[F]$ as above, and then the principal bundle $P(\xi[F])$, we get a principal bundle equivalent to ξ. Therefore, the maps

$$\xi \mapsto \xi[F] \quad \text{and} \quad \xi \mapsto P(\xi)$$

are mutual inverses, and they set up a bijection between equivalence classes of principal G-bundles over B and equivalence classes of fiber bundles over B (with structure group G). Moreover, this map extends to morphisms, so it is functorial (see Steenrod [101], Part I, Section 2, Lemma 2.6–Lemma 2.10).

As a consequence, in order to "classify" equivalence classes of fiber bundles (assuming B and G fixed), it is enough to know how to classify principal G-bundles over B. Given some reasonable conditions on the coverings of B, Milnor solved this classification problem, but this is taking us way beyond the scope of these notes!

The classical reference on fiber bundles, vector bundles, and principal bundles is Steenrod [101]. More recent references include Bott and Tu [12], Madsen and Tornehave [75], Morita [82], Griffith and Harris [52], Wells [110], Hirzebruch [57], Milnor and Stasheff [78], Davis and Kirk [28], Atiyah [4], Chern [22], Choquet-Bruhat, DeWitt-Morette, and Dillard-Bleick [26], Hirsh [56], Sato [95], Narasimham [83], Sharpe [100], and also Husemoller [60], which covers more, including characteristic classes.

Proposition 9.21 shows that principal bundles are induced by suitable right actions, but we still need sufficient conditions to guarantee Conditions (a), (b), and (c). The special situation of homogeneous spaces is considered in the next section.

9.10   Proper and Free Actions and Homogeneous Spaces

Now that we have introduced the notion of principal bundle, we can state various results about homogeneous spaces. These results are stronger than those stated in Gallier and Quaintance [47] which apply to groups and sets without any topology or differentiable structure. We need to review the notion of proper map and proper action.

Definition 9.29. If X and Y are two Hausdorff topological spaces,[1] a function a $\varphi\colon X \to Y$ is *proper* iff it is continuous and for every topological space Z, the map $\varphi \times \mathrm{id}\colon X \times Z \to Y \times Z$ is a *closed map* (recall that f is a closed map iff the image of any closed set by f is a closed set).

If we let Z be a one-point space, we see that *a proper map is closed.*

At first glance, it is not obvious how to check that a map is proper just from Definition 9.29. Proposition 9.27 gives a more palatable criterion.

The following proposition is easy to prove (see Bourbaki, General Topology [16], Chapter 1, Section 10).

Proposition 9.26. *If $\varphi\colon X \to Y$ is any proper map, then for any closed subset F of X, the restriction of φ to F is proper.*

The following result providing a "good" criterion for checking that a map is proper can be shown (see Bourbaki, General Topology [16], Chapter 1, Section 10).

Proposition 9.27. *A continuous map $\varphi\colon X \to Y$ is proper iff φ is closed and if $\varphi^{-1}(y)$ is compact for every $y \in Y$.*

Proposition 9.27 shows that a homeomorphism (or a diffeomorphism) is proper.

If φ is proper, it is easy to show that $\varphi^{-1}(K)$ is compact in X whenever K is compact in Y. Moreover, if Y is also locally compact, then we have the following result (see Bourbaki, General Topology [16], Chapter 1, Section 10).

Proposition 9.28. *If Y is locally compact, a continuous map $\varphi\colon X \to Y$ is a proper map iff $\varphi^{-1}(K)$ is compact in X whenever K is compact in Y*

---

[1]It is not necessary to assume that X and Y are Hausdorff but, if X and/or Y are not Hausdorff, we have to replace "compact" by "quasi-compact." We have no need for this extra generality.

In particular, this is true if Y is a manifold since manifolds are locally compact. This explains why Lee [73] (Chapter 9) takes the property stated in Proposition 9.28 as the definition of a proper map (because he only deals with manifolds).[2]

Proper actions are defined as follows.

Definition 9.30. Given a Hausdorff topological group G and a topological space M, a left action $\cdot: G \times M \to M$ is *proper* if it is continuous and if the map

$$\theta: G \times M \longrightarrow M \times M, \quad (g, x) \mapsto (g \cdot x, x)$$

is proper.

Proposition 9.29. *The action* $\cdot: H \times G \to G$ *of a closed subgroup* H *of a group* G *on* G *(given by* $(h, g) \mapsto hg$*) is proper. The same is true for the right action of* H *on* G.

Definition 9.31. An action $\cdot: G \times M \to M$ is *free* if for all $g \in G$ and all $x \in M$, if $g \neq 1$, then $g \cdot x \neq x$.

An equivalent way to state that an action $\cdot: G \times M \to M$ is free is as follows. For every $g \in G$, let $\tau_g: M \to M$ be the diffeomorphism of M given by

$$\tau_g(x) = g \cdot x, \quad x \in M.$$

Then the action $\cdot: G \times M \to M$ is free iff for all $g \in G$, if $g \neq 1$, then τ_g has no fixed point.

Consequently, an action $\cdot: G \times M \to M$ is free iff for every $x \in M$, the stabilizer G_x of x is reduced to the trivial group $\{1\}$.

If H is a subgroup of G, obviously H acts freely on G (by multiplication on the left or on the right). This fact together with Proposition 9.29 yields the following corollary which provides a large supply of free and proper actions.

Corollary 9.30. *The action* $\cdot: H \times G \to G$ *of a closed subgroup* H *of a group* G *on* G *(given by* $(h, g) \mapsto hg$*) is free and proper. The same is true for the right action of* H *on* G.

Before stating the main results of this section, observe that in the definition of a fiber bundle (Definition 9.1), the local trivialization maps are of the form

$$\varphi_\alpha: \pi^{-1}(U_\alpha) \to U_\alpha \times F,$$

where the fiber F appears on the right. In particular, for a principal fiber bundle ξ, the fiber F is equal to the structure group G, and this is the reason why G acts *on the right* on the total space E of ξ (see Proposition 9.20).

---

[2]However, Duistermaat and Kolk [38] seem to have overlooked the fact that a condition on Y (such as local compactness) is needed in their remark on lines 5–6, page 53, just before Lemma 1.11.3.

To be more precise, we call a *right bundle* a bundle $\xi = (E, \pi, B, F, G)$ where the group G acts effectively on the left on the fiber F and where the local trivialization maps are of the form

$$\varphi_\alpha : \pi^{-1}(U_\alpha) \to U_\alpha \times F.$$

If ξ is a right principal bundle, the group G acts on E *on the right* . We call a *left bundle* a bundle $\xi = (E, \pi, B, F, G)$ where the group G acts effectively on the right on the fiber F and the local trivialization maps are of the form

$$\varphi_\alpha : \pi^{-1}(U_\alpha) \to F \times U_\alpha.$$

Then if ξ is a left principal bundle, the group G acts on E *on the left*.

Duistermaat and Kolk [38] address this issue at the end of their Appendix A, and prove the theorem stated below (Chapter 1, Section 11). Beware that in Duistermaat and Kolk [38], this theorem is stated for *right bundles*. However, the weaker version that does not mention principal bundles is usually stated for left actions; for instance, see Lee [73] (Chapter 9, Theorem 9.16). We formulate both versions at the same time.

Theorem 9.31. *Let M be a smooth manifold, G be a Lie group, and let $\cdot : M \times G \to M$ be a right smooth action (resp. $\cdot : G \times M \to M$ a left smooth action) which is proper and free. Then, M/G is a principal right G-bundle (resp. left G-bundle) of dimension $\dim M - \dim G$. Moreover, the canonical projection $\pi : M \to M/G$ is a submersion,[3] and there is a unique manifold structure on M/G with this property.*

Theorem 9.31 has some interesting corollaries. Because a closed subgroup H of a Lie group G is a Lie group, and because the action of a closed subgroup is free and proper, we get the following result (proofs can also be found in Bröcker and tom Dieck [18] (Chapter I, Section 4) and in Duistermaat and Kolk [38] (Chapter 1, Section 11)).

Theorem 9.32. *If G is a Lie group and H is a closed subgroup of G, then the right action of H on G defines a principal (right) H-bundle $\xi = (G, \pi, G/H, H)$, where $\pi : G \to G/H$ is the canonical projection. Moreover, π is a submersion, and there is a unique manifold structure on G/H with this property.*

In the special case where G acts transitively on M, for any $x \in M$, if G_x is the stabilizer of x, then with $H = G_x$, we get Proposition 9.33 below. Recall the definition of a homogeneous space.

Definition 9.32. A *homogeneous space* is a smooth manifold M together with a smooth transitive action $\cdot : G \times M \to M$, of a Lie group G on M.

---

[3]Recall that this means that the derivative $d\pi_p : T_p M \to T_{\pi(p)} M/G$ is surjective for every $p \in$ M.

The following result can be shown as a corollary of Theorem 9.32 (see Lee [73], Chapter 9, Theorem 9.24). It is also mostly proved in Bröcker and tom Dieck [18], Chapter I, Section 4):

Proposition 9.33. *Let* $\cdot\colon G \times M \to M$ *be smooth transitive action of a Lie group* G *on a manifold* M. *Then,* G/G_x *and* M *are diffeomorphic, and* G *is the total space of a principal bundle* $\xi = (G, \pi, M, G_x)$, *where* G_x *is the stabilizer of any element* $x \in M$. *Furthermore, the projection* $\pi\colon G \to G/G_x$ *is a submersion.*

Thus, we finally see that homogeneous spaces induce principal bundles. Going back to some of the examples mentioned earlier (also, see Gallier and Quaintance [47]), we see that

(1) $SO(n + 1)$ is a principal $SO(n)$-bundle over the sphere S^n (for $n \geq 0$).
(2) $SU(n + 1)$ is a principal $SU(n)$-bundle over the sphere S^{2n+1} (for $n \geq 0$).
(3) $SL(2, \mathbb{R})$ is a principal $SO(2)$-bundle over the upper-half space H.
(4) $GL(n, \mathbb{R})$ is a principal $O(n)$-bundle over the space $SPD(n)$ of symmetric, positive definite matrices.
(5) $GL^+(n, \mathbb{R})$, is a principal $SO(n)$-bundle over the space $SPD(n)$ of symmetric, positive definite matrices, with fiber $SO(n)$.
(6) $SO(n + 1)$ is a principal $O(n)$-bundle over the real projective space $\mathbb{RP}^n$ (for $n \geq 0$).
(7) $SU(n + 1)$ is a principal $U(n)$-bundle over the complex projective space $\mathbb{CP}^n$ (for $n \geq 0$).
(8) $O(n)$ is a principal $O(k) \times O(n - k)$-bundle over the Grassmannian $G(k, n)$.
(9) $SO(n)$ is a principal $S(O(k) \times O(n - k))$-bundle over the Grassmannian $G(k, n)$.
(10) $SO(n)$ is a principal $SO(n - k)$-bundle over the Stiefel manifold $S(k, n)$, with $1 \leq k \leq n - 1$.
(11) The Lorentz group $SO_0(n, 1)$ is a principal $SO(n)$-bundle over the space $\mathcal{H}_n^+(1)$, consisting of one sheet of the hyperbolic paraboloid $\mathcal{H}_n(1)$.

Thus, we see that both $SO(n+1)$ and $SO_0(n, 1)$ are principal $SO(n)$-bundles, the difference being that the base space for $SO(n+1)$ is the sphere S^n, which is compact, whereas the base space for $SO_0(n, 1)$ is the (connected) surface $\mathcal{H}_n^+(1)$, which is not compact. Many more examples can be given, for instance, see Arvanitoyeorgos [3].

9.11   Problems

Problem 9.1. Show that a Klein bottle is a fiber bundle with $B = F = S^1$ and $G = \{-1, 1\}$.

Problem 9.2. Adjust the proof of Proposition 9.1 to prove the following: If $\xi_1 = (E_1, \pi_1, B_1, F, G)$ and $\xi_2 = (E_2, \pi_2, B_2, F, G)$ are two bundles over different bases and $f\colon \xi_1 \to \xi_2$ is a bundle isomorphism, with $f = (f_B, f_E)$, then f_E

and f_B are diffeomorphisms, and

$$g'_{\alpha\beta}(f_B(b)) = \rho_\alpha(b)g_{\alpha\beta}(b)\rho_\beta(b)^{-1}, \qquad \text{for all } b \in U_\alpha \cap U_\beta.$$

Problem 9.3. Adjust the proof of Proposition 9.2 to prove the following: If $\xi_1 = (E_1, \pi_1, B_1, F, G)$ and $\xi_2 = (E_2, \pi_2, B_2, F, G)$ are two bundles over different bases and if there is a diffeomorphism $f_B : B_1 \to B_2$, and the conditions

$$g'_{\alpha\beta}(f_B(b)) = \rho_\alpha(b)g_{\alpha\beta}(b)\rho_\beta(b)^{-1}, \qquad \text{for all } b \in U_\alpha \cap U_\beta$$

hold, then there is a bundle isomorphism (f_B, f_E) between ξ_1 and ξ_2.

Problem 9.4. Complete the proof details of Theorem 9.4. In particular check that the cocycle condition produces an equivalence relation on $Z \times Z$. Also verify that the corresponding transition functions are the original $g_{\alpha\beta}$. Finally prove that ξ_g and $\xi_{g'}$ are isomorphic when g and g' are equivalent cocycles.
Hint. See Steenrod [101], Part I, Section 3, Theorem 3.2. Also see Steenrod [101], Part I, Section 2, Lemma 2.10.

Problem 9.5. Show that the transition functions of the pullback bundle $f^*\xi$ can be constructed as follows: Pick any open cover (U_α) of B, then $(f^{-1}(U_\alpha))$ is an open cover of N, and check that if $(g_{\alpha\beta})$ is a cocycle for ξ, then the maps $g_{\alpha\beta} \circ f : f^{-1}(U_\alpha) \cap f^{-1}(U_\beta) \to G$ satisfy the cocycle conditions.

Problem 9.6. Show that the pullback bundle $f^*\xi$ of Definition 9.11 can be defined explicitly as follows. Set

$$f^*E = \{(n, e) \in N \times E \mid f(n) = \pi(e)\},$$

$\pi^* = pr_1$ and $f^* = pr_2$. For any trivialization $\varphi_\alpha : \pi^{-1}(U_\alpha) \to U_\alpha \times F$ of ξ, we have

$$(\pi^*)^{-1}(f^{-1}(U_\alpha)) = \{(n, e) \in N \times E \mid n \in f^{-1}(U_\alpha), e \in \pi^{-1}(f(n))\},$$

and so we have a bijection $\widetilde{\varphi}_\alpha : (\pi^*)^{-1}(f^{-1}(U_\alpha)) \to f^{-1}(U_\alpha) \times F$, given by

$$\widetilde{\varphi}_\alpha(n, e) = (n, pr_2(\varphi_\alpha(e))).$$

By giving f^*E the smallest topology that makes each $\widetilde{\varphi}_\alpha$ a diffeomorphism, prove that each $\widetilde{\varphi}_\alpha$ is a trivialization of $f^*\xi$ over $f^{-1}(U_\alpha)$.

Problem 9.7. If $g : M \to N$ is another smooth map of manifolds, show that that

$$(f \circ g)^*\xi = g^*(f^*\xi).$$

Problem 9.8. Let $H_n^{\mathbb{R}} \subseteq \mathbb{RP}^n \times \mathbb{R}^{n+1}$ be the subset

$$H_n^{\mathbb{R}} = \{(L, v) \in \mathbb{RP}^n \times \mathbb{R}^{n+1} \mid v \in L\},$$

where $\mathbb{RP}^n$ is viewed as the set of lines L in $\mathbb{R}^{n+1}$ through 0, or more explicitly,

$$H_n^{\mathbb{R}} = \{((x_0 : \cdots : x_n), \lambda(x_0, \ldots, x_n)) \mid (x_0 : \cdots : x_n) \in \mathbb{RP}^n, \lambda \in \mathbb{R}\}.$$

Show that $H_n^{\mathbb{R}}$ is a manifold of dimension $n + 1$.

Problem 9.9. For rank 1 vector bundles, show that the set of equivalence classes of line bundles over a base B forms a group, where the group operation is $\otimes$, the inverse is $*$ (dual), and the identity element is the trivial bundle.

Problem 9.10. Given a real vector bundle, ξ, recall that the *complexification* $\xi_{\mathbb{C}}$ *of* ξ is the complex vector bundle defined by

$$\xi_{\mathbb{C}} = \xi \otimes_{\mathbb{R}} \epsilon_{\mathbb{C}},$$

where $\epsilon_{\mathbb{C}} = B \times \mathbb{C}$ is the trivial complex line bundle. Given a complex vector bundle ξ, recall that by viewing its fiber as a real vector space we obtain the real vector bundle $\xi_{\mathbb{R}}$. Prove the following.

(1) For every real vector bundle ξ,

$$(\xi_{\mathbb{C}})_{\mathbb{R}} \cong \xi \oplus \xi.$$

(2) For every complex vector bundle ξ,

$$(\xi_{\mathbb{R}})_{\mathbb{C}} \cong \xi \oplus \xi^*.$$

Problem 9.11. Given ξ, a subbundle of ξ', we can form the *quotient bundle* ξ'/ξ as the bundle over B whose fiber at $b \in B$ is the quotient space $(\pi')^{-1}(b)/\pi^{-1}(b)$. Define the trivializations and transition maps for ξ'/ξ.

Problem 9.12. Prove Proposition 9.13.
Hint. See O'Neill [84], Chapter 2, Proposition 13 and Theorem 15.

Problem 9.13. Prove the following: If $\xi = (E, \pi, B, V)$ is an orientable vector bundle and its base B is an orientable manifold, then E is orientable too.
Hint. Assume that B is a manifold of dimension m, ξ is a rank n vector bundle with fiber V, let $((U_\alpha, \psi_\alpha))_\alpha$ be an atlas for B, let $\varphi_\alpha \colon \pi^{-1}(U_\alpha) \to U_\alpha \times V$ be a collection of trivializing maps for ξ, and pick any isomorphism, $\iota \colon V \to \mathbb{R}^n$.

(a) Show that the maps

$$(\psi_\alpha \times \iota) \circ \varphi_\alpha \colon \pi^{-1}(U_\alpha) \longrightarrow \mathbb{R}^m \times \mathbb{R}^n.$$

form an atlas for E.

(b) Check that the corresponding transition maps for E are of the form

$$(x, y) \mapsto (\psi_\beta \circ \psi_\alpha^{-1}(x), g_{\alpha\beta}(\psi_\alpha^{-1}(x))y).$$

(c) Since B and ξ are orientable, check that these transition maps have positive Jacobian.

Problem 9.14. Prove Proposition 9.19.
Hint. See Milnor and Stasheff [78], Chapter 3.

Problem 9.15. Prove Proposition 9.11.
Hint. See Madsen and Tornehave [75], Chapter 16.

Problem 9.16. Complete the proof of Proposition 9.20. Recall that right action is defined as follows: Let $\{(U_\alpha, \varphi_\alpha)\}$ be some trivializing cover defining ξ. For every $z \in E$, pick some U_α so that $\pi(z) \in U_\alpha$, and let $\varphi_\alpha(z) = (b, h)$, where $b = \pi(z)$ and $h \in G$. For any $g \in G$, we set

$$z \cdot g = \varphi_\alpha^{-1}(b, hg).$$

Show this action takes each fiber to itself. Also show that E/G is diffeomorphic to B.

Problem 9.17. Complete the proof details of Proposition 9.25. In particular show that the map $\theta : (U_\alpha \times G) \times_G F \to U_\alpha \times F$ given by

$$\theta([((b_1, g_2), g_2^{-1}g_1 \cdot f_1)]) = (b_1, g_1 \cdot f_1)$$

is well defined on the equivalence class $[((b_1, g_1), f_1)]$ and that it is a bijection. Also show that the transition functions of $\xi[F]$ are the same as the transition functions of ξ.

Chapter 10
Connections and Curvature in Vector Bundles

10.1 Introduction to Connections in Vector Bundles

A connection on a manifold B is a means of relating different tangent spaces. In particular, a *connection* on B is a $\mathbb{R}$-bilinear map

$$\nabla \colon \mathfrak{X}(B) \times \mathfrak{X}(B) \to \mathfrak{X}(B), \qquad (\dagger)$$

such that the following two conditions hold:

$$\nabla_{fX} Y = f \nabla_X Y$$
$$\nabla_X (fY) = X[f]Y + f \nabla_X Y,$$

for all smooth vector fields $X, Y \in \mathfrak{X}(B)$ and all $f \in C^\infty(B)$; see Gallot, Hulin, and Lafontaine [48], Do Carmo [37], or Gallier and Quaintance [47].

Given $p \in B$ and $X, Y \in \mathfrak{X}(B)$, we know that Equation $(\dagger)$ is related to the directional derivative $D_X Y(p)$ of Y with respect to X, namely

$$D_X Y(p) = \lim_{t \to 0} \frac{Y(p + tX(p)) - Y(p)}{t},$$

since

$$D_X Y(p) = \nabla_X Y(p) + (D_n)_X Y(p),$$

where its horizontal (or tangential) component is $\nabla_X Y(p) \in T_p M$, and its normal component is $(D_n)_X Y(p)$. A natural question is to wonder whether we can generalize this notion of directional derivative to the case of a vector bundle $\xi = (E, \pi, B, V)$. The answer is yes if we let Y be a smooth global vector field of

© Springer Nature Switzerland AG 2020, corrected publication 2020
J. Gallier, J. Quaintance, *Differential Geometry and Lie Groups*, Geometry and Computing 13, https://doi.org/10.1007/978-3-030-46047-1_10

V instead of a smooth global vector field of tangent vectors. In other words, since $\mathfrak{X}(B)$ is the set of smooth sections of the tangent bundle TB, we may rewrite (†) as

$$\nabla \colon \mathfrak{X}(B) \times \Gamma(TB) \to \Gamma(TB), \tag{††}$$

replace the two occurrence of $\Gamma(TB)$ with $\Gamma(\xi)$ and say a connection on ξ is an $\mathbb{R}$-bilinear map

$$\nabla \colon \mathfrak{X}(B) \times \Gamma(\xi) \to \Gamma(\xi),$$

such that the following two conditions hold:

$$\nabla_{fX} s = f \nabla_X s$$
$$\nabla_X(fs) = X[f]s + f \nabla_X s,$$

for all $s \in \Gamma(\xi)$, all $X \in \mathfrak{X}(B)$ and all $f \in C^\infty(B)$. We refer to $\nabla_X s$ the covariant derivative of s relative to X.

This definition of a connection on a vector bundle has the advantage in that it readily allows us to transfer all the concepts of connections on a manifold to the context of connections in vector bundles. In particular, we will show that connections in vector bundles exist and are local operators; see Propositions 10.4 and 10.1 respectively. We will be able to define the notion of parallel transport along a curve of B in terms of the $\mathbb{R}$-linear map $\frac{D}{dt}$ where

$$\frac{DX}{dt}(t_0) = (\nabla_{\gamma'(t_0)} s)_{\gamma(t_0)},$$

whenever X is induced by a global section $s \in \Gamma(\xi)$, i.e., $X(t_0) = s(\gamma(t_0))$ for all $t_0 \in [a, b]$; see Proposition 10.6 and Definition 10.7. We will also be able to define the notion of a metric connection in a vector bundle as follows. Given any metric $\langle -, - \rangle$ on a vector bundle ξ, a connection ∇ on ξ is compatible with the metric if and only if

$$X(\langle s_1, s_2 \rangle) = \langle \nabla_X s_1, s_2 \rangle + \langle s_1, \nabla_X s_2 \rangle,$$

for every vector field $X \in \mathfrak{X}(B)$ and sections $s_1, s_2 \in \Gamma(\xi)$; see Definition 10.17.

We can also generalize the notion of curvature in a Riemannian manifold to the context of vector bundles if we define the curvature tensor of $\Gamma(\xi)$ as

$$R(X, Y) = \nabla_X \nabla_Y - \nabla_Y \nabla_X - \nabla_{[X,Y]},$$

where $X, Y \in \mathfrak{X}(B)$. Note that this definition of curvature implies that

$$R \colon \mathfrak{X}(B) \times \mathfrak{X}(B) \times \Gamma(\xi) \longrightarrow \Gamma(\xi)$$

is a $\mathbb{R}$-trilinear map where

$$R(X, Y)s = \nabla_X \nabla_Y s - \nabla_Y \nabla_X s - \nabla_{[X,Y]}s,$$

whenever $X, Y \in \mathfrak{X}(B)$ and $s \in \Gamma(\xi)$.

The reason we are interested in having a definition of curvature on a vector bundle $\xi = (E, \pi, B, V)$ is that it allows us to define global invariants on ξ called the Pontrjagin and Chern classes; see Section 10.11. However, in order to define the Pontrjagin and Chern classes in an accessible manner, we will need to associate $R(X, Y)$ with a vector valued two-form R^∇. We are able to make this association if we realize that a connection on $\xi = (E, \pi, B, V)$ is actually a vector valued one-form with the vector values taken from $\Gamma(\xi)$. Therefore, following the lead of Appendix C in Milnor and Stasheff [78], we will rephrase the notions of connection, metric connection, and curvature in terms of vector valued forms. This vector valued form approach has another advantage in that it allows for elegant proofs of the essential properties of connections on vector bundles.

In Section 10.2 we define connections on a vector bundle. This can be done in two equivalent ways. One of the two definitions is more abstract than the other because it involves a tensor product, but it is technically more convenient. This definition states that a connection on a vector bundle ξ, as an $\mathbb{R}$-linear map

$$\nabla \colon \Gamma(\xi) \to \mathcal{A}^1(B) \otimes_{C^\infty(B)} \Gamma(\xi) \tag{$*$}$$

that satisfies the "Leibniz rule"

$$\nabla(fs) = df \otimes s + f \nabla s,$$

with $s \in \Gamma(\xi)$ and $f \in C^\infty(B)$, where $\Gamma(\xi)$ and $\mathcal{A}^1(B)$ are treated as $C^\infty(B)$-modules. Here, $\mathcal{A}^1(B) = \Gamma(T^*B)$ is the space of 1-forms on B. Since there is an isomorphism

$$\mathcal{A}^1(B) \otimes_{C^\infty(B)} \Gamma(\xi) \cong \Gamma(T^*B \otimes \xi),$$

a connection can be defined equivalently as an $\mathbb{R}$-linear map

$$\nabla \colon \Gamma(\xi) \to \Gamma(T^*B \otimes \xi)$$

satisfying the Leibniz rule. Milnor and Stasheff [78] (Appendix C) use this second version, and Madsen and Tornehave [75] (Chapter 17) use the equivalent version stated in $(*)$. We show that a connection is a local operator.

In Section 10.3, we show how a connection can be represented in a chart in terms of a certain matrix called a *connection matrix*. We prove that every vector bundle possesses a connection, and we give a formula describing how a connection matrix changes if we switch from one chart to another.

In Section 10.4 we define the notion of covariant derivative along a curve and parallel transport.

Section 10.5 is devoted to the very important concept of *curvature form* R^∇ of a connection ∇ on a vector bundle ξ. We show that the curvature form is a vector-valued two-form with values in $\Gamma(\mathcal{H}om(\xi, \xi))$. We also establish the relationship between R^∇ and the more familiar definition of the Riemannian curvature in terms of vector fields.

In Section 10.6 we show how the curvature form can be expressed in a chart in terms of a matrix of two-forms called a *curvature matrix*. The connection matrix and the curvature matrix are related by the *structure equation*. We also give a formula describing how a curvature matrix changes if we switch from one chart to another. Bianchi's identity gives an expression for the exterior derivative of the curvature matrix in terms of the curvature matrix itself and the connection matrix.

Section 10.8 deals with connections compatible with a metric, and the Levi-Civita connection, which arises in the Riemannian geometry of manifolds. One way of characterizing the Levi-Civita connection involves defining the notion of connection on the dual bundle. This is achieved in Section 10.9.

Levi-Civita connections on the tangent bundle of a manifold are investigated in Section 10.10.

The purpose of Section 10.11 is to introduce the reader to *Pontrjagin Classes and Chern Classes*, which are fundamental invariants of real (resp. complex) vector bundles. Here we are dealing with one of the most sophisticated and beautiful parts of differential geometry.

Pontrjagin, Stiefel, and Chern (starting from the late 1930s) discovered that invariants with "good" properties could be defined if we took these invariants to belong to various cohomology groups associated with B. Such invariants are usually called *characteristic classes*. Roughly, there are two main methods for defining characteristic classes: one using topology, and the other due to Chern and Weil, using differential forms.

A masterly exposition of these methods is given in the classic book by Milnor and Stasheff [78]. Amazingly, the method of Chern and Weil using differential forms is quite accessible for someone who has reasonably good knowledge of differential forms and de Rham cohomology, as long as one is willing to gloss over various technical details. We give an introduction to characteristic classes using the method of Chern and Weil.

If ξ is a real orientable vector bundle of rank $2m$, and if ∇ is a metric connection on ξ, then it is possible to define a closed global form $\mathrm{eu}(R^\nabla)$, and its cohomology class $e(\xi)$ is called the *Euler class* of ξ. This is shown in Section 10.13. The Euler class $e(\xi)$ turns out to be a square root of the top Pontrjagin class $p_m(\xi)$ of ξ. A complex rank m vector bundle can be viewed as a real vector bundle of rank $2m$, which is always orientable. The Euler class $e(\xi)$ of this real vector bundle is equal to the top Chern class $c_m(\xi)$ of the complex vector bundle ξ.

The global form $\mathrm{eu}(R^\nabla)$ is defined in terms of a certain polynomial $\mathrm{Pf}(A)$ associated with a real skew-symmetric matrix A, which is a kind of square root of the determinant $\det(A)$. The polynomial $\mathrm{Pf}(A)$, called the *Pfaffian*, is defined in Section 10.12.

The culmination of this chapter is a statement of the generalization due to Chern of a classical theorem of Gauss and Bonnet. This theorem known as the *generalized Gauss–Bonnet formula* expresses the Euler characteristic $\chi(M)$ of an orientable, compact smooth manifold M of dimension $2m$ as

$$\chi(M) = \int_M \mathrm{eu}(R^\nabla),$$

where $\mathrm{eu}(R^\nabla)$ is the Euler form associated with the curvature form R^∇ of a metric connection ∇ on M.

10.2 Connections and Connection Forms in Vector Bundles

The goal of this section is to generalize the notions of a connection to vector bundles. Among other things, this material has applications to theoretical physics. This chapter makes heavy use of differential forms (and tensor products), so the reader may want to brush up on these notions before reading it.

Given a manifold M, as $\mathfrak{X}(M) = \Gamma(M, TM) = \Gamma(TM)$, the set of smooth sections of the tangent bundle TM, it is natural that for a vector bundle $\xi = (E, \pi, B, V)$, a connection on ξ should be some kind of bilinear map,

$$\mathfrak{X}(B) \times \Gamma(\xi) \longrightarrow \Gamma(\xi),$$

that tells us how to take the covariant derivative of sections.

Technically, it turns out that it is cleaner to define a connection on a vector bundle ξ, as an $\mathbb{R}$-linear map

$$\nabla \colon \Gamma(\xi) \to \mathcal{A}^1(B) \otimes_{C^\infty(B)} \Gamma(\xi) \qquad (*)$$

that satisfies the "Leibniz rule"

$$\nabla(fs) = df \otimes s + f\nabla s,$$

with $s \in \Gamma(\xi)$ and $f \in C^\infty(B)$, where $\Gamma(\xi)$ and $\mathcal{A}^1(B)$ are treated as $C^\infty(B)$-modules. Since $\mathcal{A}^1(B) = \Gamma(B, T^*B) = \Gamma(T^*B)$ is the space of 1-forms on B, and by Proposition 9.11,

$$\mathcal{A}^1(B) \otimes_{C^\infty(B)} \Gamma(\xi) = \Gamma(T^*B) \otimes_{C^\infty(B)} \Gamma(\xi)$$

$$\cong \Gamma(T^*B \otimes \xi)$$

$$\cong \Gamma(\mathcal{H}om(TB, \xi))$$

$$\cong \mathrm{Hom}_{C^\infty(B)}(\Gamma(TB), \Gamma(\xi))$$

$$= \mathrm{Hom}_{C^\infty(B)}(\mathfrak{X}(B), \Gamma(\xi)),$$

the range of ∇ can be viewed as a space of $\Gamma(\xi)$-valued differential forms on B. Milnor and Stasheff [78] (Appendix C) use the version where

$$\nabla \colon \Gamma(\xi) \to \Gamma(T^*B \otimes \xi),$$

and Madsen and Tornehave [75] (Chapter 17) use the equivalent version stated in (∗). A thorough presentation of connections on vector bundles and the various ways to define them can be found in Postnikov [88] which also constitutes one of the most extensive references on differential geometry.

If we use the isomorphism

$$\mathcal{A}^1(B) \otimes_{C^\infty(B)} \Gamma(\xi) \cong \mathrm{Hom}_{C^\infty(B)}(\mathfrak{X}(B), \Gamma(\xi)),$$

then a connection is an $\mathbb{R}$-linear map

$$\nabla \colon \Gamma(\xi) \longrightarrow \mathrm{Hom}_{C^\infty(B)}(\mathfrak{X}(B), \Gamma(\xi))$$

satisfying a Leibniz-type rule, or equivalently, an $\mathbb{R}$-bilinear map

$$\nabla \colon \mathfrak{X}(B) \times \Gamma(\xi) \longrightarrow \Gamma(\xi)$$

such that, for any $X \in \mathfrak{X}(B)$ and $s \in \Gamma(\xi)$, if we write $\nabla_X s$ instead of $\nabla(X, s)$, then the following properties hold for all $f \in C^\infty(B)$:

$$\nabla_{fX} s = f \nabla_X s$$
$$\nabla_X(fs) = X[f]s + f \nabla_X s.$$

This second version may be considered simpler than the first since it does not involve a tensor product. Since, by Proposition 2.17,

$$\mathcal{A}^1(B) = \Gamma(T^*B) \cong \mathrm{Hom}_{C^\infty(B)}(\mathfrak{X}(B), C^\infty(B)) = (\mathfrak{X}(B))^*,$$

using Proposition 2.33, the isomorphism

$$\alpha \colon \mathcal{A}^1(B) \otimes_{C^\infty(B)} \Gamma(\xi) \cong \mathrm{Hom}_{C^\infty(B)}(\mathfrak{X}(B), \Gamma(\xi))$$

can be described in terms of the evaluation map

$$\mathrm{Ev}_X \colon \mathcal{A}^1(B) \otimes_{C^\infty(B)} \Gamma(\xi) \to \Gamma(\xi),$$

given by

$$\mathrm{Ev}_X(\omega \otimes s) = \omega(X)s, \qquad X \in \mathfrak{X}(B), \ \omega \in \mathcal{A}^1(B), \ s \in \Gamma(\xi).$$

Namely, for any $\theta \in \mathcal{A}^1(B) \otimes_{C^\infty(B)} \Gamma(\xi)$,

$$\alpha(\theta)(X) = \mathrm{Ev}_X(\theta).$$

In particular, we have

$$\mathrm{Ev}_X(df \otimes s) = df(X)s = X[f]s.$$

Then it is easy to see that we pass from the first version of ∇, where

$$\nabla \colon \Gamma(\xi) \to \mathcal{A}^1(B) \otimes_{C^\infty(B)} \Gamma(\xi) \tag{$*$}$$

with the Leibniz rule

$$\nabla(fs) = df \otimes s + f\nabla s,$$

to the second version of ∇, denoted ∇', where

$$\nabla' \colon \mathfrak{X}(B) \times \Gamma(\xi) \to \Gamma(\xi) \tag{$**$}$$

is $\mathbb{R}$-bilinear and where the two conditions

$$\nabla'_{fX}s = f\nabla'_X s$$
$$\nabla'_X(fs) = X[f]s + f\nabla'_X s$$

hold, *via* the equation

$$\nabla'_X = \mathrm{Ev}_X \circ \nabla.$$

From now on, we will simply write $\nabla_X s$ instead of $\nabla'_X s$, unless confusion arise. As summary of the above discussion, we make the following definition.

Definition 10.1. Let $\xi = (E, \pi, B, V)$ be a smooth real vector bundle. A *connection* on ξ is an $\mathbb{R}$-linear map

$$\nabla \colon \Gamma(\xi) \to \mathcal{A}^1(B) \otimes_{C^\infty(B)} \Gamma(\xi) \tag{$*$}$$

such that the *Leibniz rule*

$$\nabla(fs) = df \otimes s + f\nabla s$$

holds, for all $s \in \Gamma(\xi)$ and all $f \in C^\infty(B)$. For every $X \in \mathfrak{X}(B)$, we let

$$\nabla_X = \mathrm{Ev}_X \circ \nabla$$

where the evaluation map

$$\mathrm{Ev}_X \colon \mathcal{A}^1(B) \otimes_{C^\infty(B)} \Gamma(\xi) \to \Gamma(\xi),$$

is given by

$$\mathrm{Ev}_X(\omega \otimes s) = \omega(X)s, \qquad X \in \mathfrak{X}(B),\ \omega \in \mathcal{A}^1(B),\ s \in \Gamma(\xi),$$

and for every $s \in \Gamma(\xi)$, we call $\nabla_X s$ the *covariant derivative of s relative to X*. Then the family (∇_X) induces a $\mathbb{R}$-bilinear map also denoted ∇,

$$\nabla \colon \mathfrak{X}(B) \times \Gamma(\xi) \to \Gamma(\xi), \qquad\qquad\qquad (**)$$

such that the following two conditions hold:

$$\nabla_{fX} s = f \nabla_X s$$
$$\nabla_X(fs) = X[f]s + f \nabla_X s,$$

for all $s \in \Gamma(\xi)$, all $X \in \mathfrak{X}(B)$ and all $f \in C^\infty(B)$. We refer to $(*)$ as the *first version* of a connection and to $(**)$ as the *second version* of a connection.

Every vector bundle admits a connection. We need some technical tools to prove this, so we postpone the proof until Proposition 10.4.

Remark. Given two connections, ∇^1 and ∇^2, we have

$$\nabla^1(fs) - \nabla^2(fs) = df \otimes s + f\nabla^1 s - df \otimes s - f\nabla^2 s = f(\nabla^1 s - \nabla^2 s),$$

which shows that $\nabla^1 - \nabla^2$ is a $C^\infty(B)$-linear map from $\Gamma(\xi)$ to $\mathcal{A}^1(B) \otimes_{C^\infty(B)} \Gamma(\xi)$. However

$$\mathrm{Hom}_{C^\infty(B)}(\Gamma(\xi), \mathcal{A}^i(B) \otimes_{C^\infty(B)} \Gamma(\xi)) \cong (\Gamma(\xi))^* \otimes_{C^\infty(B)} (\mathcal{A}^i(B) \otimes_{C^\infty(B)} \Gamma(\xi))$$

$$\cong \mathcal{A}^i(B) \otimes_{C^\infty(B)} ((\Gamma(\xi))^* \otimes_{C^\infty(B)} \Gamma(\xi))$$

$$\cong \mathcal{A}^i(B) \otimes_{C^\infty(B)} \mathrm{Hom}_{C^\infty(B)}(\Gamma(\xi), \Gamma(\xi))$$

$$\cong \mathcal{A}^i(B) \otimes_{C^\infty(B)} \Gamma(\mathcal{H}om(\xi, \xi)).$$

Therefore, $\nabla^1 - \nabla^2$ is a one-form with values in $\Gamma(\mathcal{H}om(\xi, \xi))$. But then, the vector space $\Gamma(\mathcal{H}om(\xi, \xi))$ acts on the space of connections (by addition) and makes the space of connections into an affine space. Given any connection, ∇ and any one-form $\omega \in \Gamma(\mathcal{H}om(\xi, \xi))$, the expression $\nabla + \omega$ is also a connection. Equivalently, any affine combination of connections is also a connection.

If $\xi = TM$, the tangent bundle of some smooth manifold M, then a connection on TM, also called a *connection on M*, is a linear map

$$\nabla: \mathfrak{X}(M) \longrightarrow \mathcal{A}^1(M) \otimes_{C^\infty(M)} \mathfrak{X}(M) \cong \mathrm{Hom}_{C^\infty(M)}(\mathfrak{X}(M), \mathfrak{X}(M)),$$

since $\Gamma(TM) = \mathfrak{X}(M)$. Then for fixed $Y \in \mathfrak{X}(M)$, the map ∇Y is $C^\infty(M)$-linear, which implies that ∇Y is a $(1, 1)$ tensor. In a local chart, (U, φ), we have

$$\nabla_{\frac{\partial}{\partial x_i}}\left(\frac{\partial}{\partial x_j}\right) = \sum_{k=1}^n \Gamma_{ij}^k \frac{\partial}{\partial x_k},$$

where the Γ_{ij}^k are Christoffel symbols.

A basic property of ∇ is that it is a local operator.

Proposition 10.1. *Let $\xi = (E, \pi, B, V)$ be a smooth real vector bundle and let ∇ be a connection on ξ. For every open subset $U \subseteq B$, for every section $s \in \Gamma(\xi)$, if $s \equiv 0$ on U, then $\nabla s \equiv 0$ on U; that is, ∇ is a local operator.*

Proof. Using a bump function applied to the constant function with value 1, for every $p \in U$, there is some open subset, $V \subseteq U$, containing p and a smooth function, $f: B \to \mathbb{R}$, such that supp $f \subseteq U$ and $f \equiv 1$ on V. Consequently, fs is a smooth section which is identically zero. By applying the Leibniz rule, we get

$$0 = \nabla(fs) = df \otimes s + f\nabla s,$$

which, evaluated at p yields $(\nabla s)(p) = 0$, since $f(p) = 1$ and $df \equiv 0$ on V. $\square$

As an immediate consequence of Proposition 10.1, if s_1 and s_2 are two sections in $\Gamma(\xi)$ that agree on U, then $s_1 - s_2$ is zero on U, so $\nabla(s_1 - s_2) = \nabla s_1 - \nabla s_2$ is zero on U, that is, ∇s_1 and ∇s_2 agree on U.

Proposition 10.1 implies the following fact.

Proposition 10.2. *A connection ∇ on ξ restricts to a connection $\nabla \restriction U$ on the vector bundle $\xi \restriction U$, for every open subset $U \subseteq B$.*

Proof. Indeed, let s be a section of ξ over U. Pick any $b \in U$ and define $(\nabla s)(b)$ as follows: Using a bump function, there is some open subset, $V_1 \subseteq U$, containing b and a smooth function, $f_1: B \to \mathbb{R}$, such that supp $f_1 \subseteq U$ and $f_1 \equiv 1$ on V_1 so, let $s_1 = f_1 s$, a global section of ξ. Clearly, $s_1 = s$ on V_1, and set

$$(\nabla s)(b) = (\nabla s_1)(b).$$

This definition does not depend on (V_1, f_1), because if we had used another pair, (V_2, f_2), as above, since $b \in V_1 \cap V_2$, we have

$$s_1 = f_1 s = s = f_2 s = s_2 \qquad \text{on} \quad V_1 \cap V_2$$

so, by Proposition 10.1,

$$(\nabla s_1)(b) = (\nabla s_2)(b).$$

$\square$

It should also be noted that $(\nabla_X s)(b)$ only depends on $X(b)$.

Proposition 10.3. *For any two vector fields* $X, Y \in \mathfrak{X}(B)$, *if* $X(b) = Y(b)$ *for some* $b \in B$, *then*

$$(\nabla_X s)(b) = (\nabla_Y s)(b), \qquad \text{for every } s \in \Gamma(\xi).$$

Proof. As above, by linearity, it is enough to prove that if $X(b) = 0$, then $(\nabla_X s)(b) = 0$ (this argument is due to O'Neill [84], Chapter 2, Lemma 3). To prove this, pick any local chart, (U, φ), with $b \in U$. We can write

$$X \upharpoonright U = \sum_{i=1}^{d} X_i \frac{\partial}{\partial x_i}.$$

However, as before, we can find a pair, (V, f), with $b \in V \subseteq U$, supp $f \subseteq U$ and $f = 1$ on V, so that $f \frac{\partial}{\partial x_i}$ is a smooth vector field on B and $f \frac{\partial}{\partial x_i}$ agrees with $\frac{\partial}{\partial x_i}$ on V, for $i = 1, \ldots, n$. Clearly, $f X_i \in C^\infty(B)$ and $f X_i$ agrees with X_i on V so if we write $\widetilde{X} = f^2 X$, then

$$\widetilde{X} = f^2 X = \sum_{i=1}^{d} f X_i \, f \frac{\partial}{\partial x_i}$$

and we have

$$f^2 \nabla_X s = \nabla_{\widetilde{X}} s = \sum_{i=1}^{d} f X_i \, \nabla_{f \frac{\partial}{\partial x_i}} s.$$

Since $X_i(b) = 0$ and $f(b) = 1$, we get $(\nabla_X s)(b) = 0$, as claimed. $\square$

Using the above property, for any point, $p \in B$, we can define the covariant derivative $(\nabla_u s)(p)$ of a section $s \in \Gamma(\xi)$, with respect to a tangent vector $u \in T_p B$.

Definition 10.2. Pick any vector field $X \in \mathfrak{X}(B)$ such that $X(p) = u$ (such a vector field exists locally over the domain of a chart, then extend it using a bump function), and define $\nabla_u s$ by $(\nabla_u s)(p) = (\nabla_X s)(p)$.

Proof. By the above property, if $X(p) = Y(p)$, then $(\nabla_X s)(p) = (\nabla_Y s)(p)$ so $(\nabla_u s)(p)$ is well defined. Since ∇ is a local operator, $(\nabla_u s)(p)$ is also well defined for any tangent vector $u \in T_p B$, and any local section $s \in \Gamma(U, \xi)$ defined in some open subset U, with $p \in U$. $\square$

From now on, we will use this property without any further justification.

Since ξ is locally trivial, it is interesting to see what $\nabla \restriction U$ looks like when (U, φ) is a local trivialization of ξ. This can be done in terms of connection matrices.

10.3 Connection Matrices

Fix once and for all some basis $(v_1, \ldots, v_n)$ of the typical fiber V ($n = \dim(V)$). To every local trivialization $\varphi \colon \pi^{-1}(U) \to U \times V$ of ξ (for some open subset, $U \subseteq B$), we associate the frame $(s_1, \ldots, s_n)$ over U, given by

$$s_i(b) = \varphi^{-1}(b, v_i), \qquad b \in U. \tag{$*$}$$

Then every section s over U can be written uniquely as $s = \sum_{i=1}^n f_i s_i$, for some functions $f_i \in C^\infty(U)$, and we have

$$\nabla s = \sum_{i=1}^n \nabla(f_i s_i) = \sum_{i=1}^n (df_i \otimes s_i + f_i \nabla s_i).$$

On the other hand, each ∇s_i can be written as

$$\nabla s_i = \sum_{j=1}^n \omega_{ji} \otimes s_j,$$

for some $n \times n$ matrix $\omega = (\omega_{ij})$ of one-forms $\omega_{ij} \in \mathcal{A}^1(U)$, which we represent in matrix form as

$$\begin{pmatrix} \nabla s_1 & \cdots & \nabla s_n \end{pmatrix} = \begin{pmatrix} s_1 & \cdots & s_n \end{pmatrix} \begin{pmatrix} \omega_{11} & \cdots & \omega_{1n} \\ \vdots & \ddots & \vdots \\ \omega_{n1} & \cdots & \omega_{nn} \end{pmatrix}.$$

Thus we get

$$\nabla s = \sum_{i=1}^n df_i \otimes s_i + \sum_{i=1}^n f_i \nabla s_i = \sum_{i=1}^n df_i \otimes s_i + \sum_{i,j=1}^n f_i \omega_{ji} \otimes s_j$$

$$= \sum_{j=1}^n (df_j + \sum_{i=1}^n f_i \omega_{ji}) \otimes s_j,$$

which we efficiently record as follows:

Definition 10.3. With respect to the frame $(s_1, \ldots, s_n)$ over the open subset U the connection ∇ has the matrix form

$$\nabla \begin{pmatrix} f_1 \\ \vdots \\ f_n \end{pmatrix} = \begin{pmatrix} df_1 \\ \vdots \\ df_n \end{pmatrix} + \omega \begin{pmatrix} f_1 \\ \vdots \\ f_n \end{pmatrix},$$

where the matrix $\omega = (\omega_{ij})$ of one-forms $\omega_{ij} \in \mathcal{A}^1(U)$ is called the *connection form* or *connection matrix* of ∇ with respect to $\varphi \colon \pi^{-1}(U) \to U \times V$.

The above computation also shows that on U, any connection is uniquely determined by a matrix of one-forms, $\omega_{ij} \in \mathcal{A}^1(U)$.

Example 10.1. Let $B = \mathbb{R}^3$ and ξ be the tangent bundle of $\mathbb{R}^3$, i.e., $\xi = (T\mathbb{R}^3, \pi, \mathbb{R}^3, \mathbb{R}^3)$. Following O'Neil, we describe the spherical frame (s_1, s_2, s_3) for the tangent space at each point of $\mathbb{R}^3$. Recall that each point of $\mathbb{R}^3$ may be parametrized via spherical coordinates as follows:

$$x = r \cos \varphi \sin \theta$$
$$y = r \sin \varphi \sin \theta$$
$$z = r \cos \theta,$$

where $r \geq 0$, $0 \leq \theta \leq \pi$, and $0 \leq \varphi < 2\pi$. For each $p \in \mathbb{R}^3$, we define the orthogonal spherical frame for $T\mathbb{R}^3_p$ as

$$s_1 = \frac{\partial}{\partial r} = (\cos \varphi \sin \theta, \sin \varphi \sin \theta, \cos \theta)$$

$$s_2 = \frac{\frac{\partial}{\partial \theta}}{\left\| \frac{\partial}{\partial \theta} \right\|} = (\cos \varphi \cos \theta, \sin \varphi \cos \theta, -\sin \theta)$$

$$s_3 = \frac{\frac{\partial}{\partial \varphi}}{\left\| \frac{\partial}{\partial \varphi} \right\|} = (-\sin \theta, \cos \theta, 0).$$

See Figure 10.1.

By utilizing an attitude matrix (see O'Neil [85], Chapter 2, Section 2.7), the connection form for (s_1, s_2, s_3) is given by

$$(\nabla s_1, \nabla s_2, \nabla s_3) = (s_1, s_2, s_3) \begin{pmatrix} 0 & -d\theta & -\sin\theta \, d\varphi \\ d\theta & 0 & -\cos\theta \, d\varphi \\ \sin\theta \, d\varphi & \cos\theta \, d\varphi & 0 \end{pmatrix}.$$

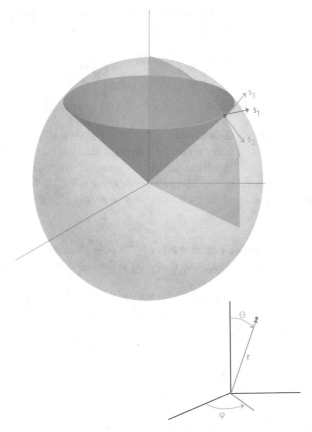

Fig. 10.1 The spherical frame (s_1, s_2, s_3) associated with the spherical coordinates (r, θ, φ). Note s_1 is normal to the sphere, s_2 is normal to the teal cone with $\theta = \theta_1$, while s_3 is normal the peach plane $\varphi = \varphi_1$.

Definition 10.4. The connection on U for which

$$\nabla s_1 = 0, \ldots, \nabla s_n = 0,$$

corresponding to the zero matrix is called the *flat connection on U* (w.r.t. $(s_1, \ldots, s_n)$).

We are following the convention in Morita [82] in expressing ∇s_i as $\nabla s_i = \sum_{j=1}^{n} \omega_{ji} \otimes s_j$, except that Morita denotes the matrix ω as (ω_j^i) where i is the row index and j is the column index, that is,

$$\nabla s_i = \sum_{j=1}^{n} \omega_i^j \otimes s_j.$$

Other authors such as Milnor and Stasheff [78] and Madsen and Tornehave [75] define ∇s_i as $\nabla s_i = \sum_{j=1}^{n} \widetilde{\omega}_{ij} \otimes s_j$, in matrix form

$$
\begin{pmatrix} \nabla s_1 \\ \vdots \\ \nabla s_n \end{pmatrix} = \begin{pmatrix} \widetilde{\omega}_{11} & \cdots & \widetilde{\omega}_{1n} \\ \vdots & \ddots & \vdots \\ \widetilde{\omega}_{n1} & \cdots & \widetilde{\omega}_{nn} \end{pmatrix} \begin{pmatrix} s_1 \\ \vdots \\ s_n \end{pmatrix},
$$

so that their matrix $\widetilde{\omega}$ *is the transpose* of our matrix ω. As a consequence, some of the results differ either by a sign (as in $\omega \wedge \omega$) or by a permutation of matrices (as in the formula for a change of frame). As we will see shortly, the advantage of Morita's convention is that it is consistent with the representation of a linear map by a matrix. This will show up in Proposition 10.5.

Remark. If $(\sigma_1, \ldots, \sigma_n)$ is a local frame of TB over U, and if $(\theta_1, \ldots, \theta_n)$ is the dual frame of $(\sigma_1, \ldots, \sigma_n)$, that is, $\theta_i \in \mathcal{A}^1(U)$ is the one-form defined so that

$$
\theta_i(b)(\sigma_j(b)) = \delta_{ij}, \qquad \text{for all} \quad b \in U, \ 1 \le i, j \le n,
$$

then we can write $\omega_{ik} = \sum_{j=1}^{n} \Gamma_{ji}^{k} \theta_j$ and so,

$$
\nabla s_i = \sum_{j,k=1}^{n} \Gamma_{ji}^{k} (\theta_j \otimes s_k),
$$

where the $\Gamma_{ji}^{k} \in C^{\infty}(U)$ are the *Christoffel symbols*.

Proposition 10.4. *Every vector bundle ξ possesses a connection.*

Proof. Since ξ is locally trivial, we can find a locally finite open cover $(U_\alpha)_\alpha$ of B such that $\pi^{-1}(U_\alpha)$ is trivial. If (f_α) is a partition of unity subordinate to the cover $(U_\alpha)_\alpha$ and if ∇^α is any flat connection on $\xi \restriction U_\alpha$, then it is immediately verified that

$$
\nabla = \sum_\alpha f_\alpha \nabla^\alpha
$$

is a connection on ξ. $\square$

If $\varphi_\alpha \colon \pi^{-1}(U_\alpha) \to U_\alpha \times V$ and $\varphi_\beta \colon \pi^{-1}(U_\beta) \to U_\beta \times V$ are two overlapping trivializations, we know that for every $b \in U_\alpha \cap U_\beta$, we have

$$
\varphi_\alpha \circ \varphi_\beta^{-1}(b, u) = (b, g_{\alpha\beta}(b)u),
$$

where $g_{\alpha\beta} \colon U_\alpha \cap U_\beta \to \mathbf{GL}(V)$ is the transition function. As

$$\varphi_\beta^{-1}(b, u) = \varphi_\alpha^{-1}(b, g_{\alpha\beta}(b)u),$$

if $(s_1, \ldots, s_n)$ is the frame over U_α associated with φ_α and $(t_1, \ldots, t_n)$ is the frame over U_β associated with φ_β, since $s_i(b) = \varphi_\alpha^{-1}(b, v_i)$ and $t_i(b) = \varphi_\beta^{-1}(b, v_i) = \varphi_\alpha^{-1}(b, g_{\alpha\beta}(b)v_i)$, if (g_{ij}) is the matrix of the linear map $g_{\alpha\beta}$ with respect to the basis $(v_1, \ldots, v_n)$, that is

$$g_{\alpha\beta}(b)v_j = \sum_{i=1}^{n} g_{ij} v_i, \tag{$**$}$$

which in matrix form is

$$\begin{pmatrix} g_{\alpha\beta}(b)v_1 & \cdots & g_{\alpha\beta}(b)v_n \end{pmatrix} = \begin{pmatrix} v_1 & \cdots & v_n \end{pmatrix} \begin{pmatrix} g_{11} & \cdots & g_{1n} \\ \vdots & \ddots & \vdots \\ g_{n1} & \cdots & g_{nn} \end{pmatrix},$$

we obtain

$$t_i(b) = \varphi_\alpha^{-1}(b, g_{\alpha\beta}(b)v_i) = \varphi_\alpha^{-1}\left(b, \sum_{j=1}^{n} g_{ji} v_j\right) = \sum_{j=1}^{n} g_{ji} \varphi_\alpha^{-1}(b, v_j) = \sum_{j=1}^{n} g_{ji} s_j(b),$$

that is

$$t_i = \sum_{j=1}^{n} g_{ji} s_j \quad \text{on } U_\alpha \cap U_\beta.$$

Proposition 10.5. *With the notations as above, the connection matrices, ω_α and ω_β respectively over U_α and U_β obey the transformation rule*

$$\omega_\beta = g_{\alpha\beta}^{-1} \omega_\alpha g_{\alpha\beta} + g_{\alpha\beta}^{-1}(dg_{\alpha\beta}),$$

where $g_{\alpha\beta}$ is viewed as the matrix function (g_{ij}) given by $g_{\alpha\beta}(b)v_j = \sum_{i=1}^{n} g_{ij} v_i$ for $j = 1, \ldots, n$ and for every $b \in U_\alpha \cap U_\beta$.

Proof. To prove the above proposition, apply ∇ to both sides of the equations

$$t_i = \sum_{j=1}^{n} g_{ji} s_j$$

on $U_\alpha \cap U_\beta$ we obtain

$$\nabla t_i = \sum_{j=1}^{n} dg_{ji} \otimes s_j + \sum_{j=1}^{n} g_{ji} \nabla s_j.$$

Since $\nabla t_i = \sum_{k=1}^{n} (\omega_\beta)_{ki} \otimes t_k$, $\nabla s_j = \sum_{k=1}^{n} (\omega_\alpha)_{kj} \otimes s_k$, and $t_k = \sum_{j=1}^{n} g_{jk} s_j$, we get

$$\nabla t_i = \sum_{j,k=1}^{n} (\omega_\beta)_{ki} g_{jk} \otimes s_j = \sum_{j=1}^{n} dg_{ji} \otimes s_j + \sum_{j,k=1}^{n} g_{ji}(\omega_\alpha)_{kj} \otimes s_k,$$

and since $(s_1, \dots, s_n)$ is a frame, the coefficients of s_j on both sides must be equal, which yields

$$\sum_{k=1}^{n} g_{jk}(\omega_\beta)_{ki} = dg_{ji} + \sum_{k=1}^{n} (\omega_\alpha)_{kj} g_{ji}$$

for all i, j, which in matrix form means that

$$g_{\alpha\beta}\omega_\beta = dg_{\alpha\beta} + \omega_\alpha g_{\alpha\beta}.$$

Since $g_{\alpha\beta}$ is invertible, we get

$$\omega_\beta = g_{\alpha\beta}^{-1} \omega_\alpha g_{\alpha\beta} + g_{\alpha\beta}^{-1}(dg_{\alpha\beta}),$$

as claimed.   □

Remark. Everything we did in this section applies to complex vector bundles by considering complex vector spaces instead of real vector spaces, $\mathbb{C}$-linear maps instead of $\mathbb{R}$-linear map, and the space of smooth complex-valued functions, $C^\infty(B; \mathbb{C}) \cong C^\infty(B) \otimes_{\mathbb{R}} \mathbb{C}$. We also use spaces of complex-valued differentials forms

$$\mathcal{A}^i(B; \mathbb{C}) = \mathcal{A}^i(B) \otimes_{\mathbb{R}} \mathbb{C},$$

and we define $\mathcal{A}^i(\xi)$ as

$$\mathcal{A}^i(\xi) = \mathcal{A}^i(B; \mathbb{C}) \otimes_{C^\infty(B;\mathbb{C})} \Gamma(\xi).$$

A connection is a $\mathbb{C}$-linear map $\nabla \colon \Gamma(\xi) \to \mathcal{A}^1(\xi)$, that satisfies the same Leibniz-type rule as before. Obviously, every differential form in $\mathcal{A}^i(B; \mathbb{C})$ can be written uniquely as $\omega + i\eta$, with $\omega, \eta \in \mathcal{A}^i(B)$. The exterior differential,

$$d \colon \mathcal{A}^i(B; \mathbb{C}) \to \mathcal{A}^{i+1}(B; \mathbb{C})$$

is defined by $d(\omega + i\eta) = d\omega + id\eta$. We obtain complex-valued de Rham cohomology groups,

$$H^i_{DR}(M; \mathbb{C}) = H^i_{DR}(M) \otimes_{\mathbb{R}} \mathbb{C}.$$

The complexification of a real vector bundle ξ is the complex vector bundle $\xi_{\mathbb{C}} = \xi \otimes_{\mathbb{R}} \epsilon^1_{\mathbb{C}}$, where $\epsilon^1_{\mathbb{C}}$ is the trivial complex line bundle $B \times \mathbb{C}$.

10.4   Parallel Transport

The notion of connection yields the notion of parallel transport in a vector bundle. First we need to define the covariant derivative of a section along a curve.

Definition 10.5. Let $\xi = (E, \pi, B, V)$ be a vector bundle and let $\gamma : [a, b] \to B$ be a smooth curve in B. A *smooth section along the curve* γ is a smooth map $X : [a, b] \to E$, such that $\pi(X(t)) = \gamma(t)$, for all $t \in [a, b]$. When $\xi = TB$, the tangent bundle of the manifold B, we use the terminology *smooth vector field along* γ.

Recall that the curve $\gamma : [a, b] \to B$ is smooth iff γ is the restriction to $[a, b]$ of a smooth curve on some open interval containing $[a, b]$. Since a section X along a curve γ does not necessarily extend to an open subset of B (for example, if the image of γ is dense in B), the covariant derivative $(\nabla_{\gamma'(t_0)} X)_{\gamma(t_0)}$ may not be defined, so we need a proposition showing that the covariant derivative of a section along a curve makes sense.

Proposition 10.6. *Let ξ be a vector bundle, ∇ be a connection on ξ, and $\gamma : [a, b] \to B$ be a smooth curve in B. There is a $\mathbb{R}$-linear map D/dt, defined on the vector space of smooth sections X along γ, which satisfies the following conditions:*

(1) For any smooth function $f : [a, b] \to \mathbb{R}$,

$$\frac{D(fX)}{dt} = \frac{df}{dt} X + f \frac{DX}{dt}$$

(2) If X is induced by a global section $s \in \Gamma(\xi)$, that is, if $X(t_0) = s(\gamma(t_0))$ for all $t_0 \in [a, b]$, then

$$\frac{DX}{dt}(t_0) = (\nabla_{\gamma'(t_0)} s)_{\gamma(t_0)}.$$

Proof. Since $\gamma([a, b])$ is compact, it can be covered by a finite number of open subsets U_α such that $(U_\alpha, \varphi_\alpha)$ is a chart for B and $(U_\alpha, \tilde{\varphi}_\alpha)$ is a local trivialization. Thus, we may assume that $\gamma : [a, b] \to U$ for some chart, (U, φ), and some local trivialization $(U, \tilde{\varphi})$. As $\varphi \circ \gamma : [a, b] \to \mathbb{R}^n$, we can write

$$\varphi \circ \gamma(t) = (u_1(t), \ldots, u_n(t)),$$

where each $u_i = pr_i \circ \varphi \circ \gamma$ is smooth. Now, for every $g \in C^\infty(B)$, as

$$d\gamma_{t_0}\left(\frac{d}{dt}\bigg|_{t_0}\right)(g) = \frac{d}{dt}(g \circ \gamma)\bigg|_{t_0} = \frac{d}{dt}((g \circ \varphi^{-1}) \circ (\varphi \circ \gamma))\bigg|_{t_0} = \sum_{i=1}^n \frac{du_i}{dt}\left(\frac{\partial}{\partial x_i}\right)_{\gamma(t_0)} g,$$

since by definition of $\gamma'(t_0)$,

$$\gamma'(t_0) = d\gamma_{t_0}\left(\frac{d}{dt}\bigg|_{t_0}\right),$$

$$\gamma'(t_0) = \sum_{i=1}^n \frac{du_i}{dt}\left(\frac{\partial}{\partial x_i}\right)_{\gamma(t_0)}.$$

If $(s_1, \ldots, s_n)$ is a frame over U determined by $(U, \tilde{\varphi})$, we can write

$$X(t) = \sum_{i=1}^n X_i(t) s_i(\gamma(t)),$$

for some smooth functions, X_i. Then Conditions (1) and (2) imply that

$$\frac{DX}{dt} = \sum_{j=1}^n \left(\frac{dX_j}{dt} s_j(\gamma(t)) + X_j(t)\nabla_{\gamma'(t)}(s_j(\gamma(t)))\right)$$

and since

$$\gamma'(t) = \sum_{i=1}^n \frac{du_i}{dt}\left(\frac{\partial}{\partial x_i}\right)_{\gamma(t)},$$

there exist some smooth functions, Γ_{ij}^k, so that

$$\nabla_{\gamma'(t)}(s_j(\gamma(t))) = \sum_{i=1}^n \frac{du_i}{dt}\nabla_{\frac{\partial}{\partial x_i}}(s_j(\gamma(t))) = \sum_{i,k}\frac{du_i}{dt}\Gamma_{ij}^k s_k(\gamma(t)).$$

It follows that

$$\frac{DX}{dt} = \sum_{k=1}^{n} \left(\frac{dX_k}{dt} + \sum_{ij} \Gamma_{ij}^k \frac{du_i}{dt} X_j \right) s_k(\gamma(t)).$$

Conversely, the above expression defines a linear operator, D/dt, and it is easy to check that it satisfies Conditions (1) and (2).   □

Definition 10.6. The operator D/dt is called the *covariant derivative along γ* and it is also denoted by $\nabla_{\gamma'(t)}$ or simply $\nabla_{\gamma'}$.

Definition 10.7. Let ξ be a vector bundle and let ∇ be a connection on ξ. For every curve $\gamma : [a, b] \to B$ in B, a section X along γ is *parallel (along γ)* iff

$$\frac{DX}{dt}(t_0) = 0 \quad \text{for all } t_0 \in [a, b].$$

If ξ was the tangent bundle of a smooth manifold M embedded in $\mathbb{R}^d$ (for some d), then to say that X is parallel along γ would mean that the directional derivative, $(D_{\gamma'}X)(\gamma(t))$, is normal to $T_{\gamma(t)}M$.

The following proposition can be shown using the existence and uniqueness of solutions of ODEs (in our case, linear ODEs).

Proposition 10.7. *Let ξ be a vector bundle and let ∇ be a connection on ξ. For every C^1 curve $\gamma : [a, b] \to B$ in B, for every $t \in [a, b]$ and every $v \in \pi^{-1}(\gamma(t))$, there is a unique parallel section X along γ such that $X(t) = v$.*

Proof. For the proof of Proposition 10.7 it is sufficient to consider the portions of the curve γ contained in some local trivialization. In such a trivialization, (U, φ), as in the proof of Proposition 10.6, using a local frame, $(s_1, \ldots, s_n)$, over U, we have

$$\frac{DX}{dt} = \sum_{k=1}^{n} \left(\frac{dX_k}{dt} + \sum_{ij} \Gamma_{ij}^k \frac{du_i}{dt} X_j \right) s_k(\gamma(t)),$$

with $u_i = pr_i \circ \varphi \circ \gamma$. Consequently, X is parallel along our portion of γ iff the system of linear ODEs in the unknowns, X_k,

$$\frac{dX_k}{dt} + \sum_{ij} \Gamma_{ij}^k \frac{du_i}{dt} X_j = 0, \qquad k = 1, \ldots, n,$$

is satisfied.   □

Remark. Proposition 10.7 can be extended to piecewise C^1 curves.

Definition 10.8. Let ξ be a vector bundle and let ∇ be a connection on ξ. For every curve $\gamma : [a, b] \to B$ in B, for every $t \in [a, b]$, the *parallel transport from $\gamma(a)$ to $\gamma(t)$ along γ* is the linear map from the fiber $\pi^{-1}(\gamma(a))$ to the fiber $\pi^{-1}(\gamma(t))$, which associates to any $v \in \pi^{-1}(\gamma(a))$ the vector $X_v(t) \in \pi^{-1}(\gamma(t))$, where X_v is the unique parallel section along γ with $X_v(a) = v$.

The following proposition is an immediate consequence of properties of linear ODEs:

Proposition 10.8. *Let* $\xi = (E, \pi, B, V)$ *be a vector bundle and let* ∇ *be a connection on* ξ. *For every* C^1 *curve* $\gamma \colon [a, b] \to B$ *in* B, *the parallel transport along* γ *defines for every* $t \in [a, b]$ *a linear isomorphism* $P_\gamma \colon \pi^{-1}(\gamma(a)) \to \pi^{-1}(\gamma(t))$ *between the fibers* $\pi^{-1}(\gamma(a))$ *and* $\pi^{-1}(\gamma(t))$.

In particular, if γ is a closed curve, that is, if $\gamma(a) = \gamma(b) = p$, we obtain a linear isomorphism P_γ of the fiber $E_p = \pi^{-1}(p)$, called the *holonomy of* γ. The *holonomy group of* ∇ *based at* p, denoted $\mathrm{Hol}_p(\nabla)$, is the subgroup of $\mathbf{GL}(V, \mathbb{R})$ (where V is the fiber of the vector bundle ξ) given by

$$\mathrm{Hol}_p(\nabla) = \{P_\gamma \in \mathbf{GL}(V, \mathbb{R}) \mid \gamma \text{ is a closed curve based at } p\}.$$

If B is connected, then $\mathrm{Hol}_p(\nabla)$ depends on the basepoint $p \in B$ up to conjugation and so $\mathrm{Hol}_p(\nabla)$ and $\mathrm{Hol}_q(\nabla)$ are isomorphic for all $p, q \in B$. In this case, it makes sense to talk about the *holonomy group of* ∇. If $\xi = TB$, the tangent bundle of a manifold, B, by abuse of language, we call $\mathrm{Hol}_p(\nabla)$ the *holonomy group of* B.

10.5   Curvature, Curvature Form, and Curvature Matrix

If $\xi = B \times V$ is the trivial bundle and ∇ is a flat connection on ξ, we obviously have

$$\nabla_X \nabla_Y - \nabla_Y \nabla_X = \nabla_{[X,Y]},$$

where $[X, Y]$ is the Lie bracket of the vector fields X and Y. However, for general bundles and arbitrary connections, the above fails. The error term

$$R(X, Y) = \nabla_X \nabla_Y - \nabla_Y \nabla_X - \nabla_{[X,Y]}$$

measures what's called the *curvature* of the connection. In order to write $R(X, Y)$ as a vector valued two-form, we need the following definition.

Definition 10.9.   Set

$$\mathcal{A}^1(\xi) = \mathcal{A}^1(B; \xi) = \mathcal{A}^1(B) \otimes_{C^\infty(B)} \Gamma(\xi),$$

and more generally, for any $i \geq 0$, set

$$\mathcal{A}^i(\xi) = \mathcal{A}^i(B; \xi) = \mathcal{A}^i(B) \otimes_{C^\infty(B)} \Gamma(\xi) \cong \Gamma\left(\left(\bigwedge^i T^*B\right) \otimes \xi\right).$$

Obviously, $\mathcal{A}^0(\xi) = \Gamma(\xi)$ (and recall that $\mathcal{A}^0(B) = C^\infty(B)$).

The space of differential forms $\mathcal{A}^i(B; \xi)$ with values in $\Gamma(\xi)$ is a generalization of the space $\mathcal{A}^i(M, F)$ of differential forms with values in F encountered in Section 4.5.

Observe that in terms of the $\mathcal{A}^i(\xi)$'s, a connection is a linear map,

$$\nabla \colon \mathcal{A}^0(\xi) \to \mathcal{A}^1(\xi),$$

satisfying the Leibniz rule. When $\xi = TB$, a connection (second version) is what is known as an *affine connection* on the manifold B.

The curvature of a connection turns up as the failure of a certain sequence involving the spaces $\mathcal{A}^i(\xi) = \mathcal{A}^i(B) \otimes_{C^\infty(B)} \Gamma(\xi)$ to be a cochain complex. Since the connection on ξ is a linear map

$$\nabla \colon \mathcal{A}^0(\xi) \to \mathcal{A}^1(\xi)$$

satisfying a Leibniz-type rule, it is natural to ask whether ∇ can be extended to a family of operators, $d^\nabla \colon \mathcal{A}^i(\xi) \to \mathcal{A}^{i+1}(\xi)$, with properties analogous to d on $\mathcal{A}^*(B)$.

This is indeed the case, and we get a sequence of maps

$$0 \longrightarrow \mathcal{A}^0(\xi) \xrightarrow{\ \nabla\ } \mathcal{A}^1(\xi) \xrightarrow{\ d^\nabla\ } \mathcal{A}^2(\xi) \longrightarrow \cdots \longrightarrow \mathcal{A}^i(\xi) \xrightarrow{\ d^\nabla\ } \mathcal{A}^{i+1}(\xi) \longrightarrow \cdots,$$

but in general, $d^\nabla \circ d^\nabla = 0$ fails. In particular, $d^\nabla \circ \nabla = 0$ generally fails.

Definition 10.10. The term $R^\nabla = d^\nabla \circ \nabla$ is the *curvature form (or curvature tensor)* of the connection ∇.

As we will see, it yields our previous curvature R, back.

Our next goal is to define d^∇. We have the notion of wedge defined for $\mathcal{A}^*(B)$. But in order to define d^∇, we require a notion of wedge that makes sense on $\mathcal{A}^*(\xi)$.

Definition 10.11. Let ξ and η be two smooth real vector bundles. We define a $C^\infty(B)$-bilinear map

$$\overline{\wedge} \colon \mathcal{A}^i(\xi) \times \mathcal{A}^j(\eta) \longrightarrow \mathcal{A}^{i+j}(\xi \otimes \eta)$$

as follows:

$$(\omega \otimes s) \overline{\wedge} (\tau \otimes t) = (\omega \wedge \tau) \otimes (s \otimes t),$$

where $\omega \in \mathcal{A}^i(B)$, $\tau \in \mathcal{A}^j(B)$, $s \in \Gamma(\xi)$, and $t \in \Gamma(\eta)$, $\omega \wedge \tau$ is the wedge defined over $\mathcal{A}^*(B)$, and where we used the fact that

$$\Gamma(\xi \otimes \eta) = \Gamma(\xi) \otimes_{C^\infty(B)} \Gamma(\eta).$$

In order to help with the calculations associated with the propositions of this section, we need to consider the special case of $\overline{\wedge}$ where $\xi = \epsilon^1 = B \times \mathbb{R}$, the trivial line bundle over B. In this case, $\mathcal{A}^i(\xi) = \mathcal{A}^i(B)$ and we have a bilinear map

$$\overline{\wedge} \colon \mathcal{A}^i(B) \times \mathcal{A}^j(\eta) \longrightarrow \mathcal{A}^{i+j}(\eta)$$

given by

$$\omega \overline{\wedge} (\tau \otimes t) = (\omega \wedge \tau) \otimes t, \qquad \tau \in \mathcal{A}^j(B), \ t \in \Gamma(\eta). \tag{1}$$

For $j = 0$, we have the bilinear map

$$\overline{\wedge} \colon \mathcal{A}^i(B) \times \Gamma(\eta) \longrightarrow \mathcal{A}^i(\eta)$$

given by

$$\omega \overline{\wedge} t = \omega \otimes t. \tag{2}$$

It can be shown that the bilinear map

$$\overline{\wedge} \colon \mathcal{A}^r(B) \times \mathcal{A}^s(\eta) \longrightarrow \mathcal{A}^{r+s}(\eta)$$

has the following properties:

$$(\omega \wedge \tau) \overline{\wedge} \theta = \omega \overline{\wedge} (\tau \overline{\wedge} \theta)$$
$$1 \overline{\wedge} \theta = \theta, \tag{3}$$

for all $\omega \in \mathcal{A}^i(B)$, $\tau \in \mathcal{A}^j(B)$ with $i + j = r$, $\theta \in \mathcal{A}^s(\xi)$, and where 1 denotes the constant function in $C^\infty(B)$ with value 1.

Proposition 10.9. *For every vector bundle ξ, for all $j \geq 0$, there is a unique $\mathbb{R}$-linear map (resp. $\mathbb{C}$-linear if ξ is a complex VB) $d^\nabla \colon \mathcal{A}^j(\xi) \to \mathcal{A}^{j+1}(\xi)$, such that*

(i) $d^\nabla = \nabla$ for $j = 0$.
(ii) $d^\nabla(\omega \overline{\wedge} t) = d\omega \overline{\wedge} t + (-1)^i \omega \overline{\wedge} d^\nabla t$, for all $\omega \in \mathcal{A}^i(B)$ and all $t \in \mathcal{A}^j(\xi)$.

Proof. Recall that $\mathcal{A}^j(\xi) = \mathcal{A}^j(B) \otimes_{C^\infty(B)} \Gamma(\xi)$, and define $\hat{d}^\nabla \colon \mathcal{A}^j(B) \times \Gamma(\xi) \to \mathcal{A}^{j+1}(\xi)$ by

$$\hat{d}^\nabla(\omega, s) = d\omega \otimes s + (-1)^j \omega \overline{\wedge} \nabla s,$$

for all $\omega \in \mathcal{A}^j(B)$ and all $s \in \Gamma(\xi)$. We claim that $\hat{d}^\nabla$ induces an $\mathbb{R}$-linear map on $\mathcal{A}^j(\xi)$, but there is a complication as $\hat{d}^\nabla$ is not $C^\infty(B)$-bilinear. The way around this problem is to use Proposition 2.34. For this we need to check that $\hat{d}^\nabla$ satisfies the condition of Proposition 2.34, where the right action of $C^\infty(B)$ on $\mathcal{A}^j(B)$ is equal to the left action, namely wedging:

$$f \wedge \omega = \omega \wedge f \qquad f \in C^\infty(B) = \mathcal{A}^0(B), \ \omega \in \mathcal{A}^j(B).$$

As $\overline{\wedge}$ and $\wedge$ are $C^\infty(B)$-bilinear, for all $\omega \in \mathcal{A}^i(B)$ and all $s \in \Gamma(\xi)$, we have

$$
\begin{aligned}
\hat{d}^\nabla(\omega f, s) &= d(\omega f) \otimes s + (-1)^j (\omega f) \, \overline{\wedge} \, \nabla s \\
&= d(\omega f) \, \overline{\wedge} \, s + (-1)^j f \omega \, \overline{\wedge} \, \nabla s, \qquad \text{by (2)} \\
&= ((d\omega) f + (-1)^j \omega \wedge df) \, \overline{\wedge} \, s + (-1)^j f \omega \, \overline{\wedge} \, \nabla s, \quad \text{by Proposition 4.12} \\
&= f d\omega \, \overline{\wedge} \, s + ((-1)^j \omega \wedge df) \, \overline{\wedge} \, s + (-1)^j f \omega \, \overline{\wedge} \, \nabla s
\end{aligned}
$$

and

$$
\begin{aligned}
\hat{d}^\nabla(\omega, fs) &= d\omega \otimes (fs) + (-1)^j \omega \, \overline{\wedge} \, \nabla(fs) \\
&= d\omega \, \overline{\wedge} \, (fs) + (-1)^j \omega \, \overline{\wedge} \, \nabla(fs), \qquad \text{by (2)} \\
&= f d\omega \, \overline{\wedge} \, s + (-1)^j \omega \, \overline{\wedge} \, (df \otimes s + f \nabla s), \qquad \text{by Definition 10.1} \\
&= f d\omega \, \overline{\wedge} \, s + (-1)^j \omega \, \overline{\wedge} \, (df \, \overline{\wedge} \, s + f \nabla s), \qquad \text{by (2)} \\
&= f d\omega \, \overline{\wedge} \, s + ((-1)^j \omega \wedge df) \, \overline{\wedge} \, s + (-1)^j f \omega \, \overline{\wedge} \, \nabla s, \qquad \text{by (3)}.
\end{aligned}
$$

Thus, $\hat{d}^\nabla(\omega f, s) = \hat{d}^\nabla(\omega, fs)$, and Proposition 2.34 shows that $d^\nabla \colon \mathcal{A}^j(\xi) \to \mathcal{A}^{j+1}(\xi)$ given by $d^\nabla(\omega \otimes s) = \hat{d}^\nabla(\omega, s)$ is a well-defined $\mathbb{R}$-linear map for all $j \geq 0$. Furthermore, it is clear that $d^\nabla = \nabla$ for $j = 0$. Now, for $\omega \in \mathcal{A}^i(B)$ and $t = \tau \otimes s \in \mathcal{A}^j(\xi)$ we have

$$
\begin{aligned}
d^\nabla(\omega \, \overline{\wedge} \, (\tau \otimes s)) &= d^\nabla((\omega \wedge \tau) \otimes s)), \qquad \text{by (1)} \\
&= d(\omega \wedge \tau) \otimes s + (-1)^{i+j} (\omega \wedge \tau) \, \overline{\wedge} \, \nabla s, \qquad \text{definition of } d^\nabla \\
&= (d\omega \wedge \tau) \otimes s + (-1)^i (\omega \wedge d\tau) \otimes s \\
&\quad + (-1)^{i+j} (\omega \wedge \tau) \, \overline{\wedge} \, \nabla s, \qquad \text{by Proposition 4.12} \\
&= d\omega \, \overline{\wedge} \, (\tau \otimes s) + (-1)^i \omega \, \overline{\wedge} \, (d\tau \otimes s) \\
&\quad + (-1)^{i+j} \omega \, \overline{\wedge} \, (\tau \, \overline{\wedge} \, \nabla s), \qquad \text{by (1) and (3)} \\
&= d\omega \, \overline{\wedge} \, (\tau \otimes s) + (-1)^i \omega \, \overline{\wedge} \, d^\nabla(\tau \otimes s), \qquad \text{definition of } d^\nabla
\end{aligned}
$$

which proves (ii). $\square$

As a consequence, we have the following sequence of linear maps

$$0 \longrightarrow \mathcal{A}^0(\xi) \xrightarrow{\nabla} \mathcal{A}^1(\xi) \xrightarrow{d^\nabla} \mathcal{A}^2(\xi) \longrightarrow \cdots \longrightarrow \mathcal{A}^i(\xi) \xrightarrow{d^\nabla} \mathcal{A}^{i+1}(\xi) \longrightarrow \cdots .$$

but in general, $d^\nabla \circ d^\nabla = 0$ fails. Although generally $d^\nabla \circ \nabla = 0$ fails, the map $d^\nabla \circ \nabla$ is $C^\infty(B)$-linear.

Proposition 10.10. *The map $d^\nabla \circ \nabla \colon \mathcal{A}^0(\xi) \to \mathcal{A}^2(\xi)$ is $C^\infty(B)$-linear.*

Proof. We have

$$
\begin{aligned}
(d^\nabla \circ \nabla)(fs) &= d^\nabla(df \otimes s + f\nabla s), && \text{by Definition 10.1} \\
&= d^\nabla(df \mathbin{\bar\wedge} s + f \mathbin{\bar\wedge} \nabla s), && \text{by (2)} \\
&= ddf \mathbin{\bar\wedge} s - df \mathbin{\bar\wedge} \nabla s + df \mathbin{\bar\wedge} \nabla s + f \mathbin{\bar\wedge} d^\nabla(\nabla s), && \text{by Proposition 10.9} \\
&= f \mathbin{\bar\wedge} d^\nabla(\nabla s), && \text{since } ddf = 0 \\
&= f((d^\nabla \circ \nabla)(s)).
\end{aligned}
$$

Therefore, $d^\nabla \circ \nabla \colon \mathcal{A}^0(\xi) \to \mathcal{A}^2(\xi)$ is a $C^\infty(B)$-linear map. $\square$

Recall that just before Proposition 10.1 we showed that

$$\mathrm{Hom}_{C^\infty(B)}(\mathcal{A}^0(\xi), \mathcal{A}^i(\xi)) \cong \mathcal{A}^i(\mathcal{H}om(\xi,\xi)),$$

therefore, $d^\nabla \circ \nabla \in \mathcal{A}^2(\mathcal{H}om(\xi,\xi)) \cong \mathcal{A}^2(B) \otimes_{C^\infty(B)} \Gamma(\mathcal{H}om(\xi,\xi))$.

Corollary 10.11. *The map $d^\nabla \circ \nabla$ is a two-form with values in $\Gamma(\mathcal{H}om(\xi,\xi))$.*
Recall from Definition 10.10 that

$$R^\nabla = d^\nabla \circ \nabla.$$

Although this is far from obvious the curvature form R^∇ is related to the curvature $R(X, Y)$ defined at the beginning of Section 10.5. To discover the relationship between R^∇ and $R(-, -)$, we need to explain how to define $R^\nabla_{X,Y}(s)$, for any two vector fields $X, Y \in \mathfrak{X}(B)$ and any section $s \in \Gamma(\xi)$. For any section $s \in \Gamma(\xi)$, the value ∇s can be written as a linear combination of elements of the form $\omega \otimes t$, with $\omega \in \mathcal{A}^1(B)$ and $t \in \Gamma(\xi)$. If $\nabla s = \omega \otimes t = \omega \mathbin{\bar\wedge} t$, as above, we have

$$
\begin{aligned}
d^\nabla(\nabla s) &= d^\nabla(\omega \mathbin{\bar\wedge} t) \\
&= d\omega \otimes t - \omega \mathbin{\bar\wedge} \nabla t, && \text{by Proposition 10.9.}
\end{aligned}
$$

But ∇t itself is a linear combination of the form

$$\nabla t = \sum_j \eta_j \otimes t_j$$

for some 1-forms $\eta_j \in \mathcal{A}^1(B)$ and some sections $t_j \in \Gamma(\xi)$, so (1) implies that

$$d^\nabla(\nabla s) = d\omega \otimes t - \sum_j (\omega \wedge \eta_j) \otimes t_j.$$

Thus it makes sense to define $R^\nabla_{X,Y}(s)$ by

$$
\begin{aligned}
R^\nabla_{X,Y}(s) &= d\omega(X,Y)t - \sum_j (\omega \wedge \eta_j)(X,Y)t_j \\
&= d\omega(X,Y)t - \sum_j (\omega(X)\eta_j(Y) - \omega(Y)\eta_j(X))t_j \\
&= d\omega(X,Y)t - \left(\omega(X) \sum_j \eta_j(Y)t_j - \omega(Y) \sum_j \eta_j(X)t_j \right) \\
&= d\omega(X,Y)t - (\omega(X)\nabla_Y t - \omega(Y)\nabla_X t), \quad\quad\quad (4)
\end{aligned}
$$

since $\nabla_X t = \sum_j \eta_j(X)t_j$ because $\nabla t = \sum_j \eta_j \otimes t_j$, and similarly for $\nabla_Y t$. We extend this formula by linearity when ∇s is a linear combinations of elements of the form $\omega \otimes t$.

The preceding discussion implies that clean way to define $R^\nabla_{X,Y}$ is to define the following evaluation map:

Definition 10.12. Let ξ be a smooth real vector bundle. Define

$$\mathrm{Ev}_{X,Y} \colon \mathcal{A}^2(\mathcal{H}om(\xi, \xi)) \to \mathcal{A}^0(\mathcal{H}om(\xi, \xi)) = \Gamma(\mathcal{H}om(\xi, \xi)) \cong \mathrm{Hom}_{C^\infty(B)}(\Gamma(\xi), \Gamma(\xi))$$

as follows: For all $X, Y \in \mathfrak{X}(B)$, all $\theta \otimes h \in \mathcal{A}^2(\mathcal{H}om(\xi, \xi)) = \mathcal{A}^2(B) \otimes_{C^\infty(B)} \Gamma(\mathcal{H}om(\xi, \xi))$, set

$$\mathrm{Ev}_{X,Y}(\theta \otimes h) = \theta(X,Y)h.$$

It is clear that this map is $C^\infty(B)$-linear and thus well defined on $\mathcal{A}^2(\mathcal{H}om(\xi, \xi))$. (Recall that $\mathcal{A}^0(\mathcal{H}om(\xi, \xi)) = \Gamma(\mathcal{H}om(\xi, \xi)) = \mathrm{Hom}_{C^\infty(B)}(\Gamma(\xi), \Gamma(\xi))$.) We write

$$R^\nabla_{X,Y} = \mathrm{Ev}_{X,Y}(R^\nabla) \in \mathrm{Hom}_{C^\infty(B)}(\Gamma(\xi), \Gamma(\xi)).$$

Since R^∇ is a linear combination of the form

$$R^\nabla = \sum_j \theta_j \otimes h_j$$

for some 2-forms $\theta_j \in \mathcal{A}^2(B)$ and some sections $h_j \in \Gamma(\mathcal{H}om(\xi,\xi))$, for any section $s \in \Gamma(\xi)$, we have

$$R^\nabla_{X,Y}(s) = \sum_j \theta_j(X,Y)h_j(s),$$

where $h_j(s)$ is some section in $\Gamma(\xi)$, and then we use the formula obtained above when ∇s is a linear combination of terms of the form $\omega \otimes s$ for some 1-forms $\mathcal{A}^1(B)$ and some sections $s \in \Gamma(\xi)$.

Proposition 10.12. *For any vector bundle ξ, and any connection ∇ on ξ, for all $X, Y \in \mathfrak{X}(B)$, if we let*

$$R(X,Y) = \nabla_X \circ \nabla_Y - \nabla_Y \circ \nabla_X - \nabla_{[X,Y]},$$

then

$$R(X,Y) = R^\nabla_{X,Y}.$$

Proof. Since for any section $s \in \Gamma(\xi)$, the value ∇s can be written as a linear combination of elements of the form $\omega \otimes t = \omega \,\bar\wedge\, t$, with $\omega \in \mathcal{A}^1(B)$ and $t \in \Gamma(\xi)$, it is sufficient to compute $R^\nabla_{X,Y}(s)$ when $\nabla s = \omega \otimes t$, and we get

$$
\begin{aligned}
R^\nabla_{X,Y}(s) &= d\omega(X,Y)t - (\omega(X)\nabla_Y t - \omega(Y)\nabla_X t), && \text{by (4)} \\
&= (X(\omega(Y)) - Y(\omega(X)) - \omega([X,Y]))t - (\omega(X)\nabla_Y t - \omega(Y)\nabla_X t), && \text{by Proposition 4.16} \\
&= \nabla_X(\omega(Y)t) - \nabla_Y(\omega(X)t) - \omega([X,Y])t, && \text{by Definition 10.1} \\
&= \nabla_X(\nabla_Y s) - \nabla_Y(\nabla_X s) - \nabla_{[X,Y]}s,
\end{aligned}
$$

since $\nabla_X s = \omega(X)t$ because $\nabla s = \omega \otimes t$ (and similarly for the other terms involving ω). $\Box$

Remark. Proposition 10.12 implies that $R(Y,X) = -R(X,Y)$ and that $R(X,Y)(s)$ is $C^\infty(B)$-linear in X, Y and s.

Definition 10.13. For any vector bundle ξ and any connection ∇ on ξ, the vector-valued two-form $R^\nabla = d^\nabla \circ \nabla \in \mathcal{A}^2(\mathcal{H}om(\xi,\xi))$ is the *curvature form* (or *curvature tensor*) of the connection ∇. We say that ∇ is a *flat connection* iff $R^\nabla = 0$.

Remark. The expression R^∇ is also denoted F^∇ or K^∇.

10.6 Structure Equations

As in the case of a connection, we can express the two-form R^∇ locally in any local trivialization $\varphi\colon \pi^{-1}(U) \to U \times V$ of ξ. Since $R^\nabla \in \mathcal{A}^2(\mathcal{H}om(\xi,\xi)) = \mathcal{A}^2(B) \otimes_{C^\infty(B)} \Gamma(\mathcal{H}om(\xi,\xi))$, if $(s_1,\ldots,s_n)$ is the frame associated with (φ,U), then

$$R^\nabla(s_i) = \sum_{j=1}^{n} \Omega_{ji} \otimes s_j,$$

for some matrix $\Omega = (\Omega_{ij})$ of two forms $\Omega_{ij} \in \mathcal{A}^2(U)$.

Definition 10.14. The matrix $\Omega = (\Omega_{ij})$ of two forms such that

$$R^\nabla(s_i) = \sum_{j=1}^{n} \Omega_{ji} \otimes s_j,$$

is called the *curvature matrix* (or *curvature form*) associated with the local trivialization.

The relationship between the connection form ω and the curvature form Ω is simple.

Proposition 10.13 (Structure Equations). *Let ξ be any vector bundle and let ∇ be any connection on ξ. For every local trivialization $\varphi\colon \pi^{-1}(U) \to U \times V$, the connection matrix $\omega = (\omega_{ij})$ and the curvature matrix $\Omega = (\Omega_{ij})$ associated with the local trivialization (φ, U), are related by the* **structure equation***:*

$$\Omega = d\omega + \omega \wedge \omega,$$

where the above formula is interpreted in an entry by entry fashion.

Proof. By definition,

$$\nabla(s_i) = \sum_{j=1}^{n} \omega_{ji} \otimes s_j,$$

so if we apply d^∇ and use Property (ii) of Proposition 10.9 we get

$$R^\nabla(s_i) = d^\nabla(\nabla(s_i)) = \sum_{k=1}^{n} \Omega_{ki} \otimes s_k$$

$$= \sum_{j=1}^{n} d^{\nabla}(\omega_{ji} \otimes s_j)$$

$$= \sum_{j=1}^{n} d\omega_{ji} \otimes s_j - \sum_{j=1}^{n} \omega_{ji} \,\bar{\wedge}\, \nabla s_j, \qquad \text{by definition of } d^{\nabla}$$

$$= \sum_{j=1}^{n} d\omega_{ji} \otimes s_j - \sum_{j=1}^{n} \omega_{ji} \,\bar{\wedge}\, \left(\sum_{k=1}^{n} \omega_{kj} \otimes s_k \right)$$

$$= \sum_{k=1}^{n} d\omega_{ki} \otimes s_k - \sum_{k=1}^{n} \left(\sum_{j=1}^{n} \omega_{ji} \wedge \omega_{kj} \right) \otimes s_k, \qquad \text{by (1)}$$

and so,

$$\Omega_{ki} = d\omega_{ki} + \sum_{j=1}^{n} \omega_{kj} \wedge \omega_{ji},$$

which means that

$$\Omega = d\omega + \omega \wedge \omega,$$

as claimed. $\square$

Some other texts including Milnor and Stasheff [78] state the structure equations as

$$\Omega = d\omega - \omega \wedge \omega.$$

Example 10.2. In Example 10.1, we showed that the connection matrix for the spherical frame of $T\mathbb{R}^3$ is given by

$$\omega = \begin{pmatrix} 0 & -d\theta & -\sin\theta d\varphi \\ d\theta & 0 & -\cos\theta d\varphi \\ \sin\theta d\varphi & \cos\theta d\varphi & 0 \end{pmatrix}.$$

Proposition 10.13 shows that the curvature matrix is

$$\Omega = d\omega - \omega \wedge \omega$$

$$= \begin{pmatrix} 0 & 0 & -\cos\theta d\theta \wedge d\varphi \\ 0 & 0 & \sin\theta d\theta \wedge d\varphi \\ \cos\theta d\theta \wedge d\varphi & -\sin\theta d\theta \wedge d\varphi & 0 \end{pmatrix}$$

$$+ \begin{pmatrix} 0 & -d\theta & -\sin\theta\, d\varphi \\ d\theta & 0 & -\cos\theta\, d\varphi \\ \sin\theta\, d\varphi & \cos\theta\, d\varphi & 0 \end{pmatrix} \wedge \begin{pmatrix} 0 & -d\theta & -\sin\theta\, d\varphi \\ d\theta & 0 & -\cos\theta\, d\varphi \\ \sin\theta\, d\varphi & \cos\theta\, d\varphi & 0 \end{pmatrix}$$

$$= \begin{pmatrix} 0 & 0 & -\cos\theta\, d\theta \wedge d\varphi \\ 0 & 0 & \sin\theta\, d\theta \wedge d\varphi \\ \cos\theta\, d\theta \wedge d\varphi & -\sin\theta\, d\theta \wedge d\varphi & 0 \end{pmatrix}$$

$$+ \begin{pmatrix} 0 & 0 & \cos\theta\, d\theta \wedge d\varphi \\ 0 & 0 & \sin\theta\, d\theta \wedge d\varphi \\ -\cos\theta\, d\theta \wedge d\varphi & -\sin\theta\, d\theta \wedge d\varphi & 0 \end{pmatrix}$$

$$= \begin{pmatrix} 0 & 0 & 0 \\ 0 & 0 & 2\sin\theta\, d\theta \wedge d\varphi \\ 0 & -2\sin\theta\, d\theta \wedge d\varphi & 0 \end{pmatrix}.$$

If $\varphi_\alpha : \pi^{-1}(U_\alpha) \to U_\alpha \times V$ and $\varphi_\beta : \pi^{-1}(U_\beta) \to U_\beta \times V$ are two overlapping trivializations, the relationship between the curvature matrices Ω_α and Ω_β, is given by the following proposition which is the counterpart of Proposition 10.5 for the curvature matrix:

Proposition 10.14. *If $\varphi_\alpha : \pi^{-1}(U_\alpha) \to U_\alpha \times V$ and $\varphi_\beta : \pi^{-1}(U_\beta) \to U_\beta \times V$ are two overlapping trivializations of a vector bundle ξ, then we have the following transformation rule for the curvature matrices Ω_α and Ω_β:*

$$\Omega_\beta = g_{\alpha\beta}^{-1} \Omega_\alpha g_{\alpha\beta},$$

where $g_{\alpha\beta}$ is viewed as the matrix function representing the linear map $g_{\alpha\beta}(b) \in$ GL(V) for every $b \in U_\alpha \cap U_\beta$.

Proof. The idea is to take the exterior derivative of the equation

$$\omega_\beta = g_{\alpha\beta}^{-1} \omega_\alpha g_{\alpha\beta} + g_{\alpha\beta}^{-1}(dg_{\alpha\beta})$$

from Proposition 10.5. To simplify notation, write g for $g_{\alpha\beta}$. Now, since g, Ω_α and Ω_β are all matrices, we apply the exterior derivative in an entry by entry fashion. Since g is a matrix of functions such that $g^{-1}g = I$, we find that

$$0 = d(g^{-1}g) = dg^{-1}\, g + g^{-1}\, dg,$$

which is equivalent to

$$dg^{-1} = -g^{-1}dg g^{-1}.$$

By recalling that

$$dd\eta = 0, \qquad d(\eta \wedge \beta) = d\eta \wedge \beta + (-1)^j \eta \wedge d\beta, \qquad \eta \in \mathcal{A}^i(B), \ \beta \in \mathcal{A}^j(B),$$

we find that

$$\begin{aligned}
d\omega_\beta &= d(g^{-1}\omega_\alpha g) + d(g^{-1}dg) \\
&= d(g^{-1}\omega_\alpha g) + dg^{-1} \wedge dg \\
&= dg^{-1} \wedge \omega_\alpha g + g^{-1} \wedge d(\omega_\alpha g) + dg^{-1} \wedge dg \\
&= -g^{-1}dgg^{-1} \wedge \omega_\alpha g + g^{-1} \wedge d(\omega_\alpha g) - g^{-1}dgg^{-1} \wedge dg \\
&= -g^{-1}dgg^{-1} \wedge \omega_\alpha g + g^{-1} \wedge (d\omega_\alpha g - \omega_\alpha \wedge dg) - g^{-1}dgg^{-1} \wedge dg \\
&= -g^{-1}dgg^{-1} \wedge \omega_\alpha g + g^{-1}d\omega_\alpha g - g^{-1}\omega_\alpha \wedge dg - g^{-1}dgg^{-1} \wedge dg,
\end{aligned}$$

so using the structure equation (Proposition 10.13) we get

$$\begin{aligned}
\Omega_\beta &= d\omega_\beta + \omega_\beta \wedge \omega_\beta \\
&= -g^{-1}dgg^{-1} \wedge \omega_\alpha g + g^{-1}d\omega_\alpha g - g^{-1}\omega_\alpha \wedge dg - g^{-1}dgg^{-1} \wedge dg \\
&\quad + (g^{-1}\omega_\alpha g + g^{-1}dg) \wedge (g^{-1}\omega_\alpha g + g^{-1}dg) \\
&= -g^{-1}dgg^{-1} \wedge \omega_\alpha g + g^{-1}d\omega_\alpha g - g^{-1}\omega_\alpha \wedge dg - g^{-1}dgg^{-1} \wedge dg \\
&\quad + g^{-1}\omega_\alpha \wedge \omega_\alpha g + g^{-1}\omega_\alpha \wedge dg + g^{-1}dg \wedge g^{-1}\omega_\alpha g + g^{-1}dg \wedge g^{-1}dg \\
&= g^{-1}d\omega_\alpha g + g^{-1}\omega_\alpha \wedge \omega_\alpha g \\
&= g^{-1}\Omega_\alpha g,
\end{aligned}$$

establishing the desired formula.   □

Proposition 10.13 also yields a formula for $d\Omega$, known as *Bianchi's identity* (in local form).

Proposition 10.15 (Bianchi's Identity). *For any vector bundle ξ and any connection ∇ on ξ, if ω and Ω are respectively the connection matrix and the curvature matrix, in some local trivialization, then*

$$d\Omega = \Omega \wedge \omega - \omega \wedge \Omega.$$

Proof. If we apply d to the structure equation, $\Omega = d\omega + \omega \wedge \omega$, we get

$$d\Omega = dd\omega + d\omega \wedge \omega - \omega \wedge d\omega$$

$$= (\Omega - \omega \wedge \omega) \wedge \omega - \omega \wedge (\Omega - \omega \wedge \omega)$$

$$= \Omega \wedge \omega - \omega \wedge \omega \wedge \omega - \omega \wedge \Omega + \omega \wedge \omega \wedge \omega$$

$$= \Omega \wedge \omega - \omega \wedge \Omega,$$

as claimed. $\square$

We conclude this section by giving a formula for $d^\nabla \circ d^\nabla (t)$, for any $t \in \mathcal{A}^i(\xi)$.

10.7   A Formula for $d^\nabla \circ d^\nabla$

Consider the special case of the bilinear map

$$\bar{\wedge} \colon \mathcal{A}^i(\xi) \times \mathcal{A}^j(\eta) \longrightarrow \mathcal{A}^{i+j}(\xi \otimes \eta)$$

given in Definition 10.11 with $j = 2$ and $\eta = \mathcal{H}om(\xi, \xi)$. This is the $C^\infty(B)$-bilinear map

$$\bar{\wedge} \colon \mathcal{A}^i(\xi) \times \mathcal{A}^2(\mathcal{H}om(\xi, \xi)) \longrightarrow \mathcal{A}^{i+2}(\xi \otimes \mathcal{H}om(\xi, \xi)).$$

Two applications of Proposition 9.11 show that

$$\Gamma(\xi \otimes \mathcal{H}om(\xi, \xi)) \cong \Gamma(\xi) \otimes_{C^\infty(B)} \Gamma(\mathcal{H}om(\xi, \xi)) \cong \Gamma(\xi) \otimes_{C^\infty(B)} \mathrm{Hom}_{C^\infty(B)}(\Gamma(\xi), \Gamma(\xi)).$$

We then have the evaluation map

$$\mathrm{ev} \colon \mathcal{A}^j(\xi \otimes \mathcal{H}om(\xi, \xi)) \cong \mathcal{A}^j(B) \otimes_{C^\infty(B)} \Gamma(\xi) \otimes_{C^\infty(B)} \mathrm{Hom}_{C^\infty(B)}(\Gamma(\xi), \Gamma(\xi))$$

$$\longrightarrow \mathcal{A}^j(B) \otimes_{C^\infty(B)} \Gamma(\xi) = \mathcal{A}^j(\xi),$$

given by

$$\mathrm{ev}(\omega \otimes s \otimes h) = \omega \otimes h(s),$$

with $\omega \in \mathcal{A}^j(B)$, $s \in \Gamma(\xi)$ and $h \in \mathrm{Hom}_{C^\infty(B)}(\Gamma(\xi), \Gamma(\xi))$.

Definition 10.15. Let

$$\bar{\bar{\wedge}} \colon \mathcal{A}^i(\xi) \times \mathcal{A}^2(\mathcal{H}om(\xi, \xi)) \longrightarrow \mathcal{A}^{i+2}(\xi)$$

be the composition

$$\mathcal{A}^i(\xi) \times \mathcal{A}^2(\mathcal{H}om(\xi, \xi)) \xrightarrow{\bar{\wedge}} \mathcal{A}^{i+2}(\xi \otimes \mathcal{H}om(\xi, \xi)) \xrightarrow{\mathrm{ev}} \mathcal{A}^{i+2}(\xi).$$

More explicitly, the above map is given (on generators) by

$$(\omega \otimes s) \,\overline{\overline{\wedge}}\, H = \omega \,\overline{\wedge}\, H(s), \tag{10.1}$$

where $\omega \in \mathcal{A}^i(B)$, $s \in \Gamma(\xi)$ and $H \in \mathrm{Hom}_{C^\infty(B)}(\Gamma(\xi), \mathcal{A}^2(\xi)) \cong \mathcal{A}^2(\mathcal{H}om(\xi, \xi))$.

Proposition 10.16. *For any vector bundle ξ and any connection ∇ on ξ, the composition $d^\nabla \circ d^\nabla : \mathcal{A}^i(\xi) \to \mathcal{A}^{i+2}(\xi)$ maps t to $t \,\overline{\overline{\wedge}}\, R^\nabla$, for any $t \in \mathcal{A}^i(\xi)$.*

Proof. Any $t \in \mathcal{A}^i(\xi)$ is some linear combination of elements $\omega \otimes s \in \mathcal{A}^i(B) \otimes_{C^\infty(B)} \Gamma(\xi)$ and by Proposition 10.9, we have

$$d^\nabla \circ d^\nabla(\omega \otimes s) = d^\nabla(d\omega \otimes s + (-1)^i \omega \,\overline{\wedge}\, \nabla s), \qquad \text{by definition of } d^\nabla$$

$$= dd\omega \otimes s + (-1)^{i+1} d\omega \,\overline{\wedge}\, \nabla s + (-1)^i d\omega \,\overline{\wedge}\, \nabla s$$

$$+ (-1)^i (-1)^i \omega \,\overline{\wedge}\, d^\nabla \circ \nabla s$$

$$= \omega \,\overline{\wedge}\, (d^\nabla \circ \nabla s)$$

$$= (\omega \otimes s) \,\overline{\overline{\wedge}}\, R^\nabla, \qquad \text{by (10.1)}$$

as claimed.   □

Proposition 10.16 shows that $d^\nabla \circ d^\nabla = 0$ iff $R^\nabla = d^\nabla \circ \nabla = 0$, that is, iff the connection ∇ is flat. Thus, the sequence

$$0 \longrightarrow \mathcal{A}^0(\xi) \xrightarrow{\nabla} \mathcal{A}^1(\xi) \xrightarrow{d^\nabla} \mathcal{A}^2(\xi) \longrightarrow \cdots \longrightarrow \mathcal{A}^i(\xi) \xrightarrow{d^\nabla} \mathcal{A}^{i+1}(\xi) \longrightarrow \cdots,$$

is a cochain complex iff ∇ is flat.

Remark. Again everything we did in this section applies to complex vector bundles.

10.8   Connections Compatible with a Metric: Levi-Civita Connections

If a vector bundle (or a Riemannian manifold) ξ has a metric, then it is natural to define when a connection ∇ on ξ is compatible with the metric. This will require first defining the following three bilinear pairings.

Definition 10.16. Let ξ be a smooth real vector bundle ξ with metric $\langle -, - \rangle$. We can use this metric to define pairings

$$\mathcal{A}^1(\xi) \times \mathcal{A}^0(\xi) \longrightarrow \mathcal{A}^1(B) \quad \text{and} \quad \mathcal{A}^0(\xi) \times \mathcal{A}^1(\xi) \longrightarrow \mathcal{A}^1(B)$$

as follows: Set (on generators)

$$\langle \omega \otimes s_1, s_2 \rangle = \langle s_1, \omega \otimes s_2 \rangle = \langle s_1, s_2 \rangle \omega,$$

for all $\omega \in \mathcal{A}^1(B)$, $s_1, s_2 \in \Gamma(\xi)$ and where $\langle s_1, s_2 \rangle$ is the function in $C^\infty(B)$ given by $b \mapsto \langle s_1(b), s_2(b) \rangle$, for all $b \in B$. More generally, we define a pairing

$$\mathcal{A}^i(\xi) \times \mathcal{A}^j(\xi) \longrightarrow \mathcal{A}^{i+j}(B),$$

by

$$\langle \omega \otimes s_1, \eta \otimes s_2 \rangle = \langle s_1, s_2 \rangle \omega \wedge \eta,$$

for all $\omega \in \mathcal{A}^i(B)$, $\eta \in \mathcal{A}^j(B)$, $s_1, s_2 \in \Gamma(\xi)$.

Definition 10.17. Given any metric $\langle -, - \rangle$ on a vector bundle ξ, a connection ∇ on ξ is *compatible with the metric*, for short, a *metric connection* iff

$$d\langle s_1, s_2 \rangle = \langle \nabla s_1, s_2 \rangle + \langle s_1, \nabla s_2 \rangle,$$

for all $s_1, s_2 \in \Gamma(\xi)$.

In terms of version-two of a connection, ∇_X is a metric connection iff

$$X(\langle s_1, s_2 \rangle) = \langle \nabla_X s_1, s_2 \rangle + \langle s_1, \nabla_X s_2 \rangle,$$

for every vector field, $X \in \mathfrak{X}(B)$.

Remark. Definition 10.17 remains unchanged if ξ is a complex vector bundle.

It is easy to prove that metric connections exist.

Proposition 10.17. *Let ξ be a rank n vector with a metric $\langle -, - \rangle$. Then ξ possesses metric connections.*

Proof. We can pick a locally finite cover $(U_\alpha)_\alpha$ of B such that $(U_\alpha, \varphi_\alpha)$ is a local trivialization of ξ. Then for each $(U_\alpha, \varphi_\alpha)$, we use the Gram-Schmidt procedure to obtain an orthonormal frame $(s_1^\alpha, \ldots, s_n^\alpha)$ over U_α, and we let ∇^α be the trivial connection on $\pi^{-1}(U_\alpha)$. By construction, ∇^α is compatible with the metric. We finish the argument by using a partition of unity, leaving the details to the reader. $\square$

Remark. If ξ is a complex vector bundle, then we use a Hermitian metric and we call a connection compatible with this metric a *Hermitian connection*. The existence of Hermitian connections is clear.

The condition of compatibility with a metric is nicely expressed in a local trivialization. Indeed, let (U, φ) be a local trivialization of the vector bundle ξ (of rank n). Then using the Gram-Schmidt procedure, we obtain an orthonormal frame $(s_1, \ldots, s_n)$, over U.

Proposition 10.18. *Using the above notations, if $\omega = (\omega_{ij})$ is the connection matrix of ∇ w.r.t. an orthonormal frame $(s_1, \ldots, s_n)$, then ω is skew-symmetric.*

Proof. Since

$$\nabla s_i = \sum_{j=1}^{n} \omega_{ji} \otimes s_j$$

and since $\langle s_i, s_j \rangle = \delta_{ij}$ (as $(s_1, \ldots, s_n)$ is orthonormal), we have $d \langle s_i, s_j \rangle = 0$ on U. Consequently,

$$
\begin{aligned}
0 &= d \langle s_i, s_j \rangle \\
&= \langle \nabla s_i, s_j \rangle + \langle s_i, \nabla s_j \rangle \\
&= \left\langle \sum_{k=1}^{n} \omega_{ki} \otimes s_k, s_j \right\rangle + \left\langle s_i, \sum_{l=1}^{n} \omega_{lj} \otimes s_l \right\rangle \\
&= \sum_{k=1}^{n} \omega_{ki} \langle s_k, s_j \rangle + \sum_{l=1}^{n} \omega_{lj} \langle s_i, s_l \rangle \\
&= \omega_{ji} + \omega_{ij},
\end{aligned}
$$

as claimed. $\square$

Remark. In Proposition 10.18, if ξ is a complex vector bundle, then ω is skew-Hermitian. This means that

$$\overline{\omega}^{\top} = -\omega,$$

where $\overline{\omega}$ is the conjugate matrix of ω; that is, $(\overline{\omega})_{ij} = \overline{\omega_{ij}}$.

If ∇ is a metric connection, then the curvature matrices are also skew-symmetric.

Proposition 10.19. *Let ξ be a rank n vector bundle with a metric $\langle -, - \rangle$. In any local trivialization of ξ, with respect to a orthonormal frame the curvature matrix $\Omega = (\Omega_{ij})$ is skew-symmetric. If ξ is a complex vector bundle, then $\Omega = (\Omega_{ij})$ is skew-Hermitian.*

Proof. By the structure equation (Proposition 10.13),

$$\Omega = d\omega + \omega \wedge \omega,$$

that is, $\Omega_{ij} = d\omega_{ij} + \sum_{k=1}^{n} \omega_{ik} \wedge \omega_{kj}$. Using the skew symmetry of ω_{ij} and wedge,

$$\Omega_{ji} = d\omega_{ji} + \sum_{k=1}^{n} \omega_{jk} \wedge \omega_{ki}$$

$$= -d\omega_{ij} + \sum_{k=1}^{n} \omega_{kj} \wedge \omega_{ik}$$

$$= -d\omega_{ij} - \sum_{k=1}^{n} \omega_{ik} \wedge \omega_{kj}$$

$$= -\Omega_{ij},$$

as claimed. $\qquad\qquad\qquad\qquad\qquad\qquad\qquad\qquad\qquad\qquad$ □

We now restrict our attention to a Riemannian manifold; that is, to the case where our bundle ξ is the tangent bundle $\xi = TM$ of some Riemannian manifold M. We know from Proposition 10.17 that metric connections on TM exist. However, there are many metric connections on TM, and none of them seems more relevant than the others. If M is a Riemannian manifold, the metric $\langle -, - \rangle$ on M is often denoted g. In this case, for every chart (U, φ), we let $g_{ij} \in C^\infty(M)$ be the function defined by

$$g_{ij}(p) = \left\langle \left(\frac{\partial}{\partial x_i} \right)_p, \left(\frac{\partial}{\partial x_j} \right)_p \right\rangle_p.$$

(Note the unfortunate clash of notation with the transitions functions!)

The notations $g = \sum_{ij} g_{ij} dx_i \otimes dx_j$ or simply $g = \sum_{ij} g_{ij} dx_i dx_j$ are often used to denote the metric in local coordinates.

We observed immediately after stating Proposition 9.13 that the covariant differential ∇g of the Riemannian metric g on M is given by

$$\nabla_X(g)(Y, Z) = d(g(Y, Z))(X) - g(\nabla_X Y, Z) - g(Y, \nabla_X Z),$$

for all $X, Y, Z \in \mathfrak{X}(M)$. Therefore, a connection ∇ on a Riemannian manifold (M, g) is compatible with the metric iff

$$\nabla g = 0.$$

It is remarkable that if we require a certain kind of symmetry on a metric connection, then it is uniquely determined. Such a connection is known as the *Levi–Civita connection*. The Levi–Civita connection can be characterized in several equivalent ways, a rather simple way involving the notion of torsion of a connection.

Recall that one way to introduce the curvature is to view it as the "error term"

$$R(X, Y) = \nabla_X \nabla_Y - \nabla_Y \nabla_X - \nabla_{[X,Y]}.$$

Another natural error term is the *torsion* $T(X, Y)$, of the connection ∇, given by

$$T(X, Y) = \nabla_X Y - \nabla_Y X - [X, Y],$$

which measures the failure of the connection to behave like the Lie bracket. Then the Levi–Civita connection is the unique metric and torsion-free connection $(T(X, Y) = 0)$ on the Riemannian manifold. The first characterization of the Levi–Civita connection is given by the following proposition.

Proposition 10.20 (Levi-Civita, Version 1). *Let M be any Riemannian manifold. There is a unique, metric, torsion-free connection ∇ on M; that is, a connection satisfying the conditions:*

$$X(\langle Y, Z \rangle) = \langle \nabla_X Y, Z \rangle + \langle Y, \nabla_X Z \rangle$$

$$\nabla_X Y - \nabla_Y X = [X, Y],$$

for all vector fields, $X, Y, Z \in \mathfrak{X}(M)$. This connection is called the Levi-Civita connection (or canonical connection) on M. Furthermore, this connection is determined by the Koszul formula

$$2\langle \nabla_X Y, Z \rangle = X(\langle Y, Z \rangle) + Y(\langle X, Z \rangle) - Z(\langle X, Y \rangle)$$
$$- \langle Y, [X, Z] \rangle - \langle X, [Y, Z] \rangle - \langle Z, [Y, X] \rangle.$$

The proof of Proposition 10.20 can be found in Gallot, Hulin, and Lafontaine [48], Do Carmo [37], Morita [82], or Gallier and Quaintance [47].

Another way to characterize the Levi-Civita connection uses the cotangent bundle T^*M. It turns out that a connection ∇ on a vector bundle (metric or not) ξ naturally induces a connection ∇^* on the dual bundle ξ^*. If ∇ is a connection on TM, then ∇^* is a connection on T^*M, namely, a linear map, $\nabla^* \colon \Gamma(T^*M) \to \mathcal{A}^1(M) \otimes_{C^\infty(B)} \Gamma(T^*M)$; that is

$$\nabla^* \colon \mathcal{A}^1(M) \to \mathcal{A}^1(M) \otimes_{C^\infty(B)} \mathcal{A}^1(M) \cong \Gamma(T^*M \otimes T^*M),$$

since $\Gamma(T^*M) = \mathcal{A}^1(M)$. With a slight abuse of notation, we denote by $\wedge$ the map $\wedge_\otimes \colon \mathcal{A}^1(M) \otimes_{C^\infty(B)} \mathcal{A}^1(M) \longrightarrow \mathcal{A}^2(M)$ induced by the $C^\infty(B)$-bilinear map $\wedge \colon \mathcal{A}^1(M) \times \mathcal{A}^1(M) \longrightarrow \mathcal{A}^2(M)$. By composition we get the map

$$\mathcal{A}^1(M) \xrightarrow{\nabla^*} \mathcal{A}^1(M) \otimes_{C^\infty(B)} \mathcal{A}^1(M) \xrightarrow{\wedge} \mathcal{A}^2(M).$$

Then miracle, a metric connection is the Levi-Civita connection iff

$$d = \wedge \circ \nabla^*,$$

where $d \colon \mathcal{A}^1(M) \to \mathcal{A}^2(M)$ is exterior differentiation. There is also a nice local expression of the above equation.

Let us now consider the second approach to torsion-freeness. For this, we have to explain how a connection ∇ on a vector bundle $\xi = (E, \pi, B, V)$ induces a connection ∇^* on the dual bundle ξ^*.

10.9 Connections on the Dual Bundle

Let $\xi = (E, \pi, B, V)$ be a vector bundle. First, there is an evaluation map $\Gamma(\xi \otimes \xi^*) \longrightarrow \Gamma(\epsilon^1)$ (where $\epsilon^1 = B \times \mathbb{R}$, the trivial line bundle over B), or equivalently

$$\langle\langle -, - \rangle\rangle \colon \Gamma(\xi) \otimes_{C^\infty(B)} \mathrm{Hom}_{C^\infty(B)}(\Gamma(\xi), C^\infty(B)) \longrightarrow C^\infty(B),$$

given by

$$\langle\langle s_1, s_2^* \rangle\rangle = s_2^*(s_1), \qquad s_1 \in \Gamma(\xi), \ s_2^* \in \mathrm{Hom}_{C^\infty(B)}(\Gamma(\xi), C^\infty(B)),$$

and thus a map

$$\mathcal{A}^k(\xi \otimes \xi^*) = \mathcal{A}^k(B) \otimes_{C^\infty(B)} \Gamma(\xi \otimes \xi^*) \overset{\mathrm{id} \otimes \langle\langle -, - \rangle\rangle}{\longrightarrow} \mathcal{A}^k(B) \otimes_{C^\infty(B)} C^\infty(B) \cong \mathcal{A}^k(B).$$

Using this map, we obtain a pairing

$$(-, -) \colon \mathcal{A}^i(\xi) \otimes \mathcal{A}^j(\xi^*) \overset{\overline{\wedge}}{\longrightarrow} \mathcal{A}^{i+j}(\xi \otimes \xi^*) \longrightarrow \mathcal{A}^{i+j}(B)$$

given by

$$(\omega \otimes s_1, \eta \otimes s_2^*) = (\omega \wedge \eta) \otimes \langle\langle s_1, s_2^* \rangle\rangle,$$

where $\omega \in \mathcal{A}^i(B)$, $\eta \in \mathcal{A}^j(B)$, $s_1 \in \Gamma(\xi)$, and $s_2^* \in \Gamma(\xi^*)$. It is easy to check that this pairing is nondegenerate. Then given a connection ∇ on a rank n vector bundle ξ, we define ∇^* on ξ^* by

$$d\langle\langle s_1, s_2^* \rangle\rangle = \big(\nabla(s_1), s_2^*\big) + \big(s_1, \nabla^*(s_2^*)\big),$$

where $s_1 \in \Gamma(\xi)$ and $s_2^* \in \Gamma(\xi^*)$. Because the pairing $(-, -)$ is nondegenerate, ∇^* is well defined, and it is immediately that it is a connection on ξ^*. Let us see how it is expressed locally.

If (U, φ) is a local trivialization and $(s_1, \ldots, s_n)$ is a frame over U, then let $(\theta_1, \ldots, \theta_n)$ be the dual frame (called a *coframe*). We have

$$\langle\langle s_j, \theta_i \rangle\rangle = \theta_i(s_j) = \delta_{ij}, \qquad 1 \le i, j \le n.$$

Recall that

$$\nabla s_j = \sum_{k=1}^{n} \omega_{kj} \otimes s_k,$$

and write

$$\nabla^* \theta_i = \sum_{k=1}^{n} \omega_{ki}^* \otimes \theta_k.$$

Applying d to the equation $\langle\langle s_j, \theta_i \rangle\rangle = \delta_{ij}$ and using the equation defining ∇^*, we get

$$
\begin{aligned}
0 &= d\langle\langle s_j, \theta_i \rangle\rangle \\
&= \left(\nabla(s_j), \theta_i\right) + \left(s_j, \nabla^*(\theta_i)\right) \\
&= \left(\sum_{k=1}^{n} \omega_{kj} \otimes s_k, \theta_i\right) + \left(s_j, \sum_{l=1}^{n} \omega_{li}^* \otimes \theta_l\right) \\
&= \sum_{k=1}^{n} \omega_{kj} \langle\langle s_k, \theta_i \rangle\rangle + \sum_{l=1}^{n} \omega_{li}^* \langle\langle s_j, \theta_l \rangle\rangle \\
&= \omega_{ij} + \omega_{ji}^*.
\end{aligned}
$$

Proposition 10.21. *If we write* $\omega^* = (\omega_{ij}^*)$, *then we have*

$$\omega^* = -\omega^{\top}.$$

If ∇ is a metric connection and $(s_1, \ldots, s_n)$ is an orthonormal frame over U, then ω is skew-symmetric; that is, $\omega^{\top} = -\omega$.

Corollary 10.22. *If* ∇ *is a metric connection on* ξ, *then* $\omega^* = -\omega^{\top} = \omega$.

Remark. If ξ is a complex vector bundle, then there is a problem because if $(s_1, \ldots, s_n)$ is a frame over U, then the $\theta_j(b)$'s defined by

$$\langle\langle s_i(b), \theta_j(b) \rangle\rangle = \delta_{ij}$$

are *not* linear, but instead conjugate-linear. (Recall that a linear form θ is *conjugate linear* (or *semi-linear*) iff $\theta(\lambda u) = \overline{\lambda}\theta(u)$, for all $\lambda \in \mathbb{C}$.)

Instead of ξ^*, we need to consider the bundle $\overline{\xi}^*$, which is the bundle whose fiber over $b \in B$ consists of all conjugate-linear forms over $\pi^{-1}(b)$. In this case, the evaluation pairing $\langle\langle s, \theta \rangle\rangle$ is conjugate-linear in s, and we find that $\omega^* = -\overline{\omega}^{\top}$, where ω^* is the connection matrix of $\overline{\xi}^*$ over U.

If ξ is a Hermitian bundle, as ω is skew-Hermitian, we find that $\omega^* = \omega$, which makes sense since ξ and $\overline{\xi}^*$ are canonically isomorphic. However, this does not give any information on ξ^*. For this, we consider the *conjugate bundle* $\overline{\xi}$. This is the bundle obtained from ξ by redefining the vector space structure on each fiber $\pi^{-1}(b)$, with $b \in B$, so that

$$(x + iy)v = (x - iy)v,$$

for every $v \in \pi^{-1}(b)$. If ω is the connection matrix of ξ over U, then $\overline{\omega}$ is the connection matrix of $\overline{\xi}$ over U. If ξ has a Hermitian metric, it is easy to prove that ξ^* and $\overline{\xi}$ are canonically isomorphic (see Proposition 10.33). In fact, the Hermitian product $\langle -, - \rangle$ establishes a pairing between $\overline{\xi}$ and ξ^*, and basically as above, we can show that if $\overline{\omega}$ is the connection matrix of $\overline{\xi}$ over U, then $\omega^* = -\omega^\top$ is the connection matrix of ξ^* over U. As ω is skew-Hermitian, $\omega^* = \overline{\omega}$.

10.10   The Levi-Civita Connection on TM Revisited

If ∇ is the Levi-Civita connection of some Riemannian manifold M, for every chart (U, φ), in an orthonormal frame we have $\omega^* = \omega$, where ω is the connection matrix of ∇ over U and ω^* is the connection matrix of the dual connection ∇^*. This implies that the Christoffel symbols of ∇ and ∇^* over U are identical. Furthermore, ∇^* is a linear map

$$\nabla^* \colon \mathcal{A}^1(M) \longrightarrow \Gamma(T^*M \otimes T^*M).$$

Thus, locally in a chart (U, φ), if (as usual) we let $x_i = pr_i \circ \varphi$, then we can write

$$\nabla^*(dx_k) = \sum_{ij} \Gamma_{ik}^j dx_i \otimes dx_j.$$

Now, if we want $\wedge \circ \nabla^* = d$, we must have

$$\wedge \nabla^*(dx_k) = \sum_{ij} \Gamma_{ik}^j dx_i \wedge dx_j = ddx_k = 0;$$

that is

$$\Gamma_{ik}^j = \Gamma_{ki}^j,$$

for all i, k. It is known that this condition on the Christoffel symbols is equivalent to torsion-freeness (see Gallot, Hulin, and Lafontaine [48], or Do Carmo [37]). We record this as follows.

Proposition 10.23. *Let M be a manifold with connection* ∇. *Then* ∇ *is torsion-free* *(i.e.,* $T(X, Y) = \nabla_X Y - \nabla_Y X - [X, Y] = 0$, *for all* $X, Y \in \mathfrak{X}(M)$) *iff*

$$\wedge \circ \nabla^* = d,$$

where $d \colon \mathcal{A}^1(M) \to \mathcal{A}^2(M)$ *is exterior differentiation.*

Proposition 10.23 together with Proposition 10.20 yield a second version of the Levi-Civita theorem:

Proposition 10.24 (Levi-Civita, Version 2). *Let M be any Riemannian manifold. There is a unique, metric connection* ∇ *on M, such that*

$$\wedge \circ \nabla^* = d,$$

where $d \colon \mathcal{A}^1(M) \to \mathcal{A}^2(M)$ *is exterior differentiation. This connection is equal to the Levi-Civita connection in Proposition 10.20.*

Our third version of the Levi-Civita connection is a local version due to Élie Cartan. Recall that locally with respect to a (orthonormal) frame over a chart (U, φ), the connection ∇^* is given by the matrix, ω^*, such that $\omega^* = -\omega^\top$, where ω is the connection matrix of TM over U. That is, we have

$$\nabla^* \theta_i = \sum_{j=1}^{n} -\omega_{ij} \otimes \theta_j,$$

for some one-forms $\omega_{ij} \in \mathcal{A}^1(M)$. Then,

$$\wedge \circ \nabla^* \theta_i = -\sum_{j=1}^{n} \omega_{ij} \wedge \theta_j$$

so the requirement that $d = \wedge \circ \nabla^*$ is expressed locally by

$$d\theta_i = -\sum_{j=1}^{n} \omega_{ij} \wedge \theta_j.$$

In addition, since our connection is metric, ω is skew-symmetric, and so $\omega^* = \omega$. Then it is not too surprising that the following proposition holds:

Proposition 10.25. *Let M be a Riemannian manifold with metric* $\langle -, - \rangle$. *For every chart* (U, φ), *if* $(s_1, \ldots, s_n)$ *is an orthonormal frame over over U and* $(\theta_1, \ldots, \theta_n)$ *is the corresponding coframe (dual frame), then there is a unique matrix* $\omega = (\omega_{ij})$ *of one-forms* $\omega_{ij} \in \mathcal{A}^1(M)$, *so that the following conditions hold:*

(i) $\omega_{ji} = -\omega_{ij}$.

(ii) $d\theta_i = -\sum\limits_{j=1}^{n} \omega_{ij} \wedge \theta_j$, *or in matrix form,* $d\theta = -\omega \wedge \theta$.

Proof. There is a direct proof using a combinatorial trick. For instance, see Morita [82], Chapter 5, Proposition 5.32, or Milnor and Stasheff [78], Appendix C, Lemma 8. On the other hand, if we view $\omega = (\omega_{ij})$ as a connection matrix, then we observed that Condition (i) asserts that the connection is metric and Condition (ii) that it is torsion-free. We conclude by applying Proposition 10.24.   □

Example 10.3. In Example 10.1, we introduced the spherical frame for $T\mathbb{R}^3$ as

$$s_1 = \frac{\partial}{\partial r} = (\cos\varphi \sin\theta, \sin\varphi \sin\theta, \cos\theta)$$

$$s_2 = \frac{\frac{\partial}{\partial\theta}}{\left\|\frac{\partial}{\partial\theta}\right\|} = (\cos\varphi \cos\theta, \sin\varphi \cos\theta, -\sin\theta)$$

$$s_3 = \frac{\frac{\partial}{\partial\varphi}}{\left\|\frac{\partial}{\partial\varphi}\right\|} = (-\sin\theta, \cos\theta, 0),$$

and found that the connection matrix is

$$\omega = \begin{pmatrix} 0 & -d\theta & -\sin\theta d\varphi \\ d\theta & 0 & -\cos\theta d\varphi \\ \sin\theta d\varphi & \cos\theta d\varphi & 0 \end{pmatrix}.$$

The dual coframe is then given by

$$\theta_1 = dr$$
$$\theta_2 = r\, d\theta$$
$$\theta_3 = r \sin\theta\, d\varphi.$$

Observe that

$$d\theta_1 = d\, dr = 0 = r\, d\theta \wedge d\theta + r \sin\varphi\, d\varphi \wedge r \sin\theta\, d\varphi = -\omega_{12} \wedge \theta_2 - \omega_{13} \wedge \theta_3$$

$$d\theta_2 = dr \wedge d\theta = -d\theta \wedge dr + \cos\theta\, d\varphi \wedge r \sin\theta\, d\varphi = -\omega_{21} \wedge \theta_1 - \omega_{23} \wedge \theta_3$$

$$d\theta_3 = \sin\theta\, dr \wedge d\varphi + r \cos\theta d\theta \wedge d\varphi = -\sin\theta\, d\varphi \wedge dr - \cos\theta\, d\varphi \wedge r\, d\theta$$

$$= -\omega_{31} \wedge \theta_1 - \omega_{32} \wedge \theta_2,$$

which shows that the connection form obeys Condition (ii) of Proposition 10.25, and hence is the Levi-Civita connection for $\mathbb{R}^3$ with the induced Euclidean metric.

For another example of Proposition 10.25 consider an orientable (compact) surface M, with a Riemannian metric. Pick any chart (U, φ), and choose an orthonormal coframe of one-forms (θ_1, θ_2), such that $\mathrm{Vol}_M = \theta_1 \wedge \theta_2$ on U. Then we have

$$d\theta_1 = a_1\theta_1 \wedge \theta_2$$
$$d\theta_2 = a_2\theta_1 \wedge \theta_2$$

for some functions, a_1, a_2, and we let

$$\omega_{12} = a_1\theta_1 + a_2\theta_2.$$

Clearly,

$$\begin{pmatrix} 0 & \omega_{12} \\ -\omega_{12} & 0 \end{pmatrix}\begin{pmatrix} \theta_1 \\ \theta_2 \end{pmatrix} = \begin{pmatrix} 0 & a_1\theta_1 + a_2\theta_2 \\ -(a_1\theta_1 + a_2\theta_2) & 0 \end{pmatrix}\begin{pmatrix} \theta_1 \\ \theta_2 \end{pmatrix} = \begin{pmatrix} d\theta_1 \\ d\theta_2 \end{pmatrix}$$

which shows that

$$\omega = \omega^* = \begin{pmatrix} 0 & \omega_{12} \\ -\omega_{12} & 0 \end{pmatrix}$$

corresponds to the Levi-Civita connection on M. Since $\Omega = d\omega + \omega \wedge \omega$, we see that

$$\Omega = \begin{pmatrix} 0 & d\omega_{12} \\ -d\omega_{12} & 0 \end{pmatrix}.$$

As M is oriented and as M has a metric, the transition functions are in $SO(2)$. We easily check that

$$\begin{pmatrix} \cos t & \sin t \\ -\sin t & \cos t \end{pmatrix}\begin{pmatrix} 0 & d\omega_{12} \\ -d\omega_{12} & 0 \end{pmatrix}\begin{pmatrix} \cos t & -\sin t \\ \sin t & \cos t \end{pmatrix} = \begin{pmatrix} 0 & d\omega_{12} \\ -d\omega_{12} & 0 \end{pmatrix},$$

which shows that Ω is a global two-form called the *Gauss-Bonnet* 2-form of M. There is a function κ, the *Gaussian curvature of* M, such that

$$d\omega_{12} = -\kappa \mathrm{Vol}_M,$$

where $\mathrm{Vol}_M = \theta_1 \wedge \theta_2$ is the oriented volume form on M. It should be noted that Milnor and Stasheff define the volume form as $\mathrm{Vol}_M = -\theta_1 \wedge \theta_2$, so in their work the curvature κ should be replaced by $-\kappa$. The reason for such a choice is explained on page 304 of Milnor and Stasheff [78]. Many other authors (including Warner and Bott and Chern) use the definition $\mathrm{Vol}_M = +\theta_1 \wedge \theta_2$ that has been adopted here.

The *Gauss-Bonnet theorem* for orientable surfaces asserts that

$$\int_M d\omega_{12} = 2\pi \chi(M),$$

where $\chi(M)$ is the *Euler characteristic* of M.

Remark. The Levi-Civita connection induced by a Riemannian metric g can also be defined in terms of the Lie derivative of the metric g. This is the approach followed in Petersen [86] (Chapter 2). If θ_X is the one-form given by

$$\theta_X = i_X g;$$

that is, $(i_X g)(Y) = g(X, Y)$ for all $X, Y \in \mathfrak{X}(M)$, and if $L_X g$ is the Lie derivative of the symmetric $(0, 2)$ tensor g, defined so that

$$(L_X g)(Y, Z) = X(g(Y, Z)) - g(L_X Y, Z) - g(Y, L_X Z)$$

(see Proposition 4.23), then it is proved in Petersen [86] (Chapter 2, Theorem 1) that the Levi-Civita connection is defined implicitly by the formula

$$2g(\nabla_X Y, Z) = (L_Y g)(X, Z) + (d\theta_Y)(X, Z).$$

10.11   Pontrjagin Classes and Chern Classes: a Glimpse

The purpose of this section is to introduce the reader to Pontrjagin Classes and Chern Classes, which are fundamental invariants of real (resp. complex) vector bundles. We focus on motivations and intuitions and omit most proofs, but we give precise pointers to the literature for proofs.

Given a real (resp. complex) rank n vector bundle $\xi = (E, \pi, B, V)$, we know that locally, ξ "looks like" a trivial bundle $U \times V$, for some open subset U of the base space B. Globally, ξ can be very twisted, and one of the main issues is to understand and quantify "how twisted" ξ really is. Now we know that every vector bundle admits a connection, say ∇, and the curvature R^∇ of this connection is some measure of the twisting of ξ. However, R^∇ depends on ∇, so curvature is not intrinsic to ξ, which is unsatisfactory as we seek invariants that depend only on ξ.

Pontrjagin, Stiefel, and Chern (starting from the late 1930s) discovered that invariants with "good" properties could be defined if we took these invariants to belong to various cohomology groups associated with B. Such invariants are usually called *characteristic classes*. Roughly, there are two main methods for defining characteristic classes: one using topology, and the other due to Chern and Weil, using differential forms.

A masterly exposition of these methods is given in the classic book by Milnor and Stasheff [78]. Amazingly, the method of Chern and Weil using differential forms is quite accessible for someone who has reasonably good knowledge of differential forms and de Rham cohomology, as long as one is willing to gloss over various technical details.

As we said earlier, one of the problems with curvature is that it depends on a connection. The way to circumvent this difficulty rests on the simple, yet subtle observation, that locally, given any two overlapping local trivializations $(U_\alpha, \varphi_\alpha)$ and (U_β, φ_β), the transformation rule for the curvature matrices Ω_α and Ω_β is

$$\Omega_\beta = g_{\alpha\beta}^{-1} \Omega_\alpha g_{\alpha\beta},$$

where $g_{\alpha\beta} \colon U_\alpha \cap U_\beta \to \mathbf{GL}(V)$ is the transition function. The matrices of two-forms Ω_α and Ω_α are local, but the stroke of genius is to glue them together to form a global form using *invariant polynomials*.

Indeed, the Ω_α are $n \times n$ matrices, so consider the algebra of polynomials $\mathbb{R}[X_1, \ldots, X_{n^2}]$ (or $\mathbb{C}[X_1, \ldots, X_{n^2}]$ in the complex case) in n^2 variables $X_1, \ldots, X_{n^2}$, considered as the entries of an $n \times n$ matrix. It is more convenient to use the set of variables $\{X_{ij} \mid 1 \leq i, j \leq n\}$, and to let X be the $n \times n$ matrix $X = (X_{ij})$.

Definition 10.18. A polynomial $P \in \mathbb{R}[\{X_{ij} \mid 1 \leq i, j \leq n\}]$ (or $P \in \mathbb{C}[\{X_{ij} \mid 1 \leq i, j \leq n\}]$) is *invariant* iff

$$P(AXA^{-1}) = P(X),$$

for all $A \in \mathbf{GL}(n, \mathbb{R})$ (resp. $A \in \mathbf{GL}(n, \mathbb{C})$). The *algebra of invariant polynomials* over $n \times n$ matrices is denoted by I_n.

Examples of invariant polynomials are the trace $\mathrm{tr}(X)$ and the determinant $\det(X)$ of the matrix X. We will characterize shortly the algebra I_n.

Now comes the punch line: For any homogeneous invariant polynomial $P \in I_n$ of degree k, we can substitute Ω_α for X; that is, substitute ω_{ij} for X_{ij}, and evaluate $P(\Omega_\alpha)$. This is because Ω is a matrix of two-forms, and the wedge product is commutative for forms of even degree. Therefore, $P(\Omega_\alpha) \in \mathcal{A}^{2k}(U_\alpha)$. But the formula for a change of trivialization yields

$$P(\Omega_\alpha) = P(g_{\alpha\beta}^{-1} \Omega_\alpha g_{\alpha\beta}) = P(\Omega_\beta),$$

so the forms $P(\Omega_\alpha)$ and $P(\Omega_\beta)$ agree on overlaps, and thus they define a *global form* denoted $P(R^\nabla) \in \mathcal{A}^{2k}(B)$.

Definition 10.19. For any vector bundle $\xi = (E, \pi, B, V)$, for any homogeneous invariant polynomial $P \in I_n$ of degree k, the global form $P(R^\nabla) \in \mathcal{A}^{2k}(B)$ defined above is called the *global curvature form* on the vector bundle ξ.

Now we know how to obtain global $2k$-forms $P(R^\nabla) \in \mathcal{A}^{2k}(B)$, but they still seem to depend on the connection, and how do they define a cohomology class? Both problems are settled, thanks to the following theorems:

Theorem 10.26. *For every real rank n vector bundle ξ, for every connection ∇ on ξ, for every invariant homogeneous polynomial P of degree k, the $2k$-form $P(R^\nabla) \in \mathcal{A}^{2k}(B)$ is closed. If ξ is a complex vector bundle, then the $2k$-form $P(R^\nabla) \in \mathcal{A}^{2k}(B; \mathbb{C})$ is closed.*

Theorem 10.26 implies that the $2k$-form $P(R^\nabla) \in \mathcal{A}^{2k}(B)$ defines a cohomology class $[P(R^\nabla)] \in H^{2k}_{\mathrm{DR}}(B)$. We will come back to the proof of Theorem 10.26 later.

Theorem 10.27. *For every real (resp. complex) rank n vector bundle ξ, for every invariant homogeneous polynomial P of degree k, the cohomology class $[P(R^\nabla)] \in H^{2k}_{\mathrm{DR}}(B)$ (resp. $[P(R^\nabla)] \in H^{2k}_{\mathrm{DR}}(B; \mathbb{C})$) is independent of the choice of the connection ∇.*

Definition 10.20. The cohomology class $[P(R^\nabla)]$, which does not depend on ∇, is denoted $P(R_\xi)$ (or $P(K_\xi)$) and is called the *characteristic class* of ξ corresponding to P.

Remark. Milnor and Stasheff [78] use the notation $P(K)$, Madsen and Tornehave [75] use the notation $P(F^\nabla)$, and Morita [82] use the notation $f(E)$ (where E is the total space of the vector bundle ξ).

The proof of Theorem 10.27 involves a kind of homotopy argument; see Madsen and Tornehave [75] (Lemma 18.2), Morita [82] (Proposition 5.28), or Milnor and Stasheff [78] (Appendix C).

The upshot is that Theorems 10.26 and 10.27 give us a method for producing invariants of a vector bundle that somehow reflect how curved (or twisted) the bundle is. However, it appears that we need to consider infinitely many invariants. Fortunately, we can do better because the algebra I_n of invariant polynomials is finitely generated, and in fact, has very nice sets of generators. For this, we recall the *elementary symmetric functions* in n variables.

Definition 10.21. Given n variables $\lambda_1, \ldots, \lambda_n$, we can write

$$\prod_{i=1}^{n}(1 + t\lambda_i) = 1 + \sigma_1 t + \sigma_2 t^2 + \cdots + \sigma_n t^n,$$

where the σ_i are symmetric, homogeneous polynomials of degree i in $\lambda_1, \ldots, \lambda_n$, called *elementary symmetric functions* in n variables.

For example,

$$\sigma_1 = \sum_{i=1}^n \lambda_i, \quad \sigma_2 = \sum_{1 \le i < j \le n} \lambda_i \lambda_j, \quad \sigma_n = \lambda_1 \cdots \lambda_n.$$

To be more precise, we write $\sigma_i(\lambda_1, \ldots, \lambda_n)$ instead of σ_i.

Definition 10.22. Given any $n \times n$ matrix $X = (X_{ij})$, we define $\sigma_i(X)$ by the formula

$$\det(I + tX) = 1 + \sigma_1(X)t + \sigma_2(X)t^2 + \cdots + \sigma_n(X)t^n.$$

Proposition 10.28. *Let X be an $n \times n$ matrix. Then*

$$\sigma_i(X) = \sigma_i(\lambda_1, \ldots, \lambda_n),$$

where $\lambda_1, \ldots, \lambda_n$ are the eigenvalues of X.

Proof. Indeed, $\lambda_1, \ldots, \lambda_n$ are the roots the polynomial $\det(\lambda I - X) = 0$, and as

$$\det(\lambda I - X) = \prod_{i=1}^n (\lambda - \lambda_i) = \lambda^n \prod_{i=1}^n \left(1 - \frac{\lambda_i}{\lambda}\right) = \lambda^n + \sum_{i=1}^n (-1)^i \sigma_i(\lambda_1, \ldots, \lambda_n)\lambda^{n-i},$$

by factoring λ^n and replacing λ^{-1} by $-\lambda^{-1}$, we get

$$\det(I + (-\lambda^{-1})X) = 1 + \sum_{i=1}^n = \sigma_i(\lambda_1, \ldots, \lambda_n)(-\lambda^{-1})^i,$$

which proves our claim. $\square$

Observe that

$$\sigma_1(X) = \text{tr}(X), \qquad \sigma_n(X) = \det(X).$$

Also, $\sigma_k(X^\top) = \sigma_k(X)$, since $\det(I + tX) = \det((I + tX)^\top) = \det(I + tX^\top)$. It is not very difficult to prove the following theorem.

Theorem 10.29. *The algebra I_n of invariant polynomials in n^2 variables is generated by $\sigma_1(X), \ldots, \sigma_n(X)$; that is,*

$$I_n \cong \mathbb{R}[\sigma_1(X), \ldots, \sigma_n(X)] \qquad (resp. \quad I_n \cong \mathbb{C}[\sigma_1(X), \ldots, \sigma_n(X)]).$$

Proof Sketch. For a proof of Theorem 10.29, see Milnor and Stasheff [78] (Appendix C, Lemma 6), Madsen and Tornehave [75] (Appendix B), or Morita [82] (Theorem 5.26). The proof uses the fact that for every matrix X, there is an upper-triangular matrix T, and an invertible matrix B, so that

$$X = BTB^{-1}.$$

Then we can replace B by the matrix $\mathrm{diag}(\epsilon, \epsilon^2, \dots, \epsilon^n)B$, where ϵ is very small, and make the off diagonal entries arbitrarily small. By continuity, it follows that $P(X)$ depends only on the diagonal entries of BTB^{-1}, that is, on the eigenvalues of X. So, $P(X)$ must be a symmetric function of these eigenvalues, and the classical theory of symmetric functions completes the proof. □

It turns out that there are situations where it is more convenient to use another set of generators instead of $\sigma_1, \dots, \sigma_n$. Define $s_i(X)$ by

$$s_i(X) = \mathrm{tr}(X^i).$$

Of course,

$$s_i(X) = \lambda_1^i + \cdots + \lambda_n^i,$$

where $\lambda_1, \dots, \lambda_n$ are the eigenvalues of X. Now the $\sigma_i(X)$ and $s_i(X)$ are related to each other by *Newton's formula*, namely:

$$s_i(X) - \sigma_1(X)s_{i-1}(X) + \sigma_2(X)s_{i-2}(X) + \cdots + (-1)^{i-1}\sigma_{i-1}(X)s_1(X)$$
$$+ (-1)^i i\sigma_i(X) = 0,$$

with $1 \le i \le n$. A "cute" proof of the Newton formulae is obtained by computing the derivative of $\log(h(t))$, where

$$h(t) = \prod_{i=1}^{n}(1 + t\lambda_i) = 1 + \sigma_1 t + \sigma_2 t^2 + \cdots + \sigma_n t^n,$$

see Madsen and Tornehave [75] (Appendix B) or Morita [82] (Exercise 5.7).

Consequently, we can inductively compute s_i in terms of $\sigma_1, \dots, \sigma_i$, and conversely σ_i in terms of $s_1, \dots, s_i$. For example,

$$s_1 = \sigma_1, \quad s_2 = \sigma_1^2 - 2\sigma_2, \quad s_3 = \sigma_1^3 - 3\sigma_1\sigma_2 + 3\sigma_3.$$

It follows that

$$I_n \cong \mathbb{R}[s_1(X), \dots, s_n(X)] \qquad (\text{resp. } I_n \cong \mathbb{C}[s_1(X), \dots, s_n(X)]).$$

Using the above, we can give a simple proof of Theorem 10.26, using Theorem 10.29.

Proof of Theorem 10.26. Since $s_1, \ldots, s_n$ generate I_n, it is enough to prove that $s_i(R^\nabla)$ is closed. We need to prove that $ds_i(R^\nabla) = 0$, and for this, it is enough to prove it in every local trivialization $(U_\alpha, \varphi_\alpha)$. To simplify notation, we write Ω for Ω_α. Now, $s_i(\Omega) = \mathrm{tr}(\Omega^i)$, so

$$ds_i(\Omega) = d\mathrm{tr}(\Omega^i) = \mathrm{tr}(d\Omega^i),$$

and we use Bianchi's identity (Proposition 10.15),

$$d\Omega = \Omega \wedge \omega - \omega \wedge \Omega.$$

We have

$$
\begin{aligned}
d\Omega^i =& d\Omega \wedge \Omega^{i-1} + \Omega \wedge d\Omega \wedge \Omega^{i-2} + \cdots + \Omega^k \wedge d\Omega \wedge \Omega^{i-k-1} + \cdots \\
& + \Omega^{i-1} \wedge d\Omega \\
=& (\Omega \wedge \omega - \omega \wedge \Omega) \wedge \Omega^{i-1} + \Omega \wedge (\Omega \wedge \omega - \omega \wedge \Omega) \wedge \Omega^{i-2} + \cdots \\
& + \Omega^k \wedge (\Omega \wedge \omega - \omega \wedge \Omega) \wedge \Omega^{i-k-1} + \Omega^{k+1} \wedge (\Omega \wedge \omega - \omega \wedge \Omega) \wedge \Omega^{i-k-2} \\
& + \cdots + \Omega^{i-1} \wedge (\Omega \wedge \omega - \omega \wedge \Omega) \\
=& -\omega \wedge \Omega^i + \Omega \wedge \omega \wedge \Omega^{i-1} - \Omega \wedge \omega \wedge \Omega^{i-1} + \Omega^2 \wedge \omega \wedge \Omega^{i-2} + \cdots + \\
& - \Omega^k \wedge \omega \wedge \Omega^{i-k} + \Omega^{k+1} \wedge \omega \wedge \Omega^{i-k-1} - \Omega^{k+1} \wedge \omega \wedge \Omega^{i-k-1} \\
& + \Omega^{k+2} \wedge \omega \wedge \Omega^{i-k-2} + \cdots - \Omega^{i-1} \wedge \omega \wedge \Omega + \Omega^i \wedge \omega \\
=& \ \Omega^i \wedge \omega - \omega \wedge \Omega^i.
\end{aligned}
$$

However, the entries in ω are one-forms, the entries in Ω are two-forms, and since

$$\eta \wedge \theta = \theta \wedge \eta$$

for all $\eta \in \mathcal{A}^1(B)$ and all $\theta \in \mathcal{A}^2(B)$ and $\mathrm{tr}(XY) = \mathrm{tr}(YX)$ for all matrices X and Y with commuting entries, we get

$$\mathrm{tr}(d\Omega^i) = \mathrm{tr}(\omega \wedge \Omega^i - \Omega^i \wedge \omega) = \mathrm{tr}(\Omega^i \wedge \omega) - \mathrm{tr}(\omega \wedge \Omega^i) = 0,$$

as required.   □

A more elegant proof (also using Bianchi's identity) can be found in Milnor and Stasheff [78] (Appendix C, page 296-298).

For real vector bundles, only invariant polynomials of even degrees matter.

Proposition 10.30. *If ξ is a real vector bundle, then for every homogeneous invariant polynomial P of odd degree k, we have $P(R_\xi) = 0 \in H^{2k}_{\mathrm{DR}}(B)$.*

Proof. As $I_n \cong \mathbb{R}[s_1(X), \ldots, s_n(X)]$ and $s_i(X)$ is homogeneous of degree i, it is enough to prove Proposition 10.30 for $s_i(X)$ with i odd. By Theorem 10.27, we may assume that we pick a metric connection on ξ, so that Ω_α is skew-symmetric in every local trivialization. Then, Ω_α^i is also skew symmetric and

$$\operatorname{tr}(\Omega_\alpha^i) = 0,$$

since the diagonal entries of a real skew-symmetric matrix are all zero. It follows that $s_i(\Omega_\alpha) = \operatorname{tr}(\Omega_\alpha^i) = 0$.   □

Proposition 10.30 implies that for a real vector bundle ξ, nonzero characteristic classes can only live in the cohomology groups $H_{\mathrm{DR}}^{4k}(B)$ of dimension $4k$. This property is specific to real vector bundles and generally fails for complex vector bundles.

Before defining Pontrjagin and Chern classes, we state another important property of the homology classes $P(R_\xi)$; see Madsen and Tornehave [75] (Chapter 18, Theorem 18.5).

Proposition 10.31. *If $\xi = (E, \pi, B, V)$ and $\xi' = (E', \pi', B', V)$ are real (resp. complex) vector bundles, for every bundle map*

$$
\begin{array}{ccc}
E & \xrightarrow{\;f_E\;} & E' \\
{\scriptstyle \pi}\downarrow & & \downarrow{\scriptstyle \pi'} \\
B & \xrightarrow[\;f\;]{} & B',
\end{array}
$$

for every homogeneous invariant polynomial P of degree k, we have

$$P(R_\xi) = f^*(P(R_{\xi'})) \in H_{\mathrm{DR}}^{2k}(B) \quad (resp. \quad P(R_\xi) = f^*(P(R_{\xi'})) \in H_{\mathrm{DR}}^{2k}(B; \mathbb{C})).$$

In particular, for every smooth map $g \colon N \to B$, we have

$$P(R_{g^*\xi}) = g^*(P(R_\xi)) \in H_{\mathrm{DR}}^{2k}(N) \quad (resp. \quad P(R_{g^*\xi}) = g^*(P(R_\xi)) \in H_{\mathrm{DR}}^{2k}(N; \mathbb{C})),$$

where $g^*\xi$ is the pullback bundle of ξ along g.

The above proposition implies that if $(f_E, f) \colon \xi \to \xi'$ is an isomorphism of vector bundles, then the pullback map f^* maps the characteristic classes of ξ' to the characteristic classes of ξ bijectively.

We finally define Pontrjagin classes and Chern classes.

Definition 10.23. If ξ be a real rank n vector bundle, then the kth *Pontrjagin class* of ξ, denoted $p_k(\xi)$, where $1 \leq 2k \leq n$, is the cohomology class

$$p_k(\xi) = \left[\frac{1}{(2\pi)^{2k}} \, \sigma_{2k}(R^\nabla) \right] \in H_{\mathrm{DR}}^{4k}(B),$$

for any connection ∇ on ξ.

If ξ be a complex rank n vector bundle, then the kth *Chern class* of ξ, denoted $c_k(\xi)$, where $1 \leq k \leq n$, is the cohomology class

$$c_k(\xi) = \left[\left(\frac{-1}{2\pi i} \right)^k \sigma_k(R^\nabla) \right] \in H_{\text{DR}}^{2k}(B),$$

for any connection ∇ on ξ. We also set $p_0(\xi) = 1$, and $c_0(\xi) = 1$ in the complex case.

The strange coefficient in $p_k(\xi)$ is present so that our expression matches the topological definition of Pontrjagin classes. The equally strange coefficient in $c_k(\xi)$ is there to ensure that $c_k(\xi)$ actually belongs to the *real* cohomology group $H_{\text{DR}}^{2k}(B)$, as stated (from the definition, we can only claim that $c_k(\xi) \in H_{\text{DR}}^{2k}(B; \mathbb{C})$).

This requires a proof which can be found in Morita [82] (Proposition 5.30), or in Madsen and Tornehave [75] (Chapter 18). One can use the fact that every complex vector bundle admits a Hermitian connection. Locally, the curvature matrices are skew-Hermitian and this easily implies that the Chern classes are real, since if Ω is skew-Hermitian, then $i\Omega$ is Hermitian. (Actually, the topological version of Chern classes shows that $c_k(\xi) \in H^{2k}(B; \mathbb{Z})$.)

If ξ is a real rank n vector bundle and n is odd, say $n = 2m + 1$, then the "top" Pontrjagin class $p_m(\xi)$ corresponds to $\sigma_{2m}(R^\nabla)$, which is not $\det(R^\nabla)$. However, if n is even, say $n = 2m$, then the "top" Pontrjagin class $p_m(\xi)$ corresponds to $\det(R^\nabla)$.

Definition 10.24. The *Pontrjagin polynomial* $p(\xi)(t) \in H_{\text{DR}}^\bullet(B)[t]$, given by

$$p(\xi)(t) = \left[\det \left(I + \frac{t}{2\pi} R^\nabla \right) \right] = 1 + p_1(\xi)t + p_2(\xi)t^2 + \cdots + p_{\lfloor \frac{n}{2} \rfloor}(\xi) t^{\lfloor \frac{n}{2} \rfloor}$$

and the *Chern polynomial* $c(\xi)(t) \in H_{\text{DR}}^\bullet(B)[t]$, given by

$$c(\xi)(t) = \left[\det \left(I - \frac{t}{2\pi i} R^\nabla \right) \right] = 1 + c_1(\xi)t + c_2(\xi)t^2 + \cdots + c_n(\xi)t^n.$$

If a vector bundle is trivial, then all its Pontrjagin classes (or Chern classes) are zero for all $k \geq 1$. If ξ is the real tangent bundle $\xi = TB$ of some manifold B of dimension n, then the $\lfloor \frac{n}{4} \rfloor$ Pontrjagin classes of TB are denoted $p_1(B), \ldots, p_{\lfloor \frac{n}{4} \rfloor}(B)$.

For complex vector bundles, the manifold B is often the real manifold corresponding to a complex manifold. If B has complex dimension n, then B has real dimension $2n$. In this case, the tangent bundle TB is a rank n complex vector bundle over the real manifold of dimension $2n$, and thus, it has n Chern classes, denoted $c_1(B), \ldots, c_n(B)$.

The determination of the Pontrjagin classes (or Chern classes) of a manifold is an important step for the study of the geometric/topological structure of the manifold. For example, it is possible to compute the Chern classes of complex projective space $\mathbb{CP}^n$ (as a complex manifold).

The Pontrjagin classes of a real vector bundle ξ are related to the Chern classes of its complexification $\xi_{\mathbb{C}} = \xi \otimes_{\mathbb{R}} \epsilon_{\mathbb{C}}^1$ (where $\epsilon_{\mathbb{C}}^1$ is the trivial complex line bundle $B \times \mathbb{C}$).

Proposition 10.32. *For every real rank n vector bundle $\xi = (E, \pi, B, V)$, if $\xi_{\mathbb{C}} = \xi \otimes_{\mathbb{R}} \epsilon_{\mathbb{C}}^1$ is the complexification of ξ, then*

$$p_k(\xi) = (-1)^k c_{2k}(\xi_{\mathbb{C}}) \in H_{\mathrm{DR}}^{4k}(B) \qquad k \geq 0.$$

Basically, the reason why Proposition 10.32 holds is that

$$\frac{1}{(2\pi)^{2k}} = (-1)^k \left(\frac{-1}{2\pi i}\right)^{2k}.$$

For details, see Morita [82] (Chapter 5, Section 5, Proposition 5.38).

We conclude this section by stating a few more properties of Chern classes.

Proposition 10.33. *For every complex rank n vector bundle ξ, the following properties hold:*

(1) If ξ has a Hermitian metric, then we have a canonical isomorphism $\xi^* \cong \overline{\xi}$.

(2) The Chern classes of ξ, ξ^* and $\overline{\xi}$ satisfy:

$$c_k(\xi^*) = c_k(\overline{\xi}) = (-1)^k c_k(\xi).$$

(3) For any complex vector bundles ξ and η,

$$c_k(\xi \oplus \eta) = \sum_{i=0}^{k} c_i(\xi) c_{k-i}(\eta),$$

or equivalently

$$c(\xi \oplus \eta)(t) = c(\xi)(t) c(\eta)(t),$$

and similarly for Pontrjagin classes when ξ and η are real vector bundles.

To prove (2), we can use the fact that ξ can be given a Hermitian metric. Then we saw earlier that if ω is the connection matrix of ξ over U, then $\overline{\omega} = -\omega^\top$ is the connection matrix of $\overline{\xi}$ over U. However, it is clear that $\sigma_k(-\Omega_\alpha^\top) = (-1)^k \sigma_k(\Omega_\alpha)$, and so $c_k(\overline{\xi}) = (-1)^k c_k(\xi)$. For details, see Morita [82] (Chapter 5, Section 5, Theorem 5.37 and Proposition 5.40).

Remark. For a real vector bundle ξ, it is easy to see that $(\xi_{\mathbb{C}})^* = (\xi^*)_{\mathbb{C}}$, which implies that $c_k((\xi_{\mathbb{C}})^*) = c_k(\xi_{\mathbb{C}})$ (as $\xi \cong \xi^*$) and (2) implies that $c_k(\xi_{\mathbb{C}}) = 0$ for k odd. This proves again that the Pontrjagin classes exit only in dimension $4k$.

A complex rank n vector bundle ξ can also be viewed as a rank $2n$ vector bundle $\xi_{\mathbb{R}}$ and we have:

Proposition 10.34. *For every rank n complex vector bundle ξ, if $p_k = p_k(\xi_{\mathbb{R}})$ and $c_k = c_k(\xi)$, then we have*

$$1 - p_1 + p_2 + \cdots + (-1)^n p_n = (1 + c_1 + c_2 + \cdots + c_n)(1 - c_1 + c_2 + \cdots + (-1)^n c_n).$$

For a proof, see Morita [82] (Chapter 5, Section 5, Proposition 5.41).

Besides defining the Chern and Pontrjagin classes, the curvature form R^∇ also defines an Euler class. But in order to efficiently define the Euler class, we need a technical tool, the Pfaffian polynomial.

10.12 The Pfaffian Polynomial

The results of this section will be needed to define the Euler class of a real orientable rank $2n$ vector bundle; see Section 10.13.

Let $\mathfrak{so}(2n)$ denote the vector space (actually, Lie algebra) of $2n \times 2n$ real skew-symmetric matrices. It is well known that every matrix $A \in \mathfrak{so}(2n)$ can be written as

$$A = PDP^\top,$$

where P is an orthogonal matrix and where D is a block diagonal matrix

$$D = \begin{pmatrix} D_1 & & & \\ & D_2 & & \\ & & \ddots & \\ & & & D_n \end{pmatrix}$$

consisting of 2×2 blocks of the form

$$D_i = \begin{pmatrix} 0 & -a_i \\ a_i & 0 \end{pmatrix}.$$

For a proof, see Horn and Johnson [59] (Corollary 2.5.14), Gantmacher [49] (Chapter IX), or Gallier [46] (Chapter 11).

Since $\det(D_i) = a_i^2$ and $\det(A) = \det(PDP^\top) = \det(D) = \det(D_1) \cdots \det(D_n)$, we get

$$\det(A) = (a_1 \cdots a_n)^2.$$

The Pfaffian is a polynomial function $\mathrm{Pf}(A)$ in skew-symmetric $2n \times 2n$ matrices A (a polynomial in $(2n-1)n$ variables) such that

$$\mathrm{Pf}(A)^2 = \det(A),$$

and for every arbitrary matrix B,

$$\mathrm{Pf}(BAB^\top) = \mathrm{Pf}(A)\det(B).$$

The Pfaffian shows up in the definition of the Euler class of a vector bundle. There is a simple alternative way to define the Pfaffian using some exterior algebra. Let $(e_1, \ldots, e_{2n})$ be any basis of $\mathbb{R}^{2n}$. For any matrix $A \in \mathfrak{so}(2n)$, let

$$\omega(A) = \sum_{i<j} a_{ij}\, e_i \wedge e_j,$$

where $A = (a_{ij})$. Then $\bigwedge^n \omega(A)$ is of the form $Ce_1 \wedge e_2 \wedge \cdots \wedge e_{2n}$ for some constant $C \in \mathbb{R}$.

Definition 10.25. For every skew symmetric matrix $A \in \mathfrak{so}(2n)$, the *Pfaffian polynomial* or *Pfaffian*, is the degree n polynomial $\mathrm{Pf}(A)$ defined by

$$\bigwedge^n \omega(A) = n!\,\mathrm{Pf}(A)\, e_1 \wedge e_2 \wedge \cdots \wedge e_{2n}.$$

Clearly, $\mathrm{Pf}(A)$ is independent of the basis chosen. If A is the block diagonal matrix D, a simple calculation shows that

$$\omega(D) = -(a_1 e_1 \wedge e_2 + a_2 e_3 \wedge e_4 + \cdots + a_n e_{2n-1} \wedge e_{2n})$$

and that

$$\bigwedge^n \omega(D) = (-1)^n n!\, a_1 \cdots a_n\, e_1 \wedge e_2 \wedge \cdots \wedge e_{2n},$$

and so

$$\mathrm{Pf}(D) = (-1)^n a_1 \cdots a_n.$$

Since $\mathrm{Pf}(D)^2 = (a_1 \cdots a_n)^2 = \det(A)$, we seem to be on the right track.

Proposition 10.35. *For every skew symmetric matrix $A \in \mathfrak{so}(2n)$ and every arbitrary matrix B, we have:*

(i) $\text{Pf}(A)^2 = \det(A)$
(ii) $\text{Pf}(BAB^{\top}) = \text{Pf}(A) \det(B)$.

Proof. If we assume that (ii) is proved, then since we can write $A = PDP^{\top}$ for some orthogonal matrix P and some block diagonal matrix D as above, as $\det(P) = \pm 1$ and $\text{Pf}(D)^2 = \det(A)$, we get

$$\text{Pf}(A)^2 = \text{Pf}(PDP^{\top})^2 = \text{Pf}(D)^2 \det(P)^2 = \det(A),$$

which is (i). Therefore, it remains to prove (ii).

Let $f_i = Be_i$ for $i = 1, \ldots, 2n$, where $(e_1, \ldots, e_{2n})$ is any basis of $\mathbb{R}^{2n}$. Since $f_i = \sum_k b_{ki} e_k$, we have

$$\tau = \sum_{i,j} a_{ij} \, f_i \wedge f_j = \sum_{i,j} \sum_{k,l} b_{ki} a_{ij} b_{lj} \, e_k \wedge e_l = \sum_{k,l} (BAB^{\top})_{kl} \, e_k \wedge e_l,$$

and so, as $BAB^{\top}$ is skew symmetric and $e_k \wedge e_l = -e_l \wedge e_k$, we get

$$\tau = 2\omega(BAB^{\top}).$$

Consequently,

$$\bigwedge^n \tau = 2^n \bigwedge^n \omega(BAB^{\top}) = 2^n n! \, \text{Pf}(BAB^{\top}) \, e_1 \wedge e_2 \wedge \cdots \wedge e_{2n}.$$

Now,

$$\bigwedge^n \tau = C \, f_1 \wedge f_2 \wedge \cdots \wedge f_{2n},$$

for some $C \in \mathbb{R}$. If B is singular, then the f_i are linearly dependent, which implies that $f_1 \wedge f_2 \wedge \cdots \wedge f_{2n} = 0$, in which case

$$\text{Pf}(BAB^{\top}) = 0,$$

as $e_1 \wedge e_2 \wedge \cdots \wedge e_{2n} \neq 0$. Therefore, if B is singular, $\det(B) = 0$ and

$$\text{Pf}(BAB^{\top}) = 0 = \text{Pf}(A) \det(B).$$

If B is invertible, as $\tau = \sum_{i,j} a_{ij} \, f_i \wedge f_j = 2 \sum_{i<j} a_{ij} \, f_i \wedge f_j$, we have

$$\bigwedge^n \tau = 2^n n! \, \text{Pf}(A) \, f_1 \wedge f_2 \wedge \cdots \wedge f_{2n}.$$

However, as $f_i = Be_i$, we have

$$f_1 \wedge f_2 \wedge \cdots \wedge f_{2n} = \det(B)\, e_1 \wedge e_2 \wedge \cdots \wedge e_{2n},$$

so

$$\bigwedge^{n} \tau = 2^n n!\, \mathrm{Pf}(A) \det(B)\, e_1 \wedge e_2 \wedge \cdots \wedge e_{2n}$$

and as

$$\bigwedge^{n} \tau = 2^n n!\, \mathrm{Pf}(BAB^\top)\, e_1 \wedge e_2 \wedge \cdots \wedge e_{2n},$$

we get

$$\mathrm{Pf}(BAB^\top) = \mathrm{Pf}(A) \det(B),$$

as claimed.   □

Remark. It can be shown that the polynomial $\mathrm{Pf}(A)$ is the unique polynomial with integer coefficients such that $\mathrm{Pf}(A)^2 = \det(A)$ and $\mathrm{Pf}(\mathrm{diag}(S, \ldots, S)) = +1$, where

$$S = \begin{pmatrix} 0 & 1 \\ -1 & 0 \end{pmatrix};$$

see Milnor and Stasheff [78] (Appendix C, Lemma 9). There is also an explicit formula for $\mathrm{Pf}(A)$, namely

$$\mathrm{Pf}(A) = \frac{1}{2^n n!} \sum_{\sigma \in \mathfrak{S}_{2n}} \mathrm{sgn}(\sigma) \prod_{i=1}^{n} a_{\sigma(2i-1)\,\sigma(2i)}.$$

For example, if

$$A = \begin{pmatrix} 0 & -a \\ a & 0 \end{pmatrix},$$

then $\mathrm{Pf}(A) = -a$, and if

$$A = \begin{pmatrix} 0 & a_1 & a_2 & a_3 \\ -a_1 & 0 & a_4 & a_5 \\ -a_2 & -a_4 & 0 & a_6 \\ -a_3 & -a_5 & -a_6 & 0 \end{pmatrix},$$

then

$$\mathrm{Pf}(A) = a_1 a_6 - a_2 a_5 + a_4 a_3.$$

It is easily checked that

$$\det(A) = (\mathrm{Pf}(A))^2 = (a_1a_6 - a_2a_5 + a_4a_3)^2.$$

 Beware, some authors use a different sign convention and require the Pfaffian to have the value $+1$ on the matrix $\mathrm{diag}(S', \ldots, S')$, where

$$S' = \begin{pmatrix} 0 & -1 \\ 1 & 0 \end{pmatrix}.$$

For example, if $\mathbb{R}^{2n}$ is equipped with an inner product $\langle -, - \rangle$, then some authors define $\omega(A)$ as

$$\omega(A) = \sum_{i<j} \langle Ae_i, e_j \rangle \, e_i \wedge e_j,$$

where $A = (a_{ij})$. But then, $\langle Ae_i, e_j \rangle = a_{ji}$ and **not** a_{ij}, and this Pfaffian takes the value $+1$ on the matrix $\mathrm{diag}(S', \ldots, S')$. This version of the Pfaffian differs from our version by the factor $(-1)^n$. In this respect, Madsen and Tornehave [75] seem to have an incorrect sign in Proposition B6 of Appendix C.

We will also need another property of Pfaffians. Recall that the ring $M_n(\mathbb{C})$ of $n \times n$ matrices over $\mathbb{C}$ is embedded in the ring $M_{2n}(\mathbb{R})$ of $2n \times 2n$ matrices with real coefficients, using the injective homomorphism that maps every entry $z = a + ib \in \mathbb{C}$ to the 2×2 matrix

$$\begin{pmatrix} a & -b \\ b & a \end{pmatrix}.$$

If $A \in M_n(\mathbb{C})$, let $A_{\mathbb{R}} \in M_{2n}(\mathbb{R})$ denote the real matrix obtained by the above process. Observe that every skew Hermitian matrix $A \in \mathfrak{u}(n)$ (*i.e.*, with $A^* = \overline{A}^\top = -A$) yields a matrix $A_{\mathbb{R}} \in \mathfrak{so}(2n)$.

Proposition 10.36. *For every skew Hermitian matrix* $A \in \mathfrak{u}(n)$, *we have*

$$\mathrm{Pf}(A_{\mathbb{R}}) = i^n \det(A).$$

Proof. It is well known that a skew Hermitian matrix can be diagonalized with respect to a unitary matrix U and that the eigenvalues are pure imaginary or zero, so we can write

$$A = U \, \mathrm{diag}(ia_1, \ldots, ia_n)U^*,$$

for some reals $a_j \in \mathbb{R}$. Consequently, we get

$$A_{\mathbb{R}} = U_{\mathbb{R}} \operatorname{diag}(D_1, \ldots, D_n) U_{\mathbb{R}}^{\top},$$

where

$$D_j = \begin{pmatrix} 0 & -a_j \\ a_j & 0 \end{pmatrix}$$

and

$$\operatorname{Pf}(A_{\mathbb{R}}) = \operatorname{Pf}(\operatorname{diag}(D_1, \ldots, D_n)) = (-1)^n \, a_1 \cdots a_n,$$

as we saw before. On the other hand,

$$\det(A) = \det(\operatorname{diag}(ia_1, \ldots, ia_n)) = i^n \, a_1 \cdots a_n,$$

and as $(-1)^n = i^n i^n$, we get

$$\operatorname{Pf}(A_{\mathbb{R}}) = i^n \, \det(A),$$

as claimed. $\square$

Madsen and Tornehave [75] state Proposition 10.36 using the factor $(-i)^n$, which is wrong.

10.13   Euler Classes and the Generalized Gauss-Bonnet Theorem

Let $\xi = (E, \pi, B, V)$ be a real vector bundle of rank $n = 2m$ and let ∇ be any metric connection on ξ. Then if ξ is orientable (as defined in Section 9.8, see Definition 9.21 and the paragraph following it), it is possible to define a global form $\operatorname{eu}(R^{\nabla}) \in \mathcal{A}^{2m}(B)$, which turns out to be closed. Furthermore, the cohomology class $[\operatorname{eu}(R^{\nabla})] \in H_{\mathrm{DR}}^{2m}(B)$ is independent of the choice of ∇. This cohomology class, denoted $e(\xi)$, is called the *Euler class* of ξ and has some very interesting properties. For example, $p_m(\xi) = e(\xi)^2$.

As ∇ is a metric connection, in a trivialization $(U_{\alpha}, \varphi_{\alpha})$, the curvature matrix Ω_{α} is a skew symmetric $2m \times 2m$ matrix of 2-forms. Therefore, we can substitute the 2-forms in Ω_{α} for the variables of the Pfaffian of degree m (see Section 10.12), and we obtain the $2m$-form, $\operatorname{Pf}(\Omega_{\alpha}) \in \mathcal{A}^{2m}(B)$. Now as ξ is orientable, the transition functions take values in $\mathbf{SO}(2m)$, so by Proposition 10.14, since

$$\Omega_{\beta} = g_{\alpha\beta}^{-1} \Omega_{\alpha} g_{\alpha\beta},$$

we conclude from Proposition 10.35 (ii) that

$$\mathrm{Pf}(\Omega_\alpha) = \mathrm{Pf}(\Omega_\beta).$$

Therefore, the local $2m$-forms $\mathrm{Pf}(\Omega_\alpha)$ patch and define a global form $\mathrm{Pf}(R^\nabla) \in \mathcal{A}^{2m}(B)$.

The following propositions can be shown.

Proposition 10.37. *For every real, orientable, rank $2m$ vector bundle ξ, for every metric connection ∇ on ξ, the $2m$-form $\mathrm{Pf}(R^\nabla) \in \mathcal{A}^{2m}(B)$ is closed.*

Proposition 10.38. *For every real, orientable, rank $2m$ vector bundle ξ, the cohomology class $[\mathrm{Pf}(R^\nabla)] \in H_{\mathrm{DR}}^{2m}(B)$ is independent of the metric connection ∇ on ξ.*

Proofs of Propositions 10.37 and 10.38 can be found in Madsen and Tornehave [75] (Chapter 19) or Milnor and Stasheff [78] (Appendix C) (also see Morita [82], Chapters 5 and 6).

Definition 10.26. Let $\xi = (E, \pi, B, V)$ be any real, orientable, rank $2m$ vector bundle. For any metric connection ∇ on ξ, the *Euler form* associated with ∇ is the closed form

$$\mathrm{eu}(R^\nabla) = \frac{1}{(2\pi)^m} \mathrm{Pf}(R^\nabla) \in \mathcal{A}^{2m}(B),$$

and the *Euler class* of ξ is the cohomology class

$$e(\xi) = \left[\mathrm{eu}(R^\nabla)\right] \in H_{\mathrm{DR}}^{2m}(B),$$

which does not depend on ∇.

 Some authors, including Madsen and Tornehave [75], have a negative sign in front of R^∇ in their definition of the Euler form; that is, they define $\mathrm{eu}(R^\nabla)$ by

$$\mathrm{eu}(R^\nabla) = \frac{1}{(2\pi)^m} \mathrm{Pf}(-R^\nabla).$$

However these authors use a Pfaffian with the opposite sign convention from ours and this Pfaffian differs from ours by the factor $(-1)^n$ (see the warning in Section 10.12). Madsen and Tornehave [75] seem to have overlooked this point and with their definition of the Pfaffian (which is the one we have adopted) Proposition 10.41 is incorrect.

Here is the relationship between the Euler class $e(\xi)$, and the top Pontrjagin class $p_m(\xi)$:

Proposition 10.39. *For every real, orientable, rank $2m$ vector bundle $\xi = (E, \pi, B, V)$, we have*

$$p_m(\xi) = e(\xi)^2 \in H_{\mathrm{DR}}^{4m}(B).$$

Proof. The top Pontrjagin class $p_m(\xi)$ is given by

$$p_m(\xi) = \left[\frac{1}{(2\pi)^{2m}} \det(R^{\nabla}) \right],$$

for any (metric) connection ∇, and

$$e(\xi) = \left[\mathrm{eu}(R^{\nabla}) \right],$$

with

$$\mathrm{eu}(R^{\nabla}) = \frac{1}{(2\pi)^m} \mathrm{Pf}(R^{\nabla}).$$

From Proposition 10.35 (i), we have

$$\det(R^{\nabla}) = \mathrm{Pf}(R^{\nabla})^2,$$

which yields the desired result. $\qquad\square$

A rank m complex vector bundle $\xi = (E, \pi, B, V)$ can be viewed as a real rank $2m$ vector bundle $\xi_{\mathbb{R}}$, by viewing V as a $2m$-dimensional real vector space.

Proposition 10.40. *For any complex vector bundle $\xi = (E, \pi, B, V)$, the real vector bundle $\xi_{\mathbb{R}}$ is naturally orientable.*

Proof. For any basis, $(e_1, \ldots, e_m)$, of V over $\mathbb{C}$, observe that $(e_1, ie_1, \ldots, e_m, ie_m)$ is a basis of V over $\mathbb{R}$ (since $v = \sum_{i=1}^m (\lambda_i + i\mu_i)e_i = \sum_{i=1}^m \lambda_i e_i + \sum_{i=1}^m \mu_i ie_i$). But, any $m \times m$ invertible matrix A, over $\mathbb{C}$ becomes a real $2m \times 2m$ invertible matrix $A_{\mathbb{R}}$, obtained by replacing the entry $a_{jk} + ib_{jk}$ in A by the real 2×2 matrix

$$\begin{pmatrix} a_{jk} & -b_{jk} \\ b_{jk} & a_{jk}. \end{pmatrix}$$

Indeed, if $v_k = \sum_{j=1}^m a_{jk}e_j + \sum_{j=1}^m b_{jk}ie_j$, then $iv_k = \sum_{j=1}^m -b_{jk}e_j + \sum_{j=1}^m a_{jk}ie_j$ and when we express v_k and iv_k over the basis $(e_1, ie_1, \ldots, e_m, ie_m)$, we get a matrix $A_{\mathbb{R}}$ consisting of 2×2 blocks as above. Clearly, the map $r \colon A \mapsto A_{\mathbb{R}}$ is a continuous injective homomorphism from $\mathbf{GL}(m, \mathbb{C})$ to $\mathbf{GL}(2m, \mathbb{R})$. Now, it is known that $\mathbf{GL}(m, \mathbb{C})$ is connected, thus $\mathrm{Im}(r) = r(\mathbf{GL}(m, \mathbb{C}))$ is connected, and as $\det(I_{2m}) = 1$, we conclude that all matrices in $\mathrm{Im}(r)$ have positive determinant.[1] Therefore, the transition functions of $\xi_{\mathbb{R}}$ which take values in $\mathrm{Im}(r)$ have positive determinant, and $\xi_{\mathbb{R}}$ is orientable. $\qquad\square$

We can give $\xi_{\mathbb{R}}$ an orientation by fixing some basis of V over $\mathbb{R}$. We have the following relationship between $e(\xi_{\mathbb{R}})$ and the top Chern class, $c_m(\xi)$.

Proposition 10.41. *For every complex, rank m vector bundle $\xi = (E, \pi, B, V)$, we have*

$$c_m(\xi) = e(\xi) \in H^{2m}_{\mathrm{DR}}(B).$$

Proof. Pick some metric connection ∇ on the complex vector bundle ξ. Recall that

$$c_m(\xi) = \left[\left(\frac{-1}{2\pi i}\right)^m \det(R^\nabla)\right] = i^m \left[\left(\frac{1}{2\pi}\right)^m \det(R^\nabla)\right].$$

On the other hand,

$$e(\xi) = \left[\frac{1}{(2\pi)^m} \mathrm{Pf}(R^\nabla_{\mathbb{R}})\right].$$

Here, $R^\nabla_{\mathbb{R}}$ denotes the global $2m$-form, which locally is equal to $\Omega_{\mathbb{R}}$, where Ω is the $m \times m$ curvature matrix of ξ over some trivialization. By Proposition 10.36,

$$\mathrm{Pf}(\Omega_{\mathbb{R}}) = i^m \det(\Omega),$$

so $c_m(\xi) = e(\xi)$, as claimed. $\qquad\square$

The Euler class enjoys many other nice properties. For example, if $f: \xi_1 \to \xi_2$ is an orientation preserving bundle map, then

$$e(f^*\xi_2) = f^*(e(\xi_2)),$$

where $f^*\xi_2$ is given the orientation induced by ξ_2. Also, the Euler class can be defined by topological means and it belongs to the integral cohomology group $H^{2m}(B; \mathbb{Z})$.

Although this result lies beyond the scope of these notes, we cannot resist stating one of the most important and most beautiful theorems of differential geometry usually called the *Generalized Gauss-Bonnet theorem* or *Gauss-Bonnet-Chern theorem*.

---

[1] One can also prove directly that every matrix in $\mathrm{Im}(r)$ has positive determinant by expressing $r(A)$ as a product of simple matrices whose determinants are easily computed.

For this we need the notion of Euler characteristic. Since we haven't discussed triangulations of manifolds, we will use a definition in terms of cohomology. Although concise, this definition is hard to motivate, and we apologize for this. Given a smooth n-dimensional manifold M, we define its *Euler characteristic* $\chi(M)$, as

$$\chi(M) = \sum_{i=0}^{n} (-1)^i \dim(H_{DR}^i(M)).$$

The integers $b_i = \dim(H_{DR}^i(M))$ are known as the *Betti numbers* of M. For example, $\chi(S^2) = 2$.

It turns out that if M is an odd dimensional manifold, then $\chi(M) = 0$. This explains partially why the Euler class is only defined for even dimensional bundles.

The Generalized Gauss-Bonnet theorem (or Gauss-Bonnet-Chern theorem) is a generalization of the Gauss-Bonnet theorem for surfaces. In the general form stated below it was first proved by Allendoerfer and Weil (1943), and Chern (1944).

Theorem 10.42 (Generalized Gauss-Bonnet Formula). *Let M be an orientable, smooth, compact manifold of dimension $2m$. For every metric connection ∇ on TM, (in particular, the Levi-Civita connection for a Riemannian manifold), we have*

$$\int_M \mathrm{eu}(R^\nabla) = \chi(M).$$

A proof of Theorem 10.42 can be found in Madsen and Tornehave [75] (Chapter 21), but beware of some sign problems. The proof uses another famous theorem of differential topology, the *Poincaré-Hopf theorem*. A sketch of the proof is also given in Morita [82], Chapter 5.

Theorem 10.42 is remarkable because it establishes a relationship between the geometry of the manifold (its curvature) and the topology of the manifold (the number of "holes"), somehow encoded in its Euler characteristic.

Characteristic classes are a rich and important topic and we've only scratched the surface. We refer the reader to the texts mentioned earlier in this section as well as to Bott and Tu [12] for comprehensive expositions.

10.14   Problems

Problem 10.1. Complete the proof of Proposition 10.4. In particular show that

$$\nabla = \sum_\alpha f_\alpha \nabla^\alpha$$

is a connection on ξ.

Problem 10.2. Prove Proposition 10.8.

Problem 10.3. Show that the bilinear map

$$\overline{\wedge} \colon \mathcal{A}^r(B) \times \mathcal{A}^s(\eta) \longrightarrow \mathcal{A}^{r+s}(\eta),$$

as defined in Definition 10.11 (with $\xi = B \times \mathbb{R}$), satisfies the following properties:

$$(\omega \wedge \tau) \overline{\wedge} \theta = \omega \overline{\wedge} (\tau \overline{\wedge} \theta) \tag{10.3}$$

$$1 \overline{\wedge} \theta = \theta,$$

for all $\omega \in \mathcal{A}^i(B)$, $\tau \in \mathcal{A}^j(B)$ with $i + j = r$, $\theta \in \mathcal{A}^s(\xi)$, and where 1 denotes the constant function in $C^\infty(B)$ with value 1.

Problem 10.4. Complete the partition of unity argument for Proposition 10.17.

Problem 10.5. Show that the pairing

$$(-, -) \colon \mathcal{A}^i(\xi) \otimes \mathcal{A}^j(\xi^*) \xrightarrow{\overline{\wedge}} \mathcal{A}^{i+j}(\xi \otimes \xi^*) \longrightarrow \mathcal{A}^{i+j}(B)$$

given by

$$(\omega \otimes s_1, \eta \otimes s_2^*) = (\omega \wedge \eta) \otimes \langle\langle s_1, s_2^* \rangle\rangle,$$

where $\omega \in \mathcal{A}^i(B)$, $\eta \in \mathcal{A}^j(B)$, $s_1 \in \Gamma(\xi)$, and $s_2^* \in \Gamma(\xi^*)$, is nondegenerate.

Problem 10.6. Let ξ is a complex vector bundle with connection matrix ω. Consider the bundle $\overline{\xi}^*$, which is the bundle whose fiber over $b \in B$ consists of all conjugate-linear forms over $\pi^{-1}(b)$.

(i) Show that the evaluation pairing $\langle\langle s, \theta\rangle\rangle$ is conjugate-linear in s.
(ii) Show that $\omega^* = -\overline{\omega}^\top$, where ω^* is the connection matrix of $\overline{\xi}^*$ over U.

Problem 10.7. Let ξ is a complex vector bundle. Consider the *conjugate bundle* $\overline{\xi}$, which is obtained from ξ by redefining the vector space structure on each fiber $\pi^{-1}(b)$, with $b \in B$, so that

$$(x + iy)v = (x - iy)v,$$

for every $v \in \pi^{-1}(b)$. If ω is the connection matrix of ξ over U, prove that $\overline{\omega}$ is the connection matrix of $\overline{\xi}$ over U.

Problem 10.8. Let θ_X is the one-form given by

$$\theta_X = i_X g;$$

that is, $(i_X g)(Y) = g(X, Y)$ for all $X, Y \in \mathfrak{X}(M)$. If $L_X g$ is the Lie derivative of the symmetric $(0, 2)$ tensor g, defined so that

$$(L_X g)(Y, Z) = X(g(Y, Z)) - g(L_X Y, Z) - g(Y, L_X Z),$$

show that the Levi-Civita connection is defined implicitly by the formula

$$2g(\nabla_X Y, Z) = (L_Y g)(X, Z) + (d\theta_Y)(X, Z).$$

Hint. See Petersen [86], Chapter 2, Theorem 1.

Problem 10.9. Investigate the following sources, namely Madsen and Tornehave [75] (Lemma 18.2), Morita [82] (Proposition 5.28), and Milnor and Stasheff [78] (Appendix C), and prove Theorem 10.27.

Problem 10.10. Complete the proof sketch of Theorem 10.29.
Hint. See Milnor and Stasheff [78], Appendix C, Lemma 6, Madsen and Tornehave [75], Appendix B, or Morita [82], Theorem 5.26.

Problem 10.11. Let X be an $n \times n$ matrix. Recall that $s_i(X) = \mathrm{tr}(X^i)$. Prove Newton's formula for symmetric polynomial namely:

$$s_i(X) - \sigma_1(X)s_{i-1}(X) + \sigma_2(X)s_{i-2}(X) + \cdots + (-1)^{i-1}\sigma_{i-1}(X)s_1(X) + (-1)^i i\sigma_i(X) = 0,$$

with $1 \leq i \leq n$.

See Madsen and Tornehave [75], Appendix B, or Morita [82], Exercise 5.7.

Problem 10.12. Prove Proposition 10.31.

Problem 10.13. Complete the proof details of Proposition 10.33.

Problem 10.14.

(i) Show that $\mathrm{Pf}(A)$ is independent of the basis chosen.
(ii) If A is the block diagonal matrix D, a show that

$$\omega(D) = -(a_1 e_1 \wedge e_2 + a_2 e_3 \wedge e_4 + \cdots + a_n e_{2n-1} \wedge e_{2n})$$

and that

$$\bigwedge^n \omega(D) = (-1)^n n! \, a_1 \cdots a_n \, e_1 \wedge e_2 \wedge \cdots \wedge e_{2n},$$

and so

$$\mathrm{Pf}(D) = (-1)^n a_1 \cdots a_n.$$

Problem 10.15. Use Definition 10.25 to that $\mathrm{Pf}(A)$ is the unique polynomial with integer coefficients such that $\mathrm{Pf}(A)^2 = \det(A)$ and $\mathrm{Pf}(\mathrm{diag}(S, \ldots, S)) = +1$, where

$$S = \begin{pmatrix} 0 & 1 \\ -1 & 0 \end{pmatrix}.$$

Hint. See Milnor and Stasheff [78], Appendix C, Lemma 9.

Problem 10.16. Show that $\mathrm{Pf}(A)$ is explicitly calculated as

$$\mathrm{Pf}(A) = \frac{1}{2^n n!} \sum_{\sigma \in \mathfrak{S}_{2n}} \mathrm{sgn}(\sigma) \prod_{i=1}^{n} a_{\sigma(2i-1)\,\sigma(2i)}.$$

Problem 10.17. Prove Propositions 10.37 and 10.38.
Hint. See Madsen and Tornehave [75], Chapter 19, or Milnor and Stasheff [78], Appendix C.

Chapter 11
Clifford Algebras, Clifford Groups, and the Groups Pin(*n*) and Spin(*n*)

11.1 Introduction: Rotations as Group Actions

The main goal of this chapter is to explain how rotations in $\mathbb{R}^n$ are induced by the action of a certain group $\mathbf{Spin}(n)$ on $\mathbb{R}^n$, in a way that generalizes the action of the unit complex numbers $\mathbf{U}(1)$ on $\mathbb{R}^2$, and the action of the unit quaternions $\mathbf{SU}(2)$ on $\mathbb{R}^3$ (*i.e.*, the action is defined in terms of multiplication in a larger algebra containing both the group $\mathbf{Spin}(n)$ and $\mathbb{R}^n$). The group $\mathbf{Spin}(n)$, called a *spinor group*, is defined as a certain subgroup of units of an algebra Cl_n, the *Clifford algebra* associated with $\mathbb{R}^n$. Furthermore, for $n \geq 3$, we are lucky, because the group $\mathbf{Spin}(n)$ is topologically simpler than the group $\mathbf{SO}(n)$. Indeed, for $n \geq 3$, the group $\mathbf{Spin}(n)$ is simply connected (a fact that it is not so easy to prove without some machinery), whereas $\mathbf{SO}(n)$ is not simply connected. Intuitively speaking, $\mathbf{SO}(n)$ is more twisted than $\mathbf{Spin}(n)$. In fact, we will see that $\mathbf{Spin}(n)$ is a double cover of $\mathbf{SO}(n)$.

Since the spinor groups are certain well-chosen subgroups of units of Clifford algebras, it is necessary to investigate Clifford algebras to get a firm understanding of spinor groups. This chapter provides a tutorial on Clifford algebra and the groups **Spin** and **Pin**, including a study of the structure of the Clifford algebra $\mathrm{Cl}_{p,q}$ associated with a nondegenerate symmetric bilinear form of signature (p, q) and culminating in the beautiful "8-periodicity theorem" of Élie Cartan and Raoul Bott (with proofs). We also explain when $\mathbf{Spin}(p, q)$ is a double-cover of $\mathbf{SO}(p, q)$. The reader should be warned that a certain amount of algebraic (and topological) background is expected. This being said, perseverant readers will be rewarded by being exposed to some beautiful and nontrivial concepts and results, including Élie Cartan and Raoul Bott "8-periodicity theorem."

Going back to rotations as transformations induced by group actions, recall that if V is a vector space, a *linear action (on the left) of a group G on V* is a map $\alpha : G \times V \to V$ satisfying the following conditions, where, for simplicity of notation, we denote $\alpha(g, v)$ by $g \cdot v$:

© Springer Nature Switzerland AG 2020
J. Gallier, J. Quaintance, *Differential Geometry and Lie Groups*, Geometry and Computing 13, https://doi.org/10.1007/978-3-030-46047-1_11

(1) $g \cdot (h \cdot v) = (gh) \cdot v$, for all $g, h \in G$ and $v \in V$;
(2) $1 \cdot v = v$, for all $v \in V$, where 1 is the identity of the group G;
(3) The map $v \mapsto g \cdot v$ is a linear isomorphism of V for every $g \in G$.

For example, the (multiplicative) group $\mathbf{U}(1)$ of unit complex numbers acts on $\mathbb{R}^2$ (by identifying $\mathbb{R}^2$ and $\mathbb{C}$) *via* complex multiplication: For every $z = a + ib$ (with $a^2 + b^2 = 1$), for every $(x, y) \in \mathbb{R}^2$ (viewing (x, y) as the complex number $x + iy$),

$$ z \cdot (x, y) = (ax - by, ay + bx). $$

Now every unit complex number is of the form $\cos\theta + i\sin\theta$, and thus the above action of $z = \cos\theta + i\sin\theta$ on $\mathbb{R}^2$ corresponds to the rotation of angle θ around the origin. In the case $n = 2$, the groups $\mathbf{U}(1)$ and $\mathbf{SO}(2)$ are isomorphic, but this is an exception.

To represent rotations in $\mathbb{R}^3$ and $\mathbb{R}^4$, we need the quaternions. For our purposes, it is convenient to define the quaternions as certain 2×2 complex matrices. Let $\mathbf{1}, \mathbf{i}, \mathbf{j}, \mathbf{k}$ be the matrices

$$ \mathbf{1} = \begin{pmatrix} 1 & 0 \\ 0 & 1 \end{pmatrix}, \qquad \mathbf{i} = \begin{pmatrix} i & 0 \\ 0 & -i \end{pmatrix}, \qquad \mathbf{j} = \begin{pmatrix} 0 & 1 \\ -1 & 0 \end{pmatrix}, \qquad \mathbf{k} = \begin{pmatrix} 0 & i \\ i & 0 \end{pmatrix}, $$

and let $\mathbb{H}$ be the set of all matrices of the form

$$ X = a\mathbf{1} + b\mathbf{i} + c\mathbf{j} + d\mathbf{k}, \quad a, b, c, d \in \mathbb{R}. $$

Thus, every matrix in $\mathbb{H}$ is of the form

$$ X = \begin{pmatrix} a + ib & c + id \\ -(c - id) & a - ib \end{pmatrix}, \quad a, b, c, d \in \mathbb{R}. $$

The quaternions $\mathbf{1}, \mathbf{i}, \mathbf{j}, \mathbf{k}$ satisfy the famous identities discovered by Hamilton:

$$ \mathbf{i}^2 = \mathbf{j}^2 = \mathbf{k}^2 = \mathbf{ijk} = -\mathbf{1}, $$

$$ \mathbf{ij} = -\mathbf{ji} = \mathbf{k}, $$

$$ \mathbf{jk} = -\mathbf{kj} = \mathbf{i}, $$

$$ \mathbf{ki} = -\mathbf{ik} = \mathbf{j}. $$

As a consequence, it can be verified that $\mathbb{H}$ is a skew field (a noncommutative field) called the *quaternions*. It is also a real vector space of dimension 4 with basis $(\mathbf{1}, \mathbf{i}, \mathbf{j}, \mathbf{k})$; thus as a vector space, $\mathbb{H}$ is isomorphic to $\mathbb{R}^4$. The *unit quaternions* are the quaternions such that

$$\det(X) = a^2 + b^2 + c^2 + d^2 = 1.$$

Given any quaternion $X = a\mathbf{1} + b\mathbf{i} + c\mathbf{j} + d\mathbf{k}$, the *conjugate* $\overline{X}$ of X is given by

$$\overline{X} = a\mathbf{1} - b\mathbf{i} - c\mathbf{j} - d\mathbf{k}.$$

It is easy to check that the matrices associated with the unit quaternions are exactly the matrices in $\mathbf{SU}(2)$. Thus, we call $\mathbf{SU}(2)$ the group of unit quaternions.

Now we can define an action of the group of unit quaternions $\mathbf{SU}(2)$ on $\mathbb{R}^3$. For this, we use the fact that $\mathbb{R}^3$ can be identified with the pure quaternions in $\mathbb{H}$, namely, the quaternions of the form $x_1\mathbf{i} + x_2\mathbf{j} + x_3\mathbf{k}$, where $(x_1, x_2, x_3) \in \mathbb{R}^3$. Then we define the action of $\mathbf{SU}(2)$ over $\mathbb{R}^3$ by

$$Z \cdot X = ZXZ^{-1} = ZX\overline{Z},$$

where $Z \in \mathbf{SU}(2)$ and X is any pure quaternion. Now it turns out that the map ρ_Z (where $\rho_Z(X) = ZX\overline{Z}$) is indeed a rotation, and that the map $\rho\colon Z \mapsto \rho_Z$ is a surjective homomorphism $\rho\colon \mathbf{SU}(2) \to \mathbf{SO}(3)$ whose kernel is $\{-\mathbf{1}, \mathbf{1}\}$, where $\mathbf{1}$ denotes the multiplicative unit quaternion. (For details, see Gallier [46], Chapter 8).

We can also define an action of the group $\mathbf{SU}(2) \times \mathbf{SU}(2)$ over $\mathbb{R}^4$, by identifying $\mathbb{R}^4$ with the quaternions. In this case, the action is

$$(Y, Z) \cdot X = YX\overline{Z},$$

where $(Y, Z) \in \mathbf{SU}(2) \times \mathbf{SU}(2)$ and $X \in \mathbb{H}$ is any quaternion. Then the map $\rho_{Y,\overline{Z}}$ is a rotation (where $\rho_{Y,\overline{Z}}(X) = YX\overline{Z}$), and the map $\rho\colon (Y, Z) \mapsto \rho_{Y,\overline{Z}}$ is a surjective homomorphism $\rho\colon \mathbf{SU}(2) \times \mathbf{SU}(2) \to \mathbf{SO}(4)$ whose kernel is $\{(\mathbf{1}, \mathbf{1}), (-\mathbf{1}, -\mathbf{1})\}$. (For details, see Gallier [46], Chapter 8.)

Thus, we observe that for $n = 2, 3, 4$, the rotations in $\mathbf{SO}(n)$ can be realized *via* the linear action of some group (the case $n = 1$ is trivial, since $\mathbf{SO}(1) = \{1, -1\}$). It is also the case that the action of each group can be somehow be described in terms of multiplication in some larger algebra "containing" the original vector space $\mathbb{R}^n$ ($\mathbb{C}$ for $n = 2$, $\mathbb{H}$ for $n = 3, 4$). However, these groups appear to have been discovered in an ad hoc fashion, and there does not appear to be any universal way to define the action of these groups on $\mathbb{R}^n$. It would certainly be nice if the action was always of the form

$$Z \cdot X = ZXZ^{-1} (= ZX\overline{Z}).$$

A systematic way of constructing groups realizing rotations in terms of linear action, using a uniform notion of action, does exist. Such groups are the spin groups.

We just observed that the rotations in $\mathbf{SO}(3)$ can be realized by the linear action of the group of unit quaternions $\mathbf{SU}(2)$ on $\mathbb{R}^3$, and how the rotations in $\mathbf{SO}(4)$ can be realized by the linear action of the group $\mathbf{SU}(2) \times \mathbf{SU}(2)$ on $\mathbb{R}^4$.

The main reasons why the rotations in $\mathbf{SO}(3)$ can be represented by unit quaternions are the following:

(1) For every nonzero vector $u \in \mathbb{R}^3$, the reflection s_u about the hyperplane perpendicular to u is represented by the map

$$v \mapsto -uvu^{-1},$$

where u and v are viewed as pure quaternions in $\mathbb{H}$ (*i.e.*, if $u = (u_1, u_2, u_2)$, then view u as $u_1\mathbf{i} + u_2\mathbf{j} + u_3\mathbf{k}$, and similarly for v).
(2) The group $\mathbf{SO}(3)$ is generated by the reflections.

As one can imagine, a successful generalization of the quaternions, *i.e.*, the discovery of a group G inducing the rotations in $\mathbf{SO}(n)$ *via* a linear action, depends on the ability to generalize Properties (1) and (2) above. Fortunately, it is true that the group $\mathbf{SO}(n)$ is generated by the hyperplane reflections. In fact, this is also true for the orthogonal group $\mathbf{O}(n)$, and more generally for the group of isometries $\mathbf{O}(\Phi)$ of any nondegenerate quadratic form Φ, by the *Cartan-Dieudonné theorem* (for instance, see Bourbaki [13], or Gallier [46], Chapter 7, Theorem 7.2.1).

In order to generalize (1), we need to understand how the group G acts on $\mathbb{R}^n$. The case $n = 3$ is special, because the underlying space $\mathbb{R}^3$ on which the rotations act can be embedded as the pure quaternions in $\mathbb{H}$. The case $n = 4$ is also special, because $\mathbb{R}^4$ is the underlying space of $\mathbb{H}$. The generalization to $n \geq 5$ requires more machinery, namely, the notions of Clifford groups and Clifford algebras.

As we will see, for every $n \geq 2$, there is a compact, connected (and simply connected when $n \geq 3$) group $\mathbf{Spin}(n)$, the "spinor group," and a surjective homomorphism $\rho\colon \mathbf{Spin}(n) \to \mathbf{SO}(n)$ whose kernel is $\{-1, 1\}$, where $\mathbf{1}$ denotes the multiplicative unit of $\mathbf{Spin}(n)$. This time, $\mathbf{Spin}(n)$ acts directly on $\mathbb{R}^n$, because $\mathbf{Spin}(n)$ is a certain subgroup of the group of units of the *Clifford algebra* Cl_n, and $\mathbb{R}^n$ is naturally a subspace of Cl_n.

The group of unit quaternions $\mathbf{SU}(2)$ turns out to be isomorphic to the spinor group $\mathbf{Spin}(3)$. Because $\mathbf{Spin}(3)$ acts directly on $\mathbb{R}^3$, the representation of rotations in $\mathbf{SO}(3)$ by elements of $\mathbf{Spin}(3)$ may be viewed as more natural than the representation by unit quaternions. The group $\mathbf{SU}(2) \times \mathbf{SU}(2)$ turns out to be isomorphic to the spinor group $\mathbf{Spin}(4)$, but this isomorphism is less obvious.

In summary, we are going to define a group $\mathbf{Spin}(n)$ representing the rotations in $\mathbf{SO}(n)$, for any $n \geq 1$, in the sense that there is a linear action of $\mathbf{Spin}(n)$ on $\mathbb{R}^n$ which induces a surjective homomorphism $\rho\colon \mathbf{Spin}(n) \to \mathbf{SO}(n)$ whose kernel is $\{-1, 1\}$. Furthermore, the action of $\mathbf{Spin}(n)$ on $\mathbb{R}^n$ is given in terms of multiplication in an algebra Cl_n containing $\mathbf{Spin}(n)$, and in which $\mathbb{R}^n$ is also embedded.

It turns out that as a bonus, for $n \geq 3$, the group $\mathbf{Spin}(n)$ is topologically simpler than $\mathbf{SO}(n)$, since $\mathbf{Spin}(n)$ is simply connected, but $\mathbf{SO}(n)$ is not. By being astute, we can also construct a group $\mathbf{Pin}(n)$ and a linear action of $\mathbf{Pin}(n)$ on $\mathbb{R}^n$ that induces a surjective homomorphism $\rho\colon \mathbf{Pin}(n) \to \mathbf{O}(n)$ whose kernel is $\{-1, 1\}$. The difficulty here is the presence of the negative sign in (1). We will see how

Atiyah, Bott, and Shapiro circumvent this problem by using a "twisted adjoint action," as opposed to the usual adjoint action (where $v \mapsto uvu^{-1}$).

Let us now outline in more detail the contents of this chapter. The first step for generalizing the quaternions is to define the notion of a Clifford algebra. Let K be any field of characteristic different from 2. Let V be a finite-dimensional vector space over a field K of characteristic $\neq 2$, let $\varphi\colon V \times V \to K$ be a possibly degenerate symmetric bilinear form, and let $\Phi(v) = \varphi(v, v)$ be the corresponding quadratic form. Roughly speaking, a *Clifford algebra associated with V and Φ* is a K-algebra $\mathrm{Cl}(V, \Phi)$ satisfying the condition

$$v^2 = v \cdot v = \Phi(v) \cdot \mathbf{1} \quad \text{for all } v \in V,$$

where $\mathbf{1}$ is the multiplicative unit of the algebra. In all rigor, V is not contained in $\mathrm{Cl}(V, \Phi)$ but there is an injection of V into $\mathrm{Cl}(V, \Phi)$.

The algebra $\mathrm{Cl}(V, \Phi)$ is the quotient $T(V)/\mathfrak{A}$ of the tensor algebra $T(V)$ over V modulo the ideal $\mathfrak{A}$ of $T(V)$ generated by all elements of the form $v \otimes v - \Phi(v) \cdot \mathbf{1}$, where $v \in V$.

If V is finite dimensional and if $(e_1, \ldots, e_n)$ is a basis of V, then $\mathrm{Cl}(V, \Phi)$ has a basis consisting of the $2^n - 1$ products

$$e_{i_1} e_{i_2} \cdots e_{i_k}, \quad 1 \leq i_1 < i_2 < \ldots < i_k \leq n,$$

and $\mathbf{1}$. Thus $\mathrm{Cl}(V, \Phi)$, also denoted by $\mathrm{Cl}(\Phi)$, has dimension 2^n. If $(e_1, \ldots, e_n)$ is an orthogonal basis of V with respect to Φ, then we can view $\mathrm{Cl}(\Phi)$ as the algebra presented by the generators $(e_1, \ldots, e_n)$ and the relations

$$e_j^2 = \Phi(e_j) \cdot \mathbf{1}, \quad 1 \leq j \leq n, \quad \text{and}$$

$$e_j e_k = -e_k e_j, \quad 1 \leq j, k \leq n, \ j \neq k.$$

If V has finite dimension n and $(e_1, \ldots, e_n)$ is a basis of V, we can define two maps t and α as follows. The map t is defined on basis elements by

$$t(e_i) = e_i$$

$$t(e_{i_1} e_{i_2} \cdots e_{i_k}) = e_{i_k} e_{i_{k-1}} \cdots e_{i_1},$$

where $1 \leq i_1 < i_2 \cdots < i_k \leq n$, and of course, $t(\mathbf{1}) = \mathbf{1}$. The map α is defined on basis elements by

$$\alpha(e_i) = -e_i$$

$$\alpha(e_{i_1} e_{i_2} \cdots e_{i_k}) = (-1)^k e_{i_1} e_{i_2} \cdots e_{i_k}$$

where $1 \leq i_1 < i_2 < \cdots < i_k \leq n$, and of course, $\alpha(\mathbf{1}) = \mathbf{1}$. The even-graded elements (the elements of $\mathrm{Cl}^0(\Phi)$) are those generated by $\mathbf{1}$ and the basis elements consisting of an even number of factors $e_{i_1} e_{i_2} \cdots e_{i_{2k}}$, and the odd-graded elements (the elements of $\mathrm{Cl}^1(\Phi)$) are those generated by the basis elements consisting of an odd number of factors $e_{i_1} e_{i_2} \cdots e_{i_{2k+1}}$.

The second step is to define the Clifford group, which is a subgroup of the group $\mathrm{Cl}(\Phi)^*$ of units of $\mathrm{Cl}(\Phi)$.

The *Clifford group of* Φ is the group

$$\Gamma(\Phi) = \{x \in \mathrm{Cl}(\Phi)^* \mid \alpha(x)vx^{-1} \in V \quad \text{for all } v \in V\}.$$

For any $x \in \Gamma(\Phi)$, let $\rho_x \colon V \to V$ be the map defined by

$$v \mapsto \alpha(x)vx^{-1}, \quad v \in V.$$

The map $\rho \colon \Gamma(\Phi) \to \mathbf{GL}(V)$ given by $x \mapsto \rho_x$ is a linear action called the *twisted adjoint representation*. It was introduced by Atiyah, Bott, and Shapiro [6] and has technical advantages over the earlier adjoint representation ρ^0 given by

$$v \mapsto xvx^{-1}.$$

The group $\Gamma^+(\Phi)$, called the *special Clifford group*, is defined by

$$\Gamma^+(\Phi) = \Gamma(\Phi) \cap \mathrm{Cl}^0(\Phi).$$

The key property of the Clifford groups is that if the bilinear form φ is nondegenerate, then the map $\rho \colon \Gamma(\Phi) \to \mathbf{GL}(V)$ is actually a surjection $\rho \colon \Gamma(\Phi) \to \mathbf{O}(\Phi)$ onto the orthogonal group $\mathbf{O}(\Phi)$ associated with the quadratic form Φ, and the map $\rho \colon \Gamma^+(\Phi) \to \mathbf{SO}(\Phi)$ is a surjection onto the special orthogonal group $\mathbf{SO}(\Phi)$ associated with the quadratic form Φ. In both cases, the kernel of ρ is $K^* \cdot 1$.

In order to cut down on the size of the kernel of ρ, we need to define groups smaller than $\Gamma(\Phi)$ and $\Gamma^+(\Phi)$. To do so, we introduce a notion of norm on $\mathrm{Cl}(V, \Phi)$. If φ is nondegenerate, then the restriction of the norm N to $\Gamma(\Phi)$ is a map $N \colon \Gamma(\Phi) \to K^* \cdot 1$.

We can now define the groups **Pin** and **Spin** as follows.

Assume φ is a nondegenerate bilinear map on V. We define the *pinor group* **Pin**(Φ) as the group

$$\mathbf{Pin}(\Phi) = \{x \in \Gamma(\Phi) \mid N(x) = \pm 1\},$$

and the *spinor group* **Spin**(Φ) as **Pin**(Φ) $\cap \Gamma^+(\Phi)$.

If the field K is not $\mathbb{R}$ or $\mathbb{C}$, it is not obvious that the restriction of ρ to **Pin**(Φ) is surjective onto $\mathbf{O}(\Phi)$, and that the restriction of ρ to **Spin**(Φ) is surjective onto $\mathbf{SO}(\Phi)$. These maps are surjective if $K = \mathbb{R}$ and $K = \mathbb{C}$, but in general it is not surjective. In all cases the kernel of ρ is equal to $\{-1, 1\}$. When $\Phi(x_1, \ldots, x_n) = -(x_1^2 + \cdots + x_n^2)$, the group **Spin**($\Phi$), denoted **Spin**($n$), is exactly the generalization of the unit quaternions (and when $n = 3$, **Spin**(n) $\cong \mathbf{SU}(2)$, the unit quaternions).

Some preliminaries on algebras and tensor algebras are reviewed in Section 11.2. In Section 11.3, we define Clifford algebras over the field $K = \mathbb{R}$. The Clifford groups (over $K = \mathbb{R}$) are defined in Section 11.4. In the second half of this section we restrict our attention to the real quadratic form $\Phi(x_1, \ldots, x_n) = -(x_1^2 + \cdots + x_n^2)$. The corresponding Clifford algebras are denoted Cl_n and the corresponding Clifford groups as Γ_n.

In Section 11.5, we define the groups **Pin**(n) and **Spin**(n) associated with the real quadratic form $\Phi(x_1, \ldots, x_n) = -(x_1^2 + \cdots + x_n^2)$. We prove that the maps $\rho\colon \textbf{Pin}(n) \to \textbf{O}(n)$ and $\rho\colon \textbf{Spin}(n) \to \textbf{SO}(n)$ are surjective with kernel $\{-1, 1\}$. We determine the groups **Spin**(n) for $n = 2, 3, 4$.

Section 11.6 is devoted to the Spin and Pin groups associated with the real nondegenerate quadratic form

$$\Phi(x_1, \ldots, x_{p+q}) = x_1^2 + \cdots + x_p^2 - (x_{p+1}^2 + \cdots + x_{p+q}^2).$$

We obtain Clifford algebras $\mathrm{Cl}_{p,q}$, Clifford groups $\Gamma_{p,q}$, and groups **Pin**(p, q) and **Spin**(p, q). We show that the maps $\rho\colon \textbf{Pin}(p, q) \to \textbf{O}(p, q)$ and $\rho\colon \textbf{Spin}(p, q) \to \textbf{SO}(p, q)$ are surjective with kernel $\{-1, 1\}$.

In Section 11.7 we show that the Lie groups **Pin**(p, q) and **Spin**(p, q) are double covers of $\textbf{O}(p, q)$ and $\textbf{SO}(p, q)$.

In Section 11.8 we prove an amazing result due to Élie Cartan and Raoul Bott, namely the 8-periodicity of the Clifford algebras $\mathrm{Cl}_{p,q}$. This result says that: for all $n \geq 0$, we have the following isomorphisms:

$$\mathrm{Cl}_{0,n+8} \cong \mathrm{Cl}_{0,n} \otimes \mathrm{Cl}_{0,8}$$

$$\mathrm{Cl}_{n+8,0} \cong \mathrm{Cl}_{n,0} \otimes \mathrm{Cl}_{8,0}.$$

Furthermore,

$$\mathrm{Cl}_{0,8} = \mathrm{Cl}_{8,0} = \mathbb{R}(16),$$

the real algebra of 16×16 matrices.

Section 11.9 is devoted to the complex Clifford algebras $\mathrm{Cl}(n, \mathbb{C})$. In this case, we have a 2-periodicity,

$$\mathrm{Cl}(n + 2, \mathbb{C}) \cong \mathrm{Cl}(n, \mathbb{C}) \otimes_{\mathbb{C}} \mathrm{Cl}(2, \mathbb{C}),$$

with $\mathrm{Cl}(2, \mathbb{C}) = \mathbb{C}(2)$, the complex algebra of 2×2 matrices.

Finally, in the last section, Section 11.10, we outline the theory of Clifford groups and of the Pin and Spin groups over any field K of characteristic $\neq 2$.

Our presentation is heavily influenced by Bröcker and tom Dieck [18] (Chapter 1, Section 6), where most details can be found. This chapter is almost entirely taken from the first 11 pages of the beautiful and seminal paper by Atiyah, Bott, and Shapiro [6], Clifford Modules, and we highly recommend it. Another excellent (but concise) exposition can be found in Kirillov [63]. A very thorough exposition can be found in two places:

1. Lawson and Michelsohn [71], where the material on **Pin**(p, q) and **Spin**(p, q) can be found in Chapter I.
2. Lounesto's excellent book [74].

One may also want to consult Baker [8], Curtis [27], Porteous [87], Fulton and Harris (Lecture 20) [45], Choquet-Bruhat [25], Bourbaki [13], and Chevalley [24], a classic. The original source is Elie Cartan's book (1937) whose translation in English appears in [19].

11.2   Preliminaries

We begin by recalling what is an algebra over a field. Let K denote any (commutative) field, although for our purposes we may assume that $K = \mathbb{R}$ (and occasionally, $K = \mathbb{C}$). Since we will only be dealing with associative algebras with a multiplicative unit, we only define algebras of this kind.

Definition 11.1. Given a field K, a *K-algebra* is a K-vector space A together with a bilinear operation $\star\colon A \times A \to A$, called *multiplication*, which makes A into a ring with unity $\mathbf{1}$ (or $\mathbf{1}_A$, when we want to be very precise). This means that $\star$ is associative and that there is a multiplicative identity element $\mathbf{1}$ so that $\mathbf{1}\star a = a\star\mathbf{1} = a$, for all $a \in A$. Given two K-algebras A and B, a *K-algebra homomorphism* $h\colon A \to B$ is a linear map that is also a ring homomorphism, with $h(\mathbf{1}_A) = \mathbf{1}_B$.

For example, the ring $\mathrm{M}_n(K)$ of all $n \times n$ matrices over a field K is a K-algebra with multiplicative identity element $\mathbf{1} = I_n$.

There is an obvious notion of *ideal* of a K-algebra:

Definition 11.2. An ideal $\mathfrak{A} \subseteq A$ is a linear subspace of a K-algebra A that is also a two-sided ideal with respect to multiplication in A.

If the field K is understood, we usually simply say an algebra instead of a K-algebra.

We will also need tensor products. A rather detailed exposition of tensor products is given in Chapter 2 and the reader may want to review Section 2.2. For the reader's convenience, we recall the definition of the tensor product of vector spaces. The basic idea is that tensor products allow us to view multilinear maps as linear maps. The maps become simpler, but the spaces (product spaces) become more complicated (tensor products). For more details, see Section 2.2 or Atiyah and Macdonald [5].

Definition 11.3. Given two K-vector spaces E and F, a *tensor product of E and F* is a pair $(E \otimes F, \otimes)$, where $E \otimes F$ is a K-vector space and $\otimes\colon E \times F \to E \otimes F$ is a bilinear map, so that for every K-vector space G and every bilinear map $f\colon E \times F \to G$, there is a unique linear map $f_\otimes\colon E \otimes F \to G$ with

$$f(u, v) = f_\otimes(u \otimes v) \quad \text{for all } u \in E \text{ and all } v \in V,$$

as in the diagram below.

The vector space $E \otimes F$ is defined up to isomorphism. The vectors $u \otimes v$, where $u \in E$ and $v \in F$, generate $E \otimes F$.

Remark. We should really denote the tensor product of E and F by $E \otimes_K F$, since it depends on the field K. Since we usually deal with a fixed field K, we use the simpler notation $E \otimes F$.

As shown in Section 2.4, we have natural isomorphisms

$$(E \otimes F) \otimes G \cong E \otimes (F \otimes G) \quad \text{and} \quad E \otimes F \cong F \otimes E.$$

Given two linear maps $f: E \to F$ and $g: E' \to F'$, we have a unique bilinear map $f \times g: E \times E' \to F \times F'$ so that

$$(f \times g)(a, a') = (f(a), g(a')) \quad \text{for all } a \in E \text{ and all } a' \in E'.$$

Thus, we have the bilinear map $\otimes \circ (f \times g): E \times E' \to F \otimes F'$, and so, there is a unique linear map $f \otimes g: E \otimes E' \to F \otimes F'$ so that

$$(f \otimes g)(a \otimes a') = f(a) \otimes g(a') \quad \text{for all } a \in E \text{ and all } a' \in E'.$$

Let us now assume that E and F are K-algebras. We want to make $E \otimes F$ into a K-algebra. Since the multiplication operations $m_E: E \times E \to E$ and $m_F: F \times F \to F$ are bilinear, we get linear maps $m'_E: E \otimes E \to E$ and $m'_F: F \otimes F \to F$, and thus the linear map

$$m'_E \otimes m'_F: (E \otimes E) \otimes (F \otimes F) \to E \otimes F.$$

Using the isomorphism $\tau: (E \otimes E) \otimes (F \otimes F) \to (E \otimes F) \otimes (E \otimes F)$, we get a linear map

$$m_{E \otimes F}: (E \otimes F) \otimes (E \otimes F) \to E \otimes F,$$

which defines a multiplication m on $E \otimes F$ (namely, $m(\alpha, \beta) = m_{E \otimes F}(\alpha \otimes \beta)$ for all $\alpha, \beta \in E \otimes F$). It is easily checked that $E \otimes F$ is indeed a K-algebra under the multiplication m. Using the simpler notation $\cdot$ for m, we have

$$(a \otimes a') \cdot (b \otimes b') = (ab) \otimes (a'b') \qquad (*)$$

for all $a, b \in E$ and all $a', b' \in F$.

Given any vector space V over a field K, there is a special K-algebra $T(V)$ together with a linear map $i: V \to T(V)$, with the following universal mapping property: Given any K-algebra A, for any linear map $f: V \to A$, there is a unique K-algebra homomorphism $\overline{f}: T(V) \to A$ so that

$$f = \overline{f} \circ i,$$

as in the diagram below.

The algebra $T(V)$ is the *tensor algebra of* V; see Section 2.6. The algebra $T(V)$ may be constructed as the direct sum

$$T(V) = \bigoplus_{i \geq 0} V^{\otimes i},$$

where $V^0 = K$, and $V^{\otimes i}$ is the i-fold tensor product of V with itself. For every $i \geq 0$, there is a natural injection $\iota_n: V^{\otimes n} \to T(V)$, and in particular, an injection $\iota_0: K \to T(V)$. The multiplicative unit **1** of $T(V)$ is the image $\iota_0(1)$ in $T(V)$ of the unit 1 of the field K. Since every $v \in T(V)$ can be expressed as a finite sum

$$v = v_1 + \cdots + v_k,$$

where $v_i \in V^{\otimes n_i}$ and the n_i are natural numbers with $n_i \neq n_j$ if $i \neq j$, to define multiplication in $T(V)$, using bilinearity, it is enough to define the multiplication $V^{\otimes m} \times V^{\otimes n} \longrightarrow V^{\otimes(m+n)}$. Of course, this is defined by

$$(v_1 \otimes \cdots \otimes v_m) \cdot (w_1 \otimes \cdots \otimes w_n) = v_1 \otimes \cdots \otimes v_m \otimes w_1 \otimes \cdots \otimes w_n.$$

(This has to be made rigorous by using isomorphisms involving the associativity of tensor products; for details, see Jacobson [61].) The algebra $T(V)$ is an example of a *graded algebra*, where the *homogeneous elements of rank n* are the elements in $V^{\otimes n}$.

Remark. It is important to note that multiplication in $T(V)$ is **not** commutative. Also, in all rigor, the unit **1** of $T(V)$ is **not equal** to 1, the unit of the field K. The field K is embedded in $T(V)$ using the mapping $\lambda \mapsto \lambda \mathbf{1}$. More generally, in view of the injections $\iota_n : V^{\otimes n} \to T(V)$, we identify elements of $V^{\otimes n}$ with their images in $T(V)$.

Most algebras of interest arise as well-chosen quotients of the tensor algebra $T(V)$. This is true for the *exterior algebra* $\bigwedge^{\bullet} V$ (also called *Grassmann algebra*), where we take the quotient of $T(V)$ modulo the ideal generated by all elements of the form $v \otimes v$, where $v \in V$, see Section 3.4.

From now on, we assume that K is a *field of characteristic different from* 2 Given a symmetric bilinear form $\varphi \colon V \times V \to K$, recall that the *quadratic form* Φ associated with φ is given by $\Phi(v) = \varphi(v, v)$ for all $v \in V$, and that φ can be recovered from Φ by the *polarization identity*

$$\varphi(u, v) = \frac{1}{2}(\Phi(u + v) - \Phi(u) - \Phi(v)).$$

The symmetric bilinear form φ is *nondegenerate* iff for every $u \in V$, if $\varphi(u, v) = 0$ for all $v \in V$, then $u = 0$.

Definition 11.4. Let (V, φ) be a vector space equipped with a nondegenerate symmetric bilinear form φ. The the set of linear maps $f \colon V \to V$ such that

$$\varphi(f(u), f(v)) = \varphi(u, v) \quad \text{for all } u, v \in V$$

forms a group denoted $\mathbf{O}(V, \Phi)$ (or $\mathbf{O}(V, \varphi)$) which is called the group of *isometries* or *orthogonal group* of (V, φ).

The condition

$$\varphi(f(u), f(v)) = \varphi(u, v) \quad \text{for all } u, v \in V$$

is equivalent to the condition

$$\Phi(f(v)) = \Phi(v) \quad \text{for all } v \in V.$$

Definition 11.5. The subgroup of $\mathbf{O}(V, \Phi)$ denoted $\mathbf{SO}(V, \Phi)$ (or $\mathbf{SO}(V, \varphi)$) is defined by

$$\mathbf{SO}(V, \Phi) = \{f \in \mathbf{O}(V, \Phi) \mid \det(f) = 1\}$$

and is called the *special orthogonal group* or *group of rotations* of (V, φ). We often abbreviate $\mathbf{O}(V, \Phi)$ as $\mathbf{O}(\Phi)$ and $\mathbf{SO}(V, \Phi)$ as $\mathbf{SO}(\Phi)$.

Definition 11.6. If $K = \mathbb{R}$ and Φ_n is the Euclidean quadratic form $\Phi_n(x_1, \ldots, x_n) = x_1^2 + \cdots + x_n^2$, we write $\mathbf{O}(n, \mathbb{R})$ or even $\mathbf{O}(n)$ for $\mathbf{O}(\mathbb{R}^n, \Phi_n)$ and $\mathbf{SO}(n, \mathbb{R})$ or even $\mathbf{SO}(n)$ for $\mathbf{SO}(\mathbb{R}^n, \Phi_n)$. Similarly when $K = \mathbb{C}$ and

$\Phi_n(x_1, \ldots, x_n) = x_1^2 + \cdots + x_n^2$, we write $\mathbf{O}(n, \mathbb{C})$ for $\mathbf{O}(\mathbb{C}^n, \Phi_n)$ and $\mathbf{SO}(n, \mathbb{C})$ for $\mathbf{SO}(\mathbb{C}^n, \Phi_n)$.

If $K = \mathbb{R}$ and if $\Phi_{p,q}(x_1, \ldots, x_{p+q}) = x_1^2 + \cdots + x_p^2 - (x_{p+1}^2 + \cdots + x_{p+q}^2)$, with $n = p + q$, we write $\mathbf{O}(p, q)$ for $\mathbf{O}(\mathbb{R}^n, \Phi_{p,q})$ and $\mathbf{SO}(p, q)$ for $\mathbf{SO}(\mathbb{R}^n, \Phi_{p,q})$.

Observe that $\Phi_{n,0} = \Phi_n$. It is not hard to show that $\mathbf{O}(p, q)$ and $\mathbf{O}(q, p)$ are isomorphic, and similarly $\mathbf{SO}(p, q)$ and $\mathbf{SO}(q, p)$ are isomorphic. In the special cases where $p = 0$ or $q = 0$, we have

$$\Phi_{0,n}(x_1, \ldots, x_n) = -(x_1^2 + \cdots + x_n^2) = -\Phi_n(x_1, \ldots, x_n) = -\Phi_{n,0}(x_1, \ldots, x_n),$$

so for any linear map f we have $\Phi_{0,n}(f(x_1, \ldots, x_n)) = \Phi_{0,n}(x_1, \ldots, x_n)$ iff $\Phi_n(f(x_1, \ldots, x_n)) = \Phi_n(x_1, \ldots, x_n)$, which shows that $\mathbf{O}(0, n) = \mathbf{O}(n, 0) = \mathbf{O}(n)$ and $\mathbf{SO}(0, n) = \mathbf{SO}(0, n) = \mathbf{SO}(n)$.

11.3 Clifford Algebras

A Clifford algebra may be viewed as a refinement of the exterior algebra, in which we take the quotient of $T(V)$ modulo the ideal generated by all elements of the form $v \otimes v - \Phi(v) \cdot \mathbf{1}$, where Φ is the quadratic form associated with a symmetric bilinear form $\varphi \colon V \times V \to K$, and $\cdot \colon K \times T(V) \to T(V)$ denotes the scalar product of the algebra $T(V)$. For simplicity, let us assume that we are now dealing with real algebras.

Definition 11.7. Let V be a real finite-dimensional vector space together with a symmetric bilinear form $\varphi \colon V \times V \to \mathbb{R}$ and associated quadratic form $\Phi(v) = \varphi(v, v)$. A *Clifford algebra associated with V and Φ* is a real algebra $\mathrm{Cl}(V, \Phi)$ together with a linear map $i_\Phi \colon V \to \mathrm{Cl}(V, \Phi)$ satisfying the condition $(i_\Phi(v))^2 = \Phi(v) \cdot \mathbf{1}$ for all $v \in V$, and so that for every real algebra A and every linear map $f \colon V \to A$ with

$$(f(v))^2 = \Phi(v) \cdot \mathbf{1}_A \quad \text{for all } v \in V,$$

there is a unique algebra homomorphism $\overline{f} \colon \mathrm{Cl}(V, \Phi) \to A$ so that

$$f = \overline{f} \circ i_\Phi,$$

as in the diagram below.

$$V \xrightarrow{\ i_\Phi\ } \mathrm{Cl}(V, \Phi)$$
$$\searrow f \qquad \downarrow \overline{f}$$
$$A$$

We use the notation $\lambda \cdot u$ for the product of a scalar $\lambda \in \mathbb{R}$ and of an element u in the algebra $\mathrm{Cl}(V, \Phi)$, and juxtaposition uv for the multiplication of two elements u and v in the algebra $\mathrm{Cl}(V, \Phi)$.

By a familiar argument, any two Clifford algebras associated with V and Φ are isomorphic. We often denote i_Φ by i.

To show the existence of $\mathrm{Cl}(V, \Phi)$, observe that $T(V)/\mathfrak{A}$ does the job, where $\mathfrak{A}$ is the ideal of $T(V)$ generated by all elements of the form $v \otimes v - \Phi(v) \cdot \mathbf{1}$, where $v \in V$. The map $i_\Phi \colon V \to \mathrm{Cl}(V, \Phi)$ is the composition

$$V \xrightarrow{\iota_1} T(V) \xrightarrow{\pi} T(V)/\mathfrak{A},$$

where π is the natural quotient map. We often denote the Clifford algebra $\mathrm{Cl}(V, \Phi)$ simply by $\mathrm{Cl}(\Phi)$.

Remark. Observe that Definition 11.7 does not assert that i_Φ is injective or that there is an injection of $\mathbb{R}$ into $\mathrm{Cl}(V, \Phi)$, but we will prove later that both facts are true when V is finite dimensional. Also, as in the case of the tensor algebra, the unit $\mathbf{1}$ of the algebra $\mathrm{Cl}(V, \Phi)$ and the unit 1 of the field $\mathbb{R}$ are **not equal**.

Since

$$\Phi(u + v) - \Phi(u) - \Phi(v) = 2\varphi(u, v)$$

and

$$(i(u + v))^2 = (i(u))^2 + (i(v))^2 + i(u)i(v) + i(v)i(u),$$

using the fact that

$$i(u)^2 = \Phi(u) \cdot \mathbf{1},$$

we get

$$i(u)i(v) + i(v)i(u) = 2\varphi(u, v) \cdot \mathbf{1}. \tag{$*$}$$

As a consequence, if $(u_1, \ldots, u_n)$ is an orthogonal basis w.r.t. φ (which means that $\varphi(u_j, u_k) = 0$ for all $j \neq k$), we have

$$i(u_j)i(u_k) + i(u_k)i(u_j) = 0 \quad \text{for all } j \neq k.$$

Remark. Certain authors drop the unit $\mathbf{1}$ of the Clifford algebra $\mathrm{Cl}(V, \Phi)$ when writing the identities

$$i(u)^2 = \Phi(u) \cdot 1$$

and

$$2\varphi(u, v) \cdot \mathbf{1} = i(u)i(v) + i(v)i(u),$$

where the second identity is often written as

$$\varphi(u, v) = \frac{1}{2}(i(u)i(v) + i(v)i(u)).$$

This is very confusing and technically wrong, because we only have an injection of $\mathbb{R}$ into $\mathrm{Cl}(V, \Phi)$, but $\mathbb{R}$ is **not** a subset of $\mathrm{Cl}(V, \Phi)$.

 We warn the readers that Lawson and Michelsohn [71] adopt the opposite of our sign convention in defining Clifford algebras, *i.e.*, they use the condition

$$(f(v))^2 = -\Phi(v) \cdot \mathbf{1} \quad \text{for all } v \in V.$$

The most confusing consequence of this is that their $\mathrm{Cl}(p, q)$ is our $\mathrm{Cl}(q, p)$.

Observe that when $\Phi \equiv 0$ is the quadratic form identically zero everywhere, then the Clifford algebra $\mathrm{Cl}(V, 0)$ is just the exterior algebra $\bigwedge^{\bullet} V$.

Example 11.1. Let $V = \mathbb{R}$, $e_1 = 1$, and assume that $\Phi(x_1 e_1) = -x_1^2$. Then $\mathrm{Cl}(\Phi)$ is spanned by the basis $(\mathbf{1}, e_1)$. We have

$$e_1^2 = -\mathbf{1}.$$

Under the bijection

$$e_1 \mapsto i,$$

the Clifford algebra $\mathrm{Cl}(\Phi)$, also denoted by Cl_1, is isomorphic to the algebra of complex numbers $\mathbb{C}$.

Example 11.2. Now let $V = \mathbb{R}^2$, (e_1, e_2) be the canonical basis, and assume that $\Phi(x_1 e_1 + x_2 e_2) = -(x_1^2 + x_2^2)$. Then $\mathrm{Cl}(\Phi)$ is spanned by the basis $(\mathbf{1}, e_1, e_2, e_1 e_2)$. Furthermore, we have

$$e_2 e_1 = -e_1 e_2, \quad e_1^2 = -\mathbf{1}, \quad e_2^2 = -\mathbf{1}, \quad (e_1 e_2)^2 = -\mathbf{1}.$$

Under the bijection

$$e_1 \mapsto \mathbf{i}, \quad e_2 \mapsto \mathbf{j}, \quad e_1 e_2 \mapsto \mathbf{k} \quad \mathbf{1} \mapsto \mathbf{1},$$

it is easily checked that the quaternion identities

$$i^2 = j^2 = k^2 = -1,$$
$$ij = -ji = k,$$
$$jk = -kj = i,$$
$$ki = -ik = j,$$

hold, and thus the Clifford algebra $\text{Cl}(\Phi)$, also denoted by Cl_2, is isomorphic to the algebra of quaternions $\mathbb{H}$.

Our prime goal is to define an action of $\text{Cl}(\Phi)$ on V in such a way that by restricting this action to some suitably chosen multiplicative subgroups of $\text{Cl}(\Phi)$, we get surjective homomorphisms onto $\mathbf{O}(\Phi)$ and $\mathbf{SO}(\Phi)$, respectively. The key point is that a reflection in V about a hyperplane H orthogonal to a vector w can be defined by such an action, but some negative sign shows up. A correct handling of signs is a bit subtle and requires the introduction of a canonical anti-automorphism t, and of a canonical automorphism α, defined as follows:

Proposition 11.1. *Every Clifford algebra* $\text{Cl}(\Phi)$ *possesses a canonical anti-automorphism* $t : \text{Cl}(\Phi) \to \text{Cl}(\Phi)$ *satisfying the properties*

$$t(xy) = t(y)t(x), \quad t \circ t = id, \quad and \quad t(i(v)) = i(v),$$

for all $x, y \in \text{Cl}(\Phi)$ *and all* $v \in V$. *Furthermore, such an anti-automorphism is unique.*

Proof. Consider the opposite algebra $\text{Cl}(\Phi)^o$, in which the product of x and y is given by yx. It has the universal mapping property. Thus, we get a unique isomorphism t, as in the diagram below.

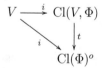

□

We also denote $t(x)$ by x^t. When V is finite dimensional, for a more palatable description of t in terms of a basis of V, see the paragraph following Theorem 11.4.

The canonical automorphism α is defined using the proposition.

Proposition 11.2. *Every Clifford algebra* $\text{Cl}(\Phi)$ *has a unique canonical automorphism* $\alpha : \text{Cl}(\Phi) \to \text{Cl}(\Phi)$ *satisfying the properties*

$$\alpha \circ \alpha = id, \quad and \quad \alpha(i(v)) = -i(v),$$

for all $v \in V$.

Proof. Consider the linear map $\alpha_0 \colon V \to \mathrm{Cl}(\Phi)$ defined by $\alpha_0(v) = -i(v)$, for all $v \in V$. We get a unique homomorphism α as in the diagram below.

$$
\begin{array}{ccc}
V & \xrightarrow{\ i\ } & \mathrm{Cl}(V, \Phi) \\
 & \alpha_0 \searrow & \downarrow \alpha \\
 & & \mathrm{Cl}(\Phi)
\end{array}
$$

Furthermore, every $x \in \mathrm{Cl}(\Phi)$ can be written as

$$ x = x_1 \cdots x_m, $$

with $x_j \in i(V)$, and since $\alpha(x_j) = -x_j$, we get $\alpha \circ \alpha = \mathrm{id}$. It is clear that α is bijective. □

When V is finite dimensional, a more palatable description of α in terms of a basis of V can be given; see the paragraph following Theorem 11.4.

If $(e_1, \ldots, e_n)$ is a basis of V, then the Clifford algebra $\mathrm{Cl}(\Phi)$ consists of certain kinds of "polynomials," linear combinations of monomials of the form $\sum_J \lambda_J e_J$, where $J = \{i_1, i_2, \ldots, i_k\}$ is any subset (possibly empty) of $\{1, \ldots, n\}$, with $1 \le i_1 < i_2 \cdots < i_k \le n$, and the monomial e_J is the "product" $e_{i_1} e_{i_2} \cdots e_{i_k}$.

We now show that if V has dimension n, then i is injective and $\mathrm{Cl}(\Phi)$ has dimension 2^n. A clever way of doing this is to introduce a graded tensor product.

First, observe that

$$ \mathrm{Cl}(\Phi) = \mathrm{Cl}^0(\Phi) \oplus \mathrm{Cl}^1(\Phi), $$

where

$$ \mathrm{Cl}^i(\Phi) = \{x \in \mathrm{Cl}(\Phi) \mid \alpha(x) = (-1)^i x\}, \quad \text{where } i = 0, 1. $$

We say that we have a $\mathbb{Z}/2$-*grading*, which means that if $x \in \mathrm{Cl}^i(\Phi)$ and $y \in \mathrm{Cl}^j(\Phi)$, then $xy \in \mathrm{Cl}^{i+j \,(\mathrm{mod}\,2)}(\Phi)$.

When V is finite dimensional, since every element of $\mathrm{Cl}(\Phi)$ is a linear combination of the form $\sum_J \lambda_J e_J$ as explained earlier, in view of the description of α given above, we see that the elements of $\mathrm{Cl}^0(\Phi)$ are those for which the monomials e_J are products of an even number of factors, and the elements of $\mathrm{Cl}^1(\Phi)$ are those for which the monomials e_J are products of an odd number of factors.

Remark. Observe that $\mathrm{Cl}^0(\Phi)$ is a subalgebra of $\mathrm{Cl}(\Phi)$, whereas $\mathrm{Cl}^1(\Phi)$ is not.

Definition 11.8. Given two $\mathbb{Z}/2$-graded algebras $A = A^0 \oplus A^1$ and $B = B^0 \oplus B^1$, their *graded tensor product* $A \mathbin{\widehat{\otimes}} B$ is defined by

$$ (A \mathbin{\widehat{\otimes}} B)^0 = (A^0 \otimes B^0) \oplus (A^1 \otimes B^1), $$

$$(A \mathbin{\widehat{\otimes}} B)^1 = (A^0 \otimes B^1) \oplus (A^1 \otimes B^0),$$

with multiplication

$$(a' \otimes b)(a \otimes b') = (-1)^{ij}(a'a) \otimes (bb'),$$

for $a \in A^i$ and $b \in B^j$.

The reader should check that $A \mathbin{\widehat{\otimes}} B$ is indeed $\mathbb{Z}/2$-graded.

Proposition 11.3. *Let V and W be finite-dimensional vector spaces with quadratic forms Φ and Ψ. Then there is a quadratic form $\Phi \oplus \Psi$ on $V \oplus W$ defined by*

$$(\Phi + \Psi)(v, w) = \Phi(v) + \Psi(w).$$

If we write $i : V \to \mathrm{Cl}(\Phi)$ and $j : W \to \mathrm{Cl}(\Psi)$, we can define a linear map

$$f : V \oplus W \to \mathrm{Cl}(\Phi) \mathbin{\widehat{\otimes}} \mathrm{Cl}(\Psi)$$

by

$$f(v, w) = i(v) \otimes \mathbf{1} + \mathbf{1} \otimes j(w).$$

Furthermore, the map f induces an isomorphism (also denoted by f)

$$f : \mathrm{Cl}(\Phi + \Psi) \to \mathrm{Cl}(\Phi) \mathbin{\widehat{\otimes}} \mathrm{Cl}(\Psi).$$

Proof. See Bröcker and tom Dieck [18], Chapter 1, Section 6, page 57. $\qquad\square$

As a corollary, we obtain the following result:

Theorem 11.4. *For every vector space V of finite dimension n, the map $i : V \to \mathrm{Cl}(\Phi)$ is injective. Given a basis $(e_1, \ldots, e_n)$ of V, the $2^n - 1$ products*

$$i(e_{i_1})i(e_{i_2}) \cdots i(e_{i_k}), \quad 1 \le i_1 < i_2 \cdots < i_k \le n,$$

and $\mathbf{1}$ form a basis of $\mathrm{Cl}(\Phi)$. Thus, $\mathrm{Cl}(\Phi)$ has dimension 2^n.

Proof. The proof is by induction on $n = \dim(V)$. For $n = 1$, the tensor algebra $T(V)$ is just the polynomial ring $\mathbb{R}[X]$, where $i(e_1) = X$. Thus, $\mathrm{Cl}(\Phi) = \mathbb{R}[X]/(X^2 - \Phi(e_1))$, and the result is obvious $((1, X)$ is a basis). Since

$$i(e_j)i(e_k) + i(e_k)i(e_j) = 2\varphi(e_i, e_j) \cdot \mathbf{1},$$

it is clear that the products

$$i(e_{i_1})i(e_{i_2}) \cdots i(e_{i_k}), \quad 1 \le i_1 < i_2 < \cdots < i_k \le n,$$

and **1** generate $\mathrm{Cl}(\Phi)$. In order to conclude that these vectors form a basis it suffices to show that the dimension of $\mathrm{Cl}(\Phi)$ is 2^n. Now there is always a basis that is orthogonal with respect to φ (for example, see Artin [2], Chapter 7, or Gallier [46], Chapter 6, Problem 6.14), and thus, we have a splitting

$$(V, \Phi) \cong \bigoplus_{k=1}^{n}(V_k, \Phi_k),$$

where V_k has dimension 1. Choosing a basis so that $e_k \in V_k$, the theorem follows by induction from Proposition 11.3. $\square$

Since i is injective, for simplicity of notation, from now on we write u for $i(u)$. Theorem 11.4 implies the following result.

Proposition 11.5. *If $(e_1, \ldots, e_n)$ is an orthogonal basis of V with respect to Φ, then $\mathrm{Cl}(\Phi)$ is the algebra presented by the generators $(e_1, \ldots, e_n)$ and the relations*

$$e_j^2 = \Phi(e_j) \cdot \mathbf{1}, \quad 1 \leq j \leq n, \quad and$$

$$e_j e_k = -e_k e_j, \quad 1 \leq j, k \leq n, \ j \neq k.$$

If V has finite dimension n and $(e_1, \ldots, e_n)$ is a basis of V, by Theorem 11.4, the maps t and α are completely determined by their action on the basis elements. Namely, t is defined by

$$t(e_i) = e_i$$

$$t(e_{i_1} e_{i_2} \cdots e_{i_k}) = e_{i_k} e_{i_{k-1}} \cdots e_{i_1},$$

where $1 \leq i_1 < i_2 \cdots < i_k \leq n$, and of course, $t(\mathbf{1}) = \mathbf{1}$. The map α is defined by

$$\alpha(e_i) = -e_i$$

$$\alpha(e_{i_1} e_{i_2} \cdots e_{i_k}) = (-1)^k e_{i_1} e_{i_2} \cdots e_{i_k}$$

where $1 \leq i_1 < i_2 < \cdots < i_k \leq n$, and of course, $\alpha(\mathbf{1}) = \mathbf{1}$. Furthermore, the even-graded elements (the elements of $\mathrm{Cl}^0(\Phi)$) are those generated by **1** and the basis elements consisting of an even number of factors $e_{i_1} e_{i_2} \cdots e_{i_{2k}}$, and the odd-graded elements (the elements of $\mathrm{Cl}^1(\Phi)$) are those generated by the basis elements consisting of an odd number of factors $e_{i_1} e_{i_2} \cdots e_{i_{2k+1}}$.

We are now ready to define the Clifford group and investigate some of its properties.

11.4 Clifford Groups

Definition 11.9. Let V be a real finite-dimensional vector space with a quadratic form Φ. Let $\mathrm{Cl}(\Phi)$ be the Clifford algebra (see Definition 11.7). The multiplicative group of invertible elements of $\mathrm{Cl}(\Phi)$ is denoted by $\mathrm{Cl}(\Phi)^*$.

Proposition 11.6. *For any $x \in V$, $\Phi(x) \neq 0$ if and only if x is invertible.*

Proof. This follows from the fact that $x^2 = \Phi(x)$ (where we abused notation and wrote $\Phi(x) \cdot \mathbf{1} = \Phi(x)$). If $\Phi(x) \neq 0$, then $x^{-1} = x(\Phi(x))^{-1}$, and if x is invertible, then $x \neq 0$ and $x = \Phi(x)x^{-1}$, so $\Phi(x) \neq 0$. $\quad\square$

We would like $\mathrm{Cl}(\Phi)^*$ to act on V *via*

$$x \cdot v = \alpha(x)vx^{-1},$$

where $x \in \mathrm{Cl}(\Phi)^*$ and $v \in V$. In general, there is no reason why $\alpha(x)vx^{-1}$ should be in V or why this action defines an automorphism of V, so we restrict this map to the subset $\Gamma(\Phi)$ of $\mathrm{Cl}(\Phi)^*$ as follows.

Definition 11.10. Given a finite-dimensional vector space V and a quadratic form Φ on V, the *Clifford group of* Φ is the group

$$\Gamma(\Phi) = \{x \in \mathrm{Cl}(\Phi)^* \mid \alpha(x)vx^{-1} \in V \quad \text{for all } v \in V\}.$$

Definition 11.11. For any $x \in \Gamma(\Phi)$, let $\rho_x : V \to V$ be the map defined by

$$v \mapsto \alpha(x)vx^{-1}, \quad v \in V.$$

It is not entirely obvious why the map $\rho \colon \Gamma(\Phi) \to \mathbf{GL}(V)$ given by $x \mapsto \rho_x$ is a linear action, and for that matter, why $\Gamma(\Phi)$ is a group. This is because V is finite dimensional and α is an automorphism.

Proposition 11.7. *The set $\Gamma(\Phi)$ is a group and ρ is a linear representation.*

Proof. For any $x \in \Gamma(\Phi)$, the map ρ_x from V to V defined by

$$v \mapsto \alpha(x)vx^{-1}$$

is clearly linear. If $\alpha(x)vx^{-1} = 0$, since by hypothesis x is invertible and since α is an automorphism $\alpha(x)$ is also invertible, so $v = 0$. Thus our linear map is injective, and since V has finite dimension, it is bijective. This proves that ρ is a linear representation.

To prove that $x^{-1} \in \Gamma(\Phi)$, pick any $v \in V$. Since the linear map ρ_x is bijective, there is some $w \in V$ such that $\rho_x(w) = v$, which means that $\alpha(x)wx^{-1} = v$. Since x is invertible and α is an automorphism, we get

$$\alpha(x^{-1})vx = w,$$

so $\alpha(x^{-1})vx \in V$; since this holds for any $v \in V$, we have $x^{-1} \in \Gamma(\Phi)$. Since α is an automorphism, if $x, y \in \Gamma(\Phi)$, for any $v \in V$ we have

$$\rho_y(\rho_x(v)) = \alpha(y)\alpha(x)vx^{-1}y^{-1} = \alpha(yx)v(yx)^{-1} = \rho_{yx}(v),$$

which shows that ρ_{yx} is a linear automorphism of V, so $yx \in \Gamma(\Phi)$ and ρ is a homomorphism. Therefore, $\Gamma(\Phi)$ is a group and ρ is a linear representation. $\square$

Definition 11.12. Given a finite-dimensional vector space V and quadratic form Φ on V, the *special Clifford group* of Φ is the group

$$\Gamma^+(\Phi) = \Gamma(\Phi) \cap \mathrm{Cl}^0(\Phi).$$

Remarks.

1. The map $\rho\colon \Gamma(\Phi) \to \mathbf{GL}(V)$ given by $x \mapsto \rho_x$ is called the *twisted adjoint representation*. It was introduced by Atiyah, Bott, and Shapiro [6]. It has the advantage of not introducing a spurious negative sign, *i.e.*, when $v \in V$ and $\Phi(v) \neq 0$, the map ρ_v is the reflection s_v about the hyperplane orthogonal to v (see Theorem 11.9). Furthermore, when Φ is nondegenerate, the kernel $\mathrm{Ker}\,(\rho)$ of the representation ρ is given by $\mathrm{Ker}\,(\rho) = \mathbb{R}^* \cdot \mathbf{1}$, where $\mathbb{R}^* = \mathbb{R} - \{0\}$. The earlier *adjoint representation* ρ^0 (used by Chevalley [24] and others) is given by

$$v \mapsto xvx^{-1}.$$

Unfortunately, in this case ρ_v^0 represents $-s_v$, where s_v is the reflection about the hyperplane orthogonal to v. Furthermore, the kernel of the representation ρ^0 is generally bigger than $\mathbb{R}^* \cdot \mathbf{1}$. This is the reason why the twisted adjoint representation is preferred (and must be used for a proper treatment of the **Pin** group).

2. According to Lounesto (in Riesz [89]), the Clifford group was actually discovered by Rudolf Lipschitz in 1880 and not by Clifford, two years after Clifford's discovery of Clifford algebras. Lounesto says (page 219): "Chevalley introduced the exterior exponential of bivectors and used it to scrutinize properties of Lipschitz's covering group of rotations (naming it unjustly a "Clifford group")."

Proposition 11.8. *The maps α and t induce an automorphism and an anti-automorphism of the Clifford group, $\Gamma(\Phi)$.*

Proof. It is not very instructive; see Bröcker and tom Dieck [18], Chapter 1, Section 6, Page 58. $\square$

The following key result shows why Clifford groups generalize the quaternions.

Theorem 11.9. *Let V be a finite-dimensional vector space and let Φ a quadratic form on V. For every element x of the Clifford group $\Gamma(\Phi)$, if $x \in V$ then $\Phi(x) \neq 0$ and the map $\rho_x \colon V \to V$ given by*

$$v \mapsto \alpha(x)vx^{-1} \quad \text{for all } v \in V$$

is the reflection about the hyperplane H orthogonal to the non-isotropic vector x.

Proof. We already observed that if $x \in V$ is an invertible element then $\Phi(x) \neq 0$. Recall that the reflection s about the hyperplane H orthogonal to the vector x is given by

$$s(u) = u - 2\frac{\varphi(u, x)}{\Phi(x)} \cdot x.$$

However, we have

$$x^2 = \Phi(x) \cdot \mathbf{1} \quad \text{and} \quad ux + xu = 2\varphi(u, x) \cdot \mathbf{1}.$$

Thus, we have

$$
\begin{aligned}
s(u) &= u - 2\frac{\varphi(u, x)}{\Phi(x)} \cdot x \\
&= u - 2\varphi(u, x) \cdot \left(\frac{1}{\Phi(x)} \cdot x \right) \\
&= u - 2\varphi(u, x) \cdot x^{-1} \\
&= u - 2\varphi(u, x) \cdot (\mathbf{1}x^{-1}) \\
&= u - (2\varphi(u, x) \cdot \mathbf{1})x^{-1} \\
&= u - (ux + xu)x^{-1} \\
&= -xux^{-1} \\
&= \alpha(x)ux^{-1},
\end{aligned}
$$

since $\alpha(x) = -x$, for $x \in V$. $\square$

Recall that the linear representation

$$\rho \colon \Gamma(\Phi) \to \mathbf{GL}(V)$$

is given by

$$\rho_x(v) = \alpha(x)vx^{-1},$$

for all $x \in \Gamma(\Phi)$ and all $v \in V$. We would like to show that ρ is a surjective homomorphism from $\Gamma(\Phi)$ onto $\mathbf{O}(\Phi)$, and a surjective homomorphism from $\Gamma^+(\Phi)$ onto $\mathbf{SO}(\Phi)$. For this, we will need to assume that φ is nondegenerate, which means that for every $v \in V$, if $\varphi(v, w) = 0$ for all $w \in V$, then $v = 0$. In order to prove that $\rho_x \in \mathbf{O}(\Phi)$ for any $x \in \Gamma(\Phi)$, we define a notion of norm on $\Gamma(\Phi)$, and for this we need to define a notion of conjugation on $\mathrm{Cl}(\Phi)$.

Definition 11.13. We define *conjugation* on a Clifford algebra $\mathrm{Cl}(\Phi)$ as the map

$$x \mapsto \overline{x} = t(\alpha(x)) \quad \text{for all } x \in \mathrm{Cl}(\Phi).$$

Observe that since $(t \circ \alpha)(v) = (\alpha \circ t)(v)$ for all $v \in V$ and since α is an automorphism and t is an anti-automorphism, we have

$$t \circ \alpha = \alpha \circ t \quad \text{on } \mathrm{Cl}(\Phi).$$

For all $x, y \in \mathrm{Cl}(\Phi)$ we also have

$$\overline{xy} = t(\alpha(xy)) = t(\alpha(x)\alpha(y)) = t(\alpha(y))t(\alpha(x)) = \overline{y}\,\overline{x}.$$

Thus we showed the following fact.

Proposition 11.10. *Conjugation is an anti-automorphism.*

If V has finite dimension n and $(e_1, \ldots, e_n)$ is a basis of V, in view of previous remarks, conjugation is defined by

$$\overline{e_i} = -e_i$$

$$\overline{e_{i_1} e_{i_2} \cdots e_{i_k}} = (-1)^k e_{i_k} e_{i_{k-1}} \cdots e_{i_1}$$

where $1 \le i_1 < i_2 \cdots < i_k \le n$, and of course, $\overline{1} = 1$.

Definition 11.14. The map $N \colon \mathrm{Cl}(\Phi) \to \mathrm{Cl}(\Phi)$ given by

$$N(x) = x\overline{x}$$

is called the *norm* of $\mathrm{Cl}(\Phi)$.

Observe that $N(v) = v\overline{v} = -v^2 = -\Phi(v) \cdot 1$ for all $v \in V$, that is,

$$N(v) = -\Phi(v) \cdot 1 \quad \text{for all } v \in V.$$

Also, if $(e_1, \ldots, e_n)$ is a basis of V, since conjugation is an anti-automorphism, we obtain

$$N(e_{i_1} e_{i_2} \cdots e_{i_k}) = e_{i_1} e_{i_2} \cdots e_{i_k} \overline{e_{i_1} e_{i_2} \cdots e_{i_k}}$$

$$= e_{i_1} e_{i_2} \cdots e_{i_k} (-1)^k e_{i_k} \cdots e_{i_2} e_{i_1}$$
$$= (-1)^k \Phi(e_{i_1}) \Phi(e_{i_2}) \cdots \Phi(e_{i_k}) \cdot \mathbf{1}.$$

In general, for an arbitrary element $x \in \text{Cl}(\Phi)$, there is no guarantee that $N(x)$ is a scalar multiple of $\mathbf{1}$. However, we will show in Proposition 11.12 that if $x \in \Gamma(\Phi)$, then $N(x) \in \mathbb{R}^* \cdot \mathbf{1}$.

For simplicity of exposition, we first assume that Φ is the quadratic form on $\mathbb{R}^n$ defined by

$$\Phi(x_1, \ldots, x_n) = \Phi_{0,n}(x_1, \ldots, x_n) = -(x_1^2 + \cdots + x_n^2).$$

Note that the isometry groups associated with $\Phi = \Phi_{0,n}$ are $\mathbf{O}(0, n)$ and $\mathbf{SO}(0, n)$, but we know that $\mathbf{O}(0, n) = \mathbf{O}(n)$ and $\mathbf{SO}(0, n) = \mathbf{SO}(n)$.

Let Cl_n denote the Clifford algebra $\text{Cl}(\Phi)$ and Γ_n denote the Clifford group $\Gamma(\Phi)$. The following lemma plays a crucial role.

Lemma 11.11. *The kernel of the map* $\rho \colon \Gamma_n \to \mathbf{GL}(n)$ *is* $\mathbb{R}^* \cdot \mathbf{1}$, *the multiplicative group of nonzero scalar multiples of* $\mathbf{1} \in \text{Cl}_n$.

Proof. If $\rho_x = \text{id}$, then

$$\alpha(x)v = vx \quad \text{for all } v \in \mathbb{R}^n. \tag{1}$$

Since $\text{Cl}_n = \text{Cl}_n^0 \oplus \text{Cl}_n^1$, we can write $x = x^0 + x^1$, with $x^i \in \text{Cl}_n^i$ for $i = 0, 1$. Then Equation (1) becomes

$$x^0 v = vx^0 \quad \text{and} \quad -x^1 v = vx^1 \quad \text{for all } v \in \mathbb{R}^n. \tag{2}$$

Using Theorem 11.4, we can express x^0 as a linear combination of monomials in the canonical basis $(e_1, \ldots, e_n)$, so that

$$x^0 = a^0 + e_1 b^1, \quad \text{with } a^0 \in \text{Cl}_n^0, \ b^1 \in \text{Cl}_n^1,$$

where neither a^0 nor b^1 contains a summand with a factor e_1. Applying the first relation in (2) to $v = e_1$, we get

$$e_1 a^0 + e_1^2 b^1 = a^0 e_1 + e_1 b^1 e_1. \tag{3}$$

Now the basis $(e_1, \ldots, e_n)$ is orthogonal w.r.t. Φ, which implies that

$$e_j e_k = -e_k e_j \quad \text{for all } j \neq k.$$

Since each monomial in a^0 is of even degree and contains no factor e_1, we get

$$a^0 e_1 = e_1 a^0.$$

Similarly, since b^1 is of odd degree and contains no factor e_1, we get

$$e_1 b^1 e_1 = -e_1^2 b^1.$$

But then from (3), we get

$$e_1 a^0 + e_1^2 b^1 = a^0 e_1 + e_1 b^1 e_1 = e_1 a^0 - e_1^2 b^1,$$

and so, $e_1^2 b^1 = 0$. However, $e_1^2 = -1$, and so, $b^1 = 0$. Therefore, x_0 contains no monomial with a factor e_1. We can apply the same argument to the other basis elements $e_2, \ldots, e_n$, and thus, we just proved that $x^0 \in \mathbb{R} \cdot \mathbf{1}$.

A similar argument applying to the second equation in (2), with $x^1 = a^1 + e_1 b^0$ and $v = e_1$ shows that $b^0 = 0$. By repeating the argument for the other basis elements, we ultimately conclude that $x^1 = \mathbf{0}$. Finally, $x = x^0 \in (\mathbb{R} \cdot \mathbf{1}) \cap \Gamma_n = \mathbb{R}^* \cdot \mathbf{1}$.

$\square$

Remark. If Φ is any nondegenerate quadratic form, we know (for instance, see Artin [2], Chapter 7, or Gallier [46], Chapter 6, Problem 6.14) that there is an orthogonal basis $(e_1, \ldots, e_n)$ with respect to φ (i.e. $\varphi(e_j, e_k) = 0$ for all $j \neq k$ and $\varphi(e_j, e_j) \neq 0$ for all j). Thus, the commutation relations

$$e_j^2 = \Phi(e_j) \cdot \mathbf{1}, \quad \text{with } \Phi(e_j) \neq 0, \quad 1 \leq j \leq n, \quad \text{and}$$

$$e_j e_k = -e_k e_j, \quad 1 \leq j, k \leq n, \ j \neq k$$

hold, and since the proof only rests on these facts, Lemma 11.11 holds for the Clifford group $\Gamma(\Phi)$ associated with any nondegenerate quadratic form.

 However, Lemma 11.11 may fail for degenerate quadratic forms. For example, if $\Phi \equiv 0$, then $Cl(V, 0) = \bigwedge^{\bullet} V$. Consider the element $x = \mathbf{1} + e_1 e_2$. Clearly, $x^{-1} = \mathbf{1} - e_1 e_2$. But now, for any $v \in V$, we have

$$\alpha(\mathbf{1} + e_1 e_2) v (\mathbf{1} + e_1 e_2)^{-1} = (\mathbf{1} + e_1 e_2) v (\mathbf{1} - e_1 e_2) = v.$$

Yet, $\mathbf{1} + e_1 e_2$ is not a scalar multiple of $\mathbf{1}$.

If instead of the twisted adjoint action we had used the action $\rho^0 \colon \Gamma^n \to \mathbf{GL}(n)$ given by

$$\rho_x^0(v) = xvx^{-1},$$

then when n is odd the kernel of ρ^0 contains other elements besides scalar multiples of $\mathbf{1}$. Indeed, if $(e_1, \ldots, e_n)$ is an orthogonal basis, we have $e_i e_j = -e_j e_i$ for all $j \neq i$ and $e_i^2 = -1$ for all i, so the element $e_1 \cdots e_n \in Cl_n^*$ commutes with all e_i (it belongs to the center of Cl_n), and thus $e_1 \cdots e_n \in \operatorname{Ker} \rho^0$. Thus, we see that another subtle consequence of the "Atiyah–Bott–Shapiro trick" of using the action

$\rho_x(v) = \alpha(x)vx$ where α takes care of the parity of $x \in \Gamma_n$ is to cut down the kernel of ρ to $\mathbb{R}^* \cdot \mathbf{1}$.

The following proposition shows that the notion of norm is well behaved.

Proposition 11.12. *If $x \in \Gamma_n$, then $N(x) \in \mathbb{R}^* \cdot \mathbf{1}$.*

Proof. The trick is to show that $N(x) = x\overline{x}$ is in the kernel of ρ. To say that $x \in \Gamma_n$ means that

$$\alpha(x)vx^{-1} \in \mathbb{R}^n \quad \text{for all } v \in \mathbb{R}^n.$$

Applying t, we get

$$t(x)^{-1}vt(\alpha(x)) = \alpha(x)vx^{-1},$$

since t is the identity on $\mathbb{R}^n$. Thus, we have

$$
\begin{aligned}
v &= t(x)\alpha(x)v(t(\alpha(x))x)^{-1} \\
&= t(x)\alpha(x)v(\overline{x}x)^{-1} \\
&= \alpha(\alpha(t(x)))\alpha(x)v(\overline{x}x)^{-1}, && \text{since } \alpha \circ \alpha = \mathrm{id} \\
&= \alpha(t(\alpha(x)))\alpha(x)v(\overline{x}x)^{-1}, && \text{since } \alpha \circ t = t \circ \alpha \\
&= \alpha(\overline{x})\alpha(x)v(\overline{x}x)^{-1} \\
&= \alpha(\overline{x}x)v(\overline{x}x)^{-1},
\end{aligned}
$$

so $\overline{x}x \in \mathrm{Ker}\,(\rho)$. By Proposition 11.8, we have $\overline{x} \in \Gamma_n$, and so, $x\overline{x} = \overline{\overline{x}}\,\overline{x} \in \mathrm{Ker}\,(\rho)$. $\qquad\square$

Remark. Again, the proof also holds for the Clifford group $\Gamma(\Phi)$ associated with any nondegenerate quadratic form Φ. When $\Phi(v) = -\|v\|^2$, where $\|v\|$ is the standard Euclidean norm of v, we have $N(v) = \|v\|^2 \cdot \mathbf{1}$ for all $v \in V$. However, for other quadratic forms, it is possible that $N(x) = \lambda \cdot \mathbf{1}$ where $\lambda < 0$, and this is a difficulty that needs to be overcome.

Proposition 11.13. *The restriction of the norm N to Γ_n is a homomorphism $N \colon \Gamma_n \to \mathbb{R}^* \cdot \mathbf{1}$, and $N(\alpha(x)) = N(x)$ for all $x \in \Gamma_n$.*

Proof. We have

$$N(xy) = xy\overline{y}\,\overline{x} = xN(y)\overline{x} = x\overline{x}N(y) = N(x)N(y),$$

where the third equality holds because $N(x) \in \mathbb{R}^* \cdot \mathbf{1}$. Next, observe that since α and t commute we have

$$\overline{\alpha(x)} = t(\alpha(\alpha(x))) = \alpha(t(\alpha(x))) = \alpha(\overline{x}),$$

so we get

$$N(\alpha(x)) = \alpha(x)\overline{\alpha(x)} = \alpha(x)\alpha(\overline{x}) = \alpha(x\overline{x}) = \alpha(N(x)) = N(x),$$

since $N(x) \in \mathbb{R}^* \cdot \mathbf{1}$.   □

Remark. The proof also holds for the Clifford group $\Gamma(\Phi)$ associated with any nondegenerate quadratic form Φ.

Proposition 11.14. *We have* $\mathbb{R}^n - \{0\} \subseteq \Gamma_n$ *and* $\rho(\Gamma_n) \subseteq \mathbf{O}(n)$.

Proof. Let $x \in \Gamma_n$ and $v \in \mathbb{R}^n$, with $v \neq 0$. We have

$$N(\rho_x(v)) = N(\alpha(x)vx^{-1}) = N(\alpha(x))N(v)N(x^{-1}) = N(x)N(v)N(x)^{-1} = N(v),$$

since $N \colon \Gamma_n \to \mathbb{R}^* \cdot \mathbf{1}$. However, for $v \in \mathbb{R}^n$, we know that

$$N(\rho_x v) = -\Phi(\rho_x v) \cdot \mathbf{1},$$

and

$$N(v) = -\Phi(v) \cdot \mathbf{1}.$$

Thus, ρ_x is norm-preserving, and so $\rho_x \in \mathbf{O}(n)$.   □

Remark. The proof that $\rho(\Gamma(\Phi)) \subseteq \mathbf{O}(\Phi)$ also holds for the Clifford group $\Gamma(\Phi)$ associated with any nondegenerate quadratic form Φ. The first statement needs to be replaced by the fact that every non-isotropic vector in $\mathbb{R}^n$ (a vector is non-isotropic if $\Phi(x) \neq 0$) belongs to $\Gamma(\Phi)$. Indeed, $x^2 = \Phi(x) \cdot \mathbf{1}$, which implies that x is invertible.

We are finally ready for the introduction of the groups **Pin**(*n*) and **Spin**(*n*).

11.5   The Groups Pin(*n*) and Spin(*n*)

Definition 11.15. We define the *pinor group* **Pin**(*n*) as the kernel Ker(N) of the homomorphism $N \colon \Gamma_n \to \mathbb{R}^* \cdot \mathbf{1}$, and the *spinor group* **Spin**(*n*) as **Pin**$(n) \cap \Gamma_n^+$.

Observe that if $N(x) = \mathbf{1}$, then x is invertible, and $x^{-1} = \overline{x}$ since $x\overline{x} = N(x) = \mathbf{1}$. Thus, we can write

$$\mathbf{Pin}(n) = \{x \in \mathrm{Cl}_n \mid \alpha(x)vx^{-1} \in \mathbb{R}^n \quad \text{for all } v \in \mathbb{R}^n, \quad N(x) = \mathbf{1}\}$$
$$= \{x \in \mathrm{Cl}_n \mid \alpha(x)v\overline{x} \in \mathbb{R}^n \quad \text{for all } v \in \mathbb{R}^n, \quad x\overline{x} = \mathbf{1}\},$$

and

$$\mathbf{Spin}(n) = \{x \in \mathbf{Cl}_n^0 \mid xvx^{-1} \in \mathbb{R}^n \quad \text{for all } v \in \mathbb{R}^n, \quad N(x) = 1\}$$
$$= \{x \in \mathbf{Cl}_n^0 \mid xv\overline{x} \in \mathbb{R}^n \quad \text{for all } v \in \mathbb{R}^n, \quad x\overline{x} = 1\}$$

Remark. According to Atiyah, Bott, and Shapiro, the use of the name **Pin**(k) is a joke due to Jean-Pierre Serre (Atiyah, Bott, and Shapiro [6], page 1).

Theorem 11.15. *The restriction of $\rho \colon \Gamma_n \to \mathbf{O}(n)$ to the pinor group **Pin**(n) is a surjective homomorphism $\rho \colon \mathbf{Pin}(n) \to \mathbf{O}(n)$ whose kernel is $\{-1, 1\}$, and the restriction of ρ to the spinor group **Spin**(n) is a surjective homomorphism $\rho \colon \mathbf{Spin}(n) \to \mathbf{SO}(n)$ whose kernel is $\{-1, 1\}$.*

Proof. By Proposition 11.14, we have a map $\rho \colon \mathbf{Pin}(n) \to \mathbf{O}(n)$. The reader can easily check that ρ is a homomorphism. By the Cartan-Dieudonné theorem (see Bourbaki [13], or Gallier [46], Chapter 7, Theorem 7.2.1), every isometry $f \in \mathbf{O}(n)$ is the composition $f = s_1 \circ \cdots \circ s_k$ of hyperplane reflections s_j. If we assume that s_j is a reflection about the hyperplane H_j orthogonal to the nonzero vector w_j, by Theorem 11.9, $\rho(w_j) = s_j$. Since $N(w_j) = \|w_j\|^2 \cdot \mathbf{1}$, we can replace w_j by $w_j / \|w_j\|$, so that $N(w_1 \cdots w_k) = \mathbf{1}$, and then

$$f = \rho(w_1 \cdots w_k),$$

and ρ is surjective. Note that

$$\mathrm{Ker}\,(\rho \mid \mathbf{Pin}(n)) = \mathrm{Ker}\,(\rho) \cap \mathrm{Ker}\,(N) = \{t \in \mathbb{R}^* \cdot \mathbf{1} \mid N(t) = \mathbf{1}\} = \{-1, 1\}.$$

As to **Spin**(n), we just need to show that the restriction of ρ to **Spin**(n) maps Γ_n into **SO**(n). If this was not the case, there would be some improper isometry $f \in \mathbf{O}(n)$ so that $\rho_x = f$, where $x \in \Gamma_n \cap \mathbf{Cl}_n^0$. However, we can express f as the composition of an odd number of reflections, say

$$f = \rho(w_1 \cdots w_{2k+1}).$$

Since

$$\rho(w_1 \cdots w_{2k+1}) = \rho_x,$$

we have $x^{-1} w_1 \cdots w_{2k+1} \in \mathrm{Ker}\,(\rho)$. By Lemma 11.11, we must have

$$x^{-1} w_1 \cdots w_{2k+1} = \lambda \cdot \mathbf{1}$$

for some $\lambda \in \mathbb{R}^*$, and thus

$$w_1 \cdots w_{2k+1} = \lambda \cdot x,$$

where x has even degree and $w_1 \cdots w_{2k+1}$ has odd degree, which is impossible. $\square$

Let us denote the set of elements $v \in \mathbb{R}^n$ with $N(v) = \mathbf{1}$ (with norm $\mathbf{1}$) by S^{n-1}. We have the following corollary of Theorem 11.15.

Corollary 11.16. *The group* **Pin**(n) *is generated by* S^{n-1}, *and every element of* **Spin**(n) *can be written as the product of an even number of elements of* S^{n-1}.

Example 11.3. In Example 11.1 we showed that Cl_1 is isomorphic to $\mathbb{C}$. The reader should verify that

$$\mathbf{Pin}(1) \cong \mathbb{Z}/4\mathbb{Z}$$

as follows. By definition

$$\mathbf{Pin}(1) = \{x \in \mathrm{Cl}_1 \mid \alpha(x)vx^{-1} \in \mathbb{R} \quad \text{for all } v \in \mathbb{R}, \ N(x) = \mathbf{1}\}.$$

A typical element in **Pin**(1) has the form $a\mathbf{1} + be_1$ where $e_1^2 = -\mathbf{1}$. Set $e_1 \mapsto i$ and $\mathbf{1} \mapsto 1$ as in Example 11.1. The condition $N(x) = \mathbf{1}$ implies that

$$N(x) = x\overline{x} = (a + bi)(a - bi) = a^2 + b^2 = 1.$$

Thus

$$x^{-1} = \frac{\overline{x}}{a^2 + b^2} = \overline{x}.$$

and $x \in \mathbf{Pin}(1)$ implies that $\alpha(x)x^{-1} \in \mathbb{R}$ where

$$\alpha(x)x^{-1} = (a - ib)(a - ib) = a^2 - b^2 - 2abi.$$

Thus either $a = 0$ or $b = 0$. This constraint, along with $a^2 + b^2 = 1$, implies that

$$\mathbf{Pin}(1) = \{1, i, -1, -i\} \cong \mathbb{Z}/4\mathbb{Z}$$

since i generates **Pin**(1) and $i^4 = 1$. Since $\mathbf{Spin}(1) = \mathbf{Pin}(1) \cap \mathrm{Cl}_n^0$, we conclude that

$$\mathbf{Spin}(1) = \{-1, 1\} \cong \mathbb{Z}/2\mathbb{Z}.$$

Example 11.4. Definition 11.15 also implies

$$\mathbf{Pin}(2) \cong \{ae_1 + be_2 \mid a^2 + b^2 = 1\} \cup \{c\mathbf{1} + de_1e_2 \mid c^2 + d^2 = 1\}, \quad \mathbf{Spin}(2) = \mathbf{U}(1).$$

We may also write $\mathbf{Pin}(2) = \mathbf{U}(1) + \mathbf{U}(1)$, where $\mathbf{U}(1)$ is the group of complex numbers of modulus 1 (the unit circle in $\mathbb{R}^2$).

Let us take a closer look at **Spin**(2). The Clifford algebra Cl$_2$ is generated by the four elements

$$\mathbf{1}, \; e_1, \; e_2, \; e_1e_2,$$

and they satisfy the relations

$$e_1^2 = -\mathbf{1}, \quad e_2^2 = -\mathbf{1}, \quad e_1e_2 = -e_2e_1.$$

We saw in Example 11.2 that Cl$_2$ is isomorphic to the algebra of quaternions $\mathbb{H}$. According to Corollary 11.16, the group **Spin**(2) consists of all products

$$\prod_{i=1}^{2k} (a_ie_1 + b_ie_2)$$

consisting of an even number of factors and such that $a_i^2 + b_i^2 = 1$. In view of the above relations, every such element can be written as

$$x = a\mathbf{1} + be_1e_2,$$

where x satisfies the conditions that $xvx^{-1} \in \mathbb{R}^2$ for all $v \in \mathbb{R}^2$, and $N(x) = \mathbf{1}$. Since

$$\overline{\mathbf{1}} = \mathbf{1}, \qquad \overline{e_1} = -e_1, \qquad \overline{e_2} = -e_2, \qquad \overline{e_1e_2} = -e_1e_2,$$

the definition of conjugation implies that

$$\overline{x} = t(\alpha(x)) = t(\alpha(a\mathbf{1}+be_1e_2)) = at(\alpha(\mathbf{1}))+bt(\alpha(e_1e_2)) = a\overline{\mathbf{1}}+b\overline{e_1e_2} = a\mathbf{1}-be_1e_2.$$

Then we get

$$N(x) = x\overline{x} = (a^2 + b^2) \cdot \mathbf{1},$$

and the condition $N(x) = \mathbf{1}$ is simply $a^2 + b^2 = 1$.

We claim that if $x \in \mathrm{Cl}_2^0$, then $xvx^{-1} \in \mathbb{R}^2$. Indeed, since $x \in \mathrm{Cl}_2^0$ and $v \in \mathrm{Cl}_2^1$, we have $xvx^{-1} \in \mathrm{Cl}_2^1$, which implies that $xvx^{-1} \in \mathbb{R}^2$, since the only elements of Cl_2^1 are those in $\mathbb{R}^2$. This observation provides an alternative proof of the fact that **Spin**(2) consists of those elements $x = a\mathbf{1} + be_1e_2$ so that $a^2 + b^2 = 1$.

If we let $\mathbf{i} = e_1e_2$, we observe that

$$\mathbf{i}^2 = -\mathbf{1},$$

$$e_1\mathbf{i} = -\mathbf{i}e_1 = -e_2,$$

$$e_2\mathbf{i} = -\mathbf{i}e_2 = e_1.$$

Thus, **Spin**(2) is isomorphic to **U**(1). Also note that

$$e_1(a\mathbf{1} + b\mathbf{i}) = (a\mathbf{1} - b\mathbf{i})e_1.$$

Let us find out explicitly what is the action of **Spin**(2) on $\mathbb{R}^2$. Given $X = a\mathbf{1} + b\mathbf{i}$, with $a^2 + b^2 = 1$, then $\overline{X} = a\mathbf{1} - b\mathbf{i}$, and for any $v = v_1 e_1 + v_2 e_2$, we have

$$\begin{aligned}
\alpha(X)vX^{-1} &= X(v_1 e_1 + v_2 e_2)X^{-1} \\
&= X(v_1 e_1 + v_2 e_2)(-e_1 e_1)\overline{X} \\
&= X(v_1 e_1 + v_2 e_2)(-e_1)(e_1 \overline{X}) \\
&= X(v_1 \mathbf{1} + v_2 \mathbf{i})Xe_1 \\
&= X^2(v_1 \mathbf{1} + v_2 \mathbf{i})e_1 \\
&= (((a^2 - b^2)v_1 - 2abv_2)\mathbf{1} + ((a^2 - b^2)v_2 + 2abv_1)\mathbf{i})e_1 \\
&= ((a^2 - b^2)v_1 - 2abv_2)e_1 + ((a^2 - b^2)v_2 + 2abv_1)e_2.
\end{aligned}$$

Since $a^2 + b^2 = 1$, we can write $X = a\mathbf{1} + b\mathbf{i} = (\cos\theta)\mathbf{1} + (\sin\theta)\mathbf{i}$, and the above derivation shows that

$$\alpha(X)vX^{-1} = (\cos 2\theta v_1 - \sin 2\theta v_2)e_1 + (\cos 2\theta v_2 + \sin 2\theta v_1)e_2.$$

This means that the rotation ρ_X induced by $X \in$ **Spin**(2) is the rotation of angle 2θ around the origin. Observe that the maps

$$v \mapsto v(-e_1), \quad X \mapsto Xe_1$$

establish bijections between $\mathbb{R}^2$ and **Spin**(2) $\cong$ **U**(1). Also, note that the action of $X = \cos\theta + i\sin\theta$ viewed as a complex number yields the rotation of angle θ, whereas the action of $X = (\cos\theta)\mathbf{1} + (\sin\theta)\mathbf{i}$ viewed as a member of **Spin**(2) yields the rotation of angle 2θ. There is nothing wrong. In general, **Spin**(n) is a two-to-one cover of **SO**(n).

Next let us take a closer look at **Spin**(3).

Example 11.5. The Clifford algebra Cl_3 is generated by the eight elements

$$\mathbf{1}, \ e_1, \ e_2, \ e_3, \ e_1 e_2, \ e_2 e_3, \ e_3 e_1, \ e_1 e_2 e_3,$$

and they satisfy the relations

$$e_i^2 = -\mathbf{1}, \quad e_i e_j = -e_j e_i, \quad 1 \le i, j \le 3, \ i \ne j.$$

It is not hard to show that Cl_3 is isomorphic to $\mathbb{H} \oplus \mathbb{H}$. By Corollary 11.16, the group
Spin(3) consists of all products

$$\prod_{i=1}^{2k} (a_i e_1 + b_i e_2 + c_i e_3)$$

consisting of an even number of factors and such that $a_i^2 + b_i^2 + c_i^2 = 1$. In view of
the above relations, every such element can be written as

$$x = a\mathbf{1} + be_2e_3 + ce_3e_1 + de_1e_2,$$

where x satisfies the conditions that $xvx^{-1} \in \mathbb{R}^3$ for all $v \in \mathbb{R}^3$, and $N(x) = \mathbf{1}$.
Since

$$\overline{e_2e_3} = -e_2e_3, \qquad \overline{e_3e_1} = -e_3e_1, \qquad \overline{e_1e_2} = -e_1e_2,$$

we observe that

$$\overline{x} = a\mathbf{1} - be_2e_3 - ce_3e_1 - de_1e_2.$$

We then get

$$N(x) = (a^2 + b^2 + c^2 + d^2) \cdot \mathbf{1},$$

and the condition $N(x) = \mathbf{1}$ is simply $a^2 + b^2 + c^2 + d^2 = 1$.

It turns out that the conditions $x \in Cl_3^0$ and $N(x) = \mathbf{1}$ imply that $xvx^{-1} \in \mathbb{R}^3$
for all $v \in \mathbb{R}^3$. To prove this, first observe that $N(x) = \mathbf{1}$ implies that $x^{-1} = \overline{x}$, and
that $\overline{v} = -v$ for any $v \in \mathbb{R}^3$, and so,

$$\overline{xvx^{-1}} = \overline{x^{-1}}\,\overline{v}\,\overline{x} = -xvx^{-1}.$$

Also, since $x \in Cl_3^0$ and $v \in Cl_3^1$, we have $xvx^{-1} \in Cl_3^1$. Thus, we can write

$$xvx^{-1} = u + \lambda e_1e_2e_3, \quad \text{for some } u \in \mathbb{R}^3 \text{ and some } \lambda \in \mathbb{R}.$$

But

$$\overline{e_1e_2e_3} = -e_3e_2e_1 = e_1e_2e_3,$$

and so,

$$\overline{xvx^{-1}} = -u + \lambda e_1e_2e_3 = -xvx^{-1} = -u - \lambda e_1e_2e_3,$$

which implies that $\lambda = 0$. Thus, $xvx^{-1} \in \mathbb{R}^3$, as claimed. By using this observation, we once again conclude that **Spin**(3) consists of those elements $x = a\mathbf{1} + be_2e_3 + ce_3e_1 + de_1e_2$ so that $a^2 + b^2 + c^2 + d^2 = 1$.

Under the bijection

$$\mathbf{i} \mapsto e_2e_3, \ \mathbf{j} \mapsto e_3e_1, \ \mathbf{k} \mapsto e_1e_2,$$

we can check that we have an isomorphism between the group **SU**(2) of unit quaternions and **Spin**(3). If $X = a\mathbf{1} + be_2e_3 + ce_3e_1 + de_1e_2 \in$ **Spin**(3), observe that

$$X^{-1} = \overline{X} = a\mathbf{1} - be_2e_3 - ce_3e_1 - de_1e_2.$$

Now using the identification

$$\mathbf{i} \mapsto e_2e_3, \ \mathbf{j} \mapsto e_3e_1, \ \mathbf{k} \mapsto e_1e_2,$$

we can easily check that

$$(e_1e_2e_3)^2 = \mathbf{1},$$
$$(e_1e_2e_3)\mathbf{i} = \mathbf{i}(e_1e_2e_3) = -e_1,$$
$$(e_1e_2e_3)\mathbf{j} = \mathbf{j}(e_1e_2e_3) = -e_2,$$
$$(e_1e_2e_3)\mathbf{k} = \mathbf{k}(e_1e_2e_3) = -e_3,$$
$$(e_1e_2e_3)e_1 = -\mathbf{i},$$
$$(e_1e_2e_3)e_2 = -\mathbf{j},$$
$$(e_1e_2e_3)e_3 = -\mathbf{k}.$$

Then if $X = a\mathbf{1} + b\mathbf{i} + c\mathbf{j} + d\mathbf{k} \in$ **Spin**(3), for every $v = v_1e_1 + v_2e_2 + v_3e_3$, we have

$$\alpha(X)vX^{-1} = X(v_1e_1 + v_2e_2 + v_3e_3)X^{-1}$$
$$= X(e_1e_2e_3)^2(v_1e_1 + v_2e_2 + v_3e_3)X^{-1}$$
$$= (e_1e_2e_3)X(e_1e_2e_3)(v_1e_1 + v_2e_2 + v_3e_3)X^{-1}$$
$$= -(e_1e_2e_3)X(v_1\mathbf{i} + v_2\mathbf{j} + v_3\mathbf{k})X^{-1}.$$

This shows that the rotation $\rho_X \in$ **SO**(3) induced by $X \in$ **Spin**(3) can be viewed as the rotation induced by the quaternion $a\mathbf{1} + b\mathbf{i} + c\mathbf{j} + d\mathbf{k}$ on the pure quaternions, using the maps

$$v \mapsto -(e_1e_2e_3)v, \quad X \mapsto -(e_1e_2e_3)X$$

to go from a vector $v = v_1 e_1 + v_2 e_2 + v_3 e_3$ to the pure quaternion $v_1 \mathbf{i} + v_2 \mathbf{j} + v_3 \mathbf{k}$, and back.

We close this section by taking a closer look at **Spin**(4).

Example 11.6. We will show in Section 11.8 that Cl_4 is isomorphic to $M_2(\mathbb{H})$, the algebra of 2×2 matrices whose entries are quaternions. According to Corollary 11.16, the group **Spin**(4) consists of all products

$$\prod_{i=1}^{2k} (a_i e_1 + b_i e_2 + c_i e_3 + d_i e_4)$$

consisting of an even number of factors and such that $a_i^2 + b_i^2 + c_i^2 + d_i^2 = 1$. Using the relations

$$e_i^2 = -1, \quad e_i e_j = -e_j e_i, \quad 1 \le i, j \le 4, \ i \ne j,$$

every element of **Spin**(4) can be written as

$$x = a_1 \mathbf{1} + a_2 e_1 e_2 + a_3 e_2 e_3 + a_4 e_3 e_1 + a_5 e_4 e_3 + a_6 e_4 e_1 + a_7 e_4 e_2 + a_8 e_1 e_2 e_3 e_4,$$

where x satisfies the conditions that $x v x^{-1} \in \mathbb{R}^4$ for all $v \in \mathbb{R}^4$, and $N(x) = \mathbf{1}$. Let

$$\mathbf{i} = e_1 e_2, \ \mathbf{j} = e_2 e_3, \ \mathbf{k} = e_3 e_1, \ \mathbf{i}' = e_4 e_3, \ \mathbf{j}' = e_4 e_1, \ \mathbf{k}' = e_4 e_2,$$

and

$\mathbb{I} = e_1 e_2 e_3 e_4$. The reader will easily verify that

$$\mathbf{i}\mathbf{j} = \mathbf{k}$$
$$\mathbf{j}\mathbf{k} = \mathbf{i}$$
$$\mathbf{k}\mathbf{i} = \mathbf{j}$$
$$\mathbf{i}^2 = -1, \quad \mathbf{j}^2 = -1, \quad \mathbf{k}^2 = -1$$
$$\mathbf{i}\mathbb{I} = \mathbb{I}\mathbf{i} = \mathbf{i}'$$
$$\mathbf{j}\mathbb{I} = \mathbb{I}\mathbf{j} = \mathbf{j}'$$
$$\mathbf{k}\mathbb{I} = \mathbb{I}\mathbf{k} = \mathbf{k}'$$
$$\mathbb{I}^2 = 1, \quad \bar{\mathbb{I}} = \mathbb{I}$$
$$\bar{\mathbf{i}} = -\mathbf{i}, \quad \bar{\mathbf{j}} = -\mathbf{j}, \quad \bar{\mathbf{k}} = -\mathbf{k}$$
$$\bar{\mathbf{i}'} = -\mathbf{i}', \quad \bar{\mathbf{j}'} = -\mathbf{j}', \quad \bar{\mathbf{k}'} = -\mathbf{k}'.$$

Then every $x \in$ **Spin**(4) can be written as

$$x = u + \mathbb{I}v, \quad \text{with} \quad u = a\mathbf{1} + b\mathbf{i} + c\mathbf{j} + d\mathbf{k} \quad \text{and} \quad v = a'\mathbf{1} + b'\mathbf{i} + c'\mathbf{j} + d'\mathbf{k},$$

with the extra conditions stated above. Using the above identities, we have

$$(u + \mathbb{I}v)(u' + \mathbb{I}v') = uu' + vv' + \mathbb{I}(uv' + vu').$$

Furthermore, the identities imply

$$\overline{u + \mathbb{I}v} = t(\alpha(u + \mathbb{I}v)) = t(\alpha(u)) + t(\alpha(\mathbb{I}v))$$
$$= \overline{u} + t(\alpha(\mathbb{I})\alpha(v)) = \overline{u} + t(\alpha(v))t(\alpha(\mathbb{I}))$$
$$= \overline{u} + \overline{v}\overline{\mathbb{I}} = \overline{u} + \overline{v}\mathbb{I}$$
$$= \overline{u} + \mathbb{I}\overline{v}.$$

As a consequence,

$$N(u + \mathbb{I}v) = (u + \mathbb{I}v)(\overline{u} + \mathbb{I}\overline{v}) = u\overline{u} + v\overline{v} + \mathbb{I}(u\overline{v} + v\overline{u}),$$

and thus, $N(u + \mathbb{I}v) = \mathbf{1}$ is equivalent to

$$u\overline{u} + v\overline{v} = \mathbf{1} \quad \text{and} \quad u\overline{v} + v\overline{u} = 0.$$

As in the case $n = 3$, it turns out that the conditions $x \in \mathrm{Cl}_4^0$ and $N(x) = \mathbf{1}$ imply that $xvx^{-1} \in \mathbb{R}^4$ for all $v \in \mathbb{R}^4$. The only change to the proof is that $xvx^{-1} \in \mathrm{Cl}_4^1$ can be written as

$$xvx^{-1} = u + \sum_{i,j,k} \lambda_{i,j,k} e_i e_j e_k, \quad \text{for some } u \in \mathbb{R}^4, \quad \text{with } \{i, j, k\} \subseteq \{1, 2, 3, 4\}.$$

As in the previous proof, we get $\lambda_{i,j,k} = 0$. So once again, **Spin**(4) consists of those elements $u + \mathbb{I}v$ so that

$$u\overline{u} + v\overline{v} = \mathbf{1} \quad \text{and} \quad u\overline{v} + v\overline{u} = 0,$$

with u and v of the form $a\mathbf{1} + b\mathbf{i} + c\mathbf{j} + d\mathbf{k}$.

Finally, we see that **Spin**(4) is isomorphic to **Spin**(3) × **Spin**(3) under the isomorphism

$$u + v\mathbb{I} \mapsto (u + v, u - v).$$

Indeed, we have

$$N(u + v) = (u + v)(\overline{u} + \overline{v}) = \mathbf{1},$$

and

$$N(u - v) = (u - v)(\overline{u} - \overline{v}) = 1,$$

since

$$u\overline{u} + v\overline{v} = 1 \quad \text{and} \quad u\overline{v} + v\overline{u} = 0,$$

and

$$(u + v, u - v)(u' + v', u' - v') = (uu' + vv' + uv' + vu', uu' + vv' - (uv' + vu')).$$

In summary, we have shown that **Spin**$(3) \cong$ **SU**(2) and **Spin**$(4) \cong$ **SU**$(2) \times$ **SU**(2). The group **Spin**(5) is isomorphic to the symplectic group **Sp**(2), and **Spin**(6) is isomorphic to **SU**(4) (see Curtis [27] or Porteous [87]).

Remark. It can be shown that the assertion if $x \in \mathbf{Cl}_n^0$ and $N(x) = 1$, then $xvx^{-1} \in \mathbb{R}^n$ for all $v \in \mathbb{R}^n$, is true up to $n = 5$ (see Porteous [87], Chapter 13, Proposition 13.58). However, this is already false for $n = 6$. For example, if $X = 1/\sqrt{2}(1 + e_1 e_2 e_3 e_4 e_5 e_6)$, it is easy to see that $N(X) = 1$, and yet, $Xe_1 X^{-1} \notin \mathbb{R}^6$.

11.6 The Groups Pin(p, q) and Spin(p, q)

For every nondegenerate quadratic form Φ over $\mathbb{R}$, there is an orthogonal basis with respect to which Φ is given by

$$\Phi(x_1, \ldots, x_{p+q}) = x_1^2 + \cdots + x_p^2 - (x_{p+1}^2 + \cdots + x_{p+q}^2),$$

where p and q only depend on Φ. The quadratic form corresponding to (p, q) is denoted $\Phi_{p,q}$ and we call (p, q) the *signature* of $\Phi_{p,q}$. Let $n = p + q$. We define the group $\mathbf{O}(p, q)$ as the group of isometries w.r.t. $\Phi_{p,q}$, i.e., the group of linear maps f so that

$$\Phi_{p,q}(f(v)) = \Phi_{p,q}(v) \quad \text{for all } v \in \mathbb{R}^n$$

and the group $\mathbf{SO}(p, q)$ as the subgroup of $\mathbf{O}(p, q)$ consisting of the isometries $f \in \mathbf{O}(p, q)$ with $\det(f) = 1$. We denote the Clifford algebra $\mathrm{Cl}(\Phi_{p,q})$ where $\Phi_{p,q}$ has signature (p, q) by $\mathrm{Cl}_{p,q}$, the corresponding Clifford group by $\Gamma_{p,q}$, and the special Clifford group $\Gamma_{p,q} \cap \mathrm{Cl}_{p,q}^0$ by $\Gamma_{p,q}^+$. Note that with this new notation, $\mathrm{Cl}_n = \mathrm{Cl}_{0,n}$.

 As we mentioned earlier, since Lawson and Michelsohn [71] adopt the opposite of our sign convention in defining Clifford algebras; their $\mathrm{Cl}(p, q)$ is our $\mathrm{Cl}(q, p)$.

As we mentioned in Section 11.4, we have the problem that $N(v) = -\Phi(v) \cdot \mathbf{1}$, but $-\Phi(v)$ is not necessarily positive (where $v \in \mathbb{R}^n$). The fix is simple: Allow elements $x \in \Gamma_{p,q}$ with $N(x) = \pm 1$.

Definition 11.16. We define the *pinor group* **Pin**(p, q) as the group

$$\mathbf{Pin}(p, q) = \{x \in \Gamma_{p,q} \mid N(x) = \pm 1\},$$

and the *spinor group* **Spin**(p, q) as **Pin**$(p, q) \cap \Gamma_{p,q}^{+}$.

Remarks.

(1) It is easily checked that the group **Spin**(p, q) is also given by

$$\mathbf{Spin}(p, q) = \{x \in \mathrm{Cl}_{p,q}^{0} \mid xv\overline{x} \in \mathbb{R}^n \quad \text{for all } v \in \mathbb{R}^n, \quad N(x) = \pm 1\}.$$

This is because **Spin**(p, q) consists of elements of even degree.

(2) One can check that if $N(x) = x\overline{x} \neq 0$, then $x^{-1} = \overline{x}(N(x))^{-1}$. This in turn implies

$$\alpha(x)vx^{-1} = \alpha(x)v\overline{x}(N(x))^{-1}$$
$$= \alpha(x)v\alpha(t(x))(N(x))^{-1}$$
$$= \alpha(x)\alpha(-v)\alpha(t(x))(N(x))^{-1}, \qquad \text{since } \alpha(v) = -v$$
$$= \alpha(-xvt(x))(N(x))^{-1}$$
$$= xvt(x)(N(x))^{-1}.$$

Thus, we have

$$\mathbf{Pin}(p, q) = \{x \in \mathrm{Cl}_{p,q} \mid xvt(x) \in \mathbb{R}^n \quad \text{for all } v \in \mathbb{R}^n, \quad N(x) = \pm 1\}$$
$$= \{x \in \mathrm{Cl}_{p,q} \mid xv\overline{x} \in \mathbb{R}^n \quad \text{for all } v \in \mathbb{R}^n, \quad N(x) = \pm 1\}.$$

Theorem 11.15 generalizes as follows:

Theorem 11.17. *The restriction of $\rho \colon \Gamma_{p,q} \to \mathbf{GL}(n)$ to the pinor group **Pin**(p, q) is a surjective homomorphism $\rho \colon \mathbf{Pin}(p, q) \to \mathbf{O}(p, q)$ whose kernel is $\{-1, 1\}$, and the restriction of ρ to the spinor group **Spin**(p, q) is a surjective homomorphism $\rho \colon \mathbf{Spin}(p, q) \to \mathbf{SO}(p, q)$ whose kernel is $\{-1, 1\}$.*

Proof. The Cartan-Dieudonné also holds for any nondegenerate quadratic form Φ, in the sense that every isometry in $\mathbf{O}(\Phi)$ is the composition of reflections defined by hyperplanes orthogonal to non-isotropic vectors (see Dieudonné [30], Chevalley [24], Bourbaki [13], or Gallier [46], Chapter 7, Problem 7.14). Thus, Theorem 11.15 also holds for any nondegenerate quadratic form Φ. The only change to the proof is the following: Since $N(w_j) = -\Phi(w_j) \cdot \mathbf{1}$, we can replace w_j by $w_j/\sqrt{|\Phi(w_j)|}$, so that $N(w_1 \cdots w_k) = \pm\mathbf{1}$, and then

$$f = \rho(w_1 \cdots w_k),$$

and ρ is surjective. If $f \in \mathbf{SO}(p, q)$, then k is even and $w_1 \cdots w_k \in \Gamma_{p,q}^+$ and by replacing w_j by $w_j/\sqrt{|\Phi(w_j)|}$ we obtain $w_1 \cdots w_k \in \mathbf{Spin}(p, q)$.   □

If we consider $\mathbb{R}^n$ equipped with the quadratic form $\Phi_{p,q}$ (with $n = p + q$), we denote the set of elements $v \in \mathbb{R}^n$ with $N(v) = \pm\mathbf{1}$ by $S_{p,q}^{n-1}$. We have the following corollary of Theorem 11.17 (generalizing Corollary 11.16).

Corollary 11.18. *The group* $\mathbf{Pin}(p, q)$ *is generated by* $S_{p,q}^{n-1}$, *and every element of* $\mathbf{Spin}(p, q)$ *can be written as the product of an even number of elements of* $S_{p,q}^{n-1}$.

Example 11.7. In Example 11.1 we showed that

$$\mathrm{Cl}_{0,1} \cong \mathbb{C}.$$

We use a similar argument to calculate $\mathrm{Cl}_{1,0}$. The basis for $\mathrm{Cl}_{1,0}$ is $\mathbf{1}, e_1$ where

$$e_1^2 = \mathbf{1}.$$

By using the bijection

$$\mathbf{1} \mapsto 1 + 0, \qquad e_1 \mapsto 0 + 1$$

we find that

$$\mathrm{Cl}_{1,0} \cong \mathbb{R} \oplus \mathbb{R},$$

where the multiplication on $\mathbb{R} \oplus \mathbb{R}$ is given by

$$(a_1 + b_1)(a_2 + b_2) = (a_1 a_2 + b_1 b_2) + (a_1 b_2 + a_2 b_1)$$
$$\cong (a_1 \mathbf{1} + b_1 e_1)(a_2 \mathbf{1} + b_2 e_1) = (a_1 a_2 + b_1 b_1)\mathbf{1} + (a_1 b_2 + a_2 b_1)e_1.$$

From Example 11.3 we have

$$\mathbf{Pin}(0, 1) \cong \mathbb{Z}/4\mathbb{Z}.$$

To calculate

$$\mathbf{Pin}(1, 0) = \{x \in \mathrm{Cl}_{1,0} \mid \alpha(x)vx^{-1} \in \mathbb{R} \quad \text{for all } v \in \mathbb{R}, \qquad N(x) = \pm\mathbf{1}\},$$

we first observe that a typical element of $Cl_{1,0}$ has the form $x = a\mathbf{1} + be_1$, where $e_1^2 = \mathbf{1}$. Then

$$N(x) = x\overline{x} = (a\mathbf{1} + be_1)(a\mathbf{1} - be_1) = (a^2 - b^2)\mathbf{1} = \pm\mathbf{1},$$

which in turn implies

$$a^2 - b^2 = \pm 1,$$

and that

$$x^{-1} = \overline{x}N(x)^{-1} = \frac{a\mathbf{1} - be_1}{a^2 - b^2}.$$

Since $x \in \mathbf{Pin}(1, 0)$, we have $\alpha(x)x^{-1} \in \mathbb{R}$, or equivalently

$$(a\mathbf{1} - be_1)\frac{a\mathbf{1} - be_1}{a^2 - b^2} = \frac{1}{a^2 - b^2}[(a^2 + b^2)\mathbf{1} - 2abe_1] \in \mathbb{R}.$$

This implies that $a = 0$ or $b = 0$. If $a = 0$, we set $a^2 - b^2 = -1$ to obtain $b = \pm 1$. If $b = 0$, we set $a^2 - b^2 = 1$ to obtain $a = \pm 1$. Thus

$$\mathbf{Pin}(1, 0) = \{\mathbf{1}, e_1, -e_1, -\mathbf{1}\} \cong \mathbb{Z}/2\mathbb{Z} \times \mathbb{Z}/2\mathbb{Z},$$

since $\mathbf{1}^2 = e_1^2 = -e_1^2$. Since $\mathbf{Spin}(1, 0) = \mathbf{Pin}(1, 0) \cap \Gamma_{1,0}^+$, we deduce that

$$\mathbf{Spin}(1, 0) = \{\mathbf{1}, -\mathbf{1}\} \cong \mathbb{Z}/2\mathbb{Z}.$$

Example 11.8. We now turn our attention to Clifford algebras over $\mathbb{R}^2$. In Example 11.2 we showed that

$$Cl_{0,2} \cong \mathbb{H}.$$

To calculate $Cl_{2,0}$ we first observe that $Cl_{2,0}$ is spanned by the basis $\{\mathbf{1}, e_1, e_2, e_1e_2\}$, where

$$e_1^2 = \mathbf{1}, \qquad e_2^2 = \mathbf{1}, \qquad e_1e_2 = -e_2e_1.$$

Define the bijection

$$\mathbf{1} \mapsto \begin{pmatrix} 1 & 0 \\ 0 & 1 \end{pmatrix}, \qquad e_1 \mapsto \begin{pmatrix} 1 & 0 \\ 0 & -1 \end{pmatrix}, \qquad e_2 \mapsto \begin{pmatrix} 0 & 1 \\ 1 & 0 \end{pmatrix}.$$

Then

$$e_1 e_2 = \begin{pmatrix} 1 & 0 \\ 0 & -1 \end{pmatrix} \begin{pmatrix} 0 & 1 \\ 1 & 0 \end{pmatrix} = \begin{pmatrix} 0 & 1 \\ -1 & 0 \end{pmatrix}.$$

A few basic computations show that $\begin{pmatrix} 1 & 0 \\ 0 & 1 \end{pmatrix}, \begin{pmatrix} 1 & 0 \\ 0 & -1 \end{pmatrix}, \begin{pmatrix} 0 & 1 \\ 1 & 0 \end{pmatrix}, \begin{pmatrix} 0 & 1 \\ -1 & 0 \end{pmatrix}$ form a basis for $M_2(\mathbb{R})$. Furthermore

$$\begin{pmatrix} 1 & 0 \\ 0 & -1 \end{pmatrix}^2 = \begin{pmatrix} 1 & 0 \\ 0 & 1 \end{pmatrix}$$

$$\begin{pmatrix} 0 & 1 \\ 1 & 0 \end{pmatrix}^2 = \begin{pmatrix} 1 & 0 \\ 0 & 1 \end{pmatrix}$$

$$\begin{pmatrix} 0 & 1 \\ 1 & 0 \end{pmatrix} \begin{pmatrix} 1 & 0 \\ 0 & -1 \end{pmatrix} = \begin{pmatrix} 0 & -1 \\ 1 & 0 \end{pmatrix}.$$

From this bijection we conclude that

$$Cl_{2,0} \cong M_2(\mathbb{R}).$$

A similar calculation shows that

$$Cl_{1,1} \cong M_2(\mathbb{R}).$$

But this time

$$\mathbf{1} \mapsto \begin{pmatrix} 1 & 0 \\ 0 & 1 \end{pmatrix}, \qquad e_1 \mapsto \begin{pmatrix} 1 & 0 \\ 0 & -1 \end{pmatrix}, \qquad e_2 \mapsto \begin{pmatrix} 0 & 1 \\ -1 & 0 \end{pmatrix},$$

which implies that

$$e_1 e_2 = \begin{pmatrix} 1 & 0 \\ 0 & -1 \end{pmatrix} \begin{pmatrix} 0 & 1 \\ -1 & 0 \end{pmatrix} = \begin{pmatrix} 0 & 1 \\ 1 & 0 \end{pmatrix},$$

and that

$$e_1^2 = \mathbf{1}, \qquad e_2^2 = -\mathbf{1}, \qquad e_1 e_2 = -e_2 e_1.$$

One can also work out what are **Pin**$(2, 0)$, **Pin**$(1, 1)$, and **Pin**$(0, 2)$. See Choquet-Bruhat [25], Chapter I, Section 7, page 26, for details As far as **Spin**$(0, 2)$, we know from Example 11.3 that

$$\mathbf{Spin}(0, 2) = \mathbf{Spin}(2) \cong \mathbf{U}(1).$$

By applying the results of the following paragraph regarding the isomorphism between $\mathrm{Cl}^0_{p,q}$ and $\mathrm{Cl}^0_{q,p}$, we may deduce that

$$\mathbf{Spin}(0,2) = \mathbf{Spin}(2,0) \cong \mathbf{U}(1).$$

Finally an application of Corollary 11.18 implies that

$$\mathbf{Spin}(1,1) = \{a\mathbf{1} + be_1e_2 \mid a^2 - b^2 = \pm 1\},$$

and

$$\mathbf{Pin}(1,1) = \{a\mathbf{1} + be_1e_2 \mid a^2 - b^2 = \pm 1\} \cup \{ae_1 + be_2 \mid a^2 - b^2 = \pm 1\}.$$

Observe that $\mathbf{Spin}(1,1)$ has four connected components and $\mathbf{Pin}(1,1)$ has eight connected components. It is easy to show that

$$\mathbf{SO}(1,1) = \left\{ \begin{pmatrix} a & b \\ b & a \end{pmatrix} \,\middle|\, a^2 - b^2 = 1 \right\},$$

which has two connected components, and

$$\mathbf{O}(1,1) = \left\{ \begin{pmatrix} a & b \\ b & a \end{pmatrix} \,\middle|\, a^2 - b^2 = \pm 1 \right\},$$

which has four connected components.

More generally, it can be shown that $\mathrm{Cl}^0_{p,q}$ and $\mathrm{Cl}^0_{q,p}$ are isomorphic, from which it follows that $\mathbf{Spin}(p,q)$ and $\mathbf{Spin}(q,p)$ are isomorphic, but $\mathbf{Pin}(p,q)$ and $\mathbf{Pin}(q,p)$ are not isomorphic in general, and in particular, $\mathbf{Pin}(p,0)$ and $\mathbf{Pin}(0,p)$ are not isomorphic in general (see Choquet-Bruhat [25], Chapter I, Section 7). However, due to the "8-periodicity" of the Clifford algebras (to be discussed in Section 11.8), it follows that $\mathrm{Cl}_{p,q}$ and $\mathrm{Cl}_{q,p}$ are isomorphic when $|p-q| = 0 \bmod 4$.

Remark. We can also define the group $\mathbf{Spin}^+(p,q)$ as

$$\mathbf{Spin}^+(p,q) = \{x \in \mathrm{Cl}^0_{p,q} \mid xv\overline{x} \in \mathbb{R}^n \quad \text{for all } v \in \mathbb{R}^n, \quad N(x) = \mathbf{1}\},$$

and $\mathbf{SO}_0(p,q)$ as the connected component of $\mathbf{SO}(p,q)$ containing the identity. Then it can be shown that the map $\rho\colon \mathbf{Spin}^+(p,q) \to \mathbf{SO}_0(p,q)$ is a surjective homomorphism with kernel $\{-\mathbf{1}, \mathbf{1}\}$; see Lounesto [74] (Chapter 17, Section 17.2). In particular,

$$\mathbf{Spin}^+(1,1) = \{a\mathbf{1} + be_1e_2 \mid a^2 - b^2 = 1\}.$$

This group has two connected components, but it can be shown that for $p + q \geq 2$ and $(p, q) \neq (1, 1)$ the groups **Spin**$^+(p, q)$ are connected; see Lounesto [74] (Chapter 17, Section 17.2).

11.7 The Groups Pin(p, q) and Spin(p, q) as Double Covers of O(p, q) and SO(p, q)

It turns out that the groups **Pin**(p, q) and **Spin**(p, q) have nice topological properties w.r.t. the groups **O**(p, q) and **SO**(p, q). To explain this, we review the definition of covering maps and covering spaces (for details, see Fulton [44], Chapter 11). Another interesting source is Chevalley [23], where it is proved that **Spin**(n) is a universal double cover of **SO**(n) for all $n \geq 3$.

Since $\mathrm{Cl}_{p,q}$ is an algebra of dimension 2^{p+q}, it is a vector space isomorphic to $V = \mathbb{R}^{2^{p+q}}$.

Proposition 11.19. *The spaces* $\mathrm{Cl}^*_{p,q}$, **Pin**(p, q), *and* **Spin**(p, q) *are Lie groups.*

Proof. The group $\mathrm{Cl}^*_{p,q}$ of units of $\mathrm{Cl}_{p,q}$ is open in $\mathrm{Cl}_{p,q}$, because $x \in \mathrm{Cl}_{p,q}$ is a unit if the linear map L_x is a bijection iff $\det(L_x) \neq 0$ (where L_x is defined by $L_x(y) = xy$ for all $y \in \mathrm{Cl}_{p,q}$). Thus we have a continuous map $\mathcal{L} \colon \mathrm{Cl}_{p,q} \to \mathbb{R}$ given by $\mathcal{L}(x) = \det(L_x)$ and since $\mathrm{Cl}^*_{p,q} = \mathcal{L}^{-1}(\mathbb{R} - \{0\})$ and $\mathbb{R} - \{0\}$ is open, $\mathrm{Cl}^*_{p,q}$ is open. Thus, $\mathrm{Cl}^*_{p,q}$ is a Lie group, and since **Pin**(p, q) and **Spin**(p, q) are clearly closed subgroups of $\mathrm{Cl}^*_{p,q}$, they are also Lie groups. $\square$

The definition below is analogous to the definition of a covering map (see Gallot, Hulin, Lafontaine [48] or Gallier and Quaintance [47]), except that now, we are only dealing with topological spaces and not manifolds.

Definition 11.17. Given two topological spaces X and Y, a *covering map* is a continuous surjective map $p \colon Y \to X$ with the property that for every $x \in X$, there is some open subset $U \subseteq X$ with $x \in U$, so that $p^{-1}(U)$ is the disjoint union of open subsets $V_\alpha \subseteq Y$, and the restriction of p to each V_α is a homeomorphism onto U. We say that U is *evenly covered by* p. We also say that Y is a *covering space* of X. A covering map $p \colon Y \to X$ is called *trivial* if X itself is evenly covered by p (*i.e.*, Y is the disjoint union of open subsets Y_α each homeomorphic to X), and *nontrivial* otherwise. When each fiber $p^{-1}(x)$ has the same finite cardinality n for all $x \in X$, we say that p is an *n-covering* (or *n-sheeted covering*). See Figure 11.1.

Note that a covering map $p \colon Y \to X$ is not always trivial, but always *locally trivial* (*i.e.*, for every $x \in X$, it is trivial in some open neighborhood of x). A covering is trivial iff Y is isomorphic to a product space of $X \times T$, where T is any set with the discrete topology. See Figure 11.1. Also, if Y is connected, then the covering map is nontrivial.

Fig. 11.1 Two coverings of
S^1. The left illustration is
$p: \mathbb{R} \to S^1$ with
$\pi(t) = (\cos(2\pi t), \sin(2\pi t))$,
while the right illustration is
the trivial 3-fold covering.

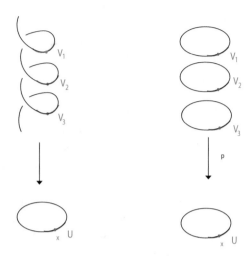

Definition 11.18. An *isomorphism* φ between covering maps $p: Y \to X$ and $p': Y' \to X$ is a homeomorphism $\varphi: Y \to Y'$ so that $p = p' \circ \varphi$.

Typically, the space X is connected, in which case it can be shown that all the fibers $p^{-1}(x)$ have the same cardinality.

One of the most important properties of covering spaces is the path-lifting property, a property that we will use to show that **Spin**(n) is path-connected.

Proposition 11.20 (Path Lifting). *Let $p: Y \to X$ be a covering map, and let $\gamma: [a, b] \to X$ be any continuous curve from $x_a = \gamma(a)$ to $x_b = \gamma(b)$ in X. If $y \in Y$ is any point so that $p(y) = x_a$, then there is a unique curve $\widetilde{\gamma}: [a, b] \to Y$ so that $y = \widetilde{\gamma}(a)$ and*

$$p \circ \widetilde{\gamma}(t) = \gamma(t) \quad for\ all\ t \in [a, b].$$

See Figure 11.2.

Proof. See Fulton [45], Chapter 11, Lemma 11.6.   □

Many important covering maps arise from the action of a group G on a space Y. If Y is a topological space, recall that an *action (on the left) of a group G on Y* is a map $\alpha: G \times Y \to Y$ satisfying the following conditions, where for simplicity of notation, we denote $\alpha(g, y)$ by $g \cdot y$:

(1) $g \cdot (h \cdot y) = (gh) \cdot y$, for all $g, h \in G$ and $y \in Y$;
(2) $1 \cdot y = y$, for all $\in Y$, where 1 is the identity of the group G;
(3) The map $y \mapsto g \cdot y$ is a homeomorphism of Y for every $g \in G$.

We define an equivalence relation on Y as follows: $x \equiv y$ iff $y = g \cdot x$ for some $g \in G$ (check that this is an equivalence relation). The equivalence class $G \cdot x = \{g \cdot x \mid g \in G\}$ of any $x \in Y$ is called the *orbit of* x. We obtain the

Fig. 11.2 The lift of a curve
γ when $\pi : \mathbb{R} \to S^1$ is
$\pi(t) = (\cos(2\pi t), \sin(2\pi t))$.

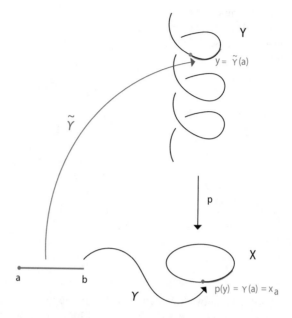

quotient space Y/G and the projection map $p : Y \to Y/G$ sending every $y \in Y$ to
its orbit. The space Y/G is given the quotient topology (a subset U of Y/G is open
iff $p^{-1}(U)$ is open in Y).

Given a subset V of Y and any $g \in G$, we let

$$g \cdot V = \{g \cdot y \mid y \in V\}.$$

Definition 11.19. We say that G *acts evenly on* Y if for every $y \in Y$, there is an
open subset V containing y so that $g \cdot V$ and $h \cdot V$ are disjoint for any two distinct
elements $g, h \in G$.

The importance of the notion a group acting evenly is that such actions induce a
covering map. See Figure 11.3.

Proposition 11.21. *If G is a group acting evenly on a space Y, then the projection
map $p : Y \to Y/G$ is a covering map.*

Proof. See Fulton [45], Chapter 11, Lemma 11.17.   □

The following proposition shows that **Pin**(p, q) and **Spin**(p, q) are nontrivial
covering spaces, unless $p = q = 1$.

Proposition 11.22. *For all $p, q \geq 0$, the groups* **Pin**(p, q) *and* **Spin**(p, q)
are double covers of **O**(p, q) *and* **SO**(p, q), *respectively. Furthermore, they are
nontrivial covers unless $p = q = 1$.*

Fig. 11.3 The 2-fold antipodal covering of $\mathbb{RP}^2$ induced by $\{-1, 1\}$ acting evenly on S^2.

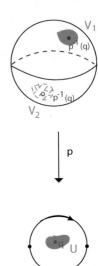

Proof. We know that kernel of the homomorphism $\rho\colon \mathbf{Pin}(p, q) \to \mathbf{O}(p, q)$ is $\mathbb{Z}_2 = \{-1, 1\}$. If we let $\mathbb{Z}_2$ act on $\mathbf{Pin}(p, q)$ in the natural way, then $\mathbf{O}(p, q) \cong \mathbf{Pin}(p, q)/\mathbb{Z}_2$, and the reader can easily check that $\mathbb{Z}_2$ acts evenly. By Proposition 11.21, we get a double cover. The argument for $\rho\colon \mathbf{Spin}(p, q) \to \mathbf{SO}(p, q)$ is similar.

Since

$$\mathbf{Pin}(1, 1) = \{a\mathbf{1} + be_1e_2 \mid a^2 - b^2 = \pm 1\} \cup \{ae_1 + be_2 \mid a^2 - b^2 = \pm 1\}$$

and

$$\mathbf{O}(1, 1) = \left\{ \begin{pmatrix} a & b \\ b & a \end{pmatrix} \,\middle|\, a^2 - b^2 = \pm 1 \right\},$$

we see that $\mathbf{Pin}(1, 1)$ is the disjoint union of two open subsets each homeomorphic with $\mathbf{O}(1, 1)$, and so the covering is trivial. Similarly, since

$$\mathbf{Spin}(1, 1) = \{a\mathbf{1} + be_1e_2 \mid a^2 - b^2 = \pm 1\},$$

and

$$\mathbf{SO}(1, 1) = \left\{ \begin{pmatrix} a & b \\ b & a \end{pmatrix} \,\middle|\, a^2 - b^2 = 1 \right\},$$

$\mathbf{Spin}(1, 1)$ is the disjoint union of two open subsets each homeomorphic with $\mathbf{SO}(1, 1)$, so the covering is also trivial.

Let us now assume that $p \neq 1$ or $q \neq 1$. In order to prove that we have nontrivial covers, it is enough to show that -1 and 1 are connected by a path in **Pin**(p, q) (If we had **Pin**(p, q) $= U_1 \cup U_2$ with U_1 and U_2 open, disjoint, and homeomorphic to **O**(p, q), then -1 and 1 would not be in the same U_i, and so, they would be in disjoint connected components. Thus, -1 and 1 can't be path-connected, and similarly with **Spin**(p, q) and **SO**(p, q).) Since $(p, q) \neq (1, 1)$, we can find two orthogonal vectors e_1 and e_2 so that either $\Phi_{p,q}(e_1) = \Phi_{p,q}(e_2) = 1$ or $\Phi_{p,q}(e_1) = \Phi_{p,q}(e_2) = -1$. Then,

$$\gamma(t) = \pm \cos(2t)\,\mathbf{1} + \sin(2t)\,e_1 e_2 = (\cos t\,e_1 + \sin t\,e_2)(\sin t\,e_2 - \cos t\,e_1),$$

for $0 \leq t \leq \pi$, defines a path in **Spin**(p, q), since

$$(\pm \cos(2t)\,\mathbf{1} + \sin(2t)\,e_1 e_2)^{-1} = \pm \cos(2t)\,\mathbf{1} - \sin(2t)\,e_1 e_2,$$

as desired.   □

In particular, if $n \geq 2$, since the group **SO**(n) is path-connected, the group **Spin**(n) is also path-connected. Indeed, given any two points x_a and x_b in **Spin**(n), there is a path γ from $\rho(x_a)$ to $\rho(x_b)$ in **SO**(n) (where $\rho\colon$ **Spin**(n) $\to$ **SO**(n) is the covering map). By Proposition 11.20, there is a path $\tilde{\gamma}$ in **Spin**(n) with origin x_a and some origin $\tilde{x}_b$ so that $\rho(\tilde{x}_b) = \rho(x_b)$. However, $\rho^{-1}(\rho(x_b)) = \{-x_b, x_b\}$, and so $\tilde{x}_b = \pm x_b$. The argument used in the proof of Proposition 11.22 shows that x_b and $-x_b$ are path-connected, and so, there is a path from x_a to x_b, and **Spin**(n) is path-connected.

In fact, for $n \geq 3$, it turns out that **Spin**(n) is simply connected. Such a covering space is called a *universal cover* (for instance, see Chevalley [23]).

This last fact requires more algebraic topology than we are willing to explain in detail, and we only sketch the proof. The notions of fiber bundle, fibration, and homotopy sequence associated with a fibration are needed in the proof. We refer the perseverant readers to Bott and Tu [12] (Chapter 1 and Chapter 3, Sections 16–17) or Rotman [91] (Chapter 11) for a detailed explanation of these concepts.

Recall that a topological space is *simply connected* if it is path connected and if $\pi_1(X) = (0)$, which means that every closed path in X is homotopic to a point. Since we just proved that **Spin**(n) is path connected for $n \geq 2$, we just need to prove that $\pi_1(\mathbf{Spin}(n)) = (0)$ for all $n \geq 3$. The following facts are needed to prove the above assertion:

(1) The sphere S^{n-1} is simply connected for all $n \geq 3$.
(2) The group **Spin**(3) $\cong$ **SU**(2) is homeomorphic to S^3, and thus, **Spin**(3) is simply connected.
(3) The group **Spin**(n) acts on S^{n-1} in such a way that we have a fiber bundle with fiber **Spin**($n - 1$):

$$\mathbf{Spin}(n - 1) \longrightarrow \mathbf{Spin}(n) \longrightarrow S^{n-1}.$$

Fact (1) is a standard proposition of algebraic topology, and a proof can be found in many books. A particularly elegant and yet simple argument consists in showing that any closed curve on S^{n-1} is homotopic to a curve that omits some point. First, it is easy to see that in $\mathbb{R}^n$, any closed curve is homotopic to a piecewise linear curve (a polygonal curve), and the radial projection of such a curve on S^{n-1} provides the desired curve. Then we use the stereographic projection of S^{n-1} from any point omitted by that curve to get another closed curve in $\mathbb{R}^{n-1}$. Since $\mathbb{R}^{n-1}$ is simply connected, that curve is homotopic to a point, and so is its preimage curve on S^{n-1}. Another simple proof uses a special version of the Seifert–van Kampen's theorem (see Gramain [50]).

Fact (2) is easy to establish directly, using (1).

To prove (3), we let $\mathbf{Spin}(n)$ act on S^{n-1} via the standard action: $x \cdot v = xvx^{-1}$. Because $\mathbf{SO}(n)$ acts transitively on S^{n-1} and there is a surjection $\mathbf{Spin}(n) \longrightarrow \mathbf{SO}(n)$, the group $\mathbf{Spin}(n)$ also acts transitively on S^{n-1}. Now we have to show that the stabilizer of any element of S^{n-1} is $\mathbf{Spin}(n-1)$. For example, we can do this for e_1. This amounts to some simple calculations taking into account the identities among basis elements. Details of this proof can be found in Mneimné and Testard [79], Chapter 4. Then by Proposition 9.33, the Lie group $\mathbf{Spin}(n)$ is a principal fiber bundle over S^{n-1} with fiber $\mathbf{Spin}(n-1)$.

Now a fiber bundle is a fibration (as defined in Bott and Tu [12], Chapter 3, Section 16, or in Rotman [91], Chapter 11). For a proof of this fact, see Rotman [91], Chapter 11, or Mneimné and Testard [79], Chapter 4. So, there is a homotopy sequence associated with the fibration (Bott and Tu [12], Chapter 3, Section 17, or Rotman [91], Chapter 11, Theorem 11.48), and in particular, we have the exact sequence

$$\pi_1(\mathbf{Spin}(n-1)) \longrightarrow \pi_1(\mathbf{Spin}(n)) \longrightarrow \pi_1(S^{n-1}).$$

Since $\pi_1(S^{n-1}) = (0)$ for $n \geq 3$, we get a surjection

$$\pi_1(\mathbf{Spin}(n-1)) \longrightarrow \pi_1(\mathbf{Spin}(n)),$$

and so, by induction and (2), we get

$$\pi_1(\mathbf{Spin}(n)) \cong \pi_1(\mathbf{Spin}(3)) = (0),$$

proving that $\mathbf{Spin}(n)$ is simply connected for $n \geq 3$.

We can also show that $\pi_1(\mathbf{SO}(n)) = \mathbb{Z}/2\mathbb{Z}$ for all $n \geq 3$. For this, we use Theorem 11.15 and Proposition 11.22, which imply that $\mathbf{Spin}(n)$ is a fiber bundle over $\mathbf{SO}(n)$ with fiber $\{-1, 1\}$, for $n \geq 2$:

$$\{-1, 1\} \longrightarrow \mathbf{Spin}(n) \longrightarrow \mathbf{SO}(n).$$

Again, the homotopy sequence of the fibration exists, and in particular we get the exact sequence

$$\pi_1(\mathbf{Spin}(n)) \longrightarrow \pi_1(\mathbf{SO}(n)) \longrightarrow \pi_0(\{-1, +1\}) \longrightarrow \pi_0(\mathbf{Spin}(n)).$$

Since $\pi_0(\{-1, +1\}) = \mathbb{Z}/2\mathbb{Z}$, $\pi_0(\mathbf{Spin}(n)) = (0)$, and $\pi_1(\mathbf{Spin}(n)) = (0)$, when $n \geq 3$, we get the exact sequence

$$(0) \longrightarrow \pi_1(\mathbf{SO}(n)) \longrightarrow \mathbb{Z}/2\mathbb{Z} \longrightarrow (0),$$

and so, $\pi_1(\mathbf{SO}(n)) = \mathbb{Z}/2\mathbb{Z}$. Therefore, $\mathbf{SO}(n)$ is not simply connected for $n \geq 3$.

Remark. Of course, we have been rather cavalier in our presentation. Given a topological space X, the group $\pi_1(X)$ is the *fundamental group of* X, *i.e.* the group of homotopy classes of closed paths in X (under composition of loops). But $\pi_0(X)$ is generally *not* a group! Instead, $\pi_0(X)$ is the set of path-connected components of X. However, when X is a Lie group, $\pi_0(X)$ is indeed a group. Also, we have to make sense of what it means for the sequence to be exact. All this can be made rigorous (see Bott and Tu [12], Chapter 3, Section 17, or Rotman [91], Chapter 11).

11.8   Periodicity of the Clifford Algebras $Cl_{p,q}$

It turns out that the real algebras $Cl_{p,q}$ can be built up as tensor products of the basic algebras $\mathbb{R}$, $\mathbb{C}$, and $\mathbb{H}$. As pointed out by Lounesto (Section 23.16 [74]), the description of the real algebras $Cl_{p,q}$ as matrix algebras and the 8-periodicity was first observed by Elie Cartan in 1908; see Cartan's article, *Nombres Complexes*, based on the original article in German by E. Study, in Molk [80], article I-5 (fasc. 3), pages 329–468. These algebras are defined in Section 36 under the name "'Systems of Clifford and Lipschitz," pages 463–466. Of course, Cartan used a very different notation; see page 464 in the article cited above. These facts were rediscovered independently by Raoul Bott in the 1960s (see Raoul Bott's comments in Volume 2 of his collected papers.).

We adopt the notation $K(n)$ for the algebra of $n \times n$ matrices over a ring or a field K; here, $K = \mathbb{R}, \mathbb{C}, \mathbb{H}$. This is the notation used in most of the literature on Clifford algebras, for instance Atiyah, Bott, and Shapiro [6], and it is a departure from the notation $M_n(K)$ that we have been using all along.

As mentioned in Examples 11.3 and 11.7, it is not hard to show that

$$Cl_{0,1} = \mathbb{C} \qquad Cl_{1,0} = \mathbb{R} \oplus \mathbb{R}$$
$$Cl_{0,2} = \mathbb{H} \qquad Cl_{2,0} = \mathbb{R}(2),$$

and

$$Cl_{1,1} = \mathbb{R}(2).$$

The key to the classification is the following lemma:

Lemma 11.23. *We have the isomorphisms*

$$Cl_{0,n+2} \cong Cl_{n,0} \otimes Cl_{0,2}$$

$$Cl_{n+2,0} \cong Cl_{0,n} \otimes Cl_{2,0}$$

$$Cl_{p+1,q+1} \cong Cl_{p,q} \otimes Cl_{1,1},$$

for all $n, p, q \geq 0$.

Proof. Let $\Phi_{0,n+2}(x) = -\|x\|^2$, where $\|x\|$ is the standard Euclidean norm on $\mathbb{R}^{n+2}$, and let $(e_1, \ldots, e_{n+2})$ be an orthonormal basis for $\mathbb{R}^{n+2}$ under the standard Euclidean inner product. We also let $(e'_1, \ldots, e'_n)$ be a set of generators for $Cl_{n,0}$ and (e''_1, e''_2) be a set of generators for $Cl_{0,2}$. We can define a linear map $f \colon \mathbb{R}^{n+2} \to Cl_{n,0} \otimes Cl_{0,2}$ by its action on the basis $(e_1, \ldots, e_{n+2})$ as follows:

$$f(e_i) = \begin{cases} e'_i \otimes e''_1 e''_2 & \text{for } 1 \leq i \leq n \\ 1 \otimes e''_{i-n} & \text{for } n+1 \leq i \leq n+2. \end{cases}$$

Observe that for $1 \leq i, j \leq n$, we have

$$f(e_i)f(e_j) + f(e_j)f(e_i) = (e'_i e'_j + e'_j e'_i) \otimes (e''_1 e''_2)^2 = -2\delta_{ij}\mathbf{1} \otimes \mathbf{1},$$

since $e''_1 e''_2 = -e''_2 e''_1$, $(e''_1)^2 = -\mathbf{1}$, and $(e''_2)^2 = -\mathbf{1}$, and $e'_i e'_j = -e'_j e'_i$, for all $i \neq j$, and $(e'_i)^2 = \mathbf{1}$, for all i with $1 \leq i \leq n$. Also, for $n+1 \leq i, j \leq n+2$, we have

$$f(e_i)f(e_j) + f(e_j)f(e_i) = \mathbf{1} \otimes (e''_{i-n} e''_{j-n} + e''_{j-n} e''_{i-n}) = -2\delta_{ij}\mathbf{1} \otimes \mathbf{1},$$

and

$$f(e_i)f(e_k) + f(e_k)f(e_i) = 2e'_i \otimes (e''_1 e''_2 e''_{k-n} + e''_{k-n} e''_1 e''_2) = 0,$$

for $1 \leq i, j \leq n$ and $n+1 \leq k \leq n+2$ (since $e''_{k-n} = e''_1$ or $e''_{k-n} = e''_2$). Thus, we have

$$(f(x))^2 = -\|x\|^2 \cdot \mathbf{1} \otimes \mathbf{1} \quad \text{for all } x \in \mathbb{R}^{n+2},$$

and by the universal mapping property of $Cl_{0,n+2}$, we get an algebra map

$$\widetilde{f} \colon Cl_{0,n+2} \to Cl_{n,0} \otimes Cl_{0,2}.$$

Since $\widetilde{f}$ maps onto a set of generators, it is surjective. However

$$\dim(\text{Cl}_{0,n+2}) = 2^{n+2} = 2^n \cdot 2 = \dim(\text{Cl}_{n,0})\dim(\text{Cl}_{0,2}) = \dim(\text{Cl}_{n,0} \otimes \text{Cl}_{0,2}),$$

and $\widetilde{f}$ is an isomorphism.

The proof of the second identity is analogous. For the third identity, we have

$$\Phi_{p,q}(x_1, \ldots, x_{p+q}) = x_1^2 + \cdots + x_p^2 - (x_{p+1}^2 + \cdots + x_{p+q}^2),$$

and let $(e_1, \ldots, e_{p+1}, \epsilon_1, \ldots, \epsilon_{q+1})$ be an orthogonal basis for $\mathbb{R}^{p+q+2}$ so that $\Phi_{p+1,q+1}(e_i) = +1$ and $\Phi_{p+1,q+1}(\epsilon_j) = -1$ for $i = 1, \ldots, p+1$ and $j = 1, \ldots, q+1$. Also, let $(e_1', \ldots, e_p', \epsilon_1', \ldots, \epsilon_q')$ be a set of generators for $\text{Cl}_{p,q}$ and (e_1'', ϵ_1'') be a set of generators for $\text{Cl}_{1,1}$. We define a linear map $f \colon \mathbb{R}^{p+q+2} \to \text{Cl}_{p,q} \otimes \text{Cl}_{1,1}$ by its action on the basis as follows:

$$f(e_i) = \begin{cases} e_i' \otimes e_1'' \epsilon_1'' & \text{for } 1 \le i \le p \\ 1 \otimes e_1'' & \text{for } i = p+1, \end{cases}$$

and

$$f(\epsilon_j) = \begin{cases} \epsilon_j' \otimes e_1'' \epsilon_1'' & \text{for } 1 \le j \le q \\ 1 \otimes \epsilon_1'' & \text{for } j = q+1. \end{cases}$$

We can check that

$$(f(x))^2 = \Phi_{p+1,q+1}(x) \cdot 1 \otimes 1 \quad \text{for all } x \in \mathbb{R}^{p+q+2},$$

and we finish the proof as in the first case. $\square$

To apply this lemma, we need some further isomorphisms among various matrix algebras.

Proposition 11.24. *The following isomorphisms hold:*

$$\mathbb{R}(m) \otimes \mathbb{R}(n) \cong \mathbb{R}(mn) \quad \text{for all } m, n \ge 0$$

$$\mathbb{R}(n) \otimes_{\mathbb{R}} K \cong K(n) \quad \text{for } K = \mathbb{C} \text{ or } K = \mathbb{H} \text{ and all } n \ge 0$$

$$\mathbb{C} \otimes_{\mathbb{R}} \mathbb{C} \cong \mathbb{C} \oplus \mathbb{C}$$

$$\mathbb{C} \otimes_{\mathbb{R}} \mathbb{H} \cong \mathbb{C}(2)$$

$$\mathbb{H} \otimes_{\mathbb{R}} \mathbb{H} \cong \mathbb{R}(4).$$

Proof. Details can be found in Lawson and Michelsohn [71]. The first two isomorphisms are quite obvious. The third isomorphism $\mathbb{C} \oplus \mathbb{C} \to \mathbb{C} \otimes_{\mathbb{R}} \mathbb{C}$ is obtained by sending

$$(1,0) \mapsto \frac{1}{2}(1 \otimes 1 + i \otimes i), \quad (0,1) \mapsto \frac{1}{2}(1 \otimes 1 - i \otimes i).$$

The field $\mathbb{C}$ is isomorphic to the subring of $\mathbb{H}$ generated by **i**. Thus, we can view $\mathbb{H}$ as a $\mathbb{C}$-vector space under left scalar multiplication. Consider the $\mathbb{R}$-bilinear map $\pi : \mathbb{C} \times \mathbb{H} \to \mathrm{Hom}_{\mathbb{C}}(\mathbb{H}, \mathbb{H})$ given by

$$\pi_{y,z}(x) = yx\overline{z},$$

where $y \in \mathbb{C}$ and $x, z \in \mathbb{H}$. Thus, we get an $\mathbb{R}$-linear map $\pi : \mathbb{C} \otimes_{\mathbb{R}} \mathbb{H} \to \mathrm{Hom}_{\mathbb{C}}(\mathbb{H}, \mathbb{H})$. However, we have $\mathrm{Hom}_{\mathbb{C}}(\mathbb{H}, \mathbb{H}) \cong \mathbb{C}(2)$. Furthermore, since

$$\pi_{y,z} \circ \pi_{y',z'} = \pi_{yy',zz'},$$

the map π is an algebra homomorphism

$$\pi : \mathbb{C} \times \mathbb{H} \to \mathbb{C}(2).$$

We can check on a basis that π is injective, and since

$$\dim_{\mathbb{R}}(\mathbb{C} \times \mathbb{H}) = \dim_{\mathbb{R}}(\mathbb{C}(2)) = 8,$$

the map π is an isomorphism. The last isomorphism is proved in a similar fashion. $\square$

We now have the main periodicity theorem.

Theorem 11.25 (Cartan/Bott). *For all $n \geq 0$, we have the following isomorphisms:*

$$Cl_{0,n+8} \cong Cl_{0,n} \otimes Cl_{0,8}$$
$$Cl_{n+8,0} \cong Cl_{n,0} \otimes Cl_{8,0}.$$

Furthermore,

$$Cl_{0,8} = Cl_{8,0} = \mathbb{R}(16).$$

Proof. By Lemma 11.23 we have the isomorphisms

$$Cl_{0,n+2} \cong Cl_{n,0} \otimes Cl_{0,2}$$
$$Cl_{n+2,0} \cong Cl_{0,n} \otimes Cl_{2,0},$$

and thus,

$$\mathrm{Cl}_{0,n+8} \cong \mathrm{Cl}_{n+6,0} \otimes \mathrm{Cl}_{0,2} \cong \mathrm{Cl}_{0,n+4} \otimes \mathrm{Cl}_{2,0} \otimes \mathrm{Cl}_{0,2} \cong \cdots$$
$$\cong \mathrm{Cl}_{0,n} \otimes \mathrm{Cl}_{2,0} \otimes \mathrm{Cl}_{0,2} \otimes \mathrm{Cl}_{2,0} \otimes \mathrm{Cl}_{0,2}.$$

Since $\mathrm{Cl}_{0,2} = \mathbb{H}$ and $\mathrm{Cl}_{2,0} = \mathbb{R}(2)$, by Proposition 11.24, we get

$$\mathrm{Cl}_{2,0} \otimes \mathrm{Cl}_{0,2} \otimes \mathrm{Cl}_{2,0} \otimes \mathrm{Cl}_{0,2} \cong \mathbb{H} \otimes \mathbb{H} \otimes \mathbb{R}(2) \otimes \mathbb{R}(2) \cong \mathbb{R}(4) \otimes \mathbb{R}(4) \cong \mathbb{R}(16).$$

The second isomorphism is proved in a similar fashion.   □

From all this, we can deduce the following table.

n	0	1	2	3	4	5	6	7	8
$\mathrm{Cl}_{0,n}$	$\mathbb{R}$	$\mathbb{C}$	$\mathbb{H}$	$\mathbb{H} \oplus \mathbb{H}$	$\mathbb{H}(2)$	$\mathbb{C}(4)$	$\mathbb{R}(8)$	$\mathbb{R}(8) \oplus \mathbb{R}(8)$	$\mathbb{R}(16)$
$\mathrm{Cl}_{n,0}$	$\mathbb{R}$	$\mathbb{R} \oplus \mathbb{R}$	$\mathbb{R}(2)$	$\mathbb{C}(2)$	$\mathbb{H}(2)$	$\mathbb{H}(2) \oplus \mathbb{H}(2)$	$\mathbb{H}(4)$	$\mathbb{C}(8)$	$\mathbb{R}(16)$

A table of the Clifford groups $\mathrm{Cl}_{p,q}$ for $0 \leq p, q \leq 7$ can be found in Kirillov [63], and for $0 \leq p, q \leq 8$, in Lawson and Michelsohn [71] (but beware that their $\mathrm{Cl}_{p,q}$ is our $\mathrm{Cl}_{q,p}$). It can also be shown that

$$\mathrm{Cl}_{p+1,q} \cong \mathrm{Cl}_{q+1,p}$$
$$\mathrm{Cl}_{p,q} \cong \mathrm{Cl}_{p-4,q+4}$$

with $p \geq 4$ in the second identity (see Lounesto [74], Chapter 16, Sections 16.3 and 16.4). Using the second identity, if $|p - q| = 4k$, it is easily shown by induction on k that $\mathrm{Cl}_{p,q} \cong \mathrm{Cl}_{q,p}$, as claimed in the previous section.

We also have the isomorphisms

$$\mathrm{Cl}_{p,q} \cong \mathrm{Cl}_{p,q+1}^0,$$

from which it follows that

$$\mathbf{Spin}(p, q) \cong \mathbf{Spin}(q, p)$$

(see Choquet-Bruhat [25], Chapter I, Sections 4 and 7). However, in general, $\mathbf{Pin}(p, q)$ and $\mathbf{Pin}(q, p)$ are not isomorphic. In fact, $\mathbf{Pin}(0, n)$ and $\mathbf{Pin}(n, 0)$ are not isomorphic if $n \neq 4k$, with $k \in \mathbb{N}$ (see Choquet-Bruhat [25], Chapter I, Section 7, page 27).

11.9   The Complex Clifford Algebras Cl(n, $\mathbb{C}$)

One can also consider Clifford algebras over the complex field $\mathbb{C}$. In this case, it is well known that every nondegenerate quadratic form can be expressed by

$$\Phi_n^{\mathbb{C}}(x_1, \ldots, x_n) = x_1^2 + \cdots + x_n^2$$

in some orthonormal basis. Also, it is easily shown that the complexification $\mathbb{C} \otimes_{\mathbb{R}}$ $\mathrm{Cl}_{p,q}$ of the real Clifford algebra $\mathrm{Cl}_{p,q}$ is isomorphic to $\mathrm{Cl}(\Phi_n^{\mathbb{C}})$. Thus, all these complex algebras are isomorphic for $p + q = n$, and we denote them by $\mathrm{Cl}(n, \mathbb{C})$. Theorem 11.23 yields the following periodicity theorem:

Theorem 11.26.   *The following isomorphisms hold:*

$$\mathrm{Cl}(n + 2, \mathbb{C}) \cong \mathrm{Cl}(n, \mathbb{C}) \otimes_{\mathbb{C}} \mathrm{Cl}(2, \mathbb{C}),$$

with $\mathrm{Cl}(2, \mathbb{C}) = \mathbb{C}(2)$.

Proof. Since $\mathrm{Cl}(n, \mathbb{C}) = \mathbb{C} \otimes_{\mathbb{R}} \mathrm{Cl}_{0,n} = \mathbb{C} \otimes_{\mathbb{R}} \mathrm{Cl}_{n,0}$, by Lemma 11.23, we have

$$\mathrm{Cl}(n{+}2, \mathbb{C}) = \mathbb{C} \otimes_{\mathbb{R}} \mathrm{Cl}_{0,n+2} \cong \mathbb{C} \otimes_{\mathbb{R}} (\mathrm{Cl}_{n,0} \otimes_{\mathbb{R}} \mathrm{Cl}_{0,2}) \cong (\mathbb{C} \otimes_{\mathbb{R}} \mathrm{Cl}_{n,0}) \otimes_{\mathbb{C}} (\mathbb{C} \otimes_{\mathbb{R}} \mathrm{Cl}_{0,2}).$$

However, $\mathrm{Cl}_{0,2} = \mathbb{H}$, $\mathrm{Cl}(n, \mathbb{C}) = \mathbb{C} \otimes_{\mathbb{R}} \mathrm{Cl}_{n,0}$, and $\mathbb{C} \otimes_{\mathbb{R}} \mathbb{H} \cong \mathbb{C}(2)$, so we get $\mathrm{Cl}(2, \mathbb{C}) = \mathbb{C}(2)$ and

$$\mathrm{Cl}(n + 2, \mathbb{C}) \cong \mathrm{Cl}(n, \mathbb{C}) \otimes_{\mathbb{C}} \mathbb{C}(2),$$

and the theorem is proven. $\square$

As a corollary of Theorem 11.26, we obtain the fact that

$$\mathrm{Cl}(2k, \mathbb{C}) \cong \mathbb{C}(2^k) \quad \text{and} \quad \mathrm{Cl}(2k + 1, \mathbb{C}) \cong \mathbb{C}(2^k) \oplus \mathbb{C}(2^k).$$

The table of the previous section can also be completed as follows:

n	0	1	2	3	4	5	6	7	8
$\mathrm{Cl}_{0,n}$	$\mathbb{R}$	$\mathbb{C}$	$\mathbb{H}$	$\mathbb{H} \oplus \mathbb{H}$	$\mathbb{H}(2)$	$\mathbb{C}(4)$	$\mathbb{R}(8)$	$\mathbb{R}(8) \oplus \mathbb{R}(8)$	$\mathbb{R}(16)$
$\mathrm{Cl}_{n,0}$	$\mathbb{R}$	$\mathbb{R} \oplus \mathbb{R}$	$\mathbb{R}(2)$	$\mathbb{C}(2)$	$\mathbb{H}(2)$	$\mathbb{H}(2) \oplus \mathbb{H}(2)$	$\mathbb{H}(4)$	$\mathbb{C}(8)$	$\mathbb{R}(16)$
$\mathrm{Cl}(n, \mathbb{C})$	$\mathbb{C}$	$2\mathbb{C}$	$\mathbb{C}(2)$	$2\mathbb{C}(2)$	$\mathbb{C}(4)$	$2\mathbb{C}(4)$	$\mathbb{C}(8)$	$2\mathbb{C}(8)$	$\mathbb{C}(16)$,

where $2\mathbb{C}(k)$ is an abbreviation for $\mathbb{C}(k) \oplus \mathbb{C}(k)$.

11.10   Clifford Groups Over a Field K

In this final section we quickly indicate which of the results about Clifford algebras, Clifford groups, and the **Pin** and **Spin** groups obtained in Sections 11.3–11.6 for vector spaces over the fields $\mathbb{R}$ and $\mathbb{C}$ generalize to nondegenerate bilinear forms on vector spaces over an arbitrary field K of *characteristic different from* 2. As we will see, most results generalize, except for some of the surjectivity results such as Theorem 11.17.

Let V be a finite-dimensional vector space over a field K of characteristic $\neq 2$, let $\varphi \colon V \times V \to K$ be a possibly degenerate symmetric bilinear form, and let $\Phi(v) = \varphi(v, v)$ be the corresponding quadratic form.

Definition 11.20. A *Clifford algebra associated with V and Φ* is a K-algebra $\mathrm{Cl}(V, \Phi)$ together with a linear map $i_\Phi \colon V \to \mathrm{Cl}(V, \Phi)$ satisfying the condition $(i_\Phi(v))^2 = \Phi(v) \cdot \mathbf{1}$ for all $v \in V$, and so that for every K-algebra A and every linear map $f \colon V \to A$ with

$$(f(v))^2 = \Phi(v) \cdot \mathbf{1}_A \quad \text{for all } v \in V,$$

there is a unique algebra homomorphism $\overline{f} \colon \mathrm{Cl}(V, \Phi) \to A$ so that

$$f = \overline{f} \circ i_\Phi,$$

as in the diagram below.

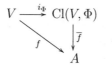

We use the notation $\lambda \cdot u$ for the product of a scalar $\lambda \in K$ and of an element u in the algebra $\mathrm{Cl}(V, \Phi)$, and juxtaposition uv for the multiplication of two elements u and v in the algebra $\mathrm{Cl}(V, \Phi)$.

By a familiar argument, any two Clifford algebras associated with V and Φ are isomorphic. We often denote i_Φ by i.

To show the existence of $\mathrm{Cl}(V, \Phi)$, since the tensor algebra $T(V)$ makes sense for a vector space V over any field K, observe that $T(V)/\mathfrak{A}$ does the job, where $\mathfrak{A}$ is the ideal of $T(V)$ generated by all elements of the form $v \otimes v - \Phi(v) \cdot \mathbf{1}$, where $v \in V$. The map $i_\Phi \colon V \to \mathrm{Cl}(V, \Phi)$ is the composition

$$V \xrightarrow{\iota_1} T(V) \xrightarrow{\pi} T(V)/\mathfrak{A},$$

where π is the natural quotient map. We often denote the Clifford algebra $\mathrm{Cl}(V, \Phi)$ simply by $\mathrm{Cl}(\Phi)$.

Proposition 11.27. *Every Clifford algebra* $\mathrm{Cl}(\Phi)$ *possesses a canonical anti-automorphism* $t \colon \mathrm{Cl}(\Phi) \to \mathrm{Cl}(\Phi)$ *satisfying the properties*

$$t(xy) = t(y)t(x), \quad t \circ t = id, \quad and \quad t(i(v)) = i(v),$$

for all $x, y \in \mathrm{Cl}(\Phi)$ *and all* $v \in V$. *Furthermore, such an anti-automorphism is unique.*

Proposition 11.28. *Every Clifford algebra* $\mathrm{Cl}(\Phi)$ *has a unique canonical automorphism* $\alpha \colon \mathrm{Cl}(\Phi) \to \mathrm{Cl}(\Phi)$ *satisfying the properties*

$$\alpha \circ \alpha = id, \quad and \quad \alpha(i(v)) = -i(v),$$

for all $v \in V$.
Write

$$\mathrm{Cl}(\Phi) = \mathrm{Cl}^0(\Phi) \oplus \mathrm{Cl}^1(\Phi),$$

where

$$\mathrm{Cl}^i(\Phi) = \{x \in \mathrm{Cl}(\Phi) \mid \alpha(x) = (-1)^i x\}, \quad \text{where } i = 0, 1.$$

We say that we have a $\mathbb{Z}/2$-*grading*.

The theorem about the existence of a nice basis of $\mathrm{Cl}(\Phi)$ only depends on the fact that there is always a basis of V that is orthogonal with respect to φ (even if φ is degenerate) so we have

Theorem 11.29. *For every vector space* V *of finite dimension* n, *the map* $i \colon V \to \mathrm{Cl}(\Phi)$ *is injective. Given a basis* $(e_1, \ldots, e_n)$ *of* V, *the* $2^n - 1$ *products*

$$i(e_{i_1})i(e_{i_2}) \cdots i(e_{i_k}), \quad 1 \le i_1 < i_2 \cdots < i_k \le n,$$

and $\mathbf{1}$ *form a basis of* $\mathrm{Cl}(\Phi)$. *Thus,* $\mathrm{Cl}(\Phi)$ *has dimension* 2^n.

Since i is injective, for simplicity of notation, from now on we write u for $i(u)$. Theorem 11.29 implies that if $(e_1, \ldots, e_n)$ is an orthogonal basis of V with respect to Φ, then $\mathrm{Cl}(\Phi)$ is the algebra presented by the generators $(e_1, \ldots, e_n)$ and the relations

$$e_j^2 = \Phi(e_j) \cdot \mathbf{1}, \quad 1 \le j \le n, \quad and$$

$$e_j e_k = -e_k e_j, \quad 1 \le j, k \le n, \ j \ne k.$$

If V has finite dimension n and $(e_1, \ldots, e_n)$ is a basis of V, by Theorem 11.29, the maps t and α are completely determined by their action on the basis elements. Namely, t is defined by

$$t(e_i) = e_i$$

$$t(e_{i_1} e_{i_2} \cdots e_{i_k}) = e_{i_k} e_{i_{k-1}} \cdots e_{i_1},$$

where $1 \leq i_1 < i_2 \cdots < i_k \leq n$, and of course, $t(\mathbf{1}) = \mathbf{1}$. The map α is defined by

$$\alpha(e_i) = -e_i$$

$$\alpha(e_{i_1} e_{i_2} \cdots e_{i_k}) = (-1)^k e_{i_1} e_{i_2} \cdots e_{i_k}$$

where $1 \leq i_1 < i_2 < \cdots < i_k \leq n$, and of course, $\alpha(\mathbf{1}) = \mathbf{1}$. Furthermore, the even-graded elements (the elements of $\mathrm{Cl}^0(\Phi)$) are those generated by $\mathbf{1}$ and the basis elements consisting of an even number of factors $e_{i_1} e_{i_2} \cdots e_{i_{2k}}$, and the odd-graded elements (the elements of $\mathrm{Cl}^1(\Phi)$) are those generated by the basis elements consisting of an odd number of factors $e_{i_1} e_{i_2} \cdots e_{i_{2k+1}}$.

The definition of the Clifford group given in Section 11.4 does not depend on the field K or on the fact that the symmetric bilinear form φ is nondegenerate.

Definition 11.21. Given a finite-dimensional vector space V over a field K and a quadratic form Φ on V, the *Clifford group of* Φ is the group

$$\Gamma(\Phi) = \{x \in \mathrm{Cl}(\Phi)^* \mid \alpha(x)vx^{-1} \in V \quad \text{for all } v \in V\}.$$

For any $x \in \Gamma(\Phi)$, let $\rho_x \colon V \to V$ be the map defined by

$$v \mapsto \alpha(x)vx^{-1}, \quad v \in V.$$

As in Section 11.4, the map $\rho \colon \Gamma(\Phi) \to \mathbf{GL}(V)$ given by $x \mapsto \rho_x$ is a linear action, and $\Gamma(\Phi)$ is a group. This is because V is finite dimensional and α is an automorphism.

We also define the group $\Gamma^+(\Phi)$, called the *special Clifford group*, by

$$\Gamma^+(\Phi) = \Gamma(\Phi) \cap \mathrm{Cl}^0(\Phi).$$

Proposition 11.30. *The maps α and t induce an automorphism and an anti-automorphism of the Clifford group, $\Gamma(\Phi)$.*

The following key result obtained in Section 11.4 still holds because its proof does not depend on the field K.

Theorem 11.31. *Let V be a finite-dimensional vector space over a field K and let Φ a quadratic form on V. For every element x of the Clifford group $\Gamma(\Phi)$, if $x \in V$ then $\Phi(x) \neq 0$ and the map $\rho_x \colon V \to V$ given by*

$$v \mapsto \alpha(x)vx^{-1} \quad \text{for all } v \in V$$

is the reflection about the hyperplane H orthogonal to the non-isotropic vector x.

We would like to show that ρ is a surjective homomorphism from $\Gamma(\Phi)$ onto $\mathbf{O}(\Phi)$, and a surjective homomorphism from $\Gamma^{+}(\Phi)$ onto $\mathbf{SO}(\Phi)$.

In order to prove that $\rho_x \in \mathbf{O}(\Phi)$ for any $x \in \Gamma(\Phi)$, we define a notion of norm on $\Gamma(\Phi)$, and for this we need to define a notion of conjugation on $\mathrm{Cl}(\Phi)$.

Definition 11.22. We define *conjugation* on a Clifford algebra $\mathrm{Cl}(\Phi)$ as the map

$$x \mapsto \overline{x} = t(\alpha(x)) \quad \text{for all } x \in \mathrm{Cl}(\Phi).$$

Conjugation is an anti-automorphism.

If V has finite dimension n and $(e_1, \ldots, e_n)$ is a basis of V, in view of previous remarks, conjugation is defined by

$$\overline{e_i} = -e_i$$

$$\overline{e_{i_1} e_{i_2} \cdots e_{i_k}} = (-1)^k e_{i_k} e_{i_{k-1}} \cdots e_{i_1}$$

where $1 \le i_1 < i_2 \cdots < i_k \le n$, and of course, $\overline{\mathbf{1}} = \mathbf{1}$.

Definition 11.23. The map $N \colon \mathrm{Cl}(\Phi) \to \mathrm{Cl}(\Phi)$ given by

$$N(x) = x\overline{x}$$

is called the *norm* of $\mathrm{Cl}(\Phi)$.

Observe that $N(v) = v\overline{v} = -v^2 = -\Phi(v) \cdot \mathbf{1}$ for all $v \in V$.

Up to this point, there is *no* assumption regarding the degeneracy of φ. Now we will need to assume that φ is nondegenerate. We observed that the proof of Lemma 11.11 goes through as long as φ is nondegenerate. Thus we have

Lemma 11.32. *Assume φ is a nondegenerate bilinear map on V. The kernel of the map $\rho \colon \Gamma(\Phi) \to \mathbf{GL}(V)$ is $K^* \cdot \mathbf{1}$, the multiplicative group of nonzero scalar multiples of $\mathbf{1} \in \mathrm{Cl}(\Phi)$.*

We also observed that the proof of Proposition 11.12 goes through as long as φ is nondegenerate.

Proposition 11.33. *Assume φ is a nondegenerate bilinear map on V. If $x \in \Gamma(\Phi)$, then $N(x) \in K^* \cdot \mathbf{1}$.*

Similarly the following holds.

Proposition 11.34. *Assume φ is a nondegenerate bilinear map on V. The restriction of the norm N to $\Gamma(\Phi)$ is a homomorphism $N \colon \Gamma(\Phi) \to K^* \cdot \mathbf{1}$, and $N(\alpha(x)) = N(x)$ for all $x \in \Gamma(\Phi)$.*

Finally we obtain the following result.

Proposition 11.35. *Assume φ is a nondegenerate bilinear map on V. The set of non-isotropic vectors in V (those $x \in V$ such that $\Phi(x) \neq 0$) is a subset of $\Gamma(\Phi)$, and $\rho(\Gamma(\Phi)) \subseteq \mathbf{O}(\Phi)$.*

We have the following theorem.

Theorem 11.36. *Assume φ is a nondegenerate bilinear map on V. The map $\rho \colon \Gamma(\Phi) \to \mathbf{O}(\Phi)$ is a surjective homomorphism whose kernel is $K^* \cdot \mathbf{1}$, and the map $\rho \colon \Gamma^+(\Phi) \to \mathbf{SO}(\Phi)$ is a surjective homomorphism whose kernel is $K^* \cdot \mathbf{1}$.*

Proof. The Cartan-Dieudonné holds for any nondegenerate quadratic form Φ over a field of characteristic $\neq 2$, in the sense that every isometry $f \in \mathbf{O}(\Phi)$ is the composition $f = s_1 \circ \cdots \circ s_k$ of reflections s_j defined by hyperplanes orthogonal to non-isotropic vectors $w_j \in V$ (see Dieudonné [30], Chevalley [24], Bourbaki [13], or Gallier [46], Chapter 7, Problem 7.14). Then we have

$$f = \rho(w_1 \cdots w_k),$$

and since the w_j are non-isotropic $\Phi(w_j) \neq 0$, so $w_j \in \Gamma(\Phi)$ and we have $w_1 \cdots w_k \in \Gamma(\Phi)$.

For the second statement, we need to show that ρ maps $\Gamma^+(\Phi)$ into $\mathbf{SO}(\Phi)$. If this was not the case, there would be some improper isometry $f \in \mathbf{O}(\Phi)$ so that $\rho_x = f$, where $x \in \Gamma(\Phi) \cap \mathrm{Cl}^0(\Phi)$. However, we can express f as the composition of an odd number of reflections, say

$$f = \rho(w_1 \cdots w_{2k+1}).$$

Since

$$\rho(w_1 \cdots w_{2k+1}) = \rho_x,$$

we have $x^{-1} w_1 \cdots w_{2k+1} \in \mathrm{Ker}\,(\rho)$. By Lemma 11.32, we must have

$$x^{-1} w_1 \cdots w_{2k+1} = \lambda \cdot \mathbf{1}$$

for some $\lambda \in K^*$, and thus

$$w_1 \cdots w_{2k+1} = \lambda \cdot x,$$

where x has even degree and $w_1 \cdots w_{2k+1}$ has odd degree, which is impossible. □

The groups **Pin** and **Spin** are defined as follows.

Definition 11.24. Assume φ is a nondegenerate bilinear map on V. We define the *pinor group* **Pin**(Φ) as the group

$$\mathbf{Pin}(\Phi) = \{x \in \Gamma(\Phi) \mid N(x) = \pm 1\},$$

equivalently

$$\mathbf{Pin}(\Phi) = \{x \in \mathrm{Cl}(\Phi) \mid xv\overline{x} \in V \quad \text{for all } v \in V, \ N(x) = \pm 1\},$$

and the *spinor group* $\mathbf{Spin}(\Phi)$ as $\mathbf{Pin}(\Phi) \cap \Gamma^+(\Phi)$. Equivalently,

$$\mathbf{Spin}(\Phi) = \{x \in \mathrm{Cl}^0(\Phi) \mid xv\overline{x} \in V \quad \text{for all } v \in V, \ N(x) = \pm 1\}.$$

This time, if the field K is not $\mathbb{R}$ or $\mathbb{C}$, it is not obvious that the restriction of ρ to $\mathbf{Pin}(\Phi)$ is surjective onto $\mathbf{O}(\Phi)$ and that the restriction of ρ to $\mathbf{Spin}(\Phi)$ is surjective onto $\mathbf{SO}(\Phi)$.

To understand this better, assume that

$$\rho_x = \rho(y) = f$$

for some $x, y \in \Gamma(\Phi)$ and some $f \in \mathbf{O}(\Phi)$. Then $\rho(yx^{-1}) = \mathrm{id}$, which by Lemma 11.32 implies that $yx^{-1} = \lambda \mathbf{1}$ for some $\lambda \in K^*$, that is,

$$y = \lambda x.$$

Then we obtain

$$N(y) = y\overline{y} = \lambda x \, \overline{\lambda x} = \lambda^2 x\overline{x} = \lambda^2 N(x).$$

This suggests defining a map σ from $\mathbf{O}(\Phi)$ to the group $K^*/(\pm(K^*)^2)$ by

$$\sigma(f) = [N(x)] \quad \text{for any } x \in \Gamma(\Phi) \text{ with } \rho_x = f,$$

where $\pm(K^*)^2$ denotes the multiplicative subgroup of K^* consisting of all elements of the form $\pm\lambda^2$, with $\lambda \in K^*$, and $[N(x)]$ denotes the equivalence class of $N(x)$ in $K^*/(\pm(K^*)^2)$. Then we have the following result.

Proposition 11.37. *Assume φ is a nondegenerate bilinear map on V. We have the exact sequences*

$$(1) \longrightarrow \{-1, 1\} \longrightarrow \mathbf{Pin}(\Phi) \xrightarrow{\ \rho\ } \mathbf{O}(\Phi) \xrightarrow{\ \sigma\ } \mathrm{Im}\,\sigma \longrightarrow (1)$$

$$(1) \longrightarrow \{-1, 1\} \longrightarrow \mathbf{Spin}(\Phi) \xrightarrow{\ \rho\ } \mathbf{SO}(\Phi) \xrightarrow{\ \sigma\ } \mathrm{Im}\,\sigma \longrightarrow (1).$$

and

$$(1) \longrightarrow \{-1, 1\} \longrightarrow \mathbf{Spin}(\Phi) \xrightarrow{\ \rho\ } \mathbf{SO}(\Phi) \xrightarrow{\ \sigma\ } \mathrm{Im}\,\sigma \longrightarrow (1).$$

Proof. Since by Lemma 11.32 the kernel of the map $\rho: \Gamma(\Phi) \to \mathbf{GL}(V)$ is $K^* \cdot \mathbf{1}$, and since $N(x) = \pm\mathbf{1}$ if $x \in \mathbf{Pin}(\Phi)$, the sequence is exact at $\mathbf{Pin}(\Phi)$. For any $x \in \mathbf{Pin}(\Phi)$, since $N(x) = \pm\mathbf{1}$, we have $\sigma(\rho_x) = 1$, which means that $\mathrm{Im}\,\rho \subseteq \mathrm{Ker}\,\sigma$. Assume that $f \in \mathrm{Ker}\,\sigma$, which means that $\rho_x = f$ some $x \in \Gamma(\Phi)$ such that $N(x) = \pm\lambda^2$ for some $\lambda \in K^*$. Then $N(\lambda^{-1}x) = \pm\mathbf{1}$ so $\lambda^{-1}x \in \mathbf{Pin}(\Phi)$ and since ρ is a homomorphism, $\rho(\lambda^{-1}x) = \rho(\lambda^{-1})\rho_x = \mathrm{id} \circ f = f$, which shows that $\mathrm{Ker}\,\sigma \subseteq \mathrm{Im}\,\rho$. The fact that the second sequence is exact follows from the fact that the first sequence is exact and by definition of $\mathbf{Spin}(\Phi)$. $\square$

If $K = \mathbb{R}$ or $K = \mathbb{C}$, *every* element of K is of the form $\pm\lambda^2$, so ρ is surjective, which gives another proof of Theorem 11.15.

Remarks.

(1) Our definition of σ is inspired by the definition of Mnemné and Testard [79] (Chapter 4, Section 4.7) who define σ from $\mathbf{SO}(\Phi)$ to the group $K^*/(K^*)^2$ by

$$\sigma(f) = [N(x)] \quad \text{for any } x \in \Gamma(\Phi) \text{ with } \rho_x = f.$$

Allowing negative squares as well as positive squares yields the surjectivity of ρ when $K = \mathbb{R}$ or $\mathbb{C}$.

(2) We define the subgroup $\mathbf{Spin}^+(\Phi)$ of $\mathbf{Spin}(\Phi)$ by

$$\mathbf{Spin}^+(\Phi) = \{x \in \mathrm{Cl}^0(\Phi) \mid xv\overline{x} \in V \quad \text{for all } v \in V, \ N(x) = \mathbf{1}\}.$$

The image of $\mathbf{Spin}^+(\Phi)$ by ρ is a subgroup of $\mathbf{SO}(\Phi)$ denoted by $\mathbf{SO}^+(\Phi)$. For example, when $K = \mathbb{R}$ and $\Phi = \Phi_{p,q}$, we have $\mathbf{SO}^+(\Phi_{p,q}) = \mathbf{SO}_0(p, q)$, the connected component of the identity. The group $\mathbf{Spin}^+(1, 1)$ has two connected components but $\mathbf{Spin}^+(p, q)$ is connected for $p + q \geq 2$ and $(p, q) \neq (1, 1)$. The groups $\mathbf{Spin}^+(n, 1) \cong \mathbf{Spin}^+(1, n)$ are simply connected for $n \geq 3$, but in general $\mathbf{Spin}^+(p, q)$ is not simply connected; for example, $\mathbf{Spin}^+(3, 3)$ is not simply connected; see Lounesto [74] (Chapter 17).

If K is an arbitrary field, we can't expect that the periodicity results of Section 11.8 and Section 11.9 hold for the Clifford algebra $\mathrm{Cl}(\Phi)$. Still some interesting facts about the structure of the algebras $\mathrm{Cl}(\Phi)$ can be established. For this, we need to define the notion of a central simple K-algebra.

If A is an associative K-algebra over a field K with identity element $\mathbf{1}$, then there is an injection of the field K into A given by $\lambda \mapsto \lambda \cdot \mathbf{1}$, so that we can view $K \cdot \mathbf{1}$ as a copy of K in A. Observe that every element $\lambda \cdot \mathbf{1} \in K \cdot \mathbf{1}$ commutes with every element $u \in A$, since by K-bilinearity of the multiplication operation $(u, v) \mapsto uv$ on A, we have

$$(\lambda \cdot \mathbf{1})u = \lambda \cdot (\mathbf{1}u) = \lambda \cdot u$$

and

$$u(\lambda \cdot \mathbf{1}) = \lambda \cdot (u\mathbf{1}) = \lambda \cdot u.$$

The preceding calculations show that $K \cdot \mathbf{1}$ is a contained in the center of A, which is defined as follows.

Definition 11.25. Given any associative K-algebra with identity element $\mathbf{1}$ (where K is a field), the *center* $Z(A)$ of A is the subalgebra of A given by

$$Z(A) = \{u \in A \mid uv = vu \quad \text{for all } v \in A\}.$$

The K-algebra A is called a *central* algebra if $Z(A) = K \cdot \mathbf{1}$.

As we just observed, $K \cdot \mathbf{1} \subseteq Z(A)$. A typical example of a central K-algebra is the algebra $M_n(K)$ of $n \times n$ matrices over a field K.

Definition 11.26. An associative K-algebra with identity element $\mathbf{1}$ is *simple* if for any two-sided ideal $\mathfrak{A}$ in A, either $\mathfrak{A} = (0)$ or $\mathfrak{A} = A$. In other words, A contains no nonzero proper two-sided ideals.

Again, a typical example of a simple K-algebra is the algebra $M_n(K)$ of $n \times n$ matrices over a field K. By a theorem of Wedderburn, any finite-dimensional central simple K-algebra is isomorphic to the algebra $M_n(\Delta)$ of $n \times n$ matrices over some division ring (also called a skew field) Δ whose center is K, for some $n \geq 1$; see Dummit and Foote [39], Chapter 17, Section 4, and Chapter 18, Section 2, Theorem 4.

Based on the results of Chevalley [24], the following results are proved in Bourbaki [13] (§9, no 4, Theorem 2, its Corollary, and Theorem 3).

Theorem 11.38. *If $m = 2r$ is even, for any nondegenerate quadratic form Φ over a K-vector space E of dimension m, the Clifford algebra $\mathrm{Cl}(\Phi)$ is a central simple algebra of dimension 2^m. If $m > 0$, the Clifford algebra $\mathrm{Cl}^0(\Phi)$ has a center Z of dimension 2, and either $\mathrm{Cl}^0(\Phi)$ is simple if Z is a field or $\mathrm{Cl}^0(\Phi)$ is the direct sum of two simple subalgebras of dimension 2^{m-2}.*

Remark. More is true when Φ is a neutral form (which means that E is the direct sum of two totally isotropic subspaces). In this case, $\mathrm{Cl}(\Phi)$ is isomorphic to the algebra of endomorphisms $\mathrm{End}(\bigwedge N)$ of the exterior product $\bigwedge N$ of some totally isotropic subspace N of E of dimension r.

Theorem 11.39. *If $m = 2r + 1$ is odd, for any nondegenerate quadratic form Φ over a K-vector space E of dimension m (with $\mathrm{char}(K) \neq 2$), the Clifford algebra $\mathrm{Cl}^0(\Phi)$ is a central simple algebra of dimension 2^{2r}. The Clifford algebra $\mathrm{Cl}(\Phi)$ has a center Z of dimension 2, and $\mathrm{Cl}(\Phi)$ is isomorphic to $Z \otimes_K \mathrm{Cl}^0(\Phi)$; as a consequence, $\mathrm{Cl}(\Phi)$ is either simple or the direct sum of two simple subalgebras.*

A related result due to Chevalley asserts that $\mathrm{Cl}(\Phi)$ is isomorphic to a subalgebra of the algebra of endomorphisms $\mathrm{End}(\bigwedge E)$. To prove this, Chevalley introduced a product operation on $\bigwedge E$ called the *Clifford product*. The reader is referred to

Lounesto [74] (Chapter 22) for details on this construction, as well as a simpler construction due to Riesz (who introduces an exterior product in $\mathrm{Cl}(\Phi)$).

The above results have some interesting applications to representation theory. Indeed, they lead to certain irreducible representations known as *spin representations* or *half-spin representations* first discovered by Élie Cartan. The spaces that they act on are called *spinors* or *half-spinors*. Such representations play an important role in theoretical physics. The interested reader is referred to Fulton and Harris [45] (Lecture 20) or Jost [62] (Section 2.4).

11.11 Problems

Problem 11.1. The "right way" (meaning convenient and rigorous) to define the *unit quaternions* is to define them as the elements of the unitary group $\mathbf{SU}(2)$, namely the group of 2×2 complex matrices of the form

$$\begin{pmatrix} \alpha & \beta \\ -\bar{\beta} & \bar{\alpha} \end{pmatrix} \quad \alpha, \beta \in \mathbb{C}, \ \alpha\bar{\alpha} + \beta\bar{\beta} = 1.$$

Then, the *quaternions* are the elements of the real vector space $\mathbb{H} = \mathbb{R}\,\mathbf{SU}(2)$. Let $\mathbf{1}, \mathbf{i}, \mathbf{j}, \mathbf{k}$ be the matrices

$$\mathbf{1} = \begin{pmatrix} 1 & 0 \\ 0 & 1 \end{pmatrix}, \quad \mathbf{i} = \begin{pmatrix} i & 0 \\ 0 & -i \end{pmatrix}, \quad \mathbf{j} = \begin{pmatrix} 0 & 1 \\ -1 & 0 \end{pmatrix}, \quad \mathbf{k} = \begin{pmatrix} 0 & i \\ i & 0 \end{pmatrix},$$

then $\mathbb{H}$ is the set of all matrices of the form

$$X = a\mathbf{1} + b\mathbf{i} + c\mathbf{j} + d\mathbf{k}, \quad a, b, c, d \in \mathbb{R}.$$

Indeed, every matrix in $\mathbb{H}$ is of the form

$$X = \begin{pmatrix} a + ib & c + id \\ -(c - id) & a - ib \end{pmatrix}, \quad a, b, c, d \in \mathbb{R}.$$

(1) Prove that the quaternions $\mathbf{1}, \mathbf{i}, \mathbf{j}, \mathbf{k}$ satisfy the famous identities discovered by Hamilton:

$$\mathbf{i}^2 = \mathbf{j}^2 = \mathbf{k}^2 = \mathbf{ijk} = -\mathbf{1},$$
$$\mathbf{ij} = -\mathbf{ji} = \mathbf{k},$$
$$\mathbf{jk} = -\mathbf{kj} = \mathbf{i},$$
$$\mathbf{ki} = -\mathbf{ik} = \mathbf{j}.$$

Prove that $\mathbb{H}$ is a skew field (a noncommutative field) called the *quaternions*, and a real vector space of dimension 4 with basis $(\mathbf{1}, \mathbf{i}, \mathbf{j}, \mathbf{k})$; thus as a vector space, $\mathbb{H}$ is isomorphic to $\mathbb{R}^4$.

A concise notation for the quaternion X defined by $\alpha = a + ib$ and $\beta = c + id$ is

$$X = [a, (b, c, d)].$$

We call a the *scalar part* of X and (b, c, d) the *vector part* of X. With this notation, $X^* = [a, -(b, c, d)]$, which is often denoted by $\overline{X}$. The quaternion $\overline{X}$ is called the *conjugate* of q. If q is a unit quaternion, then $\overline{q}$ is the multiplicative inverse of q. A *pure quaternion* is a quaternion whose scalar part is equal to zero.

(2) Given a unit quaternion

$$q = \begin{pmatrix} \alpha & \beta \\ -\overline{\beta} & \overline{\alpha} \end{pmatrix} \in \mathbf{SU}(2),$$

the usual way to define the rotation ρ_q (of $\mathbb{R}^3$) induced by q is to embed $\mathbb{R}^3$ into $\mathbb{H}$ as the pure quaternions, by

$$\psi(x, y, z) = \begin{pmatrix} ix & y + iz \\ -y + iz & -ix \end{pmatrix}, \quad (x, y, z) \in \mathbb{R}^3.$$

Observe that the above matrix is skew-Hermitian $(\psi(x, y, z)^* = -\psi(x, y, z))$. But, the space of skew-Hermitian matrices is the Lie algebra $\mathfrak{su}(2)$ of $\mathbf{SU}(2)$, so $\psi(x, y, z) \in \mathfrak{su}(2)$. Then, q defines the map ρ_q (on $\mathbb{R}^3$) given by

$$\rho_q(x, y, z) = \psi^{-1}(q\psi(x, y, z)q^*),$$

where q^* is the inverse of q (since $\mathbf{SU}(2)$ is a unitary group) and is given by

$$q^* = \begin{pmatrix} \overline{\alpha} & -\beta \\ \overline{\beta} & \alpha \end{pmatrix}.$$

Actually, the *adjoint representation* of the group $\mathbf{SU}(2)$ is the group homomorphism
Ad: $\mathbf{SU}(2) \rightarrow \mathbf{GL}(\mathfrak{su}(2))$ defined such that for every $q \in \mathbf{SU}(2)$,

$$\mathrm{Ad}_q(A) = qAq^*, \quad A \in \mathfrak{su}(2).$$

Therefore, modulo the isomorphism ψ, the linear map ρ_q is the linear isomorphism Ad_q. In fact, ρ_q is a rotation (and so is Ad_q), which you will prove shortly.

Since the matrix $\psi(x, y, z)$ is skew-Hermitian, the matrix $-i\psi(x, y, z)$ is Hermitian, and we have

$$-i\psi(x, y, z) = \begin{pmatrix} x & z - iy \\ z + iy & -x \end{pmatrix} = x\sigma_3 + y\sigma_2 + z\sigma_1,$$

where $\sigma_1, \sigma_2, \sigma_3$ are the *Pauli spin matrices*

$$\sigma_1 = \begin{pmatrix} 0 & 1 \\ 1 & 0 \end{pmatrix}, \quad \sigma_2 = \begin{pmatrix} 0 & -i \\ i & 0 \end{pmatrix}, \quad \sigma_3 = \begin{pmatrix} 1 & 0 \\ 0 & -1 \end{pmatrix}.$$

Check that $\mathbf{i} = i\sigma_3$, $\mathbf{j} = i\sigma_2$, $\mathbf{k} = i\sigma_1$. Prove that matrices of the form $x\sigma_3 + y\sigma_2 + z\sigma_1$ (with $x, y, x \in \mathbb{R}$) are exactly the 2×2 Hermitian matrix with zero trace.

(3) Prove that for every $q \in \mathbf{SU}(2)$, if A is any 2×2 Hermitian matrix with zero trace as above, then qAq^* is also a Hermitian matrix with zero trace.
   Prove that

$$\det(x\sigma_3 + y\sigma_2 + z\sigma_1) = \det(qAq^*) = -(x^2 + y^2 + z^2).$$

We can embed $\mathbb{R}^3$ into the space of Hermitian matrices with zero trace by

$$\varphi(x, y, z) = x\sigma_3 + y\sigma_2 + z\sigma_1.$$

Check that

$$\varphi = -i\psi$$

and

$$\varphi^{-1} = i\psi^{-1}.$$

Prove that every quaternion q induces a map r_q on $\mathbb{R}^3$ by

$$r_q(x, y, z) = \varphi^{-1}(q\varphi(x, y, z)q^*) = \varphi^{-1}(q(x\sigma_3 + y\sigma_2 + z\sigma_1)q^*)$$

which is clearly linear, and an isometry. Thus, $r_q \in \mathbf{O}(3)$.

(4) Find the fixed points of r_q, where $q = (a, (b, c, d))$. If $(b, c, d) \neq (0, 0, 0)$, then show that the fixed points (x, y, z) of r_q are solutions of the equations

$$-dy + cz = 0$$
$$cx - by = 0$$
$$dx - bz = 0.$$

This linear system has the nontrivial solution (b, c, d) and the matrix of this system is

$$\begin{pmatrix} 0 & -d & c \\ c & -b & 0 \\ d & 0 & -b \end{pmatrix}.$$

Prove that the above matrix has rank 2, so the fixed points of r_q form the one-dimensional space spanned by (b, c, d). Deduce from this that r_q must be a rotation.

Prove that $r \colon \mathbf{SU}(2) \to \mathbf{SO}(3)$ given by $r(q) = r_q$ is a group homomorphism whose kernel is $\{I, -I\}$.

(5) Find the matrix R_q representing r_q explicitly by computing

$$q(x\sigma_3 + y\sigma_2 + z\sigma_1)q^* = \begin{pmatrix} \alpha & \beta \\ -\overline{\beta} & \overline{\alpha} \end{pmatrix} \begin{pmatrix} x & z - iy \\ z + iy & -x \end{pmatrix} \begin{pmatrix} \overline{\alpha} & -\beta \\ \overline{\beta} & \alpha \end{pmatrix}.$$

You should find

$$R_q = \begin{pmatrix} a^2 + b^2 - c^2 - d^2 & 2bc - 2ad & 2ac + 2bd \\ 2bc + 2ad & a^2 - b^2 + c^2 - d^2 & -2ab + 2cd \\ -2ac + 2bd & 2ab + 2cd & a^2 - b^2 - c^2 + d^2 \end{pmatrix}.$$

Since $a^2 + b^2 + c^2 + d^2 = 1$, this matrix can also be written as

$$R_q = \begin{pmatrix} 2a^2 + 2b^2 - 1 & 2bc - 2ad & 2ac + 2bd \\ 2bc + 2ad & 2a^2 + 2c^2 - 1 & -2ab + 2cd \\ -2ac + 2bd & 2ab + 2cd & 2a^2 + 2d^2 - 1 \end{pmatrix}.$$

Prove that $r_q = \rho_q$.

(6) To prove the surjectivity of r algorithmically, proceed as follows. First, prove that $\operatorname{tr}(R_q) = 4a^2 - 1$, so

$$a^2 = \frac{\operatorname{tr}(R_q) + 1}{4}.$$

If $R \in \mathbf{SO}(3)$ is any rotation matrix and if we write

$$R = \begin{pmatrix} r_{11} & r_{12} & r_{13} \\ r_{21} & r_{22} & r_{23} \\ r_{31} & r_{32} & r_{33}, \end{pmatrix}$$

we are looking for a unit quaternion $q \in \mathbf{SU}(2)$ such that $r_q = R$. Therefore, we must have

$$a^2 = \frac{\operatorname{tr}(R) + 1}{4}.$$

We also know that

$$\operatorname{tr}(R) = 1 + 2\cos\theta,$$

where $\theta \in [0, \pi]$ is the angle of the rotation R. Deduce that

$$|a| = \cos\left(\frac{\theta}{2}\right) \quad (0 \leq \theta \leq \pi).$$

There are two cases.

Case 1. $\operatorname{tr}(R) \neq -1$, or equivalently $\theta \neq \pi$. In this case $a \neq 0$. Pick

$$a = \frac{\sqrt{\operatorname{tr}(R) + 1}}{2}.$$

Then, show that

$$b = \frac{r_{32} - r_{23}}{4a}, \quad c = \frac{r_{13} - r_{31}}{4a}, \quad d = \frac{r_{21} - r_{12}}{4a}.$$

Case 2. $\operatorname{tr}(R) = -1$, or equivalently $\theta = \pi$. In this case $a = 0$. Prove that

$$4bc = r_{21} + r_{12}$$
$$4bd = r_{13} + r_{31}$$
$$4cd = r_{32} + r_{23}$$

and

$$b^2 = \frac{1 + r_{11}}{2}$$

$$c^2 = \frac{1 + r_{22}}{2}$$

$$d^2 = \frac{1 + r_{33}}{2}.$$

Since $q \neq 0$ and $a = 0$, at least one of b, c, d is nonzero.
  If $b \neq 0$, let

$$b = \frac{\sqrt{1 + r_{11}}}{\sqrt{2}},$$

and determine c, d using

$$4bc = r_{21} + r_{12}$$
$$4bd = r_{13} + r_{31}.$$

If $c \neq 0$, let

$$c = \frac{\sqrt{1 + r_{22}}}{\sqrt{2}},$$

and determine b, d using

$$4bc = r_{21} + r_{12}$$
$$4cd = r_{32} + r_{23}.$$

If $d \neq 0$, let

$$d = \frac{\sqrt{1 + r_{33}}}{\sqrt{2}},$$

and determine b, c using

$$4bd = r_{13} + r_{31}$$
$$4cd = r_{32} + r_{23}.$$

(7) Given any matrix $A \in \mathfrak{su}(2)$, with

$$A = \begin{pmatrix} iu_1 & u_2 + iu_3 \\ -u_2 + iu_3 & -iu_1 \end{pmatrix},$$

write $\theta = \sqrt{u_1^2 + u_2^2 + u_3^2}$ and prove that

$$e^A = \cos\theta I + \frac{\sin\theta}{\theta} A, \quad \theta \neq 0,$$

with $e^0 = I$. Therefore, e^A is a unit quaternion representing the rotation of angle 2θ and axis (u_1, u_2, u_3) (or I when $\theta = k\pi$, $k \in \mathbb{Z}$). The above formula shows that we may assume that $0 \leq \theta \leq \pi$.

An equivalent but often more convenient formula is obtained by assuming that $u = (u_1, u_2, u_3)$ is a unit vector, equivalently $\det(A) = -1$, in which case $A^2 = -I$, so we have

$$e^{\theta A} = \cos\theta I + \sin\theta A.$$

Using the quaternion notation, this is read as

$$e^{\theta A} = [\cos\theta, \sin\theta\, u].$$

Prove that the logarithm $A \in \mathfrak{su}(2)$ of a unit quaternion

$$q = \begin{pmatrix} \alpha & \beta \\ -\overline{\beta} & \overline{\alpha} \end{pmatrix}$$

with $\alpha = a + bi$ and $\beta = c + id$ can be determined as follows:
If $q = I$ (i.e. $a = 1$), then $A = 0$. If $q = -I$ (i.e. $a = -1$), then

$$A = \pm\pi \begin{pmatrix} i & 0 \\ 0 & -i \end{pmatrix}.$$

Otherwise, $a \neq \pm 1$ and $(b, c, d) \neq (0, 0, 0)$, and we are seeking some $A = \theta B \in \mathfrak{su}(2)$ with $\det(B) = 1$ and $0 < \theta < \pi$, such that

$$q = e^{\theta B} = \cos\theta I + \sin\theta B.$$

Then,

$$\cos\theta = a \qquad (0 < \theta < \pi)$$

$$(u_1, u_2, u_3) = \frac{1}{\sin\theta}(b, c, d).$$

Since $a^2 + b^2 + c^2 + d^2 = 1$ and $a = \cos\theta$, the vector $(b, c, d)/\sin\theta$ is a unit vector. Furthermore if the quaternion q is of the form $q = [\cos\theta, \sin\theta u]$ where $u = (u_1, u_2, u_3)$ is a unit vector (with $0 < \theta < \pi$), then

$$A = \theta \begin{pmatrix} iu_1 & u_2 + iu_3 \\ -u_2 + iu_3 & -iu_1 \end{pmatrix}$$

is a logarithm of q.

Show that the exponential map $\exp: \mathfrak{su}(2) \to \mathbf{SU}(2)$ is surjective, and injective on the open ball

$$\{\theta B \in \mathfrak{su}(2) \mid \det(B) = 1, 0 \leq \theta < \pi\}.$$

(8) You are now going to derive a formula for interpolating between two quaternions. This formula is due to Ken Shoemake, once a Penn student and my TA! Since rotations in $\mathbf{SO}(3)$ can be defined by quaternions, this has applications to computer graphics, robotics, and computer vision.

First, we observe that multiplication of quaternions can be expressed in terms of the inner product and the cross-product in $\mathbb{R}^3$. Indeed, if $q_1 = [a, u_1]$ and $q_2 = [a_2, u_2]$, then check that

$$q_1 q_2 = [a_1, u_1][a_2, u_2] = [a_1 a_2 - u_1 \cdot u_2, \; a_1 u_2 + a_2 u_1 + u_1 \times u_2].$$

We will also need the identity

$$u \times (u \times v) = (u \cdot v)u - (u \cdot u)v.$$

Given a quaternion q expressed as $q = [\cos \theta, \sin \theta \, u]$, where u is a unit vector, we can interpolate between I and q by finding the logs of I and q, interpolating in $\mathfrak{su}(2)$, and then exponentiating. We have

$$A = \log(I) = \begin{pmatrix} 0 & 0 \\ 0 & 0 \end{pmatrix}, \quad B = \log(q) = \theta \begin{pmatrix} i u_1 & u_2 + i u_3 \\ -u_2 + i u_3 & -i u_1 \end{pmatrix}.$$

Since $\mathbf{SU}(2)$ is a compact Lie group and since the inner product on $\mathfrak{su}(2)$ given by

$$\langle X, Y \rangle = \operatorname{tr}(X^\top Y)$$

is $\mathrm{Ad}(\mathbf{SU}(2))$-invariant, it induces a biinvariant Riemannian metric on $\mathbf{SU}(2)$, and the curve

$$\lambda \mapsto e^{\lambda B}, \quad \lambda \in [0, 1]$$

is a geodesic from I to q in $\mathbf{SU}(2)$. We write $q^\lambda = e^{\lambda B}$. Given two quaternions q_1 and q_2, because the metric is left-invariant, the curve

$$\lambda \mapsto Z(\lambda) = q_1(q_1^{-1} q_2)^\lambda, \quad \lambda \in [0, 1]$$

is a geodesic from q_1 to q_2. Remarkably, there is a closed-form formula for the interpolant $Z(\lambda)$. Say $q_1 = [\cos \theta, \sin \theta \, u]$ and $q_2 = [\cos \varphi, \sin \varphi \, v]$, and assume that $q_1 \neq q_2$ and $q_1 \neq -q_2$.

Define Ω by

$$\cos \Omega = \cos \theta \cos \varphi + \sin \theta \sin \varphi (u \cdot v).$$

Since $q_1 \neq q_2$ and $q_1 \neq -q_2$, we have $0 < \Omega < \pi$. Prove that

$$Z(\lambda) = q_1 (q_1^{-1} q_2)^\lambda = \frac{\sin(1 - \lambda)\Omega}{\sin \Omega} q_1 + \frac{\sin \lambda \Omega}{\sin \Omega} q_2.$$

(9) We conclude by discussing the problem of a consistent choice of sign for the quaternion q representing a rotation $R = \rho_q \in \mathbf{SO}(3)$. We are looking for a "nice" section $s : \mathbf{SO}(3) \to \mathbf{SU}(2)$, that is, a function s satisfying the condition

$$\rho \circ s = \mathrm{id},$$

where ρ is the surjective homomorphism $\rho : \mathbf{SU}(2) \to \mathbf{SO}(3)$.

   I claim that any section $s : \mathbf{SO}(3) \to \mathbf{SU}(2)$ of ρ is neither a homomorphism nor continuous. Intuitively, this means that there is no "nice and simple " way to pick the sign of the quaternion representing a rotation.

   To prove the above claims, let Γ be the subgroup of $\mathbf{SU}(2)$ consisting of all quaternions of the form $q = [a, (b, 0, 0)]$. Then, using the formula for the rotation matrix R_q corresponding to q (and the fact that $a^2 + b^2 = 1$), show that

$$R_q = \begin{pmatrix} 1 & 0 & 0 \\ 0 & 2a^2 - 1 & -2ab \\ 0 & 2ab & 2a^2 - 1 \end{pmatrix}.$$

Since $a^2 + b^2 = 1$, we may write $a = \cos \theta$, $b = \sin \theta$, and we see that

$$R_q = \begin{pmatrix} 1 & 0 & 0 \\ 0 & \cos 2\theta & -\sin 2\theta \\ 0 & \sin 2\theta & \cos 2\theta \end{pmatrix},$$

a rotation of angle 2θ around the x-axis. Thus, both Γ and its image are isomorphic to $\mathbf{SO}(2)$, which is also isomorphic to $\mathbf{U}(1) = \{w \in \mathbb{C} \mid |w| = 1\}$. By identifying $\mathbf{i}$ and i and identifying Γ and its image to $\mathbf{U}(1)$, if we write $w = \cos \theta + i \sin \theta \in \Gamma$, show that the restriction of the map ρ to Γ is given by $\rho(w) = w^2$.

   Prove that any section s of ρ is not a homomorphism. (Consider the restriction of s to the image $\rho(\Gamma)$.)

   Prove that any section s of ρ is not continuous.

Problem 11.2. Let $A = A^0 \oplus A^1$ and $B = B^0 \oplus B^1$ be two $\mathbb{Z}/2$-graded algebras.

(i) Show that $A \widehat{\otimes} B$ is $\mathbb{Z}/2$-graded by

$$(A \widehat{\otimes} B)^0 = (A^0 \otimes B^0) \oplus (A^1 \otimes B^1),$$
$$(A \widehat{\otimes} B)^1 = (A^0 \otimes B^1) \oplus (A^1 \otimes B^0).$$

(ii) Prove Proposition 11.3.
  Hint. See Bröcker and tom Dieck [18], Chapter 1, Section 6, page 57.

Problem 11.3. Prove Proposition 11.8.

Hint. See Bröcker and tom Dieck [18], Chapter 1, Section 6, page 58.

Problem 11.4. Recall that

$$\mathbf{Spin}^+(p,q) = \{x \in \mathbf{Cl}_{p,q}^0 \mid x v \overline{x} \in \mathbb{R}^n \quad \text{for all } v \in \mathbb{R}^n, \quad N(x) = 1\},$$

and that $\mathbf{SO}_0(p,q)$ is the connected component of $\mathbf{SO}(p,q)$ containing the identity. Show that the map $\rho\colon \mathbf{Spin}^+(p,q) \to \mathbf{SO}_0(p,q)$ is a surjective homomorphism with kernel $\{-\mathbf{1}, \mathbf{1}\}$.

Hint. See Lounesto [74], Chapter 17, Section 17.2.

Problem 11.5. Prove Proposition 11.20.

Hint. See Fulton [45], Chapter 11, Lemma 11.6.

Problem 11.6.

(i) Prove Proposition 11.21.
  Hint. See Fulton [45], Chapter 11, Lemma 11.17.
(ii) Show that $\mathbb{Z}_2$ acts evenly on $\mathbf{Pin}(p,q)$.

Problem 11.7. Show that the complexification $\mathbb{C} \otimes_{\mathbb{R}} \mathrm{Cl}_{p,q}$ of the real Clifford algebra $\mathrm{Cl}_{p,q}$ is isomorphic to $\mathrm{Cl}(\Phi_n^{\mathbb{C}})$, where

$$\Phi_n^{\mathbb{C}}(x_1, \ldots, x_n) = x_1^2 + \cdots + x_n^2$$

in some orthonormal basis.

Correction to: Differential Geometry and Lie Groups

Jean Gallier and Jocelyn Quaintance

Correction to:
J. Gallier, J. Quaintance, *Differential Geometry and Lie Groups*, Geometry and Computing 13, https://doi.org/10.1007/978-3-030-46047-1

The book was inadvertently published with some typesetting errors, which have now been corrected. The details of the corrections are as follows:

Chapters 3–5: Proposition, Theorem, and Corollary environments were incorrectly numbered with separate counters: Proposition 3.1, Theorem 3.1, Corollary 3.1, etc. The numbering has been revised to be consecutive, e.g., Proposition 3.1, Theorem 3.2, Corollary 3.3, etc.

Page 70, line 11: "mu_2" has been replaced with the correctly typeset version: "μ_2".

Page 83, lines 1 and 2: The equation has been replaced with the following corrected version:

$$f(v_1, \ldots, v_n) = \left(\sum_{\sigma \in \mathfrak{S}_n} \mathrm{sgn}(\sigma)\, a_{\sigma(1),1} \cdots a_{\sigma(n),n} \right) f(u_1, \ldots, u_n)$$
$$= \det(A)\, f(u_1, \ldots, u_n),$$

The updated online version of these chapters can be found at
https://doi.org/10.1007/978-3-030-46047-1_2
https://doi.org/10.1007/978-3-030-46047-1_3
https://doi.org/10.1007/978-3-030-46047-1_4
https://doi.org/10.1007/978-3-030-46047-1_5

Bibliography

1. G.E. Andrews, R. Askey, R. Roy, *Special Functions*, 1st edn. (Cambridge University Press, Cambridge, 2000)
2. M. Artin, *Algebra*, 1st edn. (Prentice Hall, Upper Saddle River, 1991)
3. A. Arvanitoyeorgos, *An Introduction to Lie Groups and the Geometry of Homogeneous Spaces*. SML, vol. 22 , 1st edn. (AMS, Providence, 2003)
4. M.F. Atiyah, *K-Theory*, 1st edn. (Addison Wesley, Boston, 1988)
5. M.F. Atiyah, I.G. Macdonald, *Introduction to Commutative Algebra*, 3rd edn. (Addison Wesley, Boston, 1969)
6. M.F. Atiyah, R. Bott, A. Shapiro, Clifford modules. Topology **3**(Suppl. 1), 3–38 (1964)
7. S. Axler, P. Bourdon, W. Ramey, *Harmonic Function Theory*. GTM, vol. 137, 2nd edn. (Springer, New York, 2001)
8. A. Baker, *Matrix Groups. An Introduction to Lie Group Theory*. SUMS (Springer, London, 2002)
9. R. Basri, D.W. Jacobs, Lambertian reflectance and linear subspaces. IEEE Trans. Pattern Anal. Mach. Intell. **25**(2), 228–233 (2003)
10. M. Berger, *A Panoramic View of Riemannian Geometry* (Springer, Berlin, Heidelberg, 2003)
11. J.E. Bertin, *Algèbre linéaire et géométrie classique*, 1st edn. (Masson, Paris, 1981)
12. R. Bott, L.W. Tu, *Differential Forms in Algebraic Topology*. GTM, vol. 82, 1st edn. (Springer, New York, 1986)
13. N. Bourbaki, *Algèbre, Chapitre 9*. Eléments de Mathématiques (Hermann, Paris, 1968)
14. N. Bourbaki, *Algèbre, Chapitres 1–3*. Eléments de Mathématiques (Hermann, Paris, 1970)
15. N. Bourbaki, *Espaces Vectoriels Topologiques*. Eléments de Mathématiques (Masson, Paris, 1981)
16. N. Bourbaki, *Topologie Générale, Chapitres 1–4*. Eléments de Mathématiques (Masson, Paris, 1990)
17. N. Bourbaki, *Topologie Générale, Chapitres 5–10*. Eléments de Mathématiques (CCLS, Miami, 1990)
18. T. Bröcker, T. tom Dieck, *Representations of Compact Lie Groups*. GTM, vol. 98, 1st edn. (Springer, Berlin, Heidelberg, 1985)
19. É. Cartan, *Theory of Spinors*, 1st edn. (Dover, New York, 1966)
20. H. Cartan, *Cours de Calcul Différentiel*. Collection Méthodes (Hermann, Paris, 1990)
21. H. Cartan, *Differential Forms*, 1st edn. (Dover, New York, 2006)
22. S.-S. Chern, *Complex Manifolds Without Potential Theory*. Universitext, 2nd edn. (Springer, New York, 1995)

© Springer Nature Switzerland AG 2020
J. Gallier, J. Quaintance, *Differential Geometry and Lie Groups*, Geometry and Computing 13, https://doi.org/10.1007/978-3-030-46047-1

23. C. Chevalley, *Theory of Lie Groups I*. Princeton Mathematical Series, vol. 8, 1st edn., 8th edn. (Princeton University Press, Princeton, 1946)
24. C. Chevalley, *The Algebraic Theory of Spinors and Clifford Algebras. Collected Works*, vol. 2, 1st edn. (Springer, Berlin, 1997)
25. Y. Choquet-Bruhat, C. DeWitt-Morette, *Analysis, Manifolds, and Physics, Part II: 92 Applications*, 1st edn. (North-Holland, Amsterdam, 1989)
26. Y. Choquet-Bruhat, C. DeWitt-Morette, M. Dillard-Bleick, *Analysis, Manifolds, and Physics, Part I: Basics*, 1st edn. (North-Holland, Amsterdam, 1982)
27. M.L. Curtis, *Matrix Groups*. Universitext, 2nd edn. (Springer, New York, 1984)
28. J.F. Davis, P. Kirk, *Lecture Notes in Algebraic Topology*. GSM, vol. 35, 1st edn. (AMS, Providence, 2001)
29. A. Deitmar, *A First Course in Harmonic Analysis*. UTM, 1st edn. (Springer, New York, 2002)
30. J. Dieudonné, *Sur les Groupes Classiques*, 3rd edn. (Hermann, Paris, 1967)
31. J. Dieudonné, *Special Functions and Linear Representations of Lie Groups*. Regional Conference Series in Mathematics, vol. 42, 1st edn. (AMS, Providence, 1980)
32. J. Dieudonné, *Éléments d'Analyse, Tome V. Groupes de Lie Compacts, Groupes de Lie Semi-Simples*, 1st edn. (Edition Jacques Gabay, Paris, 2003)
33. J. Dieudonné, *Éléments d'Analyse, Tome VI. Analyse Harmonique*, 1st edn. (Edition Jacques Gabay, Paris, 2003)
34. J. Dieudonné, *Éléments d'Analyse, Tome VII. Équations Fonctionnelles Linéaires. Première partie, Opérateurs Pseudo-Différentiels*, 1st edn. (Edition Jacques Gabay, Paris, 2003)
35. J. Dieudonné, *Éléments d'Analyse, Tome II. Chapitres XII à XV*, 1st edn. (Edition Jacques Gabay, Paris, 2005)
36. J. Dixmier, *General Topology*. UTM, 1st edn. (Springer, New York, 1984)
37. M.P. do Carmo, *Riemannian Geometry*, 2nd edn. (Birkhäuser, Basel, 1992)
38. J.J. Duistermaat, J.A.C. Kolk, *Lie Groups*. Universitext, 1st edn. (Springer, New York, 2000)
39. D.S. Dummit, R.M. Foote, *Abstract Algebra*, 2nd edn. (Wiley, New York, 1999)
40. A. Edelman, T.A. Arias, S.T. Smith, The geometry of algorithms with orthogonality constraints. SIAM J. Matrix Anal. Appl. **20**(2), 303–353 (1998)
41. C.H. Edwards Jr., *Advanced Calculus of Several Variables*, 1st edn. (Academic, New York, 1973)
42. G.B. Folland, *A Course in Abstract Harmonic Analysis*, 1st edn. (CRC Press, Boca Raton, 1995)
43. J. Fourier, *Théorie Analytique de la Chaleur*, 1st edn. (Edition Jacques Gabay, Paris, 1822)
44. W. Fulton, *Algebraic Topology: A First Course*. GTM, vol. 153, 1st edn. (Springer, New York, 1995)
45. W. Fulton, J. Harris, *Representation Theory: A First Course*. GTM, vol. 129, 1st edn. (Springer, Berlin, 1991)
46. J.H. Gallier, *Geometric Methods and Applications: For Computer Science and Engineering*. TAM, vol. 38, 2nd edn. (Springer, New York, 2011)
47. J.H. Gallier, J. Quaintance, *Differential Geometry and Lie Groups*, A Computational Perspective (Springer, Berlin, 2018)
48. S. Gallot, D. Hulin, J. Lafontaine, *Riemannian Geometry*. Universitext, 2nd edn. (Springer, New York, 1993)
49. F.R. Gantmacher, *The Theory of Matrices*, vol. I, 1st edn. (AMS Chelsea, Providence, 1977)
50. A. Gramain, *Topologie des Surfaces*. Collection SUP, 1st edn. (PUF, Paris, 1971)
51. R. Green, Spherical harmonic lighting: the gritty details, in *Archives of the Game Developers' Conference* (2003), pp. 1–47.
52. P. Griffiths, J. Harris, *Principles of Algebraic Geometry*, 1st edn. (Wiley Interscience, Hoboken, 1978)
53. B. Hall, *Lie Groups, Lie Algebras, and Representations. An Elementary Introduction*. GTM, vol. 222, 1st edn. (Springer, New York, 2003)
54. S. Helgason, *Geometric Analysis on Symmetric Spaces*. SURV, vol. 39, 1st edn. (AMS, Providence, 1994)

55. S. Helgason, *Groups and Geometric Analysis. Integral Geometry, Invariant Differential Operators and Spherical Functions*. MSM, vol. 83, 1st edn. (AMS, Providence, 2000)

56. M.W. Hirsch, *Differential Topology*. GTM, vol. 33, 1st edn. (Springer, New York, 1976)

57. F. Hirzebruch, *Topological Methods in Algebraic Geometry*. Springer Classics in Mathematics, 2nd edn. (Springer, Berlin, 1978)

58. H. Hochstadt, *The Functions of Mathematical Physics*, 1st edn. (Dover, New York, 1986)

59. R.A. Horn, C.R. Johnson, *Matrix Analysis*, 1st edn. (Cambridge University Press, Cambridge, 1990)

60. D. Husemoller, *Fiber Bundles*. GTM, vol. 20, 3rd edn. (Springer, New York, 1994)

61. N. Jacobson, *Basic Algebra II*, 1st edn. (Freeman, New York, 1980)

62. J. Jost, *Riemannian Geometry and Geometric Analysis*. Universitext, 4th edn. (Springer, New York, 2005)

63. A.A. Kirillov, Spinor representations of orthogonal groups. Technical report, University of Pennsylvania, Mathematics Department, Philadelphia, PA (2001). Course Notes for Mathematics, vol. 654

64. A.A. Kirillov (ed.) *Representation Theory and Noncommutative Harmonic Analysis*. Encyclopaedia of Mathematical Sciences, vol. 22, 1st edn. (Springer, Berlin, 1994)

65. A.W. Knapp, *Representation Theory of Semisimple Groups*, 1st edn. Princeton Landmarks in Mathematics (Princeton University Press, Princeton, 2001)

66. A.W. Knapp, *Lie Groups Beyond an Introduction*. Progress in Mathematics, vol. 140, 2nd edn. (Birkhäuser, Basel, 2002)

67. S. Lang, *Algebra*, 3rd edn. (Addison Wesley, Boston, 1993)

68. S. Lang, *Real and Functional Analysis*. GTM, vol. 142, 3rd edn. (Springer, New York, 1996)

69. S. Lang, *Undergraduate Analysis*. UTM, 2nd edn. (Springer, New York, 1997)

70. S. Lang, *Fundamentals of Differential Geometry*. GTM, vol. 191, 1st edn. (Springer, New York, 1999)

71. B.H. Lawson, M.-L. Michelsohn, *Spin Geometry*. Princeton Mathematical Series, vol. 38 (Princeton University Press, Princeton, 1989)

72. N.N. Lebedev, *Special Functions and Their Applications*, 1st edn. (Dover, New York, 1972)

73. J.M. Lee, *Introduction to Smooth Manifolds*. GTM, vol. 218, 1st edn. (Springer, New York, 2006)

74. P. Lounesto, *Clifford Algebras and Spinors*. LMS, vol. 286, 2nd edn. (Cambridge University Press, Cambridge, 2001)

75. I. Madsen, J. Tornehave, *From Calculus to Cohomology. De Rham Cohomology and Characteristic Classes*, 1st edn. (Cambridge University Press, Cambridge, 1998)

76. P. Malliavin, *Géométrie Différentielle Intrinsèque*. Enseignement des Sciences, vol. 14, 1st edn. (Hermann, Paris, 1972)

77. J.E. Marsden, T.S. Ratiu, *Introduction to Mechanics and Symmetry*. TAM, vol. 17, 1st edn. (Springer, New York, 1994)

78. J.W. Milnor, J.D. Stasheff, *Characteristic Classes*. Annals of Mathematics Series, vol. 76, 1st edn. (Princeton University Press, Princeton, 1974)

79. R. Mneimné, F. Testard, *Introduction à la Théorie des Groupes de Lie Classiques*, 1st edn. (Hermann, Paris, 1997)

80. J. Molk, *Encyclopédie des Sciences Mathématiques Pures et Appliquées. Tome I (premier volume), Arithmétique*, 1st edn. (Gauthier-Villars, Paris, 1916)

81. M. Morimoto, *Analytic Functionals on the Sphere*. Translations of Mathematical Monographs, vol. 178, 1st edn. (AMS, Providence, 1998)

82. S. Morita, *Geometry of Differential Forms*. Translations of Mathematical Monographs, vol. 201, 1st edn. (AMS, Providence, 2001)

83. R. Narasimham, *Compact Riemann Surfaces*. Lecture in Mathematics. ETH Zürich, 1st edn. (Birkhäuser, Basel, 1992)

84. B. O'Neill, *Semi-Riemannian Geometry with Applications to Relativity*. Pure and Applies Mathematics, vol. 103, 1st edn. (Academic, New York, 1983)

85. B. O'Neill, *Elementary Differential Geometry*, 2nd edn. (Academic, Cambridge, 2006)

86. P. Petersen, *Riemannian Geometry*. GTM, vol. 171, 2nd edn. (Springer, New York, 2006)
87. I.R. Porteous, *Topological Geometry*, 2nd edn. (Cambridge University Press, Cambridge, 1981)
88. M.M. Postnikov, *Geometry VI. Riemannian Geometry*. Encyclopaedia of Mathematical Sciences, vol. 91, 1st edn. (Springer, Berlin, 2001)
89. M. Riesz, *Clifford Numbers and Spinors*, 1st edn., ed. by E. Folke Bolinder, P. Lounesto (Kluwer Academic Press, Dordrecht, 1993)
90. S. Rosenberg, *The Laplacian on a Riemannian Manifold*, 1st edn. (Cambridge University Press, Cambridge, 1997)
91. J.J. Rotman, *Introduction to Algebraic Topology*. GTM, vol. 119, 1st edn. (Springer, New York, 1988)
92. W. Rudin, *Real and Complex Analysis*, 3rd edn. (McGraw Hill, New York, 1987)
93. T. Sakai, *Riemannian Geometry*. Mathematical Monographs, vol. 149, 1st edn. (AMS, Providence, 1996)
94. G. Sansone, *Orthogonal Functions*, 1st edn. (Dover, New York, 1991)
95. H. Sato, *Algebraic Topology: An Intuitive Approach*. Mathematical Monographs, vol. 183, 1st edn. (AMS, Providence, 1999)
96. L. Schwartz, *Analyse I. Théorie des Ensembles et Topologie*. Collection Enseignement des Sciences (Hermann, Paris, 1991)
97. L. Schwartz, *Analyse III. Calcul Intégral*. Collection Enseignement des Sciences (Hermann, Paris, 1993)
98. L. Schwartz, *Analyse IV. Applications à la Théorie de la Mesure*. Collection Enseignement des Sciences (Hermann, Paris, 1993)
99. I.R. Shafarevich, *Basic Algebraic Geometry 1*, 2nd edn. (Springer, New York, 1994)
100. R.W. Sharpe, *Differential Geometry. Cartan's Generalization of Klein's Erlangen Program*. GTM, vol. 166, 1st edn. (Springer, New York, 1997)
101. N. Steenrod, *The Topology of Fibre Bundles*. Princeton Mathematical Series, vol. 1 (Princeton University Press, Princeton, 1956)
102. E.M. Stein, G. Weiss, *Introduction to Fourier Analysis on Euclidean Spaces*. Princeton Mathematical Series, vol. 32 (Princeton University Press, Princeton, 1971)
103. M.E. Taylor, *Noncommutative Harmonic Analysis*. Mathematical Surveys and Monographs, vol. 22, 1st edn. (AMS, Providence, 1986)
104. A. Terras, *Fourier Analysis on Finite Groups and Applications*. Student Text, vol. 43, 1st edn. (Cambridge University Press, Cambridge, 1999)
105. L.W. Tu, *An Introduction to Manifolds*. Universitext, 1st edn. (Springer, New York, 2008)
106. V.S. Varadarajan, *An Introduction to Harmonic Analysis on Semisimple Lie Groups*. Cambridge Studies in Advanced Mathematics, 1st edn. (Cambridge University Press, Cambridge, 1999)
107. N.J. Vilenkin, *Special Functions and the Theory of Group Representations*. Translations of Mathematical Monographs, vol. 22, 1st edn. (AMS, Providence, 1968)
108. R. Vinroot, The Haar measure. Notes for math 519. Technical report, 2013. www.math.wm.edu/~vinroot/PadicGroups/haar.pdf
109. F. Warner, *Foundations of Differentiable Manifolds and Lie Groups*. GTM, vol. 94, 1st edn. (Springer, Berlin, 1983)
110. R.O. Wells, *Differential Analysis on Complex Manifolds*. GTM, vol. 65, 2nd edn. (Springer, New York, 1980)

Symbol Index

© Springer Nature Switzerland AG 2020

J. Gallier, J. Quaintance, *Differential Geometry and Lie Groups*, Geometry and Computing 13, https://doi.org/10.1007/978-3-030-46047-1

Index

© Springer Nature Switzerland AG 2020

J. Gallier, J. Quaintance, *Differential Geometry and Lie Groups*, Geometry and Computing 13, https://doi.org/10.1007/978-3-030-46047-1

Printed in the United States
by Baker & Taylor Publisher Services